VOLUME 2

MODERN ELECTROCHEMISTRY

VOLUME 2
MODERN ELECTROCHEMISTRY

An Introduction to an Interdisciplinary Area

John O'M. Bockris
Professor of Electrochemistry
University of Pennsylvania, Philadelphia, Pennsylvania
and
Amulya K. N. Reddy
Professor of Electrochemistry
Indian Institute of Science, Bangalore, India

A Plenum/Rosetta Edition

Library of Congress Cataloging in Publication Data

Bockris, John O'M
  Modern electrochemistry.

  "A Plenum/Rosetta edition."
  1. Electrochemistry. I. Reddy, Amulya K. N., joint author. II. Title.
QD553.B63   1973   541'.37   73-13712
ISBN 0-306-25002-0 (v. 2)

First Paperback Printing — October 1973
Second Printing — March 1976
Third Printing — July 1977

A Plenum/Rosetta Edition
Published by Plenum Publishing Corporation
227 West 17th Street, New York, N.Y. 10011

© 1970 Plenum Press, New York
A Division of Plenum Publishing Corporation

United Kingdom edition published by Plenum Press, London
A Division of Plenum Publishing Company, Ltd.
Davis House (4th Floor), 8 Scrubs Lane, Harlesden, London, NW10 6SE, England

All rights reserved

No part of this publication may be reproduced in any form
without written permission from the publisher

Printed in the United States of America

CONTENTS

VOLUME 2

CHAPTER 7

The Electrified Interface

7.1	Electrification of an Interface	623
7.1.1	The Electrode–Electrolyte Interface: The Basis of Electrodics	623
7.1.2	New Forces at the Boundary of an Electrolyte	623
7.1.3	The Interphase Region Has New Properties and New Structures	626
7.1.4	An Electrode Is Like a Giant Central Ion	626
7.1.5	The Consequences of Compromise Arrangements: The Electrolyte Side of the Boundary Acquires a Charge	627
7.1.6	Both Sides of the Interface Become Electrified: The So-Called "Electrical Double Layer"	629
7.1.7	Double Layers Are Characteristic of All Phase Boundaries	630
7.1.8	A Look into an Electrified Interface	632
Further Reading		639
7.2	Some Problems in Understanding an Electrified Interface	640
7.2.1	What Knowledge Is Required before an Electrified Interface Can Be Regarded as Understood?	640
7.2.2	Predicting the Interphase Properties from the Bulk Properties of the Phases	641
7.2.3	Why Bother about Electrified Interfaces?	642
7.2.4	The Need to Clarify Some Concepts	643
7.2.5	The Potential Difference across Electrified Interfaces	644
	7.2.5a What Happens when One Tries to Measure the Absolute Potential Difference across a Single Electrode–Electrolyte Interface	644

	7.2.5b	The Absolute Potential Difference across a Single Electrified Interface Cannot Be Measured	648
	7.2.5c	Can One Measure Changes in the Metal–Solution Potential Difference?	650
	7.2.5d	The Extreme Cases of Ideally Nonpolarizable and Polarizable Interfaces	653
	7.2.5e	The Development of a Scale of Relative Potential Differences	655
	7.2.5f	Can One Meaningfully Analyze an Electrode–Electrolyte Potential Difference?	659
	7.2.5g	A Thought Experiment Involving a Charged Electrode in Vacuum	660
	7.2.5h	The Test Charge Must Avoid Image Interactions with the Charged Electrode	660
	7.2.5i	The Outer Potential ψ of a Material Phase in Vacuum	663
	7.2.5j	What is the Relevance of the Outer Potential to Double-Layer Studies?	665
	7.2.5k	Another Thought Experiment Involving an Uncharged, Dipole-Covered Phase	667
	7.2.5l	The Dipole Potential Difference $^M\Delta^s\chi$ across an Electrode–Electrolyte Interface	670
	7.2.5m	The Sum of the Potential Differences Due to Charges and Dipoles: The Absolute Electrode–Electrolyte (or Galvani) Potential Difference	670
	7.2.5n	The Outer, Surface, and Inner Potential Differences	673
	7.2.5o	An Apparent Contradiction: The Sum of the $\Delta\phi$'s across a System of Interfaces Can and the $\Delta\phi$ across One Interface Cannot Be Measured	674
	7.2.5p	What Deeper Understanding Has Been Hitherto Gained Regarding the Absolute Potential Difference Across an Electrified Interface?	677
7.2.6		The Accumulation and Depletion of Substances at an Interface	679
	7.2.6a	What Would Represent Complete Structural Information Regarding an Electrified Interface?	679
	7.2.6b	The Concept of Surface Excess	680
	7.2.6c	Does Knowledge of the Surface Excess Contribute to Knowledge of the Distribution of Species in the Interphase Region?	683
	7.2.6d	Is the Surface Excess Equivalent to the Amount Adsorbed?	684
	7.2.6e	Is the Surface Excess Measurable?	685
	7.2.6f	The Special Position of Mercury in Double-Layer Studies	687
Further Reading			687

7.3 The Thermodynamics of Electrified Interfaces — 688

7.3.1		The Measurement of Interfacial Tension as a Function of the Potential Difference across the Interface	688
7.3.2		Some Basic Facts about Electrocapillary Curves	690
7.3.3		A Digression on the Electrochemical Potential	693
	7.3.3a	Definition of Electrochemical Potential	693
	7.3.3b	Can the Chemical and Electrical Work Be Determined Separately?	695
	7.3.3c	A Criterion of Thermodynamic Equilibrium between Two Phases: Equality of Electrochemical Potentials	696
	7.3.3d	Nonpolarizable Interfaces and Thermodynamic Equilibrium	697
7.3.4		Some Thermodynamic Thoughts on Electrified Interfaces	698

7.3.5	Interfacial Tension Varies with Applied Potential: Determination of the Charge Density on the Electrode	701
7.3.6	Electrode Charge Varies with Applied Potential: Determination of the Electrical Capacitance of the Interface	703
7.3.7	The Potential at Which an Electrode Has a Zero Charge	706
7.3.8	Surface Tension Varies with Solution Composition: Determination of the Surface Excess	707
7.3.9	Reflections on Electrocapillary Thermodynamics	714
7.3.10	Retrospect and Prospect in the Study of Electrified Interfaces	715
Further Reading		717
7.4	**The Structure of Electrified Interfaces**	**718**
7.4.1	The Parallel-Plate Condenser Model: The Helmholtz–Perrin Theory	718
7.4.2	The Double Layer in Trouble: Neither Perfect Parabolas nor Constant Capacities	719
7.4.3	The Ionic Cloud: The Gouy–Chapman Diffuse-Charge Model of the Double Layer	722
7.4.4	Ions under Thermal and Electric Forces near an Electrode	724
7.4.5	A Picture of the Potential Drop in the Diffuse Layer	728
7.4.6	An Experimental Test of the Gouy–Chapman Model: Potential Dependence of the Capacitance, but at What Cost?	732
7.4.7	Some Ions Stuck to the Electrode, Others Scattered in Thermal Disarray: The Stern Model	733
7.4.8	A Consequence of the Stern Picture: Two Potential Drops across an Electrified Interface	734
7.4.9	Another Consequence of the Stern Model: An Electrified Interface Is Equivalent to Two Capacitors in Series	735
7.4.10	The Relative Contributions of the Helmholtz–Perrin and Gouy–Chapman Capacities	737
7.4.11	Some Questions Regarding the Sticking of Ions to the Electrode	738
7.4.12	An Electrode Is Largely Covered with Adsorbed Water Molecules	739
7.4.13	Metal–Water Interactions	740
7.4.14	The Orientation of Water Molecules on Charged Electrodes	741
7.4.15	How Close Can Hydrated Ions Come to a Hydrated Electrode?	741
7.4.16	Is It Only Desolvated Ions which Contact-Adsorb on the Electrode?	742
7.4.17	The Free-Energy Change for Contact Adsorption	742
7.4.18	What Determines the Degree of Contact Adsorption?	743
7.4.19	How Is Contact Adsorption Measured?	745
7.4.20	Contact Adsorption, Specific Adsorption, or Superequivalent Adsorption	748
7.4.21	Contact Adsorption: Its Influence of the Capacity of the Interface	749
7.4.22	Looking Back to Look Forward	752
7.4.23	The Complete Capacity–Potential Curve	753
7.4.24	The Constant-Capacity Region	753
	7.4.24a The So-Called "Double Layer" Is a Double Layer	753
	7.4.24b The Dielectric Constant of the Water between the Metal and the Outer Helmholtz Plane	756
	7.4.24c The Position of the Outer Helmholtz Plane and an Interpretation of the Constant Capacity	757
7.4.25	The Capacitance Hump	761
7.4.26	How Does the Population of Contact-Adsorbed Ions Change with Electrode Charge?	762
7.4.27	The Test of the Population Law for Contact-Adsorbed Ions	769
7.4.28	The Lateral-Repulsion Model for Contact Adsorption	776

viii CONTENTS

7.4.29	Flip-Flop Water on Electrodes	779
7.4.30	Calculation of the Potential Difference Due to Water Dipoles	781
7.4.31	The Excess of Flipped Water Dipoles over Flopped Water Dipoles	782
7.4.32	The Contribution of Adsorbed Water Dipoles to the Capacity of the Interface	789
Further Reading		790
7.5	The Competition between Water and Organic Molecules at the Electrified Interfaces	791
7.5.1	The Relevance of Organic Adsorption	791
7.5.2	The Forces Involved in Organic Adsorption	792
7.5.3	Does Organic Adsorption Depend on Electrode Charge?	793
7.5.4	The Examination of the Water Flip-Flop Model for Simple Cases of Organic Adsorption	797
7.5.5	At What Potential Does Maximum Organic Adsorption Occur?	798
Further Reading		801
7.6	Electrified Interfaces at Metals Other than Mercury	801
Further Reading		803
7.7	The Structure of the Semiconductor–Electrolyte Interface	803
7.7.1	How Is the Charge Distributed inside a Solid Electrode?	803
7.7.2	The Band Theory of Crystalline Solids	804
7.7.3	Conductors, Insulators, and Semiconductors	806
7.7.4	Some Analogies between Semiconductors and Electrolytic Solutions	811
7.7.5	The Diffuse-Charge Region inside an Intrinsic Semiconductor: The Garrett–Brattain Space Charge	813
7.7.6	The Differential Capacity Due to the Space Charge	816
7.7.7	Impurity Semiconductors, n Type and p Type	818
7.7.8	Surface States: The Semiconductor Analogue of Contact Adsorption	821
7.7.9	Semiconductor Electrochemistry: The Beginnings of the Electrochemistry of Nonmetallic Materials	823
Further Reading		823
7.8	A Bird's-Eye View of the Structure of Charged Interfaces	824
7.9	Double Layers between Phases Moving Relative to Each Other	826
7.9.1	The Phenomenology of Mobile Electrified Interfaces: Electrokinetic Properties	826
7.9.2	The Relative Motion of One of the Phases Constituting an Electrified Interface Produces a Streaming Current	829
7.9.3	A Potential Difference Applied Parallel to an Electrified Interface Produces an Electro-osmotic Motion of One of the Phases Relative to the Other	831
7.9.4	Electrophoresis: Moving Solid Particles in a Stationary Electrolyte	832
Further Reading		835
7.10	Colloid Chemistry	835
7.10.1	Colloids: The Thickness of the Double Layer and the Bulk Dimensions Are of the Same Order	835
7.10.2	The Interaction of Double Layers and the Stability of Colloids	836
7.10.3	Sols and Gels	839
Further Reading		841
Appendix 7.1	Measurement of the Electrode–Solution Volta Potential Difference	841

CHAPTER 8

Electrodics

8.1	Introduction	845
8.1.1	The Situation Thus Far	845
8.1.2	Charge Transfer: Its Chemical and Electrical Implications	846
8.1.3	Can an Isolated Electrode–Solution Interface Be Used as a Device?	849
8.1.4	Electrochemical Systems Can Be Used as Devices	851
8.1.5	An Electrochemical Device: The Substance Producer	851
8.1.6	Another Electrochemical Device: The Energy Producer	855
8.1.7	The Electrochemical Undevice: The Substance Destroyer and Energy Waster	859
8.1.8	Some Basic Questions	861
8.2	The Basic Electrodic Equation: The Butler–Volmer Equation	862
8.2.1	The Instant of Immersion of a Metal in an Electrolytic Solution	862
8.2.2	The Rate of Charge-Transfer Reactions under Zero Field: The Chemical Rate Constant	865
8.2.3	Some Consequences of Electron Transfer at an Interface	868
8.2.4	What Is the Rate of an Electron-Transfer Reaction under the Influence of an Electric Field?	869
8.2.5	The Two-Way Electron Traffic across the Interface	873
8.2.6	The Interface at Equilibrium: The Equilibrium Exchange-Current Density i_0	876
8.2.7	The Interface Departs from Equilibrium: The Nonequilibrium Drift-Current Density i	879
8.2.8	The Current-Producing (or Current-Produced) Potential Difference: The Overpotential η	880
8.2.9	The Basic Electrodic (Butler–Volmer) Equation: Some General and Special Cases	883
8.2.10	The High-Field Approximation: The Exponential i versus η Law	888
8.2.11	The Low-Field Approximation: The Linear i versus η Law	892
8.2.12	Nonpolarizable and Polarizable Interfaces	894
8.2.13	Zero Net Current and the Classical Law of Nernst	897
8.2.14	The Nernst Equation	901
8.2.15	The Nernst Equation: Its Sphere of Relevance	906
8.2.16	Looking Back	908
Further Reading		909
8.3	The Butler–Volmer Equation: Further Details	910
8.3.1	The Need for a Careful Look at Some Quantities in the Butler–Volmer Equation	910
8.3.2	The Relation between Structure at the Electrified Interface and the Rate of Charge-transfer Reactions	911
8.3.3	The Interfacial Concentrations May Depend on Ionic Transport in the Electrolyte	916
8.3.4	What Is the Physical Meaning of the Symmetry factor β?	917
	8.3.4a The Factor β Is at the Center of Electrode Kinetics	917
	8.3.4b A Preliminary to a Second Theory of β: Potential-Energy-Distance Relations of Particles Undergoing Charge Transfer	918
	8.3.4c A Simple Picture of the Symmetry Factor	922

x CONTENTS

	8.3.4d Is the β in the Butler–Volmer Equation Independent of Overpotential?	926
8.3.5	Summing-up of Further Details on the Butler–Volmer Equation	928
Further Reading		929
8.4	The Current–Potential Laws at Other Types of Charged Interfaces	930
8.4.1	Semiconductor n–p Junctions	930
8.4.2	The Current across Biological Membranes	937
8.4.3	The Hot Emission of Electrons from a Metal into Vacuum	942
8.4.4	The Cold Emission of Electrons from a Metal into Vacuum	944
Further Reading		946
8.5	The Quantum Aspects of Charge-Transfer Reactions at Electrode–Solution Interfaces	947
8.5.1	A Few Words on the Mechanics of Electrons	947
8.5.2	The Penetration of Electrons into Classically Forbidden Regions	950
8.5.3	The Probability of Electron Tunneling through Barriers	953
8.5.4	The Distribution of Electrons among the Energy Levels in a Metal	956
8.5.5	Under What Conditions Do Electrons Tunnel between the Electrode and Ions in Solution?	959
8.5.6	The Tunneling Condition and the Proton-Transfer Curve	966
8.5.7	Electron Tunneling and the De-electronation Reaction	973
8.5.8	A Perspective View of Charge-Transfer Reactions at an Electrode	974
8.5.9	The Symmetry Factor β: A Better View	975
8.5.10	Quantifying the Charge-Transfer Picture	977
8.5.11	Some Desirable Refinements and Generalizations	981
8.5.12	Surveying the Progress	983
Further Reading		986
8.6	Electrodic Reactions and Chemical Reactions	986
Further Reading		989
Appendix 8.1	The Number of Electrons Having Energy E_F Striking the Surface of a Metal from the Inside	990

CHAPTER 9

Electrodics: More Fundamentals

9.1	Multistep Reactions	991
9.1.1	The Question of Multistep Reactions	991
9.1.2	Some Ideas on Queues, or Waiting Lines	992
9.1.3	The Overpotential η Is Related to the Electron Queue at an Interface	993
9.1.4	A Near-Equilibrium Relation between the Current Density and Overpotential for a Multistep Reaction	994
9.1.5	The Concept of a Rate-Determining Step	997
9.1.6	Rate-Determining Steps and Energy Barriers for Multistep Reactions	1002
9.1.7	How Many Times Must the Rate-Determining Step Take Place for the Overall Reaction to Occur Once? The Stoichiometric Number ν	1004
9.1.8	The Order of an Electrodic Reaction	1008

9.1.9	Blockage of the Electrode Surface during Charge Transfer: The Surface-Coverage Factor	1014
Further Reading		1017
9.2	**The Transient Behavior of Interfaces**	1017
9.2.1	The Interface under Equilibrium, Transient, and Steady-State Conditions	1017
9.2.2	How an Interface Is Stimulated to Show Time Variations	1019
9.2.3	Some Ideas on the Understanding of Transients	1020
9.2.4	Intermediates in Electrodic Reactions and Their Effects on Potential–Time Transients	1026
9.2.5	Experimental Methods for the Determination of Partial Coverage, with Adsorbed Entities, of the Surface of Electrocatalysts	1029
	9.2.5a Radiotracer Method	1030
	9.2.5b Galvanostatic Transient Method	1030
	9.2.5c Potentiostatic Transients	1032
	9.2.5d The Potential-Sweep, or Potentiodynamic, Method	1033
Further Reading		1035
9.3	**Transport in the Electrolyte Effects Charge Transfer at the Interface**	1036
9.3.1	Ionics Looks after the Material Needs of the Interface	1036
9.3.2	How the Transport Flux Is Linked to the Charge-Transfer Flux: The Flux-Equality Condition	1038
9.3.3	Appropriations from the Theory of Heat Transfer	1040
9.3.4	A Qualitative Study of How Diffusion Affects the Response of an Interface to a Constant Current	1041
9.3.5	A Quantitative Treatment of How Diffusion to an Electrode Affects the Response with Time of an Interface to a Constant Current	1044
9.3.6	The Concept of Transition Time	1047
9.3.7	Convection Can Maintain Steady Interfacial Concentrations	1050
9.3.8	The Origin of Concentration Overpotential	1052
9.3.9	The Diffusion Layer	1055
9.3.10	The Limiting Current Density and Its Practical Importance	1059
	9.3.10a Polarography: The Dropping-Mercury Electrode	1060
	9.3.10b The Rotating-Disc Electrode	1070
9.3.11	The Steady-State Current–Potential Relation under Conditions of Transport Control	1072
9.3.12	Transport-Controlled De-electronation Reactions	1074
9.3.13	What Is the Effect of Electrical Migration on the Limiting Diffusion-Current Density?	1075
9.3.14	Some Summarizing Remarks on the Transport Aspects of Electrodics	1076
Further Reading		1079
9.4	**Determining the Stepwise Mechanism of an Electrodic Reaction**	1080
9.4.1	How One Tries to Determine the Reaction Mechanism	1080
9.4.2	Which Is the Rate-Determining Step in the Iron Deposition and Dissolution Reaction?	1083
9.4.3	The Transfer Coefficient α and Reaction Mechanisms	1089
9.4.4	Summarizing Remarks Concerning Mechanistic Studies	1090
Further Reading		1093
9.5	**More on Mechanism Determination**	1093
9.5.1	Why Review Mechanism Determination?	1093

xii CONTENTS

9.5.2	What Is Mechanism Determination?	1094
9.5.3	Stages in the Elucidation of a Reaction Mechanism	1095
9.5.4	The Elucidation of the Overall Reaction, the Entities in Solution, and the Surface Coverage	1096
9.5.5	Some Techniques for Mechanism Determination	1099
	9.5.5a The Determination of Reaction Order	1099
	9.5.5b The Determination of the Transfer Coefficients	1100
	9.5.5c The Determination of the Stoichiometric Number	1101
	9.5.5d Auxiliary Methods	1104
9.5.6	Mechanism Determination for Saturated Hydrocarbons	1107
Further Reading		1110

9.6	Current–Potential Laws for Electrochemical Systems	1111
9.6.1	The Potential Difference across an Electrochemical System	1111
9.6.2	The Equilibrium Potential Difference across an Electrochemical Cell	1114
9.6.3	The Problem with Tables of Standard Electrode Potentials	1115
9.6.4	The pH–Potential Diagrams: A General Representation of Equilibrium Potential Differences across Cells	1121
9.6.5	Are Equilibrium Cell Potential Differences Useful?	1124
9.6.6	Electrochemical Cells: A Qualitative Discussion of the Variation of Cell Potential with Current	1129
9.6.7	Electrochemical Cells in Action: Some Quantitative Relations between Cell Current and Cell Potential	1132
Further Reading		1137

| 9.7 | The Grand Divide | 1138 |

CHAPTER 10
Electrodic Reactions of Special Interest

10.1	Electrocatalysis	1141
10.1.1	A Chemical Catalyst and an Electrocatalyst	1141
10.1.2	At What Potential Should Electrocatalysts Be Compared?	1143
10.1.3	Electrocatalysis in Simple Redox Reactions	1146
	10.1.3a How Does the Electrocatalytic Rate Depend upon the Substrate Work Function at the Reversible Potential?	1146
	10.1.3b Can the Exchange-Current Density Depend upon the Work Function?	1149
10.1.4	Electrocatalysis in Reactions Involving Adsorbed Species	1153
	10.1.4a Electrocatalysis in the Hydrogen-Evolution and -Dissolution Reaction	1153
	10.1.4b The Electrocatalytic De-electronation of Hydrocarbons	1156
	10.1.4c The Dependence of the Rate upon Substrate for the Oxidation of Ethylene	1161
	10.1.4d The Special Position of Platinum as an Electrocatalyst	1166
10.1.5	Special Features of Electrocatalysis	1168
	10.1.5a The Effect of the Electric Field	1168
	10.1.5b Reactivity at Low Temperatures	1169
	10.1.5c The Activation of an Electrocatalyst	1170
	10.1.5d Increasing the Power Output by Changing the Reaction Path	1171
	10.1.5e The Use of Porous Electrodes	1171
Further Reading		1172

10.2	The Electrogrowth of Metals on Electrodes	1173
10.2.1	The Two Aspects of Electrogrowth	1173
10.2.2	The Reaction Pathway for Electrodeposition	1175
10.2.3	Stepwise Dehydration of an Ion; the Surface Diffusion of Adions	1177
10.2.4	Mechanism Determination on Surfaces Which Change with Time	1182
10.2.5	The Time Variation of the Average Adion Concentration in Response to the Switching on of a Constant Current	1185
10.2.6	The Contributions of Double-Layer Charging and Faradaic Reaction to the Total Deposition-Current Density	1190
10.2.7	The Time Variation of the Overpotential and the Rate-Determining Step in Electrodeposition	1192
10.2.8	The Contribution of Charge Transfer and Surface Diffusion to the Total Overpotential for Electrodeposition at Steady State	1199
10.2.9	From Deposition to Crystallization	1202
10.2.10	Some Devices for Building Lattices from Adions: Screw Dislocations and Spiral Growths	1203
10.2.11	Microsteps and Macrosteps	1207
10.2.12	How Steps from a Pair of Screw Dislocations Interact	1210
10.2.13	Crystal Facets Form	1212
10.2.14	Deposition on Single-Crystal and Polycrystal Substrates	1218
10.2.15	How the Diffusion of Ions in Solution May Affect Electrogrowth	1218
10.2.16	Organic Additives and Electrodeposits	1221
10.2.17	The Simultaneous Deposition of More Than One Metal: Alloy Deposition	1223
10.2.18	The Sometimes Unavoidable Complication: Hydrogen Codeposition	1227
Further Reading		1230

10.3	The Hydrogen-Evolution Reaction	1231
10.3.1	A Reaction with a Special History	1231
10.3.2	What Are the Possible Paths for the Hydrogen-Evolution Reaction?	1233
10.3.3	What Mechanisms Are Possible in Hydrogen Evolution?	1235
10.3.4	How One Determines the Path and Rate-Determining Step of the Hydrogen-Evolution Reaction	1237
	10.3.4a The Determination of the Exchange-Current Density	1238
	10.3.4b The Determination of the Transfer Coefficient	1238
	10.3.4c The Determination of Reaction Order with Respect to Hydrogen Ions in Solution	1243
	10.3.4d The Stoichiometric Number v	1244
	10.3.4e The Determination of Hydrogen Coverage	1245
	10.3.4f The Heat of Adsorption of Atomic Hydrogen on the Electrode	1247
	10.3.4g Isotopic Separation Factors	1248
	10.3.4h What Are the Probable Mechanisms for Hydrogen Evolution?	1250
Further Reading		1250

10.4	The Electronation of Oxygen	1251
10.4.1	The Importance of the Oxygen-Electronation Reaction	1251
10.4.2	The Evaluation of One of the Mechanisms of Oxygen Electronation	1253
10.4.3	Catalysis and the Oxygen Reaction	1256
10.4.4	Some Special Difficulties with Electrodic Reactions Having Small Exchange-Current Densities	1259
10.4.5	An Electrodic Method of Purifying Solutions	1261
10.4.6	Observing Very Slow Reactions near Equilibrium	1263
Further Reading		1263

CHAPTER 11

Some Electrochemical Systems of Technological Interest

11.1	Technological Aspects of Electrochemistry	1265
11.2	Corrosion and the Stability of Metals	1267
11.2.1	Civilization and Surfaces	1267
11.2.2	Charge-Transfer Reactions Are the Origin of the Instability of a Surface	1268
11.2.3	A Corroding Metal is Analogous to a Short-Circuited Energy-Producing Cell	1269
11.2.4	The Mechanism of the Corrosion of Ultrapure Metals	1273
11.2.5	What Is the Electronation Reaction in Corrosion?	1275
11.2.6	Thermodynamics and the Stability of Metals	1277
11.2.7	Potential–pH (or Pourbaix) Diagrams: Uses and Abuses	1281
11.2.8	The Corrosion Current and the Corrosion Potential	1285
11.2.9	The Basic Electrodics of Corrosion in the Absence of Oxide Films ...	1287
11.2.10	An Understanding of Corrosion in Terms of Evans Diagrams	1291
11.2.11	Which Step in the Corrosion Process Controls the Corrosion Current?	1296
11.2.12	Metals, pH, and Air	1297
11.2.13	Some Common Examples of Corrosion	1301
11.2.14	Electrodic Approaches to Increasing the Stability of Metals	1306
	11.2.14a Corrosion Inhibition by the Addition of Substances to the Electrolytic Environment of a Corroding Metal	1306
	11.2.14b Corrosion Prevention by Charging the Corroding Metal with Electrons from an External Source	1309
11.2.15	Passivation: The Transformation from a Corroding and Unstable Surface to a Passive and Stable Surface	1315
11.2.16	The Mechanism of Passivation	1318
11.2.17	The Dissolution–Precipitation Model for Film Formation	1321
11.2.18	Spontaneous Passivation: Nature's Method of Stabilizing Surfaces ...	1323
11.2.19	A Competition in Models for Passivation?	1324
11.2.20	The Thermodynamics of Passivation	1326
11.2.21	Hydrogen Diffusion into a Metal	1328
11.2.22	The Preferential Diffusion of Absorbed Hydrogen to Regions of Stress in a Metal	1330
11.2.23	Interstitial Hydrogen Can Crack Open a Metal Surface	1333
11.2.24	Surface Instability and the Internal Decay of Metals: Stress-Corrosion Cracking	1335
11.2.25	Surface Instability and Internal Decay of Metals: Hydrogen Embrittlement	1338
11.2.26	Charge Transfer and the Stability of Metals	1345
11.2.27	The Cost of Corrosion	1346
11.2.28	A Bird's-Eye View of Corrosion	1347
Further Reading		1349
11.3	Electrochemical Energy Conversion	1350
11.3.1	The Present Situation in Energy Consumption	1350
11.3.2	How Are the Hydrocarbon Fuels Used at Present?	1352
11.3.3	The Pollution of the Atmosphere with Products from Internal-Combustion Reactions and Its Possible Effect on World Temperature and Sea Levels	1353
	11.3.3a Products of Combustion Other than Carbon Dioxide	1353
	11.3.3b Carbon Dioxide	1355

	11.3.3c Uncertainties in Predicting the Future Pollution of the Atmosphere	1357
11.3.4	Thermal-Combustion Engines Waste the Chemical Energy Available from Burning Hydrocarbons in Air	1357
11.3.5	Direct Energy Conversion	1358
11.3.6	Direct Energy Conversion by Electrochemical Means	1361
11.3.7	The Maximum Intrinsic Efficiency in Electrochemical Conversion of the Energy of a Chemical Reaction to Electric Energy	1361
11.3.8	The Actual Efficiency of an Electrochemical Energy Converter	1366
11.3.9	The Physical Interpretation of the Absence of the Carnot Efficiency Factor in Electrochemical Energy Conversion	1366
11.3.10	Cold Combustion	1369
11.3.11	Making V near V_e Is the Central Problem of Electrochemical Energy Conversion	1369
11.3.12	The Electrochemical Quantities Which Must Be Optimized for Good Energy Conversion	1374
11.3.13	The Power Output of an Electrochemical Energy Converter	1376
11.3.14	The Electrochemical Engine	1378
11.3.15	Was the Wrong Path Taken in the Development of Power Sources at the End of the Nineteenth Century?	1379
11.3.16	Electrodes Burning Oxygen from Air	1382
11.3.17	The Special Configurations of Electrodes in Electrochemical Reactors	1382
11.3.18	Electrochemical Electricity Producers: The Two Basic Types	1385
11.3.19	Examples of Electrochemical Generators	1386
	11.3.19a The Hydrogen–Oxygen Cell	1388
	11.3.19b Reformer-Supplied Hydrogen–Air Cells	1389
	11.3.19c Hydrocarbon–Air Cells	1391
	11.3.19d Dissolved-Fuel Fuel Cells	1393
	11.3.19e Natural Gas and CO–Air Cells	1393
11.3.20	The Relations between Electrochemical Energy Conversion and the Future Dominance of Atomic Energy as the Source of Power	1395
	11.3.20a Will Atomic Power Sources Compete for Any of the Uses Foreseen for Electrochemical Power Sources?	1395
	11.3.20b Will Electrochemical Means Be Used to Convert Nuclear Power to Electricity?	1396
	11.3.20c What Is the Relation between Electricity Storage and Atomic Energy?	1397
11.3.21	A Summary of the Direct Conversion of Chemical Energy to Electricity	1398
Further Reading		1400
11.4	**Electricity Storage**	**1401**
11.4.1	Conventional and Descriptive Terminology in Energy Conversion and Storage	1403
11.4.2	The Important Quantities in Electricity Storage	1404
	11.4.2a Electricity Storage Density	1404
	11.4.2b Energy Density	1407
	11.4.2c Power	1410
	11.4.2d Desirable Trends	1412
11.4.3	Classical Electricity Storers	1413
	11.4.3a The Lead–Acid Storage Battery	1413
	11.4.3b A Dry Cell	1415
	11.4.3c Two Relatively New Electricity Storers	1416

11.4.4	The Large Gap between the Maximum Feasible and the Present Actual Energy Densities of Electricity Storers		1418
11.4.5	Outlines of Some Possible Future Electricity Storers		1420
	11.4.5*a*	Electricity Storage in Hydrogen	1420
	11.4.5*b*	Storage by Using Alkali Metals	1422
	11.4.5*c*	Storers Involving Nonaqueous Solutions	1424
	11.4.5*d*	Storers with Zinc in Combination with an Air Electrode	1427
11.4.6	The Respective Realms of Applicability of Electrochemical Energy Converters and Electricity Storers		1428
11.4.7	Electrochemical Electricity Storage in a Nutshell		1430
Further Reading			1432
Index			xxix

CONTENTS

VOLUME 1

CHAPTER 1

Electrochemistry

1.1	Introduction	1
1.2	Electrons at and across Interfaces	3
1.2.1	Many Properties of Materials Depend upon Events Occurring at Their Surfaces	3
1.2.2	Almost All Interfaces Are Electrified	3
1.2.3	The Continuous Flow of Electrons across an Interface: Electrochemical Reactions	7
1.2.4	Electrochemical and Chemical Reactions	8
1.3	Basic Electrochemistry	12
1.3.1	Electrochemistry before 1950	12
1.3.2	The Treatment of Interfacial Electron Transfer as a Rate Process: The 1950's	17
1.3.3	Quantum Electrochemistry: The 1960's	19
1.3.4	Ions in Solution, as well as Electron Transfer across Interfaces	22
1.4	The Relation of Electrochemistry to Other Sciences	26
1.4.1	Some Diagrammatic Presentations	26
1.4.2	Some Examples of the Involvement of Electrochemistry in Other Sciences	28
1.4.3	Electrochemistry as an Interdisciplinary Field, Apart from Chemistry?	29
1.5	Electrodics and Electronics	31
1.6	Transients	32

xviii   CONTENTS

1.7	Electrodes are Catalysts	34
1.8	The Electromagnetic Theory of Light and the Examination of Electrode Surfaces	35
1.9	Science, Technology, Electrochemistry, and Time	38
1.9.1	Do Interfacial Charge-Transfer Reactions Have a Wider Significance Than Has Hitherto Been Realized?	38
1.9.2	The Relation between Three Major Advances in Science, and the Place of Electrochemistry in the Developing World	39

CHAPTER 2

Ion–Solvent Interactions

2.1	Introduction	45
2.2	The Nonstructural Treatment of Ion–Solvent Interactions	48
2.2.1	A Quantitative Measure of Ion–Solvent Interactions	48
2.2.2	The Born Model: A Charged Sphere in a Continuum	49
2.2.3	The Electrostatic Potential at the Surface of a Charged Sphere	52
2.2.4	On the Electrostatics of Charging (or Discharging) Spheres	54
2.2.5	The Born Expression for the Free Energy of Ion–Solvent Interactions	56
2.2.6	The Enthalpy and Entropy of Ion–Solvent Interactions	59
2.2.7	Can One Experimentally Study the Interactions of a Single Ionic Species with the Solvent?	61
2.2.8	The Experimental Evaluation of the Heat of Interaction of a Salt and Solvent	64
2.2.9	How Good Is the Born Theory?	68
Further Reading		72
2.3	Structural Treatment of the Ion–Solvent Interactions	72
2.3.1	The Structure of the Most Common Solvent, Water	72
2.3.2	The Structure of Water near an Ion	76
2.3.3	The Ion–Dipole Model of Ion–Solvent Interactions	80
2.3.4	Evaluation of the Terms in the Ion–Dipole Approach to the Heat of Solvation	88
2.3.5	How Good Is the Ion–Dipole Theory of Solvation?	93
2.3.6	The Relative Heats of Solvation of Ions on the Hydrogen Scale	95
2.3.7	Do Oppositely Charged Ions of Equal Radii Have Equal Heats of Solvation?	96
2.3.8	The Water Molecule Can Be Viewed as an Electrical Quadrupole	98
2.3.9	The Ion–Quadrupole Model of Ion–Solvent Interactions	99
2.3.10	Ion-Induced-Dipole Interactions in the Primary Solvation Sheath	102
2.3.11	How Good Is the Ion–Quadrupole Theory of Solvation?	103
2.3.12	The Special Case of Interactions of the Transition-Metal Ions with Water	108
2.3.13	Some Summarizing Remarks on the Energetics of Ion–Solvent Interactions	113
Further Reading		116
2.4	The Solvation Number	117
2.4.1	How Many Water Molecules Are Involved in the Solvation of an Ion?	117

2.4.2	Static and Dynamic Pictures of the Ion–Solvent Molecule Interaction	120
2.4.3	The Meaning of Hydration Numbers	123
2.4.4	Why Is the Concept of Solvation Numbers Useful?	124
2.4.5	On the Determination of Solvation Numbers	125
Further Reading		132
2.5	The Dielectric Constant of Water and Ionic Solutions	132
2.5.1	An Externally Applied Electric Field Is Opposed by Counterfields Developed within the Medium	132
2.5.2	The Relation between the Dielectric Constant and Internal Counterfields	136
2.5.3	The Average Dipole Moment of a Gas-Phase Dipole Subject to Electrical and Thermal Forces	139
2.5.4	The Debye Equation for the Dielectric Constant of a Gas of Dipoles	142
2.5.5	How the Short-Range Interactions between Dipoles Affect the Average Effective Moment of the Polar Entity Which Responds to an External Field	145
2.5.6	The Local Electric Field in a Condensed Polar Dielectric	147
2.5.7	The Dielectric Constant of Liquids Containing Associated Dipoles	152
2.5.8	The Influence of Ionic Solvation on the Dielectric Constant of Solutions	155
Further Reading		158
2.6	Ion–Solvent–Nonelectrolyte Interactions	158
2.6.1	The Problem	158
2.6.2	The Change in Solubility of a Nonelectrolyte Due to Primary Solvation	159
2.6.3	The Change in Solubility Due to Secondary Solvation	160
2.6.4	The Net Effect on Solubility of Influences from Primary and Secondary Solvation	163
2.6.5	The Case of Anomalous Salting in	164
Further Reading		168
Appendix 2.1	Free Energy Change and Work	168
Appendix 2.2	The Interaction between an Ion and a Dipole	169
Appendix 2.3	The Interaction between an Ion and a Water Quadrupole	171

CHAPTER 3

Ion–Ion Interactions

3.1	Introduction	175
3.2	True and Potential Electrolytes	176
3.2.1	Ionic Crystals Are True Electrolytes	176
3.2.2	Potential Electrolytes: Nonionic Substances Which React with the Solvent to Yield Ions	176
3.2.3	An Obsolete Classification: Strong and Weak Electrolytes	177
3.2.4	The Nature of the Electrolyte and the Relevance of Ion–Ion Interactions	180
Further Reading		180
3.3	The Debye–Hückel (or Ion-Cloud) Theory of Ion–Ion Interactions	180
3.3.1	A Strategy for a Quantitative Understanding of Ion–Ion Interactions	180

3.3.2	A Prelude to the Ionic-Cloud Theory	183
3.3.3	How the Charge Density near the Central Ion Is Determined by Electrostatics: Poisson's Equation	186
3.3.4	How the Excess Charge Density near the Central Ion Is Given by a Classical Law for the Distribution of Point Charges in a Coulombic Field	187
3.3.5	A Vital Step in the Debye–Hückel Theory of the Charge Distribution around Ions: Linearization of the Boltzmann Equation	189
3.3.6	The Linearized Poisson–Boltzmann Equation	190
3.3.7	The Solution of the Linearized P–B Equation	191
3.3.8	The Ionic Cloud around a Central Ion	193
3.3.9	How Much Does the Ionic Cloud Contribute to the Electrostatic Potential ψ_r at a Distance r from the Central Ion?	199
3.3.10	The Ionic Cloud and the Chemical-Potential Change Arising from Ion–Ion Interactions	201
Further Reading		202

3.4 Activity Coefficients and Ion–Ion Interactions 202

3.4.1	The Evolution of the Concept of Activity Coefficient	202
3.4.2	The Physical Significance of Activity Coefficients	204
3.4.3	The Activity Coefficient of a Single Ionic Species Cannot Be Measured	206
3.4.4	The Mean Ionic Activity Coefficient	207
3.4.5	The Conversion of Theoretical Activity-Coefficient Expressions into a Testable Form	209
Further Reading		212

3.5 The Triumphs and Limitations of the Debye–Hückel Theory of Activity Coefficients 212

3.5.1	How Well Does the Debye–Hückel Theoretical Expression for Activity Coefficients Predict Experimental Values?	212
3.5.2	Ions Are of Finite Size, Not Point Charges	219
3.5.3	The Theoretical Mean Ionic-Activity Coefficient in the Case of Ionic Clouds with Finite-Sized Ions	222
3.5.4	The Ion-Size Parameter a	224
3.5.5	Comparison of the Finite-Ion-Size Model with Experiment	227
3.5.6	The Debye–Hückel Theory of Ionic Solutions: An Assessment	230
3.5.7	On the Parentage of the Theory of Ion–Ion Interactions	237
Further Reading		238

3.6 Ion–Solvent Interactions and the Activity Coefficient 238

3.6.1	The Effect of Water Bound to Ions on the Theory of Deviations from Ideality	238
3.6.2	Quantitative Theory of the Activity of an Electrolyte as a Function of the Hydration Number	240
3.6.3	The Water-Removal Theory of Activity Coefficients and Its Apparent Consistency with Experiment at High Electrolytic Concentrations	243
Further Reading		246

3.7 The So-Called "Rigorous" Solutions of the Poisson–Boltzmann Equation 246

Further Reading 250

3.8 Temporary Ion Association in an Electrolytic Solution: Formation of Pairs, Triplets, etc. 251

3.8.1	Positive and Negative Ions Can Stick Together: Ion-Pair Formation	251
3.8.2	The Probability of Finding Oppositely Charged Ions near Each Other	251
3.8.3	The Fraction of Ion Pairs, According to Bjerrum	253
3.8.4	The Ion-Association Constant K_A of Bjerrum	257
3.8.5	Activity Coefficients, Bjerrum's Ion Pairs, and Debye's Free Ions	260
3.8.6	The Fuoss Approach to Ion-Pair Formation	261
3.8.7	From Ion Pairs to Triple Ions to Clusters of Ions	265
Further Reading		266
3.9	The Quasi-Lattice Approach to Concentrated Electrolytic Solutions	267
3.9.1	At What Concentration Does the Ionic-Cloud Model Break Down?	267
3.9.2	The Case for a Cube-Root Law for the Dependence of the Activity Coefficient on Electrolyte Concentration	269
3.9.3	The Beginnings of a Quasi-Lattice Theory for Concentrated Electrolytic Solutions	271
Further Reading		272
3.10	The Study of the Constitution of Electrolytic Solutions	273
3.10.1	The Temporary and Permanent Association of Ions	273
3.10.2	Electromagnetic Radiation, a Tool for the Study of Electrolytic Solutions	274
3.10.3	Visible and Ultraviolet Absorption Spectroscopy	275
3.10.4	Raman Spectroscopy	276
3.10.5	Infrared Spectroscopy	278
3.10.6	Nuclear Magnetic Resonance Spectroscopy	278
Further Reading		279
3.11	A Perspective View on the Theory of Ion–Ion Interactions	279
Appendix 3.1	Poisson's Equation for Spherically Symmetrical Charge Distribution	282
Appendix 3.2	Evaluation of the Integral $\int_{r=0}^{r\to\infty} e^{-(\varkappa r)}(\varkappa r)\,d(\varkappa r)$	283
Appendix 3.3	Derivation of the Result $f_\pm = (f_+^{\nu_+} + f_-^{\nu_-})^{1/\nu}$	284
Appendix 3.4	To Show That the Minimum in the P_r versus r Curve Occurs at $r = \lambda/2$	284
Appendix 3.5	Transformation from the Variable r to the Variable $y = \lambda/r$	285
Appendix 3.6	Relation Between Calculated and Observed Activity Coefficients	285

CHAPTER 4

Ion Transport in Solutions

4.1	Introduction	287
4.2	Ionic Drift under a Chemical-Potential Gradient: Diffusion	289
4.2.1	The Driving Force for Diffusion	291
4.2.2	The "Deduction" of an Empirical Law: Fick's First Law of Steady-State Diffusion	293

4.2.3	On the Diffusion Coefficient D	296
4.2.4	Ionic Movements: A Case of the Random Walk	299
4.2.5	The Mean Square Distance Traveled in a Time t by a Random-Walking Particle	301
4.2.6	Random-Walking Ions and Diffusion: The Einstein–Smoluchowski Equation	304
4.2.7	The Gross View of Non-Steady-State Diffusion	307
4.2.8	An Often Used Device for Solving Electrochemical Diffusion Problems: The Laplace Transformation	309
4.2.9	Laplace Transformation Converts the Partial Differential Equation Which Is Fick's Second Law into a Total Differential Equation	312
4.2.10	The Initial and Boundary Conditions for the Diffusion Process Stimulated by a Constant Current (or Flux)	313
4.2.11	The Concentration Response to a Constant Flux Switched on at $t=0$	317
4.2.12	How the Solution of the Constant-Flux Diffusion Problem Leads On to the Solution of Other Problems	323
4.2.13	Diffusion Resulting from an Instantaneous Current Pulse	328
4.2.14	What Fraction of Ions Travels the Mean Square Distance $\langle x^2 \rangle$ in the Einstein–Smoluchowski Equation?	332
4.2.15	How Can the Diffusion Coefficient Be Related to Molecular Quantities?	338
4.2.16	The Mean Jump Distance l, a Structural Question	339
4.2.17	The Jump Frequency, a Rate-Process Question	340
4.2.18	The Rate-Process Expression for the Diffusion Coefficient	342
4.2.19	Diffusion: An Overall View	342
Further Reading		345
4.3	**Ionic Drift under an Electric Field: Conduction**	**345**
4.3.1	The Creation of an Electric Field in an Electrolyte	345
4.3.2	How Do Ions Respond to the Electric Field?	349
4.3.3	The Tendency for a Conflict between Electroneutrality and Conduction	351
4.3.4	The Resolution of the Electroneutrality-versus-Conduction Dilemma: Electron-Transfer Reactions	351
4.3.5	The Quantitative Link between Electron Flow in the Electrodes and Ion Flow in the Electrolyte: Faraday's Law	353
4.3.6	The Proportionality Constant Relating the Electric Field and the Current Density: The Specific Conductivity	354
4.3.7	Molar Conductivity and Equivalent Conductivity	357
4.3.8	The Equivalent Conductivity Varies with Concentration	360
4.3.9	How the Equivalent Conductivity Changes with Concentration: Kohlrausch's Law	363
4.3.10	The Vectorial Character of Current: Kohlrausch's Law of the Independent Migration of Ions	364
Further Reading		367
4.4	**The Simple Atomistic Picture of Ionic Migration**	**367**
4.4.1	Ionic Movements under the Influence of an Applied Electric Field	367
4.4.2	What Is the Average Value of the Drift Velocity?	368
4.4.3	The Mobility of Ions	369
4.4.4	The Current Density Associated with the Directed Movement of Ions in Solution, in Terms of the Ionic Drift Velocities	371
4.4.5	The Specific and Equivalent Conductivities in Terms of the Ionic Mobilities	373
4.4.6	The Einstein Relation between the Absolute Mobility and the Diffusion Coefficient	374

4.4.7	What Is the Drag (or Viscous) Force Acting on an Ion in Solution?	377
4.4.8	The Stokes–Einstein Relation	379
4.4.9	The Nernst–Einstein Equation	381
4.4.10	Some Limitations of the Nernst–Einstein Relation	382
4.4.11	A Very Approximate Relation between Equivalent Conductivity and Viscosity: Walden's Rule	385
4.4.12	The Rate-Process Approach to Ionic Migration	387
4.4.13	The Rate-Process Expression for Equivalent Conductivity	391
4.4.14	The Total Driving Force for Ionic Transport: The Gradient of the Electrochemical Potential	394
Further Reading		399
4.5	**The Interdependence of Ionic Drifts**	**399**
4.5.1	The Drift of One Ionic Species May Influence the Drift of Another	399
4.5.2	A Consequence of the Unequal Mobilities of Cations and Anions, the Transport Numbers	400
4.5.3	The Significance of a Transport Number of Zero	402
4.5.4	The Diffusion Potential, Another Consequence of the Unequal Mobilities of Ions	406
4.5.5	Electroneutrality Coupling between the Drifts of Different Ionic Species	410
4.5.6	How Does One Represent the Interaction between Ionic Fluxes? The Onsager Phenomenological Equations	411
4.5.7	An Expression for the Diffusion Potential	413
4.5.8	The Integration of the Differential Equation for Diffusion Potentials: The Planck–Henderson Equation	417
Further Reading		420
4.6	**The Influence of Ionic Atmospheres on Ionic Migration**	**420**
4.6.1	The Concentration Dependence of the Mobility of Ions	420
4.6.2	Ionic Clouds Attempt to Catch Up with Moving Ions	422
4.6.3	An Egg-Shaped Ionic Cloud and the "Portable" Field on the Central Ion	423
4.6.4	A Second Braking Effect of the Ionic Cloud on the Central Ion: The Electrophoretic Effect	424
4.6.5	The Net Drift Velocity of an Ion Interacting with Its Atmosphere	425
4.6.6	The Electrophoretic Component of the Drift Velocity	427
4.6.7	The Procedure for Calculating the Relaxation Component of the Drift Velocity	427
4.6.8	How Long Does an Ion Atmosphere Take to Decay?	428
4.6.9	The Quantitative Measure of the Asymmetry of the Ionic Cloud Around a Moving Ion	429
4.6.10	The Magnitude of the Relaxation Force and the Relaxation Component of the Drift Velocity	430
4.6.11	The Net Drift Velocity and Mobility of an Ion Subject to Ion–Ion Interactions	432
4.6.12	The Debye–Hückel–Onsager Equation	434
4.6.13	The Theoretical Predictions of the Debye–Hückel–Onsager Equation versus the Observed Conductance Curves	435
4.6.14	A Theoretical Basis for Some Modifications of the Debye–Hückel–Onsager Equation	438
Further Reading		439
4.7	**Nonaqueous Solutions: A New Frontier in Ionics?**	**440**
4.7.1	Water Is the Most Plentiful Solvent	440

xxiv CONTENTS

4.7.2	Water Is Often Not an Ideal Solvent	441
4.7.3	The Debye–Hückel–Onsager Theory for Nonaqueous Solutions	442
4.7.4	The Solvent Effect on the Mobility at Infinite Dilution	443
4.7.5	The Slope of the Λ versus $c^{\frac{1}{2}}$ Curve as a Function of the Solvent	445
4.7.6	The Effect of the Solvent on the Concentration of Free Ions: Ion Association	447
4.7.7	The Effect of Ion Association upon Conductivity	448
4.7.8	Even Triple Ions Can Be Formed in Nonaqueous Solutions	450
4.7.9	Some Conclusions about the Conductance of Nonaqueous Solutions of True Electrolytes	452
Further Reading		452
Appendix 4.1	The Mean Square Distance Traveled by a Random-Walking Particle	453
Appendix 4.2	The Laplace Transform of a Constant	454
Appendix 4.3	A Few Elementary Ideas on the Theory of Rate Processes	455
Appendix 4.4	The Derivation of Equations (4.257) and (4.258)	458
Appendix 4.5	The Derivation of Equation (4.318)	460

CHAPTER 5

Protons in Solution

5.1	The Case of the Nonconforming Ion: The Proton	461
5.2	Proton Solvation	462
5.2.1	What Is the Condition of the Proton in Solution?	462
5.2.2	Proton Affinity	466
5.2.3	The Overall Heat of Hydration of a Proton	467
5.2.4	The Coordination Number of a Proton	468
Further Reading		470
5.3	Proton Transport	470
5.3.1	The Abnormal Mobility of a Proton	470
5.3.2	Protons Conduct by a Chain Mechanism	474
5.3.3	Classical Proton Jumps and Proton Mobility	476
5.3.4	Do Proton Jumps Obey Classical Laws?	478
5.3.5	Quantum-Mechanical Proton Jumps and Proton Mobility	480
5.3.6	Water Reorientation, a Prerequisite for Proton Jumps	481
5.3.7	The Rate of Water Reorientation and Proton Mobility	482
5.3.8	A Picture of Proton Mobility in Aqueous Solutions	484
5.3.9	The Rate-Determining Water-Rotation Model of Proton Mobility and the Other Anomalous Facts	485
5.3.10	Proton Mobility in Ice	486
5.3.11	The Existence of the Hydronium Ion from the Point of View of Proton Mobility	487
5.3.12	Why Is the Mechanism of Proton Mobility So Important?	487
Further Reading		488

5.4		Homogeneous Proton-Transfer Reactions and Potential Electrolytes	488
5.4.1		Acids Produce Hydrogen Ions and Bases Produce Hydroxyl Ions: The Initial View	488
5.4.2		Acids Are Proton Donors, and Bases Are Proton Acceptors: The Brönsted View	489
5.4.3		The Dissolution of Potential Electrolytes and Other Types of Proton-Transfer Reactions	491
5.4.4		An Important Consequence of the Brönsted View: Conjugate Acid–Base Pairs	493
5.4.5		The Absolute Strength of an Acid or a Base	494
5.4.6		The Relative Strengths of Acids and Bases	495
5.4.7		Proton Free-Energy Levels	500
5.4.8		The Primary Effect of the Solvent upon the Relative Strength of an Acid	504
5.4.9		A Secondary (Electrostatic) Effect of the Solvent on the Relative Strength of Acids	507
Further Reading			511

CHAPTER 6

Ionic Liquids

6.1		Introduction	513
6.1.1		The Limiting Case of Zero Solvent: Pure Liquid Electrolytes	513
6.1.2		The Thermal Dismantling of an Ionic Lattice	514
6.1.3		Some Features of Ionic Liquids (Pure Liquid Electrolytes)	515
6.1.4		Liquid Electrolytes Are Ionic Liquids	517
6.1.5		The Fundamental Problems in Pure Liquid Electrolytes	518
Further Reading			522
6.2		Models of Simple Ionic Liquids	522
6.2.1		The Origin of Liquid Electrolyte Models	522
6.2.2		Lattice-Oriented Models	523
	6.2.2a	The Experimental Basis for Model Building	523
	6.2.2b	The Need to Pour Empty Space into a Fused Salt	523
	6.2.2c	The Vacancy Model: A Fused Salt Is an Ionic Lattice with Numerous Vacancies	526
	6.2.2d	The Hole Model: A Fused Salt Is Full of Holes like Swiss Cheese	527
6.2.3		Gas-Oriented Models for Liquid Electrolytes	529
	6.2.3a	The Cell-Theory Approach	529
	6.2.3b	The Free Volume Belongs to the Liquid and Not to the Particles: The Liquid Free-Volume Model	530
6.2.4		A Summary of the Models for Liquid Electrolytes	532
Further Reading			533
6.3		Quantification of the Hole Model for Liquid Electrolytes	533
6.3.1		An Expression for the Probability That a Hole Has a Radius between r and $r + dr$	533
6.3.2		The Fürth Approach to the Work of Hole Formation	536
6.3.3		The Distribution Function for the Size of the Holes in a Liquid Electrolyte	537

xxvi CONTENTS

6.3.4	What Is the Average Size of a Hole?	539
Further Reading		541
6.4	**Transport Phenomena in Liquid Electrolytes**	**541**
6.4.1	Some Simplifying Features of Transport in Fused Salts	541
6.4.2	Diffusion in Fused Salts	542
	6.4.2a Self-Diffusion in Pure Liquid Electrolytes: It May Be Revealed by Introducing Isotopes	542
	6.4.2b Results of Self-Diffusion Experiments	544
6.4.3	The Viscosity of Molten Salts	547
6.4.4	What Is the Validity of the Stokes–Einstein Relation in Ionic Liquids?	550
6.4.5	The Conductivity of Pure Liquid Electrolytes	553
6.4.6	The Nernst–Einstein Relation in Ionic Liquids	555
	6.4.6a The Nernst–Einstein Relation: Its Degree of Applicability	555
	6.4.6b The Gross View of Deviations from the Nernst–Einstein Equation	557
	6.4.6c Possible Molecular Mechanisms for Nernst–Einstein Deviations	560
6.4.7	Transport Numbers in Pure Liquid Electrolytes	564
	6.4.7a Some Ideas about Transport Numbers in Fused Salts	564
	6.4.7b The Measurement of Transport Numbers in Liquid Electrolytes	566
	6.4.7c A Radiotracer Method of Calculating Transport Numbers in Molten Salts	571
	6.4.7d A Stokes' Law Approach to a Rough Estimate of Transport Numbers	572
Further Reading		573
6.5	**The Atomistic View of Transport Processes in Simple Ionic Liquids**	**574**
6.5.1	Holes and Transport Processes	574
6.5.2	What Is the Mean Lifetime of Holes in Fused Salts?	576
6.5.3	Expression for Viscosity in Terms of Holes	577
6.5.4	The Diffusion Coefficient from the Hole Model	577
6.5.5	A Critical Test of a Model for Ionic Liquids Is a Rationalization of the Heat of Activation of $3.7RT_m$ for Transport Processes	580
6.5.6	An Attempt to Rationalize $E_D = E_\eta = 3.7RT_m$	581
6.5.7	The Hole Model, the Most Consistent Present Model for Liquid Electrolytes	584
Further Reading		587
6.6	**Mixture of Simple Ionic Liquids—Complex Formation**	**587**
6.6.1	Mixtures of Simple Ionic Liquids May Not Behave Ideally	587
6.6.2	Interactions Lead to Nonideal Behavior	588
6.6.3	Can One Meaningfully Refer to Complex Ions in Fused Salts?	589
6.6.4	Raman Spectra, and Other Means of Detecting Complex Ions	590
Further Reading		593
6.7	**Mixtures of Liquid Oxide Electrolytes**	**594**
6.7.1	The Liquid Oxides	594
6.7.2	Pure Fused Nonmetallic Oxides Form Network Structures Like Liquid Water	594
6.7.3	Why Does Fused Silica Have a Much Higher Viscosity Than Do Liquid Water and the Fused Salts?	597
6.7.4	The Solvent Properties of Fused Nonmetallic Oxides	601

6.7.5	Ionic Additions to the Liquid-Silica Network: Glasses	603
6.7.6	The Extent of Structure Breaking of Three-Dimensional Network Lattices and Its Dependence on the Concentration of Metal Ions	604
6.7.7	The Molecular and Network Models of Liquid Silicate Structure	606
6.7.8	Liquid Silicates Contain Large Discrete Polyanions	610
6.7.9	The "Iceberg" Model	615
6.7.10	Fused-Oxide Systems in Metallurgy: Slags	616
Further Reading		618
Appendix 6.1	The Effective Mass of a Hole	619
Appendix 6.2	Some Properties of the Gamma Function	620
Appendix 6.3	The Kinetic Theory Expression for the Viscosity of a Fluid	621
Index		xxxiii

VOLUME 2

MODERN ELECTROCHEMISTRY

CHAPTER 7
THE ELECTRIFIED INTERFACE

7.1. ELECTRIFICATION OF AN INTERFACE

7.1.1. The Electrode–Electrolyte Interface: The Basis of Electrodics

The situation inside an electrolyte—the *ionic* aspect of electrochemistry—has been considered in the first volume. The basic phenomena involve ion–solvent interactions (Chapter 2), ion–ion interactions (Chapter 3), and the random walk of ions, which becomes a drift in a preferred direction under the influence of a concentration or a potential gradient (Chapter 4). In what way is the situation at the electrode–electrolyte interface any different from that in the bulk of the electrolyte? To answer this question, one must treat quiescent (equilibrium) and active (nonequilibrium) interfaces, the structural and electrical characteristics of the interface, the rates and mechanism of change-over from ionic to electronic conduction, etc. In short, one is led into *electrodics*, the newest and most exciting part of electrochemistry.

7.1.2. New Forces at the Boundary of an Electrolyte

It has been stressed that, so long as no irreversible transport processes occur, every particle (ion or solvent molecule) in the bulk of the electrolyte looks out upon a spherically symmetrical world. On a time average, the ions and water molecules (in aqueous solutions) experience forces which are independent of direction and independent of position in the electrolyte.

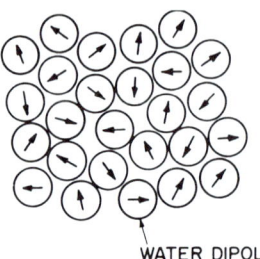

Fig. 7.1. A schematic representation of the random orientation of water dipoles in the interior of the electrolyte (the network structure of water is ignored in the diagram).

Thus, if each water dipole is represented by a vector, the vectors are completely randomized in direction[†] (Fig. 7.1). There is no *net* resultant vector, i.e., there is no alignment of the solvent dipoles in any preferred direction.

Further, the positive and negative ions are equally distributed[‡] in any given volume of electrolyte. Electroneutrality must prevail. Consider any lamina of electrolyte parallel to a planar electrode (Fig. 7.2). As long as the lamina is in the bulk of the electrolyte, the net charge on the lamina will be zero. Since the charges on any two parallel laminae are equal to zero, there will be no potential gradient inside the electrolyte *under equilibrium conditions*.

To summarize: Under equilibrium conditions, the time-average forces are the same in all directions and at all points in the bulk of the electrolyte (perfect isotropy and homogeneity), and there are no net preferentially directed electrical fields.

Every electrolyte, however, is bounded. It must ultimately contact some other material, e.g., the gas phase above the electrolyte or the metallic electrode or, for that matter, the walls of the container. The frontier is reached. What happens at such a *phase boundary*?

It will be shown further on that the phases on either side of the boundary become charged to an equal and opposite extent and this gives rise to a potential difference across the boundary. There are several ways in which this potential difference can arise. If one of the phases is an electronic conductor and the other is an ionic conductor, electron-transfer reactions

---

[†] In fact, the structure of the solvent water is a little more complicated (*cf.* Chapter 2).
[‡] One is talking here of volumes which are large compared with the dimension $\varkappa^{-1}$ of the ionic cloud of Chapter 3.

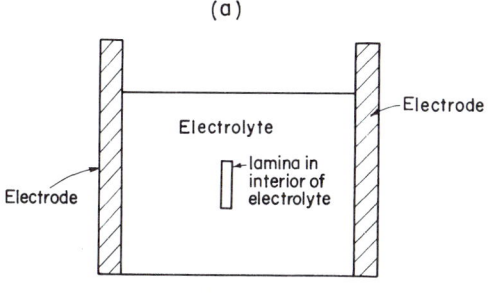

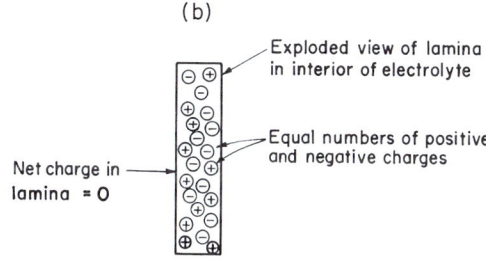

Fig. 7.2. A schematic representation of electroneutrality in the interior of an electrolyte. (a) A lamina in the bulk of the electrolyte. (b) An exploded view of the lamina showing that it contains an equal amount of positive and negative charge and therefore has a zero net charge.

can occur at the boundary and lead to the development of a potential difference. A discussion of this type of mechanism will be reserved for Chapter 8. Or, the electronic conductor can be deliberately charged by a flow of electrons from an external source of electricity. The electrolyte side of the boundary then responds with an equal and opposite charge, and, therefore, a potential difference develops across the boundary. However, even without an external connection or the occurrence of electron-transfer reactions, it is possible for a potential difference to develop across a phase boundary. How this comes about will now be described.

The electrolyte is terminated at the phase boundary by the presence of an alien material. One would expect, therefore, that the characteristics of the electrolyte, i.e., its properties, are also physically interrupted at the frontier. Now, the essential characteristics of the bulk of the electrolyte are homogeneity and isotropy. Are these uniform properties perturbed by the presence of the phase boundary?

Consider an ion near enough† to the electrode to feel its influence. (Particles sense each other through the forces they exert on each other.) This ion sees its world as quite different from that of an ion in the bulk of the electrolyte. Things are not the same in all directions. When it looks toward the bulk of the electrolyte, it feels electrolyte forces, but, when it looks across the frontier (the phase boundary), it feels new forces that it never experienced as long as it was content to stay deep inside the homogeneous electrolyte.

The forces operating on particles near the phase boundary are therefore *anisotropic*. They are different in a direction toward the boundary compared with the direction toward the electrolyte bulk. Further, the forces due to the phase (e.g., the electrode) on the other side of the phase boundary (e.g., the electrode–electrolyte boundary) should vary with distance from the boundary; the deeper the ion recedes into the bulk, the less is the frontier influence felt and the more do things become normal again.

7.1.3. The Interphase Region Has New Properties and New Structures

Now, the properties of any material are dependent on the particles present and the forces operating on the particles. Since these forces are different at the frontier compared with the forces in the bulk, the properties of the frontier region, the *interphase* region, will differ from the bulk properties. Thus, the uniform properties of the electrolyte are perturbed in the interphase region by the presence of another phase.

The arrangement of particles, however, depends on the forces operating on them. Since new forces exist near the phase boundary, new structures would tend to exist. The arrangement of particles in the interphase region is a compromise between the structures demanded by both phases. Thus, the electrode, e.g., would like the ions and water molecules of an electrolytic solution to assume a certain time-average arrangement. The solution, on the other hand, demands another arrangement. The ions and other particles, caught between contradictory demands, adopt compromise positions which are characteristic of the interphase region.

7.1.4. An Electrode Is Like a Giant Central Ion

An analogy between the situation just described and those involved in ion–solvent and ion–ion interactions can be drawn. The solvent water,

---

† The phrase *near enough* will be quantified later.

e.g., normally has a particular structure, the water network. Near an ion, however, the water dipoles are under the conflicting influences of the water network and the charged central ion. They adopt compromise positions which correspond to primary and secondary solvation (Chapter 2). Similarly, in an electrolytic solution, the presence of the central ion makes the surrounding ions redistribute themselves—an ionic cloud is formed (*cf.* Chapter 3).

Just as the central ion can perturb and cause a rearrangement of the surrounding solvent molecules and ions, the electrode itself can cause the surrounding particles to assume abnormal, compromise positions (relative to the bulk of the electrolyte). It will be seen later that an electrode also can get enveloped by a solvent sheath and an ionic cloud. There are, however, many other interesting phenomena arising from the fact that one can connect an external potential source (e.g., a battery) to the electrode by a metallic wire and thus control the electrode charge. New possibilities emerge which do not exist in the case of the central ion.

7.1.5. The Consequences of Compromise Arrangements: The Electrolyte Side of the Boundary Acquires a Charge

The new forces operating at the electrode–electrolyte interface give rise to new arrangements of solvent dipoles and charged species. At the same time, they incite the particles of the electrolyte to be governed no

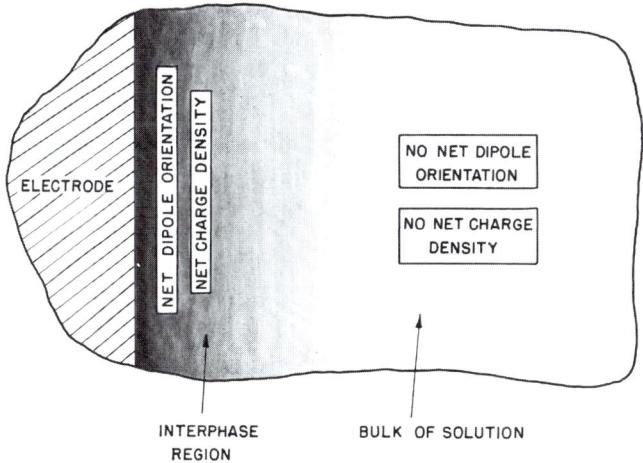

Fig. 7.3. A schematic diagram to illustrate that, in the interphase region (indicated by shading), there generally is net dipole orientation and net, or excess, charge density.

longer by the characteristics of the situation inside the electrolyte, i.e., random orientation of dipoles and equal distribution of positive and negative charges in any macroscopic lamina of the electrolyte. These two laws are not applicable in the interphase region (Fig. 7.3). Thus, there can be (and generally is) a net orientation of the solvent dipoles and a net or excess charge on a lamina parallel to the planar electrode surface (because of unequal numbers of positive and negative charges present there).

All this means that electroneutrality has broken down on the electrolyte side of the phase boundary. The electrolyte side of the frontier has become charged or electrified (Fig. 7.4).

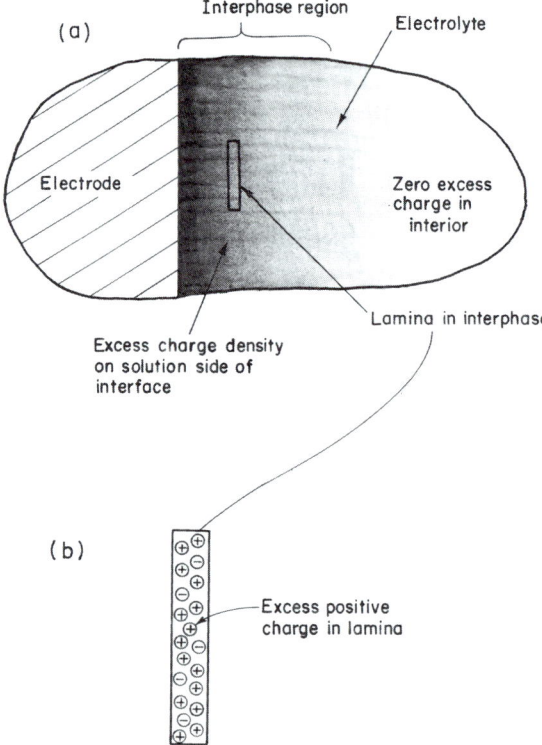

Fig. 7.4. A schematic representation of the charging of the solution side of an interface. (a) Shading indicates excess-charge density in the interphase region. (b) An exploded view shows that the positive charge in a lamina in the interphase exceeds the negative charge and there is a net, or excess, positive-charge density in the lamina.

How does this electrification affect the phase (e.g., the electrode) on the other side of the phase boundary[†]?

7.1.6. Both Sides of the Interface Become Electrified: The So-Called "Electrical Double Layer"

Once the electrolyte side of the phase boundary acquires a net, or excess, charge, an electric force, or field, operates across the boundary. All charged particles feel this field. But the other phase (e.g., the electrode) consists of charged particles. Hence, the charges of the second phase respond to the stimulus of the field arising from the charging of the electrolyte side of the boundary. The nature of their response depends on whether the nonelectrolyte phase is a conductor, a semiconductor, or an insulator. But, in any case, there *is* a response.

Consider that the other phase is a metallic conductor, i.e., an electrode. It consists of a three-dimensional, periodic network of positive ions and a communal pool of mobile electrons. The positive ions of the metallic lattice feel the field that is due to the excess charge at the boundary of the electrolyte, but they can move only with great difficulty.[‡] In contrast to these clumsy cumbersome ionic movements in a metal, the free electrons move with agility in response to the field produced by the charging of the electrolyte side of the frontier. The electrons move either toward or away from the boundary, depending on the direction of the field.

Thus, a charge is induced on the metal. This induced charge is equal and opposite to that on the electrolyte side of the phase boundary (Fig. 7.5).

What has happened as a result of this induced charge? Separation of charge has occurred across the electrode–electrolyte interface, a net charge of one sign on the electrode side of the interface and a net charge of another sign on the electrolyte side. Note, however, that the interphase region as a *whole* (not any one side, but the two sides taken together) is electrically neutral.

When charges are separated, a potential difference develops across the interface.

The electrical forces which operate between the metal and the solution

---

[†] One is discussing here a situation in which the electrode is *not* deliberately charged by connecting it up to an external source of electricity and in which electron-transfer reactions do *not* occur at the interface.

[‡] Recall the treatment (Chapter 6) of hole formation and jumping in liquid electrolytes; the energies involved in solids are at least an order of magnitude greater than those in liquids.

630 CHAPTER 7

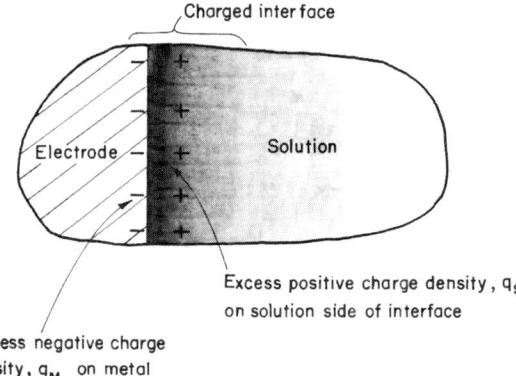

Fig. 7.5. The electrified interface. The excess-charge density q_S on the solution side of the interface is equal and opposite to that on the metal q_M.

constitute the electrical field across the electrode–electrolyte phase boundary. It will be seen that, though the potential differences across the interface are not large ($\sim$1 V), the dimensions of the interphase region are very small ($\sim$10 Å) and, thus, the field strength (gradient of potential) is enormous—it is of the order of 10^7 V cm^{-1}. *The effect of this enormous field at the electrode–electrolyte interface is, in a sense, the essence of electrochemistry.*

The term *electrical double layer*, or just *double layer*, is used to describe the arrangement of charges and oriented dipoles constituting the interphase region at the boundary of an electrolyte. The terms are a legacy from an early stage in understanding, when the interphase was pictured as always consisting of only two layers, or sheets, of charge,[†] one positive and the other negative. It is now known that the situation is more complex. Nevertheless, the term double layer is still used, not in a literal sense, but loosely, as a near-synonym for *electrified interface*.

7.1.7. Double Layers Are Characteristic of All Phase Boundaries

The argument for the formation of the double layer has proceeded simply. The existence of a boundary for the electrolyte necessarily implies a basic anisotropy in the forces operating on the particles in the interphase region. Owing to this anisotrpy, there occurs a redistribution of the mobile charges and orientable dipoles (compared with their distribution in the

---

[†] It will be shown later that, under some circumstances, the electrified interface is indeed a double layer, and the term *double layer* is literally justified.

bulk of the phases). This redistribution is the structural basis of the potential difference across the interface.

The argument is so general that its particularization for the metal–electrolyte interface was only for convenience. One could have carried out the discussion with equal validity for the gas–electrolyte or the glass (container)–electrolyte boundary of the electrolyte. Of course, one would have had to note the difference between the particles that constitute gases and glass and those that compose a metal. In all these systems, the conclusion would be reached that forces are direction dependent at the phase boundary and, therefore, new and compromise arrangements are assumed by the particles (of the two phases) in the phase boundary. If the particles are charged or are dipoles, not only is there a redistribution of particles but also an electrification of the interface and the development of a potential difference across it.

Double layers, therefore, are *not* a special feature of the electrode–electrolyte interfaces; they are a *general* consequence of the meeting of two phases at a boundary. Across almost any junction between two phases, i.e., between two materials, a potential difference will develop. If the materials contain mobile free charges (electrons or ions), the potential difference arises from the electrification of the two sides of the boundary by the mechanism described above (*cf.* Section 7.1.6), by the occurrence of charge-transfer reactions, or by connecting up the electronically conducting phase to an external source of electricity and charging it. Even if the materials consist not of free charges but of permanent dipoles or of molecules in which dipoles can be induced, a potential difference across the boundary can arise from a net orientation of the dipoles constituting it.

Some examples of double layers are shown in Fig. 7.6.

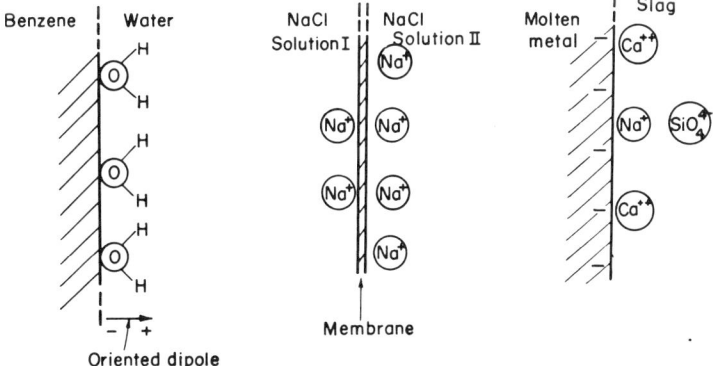

Fig. 7.6. Some examples of electrified interfaces.

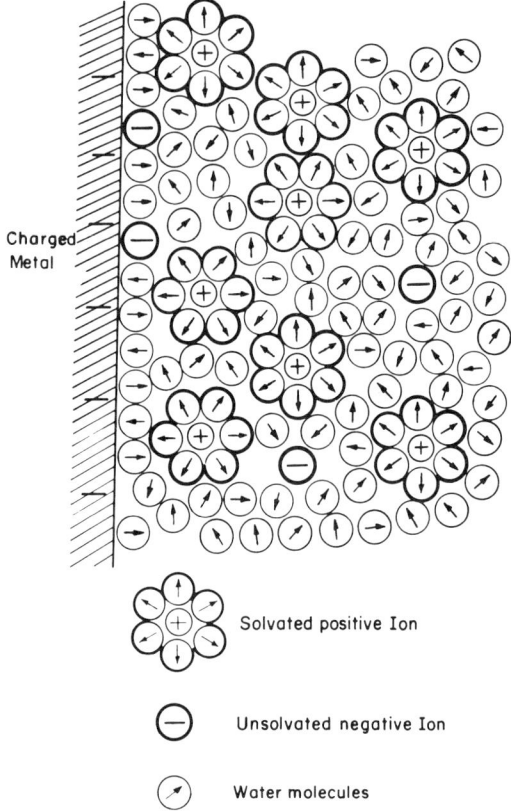

Fig. 7.7. A schematic representation of the structure of an electrified interface. The small, positive ions tend to be solvated, while the larger, negative ions are usually unsolvated (*cf.* Section 2.3.7).

7.1.8. A Look into an Electrified Interface

A very elementary and qualitative picture of the formation of a double layer has been presented. Some readers would be content with this introduction, preferring to discover more about the electrified interface as they go along. As facts pile up and theories grapple with them, these readers would like to see the structure of the interface gradually unfolding and the corresponding potential variations near the electrode surface being plotted. Other readers, however, would prefer a preview so that they might know what to expect.

Consider the picture of the metal–solution interface shown in Fig. 7.7.

THE ELECTRIFIED INTERFACE 633

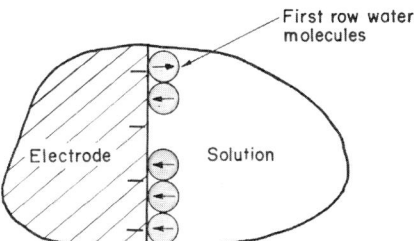

Fig. 7.8. The layer of oriented water molecules on a charged electrode. Due to excess negative charge on the electrode, there is an excess of water dipoles with their positive, hydrogen ends (the arrowheads) pointing toward the metal. For clarity, the structure of the rest of the solution is not drawn.

It looks complicated, but, actually, the picture will be seen to consist of simple elements.

The metal is made up of a lattice of positive ions and free electrons. When the metal is charged with an excess-charge density q_M, it means that there is either an excess (q_M is negative) or deficit (q_M is positive) of free electrons at the surface of the metal.

The metal surface can be compared to a stage occupied by this excess-charge density q_M. The particles of the solution constitute the audience which responds to the scene on the stage.† The *first row* is largely occupied by water dipoles (Fig. 7.8). The excess charge on the metal produces a preferential orientation of the water dipoles. This is the *hydration sheath* of the electrode (*cf.* Chapter 2). The net orientation of the dipoles varies with the charge on the metal, and the dipoles can even turn around and look away from the electrode.

The *second row* is largely reserved for solvated ions. The *locus* of centers of these solvated ions is called, for historical reasons, the *outer Helmholtz plane*, hereafter referred to as OHP (Fig. 7.9). On top of the first-row water (the primary water layer) and in between the solvated ions are other water molecules, a sort of secondary hydration sheath, feebly bound to the electrode.

---

† The relationship between the excess charge on the metal, q_M, and the excess charge on the solution side of the interface is reciprocal. The particles in solution, unevenly distributed owing to the presence of a phase boundary, may give rise to the excess-charge density q_M in the metal with which they interact, or an excess-charge density q_M generated by an external source (a battery) may effect the charge distribution on the solution side of the interface.

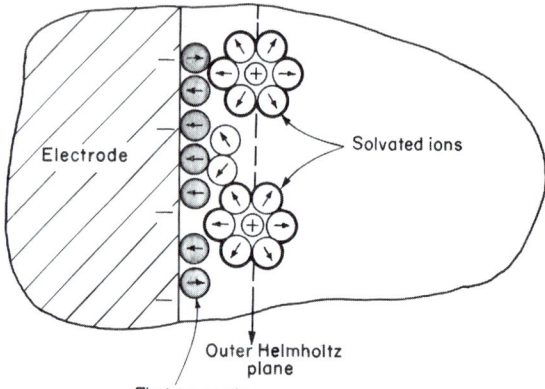

Fig. 7.9. A layer of solvated ions on the layer of first-row water. The *locus* of centers of these solvated ions defines the OHP.

In the simplest case, the excess-charge density at the OHP (due to the solvated ions) is equal and opposite to that on the metal (Fig. 7.10). This is the situation—two layers of excess charge—which gave rise to the term *double* layer. The electrical equivalent of this situation (two sheets of

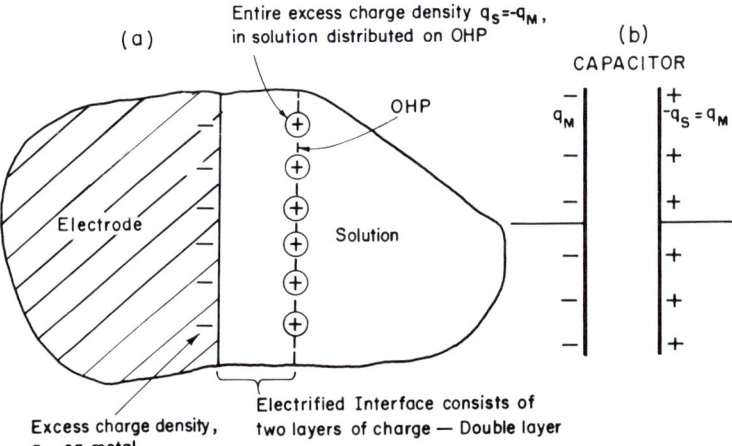

Fig. 7.10. (a) A *double* layer, a simple hypothetical type of electrified interface in which a layer of ions on the OHP constitutes the entire excess charge q_S in the solution. The solvation sheaths of these ions and the first row of water molecules on the electrode are not shown in the diagram. (b) The electrical equivalent of such a double layer is a parallel-plate condenser.

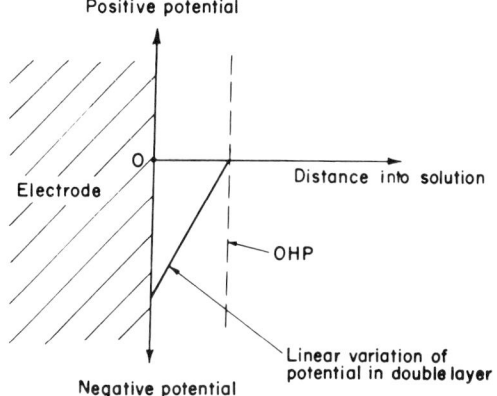

Fig. 7.11. The linear variation of potential corresponding to the double layer of Fig. 7.10(a).

smeared-out charge opposite each other) is a capacitor. The potential drop between these two layers of charge is a linear one (Fig. 7.11).

A case which is not so simple is the one in which the excess-charge density on the OHP is not equivalent to that on the metal but less (Fig. 7.12). Some of the solvated ions leave their second-row seats and random-walk about in the solution. In this case, the excess-charge density in the solution decreases with distance from the electrode. Near the metal, its

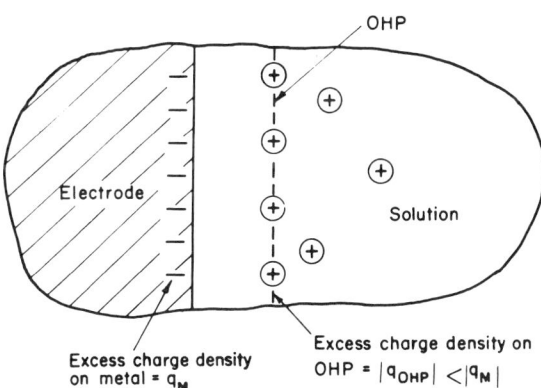

Fig. 7.12. A situation where the excess-charge density on the OHP is smaller in magnitude than the charge on the metal, $|q_S| = |q_M| > |q_{OHP}|$. The remaining charge is distributed in the solution. The solvation sheaths of the ions and the water molecules on the electrode are not shown in the diagram.

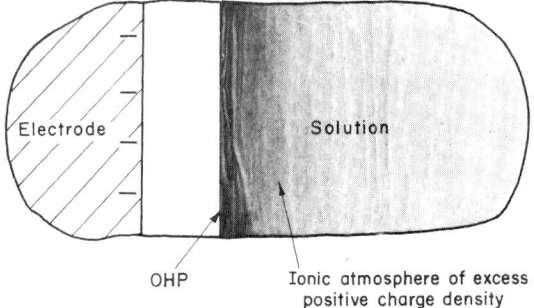

Fig. 7.13. The excess-charge density (indicated by shading) in the solution is distributed in the form of an ionic atmosphere for the electrode. The hydration sheath of the electrode is not shown in the diagram.

charge attracts the solvated ions to the second row. Further out, thermal motions are of comparable influence to the forces from the electrode. Sufficiently far into the solution, the net charge density is zero because positive and negative ions are equally likely in any region—thermal motion reigns supreme.

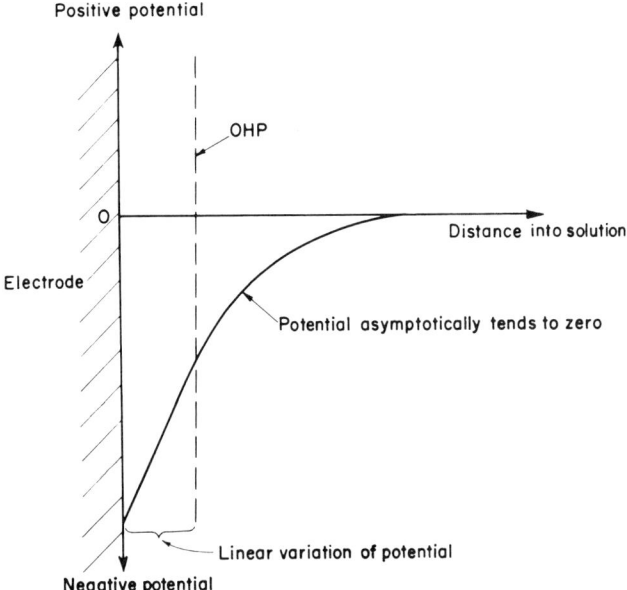

Fig. 7.14. The variation of potential corresponding to the ionic atmosphere of Fig. 7.12.

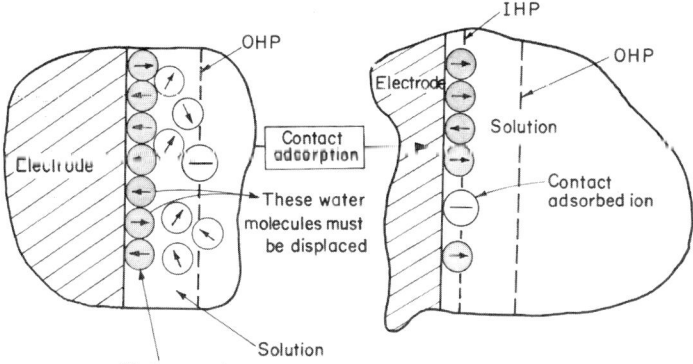

Fig. 7.15. The process of contact adsorption in which ions from the OHP displace first-row water molecules and adsorb in contact with the electrode.

Thus, the excess-charge density in the solution (equal and opposite to that on the metal) may not all be fixed compactly on the OHP, but some of it is dispersed into the solution. The electrode has a sort of ionic atmosphere (*cf*. Chapter 3 and Fig. 7.13). The potential falls off into the solution, at first sharply, and then asymptotically tends to zero (Fig. 7.14).

Although the first row is largely occupied by water molecules, there are some ionic species which find their way to the front (Fig. 7.15). But they cannot get there if obstructed by their own primary solvent sheaths. Thus, ions in contact with the electrode are those which do not have a primary hydration sheath when they are in the bulk of the solution. (In Chapter 2 it was shown that, in general, most anions and large cations do not possess primary solvation sheaths, and it happens that it is anions

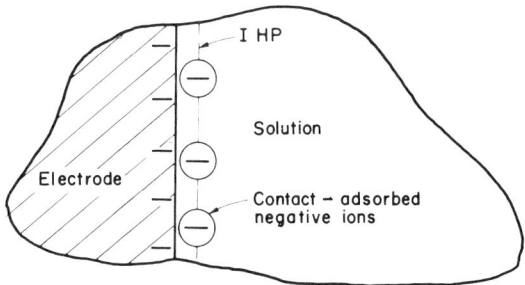

Fig. 7.16. Negative ions may contact-adsorb on a negatively charged electrode.

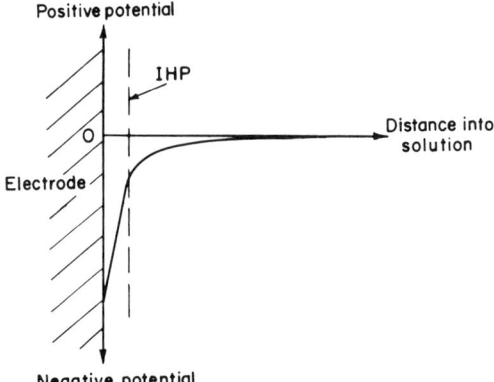

Fig. 7.17. The variation of potential corresponding to the contact adsorption shown in Fig. 7.16.

and big cations which do achieve contact with the electrode.) The *locus* of centers of these ions is known as the *inner Helmholtz plane*, often referred to as IHP.

Such ions, *in contact adsorption* with the electrode, are sometimes said to be *specifically adsorbed*. The word *specific* is used because the extent of the phenomenon seems to depend on the chemical nature of the ion (is it, e.g., a chloride or fluoride ion?) rather than on its charge (chloride and fluoride both have the same charge). The word *specific* is used also because the contact-adsorbing ion is partly oblivious to the charge on the metal; negative anions may even contact-adsorb on a negatively charged electrode (Fig. 7.16). The type of potential variation that may occur is shown in Fig. 7.17.

An interesting case is that of the specific adsorption of anions on a metal charged opposite in sign. Now, forces other than simple coulombic (like charges repel; unlike charges attract) are operative in specific adsorption. Hence, it may turn out that, if $|q_M|$ is the numerical value of the excess-charge density on the metal, a quantity[†] $|q_{CA}|$, not equivalent as required by simple Coulomb's law forces, but greater than, or *superequivalent* to, the excess-charge density on the electrode, i.e., $|q_{CA}| > |q_M|$, specifically adsorbs on the electrode (Fig. 7.18). Specific adsorption is therefore sometimes known as *superequivalent adsorption*.

The potential variation corresponding to superequivalent adsorption is interesting (Fig. 7.19). For example, going from a positively charged

---

[†] The subscript *CA* represents *contact adsorption*.

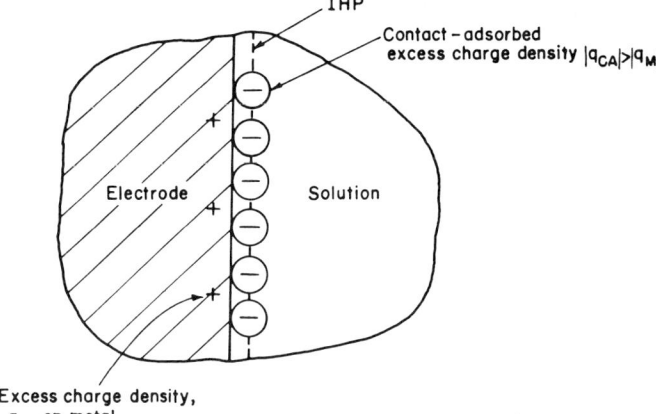

Fig. 7.18. Superequivalent adsorption, a situation where the contact-adsorbed excess charge $|q_{CA}|$ is greater in magnitude than the excess charge on the metal. The electroneutrality is made up by excess positive ion adsorption in the "solution" section (no ions shown).

electrode toward the solution, the potential falls linearly up to a plane through the *locus* of centers of the specifically adsorbed ions—the IHP—and then changes course and, turning up, goes asymptotically to the value in the bulk of the solution.

This, then, is a brief sketch of the structure and potential variation near an electrified interface.

Further Reading

1. B. E. Conway and M. Salomon, *J. Chem. Educ.*, **44**: 554 (1967).

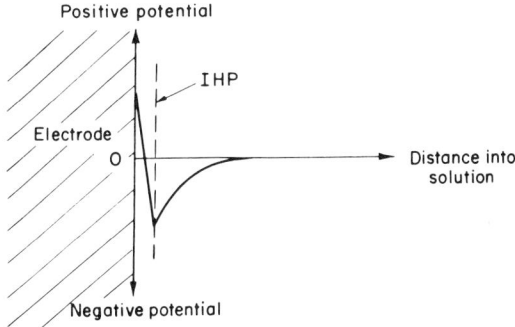

Fig. 7.19. The potential variation corresponding to superequivalent adsorption.

7.2. SOME PROBLEMS IN UNDERSTANDING AN ELECTRIFIED INTERFACE

7.2.1. What Knowledge Is Required before an Electrified Interface Can Be Regarded as Understood?

The double layer formed at a boundary between two phases containing charged entities has two fundamental aspects, the *electrical* aspect and the *structural* aspect (Fig. 7.20). The electrical aspect concerns the magnitude of the excess-charge densities on each phase. (Recall that the total excess charge on one phase is always equal in magnitude to the total excess charge on the other phase). It also concerns the variation of potential with distance from the interface. The structural aspect is a matter of knowing how the particles of the two phases (ions, electrons, dipoles, neutral molecules) are arranged in the interphase region so as to electrify the interface.

The electrical and structural aspects of the double layer are intimately related. The charge or potential difference is characteristic of the particular structure, and *vice versa*.

The formation of an electrified interface has been described in the following steps

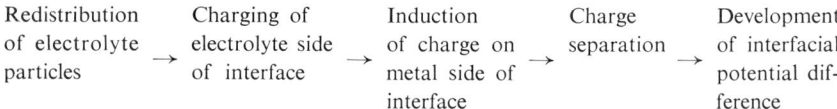

| Redistribution of electrolyte particles | → | Charging of electrolyte side of interface | → | Induction of charge on metal side of interface | → | Charge separation | → | Development of interfacial potential difference |

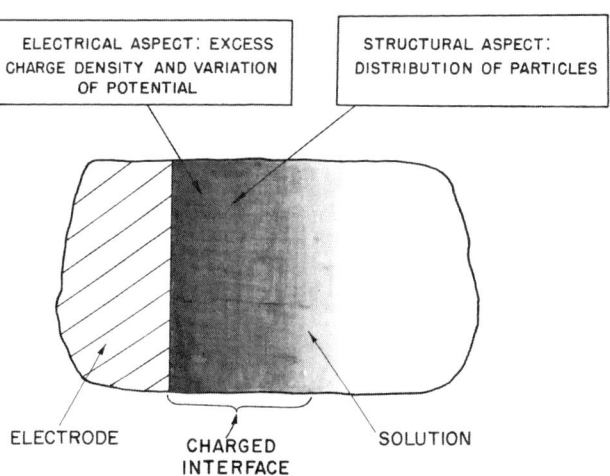

Fig. 7.20. The electrical and structural aspects of the electrified interface.

In systems in which one of the phases, e.g., a metal electrode, can be connected to an external source of charge, the formation of an electrified interface can be conceived in the following way

Charge flows from outside source into one phase (e.g., metal) → Charging of one phase → Redistribution of electrolyte particles in the interface → Development of net charge on electrolyte side of interface → Charge separation across interface → Development of potential difference across interface

In fact, these processes occur almost simultaneously.

There is a functional relationship between the charge on each phase (or the potential difference across the interface) and the structure of the interphase region. The fundamental problem of double-layer studies is to unravel this functional relationship. One has understood a particular electrified interface if, on the basis of a model, i.e., an assumed type of arrangement of the particles in the interphase, one can predict the observed distribution of charge (or variation of potential) across the interphase.

7.2.2. Predicting the Interphase Properties from the Bulk Properties of the Phases

A deeper level of understanding is gained if the double-layer structure can be predicted on the basis of the properties of the bulk phases.

Double layers are formed because, in the interphase region, particles are not distributed in the same way as in the bulk. For example, perhaps more positive ions than negative ions exist in a lamina in the interphase region. If, however, a lamina in the bulk were considered, the numbers of positive and negative ions would be equal. Evidently, there is a depletion of negative ions or an accumulation of positive ions, or both, relative to the bulk. This phenomenon of substances collecting in or departing from a phase boundary is known as *adsorption*.

Why do particles tend to accumulate or deplete from an interface? A phenomenological answer is in terms of the free-energy change associated with the adsorption process. If one knew these free energies of adsorption, one could state: Given these bulk compositions, this will be the composition of the interface and these will be the properties of the interface.

But why take the free energies of adsorption from experiment? Instead, one can attempt to calculate the values from a knowledge of the particles, the forces between them, and the effect on the particles of the electric field operating on the electrified interface.

These, then, are some of the ultimate goals of double-layer research. They may be summarized thus: From a knowledge of the bulk phases, to get at the structure of the electrified interface and finally the potential variation across the interphase region.

7.2.3. Why Bother about Electrified Interfaces?

Why is the spatial distribution of charges in the interphase region between two phases of interest and importance?

There is, of course, the philosophical reason stemming from man's search for understanding and for a coherent picture of natural phenomena. Surfaces are found almost everywhere in nature, and many of them carry a charge. A substantial part of the understanding of nature is dependent upon a satisfactory model of charged interfaces.

But there are also the reasons which arise from utilitarian needs that often inspire scientific quests. Thus, electrified interfaces are of vital importance in many aspects of everyday life. This fact can be exemplified in many ways.

Consider, e.g., colloidal particles, i.e., particles which are too small to display the properties of macroscopic objects, say, <0.01 mm, and too large to behave like atoms and small molecules, approximately >100 Å. These colloidal particles move under electric fields, and, if they are pigments, then electric fields can be used to "guide" the colloidal particles to deposit upon metals and color them. The hues formed in this way may be more permanent than paint. But why do the particles move? The answer lies in the electrified interface between a colloid and the medium. In other words, the charge separation and the resulting potential difference at the particle's interface provides a handle with which the externally applied electrical field can guide the particle along. Thus, the understanding of double layers at the surface of colloidal particles is a basis for technological improvements in the coatings of metals.

A rather unusual example of the ubiquitous role of electrified interfaces is based on the friction between two solids which, in the presence of liquid films, may depend on the double layers at their interfaces. Thus, the efficiency of a wetted rock drill depends on the double-layer structure at the metal-drill-aqueous-solution interface.

Then, electrodic reactions which underlie the processes of metal deposition, etc., cannot be understood without knowing the potential difference at the electrode–solution interface and how it varies with distance from the electrode. The ions from the solution must be electrically energized to cross the interphase region and deposit on the metal. This electrical energy must be picked up from the field at the interface, which itself depends upon the double-layer structure. Thus, control over metal deposition processes can be bettered by an increased understanding of double layers at metal–solution interfaces.

An electrodic process of vast practical significance is that resulting in the dissolution of a metal into solution or into a film of conducting moisture adhering to the metal surface. The process is corrosion. Processes connected with corrosion may lead to the breaking-off of an aircraft's wing. Many things corrode slightly; it is the corrosion *rate* which determines the significance of the corrosion. It depends partly on the structure of the double layer, i.e., on the electric field across the interface, which in turn governs the rate of metal dissolution. Thus, double layers influence the stability of metal surfaces and, hence, the strength of metals. Nor must it be thought that these remarks apply only to metals in contact with a visible solution. They apply to all substances which corrode—for this always occurs by electrodic reactions across surfaces (Section 11.1) even if the solution phase is a moisture film only a few microns thick.

Molecular mechanisms in biology, too, depend to a great extent on electrified interfaces. Thus, the mechanism by which nerves carry messages from brain to muscles is based on the potential difference across the membrane which separates a nerve cell from the environment. What are the laws which apply to this electrified interface? If this question is answered, then the mechanism by which nerves transmit messages may become agreed upon at the molecular level and thus the process concerned may become controllable.

These are only a few examples cited to stress the wide range of phenomena in which electrified interfaces play an important part. The number of such examples could be multiplied many times. They all emphasize the crucial role of double-layer studies as a basis to the understanding at a molecular level of many very practical happenings. Understanding the electrified interface is one of the most exciting aspects of electrochemistry.

7.2.4. The Need to Clarify Some Concepts

Before plunging into a discussion of the double layer, it is necessary to clear some ground. Some special terms, such as potential difference and adsorption, have already been used. They convey certain notions and suggest certain pictures. But do they represent precise quantitative concepts?

Take, e.g., the question of potential difference across the interphase. This concept calls to mind elementary ideas from electrostatics, e.g., when there is a potential difference between two points, work must be done to transport a charge from one point to the other. The concept is simple as long as the charge is moved in one phase. But what happens when the charge has to cross a phase boundary? What law will be used to calculate the work? Can one use Coulomb's law in its usual simple form? These and other questions have to be discussed.

Or consider, e.g., the questions which the term *adsorption* can arouse if a double layer is involved. How does one detect adsorption? What is its quantitative measure? Is it all the same whether the particles of one phase which accumulate in the interphase are lined up in contact with the other phase or are scattered in a haphazard, diffuse manner?

Further understanding of the double layer hinges on bringing many of the concepts mentioned so far into sharper focus. The ideas must not only acquire sharp qualitative features but also be quantitatively measurable.

7.2.5. The Potential Difference across Electrified Interfaces

7.2.5a. What Happens When One Tries to Measure the Absolute Potential Difference across a Single Electrode–Electrolyte Interface? Consider the operations necessary to measure the potential difference across a metal–solution interface. Various potential-measuring instruments can be used, potentiometers, vacuum-tube voltmeters, electrometers, etc. All these instruments have two metallic terminals which must be connected to the two points between which the potential difference is to be measured.

One terminal is connected directly to the metallic electrode. But what does one do with the other terminal (Fig. 7.21)? It must of course be connected to the other phase, the potential of which is to be measured. This is the electrolyte. The second connecting wire has therefore to be immersed in the solution (Fig. 7.22).

The connecting wires of potential-measuring devices are intended however to act as pure probes. They must be spectators of the potential scene. They must sense the potential without actively introducing any potential differences. But this is not possible because the immersion of the second connecting wire M_2 in the electrolyte inevitably produces a *new* phase boundary, i.e., the M_2–solution interface (Fig. 7.22). At this interface, there must be a second double layer and a second metal–solution potential difference, i.e., $PD_{M_2/S}$, which cannot be avoided.

Thus, the very operation of measurement (a vacuum-tube voltmeter

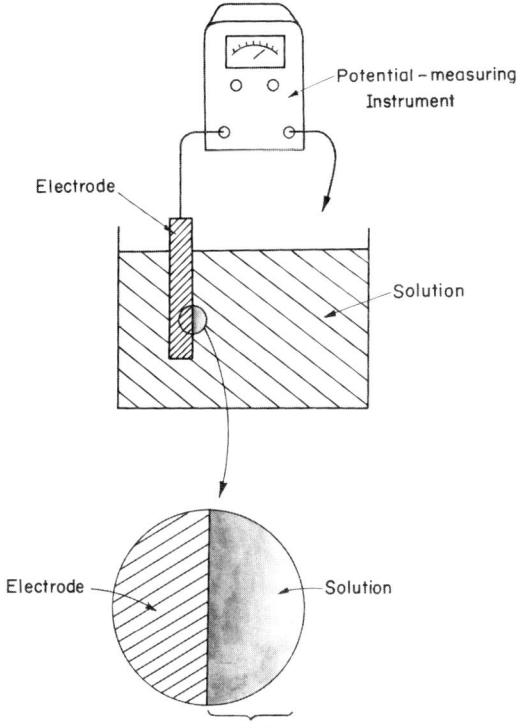

Fig. 7.21. To measure the potential difference across a metal–solution electrified interface (see exploded view), one terminal of the potential-measuring instrument is connected to the metal electrode. What is to be done with the second terminal?

must use both its terminals to grasp a potential difference) cannot but involve a second double layer and a second potential difference. This is the difficulty. One sets out to measure one potential difference $PD_{M_1/S}$ and, in the attempt, one creates *at least* one[†] additional potential difference. One ends up with the measurement of the *sum* of at least two potential differences (Fig. 7.23).

The system created by the measuring procedure is in fact an electrochemical system, or cell, consisting of *two* electronic conductors (electrodes) immersed in an ionic conductor (electrolyte). *All one can measure, in practice,*

[†] There can be more than one additional potential difference due to the act of measurement, as is shown further on.

646 CHAPTER 7

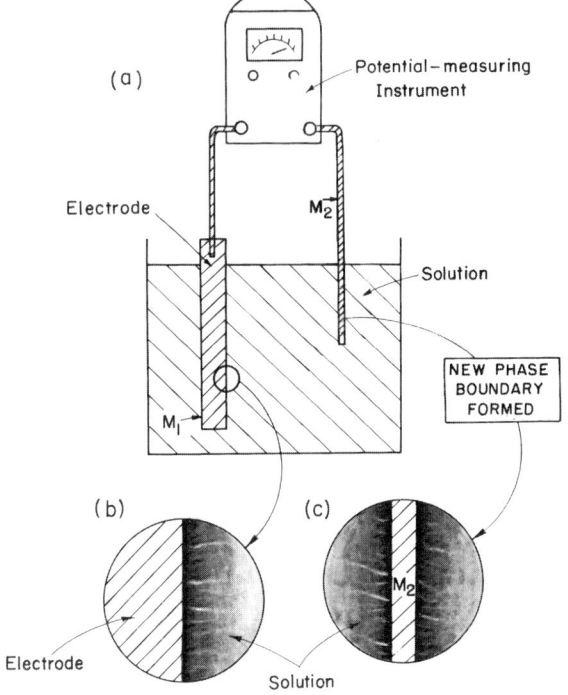

Fig. 7.22. (a) If the second terminal M_2 of potential-measuring instrument is dipped into the solution, an extra M_2–solution electrified interface is formed—see exploded view (c)—, apart from the electrified electrode–solution interface under study—see exploded view (b).

is the potential difference across a system of interfaces, or cell, not the potential difference across one electrode–electrolyte interface.

Some care must be exercised in measuring cell potentials. Simple measuring devices (e.g., voltmeters) often draw a substantial amount of current, and, since the potential drop at an interface depends on the current passing across the interface (*cf.* Section 8.2.12), a situation may arise in which the measuring instrument affects and changes the quantity being measured. To avoid this type of complication, modern instruments which draw negligibly small currents may be used (e.g., electrometers which require currents in the range of 10^{-14} amp). Alternatively, a simple Poggendorf type of compensating circuit can be employed, in which the potential of the cell is balanced against an external potential source of accurately known value, with the aid of a sensitive galvanometer used as a null detector.

In the particular cell (Fig. 7.23) generated by the measuring process, it will be only as a special case that the metal M_1 (of the electrode–electrolyte interface under study) is identical with the metal M_2 (the connecting wires of the measuring instrument). In general, M_1 and M_2 will be different metals, say, platinum and copper. The meeting of the platinum and copper phases produces another double layer and an additional potential difference across the platinum–copper interface. This additional potential difference is known as the *contact potential difference*. Thus, the procedure of using a potential-measuring device has introduced, if M_1 and M_2 are dissimilar metals, two additional and unwanted potential differences (Fig. 7.24).

One can have more-complicated cells (Fig. 7.25), and, in all of them, it can be seen that the attempted measurement of a metal–solution po-

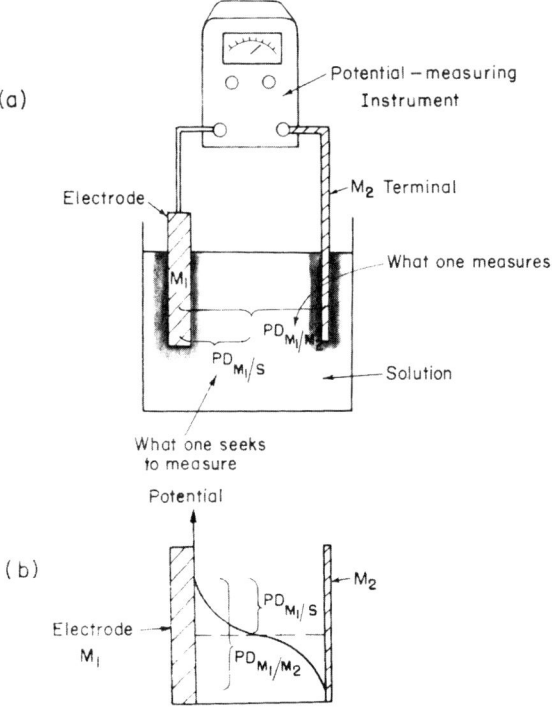

Fig. 7.23. (a) Though one seeks to measure the potential difference $PD_{M_1/S}$ across the interface under study, one actually measures the potential difference PD_{M_1/M_2} across at least two interfaces M_1/S and M_2/S. (b) A schematic representation of the potential variation from electrode M_1 to terminal M_2.

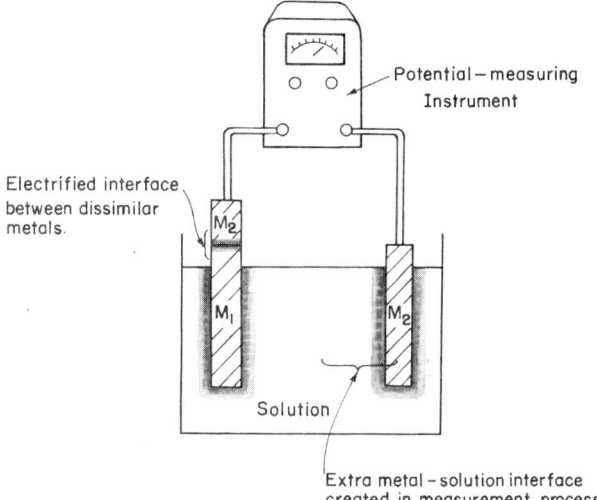

Fig. 7.24. If electrode M_1 and the connecting wires M_2 are dissimilar metals, a contact potential difference PD_{M_2/M_1} at the metal M_1–metal M_2 interface is generated in the measurement process, in addition to the extra metal-solution potential difference $PD_{M_2/S}$.

tential difference will conclude with the measurement of the *sum* of *at least* two interfacial potential differences, i.e., the desired $PD_{M_1/S}$ and as many extra potential differences as there are new phase boundaries created in the measurement. In symbolic form, therefore, the potential difference V indicated by the measuring instrument can be expressed thus

$$V = PD_{M_1/S} + \sum_{i=1}^{n} PD_i \tag{7.1}$$

where PD_i is the potential difference at the ith phase boundary created in the process of measurement, there being $n \geq 1$ such phase boundaries. In the case of the particular system shown in Fig. 7.24, one has

$$V = PD_{M_1/S} + \sum_{i=1}^{2} PD$$
$$= PD_{M_1/S} + PD_{S/M_2} + PD_{M_2/M_1} \tag{7.2}$$

7.2.5b. The Absolute Potential Difference across a Single Electrified Interface Cannot Be Measured The analysis of the process of measuring potential differences across phase boundaries demonstrates the impossibility of measuring the *absolute* value of a metal–solution potential difference.

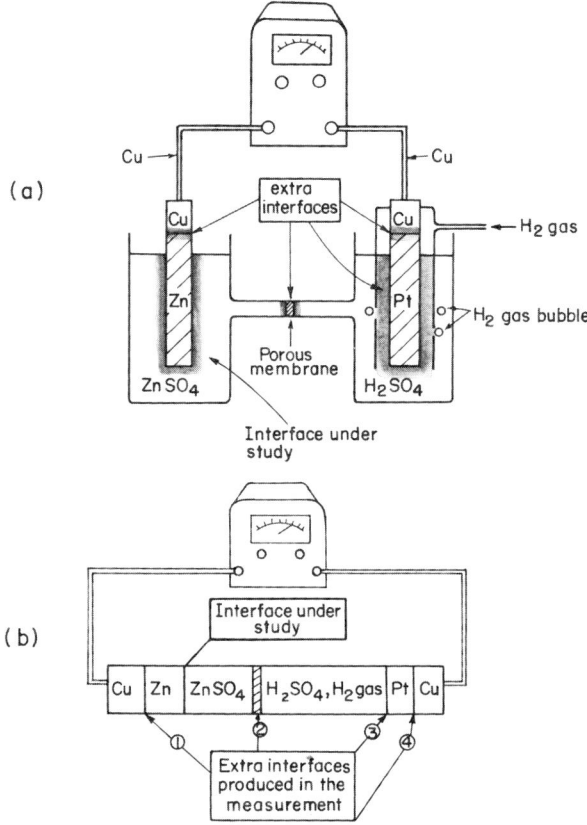

Fig. 7.25. (a) A more complex cell in which the two electrodes both differ in nature from the copper connecting wires of the potential-measuring instrument. In addition, the solutions in contact with the two electrodes are different from each other and are separated by a porous membrane. (b) Another representation of the cell.

Simple truths are sometimes accepted with great reluctance. This was indeed the case with the impossibility of making an experimental measurement of the absolute metal–solution potential difference. Many attempts of different kinds have been made to measure the absolute value of $PD_{M_1/S}$. All of them however are subject to some defect, as simply exemplified in the above analysis where it has been shown that the measurement process itself introduces additional potential differences, so that it is only the sum of several potential differences which is measured and, from this sum, one cannot extricate the desired potential difference.

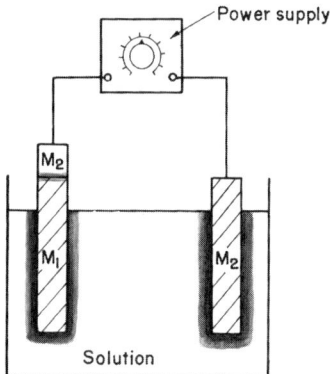

Fig. 7.26. An electrochemical cell connected to an external source of potential.

The electrochemist proceeds, therefore, somewhat humbled but not defeated. He must ask himself how much information about the potential difference across an electrode–electrolyte interface can be obtained.

7.2.5c. Can One Measure Changes in the Metal–Solution Potential Difference? In the preceding discussion, the electrochemical system has itself been the origin of the potentials discussed, these potentials being *measured* by a device.

Consider now a situation where, instead of a measuring instrument, one inserts (Fig. 7.26) into the "circuit" a *source* of potential (e.g., an electronically regulated power supply). Here, the total potential difference across the cell must equal (in magnitude) that put out by the source.[†] This is, in fact, the law of conservation of energy applied to an electrical circuit, or Kirchhoff's second law: the algebraic sum of all potential differences around a closed circuit must be equal to zero. For the simple hypothetical system shown in Fig. 7.27, one has

$$PD_{M_1/S} + PD_{S/M_2} + PD_{M_2/M_1'} + PD_{M_1'/M_1} = 0$$

or

$$PD_{M_1/S} + PD_{S/M_2} + PD_{M_2/M_1'} = -PD_{M_1'/M_1} = V \quad (7.3)$$

[†] If a current passes through the electrolyte, the sum of potential differences will include an ohmic drop in addition to those at the interfaces. This potential difference due to the ohmic drop depends on the current I and the resistance R of the electrolyte and is equal to IR.

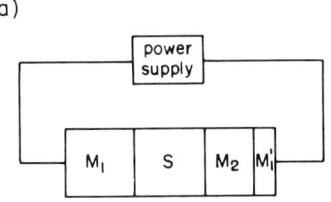

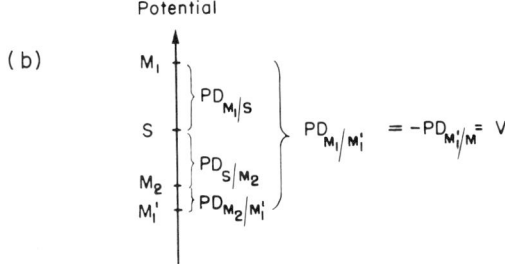

Fig. 7.27. (a) A cell consisting of two metals M_1 and M_2 in contact with a solution S. The M_1' is a metal of the same composition as M_1. (b) The potentials in M_1, S, M_2, and M_1' and the corresponding potential differences.

where, according to Kirchhoff's laws, the input potential V equals the potential difference $PD_{M_1/M_1'}$, between the two metal leads.

Now, let the input potential[†] be changed by an amount δV. Since the changes in the potential of the source must be equal to the changes in potential in the system, one can write

$$\delta V = \delta PD_{M_1/S} - \delta PD_{M_2/S} + \delta PD_{M_2/M_1'} \tag{7.4}$$

It appears that the change δV at the source is distributed over all the interfaces and produces changes in the potential differences across them.

Imagine, however, at M_2/S, a *nonpolarizable*[‡] interface (to be described further below) which is characterized by the fact that the potential across

---

[†] For convenience of exposition, it is customary to use the contraction *potential* for the term *potential difference*.

[‡] *To polarize an interface* means *to alter the potential difference across it*; *to be polarizable* means *to be susceptible to changes in potential difference*. The quantitative definition of polarizability will be given in Section 8.2.12.

it does *not* change except under extreme "duress" (i.e., a large change of input potential). Then, for small changes δV at the external source, the potential difference across the nonpolarizable interface will not depart significantly from its "fixed" value, i.e.,

$$\delta PD_{M_2/S} = 0 \qquad (7.5)$$

The way in which nonpolarizable interfaces resist changes in potential across them is important to understand. A simple picture is as follows (a detailed description is given in Section 8.2.12). The potential variation across a double layer depends on the charge separation, or distribution, at the interface. The only way that the potential can change is by changing the magnitude of the charge on each side of the interface. Suppose, however, that any charge flowing into an interface from an external source promptly leaks across the interface; then the charge separation, and thus $PD_{M_2/S}$, stays constant. The interface has resisted changes in its potential. The more easily charges leak across the interface, the more resistant is the interface to changes in potential difference. One is reminded here of a gyroscope or spinning top. The faster it spins, the more firmly it retains its spinning axis.

Further, the contact potential difference between two metals depends on the composition of the two metals and is unaffected by potential difference across the cell, and, hence,

$$\delta PD_{M_2/M_1'} = 0 \qquad (7.6)$$

One can now resort to a simple artifice. Combine the interface under study, M_1/S, with an interface which resists changes in potential, i.e., a nonpolarizable interface M_2/S (Fig. 7.28). By using this electrochemical system, or cell, all changes in the potential of the source find their way to only one interface, i.e., that under study. An excellent method of producing *changes* of potential at one interface only has thus been devised. Then

$$\delta V \approx \delta PD_{M_1/S} \qquad (7.7)$$

Of course, this argument implies that the M_1/S interface is completely polarizable. This is important. The point is that the power supply only insists that the whole cell shall change its potential difference by an amount δV. Only if one interface is completely nonpolarizable and the other one completely polarizable can the latter be argued into accepting wholly the changes of potential put out by the source.

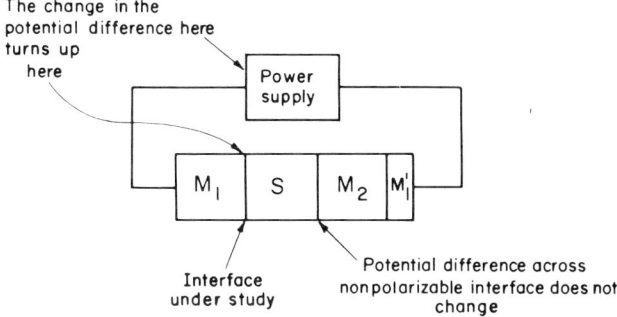

Fig. 7.28. By combining the M_1/S interface under study with a nonpolarizable M_2/S interface, changes in the potential difference across the source turn up at the interface under study.

If both interfaces are partially nonpolarizable, then the potential differences across both of them will change and one will be at a loss to know the magnitude of the individual changes at each interface.

7.2.5d. The Extreme Cases of Ideally Nonpolarizable and Polarizable Interfaces. Are nonpolarizable and polarizable interfaces fictions, or can one find them in the laboratory? The fact is that such interfaces can indeed be fabricated and have been used in double-layer studies. Of course, no interface is *ideally* nonpolarizable or *ideally* polarizable, i.e., nonpolarizable interfaces do change their potential to some extent and polarizable interfaces do resist such changes to some extent. The distinction is one of degree rather than kind.

An example of a nonpolarizable interface is the well-known calomel electrode (Fig. 7.29), and the classic example of a polarizable interface is the interface between pure mercury and an aqueous electrolyte solution (e.g., potassium chloride). Each of these interfaces possesses the characteristics required of them to such a degree of completeness that the combination of these two interfaces has dominated double-layer studies. The preoccupation with the mercury–solution interface has been so great that it has sometimes appeared that other interfaces are of no interest. This is not the case. It has been, rather, a matter of experimentalists choosing the most convenient path to understanding (*cf.* Section 7.3.7).

What makes an interface polarizable? In other words, what makes an interface decide to resist or accept potential changes? This question can be answered, but the answer has to be in terms of the rates at which charges transfer across the interface, i.e., in electrodic terms (*cf.* Section 8.2.12).

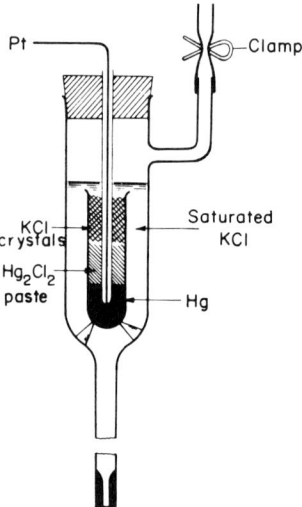

Fig. 7.29. The nonpolarizable calomel electrode.

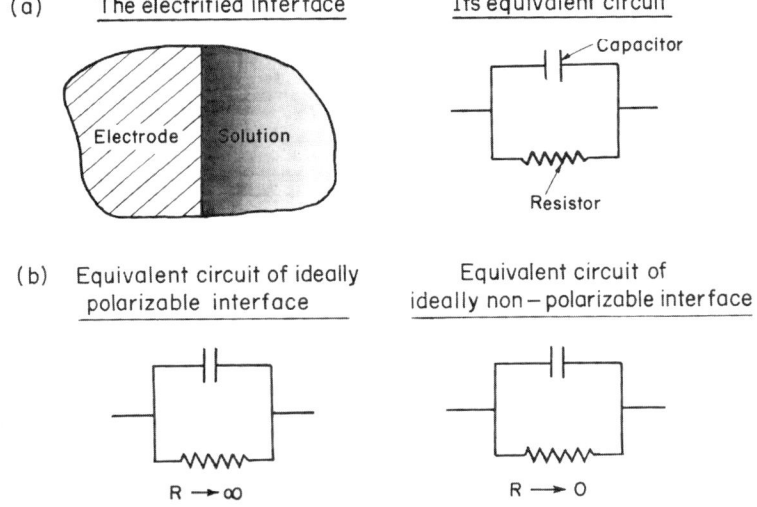

Fig. 7.30. (a) The equivalent circuit for an electrified interface is a capacitor and resistor connected in parallel. (b) In the equivalent circuit for an ideally polarizable interface, the resistance tends to infinity, and, for a nonpolarizable interface, the resistance tends to zero.

At this stage, therefore, an electrical analogy will be given. The *electrical behavior* of a metal–solution interface can be compared to that of a capacitor and resistor connected in parallel (Fig. 7.30). A circuit, consisting of electrical components, which simulates the electrical behavior of an electrified interface is referred to as an *equivalent circuit*. To understand the difference between nonpolarizable and polarizable interfaces in terms of the model consisting of a capacitor and resistor connected in parallel, consider what happens when the capacitor–resistor combination is connected to a source of potential difference. If the resistance is very high, then the capacitor charges up to the value of the potential difference put out by the source; this is the behavior of a polarizable interface. If, on the other hand, the resistance in parallel with the capacitor is low, then any attempt to change the potential difference across the capacitor is compensated by charge's leaking through the low resistance path; this is the behavior of a nonpolarizable interface.

7.2.5e. The Development of a Scale of Relative Potential Differences.
The essential feature of a nonpolarizable interface is that the potential difference across it remains effectively a constant as the potential applied to a cell which contains the nonpolarizable electrode changes. This property of nonpolarizable interfaces can be taken advantage of to develop a scale of *relative* potential differences across interfaces.

Thus, suppose that, when one wishes to measure an interfacial potential difference $PD_{M_1/S}$, one connects up the electrode with another electrode and measures the potential of this *cell*. Suppose one always keeps this second electrode constant in nature (i.e., the algebraic sum of the potentials associated with it is kept constant). Then, measurements of the potential of a cell in which the one (same) electrode and its associated solution were *always* present and the *other*, i.e., the first electrode mentioned here (M_1), and its solution were changed would clearly reflect the chang-

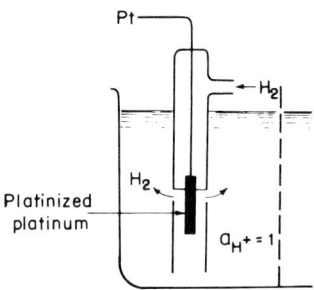

Fig. 7.31. The nonpolarizable standard hydrogen electrode.

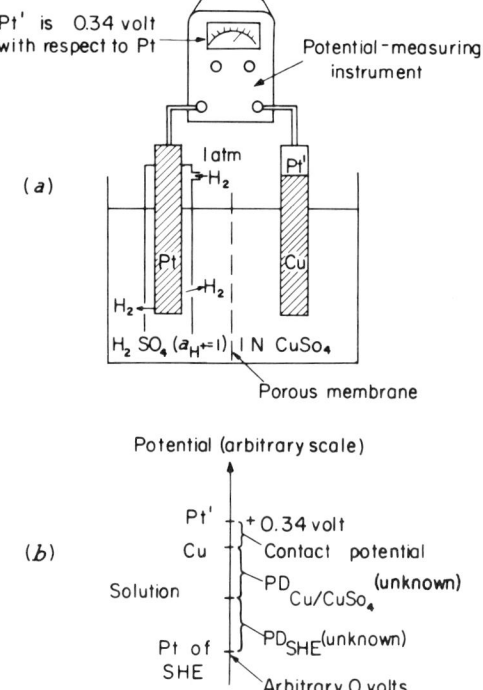

Fig. 7.32. The meaning of the relative potential difference across the $Cu/CuSO_4$ interface, i.e., the relative electrode potential. (a) The electrochemical cell corresponding to the relative potential difference. (b) The relative potential difference includes a platinum–copper contact potential and the unknown potential difference across the SHE, apart from the absolute potential difference across the $Cu/CuSO_4$ interface.

ing interfacial potential difference $PD_{M_1/S}$. This is in fact what is done to measure the *relative* values of $PD_{M_1/S}$ as M_1 or S (or both) are varied. Of course, in order that measurements of such potential differences of cells should directly give indications of the change of the nonconstant nonreference half (M_1), the reference electrode itself must always be the same one.

It has long been a convention in this field to utilize an electrode at which the potentials are controlled (see Section 8.2.14) by the reaction of the exchange of H^+ between the solution and H_2 gas, through the medium of a highly nonpolarizable interface, which exchanges electrons between

H⁺ and H_2 according to $2H^+ + 2e_0 \rightleftharpoons H_2$ [gas]. This is called the *reversible hydrogen electrode*. It will give rise to the same potential contribution to the cell (in which it is used as a reference for *comparing* the potentials of cells containing other electrodes) so long as the hydrogen-ion activity in the solution and the pressure of H_2 with which this is in equilibrium are kept always the same. The values chosen are unit activity for the hydrogen ion and unit pressure for the hydrogen gas.

Hence, if one continues to regard this matter of the relative scale of electrode potentials thermodynamically—the overall view—or phenomenologically, everything about it is very clear. Electrode potentials [a copper electrode in equilibrium with a solution consisting of a $0.07N$ solution of $Cu(NO_3)_2$, say] are measured by coupling the electrode up into a cell, the second electrode of which is a certain constant one. The one which is chosen is, as stated, a hydrogen electrode in which the solution component is H⁺ at unit activity in equilibrium with H_2 gas. It is then the *potentials of such cells* which are called *relative electrode potentials* or *potentials of electrodes on the standard hydrogen scale*. What is meant is that, by an arbitrary convention, these particular cell potentials are no longer called cell potentials but relative electrode potentials, and, indeed, the word *relative* is often dropped because those in the know realize what is meant.

Actually, what appears mostly under the heading "Standard Electrode Potentials" in classical (i.e., pre-1950) textbooks of electrochemistry are potentials arrived at in this kind of way. They have a further arbitrary aspect about them: The activity of the ions to which the *test* electrode is reversible is kept at unit activity; the temperature, at 25°C.

The presentation given above has been, as may perhaps have been gathered, a little bit with tongue in cheek; it is not quite complete (even if absolutely thermodynamically and phenomenologically—and conventionally—acceptable). The reason is that one started talking about absolute potential differences, i.e., the $PD_{M_1/S}$, and then with a slight change in phraseology slid into the presentation—one began to speak of an *electrode* potential.

Is electrode potential the same as $PD_{M_1/S}$, the potential difference of the metal–solution interface? The fact that it is not has already been implied in Eq. (7.3). An electrochemical cell consists of at least four potential differences. One is at each of the metal–solution interfaces. Another is at the junction between the two metals which constitute a part of each electrode. Another is within the electrometer, or vacuum-tube voltmeter, which opposes the sum of the other potential differences and constitutes a way of measuring a cell potential difference.

Equation (7.3) may be repeated here in a slightly different form

$$PD_{M_1/S} + PD_{S/M_{ref}} + PD_{M_{ref}/M_1'} = -PD_{M_1'/M_1}$$
$$= V \qquad (7.8)$$

The V is the vacuum-tube voltmeter reading.

How should one find, from (7.8), the meaning of a relative electrode potential? One knows now that it is not a simple absolute metal–solution potential difference, i.e., $PD_{M_1/S}$. It is the whole cell potential V, and that [cf. (7.8)] is at least three potential differences. The question to ask, therefore, is: What potentials are maintained constant when one measures a number of electrode potential in cells which always contain a standard hydrogen electrode (SHE)?

The answer to this question is relatively simple if one breaks down $PD_{M_{ref}/M_1'}$, at the junction between the two metals in the cell, into two potentials, $\phi_{M_{ref}}$ and $\phi_{M_1'}$, say. Then (7.8) becomes

$$(PD_{M_1/S} - \phi_{M_1'}) + \phi_{ref} + PD_{S/M_{ref}} = V \qquad (7.9)$$

The part which is constant in the relative scale of potentials is therefore

$$\phi_{ref} + PD_{S/M_{ref}}$$

It is the potential across the interface between the metal and the solution, together with that part of the metal–metal potential difference associated with the metal on which one carries out the exchange reaction between H^+ and H_2 (often, but by no means necessarily, platinum). This is the constant entity in cells for the measurement of relative electrode potentials. When one refers to the relative electrode potentials of a number of systems on the hydrogen scale and then equates this with the potentials of the cells made up of the systems (the electrode–solution systems concerned) in combination with a hydrogen electrode, the situation is that one is arbitrarily taking the constant

$$PD_{S/M_{ref}} + \phi_{ref} = 0 \qquad (7.10)$$

Of course, if this constant is taken as zero for all the metals with which the potentials of the other systems are compared, there will be no effect of this constant, but one may never forget that the relative electrode potential, to which reference is so often made, is in fact *not* a metal–solution potential difference.

Before we leave this rather confusing subject, it is a good idea to point out what an electrode potential *is* rather than what it is not. Unfortunately, it is not a simple thing at all. Much has been said about the fact that the absolute potential difference at a metal–solution interface cannot be measured and the fact that only *relative measurements* against another electrode can be made. But suppose one takes all this in, understands it, and realizes that the constant presence of

$$PD_{S/M_{ref}} + \phi_{ref} = 0 \qquad (7.11)$$

can be more or less neglected and taken as an arbitrary zero because it remains the same while the potential differences due to the test electrode vary. Then, when one looks at a table of standard potentials, *what* varies; what is the cause of the changes from metal to metal; what is an *electrode potential*?

The answer is essentially this: An electrode potential is not a metal–solution potential difference, not even on some arbitrary or relative scale; it is a combination of two potential differences, one at the metal–solution interface and the other *part of* (one may think of half the span of a bridge) the metal–metal potential difference in a cell (Fig. 7.32).

7.2.5f. Can One Meaningfully Analyze an Electrode–Electrolyte Potential Difference? The discussion so far can be summarized as follows (Table 7.1): The absolute value of the potential difference across an electrode–electrolyte interface cannot be *measured*. The sum of the potential differences across at least two interfaces, i.e., across an electrochemical cell, *can* be measured. Further, *changes* in the potential difference across any one interface can be measured provided the interface under study can be

TABLE 7.1

The Measurability of Various Potential Differences

1. Absolute potential difference across a single interface	Cannot be measured
2. Changes in the potential difference across a single interface	Can be measured
3. Sum of the potential differences across an electrochemical cell	Can be measured
4. Relative potential difference, i.e., potential difference across a cell which includes a standard hydrogen electrode	Can be measured

built into a system or cell, the other interface being a nonpolarizable one across which there is a *constant* potential difference.

Does the impossibility of measurement of a quantity preclude further thought about it? Discussion of a concept, even if it cannot be measured, does often lead to its better understanding. With this view, attempts will be made to probe further into the question of the absolute potential difference across an individual metal–solution interface.

Can the absolute potential difference across an interface be "structured," or resolved into contributions? This potential difference depends on the arrangement of charges, oriented dipoles, etc. Can one speak of separate contributions to the total potential difference from the excess charges on the metal and solution phases, on the one hand, and from the oriented dipoles, on the other? Perhaps these individual contributions can be measured or calculated. Thereafter, one may be able to add them together to calculate the elusive metal–solution potential difference.

7.2.5g. A Thought Experiment Involving a Charged Electrode in Vacuum.

Consider a thought experiment in which the interface is conceptually disassembled in the following manner: The two phases are separated and each placed in vacuum. The particular double layer which was present at the interface will disapear because its reason for existence, i.e., the meeting of the metal–solution interfaces, has disappeared. Let the metal and solution phases now be charged to the extent they were charged in the presence of a double layer, i.e., before the interface was disassembled (Fig. 7.33).

What one has at this point is two isolated charged phases in vacuum. The process of carrying a unit test charge from infinity in vacuum toward each charged phase will now be analyzed.

7.2.5h. The Test Charge Must Avoid Image Interactions with the Charged Electrode.

A charged metal in vacuum exerts an electric force on charges in the surrounding space (vacuum). Hence, in the process of transporting the unit charge toward the metal, work has to be done against the electric force. If the charge is moving along the x coordinate, then the work done W_x in bringing the charge from infinity up to any point x is given by

$$dW_x = -F_x\,dx \tag{7.12}$$

or

$$W_x = -\int_{\infty}^{x} F_x\,dx \tag{7.13}$$

THE ELECTRIFIED INTERFACE 661

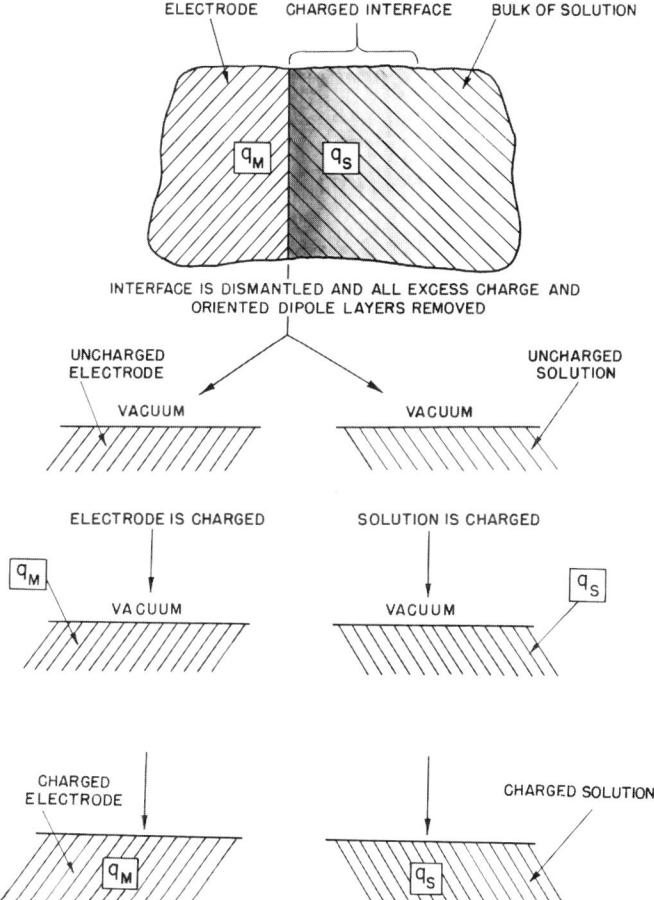

Fig. 7.33. A thought experiment involving the following steps is performed upon a charged interface: (1) The interface is dismantled, and all excess charges and dipole layers on the two phases are removed; (2) the electrode and solution phases are charged to the same extent as they were before dismantling the interface.

where F_x is the x component of the electric force per unit of charge, i.e., the electric field.

The electric forces operating on the test charge consist of two types. First there are those forces which originate in Coulomb's law. Secondly, there is what is called the *image force*, which arises as follows: When a charge approaches a material phase, it induces an equal and opposite charge on the surface of the material (Fig. 7.34). The spatial distribution of the induced charge is a complicated affair, and, therefore, the calculation of the interaction between the test charge and the metal is not simple. However, one can resort to a simple device: The metal is replaced by an "electrostatic mirror" located in the same position as the metal surface, in which case the test charge will have an image charge of equal magnitude and opposite sign as far behind the mirror as the test charge is in front of it (Fig. 7.34). The image interaction between the test charge and the metal is given by the coulombic interaction between the test charge and the fictitious image charge.[†] Thus, considering a metal sphere of radius r, the image force acting on unit charge q will be $-q^2/(2d)^2 = -q^2/4d^2$ and the corresponding potential will be $\psi_{im} = q^2/4d$, where d is the distance between the particle and the metal surface [cf. Fig. 7.35(a)]. If the metal sphere bears a charge $+Q$, the force due to coulombic interaction will be given by $Qq/(r + d)^2$, and the Coulomb potential will have a numerical value of $Qq/(r + d)$ since, for the purpose of calculation, the sphere can be substituted for by a point charge Q situated at the center of the sphere. At a large distance from the sphere, $d \gg r$, the Coulomb potential varies with $1/d$. As the test charge approaches the surface to distances small compared with the radius of the sphere, $d \ll r$, the term $Qq/(r + d)$ becomes essentially independent of d [cf. Fig. 7.35(b)]. Thus, at very short distances from the surface, the potential due to the image force is predominant. At very long distances, the coulombic force essentially determines the potential. Numerical calculation shows [cf. Fig. 7.35(c)] that there is an intermediate distance d where the contribution due to image force is negligible, yet $d \ll r$, so that the potential is essentially that due to the surface of the sphere owing to coulombic forces alone. It is in this region that a test charge would experience the constant potential which is characteristic of the surface at a given charge density and be negligibly affected by image interactions.

Now, it is very important that, in the thought experiment, the unit charge being transported only tests (or "probes") the charge on the metal. The unit test charge must not itself *interact* with the metal; it should only *sense* the charge on the metal. It must be a spectator, not an actor. From this point of view, it is clear that, when the test charge is involved in image

---

[†] The image charge does not have physical reality in the sense of a well-defined unit charge moving inside the metal in accordance with the movement of the test charge toward the metal. It is a mathematical device, proved rigorously to be valid, by which simple equations can be derived which describe correctly the complex effect of an approaching test charge on the actual charge distribution in the metal.

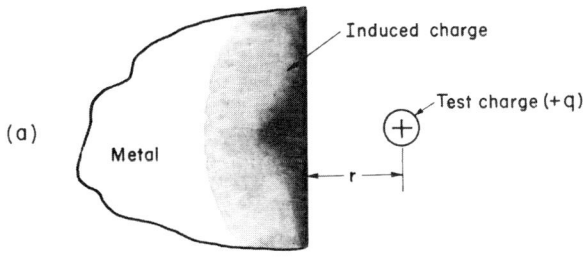

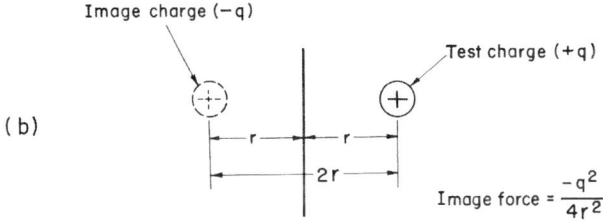

Fig. 7.34. (a) When a charge comes near a material, e.g., a metal, it induces a charge which is distributed in a complicated way. (b) The interaction between the test and induced charges can be calculated by considering that the metal is replaced by an image charge (equal in magnitude and opposite in sign to the test charge) situated as far behind the plane corresponding to the metal surface as the test charge is in front of it.

interactions, it is not taking a detached view of the charge on the electrode. Hence, one must try to ensure that the test charge experiences only the coulombic force and not the image force. Thus, as long as the test charge is sufficiently far away from the phase, the coulombic force characteristic of the sphere predominates and the test body behaves only as a noninteracting charge sensor. The closer the test charge comes to the phase, the greater becomes the magnitude of the short-range image interaction, and, close enough, such interactions begin to predominate. The test charge behaves here, not as probe, but as an interacting material particle which affects the charge on the phase under study. The distance at which the short-range image forces start to become significant is of the order of 10^{-5} cm [Fig. 7.35(a)].

7.2.5i. The Outer Potential ψ of a Material Phase in Vacuum. It may be recalled that the work done in bringing a unit test charge along the x axis from infinity up to a point P defines the potential at the point. This potential is therefore given by Eq. (7.13). In symbols,

$$\text{Potential} = \psi = -\int_{\infty}^{P} F_x\, dx \qquad (7.14)$$

Consider that the point P is a point in vacuum and *just outside* the reach of the image-force interactions arising from the presence of the electrode.

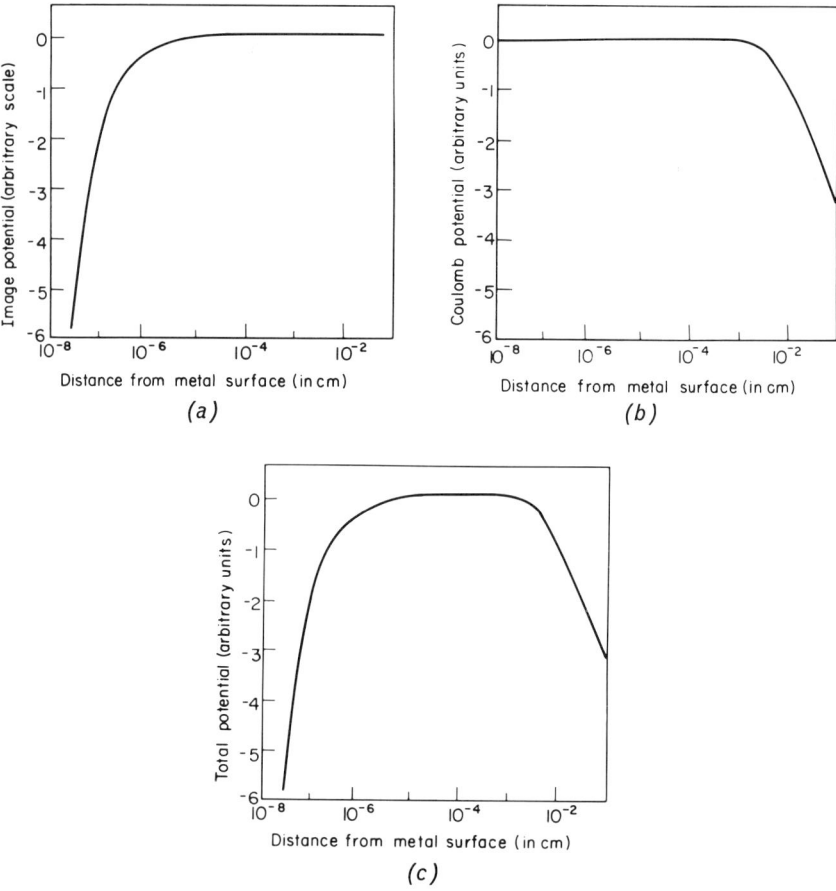

Fig. 7.35. Variation of the electrostatic potential near the surface of a charged sphere. (a) The short-range image potential contributes significantly only when the test charge comes closer than about 10^{-5} cm to the surface. (b) The long-range coulombic potential is independent of the distance d from the metal surface until this distance becomes comparable to the radius of the sphere. (c) The total potential resulting from Coulomb and image forces is independent of distance from the surface over a range of 10^{-5} to 10^{-3} cm, approximately.

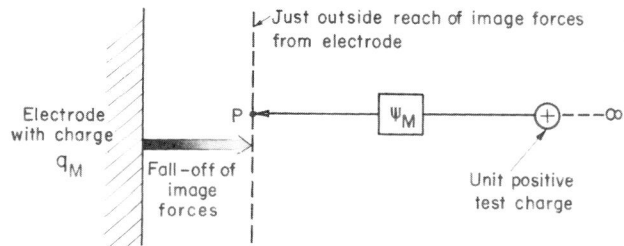

Fig. 7.36. The outer, or ψ_M, potential of the electrode is the work done to bring a unit of positive charge from infinity to a point P just outside the reach of the image forces from the electrode.

Then, the work done in bringing a unit test charge from infinity up to the point P and, hence, the potential at this point is determined purely by the *charge* on the electrode and is not influenced by any image interactions between the test charge and the electrode. At the same time [Fig. 7.35(c)], it is independent of distance. This potential just outside the charged electrode is termed, for obvious reasons, the *outer potential*. It is also referred to as the *psi* (ψ) *potential* (Fig. 7.36).

7.2.5j. What Is the Relevance of the Outer Potential to Double-Layer Studies? In Section 7.2.5g, a real electrified interface was thought of as dismantled into two parts, a solution part and a metal part. Further, *in a pure thought experiment*, these two parts were imagined as placed separately in vacuum. Then (in Section 7.2.5i), a unit test charge was brought to a point just outside the charged metal, and the outer potential, or ψ potential, defined.

But there is also the charged solution phase to be considered. Here, too, a unit test charge can be used to define the outer potential (Fig. 7.37).

Thus, by carrying out two thought experiments, one involving the electrode in vacuum and the other the electrolyte in vacuum, one obtains two outer potentials. The outer potential due to the charge on the metal electrode is termed ψ_M; and that due to the charge on the electrolytic solution, ψ_S.

At this stage, let the two conceptually separated parts of the double layer be brought together again. The interface has been reassembled. One can now refer to the *outer potential difference*, sometimes called the *Volta potential difference*, between the metal and solution. This outer potential difference is written

$$^M\Delta^S\psi = \psi_M - \psi_S \qquad (7.15)$$

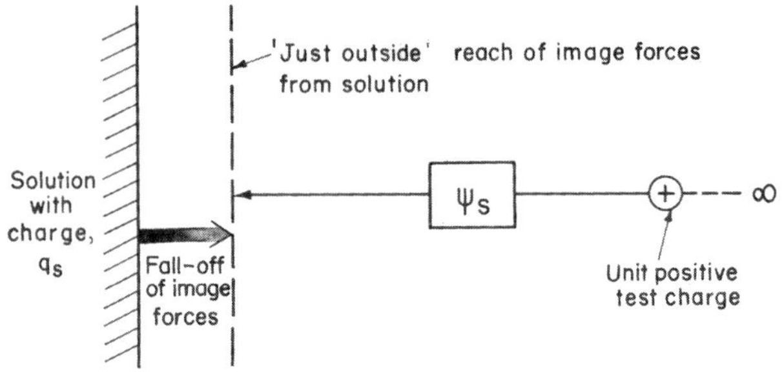

Fig. 7.37. The outer, or ψ_S, potential of the solution is the work done to bring a unit of positive charge from infinity to a point P just outside the reach of the image forces from the solution.

What is the physical significance of $^M\Delta^S\psi$? The conditions of the thought experiment may be recalled. After separating the metal and solution phases, the phases were charged (cf. Section 7.2.5g) to the same extent that they would be if the double layer existed. Hence, the ψ potentials of the metal and solution phases correspond to the charges which these phases actually have in the presence of the double layer at a metal–solution interface.

The outer potential difference $^M\Delta^S\psi$ is, therefore, the contribution to the potential difference across an electrified interface arising from the charges on the two phases.

Now the Volta (or outer) potential difference (i.e., that part of the total potential difference across a metal–solution interface which arises from the charges on the two phases) differs in an important way from the absolute and relative potential differences discussed in Sections 7.2.5b and 7.2.5e. The absolute potential difference across an interface is not a directly measurable quantity; it could only be included as a contribution to the measurable potential difference across an electrochemical cell and, then, expressed in the rather complicated way of a relative scale. And these relative potentials include a metal–metal contact potential which is an intrinsic part of the potential difference across every electrochemical cell. In contrast, the Volta potential difference is not only a straightforward quantity but a quantity *which can be measured experimentally* (cf. Appendix 7.1). Further, if one chooses a plausible model of the arrangement of charges at the interface, the Volta potential difference can be calculated by simple electrostatic reasoning [cf. Section 7.4.1, Eq. (7.79)]. It is for these reasons that the outer potential is an important aspect of the study of the electrified interface.

7.2.5k. Another Thought Experiment Involving an Uncharged, Dipole-Covered Phase. The basic picture of an electrified interface at an electrolyte boundary (Section 7.1.8) is now recalled. In general, not only is there charge separation, but there is also the possibility of a net preferential orientation of *dipoles* in the interphase region (Lange and Miscenko). When dipoles are tacked together so that more of them point one way than another, i.e., there is a net orientation, the arrangement is *equivalent to* a charge separation and, therefore, a potential difference occurs across the dipole layer. This dipole potential is an integral part of the potential difference across an electrified interface. Hence, the outer potential difference $^M\Delta^S\psi$ is *not* the only contribution to the electrode–electrolyte potential difference; the dipole contribution must also be analyzed and added to the Volta potential to give the total electrode–electrolyte potential difference.

How can this dipole potential be visualized? Once again a thought experiment can be performed (Fig. 7.38). The electrode and electrolyte phases are conceptually detached from each other and the double layer "turned off." In this process, the excess charges and oriented-dipole layers which characterized the double layer are considered eliminated. For the definition of the outer potential, the appropriate amounts of charge were then conferred on the two separated phases. Here, a layer of solvent dipoles will be fixed on the electrolyte so that the net orientation and the number per unit area of dipoles corresponds to that obtaining on the solution side of the double layer.

From this point on, the thought experiment proceeds as in Section 7.2.5*i* (Fig. 7.39). A test charge is brought in from infinity toward the dipole layer and then made just to cross it.[†] The work done in this process defines a potential. This potential has nothing to do with the excess charge on the solution phase because, during the thought experiment, the excess charge on the solution phase is maintained at a value of zero. Since the work has to do with traversing a surface layer on the electrolyte, the corresponding potential is a *surface potential*. Equally, because the potential is that associated with a dipole layer, it may be termed a *dipole potential*. It is also referred to as a *chi* (χ) *potential*.

In the case of the electrolytic solution, what has been defined here is its surface potential χ_S. Is there a χ potential for a metal electrode? This

---

[†] Once the test charge enters the phase, it will begin to interact with the atoms and molecules of the phase. Such interactions and the corresponding work terms are excluded from the dipole potential since they are considered separately in the definition of the chemical potential (Section 7.3.2).

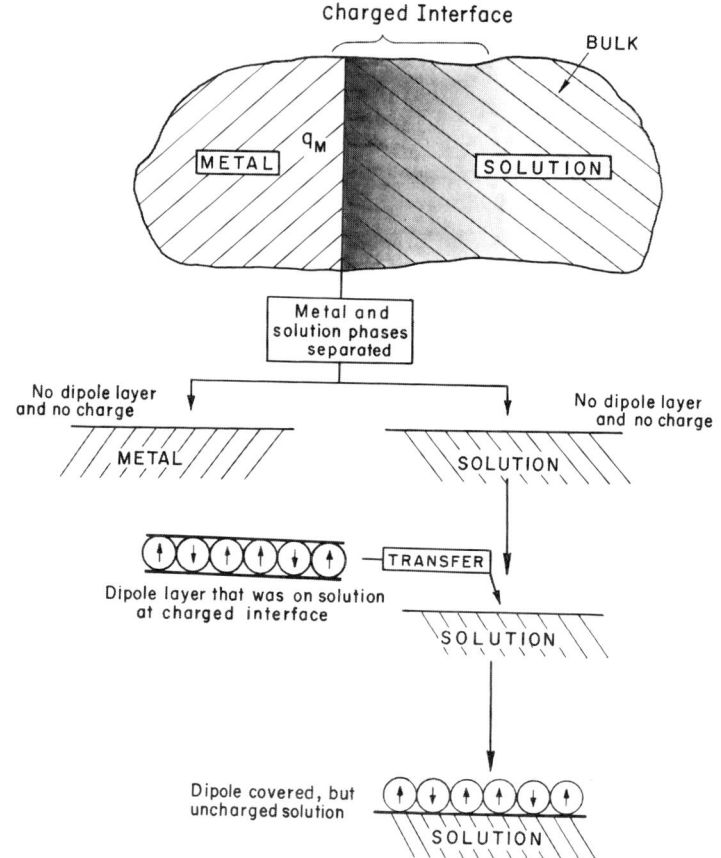

Fig. 7.38. A thought experiment involving the following steps is performed upon a charged interface: (1) The interface is dismantled, and all excess charges and dipole layers on the two phases are removed; (2) the solution phase is then covered with the same oriented-dipole layer as was present before the interface was dismantled.

question arises from the fact that, in conceptually dismantling the interface and then transferring the oriented dipoles, one has placed all the oriented solvent dipoles on the electrolyte and left none on the metal. Does this mean that the metal has no surface potential, i.e., $\chi_M = 0$, because it has no dipole layer? At first sight, this seems to be the case.

Further consideration, however, reveals that, even in the case of a metal, there is what might be termed a dipole layer. The situation is roughly as follows (Fig. 7.40): The physical surface of the metal tries to confine the free electrons inside the metal. It is as if the electrons are in "potential

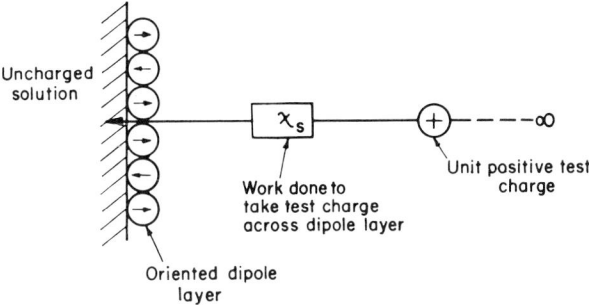

Fig. 7.39. The surface, or dipole, or χ_S, potential of the solution is the work done to bring a unit of positive charge from infinity and take it across the oriented-dipole layer of the solution.

wells." But electrons have the characteristic of being able to penetrate potential barriers (see Section 8.5.2). If they succeed, a positive charge is left behind for every electron jumping out of the metal. This is tantamount to a charge separation and a dipole layer. Thus, there is a surface potential for metals, too, and therefore a χ_M.[†]

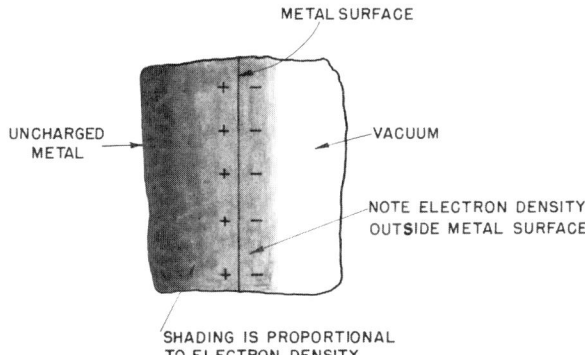

Fig. 7.40. A schematic representation of the origin of the surface, or χ, potential of a metal. Because of the finite probability of an electron's being found outside the metal surface, the electron density decays to zero outside the metal. This phenomenon is equivalent to the formation of a dipole layer across the metal surface.

[†] It will be shown in Section 8.5.2 that the probability of finding the electron in any region in space is given by the square of its wave function. Since the values of the wave function for electrons in metals do not drop abruptly to zero at the surface,

7.2.5l. The Dipole Potential Difference $^M\Delta^S\chi$ across an Electrode–Electrolyte Interface. The two surface potentials χ_S and χ_M represent the work done to carry a unit test charge from infinity in vacuum through and just across the dipole layers at the surfaces of an uncharged electrolyte and an uncharged metal. If, now, the electrode and solution phases are brought together, there will be dipole layers in the two phases. The work done to take a test charge across both these dipole layers is given by the difference of the two χ potentials. Hence, this difference

$$^M\Delta^S\chi = \chi_M - \chi_S \qquad (7.16)$$

is the *dipole* contribution to the potential difference across the interface.

By conceiving a model for the dipole layers, it is possible to make some rough calculations for the individual surface potentials (see Section 7.4.30). The χ potential for a metal has not been theoretically analyzed to the same extent as that for a semiconductor or for an electrolyte with oriented solvent dipoles. At the present time, it is customary to turn a blind eye to the existence of the χ potential for a metal.

7.2.5m. The Sum of the Potential Differences Due to Charges and Dipoles: The Absolute Electrode–Electrolyte (or Galvani) Potential Difference. The result of the two thought experiments can be summarized thus: One thought experiment described above yielded the potential difference at an electrified interface arising only from the charges, the other thought experiment yielded the potential difference arising only from the dipole layers. The former was the $\Delta\psi$ potential; and the latter, the $\Delta\chi$ potential. Thus, a *conceptual* separation of the charge and dipole contributions to the total potential has been achieved.

Since the charge separation and dipole orientation are the only two sources of a potential difference across an electrified interface, the two contributions can be summed to give the *total*, or absolute, potential across the electrode–electrolyte interface

$$\text{Absolute metal–solution potential difference} = \Delta\psi + \Delta\chi \qquad (7.17)$$

The thought experiments have therefore served the very useful purpose of permitting the conceptual synthesis of the absolute potential difference from the individual contributions—the charge-dependent contribution and

---

there is always a finite probability of finding the electron (i.e., a negative charge) outside the metal.

the dipole-dependent contribution. The term used for the total, or absolute, potential difference across an electrified interface is the *Galvani potential difference* and the symbol used is $\Delta\phi$ (delta phi).

The outer ψ and surface χ potentials were conceived in two thought experiments, one involving a charged, dipole-free phase in vacuum; and the other, an uncharged, dipole-covered phase in vacuum. What does the synthesis of these two imaginary situations represent? Obviously, it can

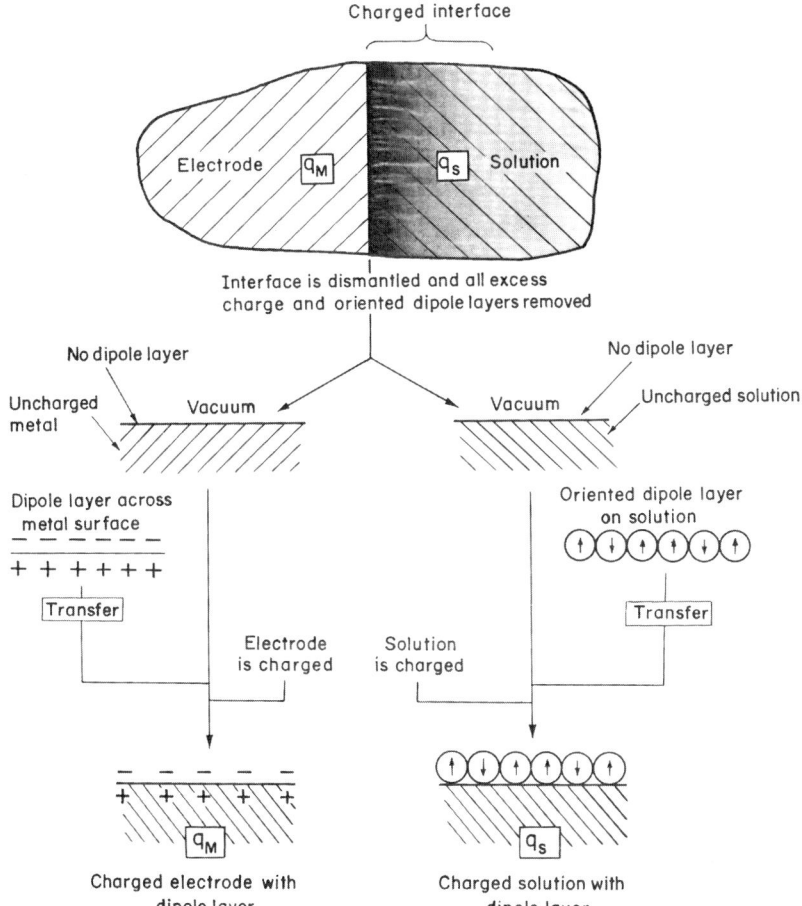

Fig. 7.41. A thought experiment to define the total potential involves the following steps performed upon an electrified interface: (1) The interface is dismantled, and all excess charges and dipole layers are removed from the two phases; (2) each phase is charged and covered with a dipole layer to reestablish its condition before the interface was dismantled.

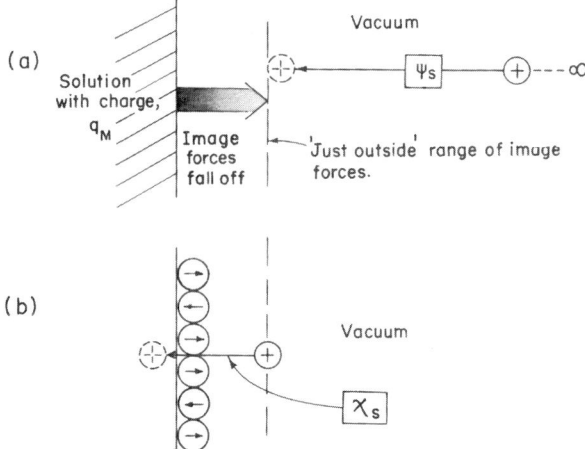

Fig. 7.42. The two stages of getting the inner, or ϕ, potential: (a) The work done to bring a unit of positive test charge from infinity to a point just outside the range of the image forces defines the outer, or ψ, potential. (b) The charge on the solution is then removed, and the solution is wrapped in an oriented-dipole layer. The work done to transport the test charge across the oriented-dipole layer defines the surface, or χ, potential. Thus, the total work to bring the test charge from infinity to a point just inside the solution is given by $\phi_S = \psi_S + \chi_S$.

be represented in terms of another thought experiment in which the phases constituting the interface are separated, the double layer turned off, and the phases charged *and* wrapped with a dipole layer (Fig. 7.41). The work done to transport a test charge from infinity to a point outside the charged phase defines the outer potential ψ; the subsequent work done in taking the charge through the surface dipole layer defines the surface potential χ.

Where is the test charge at this stage of the thought experiment? It is inside the material phase (Fig. 7.42). Thus, the work done to bring from infinity a unit test charge across the charged *surface* covered with a dipole layer to a point inside[†] the phase is the inner potential ϕ, the sum of the outer ψ and surface χ potentials; i.e.,

$$\phi = \psi + \chi \tag{7.18}$$

[†] Even though the test charge is inside the phase, its interactions with the material of the phase are not included in the definition of the ϕ potential but are reserved for the computation of the chemical potential (see Section 7.3.3).

or
$$\text{Inner potential} = \text{Outer potential} + \text{Surface potential} \quad (7.19)$$

It has been said, however, that the sum of the outer and surface potentials is the Galvani potential, i.e., the total potential difference across the interface, e.g., the metal–solution interface. Thus,

$$\begin{aligned}
\text{Absolute potential difference} &= \text{Galvani potential difference} \\
&= \Delta\psi + \Delta\chi \\
&= (\psi_M - \psi_S) + (\chi_M - \chi_S) \\
&= (\psi_M + \chi_M) - (\psi_S + \chi_S) \\
&= \phi_M - \phi_S \\
&= {}^M\Delta^S\phi
\end{aligned} \quad (7.20)$$

Thus, the absolute potential difference across an electrified interface is a difference of inner potentials.

7.2.5n. The Outer, Surface, and Inner Potential Differences. The impossibility of measuring the absolute potential difference across an electrode–electrolyte interface arises because (*cf.* Section 7.2.5*b*) the operation of using a measuring instrument necessarily generates at least one extra electrified interface. What is measured, therefore, is the total potential difference across at least *two* double layers rather than across the interface sought after. Since the absolute potential difference is in fact the difference in inner potentials, $\Delta\phi$, it means that inner potentials cannot be experimentally measured.

This conclusion can be seen from another point of view. As long as test charges stand aloof from the potential scene and only measure it, the potential is measurable. This aloofness is guaranteed as long as the test charges are outside the material phases, i.e., where they do not interact with the particles of the material phases. Only the measurement of the outer potential satisfies this criterion. Hence, it is only the ψ potential and, correspondingly, $\Delta\psi$, which can be experimentally measured.

As regards the surface potential χ, it is the result of a thought experiment involving the transport of a unit test charge across a dipole layer. The final step in its journey is to a point on the inside "fence" of the double layer. If, now, the test charge looks in a direction away from the surface and toward the interior of the phase, there lies the material medium (e.g., the electrolyte) with its particles, each one of which may interact with the test charge.

These interactions with the bulk of the phase (e.g., the electrolyte) have been tacitly ignored in the definition of the χ potential. If the test charge is an ion, e.g., all the ion–solvent (Chapter 2) and ion–ion (Chapter 3) interactions with the electrolyte bulk are switched off—an operation possible only in a thought experiment.

Hence, no direct physical operation can be prescribed for testing or probing or measuring the χ potential inside a material phase, e.g., the electrolyte. One can probe potentials inside matter only with material probes which themselves interact with matter and invalidate the whole probing process.

Since surface potentials are not measurable, any quantity which includes them is also not measurable. The inner potential ϕ is one such quantity since

$$\phi = \psi + \chi \tag{7.18}$$

Hence, the inner potential cannot be measured.

The argument just presented can be extended to the *differences* of the various potentials. The outer potential difference $\Delta\psi$ can be measured (Klein and Lange; Appendix 7.1); the surface potential difference $\Delta\chi$ cannot;[†] and, therefore, the inner potential differences $\Delta\phi = \Delta\psi + \Delta\chi$ also cannot be experimentally obtained.

7.2.5o. An Apparent Contradiction: The Sum of the $\Delta\phi$'s across a System of Interfaces Can and the $\Delta\phi$ across One Interface Cannot Be Measured. The potential difference across a phase boundary has been identified as a difference of two inner potentials. For example, the potential difference across a metal–solution interface is $^M\Delta^S\phi = \phi_M - \phi_S$. In Section 7.2.5b, however, it was shown that attempts to measure the potential difference across one interface only conclude by measuring the potential across an electrochemical system (or cell) which is a succession of interfaces. The potential difference recorded by the measuring instrument is the sum of all the potential differences across these interfaces [Eq. (7.1)]. Since each interfacial potential difference is a difference of inner potentials, Eq. (7.1)

$$V = PD_{M_1/S} + \sum_{i=1}^{n} PD_i \tag{7.1}$$

[†] While it is impossible to measure $\Delta\chi$, the change in this quantity $\delta\Delta\chi$ may under certain conditions be measured. Thus, the change in the electronic work function of metals due to adsorption of gases involves a change in the surface potential. This quantity can be obtained by observing the lowest frequency of electromagnetic radiation which will cause electron emission (the photoeletric effect).

(PD_i being the potential difference across the extra phase boundary i created by the measurement process) can be rewritten in the form

$$V = {}^{M_1}\Delta^S\phi + \sum_{i=1}^{n} (\Delta\phi_{PB})_i \tag{7.21}$$

where the symbol ${}^{M_1}\Delta^S\phi$ represents the difference of inner potentials of the phases M_1 and S, and $(\Delta\phi_{PB})_i$ is the $\Delta\phi$ across the ith extra phase boundary created by the measurement process.

Equation (7.21) shows that the potential difference recorded by the measuring instrument (equal to the potential difference across the cell) is the sum of all the inner potentials at the various interfaces *outside* the measuring instrument. Hence, a cell potential is a Galvani, or inner, potential difference.

Consider the simple system, or cell (Fig. 7.43), made up of the interfaces M_1/S, S/M_2, and M_2/M_1', with M_1''s *being the same metal as* M_1. If the potential difference across the system is written in the expanded form and terms of equal magnitude but opposite sign are canceled out, it is easy to see that

$$\begin{aligned}V = (\Delta\phi)_{\text{cell}} &= {}^{M_1}\Delta^S\phi + {}^S\Delta^{M_2}\phi + {}^{M_2}\Delta^{M_1'}\phi \\ &= (\phi_{M_1} - \phi_S) + (\phi_S - \phi_{M_2}) + (\phi_{M_2} - \phi_{M_1'}) \\ &= \phi_{M_1} - \phi_{M_1'}\end{aligned} \tag{7.22}$$

It appears, therefore, that the potential difference V registered by a measuring system, i.e., the potential difference across the system, is effectively equivalent to a difference of inner potentials between the two pieces of the same metal (or two phases of the same composition) which constitute the terminals of the measuring instrument.

It has been previously emphasized that the $\Delta\phi$ for an interface cannot be measured (Section 7.2.5n), and now it is stated that a $\Delta\phi$ can be measured between two wires of the same composition. Are these two statements contradictory?

To answer this question, let $\phi_{M_1} - \phi_{M_1'}$ be broken up into its constituent outer and surface potentials. One has

$$V = \phi_{M_1} - \phi_{M_1'} = (\psi_{M_1} - \psi_{M_1'}) + (\chi_{M_1} - \chi_{M_1'}) \tag{7.23}$$

Examine the term $\chi_{M_1} - \chi_{M_1'}$. It represents the difference of surface potentials in two wires of *identical* material, say, two copper wires. Unless their surface preparations, etc., are different, it *is to be expected that their*

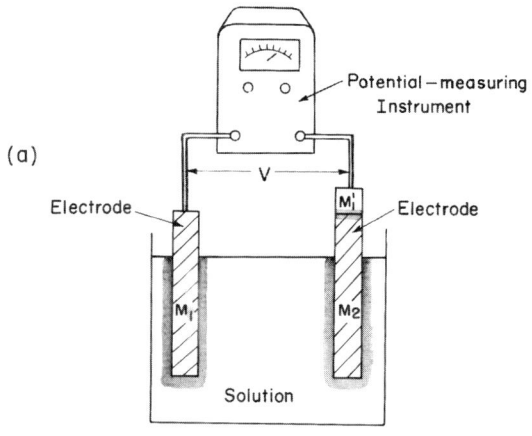

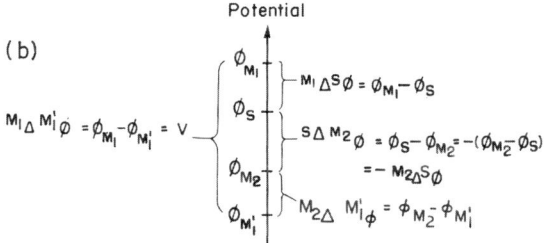

Fig. 7.43. (a) A simple electrochemical system with two electrodes M_1 and M_2 in contact with a solution S. The M_2 is joined to the potential-measuring instrument through $M_1{'}$, which is of the same composition as M_1. (b) A breakup of the potential difference V read by the potential-measuring instrument.

surface potentials are the same, i.e.,

$$\chi_{M_1} - \chi_{M_1'} = 0 \tag{7.24}$$

from which it follows that

$$\phi_{M_1} - \phi_{M_1'} = \psi_{M_1} - \psi_{M_1'} \tag{7.25}$$

Hence, when the two phases are of the same composition, their difference of inner potentials, i.e., their Galvani potential difference, reduces to a difference of outer potentials which is a *measurable* quantity (*cf.* Appendix 7.1). Thus, the potential difference V across a cell is measurable because it reduces in fact to a Volta potential difference which is a measurable quantity.†

On the other hand, the $\Delta\chi$ across an interface between two different materials, e.g., metal and solution, cannot be set equal to zero, i.e., $\chi_1 \neq \chi_2$, as it can be for an interface between two identical metals. Consequently, a $\Delta\phi$ for two *dissimilar* phases cannot be reduced to a measurable $\Delta\phi$.

Incidentally, the two wires M_1 and M_1' are really one phase. Thus, e.g., when two clean copper wires are joined, one does not form a phase boundary; one simply produces a longer copper wire.

The measurement of a $\Delta\phi$ across the whole system (or cell), i.e., the sum of all the potential differences seen by the measuring instrument, does not therefore contradict the impossibility of measuring a $\Delta\phi$ across one interface.

7.2.5p. What Deeper Understanding Has Been Hitherto Gained Regarding the Absolute Potential Difference across an Electrified Interface?
In presenting a picture of the formation of a double layer at an electrode–electrolyte interface, it became necessary to bring the nature of the interphasial potential difference into sharper focus.

The first question posed was: Can the absolute potential difference across an interface be measured? Analysis of the procedure involved in a measurement clearly showed that the potential difference across the interface under study cannot be measured; the only thing that *can* be measured is the potential difference across a cell (a system of interfaces). It looked at this stage as if the whole situation was lost and it was impossible to know the potential across the double layer and hence very difficult to investigate its structure. Fortunately, the existence of nonpolarizable interfaces saved the day. By coupling the interface under study with a nonpolarizable interface, it became possible to equate *changes* in *cell* potential (produced by an external source) with *changes* in the Galvanic potential $\Delta\phi$ across the interphase under examination.

Thus, *changes* in the phase boundary potential became measurable—i.e., the changes in potential difference across a system (or cell) in which there was one polarizable and one nonpolarizable interface.

---

† It may be said in general that the inner potential difference $\Delta\phi$ between two identical phases is measurable because the close-range interactions between the test charge and the interior of a phase, which make this quantity inaccessible in general, cancel out in this particular case. The work involved in transferring a unit test charge from one identical phase to another turns out to be simply the product of charge times the inner potential difference, $q\Delta\phi$.

A fundamental question was then posed: What contributions make up an electrode–electrolyte potential difference? To consider this question, thought experiments were analyzed. It turned out that the total potential difference across an interface could be defined as a difference of inner potentials, $\Delta\phi$, or Galvani potential difference, and a $\Delta\phi$ was contributed to by the charges at the interface (the outer potential difference, or $\Delta\psi$, portion) and also by the dipole layers at the interface (the surface potential difference, or $\Delta\chi$, portion). The $\Delta\psi$ is a *measurable* quantity, but the $\Delta\chi$ and therefore the $\Delta\phi$ (for $\Delta\phi = \Delta\psi + \Delta\chi$) *cannot* be measured.

The potential difference across an electrochemical system (or cell) is a Galvani potential difference. It is the one type of potential difference which is measurable, because the cell terminates in two wires of identical composition across which the $\Delta\phi$ reduces to the experimentally accessible $\Delta\psi$. In contrast, the potential difference across an interface (the boundary between two phases of different composition) is a $\Delta\phi$ which cannot be measured. But, as stated above, measurable changes in the $\Delta\phi$ of one interface alone can be produced by using a polarizable interface built into a system along with a nonpolarizable interface.

In the light of this discussion, it appears that what can be experimentally measured in double-layer studies are the changes in $\Delta\phi$ across an interface and the potential difference across systems of interfaces, or cells.

But the inner potential difference $\Delta\phi$ is made up of contributions from the charges and from the dipole layers at the interface. Hence, though the absolute value of $\Delta\phi$ is experimentally inaccessible, the measurable Galvani potential *changes* yield information on the distribution of charges and orientation of dipoles, i.e., *on the structure of the electrified interface*.

TABLE 7.2

The Outer ψ, Surface χ, and Inner ϕ Potentials

1. ψ potential of one phase	Can be measured
2. $\Delta\psi$ across a phase boundary	Can be measured
3. χ potential of one phase	Cannot be measured
4. $\Delta\chi$ across a phase boundary	Cannot be measured
5. ϕ potential of one phase	Cannot be measured
6. Absolute potential difference across an interface = $\Delta\phi$	Cannot be measured
7. Changes in absolute potential difference across one interface = $d\Delta\phi$	Can be measured
8. Potential difference across a cell = $\Delta\phi$ between two phases of the same composition	Can be measured

The nature of the potential difference across a phase boundary has been analyzed (Table 7.2). One's next task is to clarify the picture of adsorption of the electrolyte constituents that occur at an interface. This having been done, it will become possible to use the facts concerning adsorption to test models of the double layer.

7.2.6. The Accumulation and Depletion of Substances at an Interface

7.2.6.a. What Would Represent Complete Structural Information regarding an Electrified Interface? In discussing the second law of thermodynamics, Maxwell, the great founder of the electromagnetic theory, conceived of a hypothetical being who could, in Lilliput fashion, enter a piece of matter, determine the velocities of the particles, and segregate the fast-moving (*hot*) particles and slow-moving (*cold*) particles into different regions. In this way, the being could create a temperature difference from a material which was originally at one temperature and thus reverse the equilibrating process by which heat flows from a hot region to a cold region. *Maxwell's demon* was the name given to this mythical being.

What would one expect of such a hypothetical being—a "Gibbs' angel"[†] —in electrified interfaces? The being should be able to dive into the interphase region and quickly return with snapshots of the arrangements of ions and dipoles and neutral molecules which dwell in that area. The superposition of a sufficiently large number of snapshots would reveal the time-average *structure* of the interface: How the interface region is constituted, which ions and dipoles are in intimate contact with the electrode and in what numbers, and which ions swarm around the electrode to what extent. This is the ideal type of information that one would like to have, the time-average positions of all the particles populating the electrified interface.

A Gibbs' angel, like Maxwell's demon, unfortunately does not exist. Neither is there at present an experimental technique to achieve what could be accomplished by Gibbs' angel. Hence, one has to try to build up a picture of the structure of the interface by letting the mind play with the other types of cruder information that are available.

The position is tricky. The accumulation of species at an electrode–electrolyte interface does not advertize itself. It is not like a textile fabric

---

[†] The angel must be named after the great American physical chemist J. Willard Gibbs, who developed most of the theorems which permit the application of thermodynamics to interfaces.

which adsorbs a colored dyestuff, the amount adsorbed being determined from the intensity of the color "picked up."

In the case of the metal–solution interface, the charge on the metal is one of the signals which can be picked up from the angstroms-wide interfacial region. This electrode charge is mirrored on the solution side by an equal and opposite net charge. But, for many years, it was difficult to sort out, from this net charge in the solution, the separate contributions of the positive and negative charges, i.e., the relative concentrations of cations and anions in the interphase.

Even when this was accomplished, this "adsorption" of cations and anions did not prove to be an elementary affair. Was the adsorption *on* the metal or *near* the metal? The situation is not as trivial as in gas-phase adsorption where gas particles stick to the solid surface in partial or complete monolayer fashion. The adsorption at a metal–solution interface concerns a more extended and tenuous matter. It is simply a change of concentration near the interface which may or may not include ions in contact with the metal.

Further, in gas-phase adsorption, the "solvent" for the gas particles, is vacuum—a most inert, indifferent, and noninterfering solvent. In solution however, the ions, etc., are dissolved in a solvent, usually water. This water, too, can change its concentration near the interface compared with its concentration in the bulk of the solution. Hence, the meaning of concentration in terms of gram ions per cubic centimeter is subtly different from that which it has in the bulk of the solution because the reference phase, water, may itself undergo a concentration change as the interphase is approached.

Lastly, many techniques that can be used at the solid–gas interface become inapplicable in electrochemical systems. For example, any technique that requires high vacuum, e.g., electron microscopy, cannot work in the presence of the electrolyte.

It is clear that the adsorption of species in the metal–solution interphase region needs a subtle analysis. The unraveling of the complex situation and the building up of a basic picture of the accumulation and depletion of species at an electrified interface is one of the principal achievements of the new electrochemistry and is largely due to the American electrochemist, Grahame.

7.2.6b. The Concept of Surface Excess. Suppose that Gibbs' angel, after taking snapshots of the arrangement of particles in the interphase region, plotted graphs showing the time-average concentrations of the various species against distance x from some reference plane parallel to a planar electrode.

At the instant of immersion of the electrode in the electrolyte, i.e., at time $t = 0$, the graph for some species i, say, the positive ions M⁺, would show that the concentration was independent of x and equal to the bulk concentration c_i^0 (Fig. 7.44). This is because, at $t = 0$, the double layer has not yet been formed, i.e., the interface has not yet become electrified.

For $t > 0$, the anisotropic forces at the boundary begin to operate, and the separation and sorting-out of the various charges in the interphase takes place.

Now, what would be the nature of the concentration profile at $t \to \infty$, i.e., after the steady-state double layer is formed? It is to be expected (Fig. 7.45) that the concentration $c_i'(x)$ at $t \to \infty$ would have altered from the initial value c_i^0, i.e., from the value before the double layer was formed.

The analogy with diffusion problems is close. In the diffusion case, e.g., the constant-flux problem (Section 4.2.10), the concentration of the diffusing species i was considered to be c_i^0 everywhere at $t = 0$. Once the diffusion source or sink was created, the concentration near the source or sink departed from the bulk value. The anisotropic forces at the electrode-electrolyte boundary do the same job for the adsorption process as the diffusion source or sink does for diffusion. In each case, the concentrations are perturbed from the initial value, which is the bulk value.

It was found most convenient in the diffusion problem not to discuss the actual concentrations $c_i'(x)$ but the *perturbations*, or departures from the bulk concentrations. This was done by defining the perturbations thus

$$c_i(x) = c_i'(x) - c_i^0 \tag{7.26}$$

i.e.,

$$\begin{matrix}\text{Perturbation} \\ \text{or excess concentration}\end{matrix} = \begin{matrix}\text{Actual} \\ \text{concentration}\end{matrix} - \begin{matrix}\text{Bulk} \\ \text{concentration}\end{matrix} \tag{7.27}$$

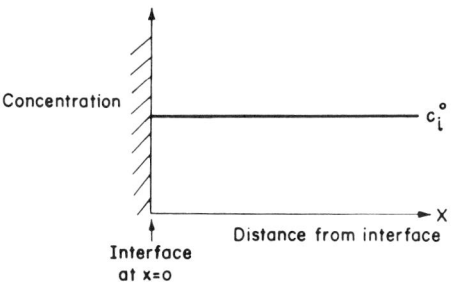

Fig. 7.44. The concentration c_i of the species i as a function of the distance x from the interface, at the instant of immersion ($t = 0$) of the electrode in the solution.

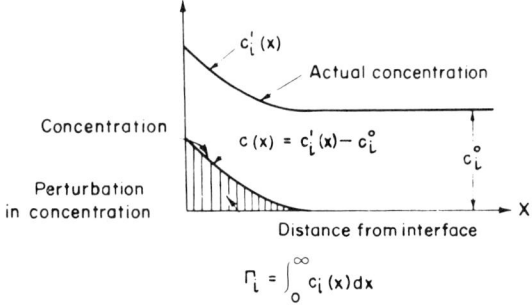

Fig. 7.45. A schematic representation of the distance variation of the concentration at $t \to \infty$, i.e., after the steady-state double layer is formed. The actual concentration $c_i'(x)$ and the concentration perturbation $c_i(x) = c_i'(x) - c_i^0$ are both shown. The surface excess Γ_i of the species i is the definite integral $\Gamma_i = \int_0^\infty c_i(x)\,dx$ indicated by the hatched area.

For the electrified interface, too, a similar procedure can be followed that results in an identical defining equation. In this case, however, the departures from the bulk value represent changes in concentration of the species resulting from the fact that the anisotropic forces at the interface have been switched on at $t = 0$. A schematic representation of these perturbations or concentration changes is shown in Fig. 7.45.

What do these perturbations in concentration represent? They are quantitative measures of the accumulation or depletion of species in the interphase region. Unfortunately, only a Gibbs' angel could provide directly the concentration–distance profile for the various cationic and anionic species in the double layer. At present, there are no techniques sensitive enough to determine experimentally the distance variation of the concentration changes of the various species in solution. One must settle for knowledge obtained by indirect argument and therefore of lesser certainty.

Gibbs conceived the idea of measuring adsorption in the interphase by using the integral of the perturbation in concentration with distance.†

---

† Note the greater complexity of defining adsorption here in studies of electric double layers than, e.g., at metal–gas systems. With electric double layers, one is concerned with the whole interphasial region. The total adsorption is the sum of the increases of concentration over a distance, which, in dilute solutions, may extend for hundreds of angstroms. Within this total adsorption, there are, as will be seen, various *types* of adsorptive situations, including one, *contact adsorption*, which counts only those ions in contact with the electronically conducting phase (and is then, like the adsorption

This is shown by the shaded area in Fig. 7.45. This *definite* integral represents the summation of the concentration perturbations, i.e., of the *excess* concentrations [Eq. (7.26)] in all the lamellae. The integration or summation is carried out from a reference plane $x = 0$ up into the bulk of the electrolyte, $x \to \infty$. In the electrolyte bulk, the anisotropic forces of the interface are negligibly small, and, hence, the perturbations tend to zero and do not contribute to the integral. The result of this summation is known as the *Gibbs surface excess* Γ, or simply the surface excess,

$$\Gamma_i = \int_0^\infty c_i(x)\, dx \tag{7.28}$$

It is easy to see that the surface excess is either a positive or a negative quantity depending on whether the departure from the bulk concentration is positive or negative, i.e., on whether there is an accumulation or depletion of the particular species i in the interface.

7.2.6c. Does Knowledge of the Surface Excess Contribute to Knowledge of the Distribution of Species in the Interphase Region? Complete knowledge regarding the structure of the interface would consist of information regarding the arrangement of all the particles or what is the analytical equivalent, the variation of the actual or perturbed concentrations of the species with distance from the reference plane. Such knowledge would be on the *microscopic* level.

The definition of surface excess, on the other hand, starts with the concentration profile but involves an integration *between limits*, i.e.,

$$\Gamma_i = \int_0^\infty c_i(x)\, dx \tag{7.28}$$

Once the integration is done and *limits are inserted*, one obtains a *number* (so many moles per square centimeter) and loses all knowledge of the function $c_i(x)$. It is a well-known mathematical fact that no unique integrand, or function, is recoverable from a definite integral. In other words, after carrying out the integration and evaluating the surface excess, the concentration profile cannot be discerned. The integration removes the possibility of recovering the microscopic detail from the surface excess.

Now, the $c_i(x)$ versus x curve used for the integral leading to the concept of a surface excess has not yet been determined experimentally. It has

---

referred to in metal–gas systems, the particles *on* the surface). Metal–gas systems deal with interfaces, one might say, whereas metal–electrolyte systems deal primarily with interphases and only secondarily with interfaces.

been based, in the discussion given here, on thought experiments illustrated with the imaginary Gibbs' angel. It is a curve imagined for the purpose of developing the concept of surface excess.

It will be shown, however, that, in interfacial chemistry, the surface excess affects many quantities, e.g., the interfacial tension at a mercury–electrolyte interface and the way in which this depends upon concentration. In fact, it will be shown (*cf.* Section 7.3.8) that surface excess can be experimentally determined frcm thermodynamic measurements. But, from these measurements, it is not possible directly to recover knowledge of how the concentrations of species vary in their distribution with distance from the electrode. Surface excess is a *macroscopic* concept, and knowledge of surface excess is macroscopic in character, attainable by thermodynamical reasoning without recourse to modelistic arguments.

What, then, is the point of a measurement of surface excess? The purpose of this measurement is that it provides material for the testing of models of the electrified interface. From these models, one can calculate the concentration variation of the surface excess of any species. The extent of agreement between the variation calculated thus and that determined from thermodynamic reasoning (*cf.* Section 7.3.8) determines the extent of validity of the model. This is why the concept of surface excess is so useful, despite its macroscopic nature.

7.2.6d. Is the Surface Excess Equivalent to the Amount Adsorbed? Often, the surface excess of a particular species has been simply assumed to be the quantity of that species adsorbed on the surface of an electrode.

To examine this point of view, consider the profile of the actual concentration. The interphase region can be said to begin from the point where

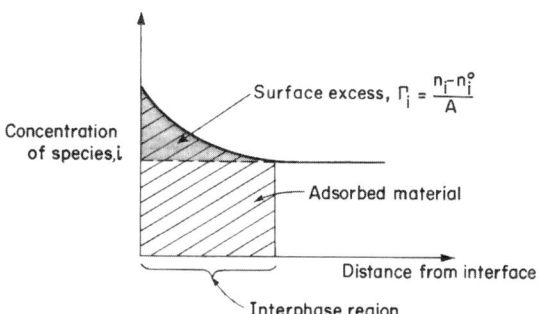

Fig. 7.46. The distinction between the amount of adsorbed material (hatched area) in the interphase region and the surface excess (shaded area).

the actual concentration departs from the bulk value. The amount of the species which can be said to have adsorbed per unit area of the interface is equal to the *total* amount of the species existing inside the interphase region divided by the area of the interface. In Fig. 7.46, the adsorbed material is indicated by the hatched area.

The surface excess, however, is the amount of material *over and above that which would have existed had there been no double layer*. This surface excess is indicated by the shaded area in Fig. 7.46. It could also be represented by the expression

$$\Gamma_i = \int_0^\infty c_i(x)\,dx = \frac{1}{A}\int_0^\infty c_i(x)\,dv = \frac{1}{A}\int_0^\infty [c_i'(x) - c_i^0]\,dv$$

$$= \frac{1}{A}\int_0^\infty d(\Delta n_i)$$

$$= \frac{n_i}{A} - \frac{n_i^0}{A} \qquad (7.29)$$

where dV is the volume of an infinitesimally thin lamina having a cross section A, n_i is the actual number of moles of species i in the interphase region, n_i^0 is the number of moles that would have been there if there had been no double layer, and A is the area of the interface.

It is now obvious that the amount of adsorbed material per unit area n_i/A, is not equal to the surface excess. In so far as the bulk concentration, or n_i^0, tends to zero, the adsorption can be taken as approximately equal to the surface excess.

7.2.6e. Is the Surface Excess Measurable?

The surface excess is measurable. There are essentially two approaches to its measurement.

One approach is to make use of the fact that the surface excess of a species is approximately equal to its adsorption on the electrode when the bulk concentration of that species is extremely low. Under these conditions, the surface excess can be approximately known by directly measuring adsorption. How can this be done?

One method of directly determining adsorption involves the use of radioactive isotopes (Fig. 7.47). The species, the adsorption of which is under study, is "tagged," and the electrode is chosen thin enough to permit a counter to be placed behind it to count the tagged particles which accumulate in the interphase region. Thus, from the number of counts per minute, it is possible to measure the adsorption and therefore the surface excess of the radioactive species.[†]

[†] Of course, corrections in the total radiation measured must be made because of the background radiation from the bulk of the solution.

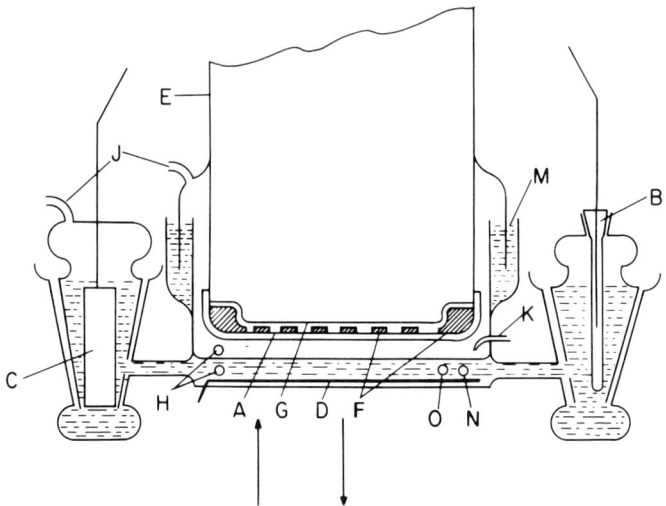

Fig. 7.47. Scheme of the apparatus for radiotracer measurements of adsorption: *A*, thin foil of adsorbent; *B*, reference electrode; *C*, counter electrode; *D*, auxiliary electrode; *E*, body of the counter; *F*, metal support; *G*, window; *H*, gas inlets; *J*, gas outlets; *K*, inlet for injecting and sampling; *M* water seal; *N*, thermistor probe for temperature control; and *O*, heating coil.

Another way of getting at the surface excess is based on the measurement of some property which depends on this surface excess.

One such property which depends on the surface excess is the surface tension, or, rigorously speaking, the interfacial tension[†]. The latter is to a surface what pressure is to a bulk volume. Work has to be done against the surface tension to increase the area of the interface, just as it has to be done against pressure to increase the volume of a bulk phase. The phenomenon of surface tension arises from the anisotropy of forces existing in an interphase. But these forces are affected by the particles in the interphase, and, therefore, the surface tension must be related to the degree to which these particles are present, i.e., to the surface excess.

The precise thermodynamic relationship between the surface tension and the surface excess can be worked out easily (see Section 7.3.8). The relationship is rigorous and accurate and provides the best method of determining surface excesses. The catch is, however, that the method is only

---

[†] The two terms *surface tension* and *interfacial tension* will be used interchangeably in the subsequent text, though the tension is strictly an interphasial tension (both phases determine it) rather than that of a surface.

suited for the interface between a *liquid* metal and a solution. This is because the surface tension of liquids can be easily determined (see Section 7.3.1); not so, the surface tension of solids.

7.2.6f. The Special Position of Mercury in Double-Layer Studies. The comparative ease and accuracy with which the surface tension of *liquid* metals can be determined has led, in the past, to an emphasis on the use of mercury as the preferred metallic electrode in studies on electrified interfaces.

There are, however, several other reasons for the preoccupation with mercury. One of the great problems in double-layer and electrode kinetic studies is how to avoid contaminating the electrode surface, i.e., the interphase region, with unwanted impurities, organics, grease, dust, etc. If a mercury-drop electrode is used (or for that matter any liquid metal), every time a drop falls and a new drop forms, the electrode presents a virgin surface to the solution. With a simple capillary connected to a reservoir, a major contamination problem is thus circumvented.

The use of a liquid metal also removes all the complications that might arise from the characteristic structure and topography of a solid surface. Liquid surfaces are nonstructured and highly reproducible.

Finally, it is important that, in acquiring the first view of an interface, one should not have the complications of charges leaking through the double layer, i.e., charge-transfer reactions. The interface should be polarizable so that it responds exactly to all the changes in the potential difference of an external source when coupled to a nonpolarizable interface (*cf.* Section 7.2.5c). The mercury–solution interface approaches closest to the ideal polarizable interface (*cf.* Section 7.2.5d) over a range of nearly 2 V. Few other interfaces present such a wide range of polarizability.

In view of all these advantages, the mercury–solution interface has occupied a unique position in double-layer research. The philosophy has been: The mercury–solution interface displays the minimum number of complications; it should therefore be investigated first, and, when understood, one must move on to other interfaces, particularly those involving the solid metals. The study of the mercury–solution interface is therefore a basis to the understanding of technologically important electrified interfaces.

Further Reading

1. W. Gibbs, *Collected Works*, Vol. I, 2nd Ed., Longmans, Green, & Co., Inc. New York, 1924.
2. E. Lange and K. P. Mishchenko, *Z. Phys. Chem.*, **A149**: 1 (1930).

688 CHAPTER 7

3. R. Parsons, "Equilibrium Properties of Electrified Interphases," in: J. O'M. Bockris and B. E. Conway, eds., *Modern Aspects of Electrochemistry*, Vol. I, Butterworth's Publications, Ltd., London, 1954.
4. R. Parsons, "The Structure of the Electric Double Layer and Its Influence on the Rates of Electrode Reactions," in: P. Delahay, ed., *Advances in Electrochemistry and Electrochemical Engineering*, Vol. I, Interscience Publications, Inc., New York, 1961.
5. P. Delahay, *Double Layer and Electrode Kinetics*, Interscience Publications, Inc., New York, 1965.
6. B. E. Conway, *Theory and Principles of Electrode Processes*, The Ronald Press Company, New York, 1965.
7. E. Gileadi, "Adsorption in Electrochemistry," in: E. Gileadi, ed., *Electrosorption*, Plenum Press, New York, 1967.

7.3. THE THERMODYNAMICS OF ELECTRIFIED INTERFACES

7.3.1. The Measurement of Interfacial Tension as a Function of the Potential Difference across the Interface

It has been emphasized that the mercury–solution interface possesses many unique advantages in the matter of understanding the electrified interface. Further, for liquid metals, the interfacial tension is a property which is easily measurable and directly relatable to the potential difference across the interface and the surface excess of various species in solution. The time has come, therefore, to indicate the nature of the basic setup for making a measurement of interfacial tension and then to relate the interfacial tension to quantities pertinent to the interface, such as the surface excess and the potential difference.

Consider the electrochemical system shown in Fig. 7.48. The essential parts are (1) a mercury–solution polarizable interface, (2) a nonpolarizable interface, (3) an external source of variable potential difference V, and (4) an arrangement to measure the surface tension of the mercury in contact with the solution.

What are the capabilities of this system? Since the system consists of a polarizable interface coupled to a nonpolarizable interface, changes in the potential of the external source are almost equal to the changes of potential only at the polarizable interface, i.e., the changes in $\Delta\phi$ across the mercury–solution interface are almost equal to changes in potential difference V across the terminals of the source. Hence, the system can be used to produce predetermined $\Delta\phi$ *changes* at the mercury–solution interface (Section 7.2.5p). Further, the measurement of the surface tension of the mercury–solution interface is possible, and, since this has been stated (Section 7.2.6e) to be

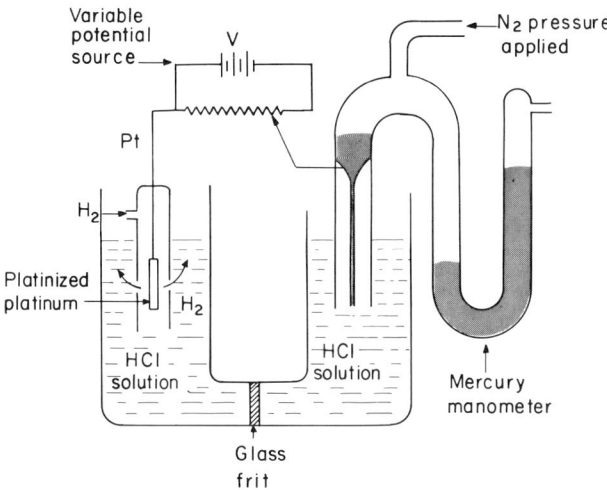

Fig. 7.48. Schematic apparatus for the measurement of the surface tension γ of mercury as a function of cell potential V.

related to the surface excess, it becomes possible to measure this quantity for a given species in the interphase.

In short, the system permits what are called *electrocapillary* measurements, i.e., the measurement of the surface tension of mercury (in contact with the solution) as a function of the electrical potential difference across the interface.

The measurements of surface tension are achieved by using a fine capillary and adjusting the height of a mercury column so that the mercury in the capillary is stationary. Under these conditions of mechanical equilibrium, the surface tension γ is obtained approximately from a simple expression (Fig. 7.49)

$$\gamma = \frac{h\varrho gr}{2} \qquad (7.30)$$

What has been described is the simplest version of the *capillary electrometer*. The system can be sophisticated by using controlled gas pressure to force the mercury in the capillary to desired distances from the tip, by using advanced optical systems in the recording of the height of the mercury column; and by connecting a sensing device of this height to the gas pressure and the applied potential. But even with the simple version of this system, it is amazing what an amount of useful information can be gained from the form of electrocapillary curves, i.e., plots of interfacial tension γ versus changes in interfacial potential difference V.

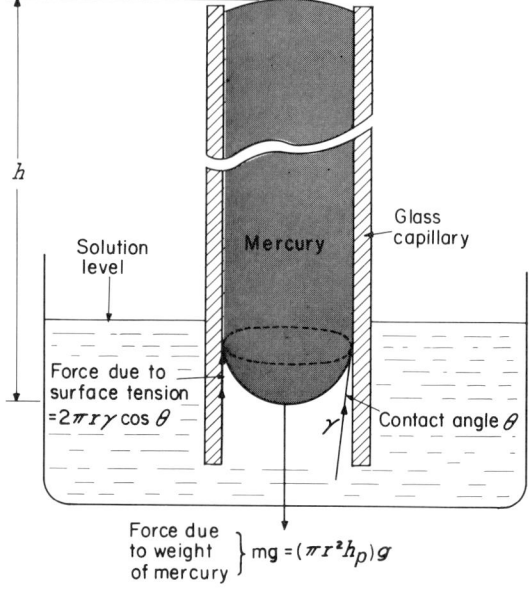

Fig. 7.49. When the mercury column in the capillary is stationary and therefore in mechanical equilibrium, its weight is exactly balanced by the total force of its surface tension. The weight of the mercury column acts downward and is equal to the density ϱ times the volume of the column, $\pi r^2 h$, times the gravitational constant g; this weight compensates the surface-tension force, which is equal to the perimeter of the contact, $2\pi r$ times the upward component $\gamma \cos \theta$ of the surface tension. Since $2\pi r \gamma \cos \theta = \pi r^2 h \varrho g$, it follows that $\gamma = rh\varrho g/2 \cos \theta$. For the mercury glass interface, $\theta \approx 0$; hence, $\gamma = rh\varrho g/2$.

7.3.2. Some Basic Facts about Electrocapillary Curves

The interfacial tension depends on the forces arising from the particles present in the interphase region. If the arrangement of these particles, i.e., the composition of the interface, is altered by varying, e.g., the potential difference across the interface, then the forces at the interface should change and thus cause a change in the interfacial tension. One would expect therefore that the surface tension γ of the metal–solution interface should vary with the potential difference V supplied by the external source.

The experimental γ versus V curves obtained by electrocapillary measurements demonstrate this variation of surface tension γ with the potential

TABLE 7.3

Variation in Tension of a Mercury–1.0N CsCl Interface with Potential Difference

Potential difference, mV vs normal calomel electrode	Interfacial tension, dynes cm^{-1}
— 0	345.0
— 100	376.4
— 200	397.1
— 300	410.2
— 400	418.8
— 500	422.6
— 600	422.9
— 700	419.9
— 800	414.0
— 900	405.6
—1000	395.1
—1100	382.9
—1200	369.2
—1300	356.6

difference V across the cell. What is informative, however, is the nature of the variation (Table 7.3). A typical γ versus V electrocapillary curve is almost a *parabola* (Fig. 7.50). The potential at which the surface tension is a maximum is known as that of the *electrocapillary maximum* (ecm).

The measurements also show that surface tension varies with the composition of the electrolyte (Table 7.4). This is easily seen by comparing electrocapillary curves obtained in solutions of different electrolyte concen-

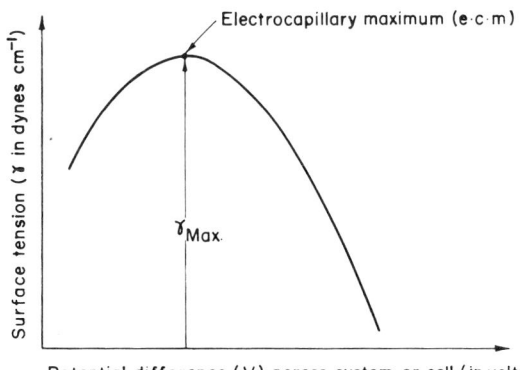

Fig. 7.50. A typical γ versus V electrocapillary curve.

TABLE 7.4
Variation in Interfacial Tension of a Hg–CsCl Interface with CsCl Concentration†

CsCl normality	Interfacial tension, dynes cm^{-1}
3.0	390.0
1.0	395.1
0.3	398.9
0.1	402.9
0.03	406.4
0.01	410.8

† The surface-tension values are at a potential of -1000 mV vs a calomel electrode in the same solution.

tration. As the solution is diluted, the maximum of surface tension rises (Fig. 7.51).

The surface tension was stated (Section 7.2.6e), on general grounds, to be related to the surface excess of species in the interphase. The surface excess in turn represents, in some way, the structure of the interface. It follows therefore that electrocapillary curves must contain many interesting messages about the double layer at the electrode–electrolyte interface. But, to understand such messages, one must learn to decode the electrocapillary data. It is necessary to derive quantitative relations among surface tension, excess charge on the metal, cell potential, surface excess, and solution composition.

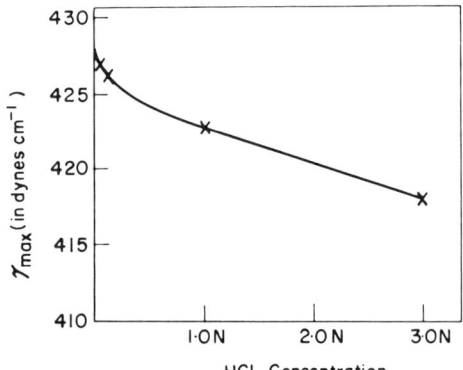

Fig. 7.51. The variation of γ_{max} with electrolyte concentration.

7.3.3. A Digression on the Electrochemical Potential

7.3.3a. Definition of Electrochemical Potential. To understand the potential difference across an electrified interface, thought experiments were used to consider the work done in moving unit test charges. The following potentials emerged from the analysis: (1) the outer, or ψ, potential arising from the work done to transport a unit test charge to a point just outside a charged but dipole-layer-free phase; (2) the surface, or χ, potential arising from the work done to carry the unit charge across the dipole layer at the surface of an uncharged phase; (3) the inner, or ϕ, potential arising from the work done to carry the test charge from infinity up to and across the dipole layer at a charged phase. In defining the χ and ϕ potentials, the test charges were prohibited from interacting with the bulk of the phases.

Now, what will happen if the material phase (e.g., the electrolytic solution) is imagined to be bereft of either surface charge or a surface dipole layer? Consider a thought experiment (Fig. 7.52) involving the transport of a test charge from infinity to a point deep inside the solution phase. The outer, or ψ, potential will be zero because the solution is uncharged. Similarly, the surface, or χ, potential will be zero because there are no surface dipole layers. Hence, the inner, or ϕ, potential will also be zero, which means that zero *electrical* work is done with the test charge.

But, once the test charge begins to be affected by all the interactions due to the particles in the solution, work will have to be done to take it inside the solution. What interactions are these? The test charge will feel, for instance, ion–solvent interactions (Section 1.2), ion–ion interactions (Section 1.3), and the repurcussions of solvent–solvent interactions. All these interactions can be lumped together and called *chemical*. In the thought experiment, therefore, one can use the chemical work, i.e., the work done against all these interactions with the particles of the material phase, to define the chemical, or μ, potential (*cf.* Chapter 3). The chemical potential μ_i of a particular species i is the work done to bring a mole of i particles from infinity into the bulk of an uncharged, dipole-layer-free material phase.

Thus, a test charge not only interacts with the charges and dipole layers on the *surface* of the phases forming the interface, but it also interacts with the *bulk* of the phases. What, therefore, is the *total* work in taking a mole of charges from infinity in vacuum into the bulk of the material phase? It is the synthesis of the result of two thought experiments, one involving a material phase without either charges or dipole layer on the surface and

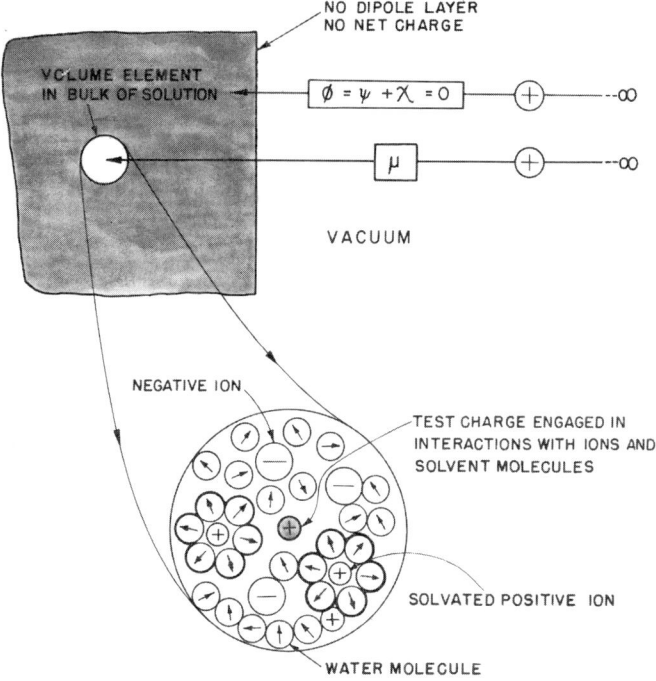

Fig. 7.52. A thought experiment for the definition of the chemical potential μ. An uncharged solution without an oriented-dipole layer on its surface is taken. The work done to transport a unit of positive test charge from infinity into the interior of the phase is the chemical potential μ of the phase. The electrical work $\phi = \psi + \chi$ is zero because there is no charge and no oriented-dipole layer on the surface of the solution.

the other involving only the charges and the dipole layer. The total work is the sum of the chemical work and the electrical work.

Total work = Chemical work + Electrical work per mole of charges

The symbol used for this catchall total work is the *electrochemical potential* $\bar{\mu}$. Hence,

$$\bar{\mu} = \mu + zF\phi = \mu + zF(\psi + \chi) \tag{7.31}$$

The factor $zF\phi$ arises because ϕ is the electrical work to bring a *unit* charge, $z_i e_0 \phi$ is the electrical work to transport one particle bearing a charge $z_i e_0$, and $N_A z_i e_0 \phi = z_i F \phi$ is the electrical work to bring an Avogadro number N_A of particles inside the material phases.

The electrochemical potential $\bar{\mu}_i$ includes all types of work involved in bringing particles[†] (of charges $z_i e_0$) into material phases. Nothing is left out. The $\bar{\mu}$ includes the chemical work μ, the charge contribution ψ, and the dipole contribution χ.

The concept of electrochemical potential has arisen frequently in earlier chapters (cf., e.g., 4.4.14). Just as the gradient of the chemical potential ($\partial \mu/\partial x$ for the x component of this gradient) acts as the driving force in pure diffusion and the gradient of the electric potential ($d\phi/dx$) acts as the driving force in pure conduction, the gradient of the electrochemical potential ($\partial \bar{\mu}/\partial x$ for the x component) can be considered the *total* driving force for the transport of a charged species, the total transport process consisting of both diffusion and conduction. In the present context, instead of approaching the concept of electrochemical potential through the facts of transport, it has been arrived at in a discussion of the work done in transporting a charge into a phase.

7.3.3b. Can the Chemical and Electrical Work Be Determined Separately? In the case of transport processes, the total driving force for the flow of a particular species j, i.e., the gradient of electrochemical potential, $\partial \bar{\mu}_j/\partial x$, was considered split up into a chemical (diffusive) driving force $\partial \mu_j/\partial x$ and an electrical driving force for conduction, $z_j F\, \partial \phi/\partial x$,

$$\frac{\partial \bar{\mu}_j}{\partial x} = \frac{\partial \mu_j}{\partial x} + z_j F \frac{\partial \phi}{\partial x}$$

It was also possible to set up experimental conditions in which either $\partial \mu_j/\partial x$ or $z_j F\, \partial \phi/\partial x$ could be reduced to zero (see Section 4.4.14). For example, by switching off the externally applied field, $\partial \phi/\partial x$ inside the electrolyte could be reduced to zero. Similarly, by avoiding a concentration gradient inside the electrolyte, $\partial \mu_j/\partial x$ can be tended to zero. Thus, the gradients of the chemical potential and the electric potential, i.e., the chemical and electrical driving forces, could be determined separately.

The separation into chemical and electrical terms is possible with the *gradients* but not with the *quantities*, i.e., μ_j and ϕ, themselves. The reason is simple. The electrochemical potential $\bar{\mu}_j$ was only *conceptually* separated into a chemical term μ_j and an electrical term $z_j F\phi$. The conceptual separation was based on thought experiments; in practice, no experimental arrangement can be devised to correspond to the thought experiment described in Section 7.3.3a. Thus, e.g., one cannot switch off the charges

---

[†] If the particles are uncharged, then $z_i = 0$ and $\bar{\mu}_i = \mu_i$.

and dipole layer at the surface of a solution as one can switch off the externally applied field in a transport experiment. Only the combined effect of μ_j and $z_j F\phi$ can be determined.

7.3.3c. A Criterion of Thermodynamic Equilibrium between Two Phases: Equality of Electrochemical Potentials. It has been stated that the total driving force responsible for the flow or transport of a species j is the gradient $d\bar{\mu}_j/dx$ of its electrochemical potential. But, when there is net flow or flux of any species, it means that the system is not at equilibrium. Conversely, for the system to be at equilibrium, it is essential that there be no drift of any species—hence, that there should be zero gradients for the electrochemical potentials of all the species. It follows, therefore, that, for an *interface* to be at equilibrium, the gradients of electrochemical potential of the various species must be zero across the phase boundary, i.e.,

$$\frac{d\bar{\mu}_j}{dx} = 0 \tag{7.32}$$

By integration, it follows that the value of the electrochemical potential of a species j must be the same on both sides of the interface, i.e.,

$$(\bar{\mu}_j)_M = (\bar{\mu}_j)_S \tag{7.33}$$

In other words, the change in electrochemical potential in transporting the species from one phase to the other must be zero, i.e.,

$$\Delta\bar{\mu}_j = 0 \tag{7.34}$$

Now, the electrochemical potential $\bar{\mu}_j$ of the species j in a particular phase is the change in free energy of the system[†] resulting from the introduction of a mole of j particles into the phase while keeping the other conditions constant, i.e.,

$$\bar{\mu}_j = \left(\frac{\partial \bar{G}}{\partial n_j}\right)_{T,p,n_{i \neq j}} \tag{7.35}$$

Hence, the equality of electrochemical potentials on either side of the phase boundary implies that the change in free energy of the system resulting from the transfer of particles from one phase to the other should be the

---

[†] More rigorously, the electrochemical free energy $\bar{G}$ should be used here. This is related to the free energy in the same way as the electrochemical potential $\bar{\mu}$ is related to the chemical potential μ.

same as that due to the transfer in the other direction. But this is only another way of stating that, when a thermodynamic system is at equilibrium, its free energy is a minimum, i.e.,

$$d\bar{G} = 0 \tag{7.36}$$

This is a well-known thermodynamic truth.

7.3.3d. Nonpolarizable Interfaces and Thermodynamic Equilibrium. It has just been shown that, for an interface to be in thermodynamic equilibrium, the electrochemical potentials of all the species must be the same in both the phases constituting the interface. Since the difference in electrochemical potential of a species i between two phases is the work done to carry a mole of this species from one phase (e.g., the electrode) to the other (e.g., the solution), it must be the same as the work in the opposite direction. This implies a free flow of species across the interface. But an interface which maintains an "open border" is none other than a nonpolarizable interface (*cf.* Section 7.2.5d).

A simple conclusion follows. *Thermodynamic equilibrium exists at a nonpolarizable interface.* Hence, one can immediately apply the criterion of thermodynamic equilibrium to a nonpolarizable interface. That is, from Eq. (7.34),

$$\begin{aligned} {}^S\!\varDelta^M\bar{\mu}_j &= {}^S\!\varDelta^M(\mu_j + z_jF\phi) \\ &= {}^S\!\varDelta^M\mu_j + z_jF{}^S\!\varDelta^M\phi \\ &= 0 \end{aligned} \tag{7.37}$$

where j is the species which is exchanged across the nonpolarizable interface.[†] Hence,

$${}^S\!\varDelta^M\phi = -\frac{1}{z_jF}{}^S\!\varDelta^M\mu_j$$

or

$$d({}^S\!\varDelta^M\phi) = -\frac{1}{z_jF}d\mu_j \tag{7.38}$$

This equation may be utilized whenever a nonpolarizable interface is treated.

[†] When j is the species which is exchanged across the nonpolarizable interface, i.e., the species which is involved in the charge-transfer reaction leading to the leakage of charge across the interface, it is customary to say that the interface, or the electrode, is *reversible* with respect to the species j.

7.3.4. Some Thermodynamic Thoughts on Electrified Interfaces

All thermodynamic thinking begins with a definition of the portion of the universe under study, i.e., the system. Here, the system is an electrode–electrolyte interface; restrictions regarding the polarizability of the interface will be introduced as and when required.

The next step is to write down the first and second laws of thermodynamics for the system. If the system is a closed one (no matter enters or leaves it), the statement of the combined first and second laws is

$$dU = T\,dS - W \tag{7.39}$$

where $T\,dS = Q$ is the heat reversibly supplied *to* the system in an infinitesimal change, and W is the work reversibly carried out *by* the system. For an open system, not only heat but also matter may be exchanged between the system and its surroundings. To introduce a mole of the species i, the chemical work done on the system is μ_i. Hence, to alter the number of moles of i in the system by dn_i, the work done by the system is $-\mu_i\,dn_i$. Hence, for an open system, this chemical work must be included and the combined first and second laws must be written

$$dU = T\,dS - W - \sum \mu_i\,dn_i \tag{7.40}$$

where $\sum \mu_i\,dn_i$ is the work done by the system in expelling dn_i moles of species i, μ_i being the work of transfer per mole.

Now, in the case of an electrode–electrolyte interface M_1/S, what are the various possible types of work? There is firstly the work of volume expansion, $p\,dV$; secondly, one might in some way increase the area of the interface by an amount dA, in which case the work of increasing the area of the interface is $\gamma\,dA$, where γ is the interfacial tension; and finally, one might, e.g., connect up the metallic phase to an external source of electricity and alter the charge on the metal by an amount dq'_M, in which case the electrical work of transferring the charge dq'_M is $^{M_1}\!\Delta^S\phi\,dq'_M$.

Introducing these work terms in place of W in Eq. (7.40), the statement of the combined first and second laws of thermodynamics applied to the above system reads

$$dU = T\,dS - p\,dV - \gamma\,dA - {}^{M_1}\!\Delta^S\phi\,dq'_M - \sum_i \mu_i\,dn_i \tag{7.41}$$

Now, each term on the right-hand side is a product of an intensive factor (one which does not depend on the amount of matter in this system) and an extensive factor (one which does depend on the amount of matter in the system). Thus,

$$dU = \Sigma \text{ Intensive factor} \times \text{Extensive factor} \tag{7.42}$$

Keeping the intensive factors $(T, p, \gamma, \Delta\phi, \mu)$ constant, let the extensive factors be increased from their differential values to their absolute values for the system concerned, S, V, A, q_M', n_i. One now has for the energy of the system

$$U = TS - PV - \gamma A - {}^M\Delta^S\phi q_M' - \sum_i \mu_i n_i \tag{7.43}$$

On differentiating this equation, the result is

$$dU = (T\, dS - p\, dV - \gamma\, dA - {}^{M_1}\Delta^S\phi\, dq_M' - \sum_i \mu_i\, dn_i)$$
$$+ [S\, dT - V\, dp - A\, d\gamma - q_M'\, d({}^{M_1}\Delta^S\phi) - \sum_i n_i\, d\mu_i] \tag{7.44}$$

The two expressions (7.44) and (7.41) must be equal to each other.

Hence, by equating them, one gets

$$0 = S\, dT - V\, dP - A\, d\gamma - q_M'\, d({}^{M_1}\Delta^S\phi) - \sum_i n_i\, d\mu_i \tag{7.45}$$

which, at constant temperature and pressure, reduces to

$$0 = -A\, d\gamma - q_M'\, d({}^{M_1}\Delta^S\phi) - \sum_i n_i\, d\mu_i$$

or

$$d\gamma = -\frac{q_M'}{A} d({}^{M_1}\Delta^S\phi) - \sum_i \frac{n_i}{A} d\mu_i \tag{7.46}$$

Thus, surface tension changes have been related to changes in the absolute potential differences across an electrode–electrolyte interface and to changes in the chemical potential of all the species, i.e., to changes in solution composition. Only one other quantity is missing, the surface excess. But this can be easily introduced by recalling the definition of surface excess [Eq. (7.29)], i.e.,

$$\Gamma_i = \frac{n_i}{A} - \frac{n_i^0}{A}$$

or

$$\frac{n_i}{A} = \Gamma_i + \frac{n_i^0}{A} \tag{7.29}$$

Hence, one can write

$$\frac{n_i}{A} d\mu_i = \Gamma_i\, d\mu_i + \frac{n_i^0}{A} d\mu_i$$

or

$$\sum_i \frac{n_i}{A} d\mu_i = \sum_i \Gamma_i d\mu_i + \sum_i \frac{n_i^0 d\mu_i}{A} \tag{7.47}$$

It is known, however, from the Gibbs–Duhem relation that

$$\sum_i n_i^0 d\mu_i = 0 \tag{7.48}$$

Introducing this relation into Eq. (7.47), one gets

$$\sum_i \frac{n_i}{A} d\mu_i = \sum_i \Gamma_i d\mu_i \tag{7.49}$$

and, by substituting this expression for $\sum_i (n_i/A) d\mu_i$ in Eq. (7.46), the result is[†]

$$d\gamma = -q_M d(^{M_1}\Delta^S\phi) - \sum_i \Gamma_i d\mu_i \tag{7.50}$$

Equation (7.50) contains the quantity $d(^{M_1}\Delta^S\phi)$, which is the change in the absolute (or Galvani) potential difference across the interface under study. It will be recalled, however (*cf.* Section 7.2.5p), that, though the absolute value of $^{M_1}\Delta^S\phi$ cannot be determined, a *change* in $^{M_1}\Delta^S\phi$, i.e., $d(^{M_1}\Delta^S\phi)$, can be measured *provided* (1) the M_1/S interface is a *polarizable* one and (2) the M_1/S interface is linked to a nonpolarizable interface M_2/S to form an electrochemical system, or cell. If such a cell is connected to an external source of electricity, one has

$$V = {}^{M_1}\Delta^S\phi + {}^S\Delta^{M_2}\phi + {}^{M_2}\Delta^{M_1'}\phi \tag{7.51}$$

since the sum of the potential drops around a circuit must be zero. The inner potential difference ${}^{M_2}\Delta^{M_1'}\phi$ does not depend upon the potential V supplied from the external source nor upon the solution composition; hence, on differentiating Eq. (7.51),

$$-d(^{M_1}\Delta^S\phi) = -dV + d(^S\Delta^{M_2}\phi) \tag{7.52}$$

Substituting this expression for $-d(^{M_1}\Delta^S\phi)$ in Eq. (7.50), it follows that

$$d\gamma = -q_M dV + q_M d(^S\Delta^{M_2}\phi) - \sum \Gamma_i d\mu_i \tag{7.53}$$

The nonpolarizable characteristics of the second interface M_2/S, which

---

[†] Note that q_M, the excess-charge *density* on the electrode, has been written instead of q'_M/A, the total excess charge on the electrode divided by its surface area.

is a necessary part of the cell and measuring setup, are now introduced. It is recalled that there is thermodynamic equilibrium at this interface, and thus

$$d(^S\Delta^{M_2}\phi) = -\frac{1}{z_j F} d\mu_j \tag{7.38}$$

where j is the particular species which is involved in the leakage of charge across the nonpolarizable interface.[†] For example, if one uses a hydrogen electrode (see Fig. 7.31), one would write (with $z_+ = 1$)

$$d(^S\Delta^{M_2}\phi) = -\frac{d\mu_{H^+}}{F} \tag{7.54}$$

Or, if one uses a calomel electrode in which Cl^- ions can be thought to do the leaking across the interface, then (with $z_- = -1$),

$$d(^S\Delta^{M_2}\phi) = +\frac{d\mu_{Cl^-}}{F} \tag{7.55}$$

By substituting Eq. (7.38) in Eq. (7.53), one obtains

$$d\gamma = -q_M\, dV - \frac{q_M}{z_j F} d\mu_j - \sum \Gamma_i\, d\mu_i \tag{7.56}$$

This is the fundamental equation for the thermodynamic treatment of polarizable interfaces. It is a relation among interfacial tension γ, surface excess Γ_i, applied potential V, charge density q_M, and solution composition. It shows that interfacial tension varies with the applied potential and with the solution composition. This is in fact the relation that was desired. Its implications will now be analyzed.

7.3.5. Interfacial Tension Varies with Applied Potential: Determination of the Charge Density on the Electrode

When an electrocapillary curve is obtained in the laboratory, a solution of a fixed composition is taken, i.e., $d\mu_i$ for all the species is zero. The

---

[†] The nonpolarizable interface has been defined above (*cf.* Section 7.2.5c) as one which, at constant solution composition, resists any change in potential due to a change in cell potential. This implies that $(\partial^S\Delta^{M_2}\phi/\partial V)_\mu = 0$. However, the inner potential difference at such an interface can change with solution composition; hence, Eq. (7.38) can be rewritten in the form of $d^{M_2}\Delta^S\phi = (RT/z_j F)\, d\ln a$, which is the Nernst equation (*cf.* Chapter 8) in differential form for a single interface.

conditions of electrocapillary-curve determinations correspond therefore to

$$\sum_i \Gamma_i \, d\mu_i = 0 \quad \text{and} \quad d\mu_j = 0 \tag{7.57}$$

from which it follows, from Eq. (7.56), that

$$\left(\frac{\partial \gamma}{\partial V} \right)_{\text{const. comp.}} = -q_M \tag{7.58}$$

This equation, known as the *Lippmann equation*, is perhaps a surprising result: The slope of the electrocapillary curve at any cell potential V is equal to the charge density on the electrode (Fig. 7.53). All that one has

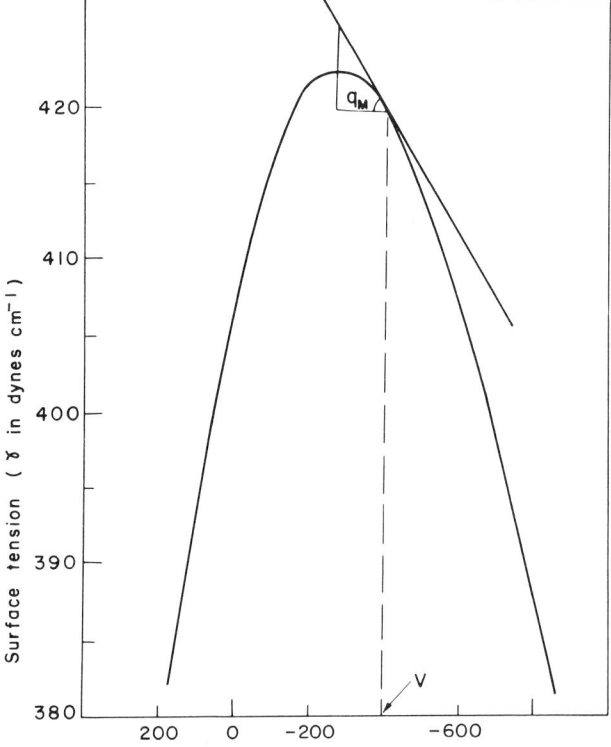

Fig. 7.53. The charge density on the electrode at a particular value of the potential difference V is given by the slope of the electrocapillary curve at that potential. The curve shown is for mercury in contact with 1.0N HCl.

TABLE 7.5

Variation of Charge Density of a Mercury Electrode in Contact with 1.0N CsCl Solution with Potential Difference

Potential difference, mV vs calomel electrode	Electrode-charge density, $\mu C\ cm^{-2}$
−100	+25.2
−200	+16.6
−300	+10.5
−400	+5.9
−500	+2.0
−600	−1.4
−700	−1.6
−800	−7.2
−900	−9.5
−1000	−11.3
−1100	−12.9
−1200	−14.4

to do to know the excess electric charge on the electrode is to find the slope of the electrocapillary curve at the corresponding value of the cell potential. A typical series of results is given in Table 7.5.

7.3.6. Electrode Charge Varies with Applied Potential: Determination of the Electrical Capacitance of the Interface

The next step is obvious. Differentiate the γ versus V electrocapillary curve at various values of cell potential, and plot these values of the slope as a function of potential. What one obtains is a plot of the electrode charge as a function of cell potential V (Fig. 7.54). If the electrocapillary curve were a perfect parabola, then the charge (strictly, excess charge density) on the electrode would vary linearly with the cell potential (Fig. 7.55).

To the extent that an electrified interphase is a *region* where charges are accumulated or depleted relative to the bulk of the electrolyte, it can be considered a system capable of *storing charge*. But the ability to store charge is the characteristic property of an electric capacitor. Hence, one can discuss the capacitance of an electrified interface in way similar to that with a *condenser*.

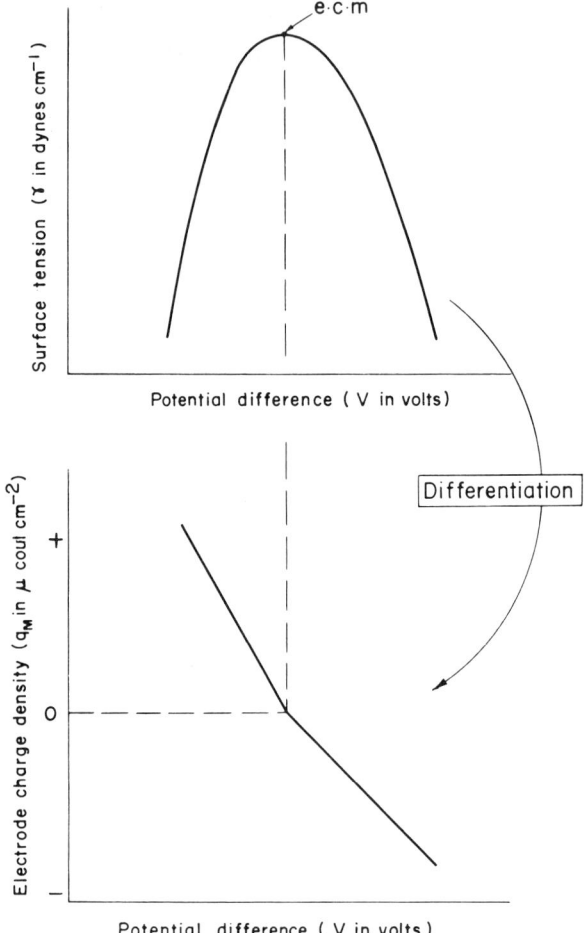

Fig. 7.54. By differentiating an electrocapillary curve, one obtains a curve for the variation of electrode-charge density versus potential difference. The maximum of the electrocapillary curve is sometimes designated by e.c.m.

What is the capacitance of a condenser? It is given by the total charge required to raise the potential difference across the condenser by 1 V

$$K = \frac{q}{V} \tag{7.59}$$

This is the *integral* capacitance, and it is generally used for electrical capacitors where the capacity is constant and independent of the potential. This constancy of capacity may not be the case with electrified interfaces, and,

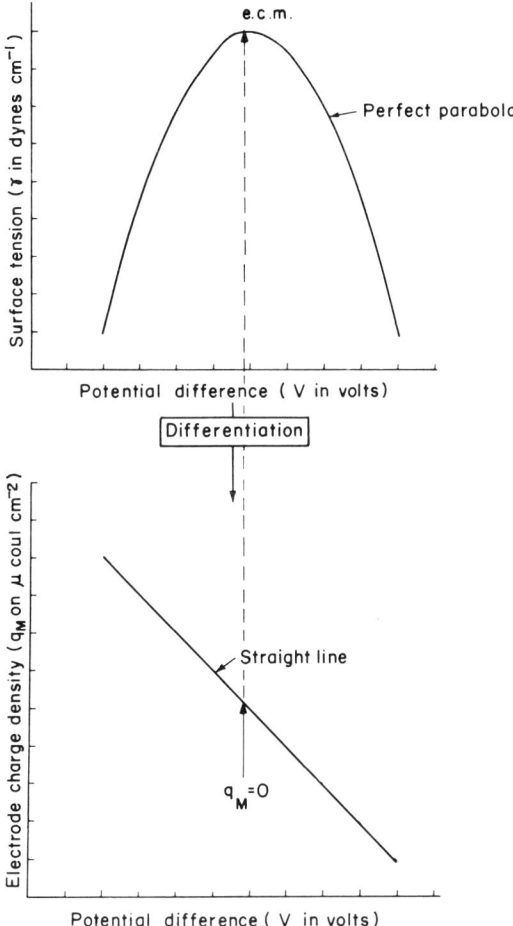

Fig. 7.55. When the electrocapillary γ versus V curve is a perfect parabola, the electrode-charge density varies linearly with potential difference.

in order to be prepared for this eventuality, it is best to define a differential capacity C thus

$$C = \left(\frac{\partial q_M}{\partial V}\right)_{\text{const. comp.}} = -\left(\frac{\partial^2 \gamma}{\partial V^2}\right)_{\text{const. comp.}} \tag{7.60}$$

What is the significance of this equation? It shows that the slope of the curve of the electrode charge versus cell potential yields the value of

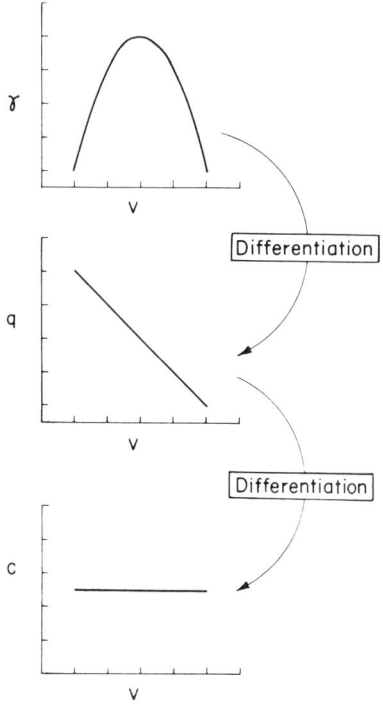

Fig. 7.56. The straight-line q_M versus V curve obtained by differentiating an ideally parabolic γ versus V electrocapillary curve yields, on further differentiation, a capacity which is potential independent.

the differential capacity of the double layer. In the case of an ideal parabolic γ versus V curve, which yields a linear q versus V curve, one obtains a constant capacitance (Fig. 7.56).

7.3.7. The Potential at Which an Electrode Has a Zero Charge

It can be seen from the q_M versus V curve (Fig. 7.54) that the charge on an electrode starts off with one sign and then changes its sign after passing through a zero-charge value. The potential difference across the system (or cell) at which the charge on the electrode is zero is known as the *potential of zero charge* (*pzc*) and given the symbol $E_{q=0}$ or E_{pzc} if this potential is measured on the hydrogen scale.

Where is the pzc on the electrocapillary γ versus V curve? The answer is easy. Since q_M is given by the slope of the curve, the pzc is defined by

$$q_M = -\left(\frac{\partial \gamma}{\partial V}\right)_{\text{const. comp.}} = 0 \qquad (7.61)$$

Hence, the pzc is the potential at which the ecm occurs (Fig. 7.54).

The E_{pzc} is a fundamental reference potential in studies of electrified interfaces. This is because, though one cannot measure the absolute potential difference $\Delta\phi$ across an electrode–electrolyte interface, one *can* measure the cell potential at which the charge on the polarizable electrode (e.g., mercury) is zero. Does this not mean that one has succeeded in knowing the absolute potential difference across the polarizable interface? No, because a $\Delta\phi$ consists (Section 7.2.5n) of a *charge* contribution $\Delta\psi$ and a *dipole* contribution $\Delta\chi$. Hence,

$$(\Delta\phi)_{q=0} = (\Delta\psi)_{q=0} + (\Delta\chi)_{q=0} \qquad (7.62)$$

At the pzc, the first term is zero, but what of the other term, i.e., $(\Delta\chi)_{q=0}$? This term, the surface potential $\Delta\chi$, cannot be measured for reasons which have been discussed (*cf.* Section 7.2.5). Hence, the absolute potential difference $\Delta\phi$ remains experimentally indeterminable even at the pzc.

With liquid metals, the most convenient method of determining the pzc is by making electrocapillary measurements. From the γ versus V curve, the q_M versus V curve can be found and thus the value of $E_{q=0}$ or E_{pzc}.

The pzc, however, is such a fundamental characteristic of the interface that there is a considerable need to know its value for interfaces involving solid electrodes. Here, surface tensions cannot be determined with capillary electrodes, and one must resort to other methods of pzc determination. Some values of the pzc for solid metals are given in Table 7.6.

7.3.8. Surface Tension Varies with Solution Composition: Determination of the Surface Excess

Equation (7.56) describes changes of surface tension both with potential and composition. In order to find out how to determine the surface excess Γ_i, one has first to eliminate one variable in this equation, *viz.*, the potential. This seems to be easy; electrocapillary curves obtained for various salt concentrations can be plotted in one diagram and a perpendicular erected to the cell-potential axis. This will then contain points relating γ to concentration at constant potential, i.e., when, in Eq. (7.53), $dV = 0$. Here, however, a very important and often not realized feature of this

TABLE 7.6

The Potential of Zero Charge on Solids

Metal	Solution		E_{pzc}, V vs SHE
Aluminum	0.01N	KCl	−0.52
Antimony	0.10	HCl	−0.19
Bismuth	0.01N	KCl	−0.36
Cadmium	0.01N	KCl	−0.92
Cobalt	0.02N	Na$_2$SO$_4$	−0.32
Copper	0.02N	Na$_2$SO$_4$	+0.03
Gold	0.02N	Na$_2$SO$_4$	+0.23
Iron	0.001N	H$_2$SO$_4$	−0.37
Lead	0.01N	KCl	−0.69
Platinum	0.003N	HClO$_4$	+0.41
Silver	0.02N	Na$_2$SO$_4$	−0.70

equation must be remembered. It contains the term $d(^S\varDelta^{M_2}\phi)$, i.e. changes in the potential difference across the reference-electrode–solution interface produced by changes in solution composition via Eq. (7.38). Thus, the V read on the potentiometer refers, for every concentration in which the γ versus V relation is determined, to a reference electrode consisting of metal immersed in a solution of *the same* concentration as that surrounding the test electrode. It is this value of V (obtained *not* with a standard reference electrode, i.e., one always immersed in the solution of the same activity, but with reference electrodes differing according to the concentration of the ions which leak charge across the interface) which has to be kept constant.

In order to remember this often-forgotten fact, it is customary to denote these potential values, referred *not* to the standard electrode but to one reversible to ions of given (varying) concentration, as V_+ or V_-, indicating at the same time whether the electrode is one at which cations or anions leak, respectively.

Now, a perpendicular erected on the axis of the cell potential V_+ or V_- (Fig. 7.57) intersects the electrocapillary curves at points for which the condition $dV = 0$ in Eq. (7.56) is satisfied, whereupon

$$d\gamma = -\frac{q_M}{z_j F}d\mu_j - \sum_i \varGamma_i\, d\mu_i \qquad (7.63)$$

This equation describes the changes of surface tension with composition at any particular cell potential, V_+ or V_- (Table 7.4).

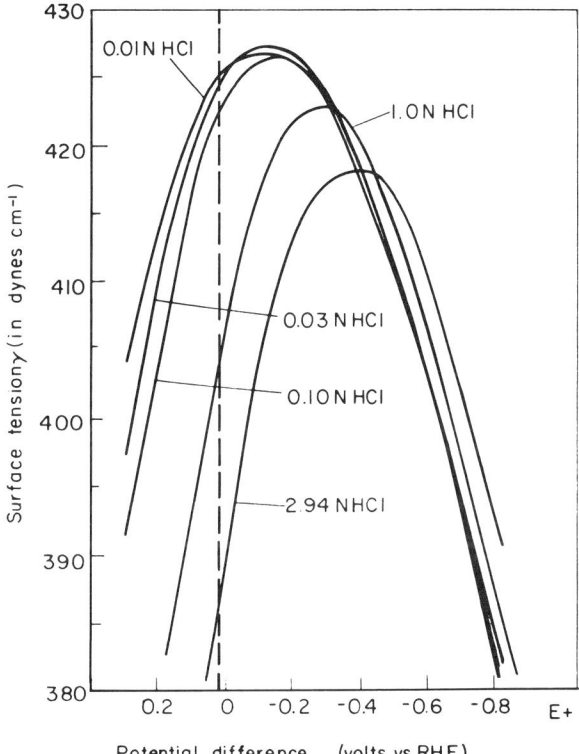

Fig. 7.57. Electrocapillary curves from solutions of different electrolyte (HCl) concentration. The symbol RHE stands for a reversible hydrogen electrode immersed, not in a standard solution, but in the same electrolyte as the electrode under study.

Now, consider a polarizable interface which consists of a metal electrode in contact with a solution of a 1:1-valent electrolyte (i.e., $z_+ = 1$ and $z_- = -1$). It will be remembered that, in order to apply electrocapillary thermodynamics to a polarizable interface M_1/S, the interface has to be assembled into a cell along with a nonpolarizable interface. Suppose that the nonpolarizable interface is one at which negative ions interchange charge with the metal surface, i.e., $z_j = -1$. Hence, Eq. (7.63) for the polarizable interface becomes

$$d\gamma = +\frac{q_M}{F} d\mu_- - \Gamma_+ d\mu_+ - \Gamma_- d\mu_- \qquad (7.64)$$

where the Σ of Eq. (7.63) has been expanded.

It would be convenient to transform this expression so that it would contain the concentration of the electrolyte used in the cell. The first step in effecting this transformation follows by noting that the chemical potential μ of the electrolyte is the sum of the chemical potentials of the ions (cf. Section 3.4), i.e.,

$$\mu = \mu_+ + \mu_-$$

or

$$d\mu = d\mu_+ + d\mu_- \tag{7.65}$$

Using equation (7.65), one can substitute for $d\mu_+$ in Eq. (7.64) to give

$$d\gamma = \frac{q_M}{F} d\mu_- - \Gamma_+ d\mu + \Gamma_+ d\mu_- - \Gamma_- d\mu_-$$

$$= -\Gamma_+ d\mu + \left(\frac{q_M + F\Gamma_+ - F\Gamma_-}{F}\right) d\mu_- \tag{7.66}$$

The second step consists in affirming that there is electroneutrality across the interface, i.e., the charge on the metal is always equal and opposite to the total charge on the solution side of the interface. Before the double layer is formed, the metal is uncharged, and, in the solution, the charge per unit area of a lamina due to positive and negative ions is zero, i.e.,

$$F\left(\frac{n_+^0}{A}\right) - F\left(\frac{n_-^0}{A}\right) = 0 \tag{7.67}$$

After the double layer is formed, electroneutrality requires that

$$F\frac{n_+}{A} - F\frac{n_-}{A} + q_M = 0 \tag{7.68}$$

Subtracting Eq. (7.67) from Eq. (7.68), one gets

$$q_M + F\left(\frac{n_+ - n_+^0}{A}\right) - F\left(\frac{n_- - n_-^0}{A}\right) = 0 \tag{7.69}$$

But, according to the definition of surface excess [see Eq. (7.29)]

$$\Gamma_+ = \frac{n_+ - n_+^0}{A} \quad \text{and} \quad \Gamma_- = \frac{n_- - n_-^0}{A} \tag{7.70}$$

Hence, according to the electroneutrality condition,

$$q_M + F\Gamma_+ - F\Gamma_- = 0 \tag{7.71}$$

and, inserting this condition into Eq. (7.66), one finds that, for a polarizable interface which is built into a cell with *a nonpolarizable interface which leaks negative ions*, the second term is zero, so that

$$d\gamma = -\Gamma_+ \, d\mu \tag{7.72}$$

or

$$\left(\frac{\partial \gamma}{\partial \mu}\right)_{\text{const.} V_-} = -\Gamma_+ \tag{7.72a}$$

Now,

$$\begin{aligned}\mu &= (\mu_+ + \mu_-) \\ &= (\mu_+^0 + \mu_-^0) + (RT \ln a_+ + RT \ln a_-) \\ &= (\mu_+^0 + \mu_-^0) + (RT \ln a_+ a_-)\end{aligned} \tag{7.73}$$

Instead of $a_+ a_-$, one can introduce the mean ionic activity $a_\pm$ obtained by multiplying Eqs. (3.29) and (3.30). Thus,

$$a_\pm = (x_\pm f_\pm) = (x_+ f_+)^{\frac{1}{2}}(x_- f_-)^{\frac{1}{2}} = (a_+ a_-)^{\frac{1}{2}}$$

or

$$a_+ a_- = a_\pm^2 \tag{7.74}$$

Hence, from (7.73) and (7.74),

$$\mu = (\mu_+^0 + \mu_-^0) + 2RT \ln a_\pm$$

or

$$d\mu = 2RT \, d\ln a_\pm \tag{7.75}$$

Hence,

$$\left(\frac{\partial \gamma}{2RT \, \partial \ln a_\pm}\right)_{\text{const.} V_-} = -\Gamma_+ \tag{7.76}$$

which shows that the slope of the surface tension *versus* $\log a_\pm$ curve at constant potential yields the surface excess. Of course, the activity of the electrolyte, $a_\pm$, is obtained by taking the bulk concentration and multiplying it by the mean activity coefficient at that concentration.

An interesting point to note is that the surface excess of the *positive* ion is determined by choosing a nonpolarizable interface which leaks *negative* ions. Following a similar argument to that given above, one could have shown that a nonpolarizable interface which leaks positive ions enables the determination of the surface excess of negative ions. Thus, in a study of the Hg–HCl polarizable interface, the surface excess Γ_- of the Cl$^-$ ions may be obtained by coupling the Hg–HCl interface with a hydrogen reference electrode which leaks hydrogen ions.

If the determination of surface excess is carried out at various cell potentials, then one can plot the surface excesses Γ_+ and Γ_- of the positive and negative ions as a function of the cell potential V. It is more useful, however, to plot $q^+ = z_+ F\Gamma_+$ and $q^- = z_- F\Gamma_-$, i.e., the excess-charge densities in the solution side of the interface due to the surface excesses of positive and negative ions (Fig. 7.58 and Table 7.7). Since the charge density $q_S = -q_M$ in the solution is made up of the algebraic sum of the excess positive and excess negative charge densities, q^+ and q^- are the *components* of the (excess) charge in the solution

$$-q_M = q_S = q^+ + q^-$$
$$= z_+ F\Gamma_+ + z_- F\Gamma_- \tag{7.77}$$

The components-of-charge curve (Fig. 7.58) must be read as follows:

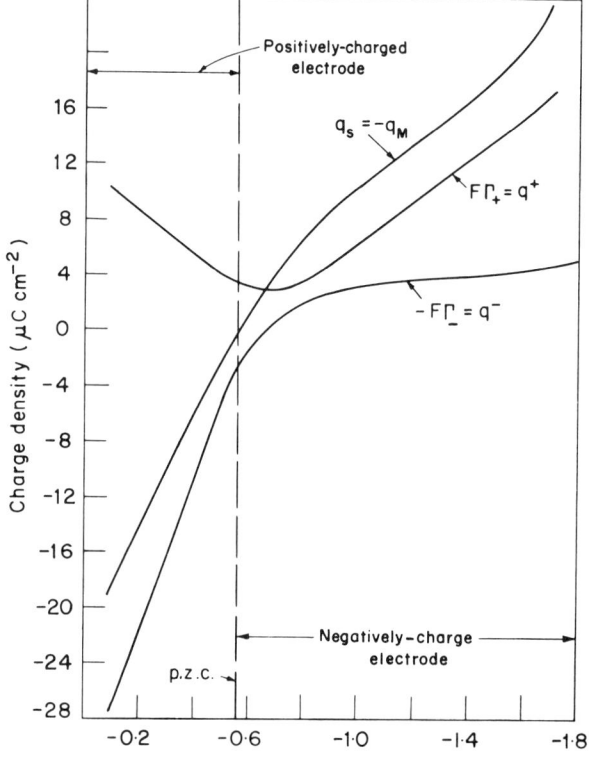

Fig. 7.58. Components of charge density as a function of the applied potential V in a 1N NaCl solution.

TABLE 7.7
Surface Excess of Cs⁺ and Cl⁻ Ions in a Mercury–1.0N CsCl Interphase as a Function of Applied Potential

Potential, mV vs calomel electrode	$q^+ = F\Gamma_+,$ $\mu C\ cm^{-2}$	$q^- = F\Gamma_-,$ $\mu C\ cm^{-2}$
−200	10.2	−26.8
−300	8.7	−19.2
−400	7.5	−13.4
−500	6.5	−8.5
−600	5.8	−4.4
−700	5.5	−0.9
−800	5.8	1.4
−900	6.9	2.6
−1000	8.5	2.8

When, e.g., q^- is negative, it means that $\Gamma_- = q^-/z_- F$ is positive, i.e., there is an accumulation of negative ions in the interphase relative to the bulk; and, when q^- is positive, Γ_- is negative, which means a depletion of negative ions in the interphase. At the ecm, $q_M = -q_S = 0$; hence, $q^- = q^+$.

7.3.8.a. A Summary of the Electrocapillary Thermodynamics

The thermodynamic equations applicable to a polarizable interface, which can be studied by means of a capillary electrometer, can now be summarized. The general equation is

$$d\gamma = -q_M\, dV - \frac{q_M}{z_j F} d\mu_j - \sum_i \Gamma_i\, d\mu_i \tag{7.56}$$

By keeping the solution composition constant, i.e., $d\mu_i = 0 = d\mu_j$, the charge density q_M on the metal is given by the slope of the interfacial tension γ versus the applied potential V curve

$$q_M = -\left(\frac{\partial \gamma}{\partial V}\right)_{\text{const. comp.}} \tag{7.58}$$

From the variation of this charge density with applied potential, the differential capacity of the interface is obtained thus

$$C = \left(\frac{\partial q_M}{\partial V}\right)_{\text{const. comp.}} = -\left(\frac{\partial^2 \gamma}{\partial V^2}\right)_{\text{const. comp.}} \qquad (7.60)$$

The surface excess of a species i is obtained from the plot of interfacial tension *versus* mean activity of the electrolyte taken under conditions of constant applied cell potential V. By considering various applied potentials, one can get the surface excess Γ_+ and Γ_- and thus the excess-charge densities $q^+ = z_+ F \Gamma_+$ and $q^- = z_- F \Gamma_-$, due to the positive and negative ions, as a function of the applied potential

$$\Gamma_\pm = -\left(\frac{\partial \gamma}{2RT\, \partial \ln a_\pm}\right)_{\text{const.}\,V_\pm} \qquad (7.76)$$

7.3.9. Reflections on Electrocapillary Thermodynamics

A remarkable piece of thermodynamic reasoning has just been presented —the thermodynamics of the ideally polarizable interface. It is a comparatively recent development, as pieces of thermodynamics go. Although the basic outline of some of the above equations was given by Gibbs in the last century, a full understanding of how to use the equations and a valid deduction of them were not obtained until the 1950's (König, Parsons, and Devanathan). The long delay in realization of a fundamental analysis of the thermodynamics of the ideally polarizable interface is one of the reasons responsible for the slow growth of understanding of the structure of electrified interfaces and, consequently, of the mechanism of electrode processes.

Once the deductions were understood, many informative results emerged. It is now possible to sit back and contemplate the achievement.

It may perhaps be thought remarkable that, from measurements of the surface tension of a liquid drop, one can obtain not only the excess electric charge on the drop but also the capacity of the interface, i.e., the ability of an angstrom-thick region to store electric charge. It is by no means obvious that measurements of dynes per centimeter can yield information on a quantity called capacity, the rate of change of charge with potential, and by thermodynamics, i.e., without the intervention of a theory involving models. Is it not remarkable that measurement of how surface tension varies with the *bulk* concentration of the components of a solution can reveal the extent to which the individual components in the *interphase region* have deviated in their concentrations from the concentrations that would be there had there been no interface?

Thus, from the *heights* of a mercury column in a capillary electrometer and *pointer readings* on an electronic voltmeter, one obtains exact thermodynamic quantities in a region which is of angstrom dimensions—the electrode charge, the electrical capacity of the interface, and the amounts of various species in the interface region. Is this achievement less spectacular than that of pointing a telescope to the stars, recording a line spectrum on a photographic plate, and determining from the lines the composition of stars?

Equilibrium thermodynamics, it may be said, tells a limited story. It cannot give us information concerning the rates of processes and, therefore, about mechanisms. But it does convert one piece of empirical, macroscopic information into another piece of macroscopic information which often cannot be got in any other way. This role is exemplified to great advantage in electrocapillary thermodynamics.

7.3.10. Retrospect and Prospect in the Study of Electrified Interfaces

What has been learned so far about electrified interfaces?

Firstly, one has learned what a double layer is. It is an interphasial region (between two homogeneous phases) in which the charged constituents have been separated so that each side of the interface is electrified. The charges on the two sides are equal in magnitude but opposite in sign, making the interphase region electrically neutral as a whole.

Secondly, the potential difference across the interface turned out to be a peculiar quantity. It is impossible to measure the absolute value of the double layer potential difference. But several things can be done. *Changes* in the potential difference across a polarizable interface *can* be measured provided the other interface is nonpolarizable. The potential difference across a cell, or system, of two electrode–electrolyte interfaces *can* be measured. The absolute potential difference $\Delta\phi$ across a double layer can however be *conceptually* resolved into a measurable contribution $\Delta\psi$ from the charges and an immeasurable contribution $\Delta\chi$ from oriented dipoles existing on one or both of the phases.

Thirdly, a curious and subtle concept was explained, the concept of surface excess. It is not to be confused with adsorption, though the surface excess may become nearly identical to the total amount adsorbed under certain limiting conditions. The surface excess of a particular species is the *excess* of that species present in the surface phase relative to the amount that would have been present had there been no double layer. The surface excess, therefore, represents the accumulation or depletion of the species in the *entire* interphase *region*. Further, electrocapillary measurements permit a direct experimental description of the surface excess of a species.

The surface excesses of the various species in an interphase do not, however, reveal the time-average locations of the ions, dipoles, etc., i.e., the structure of the interphase. The various species, e.g., the ions, are distributed in the interphase region so that their concentrations vary with distance from the electrode. Knowledge of structure hinges upon knowing these concentration variations. It is by the integrations of these concentrations that the surface excesses are obtained, but, being definite integrals, the surface excesses retain no information on that concentration variation.

What next? One must seek knowledge of the distribution of particles in the interphase region—*the structure*. For instance, from components of the charge curve (e.g., Fig. 7.58), one can get the excess-charge densities q^- and q^+ in the interphase region, but one must go further and seek the spatial distribution of the excess charge. One must also seek the variation of potential with distance. One seeks to develop the atomistic theory of the interface. To achieve these tasks, one must learn to intuit models, for these are the crutches with the aid of which one can acquire an atomistic view of an electrified interface. A brief preview of these models will be presented here.

The first one is easy to guess, a parallel-plate condenser, i.e., two sheets (or a literal double layer) of rigidly fixed charges (Fig. 7.10). This model arose from the work of Helmholtz and Perrin.

This very simple view was succeeded by that in which the effect of the thermal forces in scattering the rigid layers of Helmholtz and Perrin's was taken into account (Fig. 7.13). This second theory of the electrified interface was the creation of Gouy and Chapman. It constitutes one of the seminal theoretical treatments of physical chemistry because it was one of the first treatments of properties of condensed systems by statistical mechanics and, incidentally, the basis of Debye and Hückel's theory of electrolytes (Section 3.5.7). The diffuse-charge model was, however, an example of overreacting against the model of Helmholtz; the particles were scattered so much that it was imagined that only scattered particles existed in the interphase and none of the rigidly fixed charges of the simple double-layer model were left.

The thesis of Helmholtz and Perrin and the antithesis of Gouy and Chapman were met by the synthesis of Stern, who put the two ideas together. Thus, some of the ions exist in the Helmholtz layer in a partly rigid sheet on the solution side, but some of them are also scattered out into the solution.

What has happened, during the development of these three models, to concepts about the oriented solvent dipoles which make up the double layer and contribute, e.g., to $\Delta\chi$? Reference has already been made to the two types of potential which contribute to the total potential difference in the double layer. The first potential $\Delta\psi$ can be thought of in terms of the Helmholtz–Perrin, Gouy–Chapman, and Stern models. The $\Delta\psi$ depends only upon the charge at the interface. But the $\Delta\chi$ potential, that due partly to adsorbed solvent molecules at the electrode, has still to be included in the picture. The structure of the water in contact with the electrode represents a breakaway from the water structure in the bulk. The water molecules stand up on the electrode with their dipoles pointing toward or away from the electrode depending upon whether the electrode is charged positive or negative. Thus, there is an orientation of the water molecules in the field of the electrode (Fig. 7.8). It was many years before this role of the solvent molecules in the electrified interface was suggested and worked out.

This then is a preview of the basic models of the interface. They refer to the ions congregating in the Helmholtz–Perrin and the Gouy–Chapman regions and to the orientation of solvent molecules.

Further Reading

1. G. Lippmann, *Ann. Chim. Phys. (Paris)*, **5**: 494 (1875).
2. E. A. Guggenheim and N. K. Adam, *Proc. Roy. Soc. (London)*, **A139**: 218 (1933).
3. D. C. Grahame, *Chem. Rev.*, **41**: 441 (1947).
4. R. Parsons, "Equilibrium Properties of Electrified Interphases," in: J. O'M. Bockris, ed., *Modern Aspects of Electrochemistry*, Vol. I, Butterworth's Publications, Ltd., London, 1954.
5. L. I. Antropov, *Kinetics of Electrode Processes and Null Points of Metals*, Council of Scientific and Industrial Research, New Delhi, India, 1959.
6. P. Delahay, *Double Layer and Electrode Kinetics*, Interscience Publications, Inc., New York, 1965.
7. B. E. Conway, *Theory and Principles of Electrode Processes*, The Ronald Press Company, New York, 1965.
8. S. A. Argade and E. Gileadi, "The Potential of Zero Charge," in: E. Gileadi, ed., *Electrosorption*, Plenum Press, New York, 1967.
9. D. Mohilner, "The Electrical Double Layer," in: A. J. Bard, ed., *Electroanalytical Chemistry*, Vol. I, Chap. 4, Marcel Dekker, Inc., 1967.
10. R. S. Perkins and T. N. Andersen, "Potential of Zero Charge of Electrodes," in: J. O'M. Bockris and B. E. Conway, eds., *Modern Aspects of Electrochemistry*, Vol. V, Plenum Press, New York, 1969.

7.4. THE STRUCTURE OF ELECTRIFIED INTERFACES

7.4.1. The Parallel-Plate Condenser Model: The Helmholtz–Perrin Theory

The nonstructural, thermodynamic thinking has yielded information on the gross quantities present at the electrified interface. But, as is characteristic of thermodynamic treatments based on the first and second laws, the relations are in terms of *changes* of the various experimental quantities. The language is that of differentials.

For example, consider the Lippmann equation, i.e., the relation between changes of surface tension and changes of cell potential.

$$\left(\frac{\partial \gamma}{\partial V}\right)_{\text{const. comp.}} = -q_M \qquad (7.58)$$

From this thermodynamic expression, it is not possible to state how the surface tension changes with cell potential. Experimentally, it is known that the γ versus V curve is almost a parabola. Can this parabolic dependence of γ on V be explained in a thermodynamic framework?

One's first reaction is to answer: Why not integrate Eq. (7.58)? Let this be attempted.

$$\int d\gamma = - \int q_M \, dV \qquad (7.78)$$

But is q_M a function of V? If so, what function? One cannot integrate without knowing this function.

These questions cannot be answered by thermodynamics because they are *structural* questions that pertain to the location and arrangement of charges.

In order to obtain the relation between q_M and V, a model has to be conceived for the arrangement of charges at an electrified interface. The first such model arose from the work of Helmholtz and Perrin. The net charge on the metal will draw out from the randomly dispersed ions in solution a counterlayer of charge of opposite sign. The electrified interface consists, therefore, of two sheets of charge, one on the electrode and the other in the solution. Hence, the term *double layer*. The charge densities on the two sheets are equal in magnitude but opposite in sign, exactly as in a parallel-plate capacitor (Fig. 7.10).

Once the electrical equivalence between an electrified interface and a capacitor is postulated, the electrostatic theory of capacitors can be used for double layers. It is known, e.g., that the potential difference V across the condenser is

$$V = \frac{4\pi d}{\varepsilon} q \tag{7.79}$$

where d is the distance between the plates and ε is the dielectric constant of the material between the plates.

Based on this parallel-plate model of the double layer, one has

$$dV = \frac{4\pi d}{\varepsilon} dq_M \tag{7.80}$$

i.e., the functional relationship required for the integration of the Lippmann equation (7.58). Inserting this expression for dV in Eq. (7.78), one has

$$\int d\gamma = -\frac{4\pi d}{\varepsilon} \int q_M \, dq_M$$

or

$$\gamma + \text{Constant} = -\frac{4\pi d}{\varepsilon} \frac{1}{2} q_M^2 \tag{7.81}$$

Now, when $q_M = 0$, i.e., at the pzc, γ is at a maximum (cf. Section 7.3.7 and the electrocapillary curve, Fig. 7.50), and, therefore,

$$\text{Constant} = -\gamma_{\max} \tag{7.82}$$

Hence,

$$\gamma = \gamma_{\max} - \frac{4\pi d}{\varepsilon} \frac{q_M^2}{2}$$

or

$$\gamma = \gamma_{\max} - \frac{\varepsilon}{4\pi d} \frac{1}{2} V^2 \tag{7.83}$$

This is the equation for a parabola symmetrical about $\gamma_{\max}$, i.e., about the ecm. It appears that the Helmholtz–Perrin model would be quite satisfactory for electrocapillary curves which are perfect parabolas.

7.4.2. The Double Layer in Trouble: Neither Perfect Parabolas nor Constant Capacities

But is the electrocapillary curve a perfect parabola? Almost, *but not quite*. There is always a slight asymmetry (Fig. 7.59).

The deviations from a parabolic shape are greater with some solutions than with others. Electrocapillary curves show, for instance, a marked

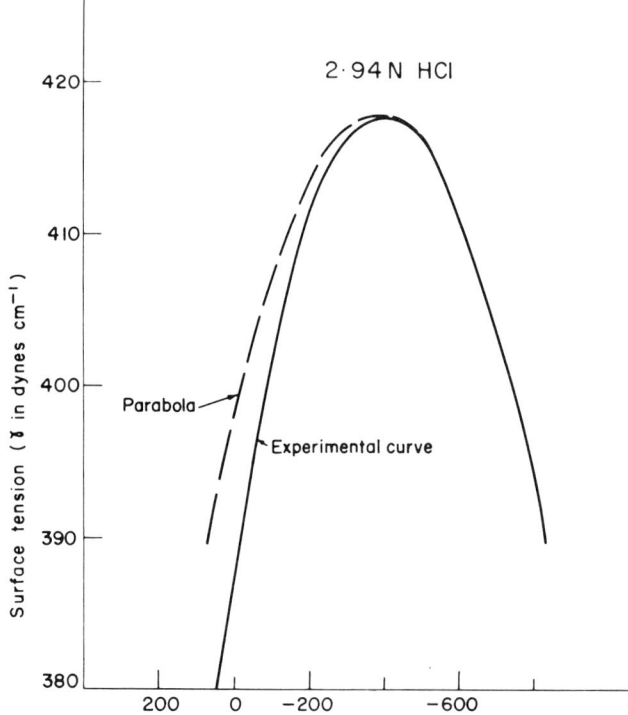

Fig. 7.59. Experimental electrocapillary curves are not perfect parabolas.

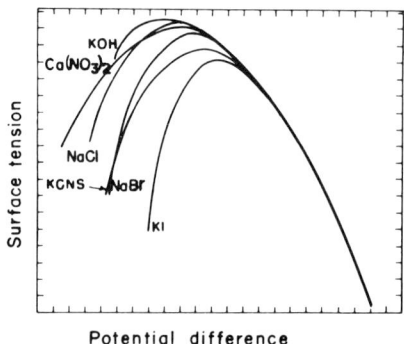

Fig. 7.60. The dependence of electrocapillary curves on the nature of the anions present in the electrolyte.

sensitivity to the nature of the anions present in the electrolyte (Fig. 7.60). In contrast, the curves do not seem to be affected significantly by the cations present, unless they are large organic cations, e.g., tetraalkylammonium ions.

The asymmetry of electrocapillary curves has important implications for the capacities of double layers.

Consider a typical electrocapillary curve which is not exactly a parabola. It yields, on differentiation [*cf.* Eq. (7.60)], a charge-density *versus* cell-potential plot which does not consist of one straight line but two straight lines meeting at $q = 0$ (Fig. 7.61). Now, the differential capacity C is given by

$$C = \frac{dq}{dV} \qquad (7.60)$$

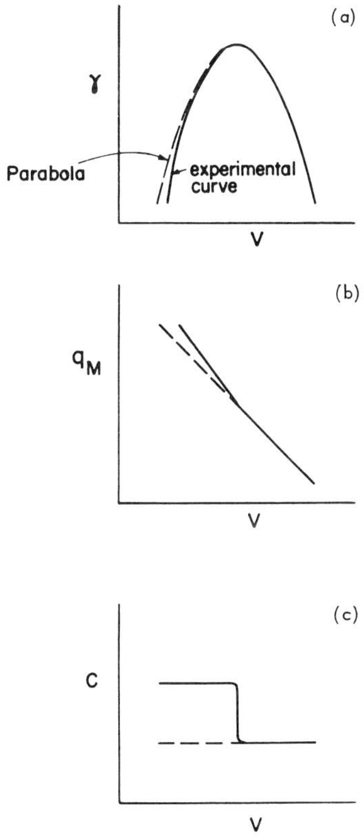

Fig. 7.61. (a) Experimental (full line) and ideally parabolic electrocapillary curves; (b) the corresponding electrode-charge density–potential and (c) capacity–potential curves.

and, therefore, a plot of C versus V can be obtained by a differentiation of the q versus V curve. It turns out that, when one starts from an asymmetric electrocapillary curve, the interface displays a differential capacity which is not constant with cell potential (*cf.* Fig. 7.61).

What has the parallel-plate model of the double layer to say regarding the capacity of the interface? A parallel-plate condenser has a specific capacity (i.e., per unit area of plates) which is given by rearranging Eq. (7.80) in the form

$$\frac{dq}{dV} = \frac{\varepsilon}{4\pi d} = C \qquad (7.84)$$

Thus, if ε and d are taken as constant, it means that the parallel-plate model predicts a *constant* capacity, i.e., one which does not change with potential. But this is not what is observed.

It appears that an electrified interface does not behave like a simple double layer. The parallel-plate condenser model is too naïve an approach. Evidently, some crucial secrets about electrified interfaces are contained in those asymmetric electrocapillary curves and the differential capacities which vary with potential. One has to think again.

7.4.3. The Ionic Cloud: The Gouy–Chapman Diffuse-Charge Model of the Double Layer

The metallic electrode has a high electronic conductivity. It cannot therefore sustain a charge separation and field inside it. The charges fly away from each other and, in doing so, flock to the surface. The electrode charge q_M is a *surface* charge.

There must be an equal amount of charge of opposite sign in the solution. But *where* in the solution; in what manner are the ions distributed out in the solution with respect to the electrode? The Helmholtz–Perrin model fixes these charges onto a sheet parallel to the metal. But this model was too rigid to explain the asymmetry of electrocapillary curves and the dependence of capacity upon potential.

Perhaps it was the lack of freedom of the charges in the solution which precipitated the inconsistency. Why not free them from their restriction to a sheet? It was Gouy and Chapman who thought of liberating the ions from a sheet parallel to the electrode. But, once the ions are free, they become exposed to the thermal buffeting from the particles of the solution. The behavior of ions in the vicinity of the electrode is affected by the electric force arising from the charge on the electrode and by thermal jostling.

Equilibrium between electric and thermal forces is attained and thus also a time-average ionic distribution.

How do the excess charges in the solution spread out? What sort of variation with distance does the potential show? What is the total amount of excess charge?

Examine these questions. Are they not precisely the type of questions asked in the theory of ion–ion interactions in solution (Chapter 3)? There, it was necessary to choose arbitrarily one ion and spotlight it as the "central ion," or source, of the field. Here, the discussion revolves on ion–electrode interactions and the electrode is the source of the field. The central ion being a sphere of charge, its field was spherically symmetrical. The response of an ion, however, does not depend on how the electric field is produced (i.e., whether the source is a central ion or a charged electrode). It depends only on the value of the field at the location of the ion. Hence, the electrostatic arguments in the problems of ion–ion interactions and ion–electrode interactions must be similar.

There are, however, differences in the *geometry* of the two problems. These differences affect the mathematical development. Thus, (Fig. 7.62) the central ion puts out a spherically symmetrical field because it is like a little sphere of charge. In contrast, the electrode is like an infinite plane (infinite *vis-à-vis* the distances at which ion–electrode interactions are considered), and its field displays a planar symmetry. Otherwise, the technique of analysis of the diffuse double layer proceeds along the same lines as in the theory of long-range ion–ion interactions (Section 3.3).

All this is another way of saying that the treatment of ion–ion interactions can be borrowed by replacing the central ion of that theory with

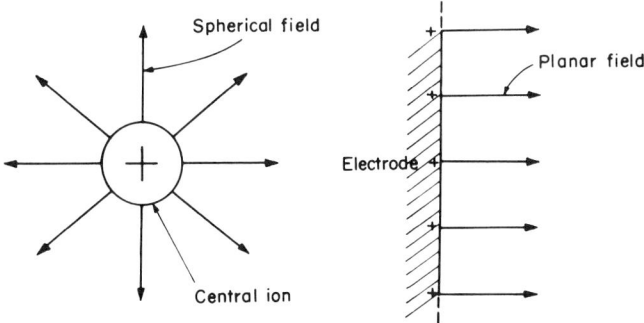

Fig. 7.62. The difference in the geometry of the ion-ion interaction and ion–electrode interaction problems.

the most gigantic central ion, namely, the electrode. Stated thus, it is easy to anticipate that the charged electrode will be enveloped by an ionic cloud (consisting of an excess of ions of opposite charge) in the same way as an ion in the bulk of the solution. This ionic atmosphere represents a falling-off, with distance from the electrode, of the net charge density in a lamina parallel to the electrode and at increasing distances out into the solution. The potential, too, will decay with distance, asymptotically settling down to a constant value (taken as zero) in the bulk of the solution.

This description of the ionic atmosphere around an electrode is qualitative. What is required is a quantitative description. Attention will now be turned to the mathematical description of the diffuse double layer.

A historical note should, however, be made here. The theory of the diffuse double layer was presented independently by Gouy and Chapman (1910) thirteen years *before* the Debye–Hückel theory of ion–ion interactions (1923). Though a historically prior treatment, the Gouy–Chapman approach to the diffuse layer is hardly known in comparison with its application by Debye and Hückel to the diffuse charge around an ion, doubtless owing to the preoccupation of the majority of scientists with bulk properties rather than those at surfaces, in particular at electrified interfaces.

7.4.4. Ions under Thermal and Electric Forces near an Electrode

Consider a lamina (in the electrolyte) parallel to the electrode and at a distance x from it (Fig. 7.63). What is the charge density ϱ_x in this lamina?

The charge density can be expressed in two ways, firstly, in terms of the Poisson equation (*cf.* Section 3.3.3), which, for the x dimension in rectangular coordinates, reads

$$\varrho_x = -\frac{\varepsilon}{4\pi}\frac{d^2\psi_x}{dx^2} \qquad (7.85)$$

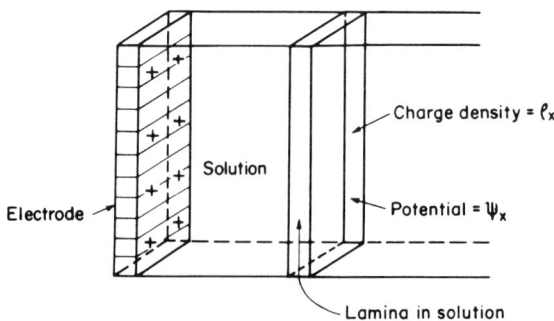

Fig. 7.63. A lamina in the solution, parallel to a plane electrode.

where ψ_x is the outer potential difference between the lamina and the bulk of the solution† (taken as $\psi_{x\to\infty} = 0$), and, secondly, in terms of the Boltzmann distribution,

$$\varrho_x = \sum_i n_i z_i e_0 = \sum_i n_i^0 z_i e_0 e^{-z_i e_0 \psi_x / kT} \tag{7.86}$$

where n_i and n_i^0 are the concentrations of the ith species in the lamina (distance x from the electrode) and in the bulk of the solution, respectively, z_i is the valence of the species i, and e_0 is the electronic charge. The factor $z_i e_0 \psi_x / kT$ represents the ratio of the electrical and thermal energies of an ion at the distance x from the electrode. From the two expressions [Eqs. (7.85) and (7.86)] for the charge density ϱ_x, one obtains the Poisson–Boltzmann equation

$$\frac{d^2\psi}{dx^2} = -\frac{4\pi}{\varepsilon} \sum_i n_i^0 z_i e_0 e^{-z_i e_0 \psi_x / kT} \tag{7.87}$$

At this stage in the Debye–Hückel theory of ion–ion interactions, it was considered that the potential was small enough to *linearize* the equation. This linearization would correspond in the present case to writing

$$e^{-z_i e_0 \psi_x / kT} = 1 - \frac{z_i e_0 \psi_x}{kT} \tag{7.88}$$

Attempts to achieve so-called "rigorous" solutions by not linearizing the Poisson–Boltzmann equation led to certain inconsistencies. Nevertheless, it has been customary,‡ in diffuse-double-layer treatments based on the Poisson–Boltzmann equation, to proceed with the solution of the unlinearized differential equation (7.87).

A simple transformation can now be used. Consider the steps

$$\frac{1}{2}\frac{d}{d\psi}\left(\frac{d\psi}{dx}\right)^2 = \frac{1}{2} 2\left(\frac{d}{d\psi}\frac{d\psi}{dx}\right)\frac{d\psi}{dx}$$

$$= \left(\frac{dx}{d\psi}\frac{d}{dx}\frac{d\psi}{dx}\right)\frac{d\psi}{dx}$$

$$= \frac{d^2\psi}{dx^2} \tag{7.89}$$

---

† If desired, one can write the inner potential difference $\phi_x - \phi_{x\to\infty}$ but this reduces to $\psi_x - \psi_{x\to\infty}$ because the $\Delta\chi$ for two points in the *same* phase (i.e., the same surface dipole layer) is zero. Note that, in the Gouy–Chapman theory—as in the Debye–Hückel theory—, one is analyzing potentials in one phase only, the solution phase.

‡ What is customary need not necessarily be right. A more consistent theory of the diffuse double layer is clearly required.

Thus, the identity
$$\frac{1}{2}\frac{d}{d\psi}\left(\frac{d\psi}{dx}\right)^2 = \frac{d^2\psi}{dx^2} \tag{7.90}$$

can be used in the differential equation (7.87) to give

$$\frac{d}{d\psi}\left(\frac{d\psi}{dx}\right)^2 = -\frac{8\pi}{\varepsilon}\sum_i n_i^0 z_i e_0 e^{-z_i e_0 \psi_x/kT} \tag{7.91}$$

which, by the following rearrangement

$$d\left(\frac{d\psi}{dx}\right)^2 = -\frac{8\pi}{\varepsilon}\sum_i n_i^0 z_i e_0 e^{-z_i e_0 \psi_x/kT}\, d\psi \tag{7.92}$$

can be integrated to give

$$\left(\frac{d\psi}{dx}\right)^2 = \frac{8\pi}{\varepsilon}\int \sum_i n_i^0 z_i e_0 e^{-z_i e_0 \psi_x/kT}\, d\psi \tag{7.93}$$

$$= \frac{8\pi}{\varepsilon}\sum_i \frac{n_i^0 z_i e_0 e^{-z_i e_0 \psi_x/kT}}{(-z_i e_0/kT)} + \text{Constant}$$

$$= -\frac{8\pi kT}{\varepsilon}\sum_i n_i^0 e^{-z_i e_0 \psi_x/kT} + \text{Constant} \tag{7.94}$$

The integration constant can be evaluated by considering that, deep in the bulk of the solution, i.e., at $x \to \infty$, not only is the Volta potential zero, $\psi_{x\to\infty} = 0$, but the field $d\psi/dx$ is also zero. Under these conditions,

$$\text{Constant} = -\frac{8\pi kT}{\varepsilon}\sum n_i^0 \tag{7.95}$$

By introducing this value of the integration constant into Eq. (7.94), the result is

$$\left(\frac{d\psi}{dx}\right)^2 = \frac{8\pi kT}{\varepsilon}\sum n_i^0(e^{-z_i e_0 \psi_x/kT} - 1) \tag{7.96}$$

Some further simplification is possible. Thus, considering a $z:z$-valent electrolyte, $|z_+| = |z_-| = z$ and $n_+^0 = n_-^0 = n^0$ and

$$\left(\frac{d\psi}{dx}\right)^2 = \frac{8\pi kT}{\varepsilon}\sum n^0(e^{-z_i e_0 \psi_x/kT} - 1)$$

$$= \frac{8\pi kT}{\varepsilon}n^0(e^{z e_0 \psi_x/kT} - 1 + e^{-z e_0 \psi_x/kT} - 1)$$

$$= \frac{8\pi kT}{\varepsilon}n^0[e^{z e_0 \psi_x/kT} - 2(e^{z e_0 \psi_x/2kT})(e^{-z e_0 \psi_x/2kT}) + e^{-z e_0 \psi_x/kT}]$$

$$= \frac{8\pi kT}{\varepsilon}n^0(e^{z e_0 \psi_x/2kT} - e^{-z e_0 \psi_x/2kT})^2 \tag{7.97}$$

But

$$e^{+x} - e^{-x} = 2 \sinh x \qquad (7.98)$$

Hence, Eq. (7.97) becomes

$$\left(\frac{d\psi}{dx}\right)^2 = \frac{32\pi kTn^0}{\varepsilon} \sinh^2 \frac{ze_0\psi_x}{2kT} \qquad (7.99)$$

From this equation, one can get the field $d\psi/dx$ in the solution by taking square roots on both sides. There is, however, a positive and a negative square root. To decide which root is to be taken, one remembers that, at the positively charged electrode, $\psi > 0$ but $d\psi/dx < 0$, while, at the negatively charged electrode, $\psi < 0$ and $d\psi/dx > 0$. Hence it is clear that only the negative root of Eq. (7.99) corresponds to the physical situation, i.e.,

$$\frac{d\psi}{dx} = -\left(\frac{32\pi kTn^0}{\varepsilon}\right)^{\frac{1}{2}} \sinh \frac{ze_0\psi_x}{2kT} \qquad (7.100)$$

What is $d\psi/dx$? It is the field (or gradient of potential) at a distance x from the electrode according to the diffuse-charge model of Gouy and Chapman. Hence, Eq. (7.100) spells out the relation between the electric field and the potential at any distance x from the electrode.

Instead of the field, it is preferable to have an expression for the total diffuse charge in the solution in terms of the potential. This diffuse charge is obtained as follows. A simple question is asked: What is the origin of the field $d\psi/dx$? One recalls Gauss's law from electrostatics: The charge contained in a closed surface, or Gaussian box, is equal to $\varepsilon/4\pi$ times the area of the closed surface (taken here as unity) times the component of the field normal to the surface

$$q = \frac{\varepsilon}{4\pi} \frac{d\psi}{dx} \qquad (7.101)$$

Hence, since $d\psi/dx$ is known from Eq. (7.100), the corresponding q can be obtained. This q is the charge enclosed within the closed volume at the surface of which the field is $d\psi/dx$.

To compute the *total* diffuse-charge density q_d, the Gaussian box chosen is (Fig. 7.64) a rectangular box with one unit-area side deep in the solution at $x \to \infty$, where ψ_x and $d\psi_x/dx = 0$, and with the other side "very close to the electrode" so as not to miss any diffuse charge. How close to the electrode? As close as possible! How close is that? If the model considered is that of *point*-charge ions (and this is the Gouy–Chapman model), then the box must be brought up to $x = 0$.

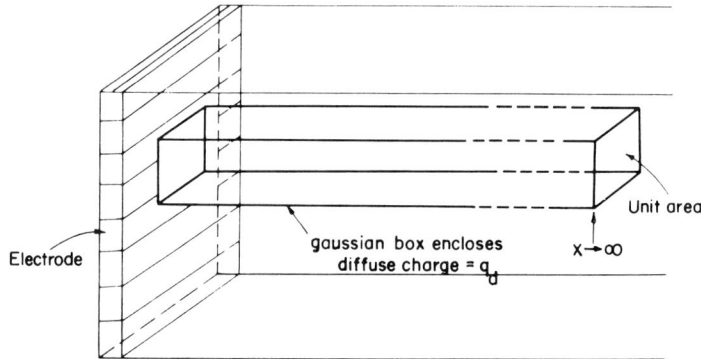

Fig. 7.64. A Gaussian box containing the entire diffuse charge q_d.

Hence, the total diffuse-charge density scattered in the solution under the interplay of thermal and electrical forces, q_d, is given from Eqs. (7.100) and (7.101) with $x = 0$, as

$$q_d = -2\left(\frac{\varepsilon n^0 kT}{2\pi}\right)^{\frac{1}{2}} \sinh \frac{ze_0\psi_0}{2kT} \tag{7.102}$$

where ψ_0 is the potential at $x = 0$ relative to the bulk of the solution where the potential is taken as zero (i.e., $\psi_\infty = 0$).

Equation (7.102) shows that the total diffuse charge varies (according to the present model) with the total potential drop in the solution according to a hyperbolic sine relation.

7.4.5. A Picture of the Potential Drop in the Diffuse Layer

The field $d\psi/dx$ is the gradient of the potential. By integrating the field, one obtains the variation of the potential with distance. Let this be done *after assuming that*[†]

$$\sinh \frac{ze_0\psi_x}{2kT} \approx \frac{ze_0\psi_x}{2kT}$$

Thus, from Eq. (7.100), one has

$$\frac{d\psi_x}{dx} \approx -\left(\frac{32n^0\pi kT}{\varepsilon}\right)^{\frac{1}{2}} \frac{ze_0\psi_x}{2kT}$$

$$\approx -\left(\frac{8\pi n^0(ze_0)^2}{\varepsilon kT}\right)^{\frac{1}{2}} \psi_x \tag{7.103}$$

[†] This assumption is made only for mathematical convenience.

Examine the constant inside the brackets. It is the familiar $\varkappa^2$ of the Debye–Hückel theory, where it was shown that $\varkappa^{-1}$ can be considered the effective thickness of the ionic cloud (cf. Section 3.3.8). In terms of $\varkappa$, Eq. (7.103) becomes

$$\frac{d\psi_x}{dx} = -\varkappa\psi_x \qquad (7.104)$$

or, by integration,

$$\ln \psi_x = -\varkappa x + \text{Constant} \qquad (7.105)$$

To evaluate the constant, the following boundary condition is used: At $x \to 0$, $\psi_x \to \psi_0$. It follows, therefore, that

$$\psi_x = \psi_0 e^{-\varkappa x} \qquad (7.106)$$

i.e., the potential decays exponentially into the solution; deep enough inside the solution, $x \to \infty$, the potential becomes zero. Further, as the solution concentration n^0 increases, $\varkappa$ increases and ψ_x falls more and more sharply. This potential–distance relation (Fig. 7.65) is an important and simple result from the Gouy–Chapman model. It forms a valuable basis for thinking about the interaction of the diffuse charges around what are called *colloidal* particles.

In the Debye–Hückel ionic-cloud model, it was found that the electrical effect of the cloud on the central ion could be simulated by placing the entire charge of the cloud, $-z_i e_0$, at the distance $\varkappa^{-1}$ from the central ion. One wonders, therefore, if the electrical effect of the diffuse-charge region could be simulated by placing the entire Gouy–Chapman charge q_d on a plane parallel to the electrode and a distance $\varkappa^{-1}$ from it. If this is done (Fig. 7.66), one has in effect a parallel-plate condenser situation, i.e., a charge of $-q_d = q_M$ at the $x = 0$ plate and the diffuse charge q_d at the $x = \varkappa^{-1}$ plate.

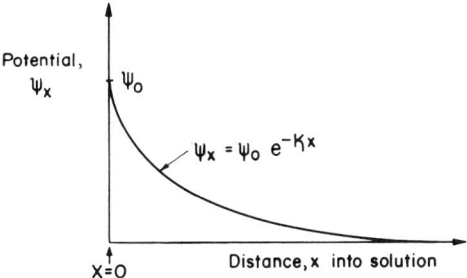

Fig. 7.65. The distance variation of the potential in the diffuse-charge region.

The potential difference across a parallel-plate condenser is

$$\Delta V = \frac{4\pi}{\varepsilon} qd \qquad (7.79)$$

where q is the charge on the plates, and d, their distance apart. By substituting [cf. Eqs. (7.102) and (3.35)]

$$q_M = -q_d = 2\left(\frac{\varepsilon n^0 kT}{2\pi}\right)^{\frac{1}{2}} \sinh \frac{ze_0\psi_0}{2kT}$$

and, for small values of the potential ψ_0 at the electrode,

$$q_M \approx 2\left(\frac{\varepsilon n^0 kT}{2\pi}\right)^{\frac{1}{2}} \frac{ze_0\psi_0}{2kT}$$

and

$$d = \varkappa^{-1} = \left(\frac{\varepsilon kT}{8\pi n^0 z^2 e_0^2}\right)^{\frac{1}{2}}$$

(a)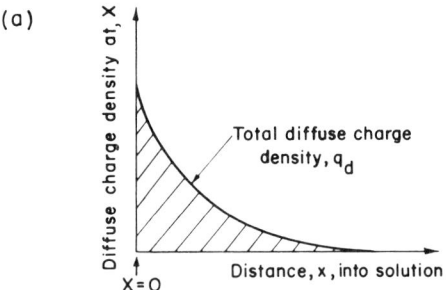

(b)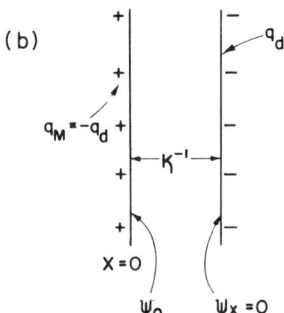

Fig. 7.66. The diffuse-charge region (a) can be simulated by (b) a sheet of charge q_d placed at a distance $\varkappa^{-1}$ from the $x = 0$ plane.

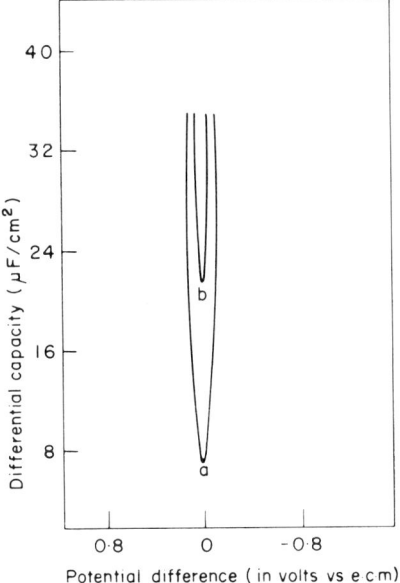

Fig. 7.67. The Gouy–Chapman theory predicts that the capacity of an interface should be a cosh function of the potential difference across it. Curve *b* is calculated for a higher concentration.

in Eq. (7.79), the result is

$$\Delta V = \psi_0 \tag{7.107}$$

which means that the $x = 0$ plate is at a potential ψ_0 relative to the $x = \varkappa^{-1}$ plate at a potential zero. This is indeed the situation.

Once the total diffuse charge q_d is known [Eq. (7.102)] and also the potential ψ_0 at $x = 0$, the differential capacity is easy to calculate. Assuming that $\psi_M = \psi_0$ (this is a reasonable approximation if one is talking of point-charge ions), one simply differentiates q_d with respect to[†] $\psi_M - \psi_B = \psi_M$ (because $\psi_B = 0$). Thus,

$$C = \frac{\partial q_M}{\partial \psi_M} = -\frac{\partial q_d}{\partial \psi_M} = \left(\frac{\varepsilon z^2 e_0^2 n^0}{2\pi kT}\right)^{\frac{1}{2}} \cosh \frac{ze_0 \psi_M}{kT} \tag{7.108}$$

Now, the cosh function gives inverted parabolas (Fig. 7.67). Hence, according to the simple diffuse-charge theory, the differential capacity of

---

[†] The symbol ψ_B is used for the potential ψ in the bulk of the solution.

an electrified interface should not be a constant. Rather, it should show an inverted-parabola dependence on the potential across the interface. This, of course, is a welcome result because the major weakness of the Helmholtz–Perrin model is that it does not predict any variation of capacity with potential, although such a variation is found experimentally (Fig. 7.68).

7.4.6. An Experimental Test of the Gouy–Chapman Model: Potential Dependence of the Capacitance, but at What Cost?

After the initial jubilation that a weakness of the parallel-plate model, namely, constant capacity with change of potential, has been overcome by the diffuse-layer model and the theory now can account for the fact that the capacitance varies with potential, one has to face a few more somber facts about the details of such dependence.

The main fact is that the experimental capacity–potential curves are just not the inverted parabolas (Fig. 7.68) which the Guoy–Chapman diffuse model predicts (Fig. 7.67). In very dilute solutions ($<0.001 M$) and at potentials near the pzc, there are *portions* of the experimental curves which suggest that the interface is behaving in a Gouy–Chapman way.

But, at potentials further away from that of zero charge and in concentrated solutions, the Gouy–Chapman model bears no relation to reality. Not only is the predicted shape of the capacity–potential curves wrong,

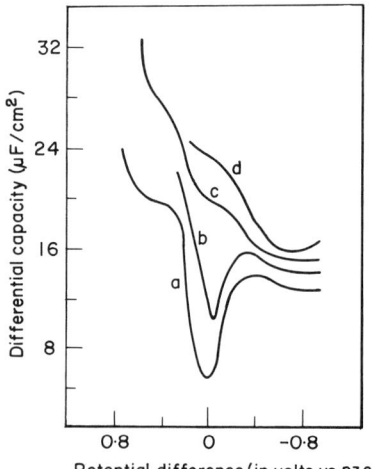

Fig. 7.68. Experimental plots of differential capacity *versus* potential at different solution concentrations. (The solution concentration increases from curve *a* to *d*.)

TABLE 7.8
Differential Capacity of a Hg–NaF Interface at the Potential of Zero Charge

	Differential capacity, $\mu F\ cm^{-2}$	
Concentration	Experimental	Calc from Eq. (7.108)
0.001N	6.0	7.2
0.01N	13.1	22.8
0.1N	20.7	72.2
1.0N	25.7	228.0

but there is also the matter of the concentration dependence. The Gouy–Chapman theory predicts that the capacity depends on the concentration to an extent that is not at all observed. At high concentrations (e.g., $1M$) the predicted capacity is almost an order of magnitude higher than the observed value (Table 7.8).

The model seems to be in sharp discord with the facts. Looked at as an attempt to represent all the essential aspects of an electrified interface, the Gouy–Chapman theory might best be described as a brilliant failure. However, as will be seen, it represents an important contribution to a truer description of the double layer; it also finds use in the understanding of the stability of colloids and, hence, of the stability of living systems (*cf.* Section 7.10) and in certain areas of electrodics (Section 8.3.2).

7.4.7. Some Ions Stuck to the Electrode, Others Scattered in Thermal Disarray: The Stern Model

The next step is fairly obvious. The Helmholtz–Perrin thesis of a *layer* of ions in contact with the electrode and the Gouy–Chapman antithesis of the ions' being scattered out in solution in thermal disarray suggest the synthesis of having some ions stuck at the electrode and the others scattered in cloudlike fashion. This synthesis was made by Stern. His theory was the result of a new inspiration, not stimulated by the arrival of new facts unavailable to Gouy and Chapman.

The simplest version of the Stern theory consists in eliminating the point-charge approximation of the diffuse-layer theory. This is done in exactly the same way (Fig. 7.69) as in the theory of ion–ion interactions (*cf.* Section 3.5.2); the ion-centers are taken as not coming closer than a

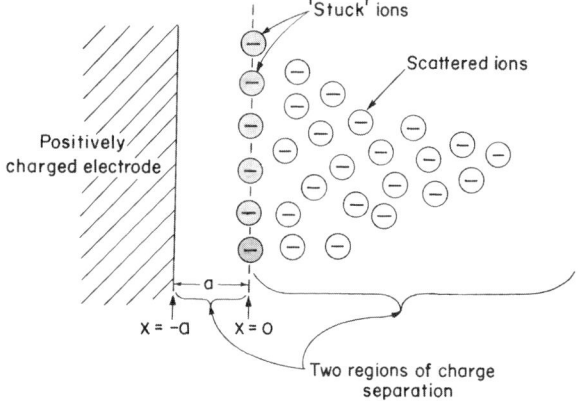

Fig. 7.69. The Stern model with a layer of (—) excess charge stuck to the electrode and the remainder scattered in cloud fashion. The *locus* of centers of the stuck ions is at a distance *a* from the electrode. (*Note*: Only the excess charges are shown in the diagram, and the water molecules are omitted. The latter sit on the electrode and separate it from the ions.)

certain critical distance *a* from the electrode. This is tantamount to applying the Gouy–Chapman theory for ions which are of finite size, not point charges. In other words, the Gouy–Chapman treatment commences not with points on the electrode, but from a layer at a certain distance *a* from the electrode. This distance *a* is the closest distance of approach of the ions to the electrode and will be specified more quantitatively as this account goes on.

7.4.8. A Consequence of the Stern Picture: Two Potential Drops across an Electrified Interface

The distribution of charge in a direction normal to the electrode can now be briefly examined from the Stern point of view.

Under all conditions, the interface as a whole (the electrode side *taken along with* the electrolyte side) is electrically neutral—the net charge density q_M on the electrode must be equal in magnitude and opposite in sign to the net charge density q_S on the solution side, i.e., $-q_M = q_S$. But, according to the Stern picture, the charge q_S on the solution is partially stuck (the Helmholtz–Perrin charge q_H) to the electrode and the remainder q_G is diffusely spread out (in Gouy–Chapman style) in the solution, i.e.,

$$q_S = q_H + q_G \qquad (7.109)$$

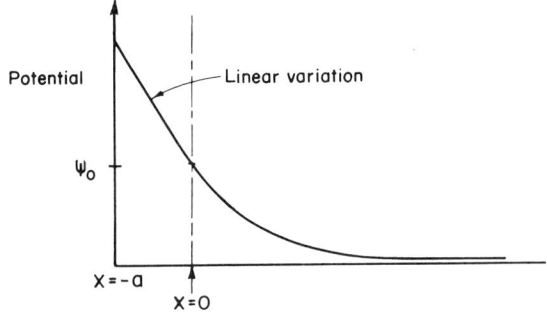

Fig. 7.70. The potential variation according to the Stern model.

There are therefore two regions of charge separation. The first region is from the electrode to the Helmholtz plane (the plane defined by the *locus* of centers of the stuck ions); and the second region is from this plane of fixed charges into the heart of the solution where the net charge density is zero. When, however, charges are separated, potential drops result. The Stern model implies, therefore, two potential drops, i.e.,

$$\phi_M - \phi_B = (\phi_M - \phi_H) + (\phi_H - \phi_B) \tag{7.110}$$

where ϕ_M and ϕ_H are the inner potentials at the metal and the Helmholtz planes, and ϕ_B is the potential in the bulk of the solution.

Why should these two potential drops, i.e., $\phi_M - \phi_H$ and $\phi_H - \phi_B$, be distinguished?

There is an important reason. The Stern synthesis of the Helmholtz–Perrin and Gouy–Chapman models also implies a synthesis of the potential-distance relations characteristic of these two models. The Helmholtz–Perrin model—it may be recalled (*cf.* Section 7.4.1)—argues for a *linear* variation of potential with distance; and the Gouy model, an approximately *exponential* potential drop (*cf.* Section 7.4.5).

Thus, the Stern model pictures (Fig. 7.70) the potential variation across an interface as consisting of *two* regions, a linear region corresponding to the ions stuck on the electrode and an exponential region corresponding to the ions which are under the combined influence of the ordering electrical and the disordering thermal forces.

7.4.9. Another Consequence of the Stern Model: An Electrified Interface Is Equivalent to Two Capacitors in Series

An interesting result emerges from the concept of two potential drops at an interface. One asks: How are the potential drops affected by small changes in the charge on the metal? In other words, what is the result of differentiating the expression for the potential difference across the interface with respect to charge on the metal? One obtains

$$\frac{\partial(\phi_M - \phi_B)}{\partial q_M} = \frac{\partial(\phi_M - \phi_H)}{\partial q_M} + \frac{\partial(\phi_H - \phi_B)}{\partial q_M} \quad (7.111)$$

In the denominator of the last term, one can replace ∂q_M with ∂q_d because the total charge on the electrode is equal to the total diffuse charge, i.e.,

$$\frac{\partial(\phi_M - \phi_B)}{\partial q_M} = \frac{\partial(\phi_M - \phi_H)}{\partial q_M} + \frac{\partial(\phi_H - \phi_B)}{\partial q_d} \quad (7.112)$$

Now examine each term in the equation. Each term is the reciprocal of a quantity which is of the form (Small change in charge/Small change in potential difference), i.e., it is the reciprocal of a differential capacity. Hence, Eq. (7.112) can be rewritten thus

$$\frac{1}{C} = \frac{1}{C_H} + \frac{1}{C_G} \quad (7.113)$$

where C is the total capacity of the interface; C_H is the Helmholtz–Perrin capacity, i.e., the capacity of the region between the metal and the Helmholtz plane to store charge; and C_G is the Gouy–Chapman, or diffuse-charge, capacity.

This result is formally identical to the expression for the total capacity displayed by two capacitors in series (Fig. 7.71). The conclusion therefore is that an electrified interface has a total differential capacity which is given by the Helmholtz and Gouy capacities in *series*.

This makes good sense. The most generalized concept of a capacitor is that of a region of space capable of storing charge. Capacitors in series

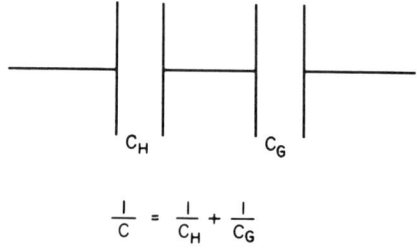

Fig. 7.71. The total differential capacity C of an electrified interface is given by the Helmholtz and Gouy capacities in series.

imply that the regions are consecutive in space, each region accounting for only a part of the total potential difference. The Helmholtz and Gouy regions do store ions and are consecutive in a direction normal to the electrode, and the total potential difference is across the whole interface.

7.4.10. The Relative Contributions of the Helmholtz–Perrin and Gouy–Chapman Capacities

It may be recalled that one of the difficulties with the diffuse-charge layer theory is that it predicted capacity values about an order of magnitude higher than those observed in the case of solutions with electrolyte concentrations of about $1M$. This is a problem which Stern's theory solves quite easily.

Thus, in Eq. (7.113), the total capacity depends on two terms

$$\frac{1}{C} = \frac{1}{C_H} + \frac{1}{C_G} \tag{7.113}$$

What happens when the concentration n^0 of the electrolyte is large? From Eq. (7.108), it can be seen that C_G becomes large, while C_H does not change. Hence, with increasing concentration, the second term in Eq. (7.113), $1/C_G$, becomes small compared with the first $1/C_H$, whereupon

$$\frac{1}{C_G} \ll \frac{1}{C_H}$$

and, for all practical purposes,

$$\frac{1}{C} \approx \frac{1}{C_H} \quad \text{or} \quad C \approx C_H$$

That is, in sufficiently concentrated solutions, the capacity of the interface is effectively equal to the capacity of the Helmholtz region, i.e., of the parallel-plate model. The analogy between the electrified interface and electrical capacitors works well here. The total capacity of two capacitors in series is effectively equal to the smaller capacity when the other one is relatively large.

What does this mean? It means that, if the Helmholtz and Gouy regions are compared at sufficiently high concentrations (C_G, high), most of the solution charge is squeezed onto the Helmholtz plane, or confined in a region very near this plane. In other words, little charge is scattered diffusely into the solution in the Gouy–Chapman disarray.

But what happens if C_G is low, that is, what happens at sufficiently low concentrations? Under these conditions [cf. Eq. (7.113)],

and
$$\frac{1}{C_H} \ll \frac{1}{C_G}$$

$$\frac{1}{C} \approx \frac{1}{C_G}$$

or
$$C \approx C_G$$

This means that the electrified interface has become in effect Gouy–Chapman-like in structure, with the solution charge scattered under the simultaneous influence of electrical and thermal forces.

It must not, however, be imagined that the Gouy and Helmholtz aspects of the double-layer structure play an equal role. Though its mathematical development is very simple, the older parallel-plate condenser approach of Helmholtz–Perrin gives much more often the essential, if very rough, picture because, for most ordinary concentrations, $C_G \gg C_H$, whereupon the capacity of the double layer is virtually equal to the capacity derived with the Helmholtz–Perrin model.

This resemblance is rather embarrassing, not only because more detailed mathematical development has been given to the Gouy–Chapman theory, but also because the simple Helmholtz–Perrin model led to several difficulties, e.g., constant capacities and strictly parabolic electrocapillary curves.

Is the Stern theory no different from and no improvement over the Helmholtz–Perrin model at practical concentrations where the diffuse charge ceases to be diffuse? A closer look at what makes up the interface is required.

7.4.11. Some Questions Regarding the Sticking of Ions to the Electrode

The description of the Stern model (Sections 7.4.7 to 7.4.10) given hitherto has been somewhat impressionist in style. Thus, one talked of ions sticking to the electrode. How does an interface look with ions stuck on the metal? What is the distance of closest approach? Are hydrated ions held on a hydrated electrode, i.e., is an electrode covered with a sheet of water molecules—arrangement *O*? Or are ions stripped of their solvent sheaths in intimate contact with a bare electrode—arrangement *I*? (See Fig. 7.72). Are *both* these arrangements *O* and *I* realized in practice? Or

THE ELECTRIFIED INTERFACE

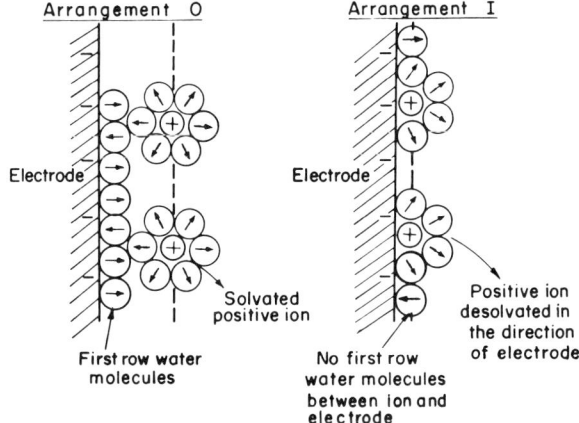

Fig. 7.72. The two types of arrangement of ions stuck on an electrode. (The compensating ions in the diffuse layer are not shown.)

does one ionic species, e.g., potassium ions, resort to arrangement O, and another ionic species, e.g., chloride ions, make use of arrangement I? If so, what determines which ion prefers one arrangement to the other? What are the forces which influence the sticking of ions to electrodes?

7.4.12. An Electrode Is Largely Covered with Adsorbed Water Molecules

A structural discussion of electrified interface must begin with the statement that the majority of sites on an electrode surface are occupied by water molecules. It is possible to show by a very rough calculation that, in the absence of directing forces, at least 70% of the metal surface must be covered with water molecules. All one has to do is to estimate the population density n_w (number per square centimeter) of water molecules on a plane in bulk water and divide this by the number of sites, n_s, per square centimeter on the metal surface. The quantity n_w is roughly estimated by taking the two-thirds power of the number of water molecules per cubic centimeter, which is equal to the Avogadro number (number of particles in a mole) divided by the molar volume V_m (i.e., the volume occupied by 1 mole). Thus,

$$n_w = \left(\frac{N_A}{V_m}\right)^{\frac{2}{3}} = \left(\frac{N_A \varrho}{M}\right)^{\frac{2}{3}} \qquad (7.114)$$

where M and ϱ are the molecular weight and density of water, respectively. The number of sites per square centimeter can be taken as equal to the number of metal atoms per square centimeter. The fraction θ_w of the surface covered turns out (by inserting the appropriate numerical values into this zeroth-approximation approach) to be about 0.7.

7.4.13. Metal–Water Interactions

There are, however, forces operating between water molecules and the metal electrode. The actual coverage with water molecules may therefore be even more than 70%. What are some of these forces?

Firstly, there are the *image forces* which have been described earlier (*cf.* Section 7.2.5*h*). What have image forces to do with metal–water interactions? A water molecule is asymmetric; it is a dipole, i.e., it has two charges separated by a certain distance (*cf.* Section 2.3.3). Each of these charges induces a charge on the metal. But, since there are two charges in a dipole, one can think of an image dipole (Fig. 7.73). Thus, the force between a dipole and a metal can be computed by replacing the metal with an *image dipole* and considering the dipole–dipole interaction.

There is another type of force between water molecules and the metal, *dispersion forces*. Dispersion forces (or London forces) can be seen classically as follows: The time-average picture of an atom shows spherical symmetry because the charge due to the electrons' orbiting around the nucleus is smoothed out in time. An instantaneous picture of, say, a hydrogen atom, would, however, show a proton here and an electron there—two charges separated by a distance. Every atom has an *instantaneous* dipole moment; of course, the time average of all these dipole moments is zero.

This instantaneous dipole will induce an instantaneous dipole in a contiguous atom, and an instantaneous dipole–dipole force arises. When these forces are averaged over all instantaneous electron configurations of the atoms, an *attractive*, nondirectional force arises, the dispersion force.

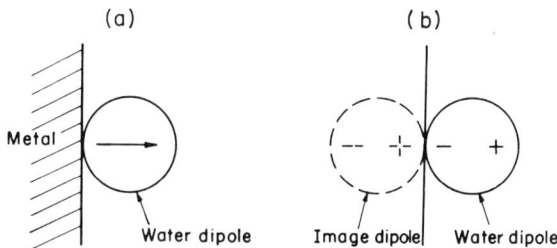

Fig. 7.73. (a) The interactions between a water molecule and a metal can be computed by (b) replacing the metal with an image dipole.

Dispersion forces are important at electrified interfaces. They contribute to adsorption at electrodes.

Apart from the image and dispersion forces which attract water to the metal, there are also the forces which induce chemical bonding. All this indicates that even an uncharged electrode has an attraction for water molecules and may overcome the forces which bind water molecules into networks in the liquid phase.

7.4.14. The Orientation of Water Molecules on Charged Electrodes

What happens to the water layer when the metal is charged? The forces of attraction between water molecules and the electrode still operate, and there will still be a water layer. But, in addition, the charge on the metal will stimulate the water molecules to orient themselves (charge implies field, and dipoles tend to align with fields). The argument is identical to that used to discuss the orientation of dipoles in the solvent sheath around an ion (*cf.* Chapter 2 on ion–solvent interactions).

The net result of all these image, dispersion, chemical, and orienting forces between an electrode and the water molecules adjacent to it is that the electrode is almost completely covered with a layer of oriented water dipoles. However, it must not be imagined that the water molecules are unaffected by the presence of their neighbors. After all, dipoles interact with dipoles. Hence, the oriented water molecules experience lateral interaction—a phenomenon which affects the net number of water molecules oriented in one direction and therefore the value of the dipole potential $\Delta\chi$. Through $\Delta\chi$, the potential difference across the interface and the structure of the interface will be affected.

7.4.15. How Close Can Hydrated Ions Come to a Hydrated Electrode?

The ions in solution have been shown to solvate with a total number n_t of water molecules (*cf.* Chapter 2). Some of these water molecules are left behind when ions random-walk and drift around. Others, however, the n_1 primary hydration molecules, resist the hydrogen bonding with the secondary water molecules and tolerate the lateral (dipole–dipole) repulsion of their like-oriented neighbors. They have a stronger attraction to the ion and follow it in its thermal, random movements.

Thus, the ion, wrapped in a primary hydration sheath, migrates up to the electrode. How close to the electrode can such a hydrated ion approach? In the first instance, it can proceed till the water molecules of the hydrated ion collide with the oriented water molecules of the hydrated electrode. The electron shells of the water molecules then start overlapping and repelling—electron-overlap interaction.

The ions have attained arrangement O (Fig. 7.72), a layer of hydrated ions in contact with a solvated electrode. The plane drawn through the *locus* of centers of these hydrated ions is the OHP (outer Helmholtz plane). The distance between the OHP and the metal surface (the plane going through the outer edge of the substrate atoms) is the closest distance of approach in this arrangement O.

7.4.16. Is It Only Desolvated Ions which Contact-Adsorb on the Electrode?

A very important question now arises. Does arrangement O—hydrated ions in contact with a *hydrated* electrode—always correspond to the configuration of lowest free energy? Why cannot ions divest themselves (at least partly) of their primary water, nudge adsorbed water molecules away from electrode sites, and come in *contact* (Fig. 7.72) with a bare electrode (arrangement I)? Defining the *locus* of centers of these contact-adsorbed ions as the IHP (inner Helmholtz plane), the question is: Do ions quit the OHP and populate the IHP.

No *a priori* answer can be given to these questions. One must calculate the free-energy change for (i.e., the work done by) an ion to move from the OHP to the IHP while at the same time displacing the appropriate number of adsorbed water molecules. If the free-energy change is negative, ions will make this move.

7.4.17. The Free-Energy Change for Contact Adsorption

Approximate calculations of the free-energy change involved in ions moving to the IHP have been made. The essence of these calculations consists in viewing the process of contact adsorption as a two-step process. In step 1, a hole of area πr_i^2 ($r_i =$ radius of contact-adsorbing ion) must be swept free of water molecules in order to make room for the ion. Thus, a certain number of water molecules must be desorbed. In step 2, the ion strips itself of part of its solvent sheath and jumps into the hole.

This two-step adsorption process can be likened to the elementary act of a transport process—hole formation followed by a jump (*cf.* Chapter 6). In a transport process, however, the conditions before and after the two-step process are identical. Not so in the case of adsorption. In both the steps, the particles involved (water molecules in step 1 and the ion plus associated

TABLE 7.9

The Thermodynamics of Contact Adsorption

Ion	Water–electrode interactions			Ion–electrode interactions			Ion–water interactions			Total
	ΔH	ΔS	ΔG	ΔH	ΔS	ΔG	ΔH	ΔS	ΔG	ΔG
Na⁺	28.1	3.6	27.0	−49.1	1.5	−49.6	39.5	11.2	36.1	+13.5
K⁺	21.1	6.7	19.1	−48.2	2.0	−48.8	33.9	5.2	32.3	+2.6
Cs⁺	20.9	10.7	17.7	−43.5	3.3	−44.5	17.7	−12.2	21.4	−5.4
F⁻	38.1	6.7	36.1	−47.1	1.4	−47.5	35.0	6.4	33.1	+21.7
Cl⁻	22.0	12.4	18.3	−49.3	2.1	−49.9	17.7	−16.6	22.7	−8.9
Br⁻	21.6	14.5	17.2	−49.2	2.8	−50.0	14.7	−22.4	21.4	−11.4
I⁻	21.9	18.0	16.5	−49.2	3.6	−50.3	10.9	−32.6	20.7	−13.1

water in step 2) break old attachments and make new ones (change of enthalpy, ΔH) and also exchange old freedoms and restrictions for freedoms and restrictions characteristic of the new neighborhood (change of entropy, ΔS).

These changes of interactions can be classified into three groups: (1) changes in water–electrode interactions; (2) changes in ion–electrode interactions (due to image, dispersion, and electron-overlap forces); and, lastly, (3) changes in ion–water interactions, i.e., changes in hydration. The enthalpy, entropy, and free-energy changes corresponding to these groups for various ions are shown in Table 7.9.

The final figures in the free-energy balance sheet are intriguing. It turns out that some ions *lose* free energy by moving to the IHP. These ions will therefore contact adsorb. Other ions would gain free energy by contact adsorbing; they will go no farther to the electrode than the OHP.

7.4.18. What Determines the Degree of Contact Adsorption?

The tabulated changes of the enthalpy, etc. (Table 7.9), reveal an important point about contact adsorption. When considered alone, the free-energy change from water–electrode interactions is positive for all species; what tip over the total free-energy change to negative values are the ion–electrode interactions (essentially dispersion and image interactions). But it is the group of ion–water interactions which gives to different ionic species different *signs* of total free-energy change and thus their differing attitudes to contact adsorption—*for* in the case of Cs^+, Cl^-, Br^-, and I^- and *against* in the case of Na^+, K^+, and F^-.

What are these ion–water interactions? They are hydration interactions (*cf.* Chapter 2). If the ions are strongly hydrated (with primary water), the changes in hydration free energy (required for contact adsorption) are too large; the transition to the IHP is not energetically worthwhile. Ions which are sufficiently strongly hydrated do not contact-adsorb.

Now, hydration is much dependent on the radius of the ions; see, e.g., a plot of free energy of hydration *versus* ionic radius (Fig. 7.74). It should be no surprise, therefore, to find a correlation between contact adsorption and ionic radius. This is indeed the case (Fig. 7.74). The large ions (Cs^+, Cl^-, Br^-, and I^-) are precisely those which have a negative free energy of contact adsorption. The smaller ions (Na^+, K^+, and F^-), tightly wrapped up in their solvent sheaths, would not be expected to contact-adsorb.

The study of the forces operating in the interphase region has shown, therefore, that there are two versions of "closest" approach of ions to the electrode and thus two corresponding arrangements O and I. In one version, *hydrated* ions on the OHP "touch" the hydrated electrode; in the other, partially hydrated ions on the IHP are contact adsorbed onto a locally

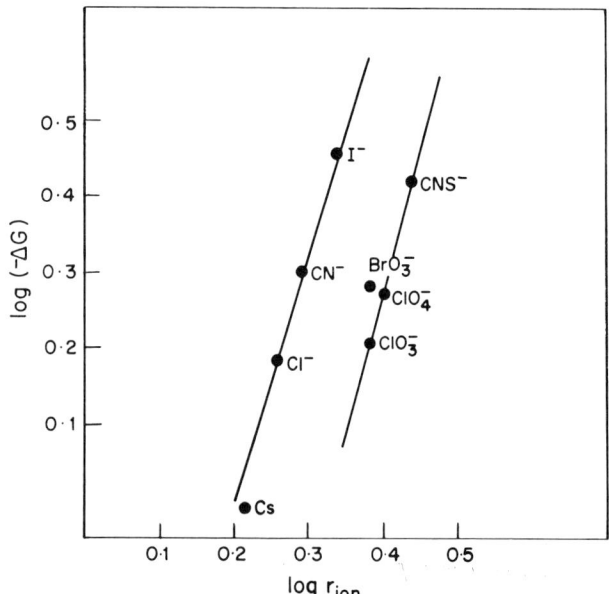

Fig. 7.74. The relation between contact adsorption and ionic radius.

dehydrated electrode. Contact adsorption, however, is not for all ionic species; some do it, others do not. The main determining fact is: How easily may ions be partly dehydrated, i.e., what is the value of their free energy of hydration?

7.4.19. How Is Contact Adsorption Measured?

The *total* surface excess Γ_i of a particular species i in an interphase, i.e., the excess which is in the diffuse region plus that which is contact adsorbed, can be obtained directly from electrocapillary measurements. If one wants, e.g., the surface excess of chloride ions, one obtains electrocapillary curves in solutions of various *hydrochloric acid* concentrations. A necessary condition, however, is to incorporate in the cell a nonpolarizable electrode which allows, *not chloride ions*, but hydrogen ions to leak across its surface. Under these conditions, it has been shown (Section 7.3.8) that, at a constant cell potential V,

$$\left(\frac{\partial \gamma}{\partial \mu_{\text{HCl}}}\right)_{\text{const.}V} = \Gamma_- \qquad (7.72)$$

Once the surface excess of the negative ions is obtained, it is easy to calculate the excess-charge density q^- due to the negative ions from the relation

$$q^- = z_- F\Gamma_- = -F\Gamma_- \qquad (7.115)$$

How is this q^- distributed in the interface? Is it completely due to the *contact* adsorption of chloride ions?

A moment's thought will show that these questions are illegitimate in the framework of the nonmodelistic thermodynamic approach which has yielded the values of surface excess. The questions pertain to the location and distribution of charges, i.e., the *structure* of the interface.

The separation of the *contact-adsorbed* negative charge from the total surface excess Γ_- must therefore involve a model of the interface, i.e., it must utilize structural thinking rather than a routine processing of thermodynamic data.

One approach is to take a close look at the table presenting the tendencies of various ionic species to contact-adsorb (Table 7.9). It appears that the small cations do not contact-adsorb. Thus, the following structural assumption could be made in the case of a hydrochloric acid solution: The H_3O^+ ions do not enter the IHP. But the OHP is the fence of the diffuse-charge region, and, therefore, all the cationic excess charge q^+ in the interphase region must be in the Gouy–Chapman region.

One must get to know the magnitude of q^+. The slope of the electrocapillary curve (at the particular potential at which q^- has been determined) furnishes the total value of the excess cationic *and* anionic charge q_S on the solution side of the interface

$$\left(\frac{\partial \gamma}{\partial V}\right)_{\text{const. comp.}} = -q_M = q_S \qquad (7.61)$$

This q_S is composed of both positive and negative ions, and, hence,

$$q^+ = q_S - q^- \qquad (7.116)$$

Having argued that q^+ is entirely in the diffuse layer, one can use the diffuse-charge theory of Gouy–Chapman (Section 7.4.4) to get the potential ψ_0 at the OHP corresponding to the given potential difference at which the

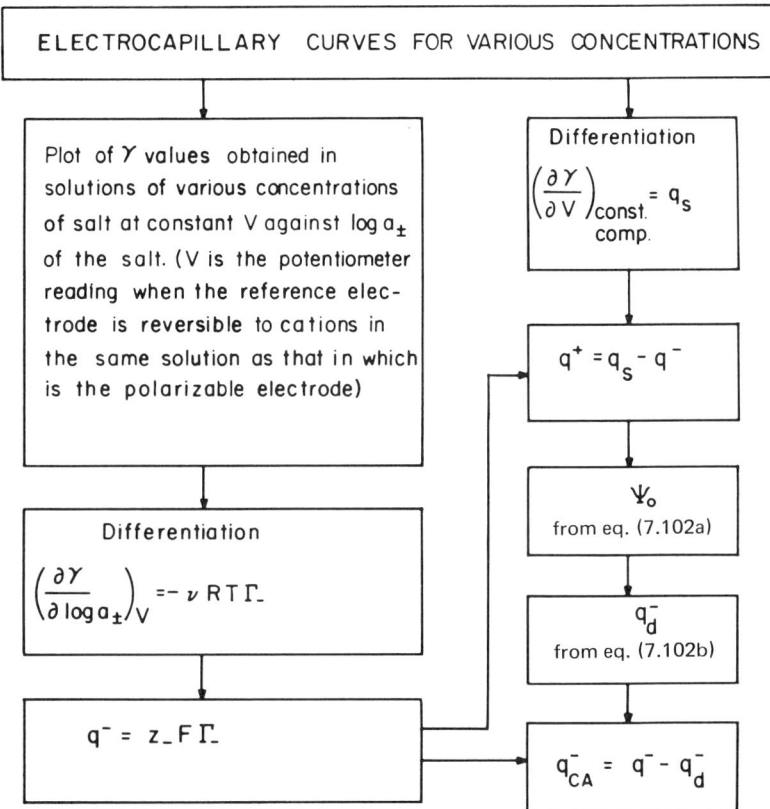

Fig. 7.75. The procedure for getting the contact-adsorbed charge density q_{CA}^-.

total surface excess of positive ions was determined. The following expression is used [cf. Eq. (7.102)]:

$$q_d^+ = \left(\frac{\varepsilon n^0 kT}{2\pi}\right)^{\frac{1}{2}} (e^{-z_+ e_0 \psi_0 /kT} - 1) \qquad (7.102a)$$

Then, the calculated value of ψ_0 for the given applied potential can be used in a similar expression for the *anionic* diffuse charge q_d^-

$$q_d^- = \left(\frac{\varepsilon n^0 kT}{2\pi}\right)^{\frac{1}{2}} (e^{-z_- e_0 \psi_0 /kT} - 1) \qquad (7.102b)$$

The last step is to subtract this anionic charge in the diffuse region from the total surface excess of negative ions, Γ_-, in the interphase expressed in form of charge, i.e., $z_- F\Gamma_- = q^-$. The result, of course, is the negative charge populating the IHP. This is the contact-adsorbed charge q_{CA}^- (*CA* means *contact-adsorbed*).

By repeating the above procedure (Fig. 7.75) at various cell potentials, i.e., various values of the charge on the electrode, q_M, one can determine the variation in the amount of contact-adsorbed charge q_{CA}^- with q_M (Table 7.10). A typical plot of q_{CA}^- versus q_M is shown in Fig. 7.76. It has very important implications, particularly for the unexpectedly varying behavior of the capacity of the interface as a function of the potential difference

TABLE 7.10

Contact Adsorption of Cl⁻ Ions at a Mercury–1.0N KCl Interface at 25°C

Electrode-charge density q_M $\mu C\ cm^{-2}$	Contact-adsorbed charge density q_{CA}^-, $\mu C\ cm^{-2}$
+20	−30.1
+16	−24.7
+12	−20.1
+8	−15.7
+4	−11.1
0	−6.5
−4	−2.6
−8	−0.5
−12	0.0
−16	+0.1
−20	+0.6

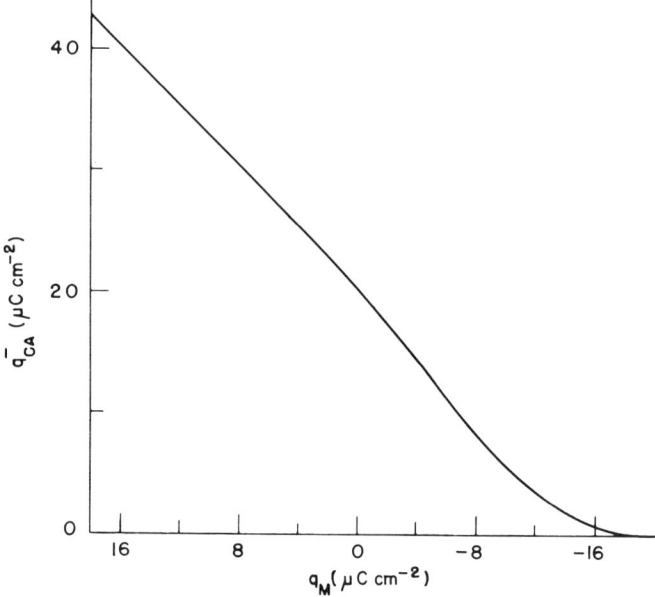

Fig. 7.76. A typical plot of the contact-adsorbed charge density $\bar{q}_{CA}$ versus the electrode-charge density q_M.

across the interface or the charge on the metal. But, more of this later (Section 7.4.23).

7.4.20. Contact Adsorption, Specific Adsorption, or Superequivalent Adsorption

Here attention is drawn to a feature of the $\bar{q}_{CA}$ versus q_M curve (Fig. 7.76), namely, the contact adsorption of *negative* ions on a *negatively* charged metal surface.

This was once an intriguing problem because it was formerly thought that only simple coulombic forces governed the distribution of charge on the solution side of the interface. Proceeding with this prejudice, one could not explain the adsorption of negative charges on a negatively charged electrode. Shelter was therefore taken under the term *specific* adsorption, which really means *adsorption by an unknown mechanism*. This term has however become an anachronism now that there is a fairly solid degree of understanding regarding the truth about the situation, i.e., concerning the forces (see Section 7.4.13) which attract ions into the IHP where they are in contact with the electrode.

Another term *superequivalent adsorption* has also been used, based on a consequence of contact adsorption. Since purely coulombic forces do not operate, there need not be an equality of charge density on the metal and IHP

$$|q_M| \neq |q_{CA}|$$

There is no equivalence, but *super*equivalence, of charge on the IHP. Since $|q_{CA}| > |q_M|$, it means that, if, e.g., negative ions are contact adsorbed on a positively charged electrode, there must be a positively charged diffuse layer to maintain overall electroneutrality across the interface, i.e.,

$$q_M = q_{CA} + q_d$$

The term *superequivalent adsorption* is thus only a *formal* description in contrast to the term *contact adsorption* which suggests a model of the structure of the interface. Based on these views, contact adsorption will be preferentially used in the following treatment.

7.4.21. Contact Adsorption: Its Influence on the Capacity of the Interface

One of the difficulties in the way of understanding the double layer (i.e., the electrified interface) is that it is often a triple layer.[†] But, in the absence of occupants in the IHP, it is indeed a double layer, a layer of excess charge on the metal and, in concentrated solutions, a layer of excess charge on the OHP. However, in the case encountered often, the IHP is populated by a layer of contact-adsorbing ions, which along with a sheet of charge on the metal and another sheet on the OHP forms a triple layer. Thus, an electrified interface in concentrated solutions comes in two models (Fig. 7.77). Model 1 is a double layer without contact-adsorbing ions, and model 2 is a triple layer with contact-adsorbing ions. Now, the type of model applicable to a given system, i.e., the presence or absence of contact adsorption, must influence the property of the interface for storing charge. In other words, contact adsorption must affect the capacity of the interface. To examine this influence, the relationship between C and q_{CA} must be derived.

The procedure is the same as in the Stern demonstration (Section 7.4.9), the total capacity of the interface being formally equivalent to two capacitors in series. That is, (1) the total potential difference is written down as the sum of the potential differences across the regions to be distin-

---

[†] Graduate students in the field often refer to it as the *trouble* layer.

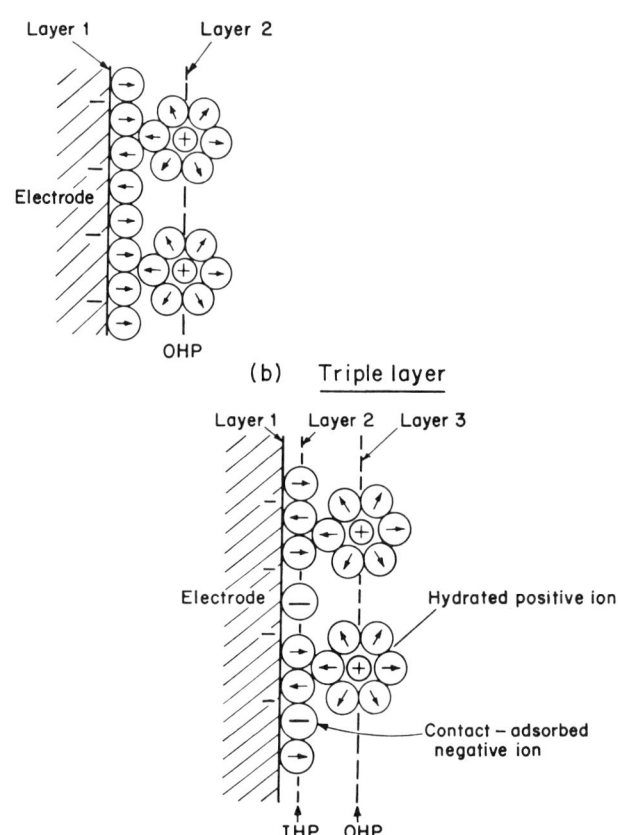

Fig. 7.77. (a) An electrical double layer and (b) an electrical triple layer. Compensating charges in the diffuse layer are not shown.

guished, and (2) the expression is differentiated with respect to the electrode charge density q_M.

It will be assumed that all the diffuse charge is perched on the OHP (the layer with the hydrated cations), i.e., the solution is concentrated. Then, the potential drop beyond the OHP into the solution tends to be negligible.

The total potential difference across an interphase which includes contact-adsorbed ions can be resolved into the component parts. Thus,

$$\phi_M - \phi_S = (\phi_M - \phi_{\text{IHP}}) + (\phi_{\text{IHP}} - \phi_{\text{OHP}}) \qquad (7.117)$$

The potential drops across the two regions, (1) the metal surface to the IHP and (2) the IHP to the OHP, will now be expressed in terms of the integral capacities of the two regions

$$\phi_M - \phi_{IHP} = \frac{q_M}{K_{M \to IHP}} \tag{7.118}$$

and

$$\phi_{IHP} - \phi_{OHP} = \frac{q_d}{K_{IHP \to OHP}} \tag{7.119}$$

Hence, one can rewrite (7.117) in the form

$$\phi_M - \phi_S = \frac{q_M}{K_{M \to IHP}} + \frac{q_d}{K_{IHP \to OHP}} \tag{7.120}$$

This expression is now differentiated with respect to q_M

$$\frac{d(\phi_M - \phi_S)}{dq_M} = \frac{1}{K_{M \to IHP}} + \frac{1}{K_{IHP \to OHP}} \frac{dq_d}{dq_M} \tag{7.121}$$

By definition, one has

$$\frac{d(\phi_M - \phi_S)}{dq_M} = \frac{1}{C} \tag{7.122}$$

and, from the condition of electroneutrality of the interface as a whole,

$$q_M = q_{CA} + q_d$$

Thus, one obtains by differentiation another relation

$$\frac{dq_d}{dq_M} = 1 - \frac{dq_{CA}}{dq_M} \tag{7.123}$$

By substituting these two relations [Eqs. (7.122) and (7.123)] in Eq. (7.121), the result is

$$\begin{aligned}\frac{1}{C} &= \frac{1}{K_{M \to IHP}} + \frac{1}{K_{IHP \to OHP}} \left(1 - \frac{dq_{CA}}{dq_M}\right) \\ &= \left(\frac{1}{K_{M \to IHP}} + \frac{1}{K_{IHP \to OHP}}\right) - \frac{1}{K_{IHP \to OHP}} \frac{dq_{CA}}{dq_M}\end{aligned} \tag{7.124}$$

Now consider the region between the metal and the OHP. The integral capacity of this region may be considered given by two capacitors in series, thus,

$$\frac{1}{K_{M \to OHP}} = \frac{1}{K_{M \to IHP}} + \frac{1}{K_{IHP \to OHP}} \tag{7.125}$$

In terms of this idea, Eq. (7.124) can be rearranged to give

$$\frac{1}{C} = \frac{1}{K_{M \to \text{OHP}}} - \left(\frac{1}{K_{M \to \text{OHP}}} - \frac{1}{K_{M \to \text{IHP}}} \right) \frac{dq_{CA}}{dq_M} \quad (7.126)$$

This is the expression for the capacitance of an interface in the presence of contact adsorption.[†] Note how the differential capacity is affected by contact-adsorbed ions populating the IHP. This situation is evidently a far cry from the situation in the electrostatic theory of capacitors, where the charges on the opposite plates must be equal in magnitude and the capacity is purely dependent on the geometry and dielectric constant of the system, never on the charge. In the electrified interface, the double-layer has become a triple layer, and the capacity varies with the contact adsorption through the quantity dq_{CA}/dq_M. This goes to show that analogies (in the present context, the analogy between an electrified interface and an electric capacitor) only serve as *starting* points for thinking.

7.4.22. Looking Back to Look Forward

It is useful at this point to review the degree of understanding.

The primitive Helmholtz–Perrin parallel-plate condenser model generated perfectly parabolic electrocapillary curves and constant double-layer capacities independent of potential—two predictions at variance with experience. The Gouy–Chapman ionic-atmosphere model argued for an inverted parabolic dependence of capacity on the electrode charge—once again, a prediction in discord with facts.

The simple Stern synthesis (of the Helmholtz–Perrin and Gouy–Chapman models) based on finite-sized ions eliminated the troublesome high Gouy–Chapman capacities. By demonstrating that the capacity of the electrified interface is formally given by the Helmholtz–Perrin and Gouy–Chapman capacitors in series, the Stern model showed that, whenever the Gouy–Chapman capacity grew too large, it tuned itself out of existence.

But one's troubles did not end. Except in very dilute solutions, the Gouy–Chapman capacitor was always tuned out. This implied that the diffuse-charge *region* got squeezed so thin that the so-called "diffuse charge" was in fact piled onto a plane, the OHP—and one was back essentially with a parallel-plate capacitor model with all its defects.

For these reasons, the consideration of the effect of noncoulombic

---

[†] The contribution of the diffuse-double-layer capacity to the total differential capacity of the interface is neglected in this treatment.

forces in the interphase region was undertaken. A new possibility emerged—contact, or specific, adsorption.

There are, therefore, two distances of closest approach in the model of an interface which includes contact-adsorbed ions (Fig. 7.77). The IHP is populated by partially desolvated ions stuck to a *bare* electrode with contact-adsorption forces. The OHP is tenanted by hydrated ions which touch a *hydrated* electrode rather than stick to a bare electrode.

Several questions arise now. Will contact adsorption solve the problem of the change of capacity with electrode charge or cell potential? What about the asymmetry of electrocapillary curves? Can this be interpreted in terms of contact adsorption? In short, how closely does the present model come to representing reality?

7.4.23. The Complete Capacity–Potential Curve

If an electrocapillary γ versus V curve is twice differentiated (*cf.* Section 7.3.6), one obtains a curve of differential capacity *versus* potential. This derived curve (Fig. 7.54) consists of two sections in which the capacity is *apparently* constant. Why apparently? Because of the limitations of the whole procedure of deriving capacity from surface-tension data. The surface tension can be determined to an accuracy within ± 0.2 dyne cm^{-1} near the pzc and up to within ± 0.8 dyne cm^{-1} far away from the pzc. The derived capacities, however, are only accurate to within $\pm 4\,\mu$F cm^{-2}.

When sufficiently accurate electrical methods of determining the capacities are used, one finds (Fig. 7.78) that the capacity–potential curve breaks out into "bumps and flats." It has a complicated fine structure, which depends upon the ions which populate the interphase region. Whereas there is a region of "constant" capacity at V more negative than the ecm, there is also a "hump" in the capacity–potential curve in a region positive to the ecm. At V much more positive than the hump region, the capacity starts shooting up. Perhaps this is because the interface is on the verge of leaking and becoming nonpolarizable, in which case the q in $C = dq/dV$ is contributed to by the transfer of charge across the double layer. The capacity–potential curve presents two basic challenges, the challenge of interpreting the constant-capacity region and that of interpreting the hump.

7.4.24. The Constant-Capacity Region

7.4.24a. The So-Called "Double Layer" Is a Double Layer. When one sees a relatively constant capacity region, e.g., on the negative side of the C versus V curve (*cf.* Fig. 7.78), one's thoughts turn naturally to a

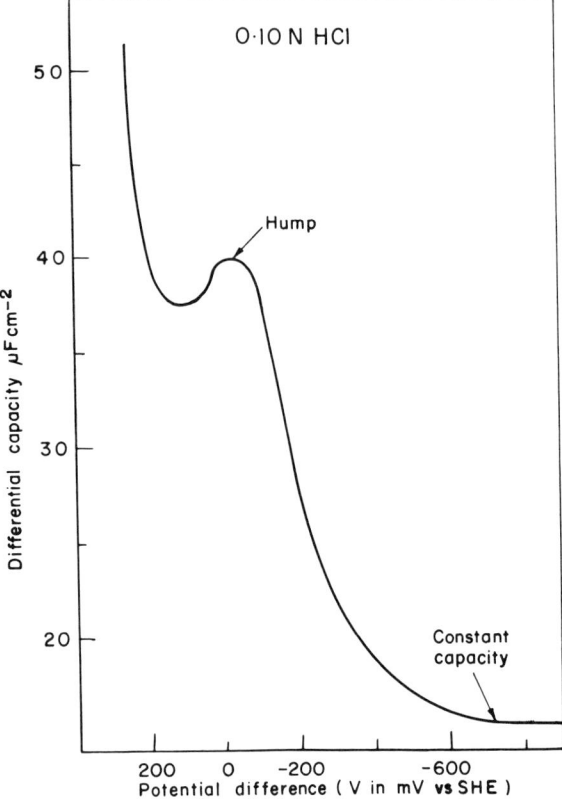

Fig. 7.78. The experimental capacity–potential curve showing the constant-capacity region and the hump.

simple parallel-plate condenser model because such a model yields a potential-independent capacity. One plate (or layer of charges) is located at the metal surface. But where is the second plate? On the IHP? On the OHP? There is a choice and, hence, the possibility of confusion.

Experiment, however, furnishes a valuable clue to the selection of a consistent model. The capacity is not only constant with potential over a region of a few hundred millivolts but also virtually independent of the nature of the ions which occupy the double layer. This point comes out clearly if a comparison is made (see Table 7.11) of the value of the capacity (in the constant-capacity region) obtained in a series of electrolytes containing ions of widely differing sizes. The capacity stays at a value of 16 to 17 μF cm^{-2} irrespective of the radii of the ions.

What is the implication of this constancy of double-layer capacity?

TABLE 7.11

Values of the Constant-Capacity 0.1N Aqueous Chloride Solutions

Ion	Unhydrated radius, Å	Estimated hydrated radius r_i, Å	Differential capacity, μF cm^{-2} at $q_M = -12\,\mu$C cm^{-2}
H$_3$O$^+$	—	—	16.6
Li$^+$	0.60	3.4	16.2
K$^+$	1.33	4.1	17.0
Rb$^+$	1.48	4.3	17.5
Mg^{++}	0.65	6.3	16.5
Sr^{++}	1.13	6.7	17.0
Al^{+++}	0.50	6.1	16.5
La^{+++}	1.15	6.8	17.1

One must think in terms of the capacity of a parallel-plate condenser, i.e., the capacity arising from the Helmholtz–Perrin model,

$$C = \frac{q}{\Delta\phi} = \frac{\varepsilon}{4\pi d} \qquad (7.127)$$

where d is the distance between the condenser plates. If the two plates constituting the condenser are the metal and the IHP, then d for this condenser would simply be the radius of the ions contact-adsorbed in the IHP. Hence, the capacitance should be very sensitive to differences in radii r_i of the ions, the capacity should be inversely proportional to the radius of the ions

$$C = \frac{\varepsilon}{4\pi r_i} \quad [\because d = r_i] \qquad (7.128)$$

Experiment does not show this radius dependence of the capacity (*cf.* Table 7.11). It seems, therefore, that there are no ions in the IHP. In other words, a constant capacity implies the absence of contact adsorption of ions.

Now, although there may not be any ions populating the IHP (in the constant-capacity region of the C versus V curve), there must be *some* ions in the double layer. These ions must therefore be in the OHP. (Note that, since it is assumed that one is dealing with fairly concentrated solutions, the diffuse charge is effectively squeezed onto the OHP). The OHP has been pictured (*cf.* Section 7.4.20) to be the *locus* of centers of hydrated ions in contact with a *hydrated* electrode. Will this model explain the constancy of capacity with the radius of the ions, as exemplified in Table 7.11? It will tend to do so because the d of Eq. (7.128) will become r_i plus other terms connected with the radii of water molecules separating the ions from the electrode. The total separation distance would be much less dependent on the radius of the ion because it consists of the water molecule around the ion and that on the electrode, and, thus, the variations in r_i would affect d to a lesser degree. However, a simple calculation shows that, even with the water between the ions and the electrode, the capacity would vary by several microfarads per square centimeter as the ionic radius varied from 0.7 Å for Li^+ to 2.1 Å for Cs^+. In fact, the capacity in this potential-independent range of the C versus V curve for systems containing these ions is constant to $\pm 0.5\,\mu F\,cm^{-2}$. It looks as though one must examine not only d but the dielectric constant as well.

The capacity of a parallel-plate condenser with charges on the metal and OHP is given by

$$C = K_{M \to OHP} = \frac{\varepsilon}{4\pi d}$$

To put numbers into this equation, one must know the dielectric constant ε of the material between the metal and the OHP. Will it be simply that of water in the bulk phase?

7.4.24b. The Dielectric Constant of the Water between the Metal and the Outer Helmholtz Plane.

The dielectric constant of water is not a constant independent of the electric field strength of the environment; this fact has been discussed in the treatment of hydration (Chapter 2).

Hydration, it may be recalled (Section 2.3), results from the interaction of ions with their environment at water dipoles. The field due to the ions orients the water molecules in the primary layer, and the orientation decreases in degree as one goes outward from the ion. The degree of orientation of the water dipoles affects the dielectric constant of this water.

Completely oriented water corresponds to a saturated dielectric. There are no more water dipoles to align; they are all aligned. Thus, such aligned dipoles make no further contribution to the dielectric constant, at least to that part of the dielectric constant which arises from orientation polarization. It has been stated that the dielectric constant of such an oriented water layer is approximately 6.

Now, there is a similarity between the situation of water molecules in contact with charged ions and that of water molecules in contact with a charged metal. On a charged electrode, too, the solvent dipoles would orient and attain saturation orientation if the charge density on the electrode is large enough. This oriented water can be termed the *primary hydration sheath* of the electrode, in analogy with the primary hydration sheath of ions. The dielectric constant of this fully oriented water in the primary hydration sheath of ions is about 6. This would then be the value to be used for the dielectric constant ε_L of the first water layer on a charged electrode.

Apart from a primary hydration sheath, ions have a secondary hydration shell in which the water molecules are somewhat uncertainly oriented toward the ion. This partial misorientation arises because the secondary hydration water has to compromise between the ion's electrostatic-orienting forces and the disorienting thermal and hydrogen-bonding forces.

Thus, near the electrode, too, one would expect a secondary water layer which would be in a state similar to that near an ion. The water molecules in this second layer are partly oriented by the field arising from the charged electrode, but they are also partly disoriented by thermal and hydrogen-bonding influences of the particles of the solution.

What is the dielectric constant ε_H of this second water layer? A quantitative answer to this question would be meaningless because the value obviously increases rapidly in the direction away from the metal and, in a few layers of the solvent, reaches the value characteristic of the bulk electrolyte.

If completely oriented, i.e., fully saturated, the dielectric constant would be about 6. Since the secondary hydration sheath is partly oriented, its dielectric constant would be between 6 and 78. Detailed theoretical calculations of the dielectric constant of this second layer indicate a mean value of about 40, about halfway between the dielectric constant of the first water layer (6) and that of the bulk (78).

7.4.24c. The Position of the Outer Helmholtz Plane and an Interpretation of the Constant Capacity.

A consideration of the dielectric constant of the material between the metal surface and the OHP (the *locus* of centers of the hydrated ions)[†] suggests, therefore, that, even when there is no contact adsorption, there are two regions, one next to the electrode with a low dielectric constant $\varepsilon_L \approx 6$ and another adjacent to the OHP

[†] The reader is once more reminded of the significance of the abbreviation OHP, i.e., the outer Helmholtz plane. Similarly, the *locus* of the contact-adsorbed ions is termed the inner Helmholtz plane, or IHP [see Fig. 7.77(b)].

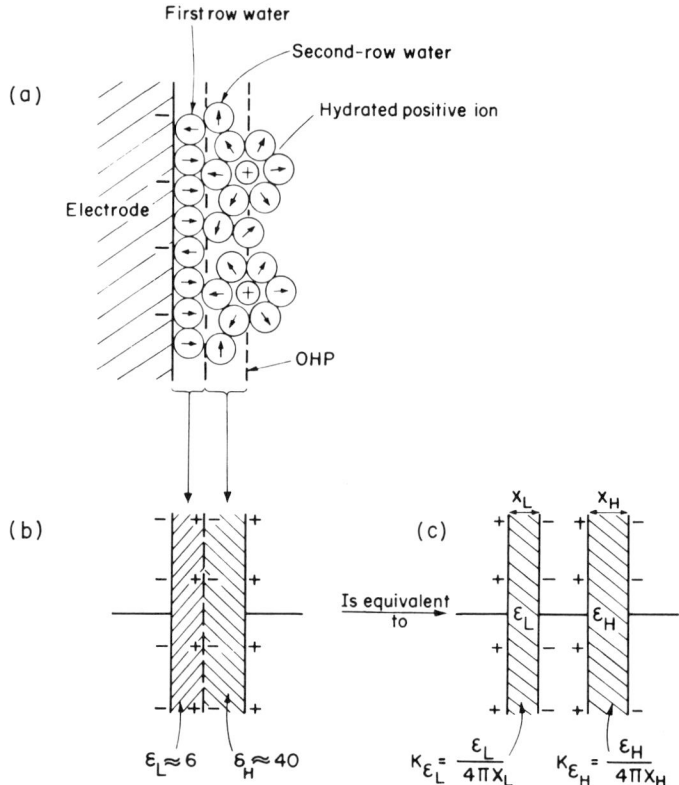

Fig. 7.79. (a) Arrangement O corresponds (b) to a parallel-plate capacitor with two dielectrics in it. Such a capacitor is equivalent to (c) two capacitors in series, each with one of the two dielectrics.

with a dielectric constant which changes as one proceeds out to the solution but can be represented by a mean value of about 40.†

This is a peculiar parallel-plate capacitor with two dielectrics inside it (Fig. 7.79). How does one calculate its capacity? One may imagine that charges arise at the boundary of the two dielectrics, charges equal in magnitude and opposite in sign to the plates facing them. Then it becomes obvious from the previous argument (Section 7.4.21) that, in effect, the total capacity‡ of the double layer is given by two capacitors in series, one

---

† The value 40 does not have any particular significance in the interpretation given here —the value varies with distance. A specific mean value is used mainly to simplify the presentation.
‡ In the part of the C versus V curve where the capacity is independent of potential, the differential capacity $C = dq/dV$ and the integral capacity $K = q/V$ must be the same.

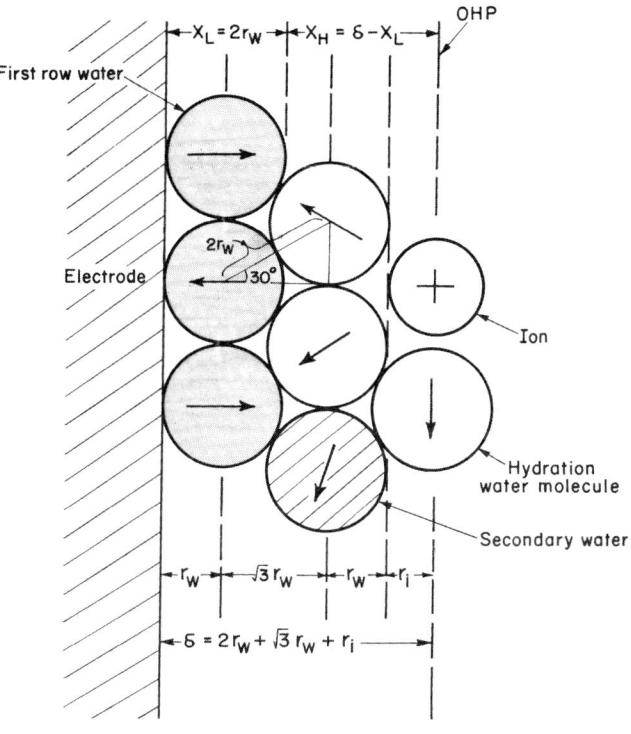

Fig. 7.80. Schematic diagram for the calculation of distances involved in the parallel-plate capacitor of Fig. 7.79.

across the first dielectric and the other across the second (Fig. 7.79). Hence,[†]

$$\frac{1}{C} = \frac{1}{K_{M \to \text{OHP}}} = 4\pi \frac{x_L}{\varepsilon_L} + 4\pi \frac{x_H}{\varepsilon_H}$$

$$= 4\pi \frac{x_L}{\varepsilon_L} + \frac{\delta - x_L}{\varepsilon_H} \quad (7.129)$$

It is obvious, however, from Fig. 7.80, that x_L is simply equal to the diameter $2r_W$ of a water molecule, and that δ, the distance between the metal and the OHP, is $2r_W + \sqrt{3}\, r_W + r_i$, where r_i is the radius of the unhydrated

---

[†] In Eq. (7.129), $K_{M \to \text{OHP}}$ means the integral capacity of the region between the metal and the OHP. One is assuming here that the solution is sufficiently concentrated for the entire diffuse charge to be located on the OHP and, consequently, here, the potential of the OHP is effectively equal to the potential in the bulk of the solution.

ion at the OHP. Hence,

$$\frac{1}{C} = \frac{1}{K_{M \to \text{OHP}}} = \frac{4\pi 2 r_W}{\varepsilon_L} + \frac{4\pi}{\varepsilon_H}\sqrt{3}\, r_W + \frac{4\pi}{\varepsilon_H} r_i \quad (7.130)$$

The capacity can now be calculated. If $\varepsilon_L \approx 6$ and $\varepsilon_H \approx 40$, it turns out that the capacity is 16 to 17 μF cm^{-2} and fairly independent of the radius of the ion (Table 7.11).

Thus, the model has attained consistency with experiment for it predicts that, in the absence of contact adsorption, the capacity of the interface does indeed become independent of the potential and of the radii of the ions in the OHP. Thus, the third term of Eq. (7.130), which contains the radii of the ions, contributes little to the total capacity because the r_i value is divided by the relatively large value of $\varepsilon_H \approx 40$, while the first term, which lacks r_i, is divided by the relatively small ε_L—the second term is small and also independent of r_i. Hence, it is the low-dielectric-constant region which contributes most to the capacity. Thus, the ionic radius scarcely affects the double-layer capacity. Simply stated, therefore, the double-layer capacity in the region of constant capacity is determined by two capacitors in series, one capacitor immediately next to the electrode containing a (saturated) water dielectric and the second one, bounded by the plane through the ion centers, having a dielectric with a much higher value of dielectric constant than that of the first layer and containing an ionic-radius term which, however—see the form of Eq. (7.130)—, contributes little to the numerical value of the capacitance.

Considering the very crude and simple nature of the present model and its neglect of factors such as the effects of the pressure generated by the electric field upon the double-layer constituents, there is satisfactory agreement between the theory and the facts of Table 7.11. Hence, one concludes that the primary hydration sheath around the electrode intervenes between the metal and the hydrated ions, the *locus* of the centers of which (Fig. 7.72) make up the OHP.

When one comes to the question of electron transfer between the hydrated ions in the OHP and the metal (Section 8.3), one wonders whether the OHP is perhaps too far away from the electrode for receipt of electrons at a rate equal to that observed. Remarkably enough, analysis of the tunneling of protons through the double layer to an electrode and a comparison of this with the rate of tunneling of deuterons give rise to a calculated answer for the ratio of the rates, which is consistent with experiment only if the ions exchanging electrons with the metal *are* as far away as the present double-layer model indicates.

The interpretation of the potential-independent capacities which are independent of the radius of the ions present in the double layer is a good example of the increase in understanding which may come by quite simple model considerations. As usual, it is the consideration of the anomalies in the previous model which lead to the next one. Models are made to be improved.

Now attention is drawn to another anomalous fact concerning double-layer capacities. This is the hump which regularly appears in curves of differential capacity C versus potential, or electrode charge. What structural information can be inferred from a consistent modelistic interpretation of the jump?

7.4.25. The Capacitance Hump

The discussion of the hump should really begin with a consideration of the question: Why does the differential capacity of the interface increase when the electrode charge becomes positive with respect to the constant-capacity region? Why does not the capacity maintain the constant value of 16 to 17 μF cm^{-2} as the potential difference across the interface changes?

Recall the general expression for the capacity of the electrode–electrolyte interface

$$\frac{1}{C} = \frac{1}{K_{M \to \text{OHP}}} - \left(\frac{1}{K_{M \to \text{OHP}}} - \frac{1}{K_{M \to \text{IHP}}}\right) \frac{dq_{CA}}{dq_M} \quad (7.126)$$

It says that C is constant when the variable second term is zero. This condition is realized when dq_{CA}/dq_M is zero, i.e., when there is no change in the amount of the contact-adsorbed ion concentration with change of the charge on the electrode, as would be so were there to be zero contact adsorption.

The expression (7.126) also indicates that $1/C$ decreases or C increases as soon as dq_{CA}/dq_M becomes nonzero. In other words, the capacity will increase if q_{CA} increases with q_M, i.e., the slope of the q_{CA} versus q_M curve is finite and increasing.

According to this argument, the capacity should keep on increasing as the potential difference across the double layer becomes more positive. In practice, C does not go on increasing. It increases up to a point and then begins to decrease; it is this rise and fall which is colloquially known as "the hump" (Fig. 7.81). The hump means, therefore, that the *slope* of the q_{CA} versus q_M curve, i.e., dq_{CA}/dq_M, should at first increase. At the peak of the hump, the dq_{CA}/dq_M slope should begin to decrease, i.e., there should

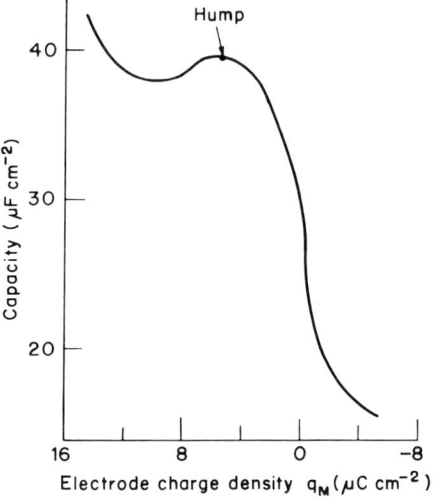

Fig. 7.81. The hump in the experimental capacity–electrode-charge-density curve.

be an inflection in the q_{CA} versus q_M curve. As to the electrode charge, the hump implies that, as the electrode charge becomes positive, the population of contact-adsorbed ions increases more and more easily, and then the *rate* of growth begins to decline.

All this is just a detailing of the implications of the general expression [Eq. (7.126)] for the capacity of an electrified interface. The content of Eq. (7.126) has been put into words. These words are a *clue* to new knowledge but not new knowledge itself. There is one term in Eq. (7.126) which has hitherto remained unilluminated; it is the term dq_{CA}/dq_M. It contains information concerning the growth of the population of ions on the IHP. In Section 7.4.18, some factors which influence whether an ionic species is contact adsorbed or not were discussed. However, the dependence of the magnitude of contact adsorption upon the electrode charge was unexamined in that section. This dependence seems to offer important information relevant to the understanding of the hump and is an important aspect of the phenomenon of contact adsorption.

7.4.26. How Does the Population of Contact-Adsorbed Ions Change with Electrode Charge?

The subject of concern here is the equilibrium value of the population density (number per square centimeter) of ions in the IHP and the sensitivity of this value to the magnitude of electrode charge.

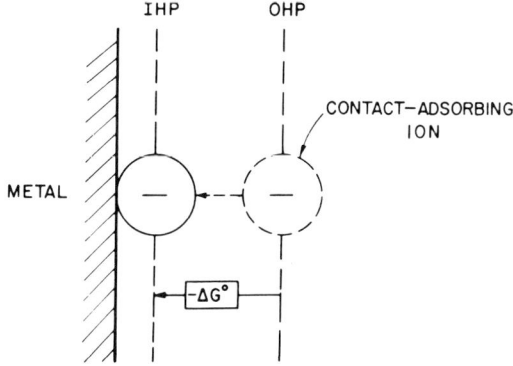

Fig. 7.82. The standard free-energy change ΔG^0 when an ion goes from the OHP to the IHP.

Since the search is for equilibrium values, the simplest approach is to consider that contact adsorption is formally similar to a chemical reaction

$$\square + \ominus \rightleftharpoons \ominus\!\!\!\!| \quad (7.131)$$

where $\ominus$ is a contact-adsorbing ion at the OHP, $\square$ is an unoccupied site on the IHP, and $\ominus\!\!\!\!|$ represents the result of a contact-adsorbing ion jumping into a vacant site, i.e., $\ominus\!\!\!\!|$ is a contact-adsorbed ion.

From the law of mass action, it follows that (writing CA for $\ominus\!\!\!\!|$)

$$\frac{n_{CA}}{n_\square} = n_\ominus e^{-\Delta G^0/RT} \quad (7.132)$$

where n_{CA} is the population density (number per square centimeter) of contact-adsorbed ions in the IHP, $n_\square$ is a number of vacant sites per square centimeter of the IHP, ΔG^0 is the standard free-energy change accompanying the migration of a mole of ions from the OHP to the IHP, and $n_\ominus$ is the concentration of the contact-adsorbing ions expressed in dimensionless units for the following reason: The terms $\Delta G^0/RT$ and $e^{-\Delta G^0/RT}$ are dimensionless quantities; so is n_{CA}/n; hence, to ensure the dimensional equality of the two sides of the equation, $n_\ominus$ must not be expressed in the usual concentration units of moles per liter. It is necessary to multiply the concentration, or activity a_i, of ions i in solution by $N_A 2r_i/1000$ to get the number of contact-adsorbing ions per square centimeter and then divide by n_T, the total number of sites (free plus occupied) per square centimeter on the IHP or parallel to it.

Thus, Eq. (7.132) becomes

$$n_{CA} = n_\square \frac{N_A 2 r_i a_i}{1000 n_T} e^{-\Delta G^0/RT}$$

$$= (1 - \theta)\frac{N_A 2 r_i}{1000} a_i e^{-\Delta G^0/RT} \qquad (7.133)$$

where θ is the fraction of the electrode surface occupied by contact-adsorbed ions.

If ΔG^0 is evaluated, the laws of growth of the population of contact-adsorbed ions have been formulated. What is the physical meaning of ΔG^0? It is the change of free energy arising from the migration of a mole of ions from the bulk of the solution to the IHP.

The problem, therefore, is to compute this work of transit, i.e., of contact adsorption. This total work when an ion is transported from the OHP to IHP can be divided into three contributions: (1) the chemical work arising from forces between the metal and the ion[†] (cf. Section 7.4.17), (2) work arising from the interaction of the ions with the electric field due to the charged electrode, and (3) work arising from the lateral interaction of the ion with its surrounding contact-adsorbed ions.

The nature of the chemical work has been discussed in Section 7.4.17. It concerns ion–electrode dispersion interaction and the changes in free energy resulting from the alteration of the hydration structure. The chemical work per mole of ions may be left with the symbol ΔG_c^0.

The electrical interaction with the field is a matter[‡] of the work of taking a charged ion through a distance $x_2 - x_1$ from the OHP to the IHP of the parallel-plate condenser consisting of the two plates, namely, the metal surface and the OHP. The field in such a condenser is $4\pi q_M/\varepsilon$, the potential difference of interest is $4\pi q_M(x_2 - x_1)/\varepsilon$, and the work of transporting one charge e_0 is $4\pi q_M e_0(x_2 - x_1)/\varepsilon$.

One must now compute the work of lateral interaction among the contact-adsorbed ions.

The first step is to visualize the spatial distribution of the ions populat-

---

[†] It must be remembered that water molecules adsorbed at the electrode will be desorbed during the process of contact adsorption. This effect is implicitly included in the chemical work $-\Delta G_c^0$.

[‡] However, it is not a simple matter, because the dielectric constant changes (Section 7.4.24c) at the boundary between the oriented water layer and the more bulk-like water. In the latter, the electrostatic work of movement is relatively small, because the dielectric constant is high. In effect, it turns out that $x_2 - x_1$ is $2r_{H_2O} - r_i$.

THE ELECTRIFIED INTERFACE   765

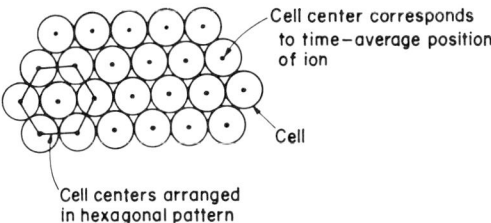

Fig. 7.83. The contact-adsorbed ions in the IHP are considered spatially distributed into cells with the centers of the cells corresponding to the time-average positions of the ions.

ing the IHP. The simplest approach is to divide a unit area in the IHP into as many cells as there are contact-adsorbed ions per unit area. Then one may consider that the time-average positions of the ions are the centers of the cells. The cell centers are assumed to be arranged according to a hexagonal pattern (Fig. 7.83). Thus, the contact-adsorbed ions are assumed to adopt time-average positions corresponding to a hexagonal array.

Now consider any reference ion (Fig. 7.84). The surrounding ions of the hexagonal array are effectively in circular rings about the central ion, the ring radii increasing thus, $1r$, $2r$, $3r$, ..., nr. Further, let the charges in a ring be smoothed out so that one can talk of a charge density, and let this charge density σ be expressed in charge per unit angle (i.e., charge per unit radian). The number of charges in the first ring can be seen (Fig. 7.84)

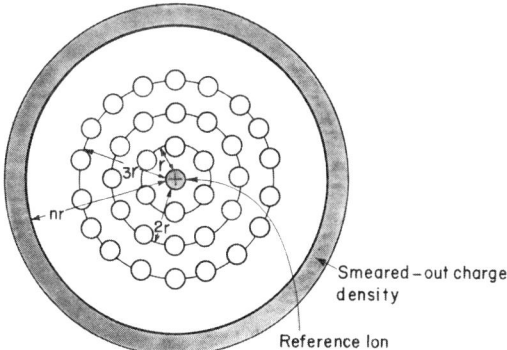

Fig. 7.84. The contact-adsorbed ions are effectively in circular rings, so that there are $6n$ ions in the nth ring of radius nr. If the charge is smoothed out, there is a charge of $6ne_0$ in the 2π radians of the nth ring, i.e., a charge density of $\sigma = 6ne_0/2\pi$.

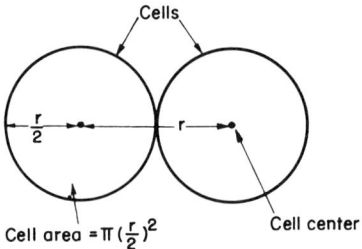

Fig. 7.85. The mean distance and area of cells.

to be 6; in the second ring, 12, ...; and, therefore, in the nth ring, $6n$. Hence, the charge per unit radian in the nth ring is

$$\sigma = \frac{6ne_0}{2\pi} \qquad (7.134)$$

The radius of the first ring is a known quantity (Fig. 7.85). If r is the mean distance between cell centers, i.e., between contact-adsorbed ions, one cell has an area $\pi(r/2)^2$ and, therefore, there are $4/\pi r^2$ cells per unit area or $4/\pi r^2$ ions per unit area since there is one ion per cell. But n_{CA} is the number of contact-adsorbed ions per unit area. Hence,[†]

$$r = \left(\frac{4}{\pi n_{CA}}\right)^{\frac{1}{2}} \qquad (7.135)$$

The potential due to the nth charged ring (on the IHP) at the site of the reference ion is equal to (charge on nth ring/$\varepsilon \times$ distance of nth ring to ion) $2\pi\sigma/\varepsilon nr$, and the interaction energy (repulsive because all contact-adsorbed ions have charges of the same sign) between the reference ion (charge e_0) and the nth charged ring is $2\pi\sigma e_0/\varepsilon nr$.

But this calculation is for conditions near a metal, i.e., for charges on the metal surface, and charges near a metal induce charges on the metal surface. This induced charge can be represented by image charges (cf. Section 7.2.5h). So, every charged ring on the IHP has associated with it a ring of image charge located (Fig. 7.86) from the reference ion at a distance $[(nr)^2 + (2r_i)^2]^{\frac{1}{2}}$. Hence, the attractive interaction between the reference ion and the nth ring of image charge is

$$-\frac{2\pi\sigma e_0}{\varepsilon} \frac{1}{[(nr)^2 + (2r_i)^2]^{\frac{1}{2}}}$$

[†] The assumption of hexagonal close packing may be shown from closer reasoning to give $r = (2\sqrt{3}/3n_{CA})^{1/2}$. The versions differ by 7%.

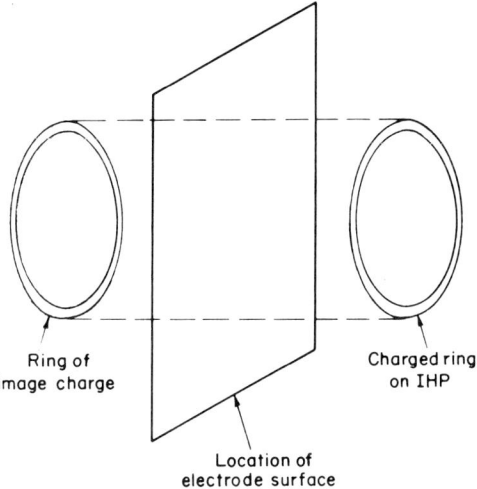

Fig. 7.86. Every charged ring on the IHP produces a ring of image charge as far behind the metal surface as the IHP is in front of it.

Thus, the interaction energy between the reference ion and the nth ring *plus* the interaction energy† of the ion with the image charge due to the nth ring is

$$\frac{2\pi\sigma e_0}{\varepsilon n r}\left[1 - \frac{1}{[1 + (2r_i/nr)^2]^{\frac{1}{2}}}\right]$$

The total lateral-interaction work W_{LI} is obtained by summing over all the rings as follows

$$W_{LI} = \sum_{n=1}^{\infty} \frac{2\pi\sigma e_0}{\varepsilon n r}\left\{1 - \frac{1}{[1 + (2r_i/nr)^2]^{\frac{1}{2}}}\right\} \quad (7.136)$$

At this stage, the binomial expansion

$$(1 + x)^{-\frac{1}{2}} = 1 - \tfrac{1}{2}x + \tfrac{3}{8}x^2 \cdots \quad (7.137)$$

---

† The image energy of an ion with the electrode can be calculated in a more complex way. Not only does the ion give an electrostatic image in the metal but also in the solution. The image *there* interacts further with the metal. Such more complex imaging processes seem to decrease, rather than increase, the agreement of predicted with observed behavior. It is possible that this unexpected discrepancy arises because the dielectric-constant change between the oriented water layer and the solution further out is not sharp but blurred.

can be used, and only the first three terms taken.† One has

$$W_{LI} = \sum_{n=1}^{\infty} \frac{4\pi\sigma e_0 r_i^2}{\varepsilon n^3 r^3} \left(1 - \frac{3r_i^2}{n^2 r^2}\right) \qquad (7.138)$$

The expressions for σ [Eq. (7.134)] and for r [Eq. (7.135)] are now inserted in Eq. (7.138) to give

$$W_{LI} = \sum_{n=1}^{\infty} \frac{3r_i^2 e_0^2 \pi^{\frac{3}{2}} n^{\frac{3}{2}}_{CA}}{2\varepsilon n^2} \left(1 - \frac{3\pi r_i^2 n_{CA}}{4} \frac{1}{n^2}\right)$$

$$= \frac{3r_i^2 e_0^2 \pi^{\frac{3}{2}} n^{\frac{3}{2}}_{CA}}{2\varepsilon} \sum_{n=1}^{\infty} \frac{1}{n^2} - \frac{9r_i^4 \pi^{\frac{5}{2}} e_0^2 n^{\frac{5}{2}}_{CA}}{8\varepsilon} \sum_{n=1}^{\infty} \frac{1}{n^4} \qquad (7.139)$$

The series $\sum_{n=1}^{\infty} 1/n^2$ and $\sum_{n=1}^{\infty} 1/n^4$ have been evaluated to be $\pi^2/6$ and $\pi^4/90$, respectively. Hence,

$$W_{LI} = \frac{(\pi^2) e_0^2 r_i^2 \pi^{\frac{3}{2}} n^{\frac{3}{2}}_{CA}}{4\varepsilon} - \frac{(\pi^2) 3 r_i^4 \pi^{\frac{5}{2}} n^{\frac{5}{2}}_{CA} e_0^2}{16\varepsilon} \frac{\pi^2}{15} \qquad (7.140)$$

The chemical, electrical, and interactional work terms‡ when summed together give $-\Delta G^0$, the free-energy change associated with the equilibrium reaction of ions contact-adsorbing on a charged electrode. When this sum of work terms is substituted in Eq. (7.133), one obtains

$$n_{CA} = (1-\theta)\frac{N_A 2r_i}{1000} a_i \exp\left[-\frac{\Delta G_c^0}{RT} + \frac{4\pi q_M e_0 (x_1 - x_2)}{\varepsilon kT}\right.$$
$$\left. - \frac{\pi^2 e_0^2 r_i^2 \pi^{\frac{3}{2}} n^{\frac{3}{2}}_{CA}}{4\varepsilon kT}\left(1 - \frac{3}{4}\frac{\pi^2}{15}\pi r_i^2 n_{CA}\right)\right] \qquad (7.141)$$

This,* then, is a relation which connects the population density of contact-adsorbed ions on the IHP, i.e., the number per unit area of contact-

---

† The number of terms taken in a binomial expression is a little arbitrary. Here, three terms (instead of two) are chosen because r_i/r is not very small, and, therefore, the series converges rather slowly.

‡ All forces are ultimately electrical in origin. The context, however, makes the classification into chemical, electrical, and interactional work terms clear: The *chemical* ones refer to the ΔG_c^0 term (dispersion interactions, etc.); the *electrical*, to the charge-dependent term; and the *interactional* terms, to the lateral repulsion of contact-adsorbed ions.

* Since the electrical and lateral interaction terms (the second and third terms in the exponent) have been computed per ion, not per mole of ions; they are divided by kT and not by RT.

adsorbed ions to the charge on the metal. It is seen that n_{CA} depends on (1) the bulk activity a_i of the adsorbing species, (2) the radius r_i of the contact-adsorbing ion, (3) the chemical term ΔG_c^0, and (4) the electrode charge q_M.

Thus, for a particular contact-adsorbing ionic species at a fixed bulk concentration, the population of contact-adsorbing ions changes as the charge on the electrode changes. This is good. If the population of contact-adsorbed ions had remained a constant, dq_{CA}/dq_M would have been zero and there would have been no hope of explaining any increases of capacity as one approached the positive region (*cf.* Fig. 7.81) by an equation such as (7.126).

7.4.27. The Test of the Population Law for Contact-Adsorbed Ions

One can extract from the population law (7.141) the characteristics of the hump (Section 7.4.23) on the differential capacity–potential curve. It should be emphasized that the hump is a peculiar and rather unexpected part of double-layer phenomenology, and its correct prediction is a good challenge for a model of the electrified interface. (Until a few years ago, it was a test which most double-layer models failed.)

One can, however, make a simpler check on the population law than that of a detailed calculation of the hump. A calculated n_{CA} versus q_M curve can be compared with a curve obtained from experimental electrocapillary curves (*cf.* Section 7.4.19). The charge q_{CA} is of course simply $n_{CA}e_0$.

The result of this check of the model for adsorbed ions in the IHP turns out to be encouraging (Fig. 7.87). Not only is there an initial increase of dq_{CA}/dq_M from its zero value at $q_M = -12$ to $-13~\mu\text{C cm}^{-2}$ (the electrode charge density at which, irrespective of the nature of the cation in the electrolyte, there is a constant capacity of 16 to 17 $\mu\text{F cm}^{-2}$), but, more significantly, there is an *inflection* in the theoretical q_{CA} versus q_M curve.

Why is this inflection of importance? It is because it has been argued (see Section 7.4.23) that the hump in the C versus q_M curve is located at a value of the electrode charge corresponding to the inflection, $d^2q_{CA}/dq_M^2 = 0$, in the q_{CA} versus q_M curve. A critical test, therefore, is to see whether the theory is able to predict at what values of q_M the capacity hump will occur. This is easily done. All one has to do is to obtain an expression for d^2q_{CA}/dq_M^2 and set it equal to zero, i.e., the procedure is to write down the mathematical condition of an inflection point.

It is convenient, however, to rewrite the expression for n_{CA} [Eq. (7.141)]

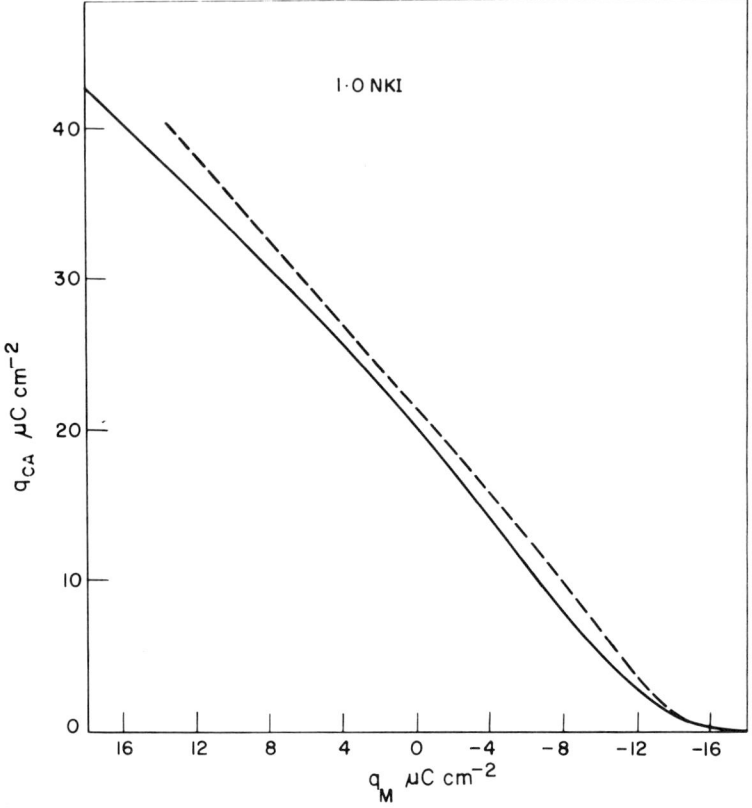

Fig. 7.87. Comparison of calculated (dotted line) and experimental (solid line) q_{CA} versus q_M curves.

in a more convenient form. One makes use of the fact that the fraction θ of the IHP occupied by contact-adsorbing ions is given by

$$\theta = \frac{n_{CA}}{n_T}$$

from which it follows that[†]

$$n_{CA} = \theta n_T = \frac{\theta n_T e_0}{e_0} = \frac{\theta q_{\max}}{e_0}$$

$$= \frac{\theta q_{\max} N_A}{F} \quad (7.142)$$

[†] The term $(q_{CA})_{\max}$, or, for simplicity, $q_{\max}$, is the *maximum* charge density arising from contact-adsorbed ions.

and

$$(n_{CA})^{\frac{3}{2}} = \left(\frac{q_{max}N_A}{F}\right)^{\frac{3}{2}} \theta^{\frac{3}{2}} \tag{7.143}$$

Relations (7.142) and (7.143) are inserted in Eq. (7.141) to give

$$n_{CA} = \frac{\theta q_{max}N_A}{F} = (1-\theta)\frac{N_A 2r_i}{1000} a_i \exp\left[-\frac{\Delta G_c^0}{RT} + \frac{4\pi q_M e_0(x_2 - x_1)}{\varepsilon kT}\right.$$
$$\left. - \frac{\pi^2 e_0^2 r_i^2 \pi^{\frac{3}{2}}}{4\varepsilon kT} n_{CA}^{\frac{3}{2}}\left(1 - \frac{3}{4}\frac{\pi^2}{15}\pi r_i^2 n_{CA}\right)\right]$$

i.e.,

$$\frac{\theta}{1-\theta} = \frac{2r_i F}{1000 q_{max}} a_\pm \exp\left[-\frac{\Delta G_c^0}{RT} + \frac{4\pi q_M e_0(x_2 - x_1)}{\varepsilon kT}\right.$$
$$\left. - \frac{\pi^{\frac{7}{2}} e_0^2 r_i^2}{4\varepsilon kT}\left(\frac{q_{max}N_A}{F}\right)^{\frac{3}{2}}\theta^{\frac{3}{2}}\left(1 - \frac{\pi^3}{20} r_i^2 \frac{q_{max}N_A}{F}\theta\right)\right] \tag{7.144}$$

Equation (7.144) can be written

$$\ln\frac{\theta}{1-\theta} = \ln\left(\frac{2r_i F}{1000 q_{max}}\right) + \ln a_\pm - \frac{\Delta G_c^0}{RT} + \frac{4\pi e_0(x_2 - x_1)}{\varepsilon kT} q_M$$
$$- \left[\frac{\pi^{\frac{7}{2}} e_0^2 r_i^2}{4\varepsilon kT}\left(\frac{q_{max}N_A}{F}\right)^{\frac{3}{2}}\right]\theta^{\frac{3}{2}} \tag{7.145}$$

where the approximation

$$1 - \frac{\pi^3 r_i^2}{20}\frac{q_{max}N_A\theta}{F} \approx 1 \tag{7.146}$$

has been made. Introducing the following notation,

$$A = \frac{4\pi e_0(x_2 - x_1)}{\varepsilon kT} \tag{7.147}$$

$$B = \frac{\pi^{\frac{7}{2}} e_0^2 r_i^2}{4\varepsilon kT}\left(\frac{q_{max}N_A}{F}\right)^{\frac{3}{2}} \tag{7.148}$$

and

$$\text{Constant} = \ln\left(\frac{2r_i F}{1000 q_{max}}\right) - \frac{\Delta G_c^0}{RT} \tag{7.149}$$

Eq. (7.145) can be rewritten in the form

$$\ln\frac{\theta}{1-\theta} = \text{Constant} + \ln a_\pm + Aq_M - B\theta^{\frac{3}{2}}$$

or
$$Aq_M = -\text{Constant} - \ln a_\pm + \ln \theta - \ln(1-\theta) + B\theta^{3/2} \quad (7.150)$$

One can now proceed rapidly to evaluate $d^2\theta/dq_M{}^2$ and locate the inflection point in the θ versus q_M curve and compare it with the position of the capacity hump, as experimentally observed.

By differentiating Eq. (7.150), keeping the bulk concentration and therefore the activity $a_\pm$ constant, one obtains

$$A\frac{dq_M}{d\theta} = \frac{1}{\theta} + \frac{1}{1-\theta} + \frac{3}{2}B\theta^{1/2}$$

or

$$\frac{d\theta}{dq_M} = \frac{A}{[1/\theta(1-\theta)] + \tfrac{3}{2}B\theta^{1/2}} \quad (7.151)$$

$$= \frac{A}{p} \quad (7.152)$$

where the symbol p is used for the denominator of the right-hand side of Eq. (7.151). To proceed further, the identity [Eq. (7.90)] used in the Gouy–Chapman theory can be resorted to

$$\frac{d^2\theta}{dq_M{}^2} = \frac{1}{2}\frac{d}{d\theta}\left(\frac{d\theta}{dq_M}\right)^2$$

$$= \frac{1}{2}\frac{d}{d\theta}\left(\frac{A}{p}\right)^2$$

$$= -\frac{A^2}{p^3}\frac{dp}{d\theta}$$

$$= -\frac{A^2}{p^3}\left[-\frac{1-2\theta}{\theta^2(1-\theta)^2} + \frac{3}{4}B\theta^{-1/2}\right]$$

Hence,

$$\frac{d^2\theta}{dq_M{}^2} = -\frac{A^2}{2p^3}\left[\frac{-2 + 4\theta + \tfrac{3}{2}B\theta^{3/2} - 3B\theta^{5/2}}{\theta^2(1-\theta)^2}\right] \quad (7.153)$$

where the $\theta^{1/2}$ term has been neglected.

The mathematical condition for an inflection point can now be used, i.e., $d^2\theta/dq_M{}^2$ is set equal to zero. Thus, from Eq. (7.153),

$$-\frac{A^2}{2p^3}\frac{1}{\theta^2(1-\theta)^2}(-2 + 4\theta + \tfrac{3}{2}B\theta^{3/2} - 3B\theta^{5/2}) = 0 \quad (7.154)$$

If either p or $\theta^2(1-\theta)^2$ is equal to zero, the left-hand side of Eq. (7.154) would tend to infinity rather than to zero. Hence, Eq. (7.154), or $d^2\theta/dq_M{}^2$, would be equal to zero only if

$$2 - 4\theta - \tfrac{3}{2}B\theta^{\frac{3}{2}} + 3B\theta^{\frac{1}{2}} = 0 \tag{7.155}$$

By solving this equation for θ, one obtains the value of θ_{infl}, i.e., the value of θ at which an inflection is expected in the θ versus q_M curve. This can be compared with θ_{hump}, i.e., the value of θ at which the hump in the experimental C versus q_M curve is observed. Figure 7.88 shows the good agreement between theory and experiment (Wroblowa and Kovac).

A further check of the theory of the population growth of contact-adsorbed ions, as contained in Eq. (7.141), (7.145), or (7.150), can be

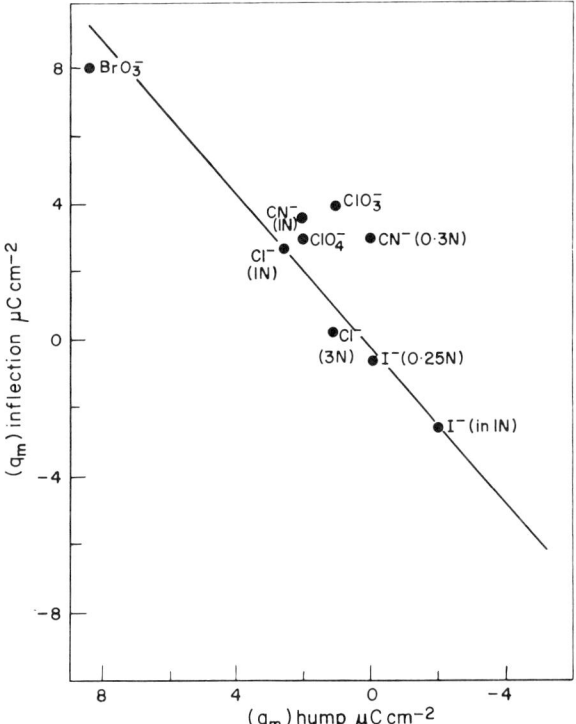

Fig. 7.88. The location of the hump on the C versus q_M curve; comparison between theory and experiment. According to theory, the hump should occur at the value of q_M corresponding to the inflection in the q_{CA} versus q_M curve. Hence, when the ordinate of a point is equal to the abscissa, there is perfect agreement between theory and experiment.

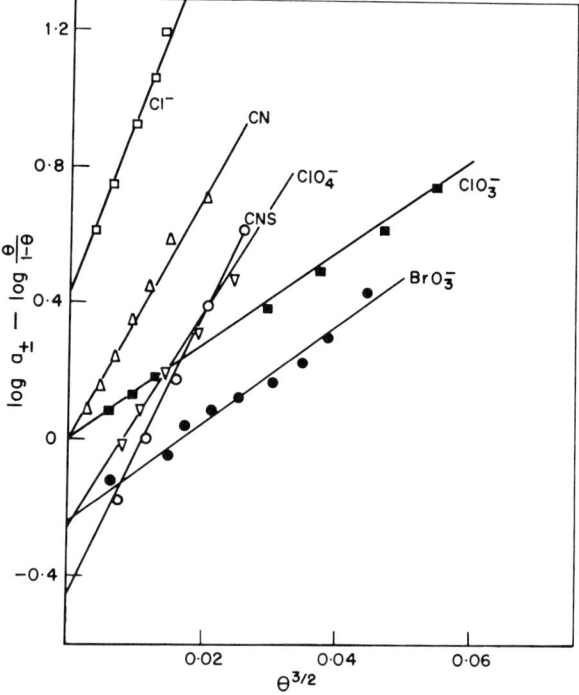

Fig. 7.89. The plot of $\log a_\pm - \log [\theta/(1-\theta)]$ versus $\theta^{3/2}$.

devised. At constant q_M, Eq. (7.150) can be rearranged in the form

$$\ln a_\pm - \ln \frac{\theta}{(1-\theta)} = \text{Constant} + B\theta^{3/2} \quad (7.156)$$

which shows that theory demands that, if $\ln a_\pm - \ln [\theta/(1-\theta)]$ is plotted against $\theta^{3/2}$ at constant q_M, a straight line should be obtained with a slope [cf. Eq. (7.148)]

$$B = \frac{\pi^{1/2} e_0^2 r_i^2}{4\varepsilon kT} \left(\frac{q_{\max} N_A}{F}\right)^{3/2} \quad (7.148)$$

The theory is in quite good agreement with experiment. The experimental plots *are* linear (Fig. 7.89), and there is reasonable agreement between the calculated and measured slopes (Table 7.12).

Yet a further test of the model can be arranged by calculating the value of dq_{CA}/dq_M (i.e., the slope of the q_{CA} versus q_M curve in Fig. 7.76) and feeding it into the expression (7.126) for the differential capacity of the interface. In this way, one emerges with a predicted curve for the capacity-

THE ELECTRIFIED INTERFACE 775

TABLE 7.12
Experimental and Calculated Slopes of the log $[a_\pm(1-\theta)/\theta]$ versus $\theta^{3/2}$ Straight Line

Ion	Ionic radius r_i, Å	$\dfrac{\log[a_\pm(1-\theta)/\theta]}{\theta^{3/2}}$	
		Calculated	$q_n = 0$ Experimental
I^-	2.19	36	15
Cl^-	1.81	43	52
CN^-	1.95	40	36
CNS^-	1.6†	49	34
BrO_3^-	1.6†	16	15
ClO_3^-	1.5†	13	14
ClO_4^-	2.5†	32	36

† Taken from scale models.

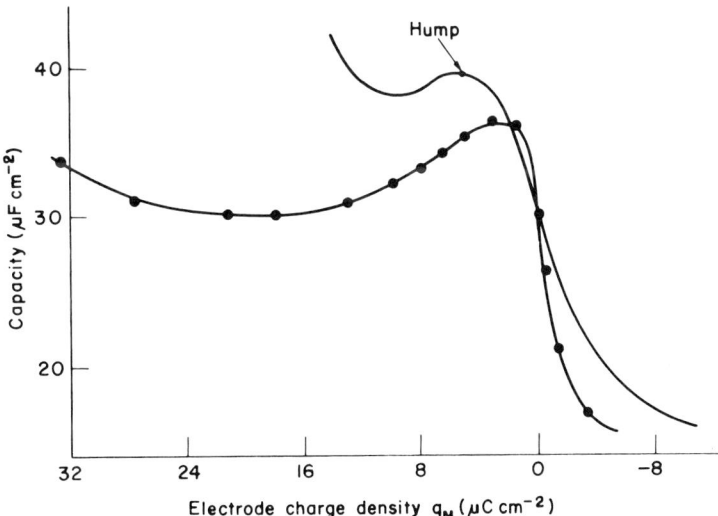

Fig. 7.90. Comparison between the predicted and observed curves of capacity *versus* electrode charge density.

electrode-charge relationship. This can be compared with the experimental curve (see Fig. 7.90). The model is able to reproduce the general shape of the capacity–potential relation.†

7.4.28. The Lateral-Repulsion Model for Contact Adsorption

The quest here has been for an understanding of the structure of electrified interfaces. Once the Stern model was evolved out of the Helmholtz–Perrin and Gouy–Chapman models (cf. Sections 7.4.1, 7.4.3, and 7.4.7), one had to face two basic problems, the explanation of the constant-capacity region and the hump in the experimental capacity–potential curve.

The importance of the C versus V or C versus q curves stems from the fact that the capacity of an electrified interface depends on the arrangement of charges in the interphase region. Thus, the variation of capacity with the potential difference across the interface is the macroscopic expression of the variation of the molecular structure of the interface, i.e., of the manner in which the IHP and OHP are populated with ions.

The interpretation of the constant-capacity region of the C versus V curve turned out to be a simple matter. If the IHP is denuded of contact-adsorbing ions, the interface becomes a literal double layer. Simple parallel-plate condenser arguments permit the description of the structure of the interface in terms of two regions. One region has a low dielectric constant corresponding to oriented water, and the second region has a high dielectric constant corresponding to partially oriented water.

The explanation of the hump was a little more sophisticated. As the electrode charge becomes positive with respect to the value (-12 to -13 μC cm^{-2}) when there is a simple double layer, ions start to contact-adsorb and populate the IHP. The dependence of differential capacity upon contact adsorption is contained in Eq. (7.126), which can be written in the form

$$\frac{1}{C} = \alpha - \beta \left(\frac{dq_{CA}}{dq_M} \right) \qquad (7.157)$$

where α and β are related to the constant integral capacities. Apparently, the capacity depends on the slope of the q_{CA} versus q_M curve. The variation of capacity with electrode charge must, therefore, be given by

$$\frac{d(1/C)}{dq_M} = -\beta \left(\frac{d^2 q_{CA}}{dq_M^2} \right) \qquad (7.158)$$

---

† The model has also been shown (Wroblowa and Müller) to be quantitatively consistent with the Essin and Markov effect, i.e., the variation of the potential of zero charge with the concentration of adsorbing ion.

If the capacity shows a hump or the reciprocal capacity $1/C$ shows an inverted hump, it means that, at the hump,

$$\frac{d(1/C)}{dq_M} = 0 \qquad (7.159)$$

or

$$\frac{d^2 q_{CA}}{dq_M^2} = 0 \qquad (7.160)$$

i.e., the inflection in the q_{CA} versus q_M curve locates the hump.

The problem reduced, therefore, to understanding the rate of growth of the population of contact-adsorbed ions and its change with the excess electric charge on the metal. A simple model was conceived. The migration of an ion from the OHP to the IHP (the contact-adsorption process) involves chemical interaction (Section 7.4.17), interaction with the electrical field arising from the electrode charge q_M, *and* lateral interaction with an already settled population of contact-adsorbed ions. The final expression [Eq. (7.141)] is of the form

$$q_{CA} = \text{Constant } e^{A q_M} e^{-B q_{CA}^{3/2}} \qquad (7.161)$$

It is seen that the electrode charge encourages the growth of the population of contact-adsorbed ions, q_{CA}, but this growth sets up and accentuates the lateral repulsion forces which try to inhibit further growth (Fig. 7.91).

It is this example of negative feedback (electrical attracting forces giving rise to lateral repulsion forces) which generates the hump. At charges in the region of constant capacitance (16 to $17 \mu\text{F cm}^{-2}$), $q_{CA} \approx 0$ and $e^{-B q_{CA}^{3/2}} \approx 1$, and at charges more positive than that of constant capacitance, q_{CA} grows at first exponentially with increasing positivity of the electrode charge. Since dq_{CA}/dq_M also increases exponentially with q_M (around $q_{CA} \approx 0$), the capacity rises. But, with the increasing departure of q_{CA} from zero, the lateral repulsion term $e^{-B q_{CA}^{3/2}}$ increases in significance. It reduces the slope dq_{CA}/dq_M, i.e., it slows down the rate of growth of the population of contact-adsorbed ions. At the inflection point $d^2 q_{CA}/dq_M^2 = 0$, the capacity goes through the hump.

The simple lateral-repulsion model not only explained the capacitance hump in qualitative terms. The shape of the q_{CA} versus q_M curve with its inflection, the location of the hump in respect to charge on the electrode, the magnitude of the capacity of the hump, the linearity of the $\{\ln a_\pm - \ln [\theta/(1-\theta)]\}$ versus $\theta^{3/2}$ plot and the magnitude of the slope of this relation, etc.—all these can be reasonably well rationalized by the lateral-repulsion model of contact-adsorbing ions in the IHP.

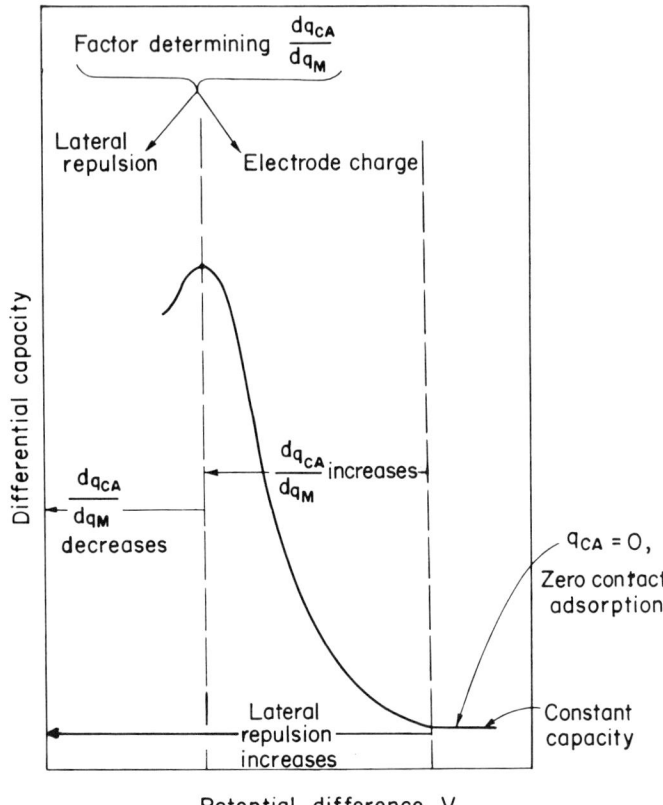

Fig. 7.91. The lateral-repulsion model for the explanation of the capacity–potential curve.

Does this mean that only more sophisticated development of the model is necessary?[†] The future of double-layer theory has plenty of problems. For example, what about the quantitative calculation of the chemical term ΔG_c^0? What about knowledge of the various distances in the electrified interface, the distances of the IHP and OHP from the metal? A basic picture has been given of these distances in explaining the constant capacity, but, e.g., the dependence of such distances on charge is not known, whereas the very high fields in the double layer make one suspect the existence of

---

† For example, electrical images of the contact-adsorbed ions occur not only in the metal but also in the solution, and reflections of such images occur many times. (But see footnote, p. 767.)

electrostriction. What are the quantitative differences between the contact adsorbability of different ions? No model contains a complete account of the phenomenology it seeks to rationalize. The lateral-repulsion model of contact adsorption increases considerably the consistency of the picture being gradually built up about the double layer with experiment.

7.4.29. Flip-Flop Water on Electrodes

The highest observed values of the coverage of a mercury electrode with contact-adsorbed ions are less than 20%. Thus, even when contact adsorption is at a maximum, water molecules cover more than three-quarters of the electrode surface and constitute the overwhelming majority of particles at the interface. It is surprising, therefore, that these adsorbed water dipoles have been ignored in all but very recent discussions of the structure of charged interfaces. The picture of the double layer has been *ioncentric* for too long.

In the present treatment, too, the contribution of water to the picture of the electrified interface has hitherto been remembered only in one context, the dielectric constant of the water. Because the adsorbed water dipoles are largely oriented, there is dielectric saturation and the dielectric constant is not 78 (as it is in the bulk), but only about 6 (*cf.* Section 7.4.24*b*). This lowering of the dielectric constant was shown to explain the potential-independent part of a typical C versus V curve.

Is this classical dielectric aspect all there is to the role of water molecules in the interface? It is difficult to accept that water dipoles, which have been the neglected masses populating the double layer even when contact-adsorbed ions intrude upon the IHP, have no effect other than lowering the dielectric constant. One must look more carefully into the matter.

Forget, for a moment, the ions, and let the viewpoint become *water-centric*. Consider a number of water dipoles adsorbed on the electrode. Then there are two limiting conditions on the relation between the charge on the electrode and the orientation of the dipoles relative to the surface of the metal.

One limiting condition arises on an electrode which has a high positive charge. The electric-field vector is pointed from the metal into the solution. In a field, dipoles reduce their potential energy by aligning themselves so that the dipole vector (which runs from the negative to the positive end of the dipole) becomes parallel to the field. In other words, the water dipoles *flip up* so that the oxygen atoms are in contact with the electrode and the hydrogen end of the water points into the solution. Let this orientation of a water molecule be called the *flip-up state* (Fig. 7.92).

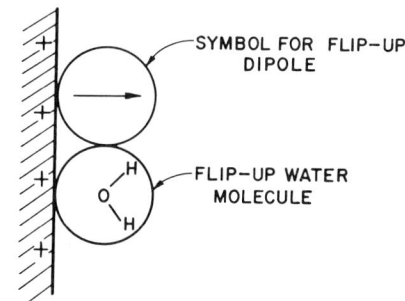

Fig. 7.92. The flip-up orientation of a water molecule on an electrode.

The other limiting condition obtains when electrons are pumped into the electrode to make it very negatively charged. What will the dipoles do? On the basis of a simple electrostatic argument, all that will happen is that the flipped-up dipoles will turn around and *flop down*. In the *flop-down state*, the hydrogens are facing the electrode and the oxygen atom is toward the solution (Fig. 7.93).

So there are, in a first and simplest model, just two states for water molecules on the electrode surface, a flip-up and a flop-down state. This flip-flop model for water turns out to be of consequence in the electrified interface and thus to electrochemistry. The model is so simple that it is surprising that the working out of its consequences was not carried out for more than the first half century of double-layer research.

The equations for the flip-flop model are simple. They just tell one about the potential difference across a dipole layer and how it affects the kinetics of charge-transfer processes at electrodes. One implication of these equations will be pursued at first, that concerning the capacity of the inter-

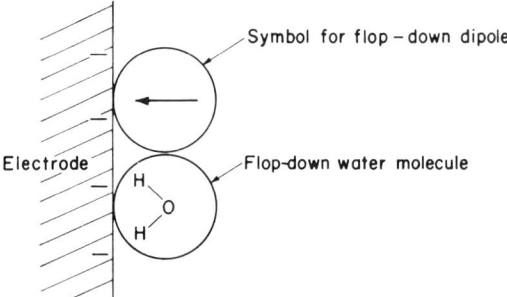

Fig. 7.93. The flop-down orientation of a water molecule on an electrode.

face. The water molecules have been on the electrode all along, and yet their presence and their flip-flop character have been hitherto neglected (apart from the dielectric-constant effect). Can one ignore the contribution of the flip-flopping water dipoles to the capacity of the interface?

7.4.30. Calculation of the Potential Difference Due to Water Dipoles

One approach to the problem is through the concept of dipoles or χ potentials. The concept will be briefly recalled. The electrical potential difference $\Delta\phi$ between the electrode and the solution depends on the charges and the dipole layer. This total $\Delta\phi$ can be conceptually synthesized from two separate thought experiments. In one experiment, the assembly of charges constituting the interfaces are isolated from the phases and the potential difference across this assembly is evaluated as the charge, or $\Delta\psi$, portion. In the second thought experiment, the dipole layer, identical to that existing at the interface, is considered in isolation from the material phases (namely, electrode and electrolyte). The dipole, or $\Delta\chi$, potential is then the potential difference across this isolated dipole layer.

What, therefore, is the expression for the $\Delta\chi$ potential due to a layer of oriented water dipoles? A dipole layer is electrically analogous to a parallel-plate condenser (Fig. 7.94), the thickness of the condenser being the thickness of the dipole and the charge density on the condenser plates being the charge e at each end of the dipole times the number N of *net oriented dipoles* per unit area. If, in a parallel-plate condenser, the plates bear a charge q and are at a distance d apart, it has been seen that the potential difference is $4\pi qd/\varepsilon$. Here, in the dipole condenser, the charge q is equal to Ne. Hence,

$$\Delta\chi = \frac{4\pi Ned}{\varepsilon} \qquad (7.162)$$

But ed is the dipole moment μ, in terms of which the $\Delta\chi$ potential becomes

$$\Delta\chi = \frac{4\pi N\mu}{\varepsilon} \qquad (7.163)$$

If any of the quantities in Eq. (7.163) depend on the electrode charge, then $d\Delta\chi/dq_M$ will not be zero. But $d\Delta\chi/dq_M$ is a reciprocal capacity; hence, if, e.g., N, the net number of oriented dipoles, changes with q_M, the dipole contribution to the capacity will be given by

$$\frac{1}{C_{\text{dipole}}} = \frac{\partial \Delta\chi}{dq_M} = \frac{4\pi\mu}{\varepsilon}\frac{\partial N}{\partial q_M} \qquad (7.164)$$

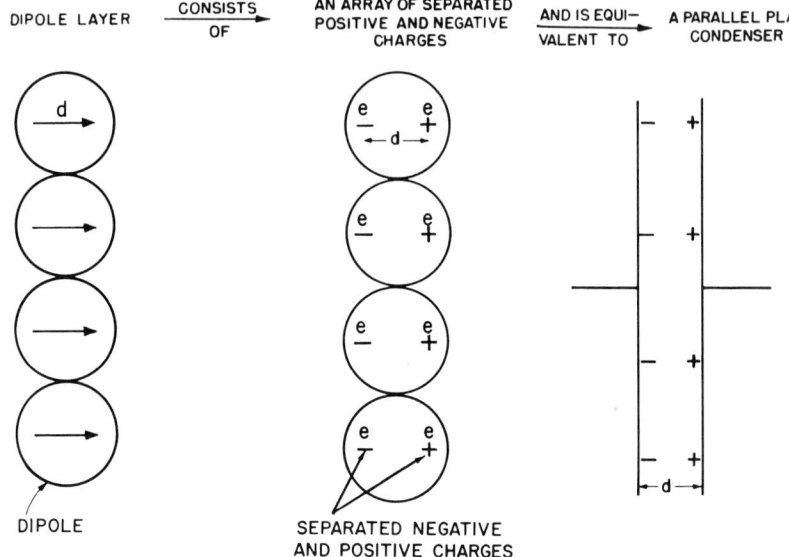

Fig. 7.94. A dipole layer is electrically equivalent to a parallel-plate condenser.

For the differentiation, one must know the functional relationship between N and q_M. This must be analyzed.

7.4.31. The Excess of Flipped Water Dipoles over Flopped Water Dipoles

It has just been seen that, if one had a unit area upon which were placed N dipoles *all pointing in the same direction*, the potential difference experienced by a unit of positive charge during a transfer across this layer is given by the equation

$$\Delta \chi = \frac{4\pi N \mu}{\varepsilon} \qquad (7.163)$$

Now consider how this electrostatic idealization could be applied to water dipoles on an electrode. Will they be *all* pointing in the same direction? In general, the answer will be no. It is true that, if the charge upon the electrode is excessively negative, the dipoles will be largely oriented with their positive (for water, this means the hydrogen side of the dipole) charges oriented toward the metal—the flop-down position. But, more normally, the dipoles will be a mixture, some with the oxygen on the metal and some with hydrogen on the metal. Let the symbol $N_\uparrow$ represent the number per unit area of the flipped dipoles with the hydrogen end toward the solution and $N_\downarrow$ be the number per unit area in the flop-down position.

Then, referring to the formula (7.163) for the potential difference across the dipole layer, one has

$$N = N_\uparrow - N_\downarrow \tag{7.165}$$

and, thus,

$$\Delta\chi = \frac{4\pi}{\varepsilon}\mu(N_\uparrow - N_\downarrow) \tag{7.166}$$

To know N, one must know the number which have flipped and how many have flopped. How can one derive an expression for $N_\uparrow$ and $N_\downarrow$? A simple approach is to consider the adsorption of water in any one position as a chemical reaction. For example,

$$\Box + H_2O \rightleftharpoons \boxed{\uparrow} \tag{7.167}$$

where $\Box$ are *free* sites on the metal and $\boxed{\uparrow}$ are sites occupied by water molecules in their *flip-up* state. Under equilibrium conditions, the law of mass action can be used. Thus,

$$\frac{N_\uparrow}{a_{H_2O}N_\Box} = e^{-\Delta G_\uparrow^0/RT} \tag{7.168}$$

where $\Delta G_\uparrow^0$ is the standard free-energy change associated with the adsorption of water in the flip-up state, $N_\Box$ is the number of free sites per unit area, and a_{H_2O} is the activity of water, which, in dilute solutions, can be taken as equal to unity. By setting $a_{H_2O} = 1$, equation (7.168) becomes

$$\frac{N_\uparrow}{N_\Box} = e^{-\Delta G_\uparrow^0/RT} \tag{7.169}$$

As in the case of the contact adsorption of ions (Section 7.4.26), $\Delta G_\uparrow^0$ can be resolved into three contributions. These arise from (1) the *chemical* work of adsorption, ΔG_c^0; (2) the *electrical*, or field, work; and (3) the work of *lateral interaction*.

The electrical work involves the free energy of a dipole in the electric field X arising from the charge q_M on the metal. This free energy has been shown (*cf.* Section 2.3.4) to be given by $-\mu X \cos \alpha$, where α is the angle between the field vector and the unit vector drawn from the negative to the positive end of the dipole (Fig. 7.95). Since the electric field is opposite in direction to the direction of positive potential gradient, the field vector runs from the solution to the electrode when the solution is charged positive (i.e., when the solution has a higher potential). In this case, the energy

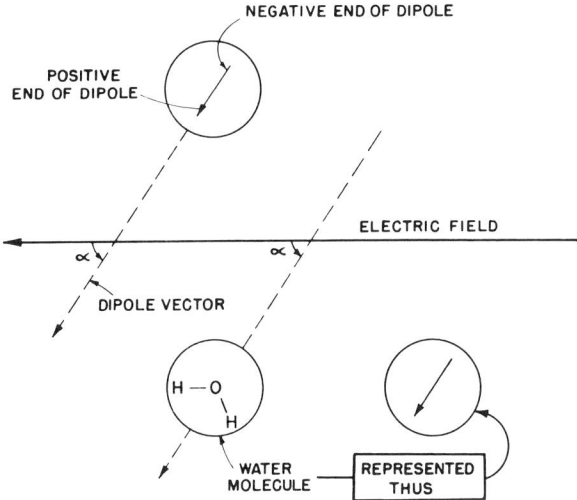

Fig. 7.95. The convention for the angle α between a dipole and the electric field.

of the flip-up dipoles is $+\mu X$ because $\alpha = \pi$ (Fig. 7.96) and that of the flop-down dipoles is $-\mu X$ because $\alpha = 0$ (Fig. 7.97).

A term remaining for consideration is that arising from lateral-interaction work. The argument here is simple and rather crude.

Consider one flip-up dipole and a dipole *in its vicinity* (Fig. 7.98). The two dipoles interact. The dipole–dipole interaction energy (see Section

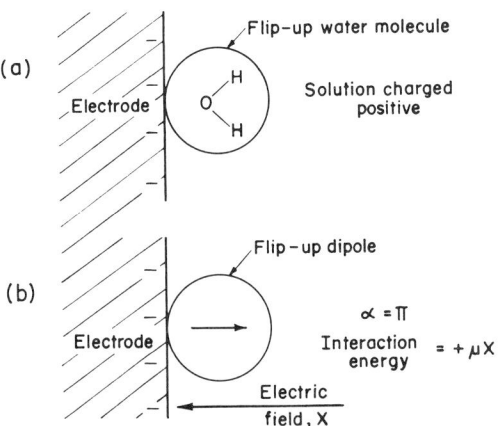

Fig. 7.96. (a) When a flip-up water molecule is on an electrode charged negative to the solution, (b) the interaction energy is $+\mu X$.

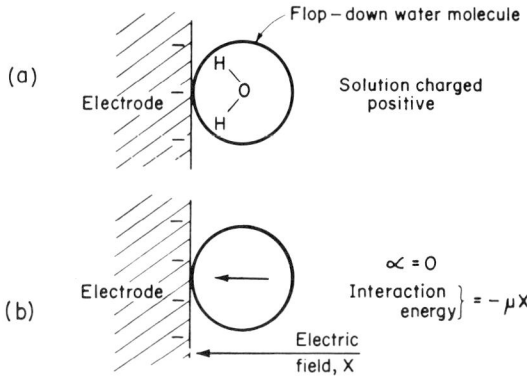

Fig. 7.97. (a) When a flop-down water molecule is on an electrode charged negative to the solution, (b) the interaction energy is $-\mu X$.

2.5.8) is minimum $(-U)$ when the second dipole is in a flop-down state and maximum $(+U)$ when both dipoles have like orientation.

Now, the field of the reference flip-up dipole dies down quite rapidly with distance, in fact, as r^{-3} (Section 2.6.3). Hence, the reference dipole can be considered to interact only with a certain number c of dipoles which surround it, i.e., with which it is coordinated.

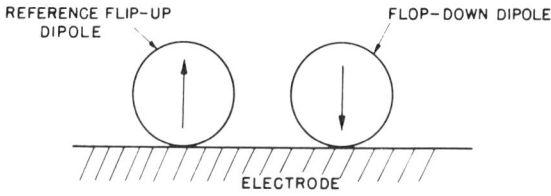

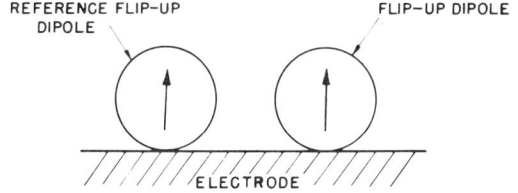

Fig. 7.98. (a) The configuration of minimum-interaction energy $-U$ between a reference flip-up dipole and a neighboring dipole; (b) The configuration of maximum-interaction energy $+U$.

But all these c dipoles do not have the same orientation; a fraction $\theta_\uparrow$ point up, and the remainder $\theta_\downarrow$ point down. These two fractions are simply related (in fact, by the definition of a fraction)

$$\theta_\uparrow = \frac{N_\uparrow}{N_T} \quad \text{and} \quad \theta_\downarrow = \frac{N_\downarrow}{N_T} \tag{7.170}$$

where N_T is the total number of water molecules on the electrode including the flip and the flop states, at the charge concerned.

It is now easy to see that $c\theta_\uparrow$ and $c\theta_\downarrow$ are the numbers of flipped and flopped dipoles, respectively, with which the reference flip-up dipole interacts. Let $+U$ be the interaction energy with one flipped-up dipole and $-U$ be that with one flopped-down dipole. The total interaction work for a flipped-up dipole is therefore $[(+U)c\theta_\uparrow + (-U)c\theta_\downarrow] = Uc(\theta_\uparrow - \theta_\downarrow)$.

If the reference dipole is in a flopped-down state, its total interaction energy is given by a similar expression. There is, however, a difference; a flopped-down dipole has a $-U$ interaction energy with a flipped-up dipole and a $+U$ interaction energy with a flopped-down dipole. Hence, its total interaction energy is $[(-U)c\theta_\uparrow + (+U)c\theta_\downarrow] = -Uc(\theta_\uparrow - \theta_\downarrow)$.

The expressions for the numbers of water dipoles which have flipped can be written by substituting $\Delta G_\uparrow^0 = (\Delta G_c^0)_\uparrow - \mu X + Uc(\theta_\uparrow - \theta_\downarrow)$ in Eq. (7.169), whereupon one has, for the flipped dipoles,

$$\frac{N_\uparrow}{N_\square} = \exp\left\{-\left[\frac{(\Delta G_c^0)_\uparrow}{RT} - \frac{\mu X}{kT} + \frac{Uc}{kT}(\theta_\uparrow - \theta_\downarrow)\right]\right\} \tag{7.171}$$

Similarly for the member of flopped dipoles, since $\Delta G_\downarrow^0 = (\Delta G_c^0)_\downarrow + \mu X - Uc(\theta_\uparrow - \theta_\downarrow)$, one can write

$$\frac{N_\downarrow}{N_\square} = \exp\left\{-\left[\frac{(\Delta G_c^0)_\downarrow}{RT} + \frac{\mu X}{kT} - \frac{Uc}{kT}(\theta_\uparrow - \theta_\downarrow)\right]\right\} \tag{7.172}$$

One can see that the free energy due to the electric field plus interaction in the case of down dipoles is simply the negative of that for up dipoles. Thus, for flipped dipoles, it is $\mu X - Uc(\theta_\downarrow - \theta_\uparrow)$, and, for flopped dipoles, it is $-\mu X + Uc(\theta_\downarrow - \theta_\uparrow)$. Using the symbol x for $(\mu X/kT) - (Uc/kT)(\theta_\downarrow - \theta_\uparrow)$ one has

$$\frac{N_\uparrow}{N_\square} = e^{[-(\Delta G_c^0)_\uparrow/RT] + x} \tag{7.173}$$

and

$$\frac{N_\downarrow}{N_\square} = e^{[-(\Delta G_c^0)_\downarrow/RT] - x} \tag{7.174}$$

Recall that the purpose of this analysis is to get $N_\uparrow - N_\downarrow$. Thus, all one has to do is take the difference of these two expressions. Hence,

$$N_\uparrow - N_\downarrow = N_\square \{e^{[-(\Delta G_c^0)\uparrow/RT]+x} - e^{[-(\Delta G_c^0)\downarrow/RT]-x}\} \tag{7.175}$$

Since $e^{+x} - e^{-x}$ is equal to $2\sinh x$, it would be *convenient* if

$$(\Delta G_c^0)_\uparrow = (\Delta G_c^0)_\downarrow = \Delta G^0 \tag{7.176}$$

in which case the chemical-work term can be taken out of the braces thus:

$$N_\uparrow - N_\downarrow = (2N_\square e^{-\Delta G_c^0/RT}) \sinh x \tag{7.177}$$

Is, however, $(\Delta G_c^0)_\uparrow = (\Delta G_c^0)_\downarrow$? Mathematical convenience must not serve as sufficient justification for an assumption. The answer is: Probably not. Nevertheless, the equality [Eq. (7.176)] of the chemical-work terms will be momentarily assumed and the consequences of the arbitrary assumption worked out. What does the assumption imply? It implies that the electrode cannot distinguish (at zero charge) between which end (the oxygen or the hydrogen end) of the water is pointing toward it. When the results of experiment show that the electrode *is* sensitive to which end of the water molecule it touches, then the assumption will be modified. Till then, as already shown,

$$N_\uparrow - N_\downarrow = (2N_\square e^{-\Delta G_c^0/RT}) \sinh x \tag{7.177}$$

The target is to substitute for $N_\uparrow - N_\downarrow$ in the expression [Eq. (7.163)] for the χ potential. To this end, it is convenient to get rid of $2N_\square$ in Eq. (7.177) by a simple trick.

The sum of $N_\uparrow$ and $N_\downarrow$ from Eqs. (7.173) and (7.174) gives

$$N_\uparrow + N_\downarrow = (2N_\square e^{-\Delta G_c^0/RT}) \cosh x \tag{7.178}$$

From (7.177) and (7.178),

$$\frac{N_\uparrow - N_\downarrow}{N_\uparrow + N_\downarrow} = \frac{\sinh x}{\cosh x} = \tanh x \tag{7.179}$$

Two symbols will be introduced, (1)

$$Y = \tanh x \tag{7.180}$$

and (2)

$$N_\uparrow + N_\downarrow = N_T \tag{7.181}$$

where N_T is the total number of adsorbed water molecules. In terms of N_T and Y, one has, from Eq. (7.179),

$$N_\uparrow - N_\downarrow = N_T Y \quad \text{or} \quad \theta_\uparrow - \theta_\downarrow = Y \tag{7.182}$$

The *excess* number N of dipoles pointing one way has been evaluated, and the dipole potential turns out to be [from (7.166) and (7.182)]

$$\Delta\chi = \frac{4\pi\mu}{\varepsilon}(N_\uparrow - N_\downarrow) = \frac{4\pi\mu N_T Y}{\varepsilon} \tag{7.183}$$

7.4.32. The Contribution of Adsorbed Water Dipoles to the Capacity of the Interface

But all this trouble was taken in order to examine whether it was reasonable—it is always done—to ignore the contribution of the water layer to the capacity of an electrified interface. It is necessary, therefore, to extract from the dipole potential the contribution of the water dipoles to the capacity. To do this, one has only to differentiate the $\Delta\chi$ potential with respect to electrode charge. The result is[†]

$$\frac{1}{C_{\text{dipole}}} = \frac{\partial(\Delta\chi)}{\partial q_M} = \frac{4\pi\mu N_T}{\varepsilon} \frac{\partial Y}{\partial q_M} \tag{7.184}$$

What is $\partial Y/\partial q_M$? One writes for the field (not the potential difference) due to the charge on the metal

$$X = \frac{4\pi}{\varepsilon} q_M \tag{7.185}$$

and, therefore, [*cf.* (7.180)] with

$$\begin{aligned} Y &= \tanh x \\ &= \tanh\left[\frac{\mu X}{kT} - \frac{Uc}{kT}(\theta_\downarrow - \theta_\uparrow)\right] \\ &= \tanh\left(\frac{4\pi\mu}{\varepsilon kT}q_M - \frac{UcY}{kT}\right) \end{aligned} \tag{7.186}$$

[†] The total number of water molecules on the surface, N_T, is taken here as a constant with respect to change of q_M. The justification of this arises from (7.178). Values of x corresponding to q_M observed in double-layer studies are $\ll 1$. In such a case, $\cosh x \simeq 1$, but $\sinh x \simeq x$.

Thus,

$$\frac{dY}{dq_M} = \text{sech}^2 x \left(\frac{4\pi\mu}{\varepsilon kT} - \frac{Uc}{kT} \frac{dY}{dq_M} \right) \quad (7.187)$$

or

$$\frac{dY}{dq_M} = \frac{4\pi\mu}{\varepsilon kT} \frac{1 - Y^2}{1 + (Uc/kT)(1 - Y^2)} \quad (7.188)$$

The dipole capacity turns out, therefore, to be [from (7.184)]

$$\frac{1}{C_{\text{dipole}}} = \frac{16\pi^2\mu^2 N_T}{\varepsilon^2 kT} \frac{1 - Y^2}{1 + (Uc/kT)(1 - Y^2)} \quad (7.189)$$

Since the parameter Y contains the field X, which depends upon the electrode charge q_M, Eq. (7.189) predicts that the dipole capacity should vary with electrode charge.

When the calculated values of C_{dipole} are plotted as a function of q_M (Fig. 7.99), it turns out that the values of the dipole capacity are extremely large compared with the experimental values of the capacity. What does

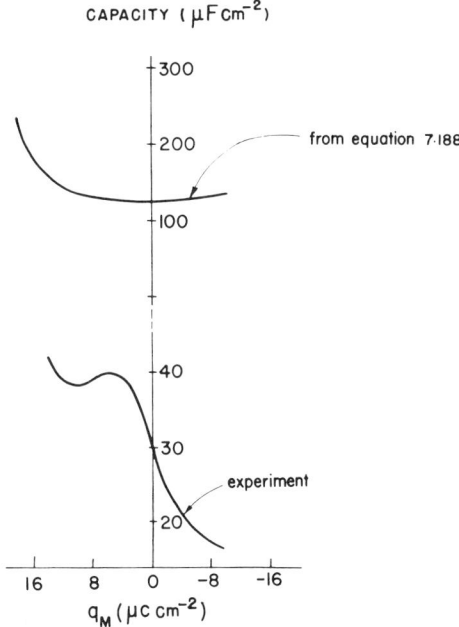

Fig. 7.99. Comparison between the calculated values of the dipole capacity and the experimental capacity.

this imply? Consider the complete expression for the differential capacity

$$\frac{1}{C} = \frac{d\Delta\phi}{dq_M} = \frac{d\Delta\psi}{dq_M} + \frac{d\Delta\chi}{dq_M} = \frac{1}{C_{\text{charge}}} + \frac{1}{C_{\text{dipole}}} \quad (7.190)$$

[The term $1/C_{\text{charge}}$ is given by Eq. (7.126) and has been the item of interest in the discussion of the constant-capacity region and the hump (cf. Sections 7.4.23 to 7.4.28).] From Eq. (7.190), it is obvious that, whenever C_{dipole} becomes too large, $1/C_{\text{dipole}}$ becomes too small to affect $1/C$ significantly. The situation is analogous to that in the Stern theory where the Gouy capacity made a negligible contribution to the total capacity whenever the magnitude of the Gouy capacity became large. Hence,

$$\frac{1}{C} \approx \frac{1}{C_{\text{charge}}} \quad (7.191)$$

and it is quite justified to neglect the contribution of the water dipoles (constituting the hydration sheath of the electrode) to the differential capacity of an electrified interface.

The flip-flop water model will be now applied to a discussion of the model for the dependence of the adsorption of organic molecules on the charge on metal electrodes, a subject which underlies, e.g., the use of organic inhibitors in metallic corrosion (Section 11.1.14a) and some aspects of whether a given electrode will act as a catalyst, or not, for a given fuel in electrochemical energy converters (see Section 10.1.4).

Further Reading

1. H. L. von Helmholtz, *Wied. Ann.*, **7**: 337 (1879).
2. G. Gouy, *J. Chim. Phys. (Paris)* **29**[7]: 145 (1903).
3. D. L. Chapman, *Phil. Mag.* **25**[6]: 475 (1913).
4. O. Stern, *Z. Elektrochem.*, **30**: 508 (1924).
5. A. N. Frumkin, *Z. Physik. Chem., (Leipzig)* **116**: 466 (1925); *Usp. Khim.*, **4** (1935).
6. D. C. Grahame, *Chem. Rev.*, **41**: 441 (1947).
7. R. Parsons, "The Structure of the Electric Double Layer and Its Influence on the Rates of Electrode Reactions," in: P. Delahay, ed., *Advances in Electrochemistry and Electrochemical Engineering*, Interscience Publishers, Inc., New York, 1961.
8. J. O'M. Bockris, M. A. V. Devanathan, and K. Muller, *Proc. Roy. Soc. (London)*, **A274**: 55 (1963).
9. T. N. Anderson, *Electrochim. Acta*, **9**: 347 (1964).
10. D. A. J. Swinkels, *J. Electrochem. Soc.*, **111**: 736 (1964).

11. P. Delahay, *Double Layer and Electrode Kinetics*, Interscience Publishers, Inc., New York, 1965.
12. B. E. Conway, *Theory and Principles of Electrode Processes*, The Ronald Press Company, New York, 1965.
13. H. Wroblowa and Z. Kovac, *Trans. Faraday Soc.*, **61**: 1523 (1965).
14. H. D. Hurwitz, in: E. Gileadi, ed., *Electrosorption*, Plenum Press, 1967.

7.5. THE COMPETITION BETWEEN WATER AND ORGANIC MOLECULES AT THE ELECTRIFIED INTERFACES

7.5.1. The Relevance of Organic Adsorption

The flip-flop water molecules which densely populate the electrode surface determine the dipole potential at the electrified interface. This is not the whole story. It will be shown that the flip-flop model may provide a basis for the understanding of the adsorption of organic molecules on an electrode.

Organic adsorption has important consequences, some of which have beneficial results. Thus, organic substances are added as brightening, leveling, and antipitting agents which help produce bright, smooth, and pit-free electrodeposits. Some organic molecules play an essential role as inhibitors in attempts to reduce corrosion. Further, an essential step in organic electrode reactions, which may generate useful electrical power from chemical reactants, is the asorption of the organic molecule.† In other cases, the consequences of organic adsorption have deleterious effects, e.g., organic impurities may "block" the electrode surface, impede charge transfer, and thus slow down the desired reactions.

There is, however, no typical organic molecule. There are many types of organics, the giant macromolecules (proteins, carbohydrates, etc.), the aromatics with their delocalization of electrons in the rings, the aliphatic substances with their C—C chains, etc. All these species do not respond in the same way to the attractions of the interface. Thus, aromatic substances may enter into actual bond formation through their π electrons, a behavior which is unlikely in the case of the aliphatic molecules. In this treatment, attention will be concentrated on the (non-charge-transfer) adsorption of aliphatic substances because it constitutes the simplest case.

---

† Such an adsorption may indeed be by means of molecules transferring from the solution to the electrode without undergoing charge transfer. In some organic reactions, however, charge transfer may occur during the adsorption. An example is C_3H_8, which adsorbs according to $C_3H_{8(solution)} \rightarrow C_3H_{7(adsorbed)} + H^+ + e$.

7.5.2. The Forces Involved in Organic Adsorption

The basic question is: Will an organic molecule adsorb onto the electrode? In principle, this question is no different from asking Will an ion contact-adsorb? The answer, therefore, can be sought with the same approach as in the calculation of the amount of contact adsorption (Section 7.4.17). Thus, the adsorption of organic molecules will be considered formally equivalent to a chemical reaction written as follows:

$$n\mathrm{H_2O_{electrode}} + \mathrm{Organic_{solution}} \rightleftharpoons \mathrm{Organic_{electrode}} + n\mathrm{H_2O_{solution}} \quad (7.192)$$

i.e., n water molecules on the electrode have to be displaced to accommodate one organic molecule.

From the law of mass action,

$$\frac{\theta_{\mathrm{org}}}{1 - \theta_{\mathrm{org}}} = \frac{c_{\mathrm{org}}}{c_W} e^{-\Delta G^0/RT} \quad (7.193)$$

where θ_{org} is the fraction of the electrode surface occupied by organic molecules, c_{org} and c_W are the bulk concentrations of organic and water, respectively, and ΔG^0 is the total standard free-energy change for the adsorption reaction [Eq. (7.192)].[†]

The standard free-energy change ΔG^0 is given by

$$\Delta G^0 = (G_{\mathrm{org}}^0)_{\mathrm{electrode}} + n\langle(G_W^{\,0})_{\mathrm{solution}}\rangle - (G_{\mathrm{org}}^0)_{\mathrm{solution}} - n\langle(G_W^{\,0})_{\mathrm{electrode}}\rangle \quad (7.194)$$

where $\langle(G_W^{\,0})_{\mathrm{solution}}\rangle$ and $\langle(G_W^{\,0})_{\mathrm{electrode}}\rangle$ are the average standard free energies of water in the solution and on the electrode and can therefore be resolved into the standard free-energy change experienced by the organic and the *average* standard free-energy change for the adsorption of water, i.e.,

$$\Delta G^0 = \Delta G_{\mathrm{org}}^0 - n\langle \Delta G_W^{\,0}\rangle \quad (7.195)$$

Introducing this expression into Eq. (7.193) results in

$$\frac{\theta_{\mathrm{org}}}{1 - \theta_{\mathrm{org}}} = \frac{c_{\mathrm{org}}}{c_W} \exp\left(-\frac{\Delta G_{\mathrm{org}}^0 - n\langle \Delta G_W^{\,0}\rangle}{RT}\right) \quad (7.196)$$

[†] Equation (7.193) is an adsorption isotherm for the simple case of $n = 1$. If the organic molecule occupies several sites on the surface, i.e., $n > 1$, a more complicated expression results which, however, yields qualitatively the same type of behavior as does the simplified analysis given here.

The thinking begins with the question: How does one get at ΔG^0_{org}? In the analysis of contact adsorption and of water adsorption, one considered a lateral interaction term, a field term, and a chemical term. In this case, the lateral interaction and field terms will be dropped out.

What does this approximation imply? It implies that either nonpolar molecules are considered, or neutral molecules with their polar groups (e.g., OH) protruding outside the range of the double-layer field and not interacting laterally with the polar groups of other adsorbed molecules (i.e., being at low coverage).

How were the forces responsible for contact adsorption analyzed? Three types of interactions were distinguished, (1) ion–electrode, (2) water–electrode, and (3) ion–water interaction. It became clear that an electrode attracts all ions, but only those turn up in the IHP which do not need to make major changes in their hydration sheath.

In the case of aliphatic molecules, there are some (rather small) chemical interactions—dispersion forces mainly and perhaps a small induced dipole moment—with the electrode. Let the resulting free energy be termed ΔG_c^0. Further, let the organic–water interactions be negligible. The predominant part of the work of adsorption therefore consists in replacing water molecules; this is the $n\langle \Delta G_W^0 \rangle$ term. In fact, since water does, and the organic molecule is assumed not to, engage in significant electrostatic interactions with the electrode, it is a case of water's surrendering its position of contact adsorption on the electrode in favor of the organic molecules, rather than a case of the organic molecules' displacing the water.

The theory of the adsorption of simple aliphatic compounds is, therefore, essentially a theory of water adsorption: insofar as this decreases, the adsorption of the organic increases.

7.5.3. Does Organic Adsorption Depend on Electrode Charge?

The arguments of the last section indicate that the dependence of organic adsorption upon electrode charge will relate to the way the electrode charge decides the "desorbability" of the water molecules upon the surface. If these are not attracted strongly, the organic molecules tend to replace them on the surface, and *vice versa*.

Consider the situation when the electrode is highly charged with excess electrons, i.e., the charge is highly negative (Fig. 7.100). Then the water dipoles are nearly all in the flopped-down position, i.e., nearly all of them are aligned and tightly held with their hydrogens on the surface of the electrode. Not much organic can adsorb under such circumstances because, as already made clear, if the water molecules are strongly held to the surface and do not come off easily, then the aliphatic organic, which interacts relatively little with the electrode, cannot land on the surface of the electrode. Hence, with an extremely negative charge on the electrode, the organic is kept out and θ_{org} is small.

What happens as the electrode charge is made less negative? An increasing fraction of the water molecules change their position from that in which the hydrogen is in contact with the electrode to that in which the oxygen is in contact with the electrode. From a flop-down position, the water molecules tend to assume a flip-up position. There will at all times be a mixture of up and down water molecules, but the excess of flopped-down dipoles tends to decrease with decreasing negative charge.

Now, the net energy of the attraction of the adsorbed water layer to the electrode depends upon the *excess* number which are aligned in the up or down state, which in turn depends on the electrode charge. Hence, as the electrode charge decreases from a highly negative value to a value of zero, the net water–electrode interaction energy decreases, the water desorbs more and more easily, and organic adsorption becomes more and more facilitated. To a first approximation, the water molecules are held most *lightly* at the pzc, and, consequently, the organic molecules will be held relatively tightly (and thus more numerously) by the electrode at this potential (Fig. 7.100). Thus, to a first approximation (particularly if the organic molecules interact little with the electrode), the maximum of organic adsorption should be reached at the pzc.

What happens on the positive side of the pzc? The tendency is for water dipoles to flip-up with hydrogens away from the electrode. Thus, the flip-up dipoles tend to be in excess, and their majority increases with increasingly positive charge on the electrode. By the same token, the net attraction of water to the surface becomes increasingly strong, and organic adsorption becomes reduced (Fig. 7.100). It will be noticed that the situation is symmetrical about the pzc. So, when the electrode charge is highly positive, virtually all the organic has been displaced by the tightly adsorbed flipped-up water dipoles.

The predictions of this very simple and qualitative picture[†] are clear-cut. The amount of adsorbed organic material will have a maximum at the pzc and will die down asymptotically to zero at extremes of positive and

---

† The crudity of the present model of flip-flop water must be stressed. A much more general—but more thermodynamic—approach to the adsorption of molecules on electrodes was formulated as early as 1926, by Frumkin. It is in terms of the work done to charge the double layer, the capacitance of which is dependent upon the constituents of water and adsorbed substances.

(a) Negatively—charged electrode with excess of flopped-down dipoles

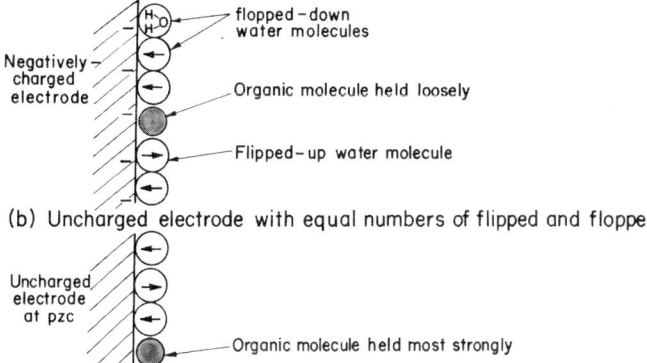

(b) Uncharged electrode with equal numbers of flipped and flopped dipole

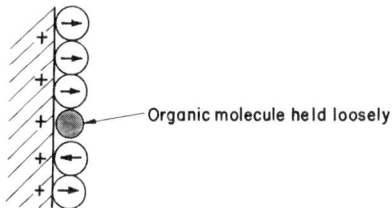

(C) Postively — charged electrode with excess of flipped-up dipoles

Fig. 7.100. The relation between the electrode charge and organic adsorption: (a) a negatively charged electrode with an excess of flopped-down dipoles has weak adsorption; (b) an uncharged electrode with equal numbers of flipped and flopped dipoles shows strong adsorption; (c) a positively charged electrode with an excess of flipped-up dipoles has weak adsorption.

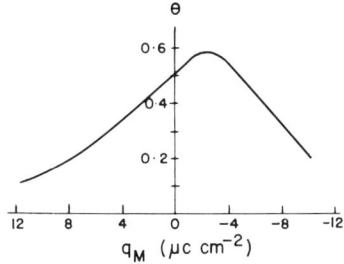

Fig. 7.101. The experimental curve for the adsorption of butanol on a mercury electrode.

negative charge. In fact, a typical experimental curve does show something near to this behavior (Fig. 7.101).

To what extent can the equations [(7.176), (7.177), (7.180), (7.181), and (7.195)] describing the orientation of water molecules and the adsorption of organic at the interface reproduce the facts?

Consider the expression for θ_{org}, the fraction of the electrode surface covered with organic,

$$\frac{\theta_{\text{org}}}{1-\theta_{\text{org}}} = \frac{c_{\text{org}}}{c_W} \exp -\left[\frac{\Delta G^0_{\text{org}} - n\langle\Delta G_W{}^0\rangle}{RT}\right] \quad (7.196)$$

The dependence of θ_{org} on the electrode charge is implicitly contained in the term $\langle\Delta G_W{}^0\rangle$, the average standard free-energy change associated with the adsorption of one water molecule. This dependence must now be made explicit.

Knowledge has already been gained about the standard free-energy change $\Delta G_\uparrow{}^0$ associated with the adsorption of a flipped-up water molecule and also the free-energy change $\Delta G_\downarrow{}^0$ arising from the adsorption of a flopped-down water molecule. If, therefore, there are $N_T = N_\uparrow + N_\downarrow$ water molecules on the electrode, the average standard free-energy per water molecule becomes

$$\langle\Delta G_W{}^0\rangle = \frac{N_\uparrow \Delta G_\uparrow{}^0 + N_\downarrow \Delta G_\downarrow{}^0}{N_T} \quad (7.197)$$

Now,

$$\frac{N_\uparrow}{N_T} = \theta_\uparrow \quad \text{and} \quad \frac{N_\downarrow}{N_T} = \theta_\downarrow \quad (7.198)$$

Hence, Eq. (7.197) can be rewritten as

$$\langle\Delta G_W{}^0\rangle = \theta_\uparrow \Delta G_\uparrow{}^0 + \theta_\downarrow \Delta G_\downarrow{}^0 \quad (7.199)$$

The individual free-energy changes $\Delta G_\uparrow{}^0$ and $\Delta G_\downarrow{}^0$ for the adsorption of flipped and flopped dipoles are not new quantities. They have already been discussed (Section 7.4.31) and shown to be

$$\Delta G_\uparrow{}^0 = (\Delta G_c^0)_\uparrow - \mu X + Uc(\theta_\uparrow - \theta_\downarrow) \quad (7.200)$$

and

$$\Delta G_\downarrow{}^0 = (\Delta G_c^0)_\downarrow + \mu X - Uc(\theta_\uparrow - \theta_\downarrow) \quad (7.201)$$

Using these expressions (7.200 and 7.201) in Eq. (7.199), one has

$$\langle\Delta G_W{}^0\rangle = [\theta_\uparrow(\Delta G_c^0)_\uparrow + \theta_\downarrow(\Delta G_c^0)_\downarrow] - \mu X(\theta_\uparrow - \theta_\downarrow) + Uc(\theta_\uparrow - \theta_\downarrow)^2 \quad (7.202)$$

or, with the notation,

$$Y = \theta_\uparrow - \theta_\downarrow \quad (7.182)$$

the result is

$$\langle \Delta G_W^0 \rangle = [\theta_\uparrow (\Delta G_c^0)_\uparrow + \theta_\downarrow (\Delta G_c^0)_\downarrow] - Y\mu X + Y^2 Uc \quad (7.203)$$

As done in the case of water adsorption, let it be assumed that the chemical interactions of water dipoles in their two states are identical. Then,

$$\Delta G_c^0 = (\Delta G_c^0)_\uparrow = (\Delta G_c^0)_\downarrow \quad (7.204)$$

and

$$[\theta_\uparrow (\Delta G_c^0)_\uparrow + \theta_\downarrow (\Delta G_c^0)_\downarrow] = \Delta G_c^0 (\theta_\uparrow + \theta_\downarrow) = \Delta G_c^0 \quad (7.205)$$

because $\theta_\uparrow + \theta_\downarrow = 1$.

The average free energy of desorbing one water molecule becomes, from Eqs. (7.203) and (7.205),

$$\langle \Delta G_W^0 \rangle = \Delta G_c^0 - Y(\mu X - YUc) \quad (7.206)$$

Inserting this evaluation of $\langle \Delta G_W^0 \rangle$ in Eq. (7.196), one obtains

$$\frac{\theta_{\text{org}}}{1 - \theta_{\text{org}}} = \frac{c_{\text{org}}}{c_W} e^{-\Delta G^0_{\text{org}}/RT} e^{n\Delta G_c^0/RT} e^{-(nY/kT)(\mu X - YUc)} \quad (7.207)$$

or, since $X = (4\pi/\varepsilon) q_M$,

$$\frac{\theta_{\text{org}}}{1 - \theta_{\text{org}}} = \frac{c_{\text{org}}}{c_W} e^{-\Delta G^0_{\text{org}}/RT} e^{n\Delta G_c^0/RT} e^{-(nY/kT)[(4\pi\mu q_M/\varepsilon) - YUc]} \quad (7.208)$$

This is the prediction that the water flip-flop model for aliphatic organic adsorption makes for the dependence of the coverage (of the electrode with organics) upon the bulk concentration c_{org} of organic and upon the electrode charge q_M. A principal assumption is that there is a negligible interaction between the organic molecule and the field due to the electrode.

7.5.4. The Examination of the Water Flip-Flop Model for Simple Cases of Organic Adsorption

When θ_{org}, calculated according to Eq. (7.208), is plotted against the electrode charge with the bulk concentration of the organic kept constant, a quasi-parabolic curve is obtained (Fig. 7.102).

This is good, because the experimental coverage–electrode-charge curves are also somewhat parabolic and symmetrical (Fig. 7.101).

The theory permits another simple test. If a series of aliphatic molecules are all assumed to land on the electrode with their hydrocarbon end in contact with it, then their chemical free energies of interaction, ΔG^0_{org}, should all be the same (the same dispersion and induced-dipole interactions).

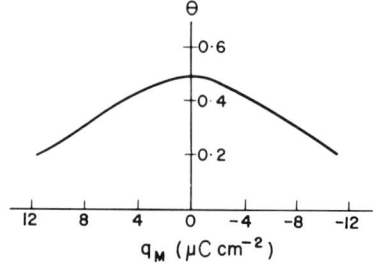

Fig. 7.102. The predicted curve for the adsorption of butanol on a mercury electrode.

Apart from this ΔG^0_{org} term, there is nothing else in the adsorption equation (7.208) which is organic dependent. All the other quantities depend only on water. Hence, shape and position on the charge axis of the θ_{org} versus q_M curve should be independent of the organic molecules. This is what is found to be the case so long as one restricts the consideration to simple aliphatic compounds, so that the assumption of limitingly low degree of interaction with the metal is likely to be applicable.

There is an interesting discrepancy between the theoretical and experimental curves. With respect to the theoretical curve, the experimental ones are clearly displaced along the charge axis. Thus, the simple theory requires maximum adsorption at the pzc ($q_M = 0$); experiment (Fig. 7.101) displays it at a negative value (about $-2\,\mu\text{C cm}^{-2}$ on a mercury electrode).

Such clashes between model and fact are good, for they immediately bring to mind that the chemical interaction between the electrode and water dipoles in their flipped and flopped states was taken [Eq. (7.176)] to be the same, i.e., $(\Delta G_c^0)_\uparrow = (\Delta G_c^0)_\downarrow$. Perhaps it is the limited applicability of this assumption which produces the sideways displacement of the experimental curve of Fig. 7.101 compared with the simple theoretical model. This is the hypothesis examined in the next section.

7.5.5. At What Potential Does Maximum Organic Adsorption Occur?

Suppose, for instance, one seeks to restrict corrosion by adding to the solution an organic molecule as a corrosion inhibitor. To know whether the added organic molecule is likely to work efficiently, one must know at what potential there is maximum adsorption. Hence, the question posed in the title of this section has important practical implications.

The adsorption of an organic has been viewed as a competition between organic and water molecules for sites on the electrode surface. When water dipoles are least attached to the electrode surface, the organic molecules push their way onto the surface.

Maximum organic adsorption occurs, therefore, when there is minimum binding of water to the electrode. But when is there least attachment of water? Water molecules are held least strongly to the electrode when the flipped and flopped dipoles have the same free energies of adsorption.

$$\Delta G_\uparrow^0 = \Delta G_\downarrow^0 \tag{7.209}$$

When this is the case, the electrode is not affected by the dipoles' having flipped or flopped. The probabilities of flipping up and flopping down are equal, and it follows that the numbers of up dipoles and down dipoles are equal

$$N_\uparrow = N_\downarrow$$

or [cf. Eq. (7.182)]:

$$Y = \theta_\uparrow - \theta_\downarrow = 0 \tag{7.210}$$

What is the potential (or electrode charge) under these conditions? This is easy to discover. Since $\Delta G_\uparrow^0 = \Delta G_\downarrow^0$, it follows, by setting $\theta_\uparrow - \theta_\downarrow = Y = 0$ in Eqs. (7.200) and (7.201), that

$$(\Delta G_c^0)_\uparrow - \mu X_{max} = (\Delta G_c^0)_\downarrow + \mu X_{max} \tag{7.211}$$

where the subscript max denotes the condition of maximum adsorption, and, therefore, that

$$[(\Delta G_c^0)_\uparrow - (\Delta G_c^0)_\downarrow] - 2\mu X_{max} = 0 \tag{7.212}$$

But the field X_{max} is simply given by

$$X_{max} = 4\pi q_{max}/\varepsilon \tag{7.213}$$

Hence, inserting Eq. (7.213) in (7.212), one obtains

$$[(\Delta G_c^0)_\uparrow - (\Delta G_c^0)_\downarrow] - \frac{8\pi\mu}{\varepsilon} q_{max} = 0$$

It may be recalled that, in the treatment of the flip-flop water model, it was assumed as a ruthless first approximation that $(\Delta G_c^0)_\uparrow = (\Delta G_c^0)_\downarrow$, i.e., that the chemical interactions of water in the flipped and flopped states are equal. What is the consequence of this approximation? The consequence is that maximum adsorption would be expected to occur when $q_{max} = 0$, i.e., at the pzc. In other words, if it is assumed that the chemical interaction between the electrode and water is the same in the two possible dipole

800 CHAPTER 7

orientations, then it follows that the pzc is also the potential of maximum adsorption.

Is this expectation borne out by experience? In general, not very well. Table 7.13 shows that, in the case of several metals, maximum adsorption occurs at values of electrode charge which are negative with respect to $q = 0$, i.e., the potential of maximum adsorption is shifted from the pzc.

The phenomenological way of explaining this shift is to argue that the chemical water–metal interactions for the flip and flop states of a water dipole are not the same:

$$(\Delta G_c^0)_\uparrow \neq (\Delta G_c^0)_\downarrow \tag{7.214}$$

This is, of course, no explanation; it is only a way of stating the experimental results. One must seek for the rationale behind the result. When one assumes that $(\Delta G_c^0)_\uparrow = (\Delta G_c^0)_\downarrow$, one is really assuming that, from the point of view of the metal, the flip and the flop states (hydrogen end away and hydrogen end toward) are indistinguishable. But the flop state is obtained by rotating the flipped dipole through 180° about an axis through the center of the dipole. When one considers an actual water molecule, not an idealized dipole with two point charges separated by a certain distance, the result of the rotation operation—a flop state—is *distinguishable* from the original flip state. The effective charge in the water molecule is farther away from the electrode when the hydrogen atoms are toward the electrode (the flopped-down state) than when the oxygen atom is toward it (the flipped-up state). When an operation (rotation, reflection, translation, etc.) produces a state distinguishable from the original state, there is no symmetry corresponding to that operation. A water molecule, therefore, has no symmetry corresponding to a rotation about an axis normal to the dipole axis through an angle of π. As a consequence, the dispersion and image inter-

TABLE 7.13

The Potential of Zero Charge, E_{pzc}, and the Potential of Maximum Adsorption, E_m

Electrode	E_{pzc}, V	E_m, V	$E_m - E_{pzc}$, V
Nickel	−0.47	−0.8	−0.33
Iron	−0.5	−0.7	−0.2
Copper	−0.2	−0.9	−0.7
Platinum, in alkali	−0.3	−0.4	−0.1

actions (see Section 7.4.13) will both vary with the state of the water molecule—has it flipped up or flopped down? But these interactions are part of the chemical interactions ΔG_c^0, which therefore turn out to be different for the two states. Thus, a shift of the potential of maximum organic adsorption away from the pzc is produced.

One can conclude from this brief survey that the model involving competition between organic molecules and flip-flop water dipoles provides a simple and fairly successful picture of organic adsorption so long as one is concerned primarily with its potential dependence and with systems in which chemical interaction of the organic with the electrode is insignificant.[†]

Further Reading

1. A. N. Frumkin, *Z. Phys.*, **35**: 792 (1926).
2. J. O'M. Bockris, M. A. V. Devanathan, and K. Muller, *Proc. Roy. Soc.* (*London*), **A274**: 55 (1963).
3. D. A. J. Swinkels, *J. Electrochem. Soc.*, **111**: 736 (1964).
4. A. N. Frumkin and B. B. Damaskin, in: J. O'M. Bockris, ed., *Modern Aspects of Electrochemistry*, Vol. III, pp. 149–223, Academic Press, Inc., New York, 1964.
5. E. Gileadi, *J. Electroanal. Chem.*, **11**: 137 (1966).
6. K. Muller, "The Role of Solvents at Electrodes," in: E. Gileadi, ed., *Electrosorption*, Plenum Press, New York, 1967.
7. B. J. Piersma, in: E. Gileadi, ed., *Electrosorption*, Plenum Press, New York, 1967.
8. E. Gileadi, in: E. Gileadi, ed., *Electrosorption*, Plenum Press, New York, 1967.
9. E. Gileadi and K. Müller, *Electrochim. Acta*, **12**: 1301 (1967).

7.6. ELECTRIFIED INTERFACES AT METALS OTHER THAN MERCURY

A discerning reader would have noticed that, despite attempts to avoid restricting the treatment of the electrified interfaces to the mercury–solution interface, the bulk of the results has concerned this metal. Mercury is a good metal to use (*cf.* Section 7.2.7). It is, at ordinary temperatures, a liquid, and, therefore, one can easily set up a dropping electrode and gain

---

[†] Of course, chemisorbed hydrogen and oxygen must also be absent for the simple dipole-competition model to have a chance of giving consistency with experiment. On metals for which hydrogen and oxygen easily deposit (e.g., platinum), there may be only a small potential region (perhaps only 0.2 to 0.3 V) in which these influences of hydrogen and oxygen can be completely neglected. On mercury, this region is about 1 V.

reproducibility because the surface is a renewable surface. It also has a known "true" surface area and no wrinkles and crumples in its microsurface to make the effective area doubtful, as solid metals have. The other advantage concerns the fact that, for a span of as much as 2 V, the mercury-solution interface does not leak, i.e., it behaves as a polarizable interface.

However, electrochemistry in practice outside the laboratory is concerned only to a very small extent with mercury electrodes. It is therefore of importance to know whether the picture of the electrified interface developed here for mercury is applicable to solid metals.

The answer is: Essentially. The electroneutrality of the interface as a whole, the contact adsorption of some ions to yield a layer of charge on the IHP, the layer of hydrated ions on the OHP, the ionic cloud diffusing out into a dilute solution, the two potential drops—one linear and the other exponentiallike—, the variation of electrode charge with potential difference across the interface, the variation of interface capacity with potential difference or electrode charge, the basic water-competition model of organic adsorptions, all these ideas can be carried over *in toto* to rationalize the electrified interface at metals other than mercury.

There are, however, three changes which must be incorporated into the general picture. One change is an obvious one but may have important consequences. The nonelectrostatic interaction between the metal (say, iron) and the constituents of the solution (say, water) may differ from the interactions when mercury is present. The relevant ΔG_c^0's must, of course, be introduced.

The second change concerns the degree of polarizability of the electrode (see Section 7.2.5d). There is often much more leakage of charge across the interface with metals other than mercury. Thus, only interfaces involving mercury and a few of the softer metals, e.g., lead, tin, gallium, and thallium have large (e.g., 1 V) ranges of potential in which the interface can be regarded as nonleaky, or polarizable. Hence, adsorbed hydrogen may interfere with adsorption or limit the range in which the electrostatic picture given above can be valid over quite large ranges of potential on many solid metals.

The really important difference between the solid metals and liquid mercury lies, however, in the fact that, on the surface of a liquid, all sites are equivalent. In short, there is homogeneity, and the surface may be treated as a featureless, structureless plane. The surface of a solid, however, has a structure. All sites are not equivalent. There are heterogeneities. (There will be more about these heterogenities in the treatment of electrodeposition in Section 10.1.) When there are different types of sites, the free energies of adsorption of ions, water molecules, and organics vary over the various parts of the surface. There will be a hierarchy of sites and a hierarchy of adsorption energies. This basic *surface heterogeneity* is a fundamental difference between the electrified interfaces at solid metals and liquid mercury. This difference will affect most considerations of *solid*–solution interfaces, but particularly those concerned with electrodics (*cf.* Chapter 8).

Further Reading

1. M. I. Temkin, *Zh. Fiz. Khim.*, **15**: 296 (1941).
2. H. Wroblowa and M. Green, *Electrochim. Acta*, **8**: 679 (1963).
3. D. A. J. Swinkels, *J. Electrochem. Soc.*, **111**: 743 (1964).
4. P. Delahay, *Double Layer and Electrode Kinetics*, Interscience Publishers, Inc., New York, 1965.
5. E. Gileadi and W. Heiland, *J. Phys. Chem.*, **69**: 3335 (1965).
6. B. E. Conway, *Theory and Principles of Electrode Processes*, The Ronald Press Company, New York, 1965.
7. M. Genshaw, in: E. Gileadi, ed., *Electrosorption*, Plenum Press, New York, 1967.
8. R. Oriani and C. Johnson, in: J. O'M. Bockris and B. E. Conway, eds., *Modern Aspects of Electrochemistry*, No. 5, Plenum Press, New York, 1969.

7.7. THE STRUCTURE OF THE SEMICONDUCTOR–ELECTROLYTE INTERFACE

7.7.1. How Is the Charge Distributed inside a Solid Electrode?

It was clearly emphasized at the outset of this chapter that phenomena which depend on electric double layers comprise a general and very widespread part of the science of surfaces. They occur wherever phases (containing charged particles or permanent or induced dipoles) meet to form an interface. Thus, double layers at air–solution, metal–solution, and solution–solution interfaces were quoted. Nevertheless, an impression might have been gained that only *metal*–solution interfaces are of consequence. If so, the impression will now be corrected by dealing in a very elementary way with another type of interface which is intellectually stimulating and technologically important, the *semiconductor*–electrolyte interface. This aspect of electrodics extends the scope of the subject beyond considerations of the *metal*–solution interface to all[†] interfaces at which electrons are

---

[†] The newly developing theory of the electrochemistry of *insulators* derives from the theory of the semiconductor–solution interface.

exchanged and thus opens up the prospect of understanding the electrochemistry of nonmetals, e.g., interfaces in biological systems and interfaces involving solid oxides.

In dealing with the metal–electrolyte interface, consideration of the metal was restricted to the statement that the charge density q_M was confined to the *surface* of the metal and to a narrow region—a few angstroms—extending into the solution, (*cf.* Section 7.2.5*l*). Thereafter, attention was turned to the solution side of the interface, i.e., toward the *ionic double layer*, and it was asked how the excess charges are arranged there as a function of distance from the metal, how the potential decays, how the concentration in the electrolyte affects the picture, etc. Now the viewpoint will be reversed. The situation in the solution will be considered somewhat understood, and the distribution of excess charge inside the electrode, i.e., the *electronic* double layer, will be scrutinized.

Consider a concentrated electrolytic solution. Then, for all intents and purposes, the entire Gouy–Chapman diffuse charge will be located on the OHP (Section 7.4.10). Further, let there be no contact adsorption, so that the IHP is unpopulated. What is being considered, therefore, is a *single* layer of charge on the solution side of the interface.

What is the situation inside the *electrode*? That depends upon whether the electrode is a metal or a semiconductor. What is the most important difference between a metal and a semiconductor? Operationally speaking, it is the order of magnitude of the conductivity. Metals have conductivities of the order of about 10^6 ohm^{-1} cm^{-1}; and semiconductors, about 10^2–10^{-9} ohm^{-1} cm^{-1}. These tremendous differences in conductivity reflect predominantly the concentration of free charge carriers. In crystalline solids, the atomic nuclei are relatively fixed, and the charge carriers which drift in response to electric fields are the electrons. So, the question is: What determines the concentration of mobile electrons? One has to take an inside look at electrons in crystalline solids.

7.7.2. The Band Theory of Crystalline Solids

Consider a crystalline solid. The atoms are arranged according to a three-dimensional pattern (or lattice) in which they have equilibrium interatomic distances.

A thought experiment is now performed. The lattice is expanded, i.e., the interatomic distances are increased. Eventually, the atoms are so far apart that they can be considered isolated and independent atoms as in a gas. The purpose of the thought experiment is to discuss the electron-energy states in gaseous atoms and then to see how these energy states are modified as the atoms are brought closer and closer together until the lattice has contracted back to its original state.

The electrons in a gaseous atom are arranged in shells. The shell structure is a result of the energy levels of the electrons' having to follow quantum rules. Thus, only a set of discrete energy states is allowed, and all other energies are forbidden. The energy states are occupied by electrons in accordance with the Pauli exclusion principle: The maximum number of electrons per energy state is two, and these two electrons must have opposite spins. The energy states fall into groups or shells.

As one shell fills up with electrons, the Pauli principle rules that any further electrons have to move to shells more removed from the nucleus. (The electrons in an incomplete outermost shell are known as *valence* electrons, and those in filled inner shells are known as *core* electrons.) The location of the various electrons can be described by talking of an *electron cloud*; the density of this cloud at any point is a measure of the probability of finding the electron at that point.

Suppose now that two gaseous atoms are made to approach each other. As long as the electron clouds of the two atoms do not overlap, the electron-energy states continue to follow the quantum rules for gaseous atoms. When, however, the electron clouds begin to overlap and the electrons interact with both atoms, the rules for electron-energy states are upset and they start changing.

They change in an interesting way; each energy state from a gaseous atom splits into two states, one with a higher energy and the other with a lower energy (Fig. 7.103). If three atoms are brought together, then each energy state of the gaseous atoms splits into three energy states; if six atoms

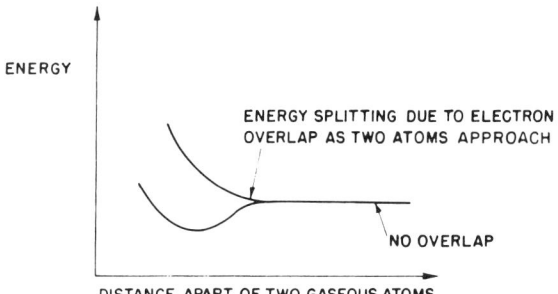

Fig. 7.103. The splitting of the energy into two energy states when two atoms approach each other.

are brought together, the splitting is into six states; and, in general, if there are N atoms, each energy state of a gaseous atom splits into N states. Some of these levels may be degenerate, i.e., they may have exactly the same energy, but none can be *lost* or *created*, i.e., the *number* of energy levels, or states, in a crystal made up of N particles must equal N times the number of electronic states in each particle to satisfy Pauli's exclusion principle. The upper and lower levels shown in Fig. 7.103 arise owing to the symmetrical and antisymmetrical linear combination of atomic orbitals.

The spacing between these N energy states depends on the value of N; the larger the value of N is, the closer together are the energy levels. In a bulky, solid electrode where there may be some 10^{22} atoms cm^{-3}, the energy levels are spaced so close together that it is more convenient to ignore the discreteness of the levels and think of a continuous *band* of allowed energies.[†]

This means that, as the expanded lattice is contracted in the thought experiment, the *discrete* energy states of the atoms are replaced by energy *bands*. Thus, when the lattice-contraction thought experiment is finally halted with atoms at their equilibrium interatomic distances, the electron-energy states show a band structure.

Now, the splitting of the energy states of gaseous atoms occurs because of the overlap between electron clouds. Obviously, therefore, atoms must come much closer before the clouds of the core electrons begin to overlap compared with the distance at which the clouds of outer (or valence) electrons overlap (Fig. 7.104). Hence, at the equilibrium interatomic distances, the energy levels of the *core* electrons (in contrast to the *valence* electrons) do not show any band structure and therefore will be neglected in the following discussion.

The above simplified picture of the band theory of solids will now be used to explain the differences in conductivity of metals, semiconductors, and insulators.

7.7.3. Conductors, Insulators, and Semiconductors

The essence of electrical conductivity is that charges must be able to move under an applied electric field. In solids, conduction requires the movement of electrons.[‡] But, for an electron to move, there must be a

---

[†] For example, if the total energy range is 10 eV and the 10^{22} levels are spaced equally apart, the distance between any two will be 10^{-21} eV, or 2.3×10^{-17} cal mole^{-1}, compared with the average thermal energy of about 6×10^2 cal mole^{-1} at room temperature.

[‡] Conduction involving the drift of ions is not under consideration here.

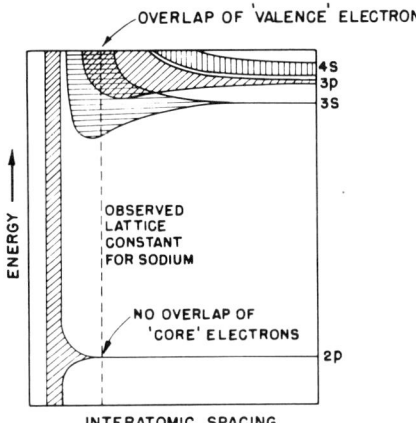

Fig. 7.104. The energy levels of the core electrons do not show any band structure. The band picture is for sodium (1s electrons not shown).

partially vacant energy band. If all energy states in a band are completely filled, then an electron cannot move for where can it move to when the Pauli principle says it cannot go into a filled state? So, differences in conductivity between different substances must be a matter of vacant, or *partially* filled, bands.

Consider the electron-energy *versus* interatomic-spacing diagram (Fig. 7.105) picturing the result of the thought experiment in which an expanded lattice is contracted. Let this contraction be stopped when the equilibrium interatomic spacing is d_M (*cf.* Fig. 7.106). Then, inside the crystal, the elec-

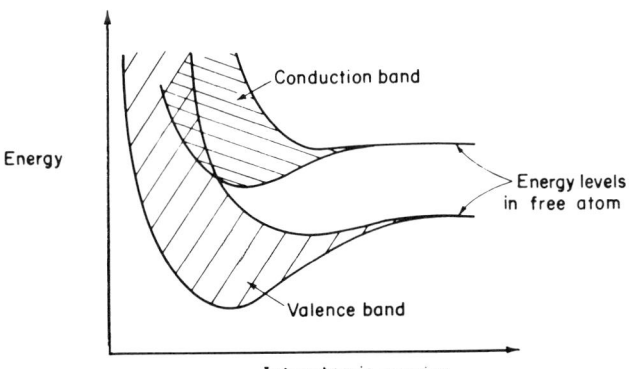

Fig. 7.105. A schematic representation of the dependence of energy upon interatomic spacing.

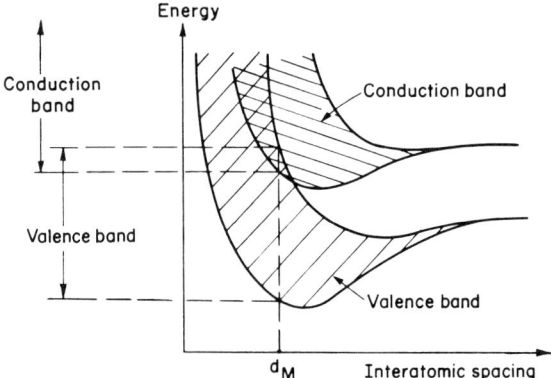

Fig. 7.106. The band picture of a metal with an interatomic spacing of d_M, showing an overlap of the valence and conduction bands.

tron-energy states would be grouped into bands (see Fig. 7.106). One can talk of the lower, or valence, band which results from the overlap of filled valence orbitals of the individual atoms and the upper, or conduction, band which results from overlap of partially filled or empty higher orbitals of the atoms involved. In such a material, mobile electrons can arise in two ways. Either the valence band containing electrons is only partially filled and thus gives rise to electron states to which electrons can migrate, or, even if this band (valence band) is completely filled, it can overlap an unfilled band (conduction band) where unoccupied energy states permit electron drift (Fig. 7.106). Since there are plenty of valence electrons (at least one valence electron per atom) and also plenty of vacant states, the concentration of mobile charge carriers is high and so will be the conductivity which depends upon this concentration. *The crystal* (e.g., copper) *will show metallic conduction.*

If, however, the equilibrium interatomic spacing in a certain solid is d_I (Fig. 7.107), it is noticed that there exists a large range of forbidden energies between the first and second bands. Now, what happens if there are just enough available valence electrons to fill the first band, the valence band (Fig. 7.107)? Then, these electrons will not be able to find any easily accessible vacant energy states in the valence band for them to move into. Further, if the energy gap E_g is large compared with the thermal energy kT of the electrons, the electrons cannot significantly be thermally excited into the conduction band. In effect, therefore, there will be no mobile electrons in either band. *The material* (e.g., diamond) *will behave like an insulator.*

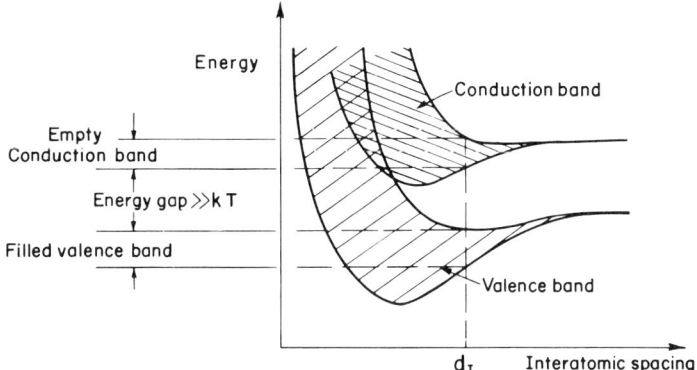

Fig. 7.107. The band picture of an insulator with an interatomic spacing of d_I, showing the filled valence band separated from the empty conduction band by a large energy gap.

A third and most interesting possibility is when the equilibrium interatomic spacing in the solid is d_{SC} (Fig. 7.108). It will be noticed that here, too, the essence of this situation is that there is an energy gap separating the valence band from the upper band. But this energy gap, in contrast to that in the case of insulators, is not much more than the thermal energy of the electrons and therefore small enough for electrons in the valence band (i.e., the electrons used for bonding atoms together) to be excited into the upper band (Table 7.14). The energy required for the excitation of electrons into the upper band may come from the thermal motions of electrons or from light shining on the material. Once they are in the upper

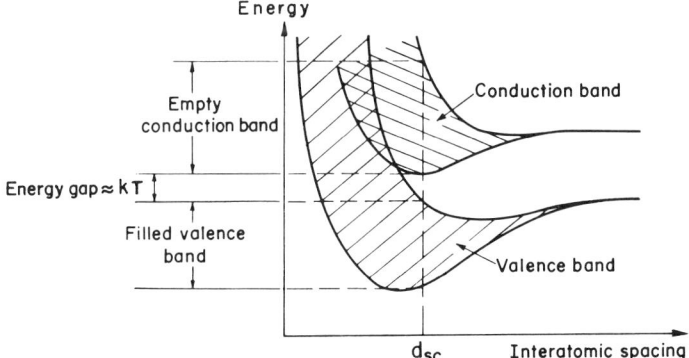

Fig. 7.108. The band picture of a semiconductor with an interatomic spacing of d_{SC}, showing the small energy gap between the valence and conduction bands.

TABLE 7.14
Room-Temperature Energy Gap for Some Materials

Substance	Energy gap, eV
Insulator	
C, diamond	5.6
Semiconductors	
Germanium	0.656
Silicon	1.089
Tellurium	0.34
GaAs	1.35
InSb	0.17
PbS	0.37

band, these electrons find plenty of unoccupied energy states into which they can move. Hence, the conduction-band electrons can conduct electricity. *This is how an intrinsic semiconductor* (e.g., silicon) *conducts electricity.*

When an electron is excited across the energy gap E_g to the conduction band in an intrinsic semiconductor, an unoccupied energy state, or *hole*, is left behind in the normally full valence band. This vacant state can be jumped into by another electron in the valence band (Fig. 7.109), but this leads to a vacant state in the place where the jumping electron was. The motion of electrons into unoccupied energy states, or holes, in the valence band is therefore equivalent to the movement of the vacant states, or holes, in the opposite direction. Since an electric field moves holes in an opposite direction to electrons, the holes may be treated as if there were positively charged (Fig. 7.110).

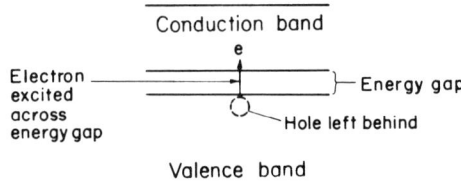

Fig. 7.109. The formation of a hole when an electron from the valence band is excited into the conduction band.

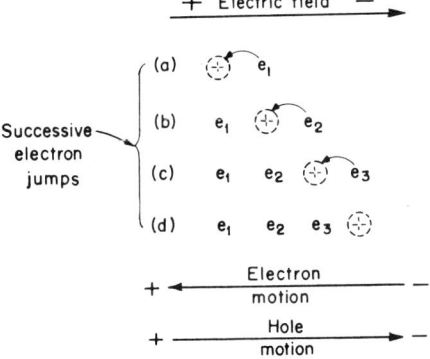

Fig. 7.110. Since hole motion is in a direction opposite to electron motion, a hole behaves as if it were positively charged.

7.7.4. Some Analogies between Semiconductors and Electrolytic Solutions

Since the electrons of the valence band are used for bonding together atoms, the removal of a valence electron by excitation into the conduction band implies the rupture of a bond in the lattice. The creation of an electron–hole pair may therefore be treated as an ionization reaction

$$\text{Lattice} \rightleftharpoons e + h$$

Viewed thus, it turns out that there are remarkable parallels between the ionization of the lattice of an intrinsic semiconductor and the ionization of water (Table 7.15)

$$\text{Water} \rightleftharpoons \text{OH}^- + \text{H}^+$$

Both equilibrium "reactions" can be treated by the law of mass action. Just as the product of the concentrations of hydroxyl ions and hydrogen ions remains a constant at a fixed temperature,

$$c_{\text{H}^+} \times c_{\text{OH}^-} = K_{\text{H}_2\text{O}} \tag{7.215}$$

the product of the electron n and hole p concentrations also remains a constant at a fixed temperature. The constant depends on the energy gap E_g across which the valence electrons must be excited into the conduction band

$$np = K_{sc} \tag{7.216}$$

Table 7.15
Some Analogies between Semiconductors and Electrolyte Solutions

This phenomenon	in aqueous solution	is paralleled by this phenomenon in a semiconductor
Dissociation of solvent	$H_2O \rightleftharpoons H^+ + OH^-$	Semiconductor lattice $\rightleftharpoons$ Electron + hole
Law of mass action	$(H^+)(OH^-) = K_w$ $n_{H^+} = n_{OH^-} \approx 10^{14}$ ions cm^{-3}	$np = K_{sc}$ $n = p \approx 10^{13} - 10^{16}$ particles cm^{-3}
Behavior of acid	$HCl \rightleftharpoons H^+ + Cl^-$ [proton donor]	$As \rightleftharpoons$ Electron + As^+ [electron donor]
Behavior of base	$NH_3 + H^+ \rightleftharpoons NH_4^+$ [proton acceptor]	$Ga + e \rightleftharpoons Ga^-$ [electron acceptor]
Common-ion effect	(a) Adding acid [proton donor] to water increases proton concentration	(a) Adding electron donor to intrinsic semiconductor increases electron concentration
	(b) Adding base [proton acceptor] to water decreases proton concentration and increases OH$^-$ concentration	(b) Adding electron acceptor to semiconductor decreases electron concentration and increases hole concentration

where K_{sc} is a constant characteristic of the intrinsic semiconductor. Further, just as, in pure water, the concentrations of hydroxyl and hydrogen ions are equal, the electron and hole concentrations in an intrinsic semiconductor are equal

$$n = p = K_{sc}^{\frac{1}{2}} \tag{7.217}$$

When one examines the value of $n = p$, it turns out that the density of charge carriers in an intrinsic semiconductor (Table 7.14) at room temperature is in the range of 10^{13} to 10^{16} cm^{-3}, compared with about 10^{22} cm^{-3} in a metal. It is this relatively low concentration of charge carriers in intrinsic semiconductors which is responsible for the most important differences between semiconductor electrodes and metal electrodes.

THE ELECTRIFIED INTERFACE   813

7.7.5. The Diffuse-Charge Region inside an Intrinsic Semiconductor: The Garrett–Brattain Space Charge

After this elementary account of the constitution of an intrinsic semiconductor, one can proceed to consider the basic question posed in Section 7.7.1: Given a layer of charge on the OHP of the electrolyte, how do the electrons and holes inside an intrinsic semiconductor distribute themselves as a function of distance from the interface?

Garrett and Brattain were the first to attack this problem by elaborating on its formal similarity to the problem of the diffuse charge in solution. Inside the electrolyte, the positive and negative ions are the charge carriers; inside the semiconductor, there are the holes and electrons. In the electrolyte bulk, the excess-charge density in any volume element is zero because the numbers per unit volume of positive and negative charges are exactly equal; similarly, deep inside the intrinsic semiconductor, the excess-charge density is zero because of the equality of the density of electrons n^0 and holes p^0. Thus, for the semiconductor bulk,

$$n^0 = p^0 \tag{7.218}$$

and

$$\varrho_{\text{bulk}} = e_0 p^0 - e_0 n^0 = 0 \tag{7.219}$$

The charged electrode exerts an electric field on the positive and negative ions in the electrolyte; similarly, the sheet of charge on the OHP exerts an electric field on the holes and electrons in the intrinsic semiconductor so that, relatively near the surface, electrons and holes are not present in equal numbers.

The charge density ϱ_x on any electrolyte lamina parallel to the electrode and a distance x from it can be obtained by the application of electrostatics (Poisson's equation) and the Boltzmann distribution. Similarly, one can write, for the intrinsic semiconductor, (1) Poisson's equation

$$\varrho_x = -\frac{\varepsilon}{4\pi} \frac{d^2 \psi_x}{dx^2} \tag{7.220}$$

and (2) the Boltzmann distribution

$$\begin{aligned}\varrho_x &= e_0(p_x - n_x) \\ &= e_0(p^0 e^{-e_0 \psi_x/kT} - n^0 e^{e_0 \psi_x/kT})\end{aligned} \tag{7.221}$$

where ψ_x is the Volta potential at a distance x from the electrode ($\psi_{x \to \infty}$

is taken as zero). Using the relation (7.218), one has

$$\varrho_x = -e_0 n_0 (e^{e_0 \psi_x / kT} - e^{-e_0 \psi_x / kT})$$
$$= -2e_0 n_0 \sinh e_0 \psi_x / kT \tag{7.222}$$

The two expressions for the charge density can be equated to give the Poisson–Boltzmann equation

$$\frac{d^2 \psi_x}{dx^2} = \frac{8\pi e_0 n^0}{\varepsilon} \sinh \frac{e_0 \psi_x}{kT} \tag{7.223}$$

This differential equation for the space variation of the potential inside the semiconductor can be easily identified with that [Eq. (7.87)] for the space variation of the potential inside the electrolyte in the Gouy–Chapman theory of the diffuse layer (Section 7.4.4). The solution can therefore be borrowed from the diffuse-layer theory. One has, from Eq. (7.100),

$$\frac{d\psi_x}{dx} = -\left(\frac{32\pi k T n^0}{\varepsilon}\right)^{\frac{1}{2}} \sinh \frac{e_0 \psi_x}{2kT} \tag{7.224}$$

and, from Eq. (7.102)

$$q_{sc} = 2 \left(\frac{\varepsilon n^0 k T}{2\pi}\right)^{\frac{1}{2}} \sinh \frac{e_0 \psi_s}{2kT} \tag{7.225}$$

where q_{sc} is total space charge, and ψ_s is the potential at the surface.

These results have very important consequences. By linearizing the hyperbolic sine function[†]

$$\sinh \frac{e_0 \psi_x}{2kT} \approx \frac{e_0 \psi_x}{2kT}$$

and solving the differential equation (cf. Section 7.4.5)

$$\frac{d\psi_x}{dx} = -\left(\frac{8\pi n^0 e_0^2}{\varepsilon kT}\right)^{\frac{1}{2}} \psi_x = -\varkappa \psi_x \tag{7.226}$$

one gets

$$\psi_x = \psi_s e^{-\varkappa x} \tag{7.227}$$

There is (Fig. 7.111) thus an exponential decay of potential due to the space charge inside the semiconductor.

[†] The system still gives an equation of the form of (7.227) even when the linearization condition is dropped.

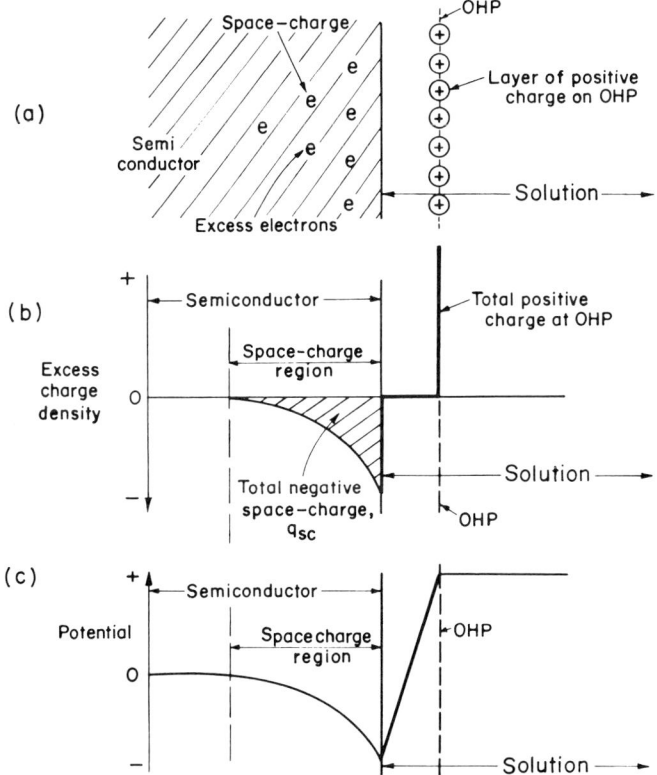

Fig. 7.111. (a) The space charge inside a semiconductor, (b) the corresponding charge-density variation, and (c) the potential variation.

This potential decay implies that there is a field inside the semiconductor and that the excess-charge density slowly decays to zero as if there were an electronic cloud analogous to the ionic cloud adjacent to an electrode in solution. It can be seen that the potential due to the atmosphere of holes and electrons is characterized by the same parameter

$$\varkappa = \left(\frac{8\pi n^0 e_0^2}{\varepsilon kT}\right)^{\frac{1}{2}}$$

as that which occurs in the Debye–Hückel ionic cloud and the Gouy–Chapman diffuse-charge treatments. The term $\varkappa^{-1}$ is the measure of the thickness of the Garrett–Brattain space charge inside a semiconductor. The value of $\varkappa^{-1}$ diminishes as the bulk concentration of charge carriers increases.

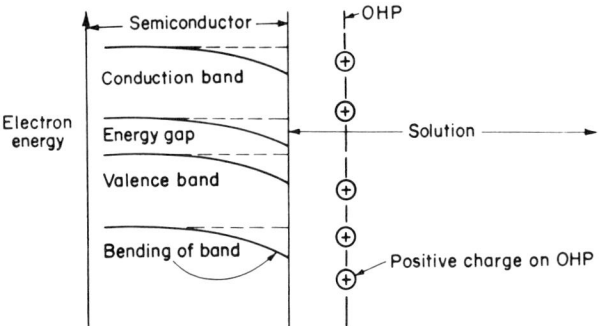

Fig. 7.112. The bending of bands near the surface of a semiconductor.

What this means is that, as the carrier concentration increases, the thickness of the space-charge region decreases. This is one way of looking at a metal; because of its high concentration of charge carriers, the space charge is all squeezed onto the surface. The situation here is analogous to the diffuse charge in solution, which gets compressed on the OHP when the electrolyte concentration is sufficiently high.

It has just been pointed out that, due to the existence of a layer of charge in the solution, there is a space charge and a potential drop inside the semiconductor. Any electron in this space-charge region will interact with the field, and its energy will either increase or decrease compared with the value in the absence of the field. The value of the electron energy in the absence of the field has been shown to be given by the band structure of the solid.

What this implies is that the energy bands near the surface of an intrinsic semiconductor are disturbed by the existence of a field. The electron energies are given by the sum of the energies due to the intrinsic-band structure and that due to the deviation of the inner potential from its zero value in the bulk. Thus, near the surface, there is a bending of the bands up or down depending upon the sign of the ionic charge populating the OHP (Fig. 7.112).

7.7.6. The Differential Capacity Due to the Space Charge

When capacity measurements are carried out on a semiconductor–electrolyte interface, one must not forget that the space-charge region inside the semiconductor has the ability to store charge. The contribution of this region to the differential capacity of the interface can easily be calculated. One simply differentiates the expression (7.225) for q_{sc} by the potential drop ψ_s inside the intrinsic semiconductor. Thus,

$$C_{sc} = \frac{dq_{sc}}{d\psi_s} = 2\left(\frac{\varepsilon n_0 kT}{2\pi}\right) \cosh \frac{e_0 \psi_x}{kT} \quad (7.228)$$

One expects that the differential capacity of a semiconductor–electrolyte interface due to the space charge inside an intrinsic semiconductor varies in a hyperbolic cosine manner with the potential. Such a variation is shown in Fig. 7.113.

It will be seen that the values of the space-charge capacities are low ($\sim$0.01–1 μF cm^{-2}) compared with the capacities ($\sim$17 μF cm^{-2}) of the region between the semiconductor surface and the OHP plane, the Helmholtz–Perrin parallel-plate region. That is why the space-charge capacities (the inverted parabolas) are noticed, for the observed capacity is given by two capacitors in series, the space-charge C_{SC} and Helmholtz–Perrin C_{HP} capacitors. Thus,

$$\frac{1}{C_{obs}} = \frac{1}{C_{SC}} + \frac{1}{C_{HP}} \quad (7.229)$$

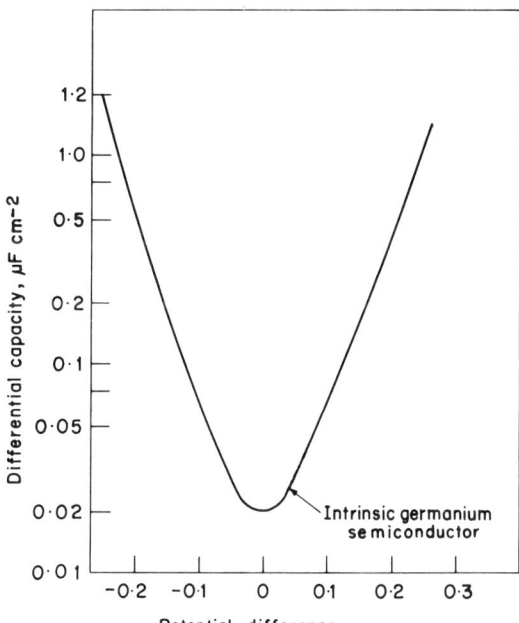

Fig. 7.113. The differential capacity of a semiconductor–electrolyte interface has a cosh dependence on the potential.

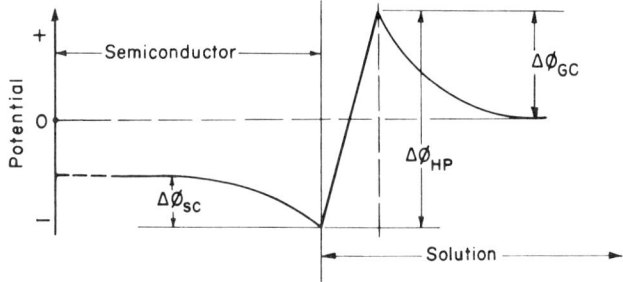

Fig. 7.114. The potential variation at a semiconductor–electrolyte interface.

but $C_{HP} \gg C_{SC}$ and, hence,

$$\frac{1}{C_{obs}} \approx \frac{1}{C_{SC}} \quad (7.229a)$$

When the electrolyte is dilute, then a diffuse-charge region will appear in the solution, too. There will be three potential drops (Fig. 7.114): one inside the semiconductor, the Garrett–Brattain drop $\Delta\phi_{SC}$; a linear Helmholtz–Perrin drop $\Delta\phi_{HP}$; and, finally, the Gouy–Chapman drop $\Delta\phi_{GC}$ in the solution. The total $\Delta\phi$ across the interface is given by

$$\Delta\phi = \Delta\phi_{SC} + \Delta\phi_{HP} + \Delta\phi_{GC} \quad (7.230)$$

and the total differential capacity, by

$$\frac{1}{C} = \frac{1}{C_{SC}} + \frac{1}{C_{HP}} + \frac{1}{C_{GC}} \quad (7.231)$$

Thus, there are three capacitors in series at a semiconductor–electrolyte interface rather than two capacitors as at a metal–solution interface. What is observed, predominantly, depends upon the electron concentration in the semiconductor (how *low* it is) and the ionic concentration in solution.

7.7.7. Impurity Semiconductors, *n* Type and *p* Type

The discussion has been restricted so far to pure intrinsic semiconductors exemplified by germanium and silicon. In these substances, there is a low concentration of charge carriers (compared with metals). Further, the hole and electron concentrations are equal, and their product is a constant given by the law of mass action

$$np = K_{sc} \quad (7.216)$$

Intrinsic semiconductors have been compared to pure water in which the OH^- and H^+ concentrations are equal and the product of these concentrations is a constant K_W

$$C_{OH^-}C_{H^+} = K_W$$

In the case of water, however, the concentration of either the OH^- or H^+ ions can be decreased by adding proton donors (acids) or proton acceptors (bases). Thus, a proton donor releases hydrogen ions into the solution (i.e., C_{H^+} increases), and the only way the product K_W (i.e., $C_{OH^-}C_{H^+}$) remains a constant is by a decrease in C_{OH^-}.

Is there an analogous situation in semiconductors? If one adds an electron donor (say, arsenic) to an intrinsic semiconductor (say, germanium), then the ionization of arsenic

$$As \rightleftharpoons As^+ + e \qquad (7.232)$$

releases electrons into the system, and the hole concentration goes down to preserve the constancy of the product np [cf. Eq. (7.216)]. In this way, the electron concentration can be made so large compared with the hole concentration that the conduction is dominantly by electrons; the substance is known as an n type of semiconductor.

There is also a parallel in semiconductors to the effect of adding a base, or proton acceptor, to water. This involves the addition of an electron acceptor (say, gallium) to an intrinsic semiconductor. The electron acceptors ionize thus

$$Ga + e \rightleftharpoons Ga^- \qquad (7.233)$$

and, by accepting electrons, force up the hole concentration in the valence band. Such a "doped" semiconductor will conduct mainly by holes; it is known as a p type of semiconductor.

What the addition of electron acceptors and donors means in the band picture can be easily understood from Figs. 7.115 and 7.116. The electron

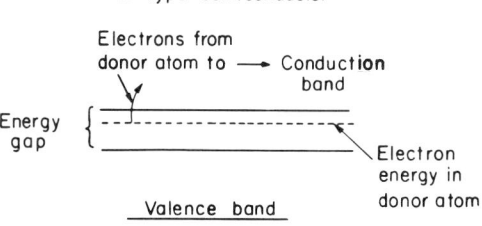

Fig. 7.115. The band picture of n-type semiconductors.

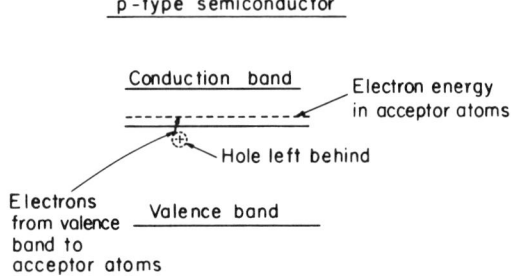

Fig. 7.116. The band picture of p-type semiconductors.

acceptors and donors enter the lattice of the semiconductor and introduce electron-energy levels in between the valence and conduction bands. Thus, with an n type of semiconductor (Fig. 7.115), only a small part of the electrons in the conduction band arise by thermal excitation from the valence band; the rest come by the ionization of electron donors. The hole concentration, however, depends only upon the number of valence electrons which are excited into the conduction band. The hole concentration can therefore be made small.

The band-picture explanation of the p type of semiconductors (Fig. 7.116) is in terms of electron-acceptor levels' appearing in the energy gap near the top of the valence band. The electron-acceptor atoms can then easily receive electrons from the valence band. Thus, holes are created in the valence band without the corresponding electrons in the conduction band. The conduction, therefore, is essentially by holes—a p type of semiconductor.

The conclusion from this description of n and p types of semiconductors is that the concentration of holes and electrons is not only low, but it can be varied by doping the material with varying amounts of electron acceptors or donors.

Apart from this variability in the choice of electron and hole concentrations, the treatment of the space charge in n and p types of semiconductor is basically the same as that of intrinsic semiconductors. One important difference, however, is that, even though the excess-charge density is zero in the bulk of the impurity semiconductor, this is not simply because the electron and hole concentrations are equal. Rather, the electroneutrality condition must be written as

$$n^0 + n_A = p^0 + n_D \qquad (7.234)$$

where n^0 and p^0 are the bulk concentrations of electrons and holes, and

n_A and n_D are the numbers per unit volume of electron acceptors and electron donors which are added to the intrinsic semiconductor. In other words, in the bulk of the impurity semiconductor, the total negative-charge density is equal to the total positive-charge density, taking into account the added donor or acceptor ions.

Another point to remember in the treatment of the diffuse charge inside a doped, or impurity, semiconductor is that, though the electron acceptors and donors affect the electroneutrality, they should be considered immobile and fixed in the lattice. Thus, their concentration remains uniform inside the doped semiconductor in contrast to that of the electron and hole concentrations, which are decided by the interplay of electrical and thermal forces.

These differences between intrinsic and doped, or impurity, semiconductors complicate the mathematics of the solution of the Poisson–Boltzmann equation, but the picture that emerges remains basically the same, a charged cloud, or space charge, and, therefore, a potential drop develops inside the semiconductor; the space charge contributes to the capacity of the interphase, etc.

7.7.8. Surface States: The Semiconductor Analogue of Contact Adsorption

In the simple theory of the space charge inside a semiconductor, it was assumed that all the electrons and holes are free to move up to the surface. Being susceptible to thermal motion, their concentrations from $x = 0$ to $x \to \infty$ were said to be given by the interplay of electrical and thermal forces only, as expressed by the Boltzmann distribution law and Poisson's equation.

What happens, however, if electrons become bound in such a way that they cannot move in a direction normal to the interface? Then, the simple theory of the space charge will have to be modified. There is charge trapped in the surface energy states (i.e., those energy levels for electrons or holes which are different from those present in the bulk and which are localized at the surface of the semiconductor). The trapped charge will have to be excluded from a space-charge analysis, in which the only charges considered were those which could distribute themselves freely under thermal and electric fields.

Surface states force a change in the picture of the double layer inside the semiconductor in the same way that contact-adsorbed ions alter the simple Gouy–Chapman picture of the diffuse charge in solution (Fig. 7.117).

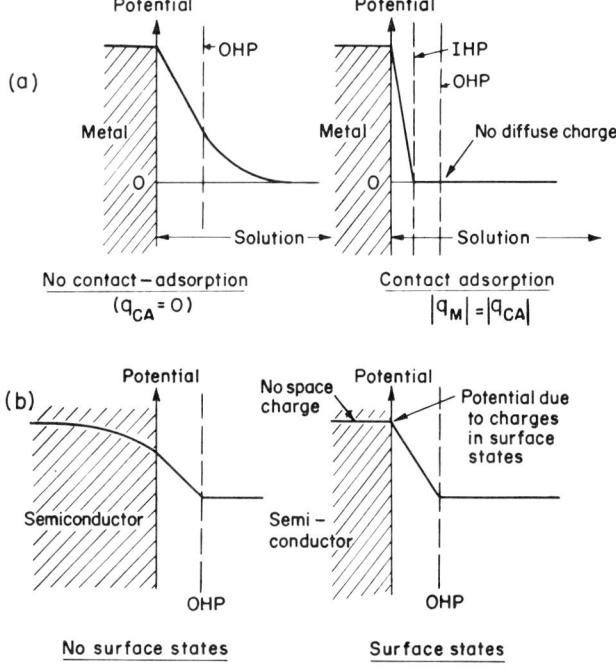

Fig. 7.117. The analogy between the role of surface states in the semiconductor and contact-adsorbed ions in the solution.

The presence of contact-adsorbed ions at a metal–solution interface means that the total Gouy–Chapman diffuse charge in solution and the potential drop across the diffuse-charge region is reduced. Similarly, the existence of surface states means that charge and the potential drop across the Garrett–Brattain space-charge region is reduced. At a high enough density of surface states, it can be assumed that there is hardly any space charge and hardly any potential drop inside the semiconductor. The semiconductor is in fact behaving like a metal—most of its charge is on the surface. This is analogous to the situation in metal–solution interfaces where the magnitude of the charge due to contact-adsorbed ions is almost equal to that on the metal. Under these circumstances, there is hardly any Gouy–Chapman diffuse charge.

But what is the atomistic nature of surface states? A complete answer to this question is not yet available, but there are strong evidences that one source of surface states is the adsorption of atoms on the semiconductor surface. Thus, adsorbed hydrogen atoms on a germanium surface apparently behave as surface states. The space charge is then reduced so drastically that the germanium–solution interface behaves like a metal–solution interface, e.g., the capacities come into the approximate 10 μF cm^{-2} range.

The study of surface states, therefore, is vital to the understanding of the semiconductor–electrolyte interface.

7.7.9. Semiconductor Electrochemistry: The Beginnings of the Electrochemistry of Nonmetallic Materials

Treatments of the semiconductor–electrolyte interface generally tend to revolve around the element semiconductors germanium and silicon. Why? For the same reason that double-layer treatments are attracted to the mercury–solution interface, namely, more is known about this metal–solution interface than any other interface. Similarly, much of the electrochemical work with semiconductors has been with the germanium and silicon electrolyte interfaces. This germanium and silicon bias has resulted mainly from the central role these elements play in semiconductor devices.

Now, there are many hundreds of other semiconducting materials, including oxide and organic semiconductors. Here lies the importance of understanding semiconductor electrochemistry. It is a way of approaching the electrochemistry of the nonmetal–solution interface and charge-transfer reactions which go on across this interface. Already, rationalizations involving oxide films and biological materials are in terms of concepts of electrochemical theories involving semiconductor electrochemistry. For example, the stability and decay of some relatively insulating refractory materials in contact with molten salts may be so discussed. The future may see many informative developments in this direction.

Further Reading

1. W. Brattain and G. Garrett, *Ann. N. Y. Acad. Sci.*, **58**: 951 (1954); and *Bell System Tech. J.*, **34**: 129 (1955).
2. M. Green, "Electrochemistry of the Semiconductor–Electrolyte Interface," in: J. O'M. Bockris, ed., *Modern Aspects of Electrochemistry*, Vol. II, Butterworths, Publications, Ltd., London, 1959.
3. H. Gerischer, "Semiconductor Electrode Reactions," in: P. Delahay, ed., *Recent Advances in Electrochemistry*, Interscience Publishers, Inc., New York, 1961.
4. P. J. Holmes, ed., *The Electrochemistry of Semiconductors*, Academic Press, Inc., New York, 1961.
5. J. F. Dewald, "Semiconductor Electrodes," in: N. B. Hannay, ed., *Semiconductors*, Reinhold Publishing Corp., New York, 1964.

6. P. J. Boddy, *J. Electroanal. Chem.* **10**: 199 (1965).
7. W. Mehl and J. M. Hale, "Insulator Electrode Reactions," in: P. Delahay and C. Tobias, eds., *Advances in Electrochemistry*, Vol. VI, pp. 399–456, Interscience Publishers, Inc., New York, 1967.

7.8. A BIRD'S-EYE VIEW OF THE STRUCTURE OF CHARGED INTERFACES

When two phases meet to form an interface, the forces in the interphase region vary with distance in a direction normal to the interface. They are anisotropic. If the phases contain charged species, these anisotropic forces at the interphase upset the normal situation of zero excess-charge density on lamellae parallel to the interface. The result of this perturbation of the distribution of charges is that the two sides of the interfaces acquire *excess*-charge densities of opposite sign. But, when there is charge separation, there is always a potential difference between the two regions of opposite charge. Thus, a potential difference develops across the interphase region.

If the solid phase is a metal electrode, its excess-charge density is concentrated at the surface and there is no potential drop extending deep inside the metal. This metal charge q_M is balanced out by a charge q_S in the solution, the other side of the interface. The metal charge also orients water dipoles at the electrode surface. Thus, the potential drops from a value in the bulk of the metal to a value in the bulk of the solution. The total potential difference $\Delta\phi$ across the interface can therefore be considered made up of two contributions, (1) $\Delta\psi$, due to the separation of charges, and (2) $\Delta\chi$, the potential difference across the oriented-dipole layer (and, perhaps, a layer of charge just inside the surface of the metal).

The water dipoles on the metal can orient in either of two ways. They can flip up with hydrogens looking into the solution; they can flop down with hydrogens facing the electrode. The ratio of flipped $N_\uparrow$ to flopped $N_\downarrow$ dipoles is determined mainly by the field at the metal surface or, what is equivalent, by the charge on the metal. At the pzc ($q_M = 0$), there would be equal numbers of flipped and flopped dipoles ($N_\uparrow = N_\downarrow$) but for the fact that the water molecule is not quite a symmetrical dipole, and, therefore, the probability of a water molecule's "landing on its oxygen feet" is slightly more (owing to dispersion interactions) than that for hydrogen-down water dipoles. This asymmetric effect shifts the conditions of minimum binding energy of water (which is the condition of $N_\uparrow = N_\downarrow$) to slightly negative values of electrode charge in the case of mercury electrodes.

The flip-flop water-molecule model is of crucial importance in two ways. Firstly, the excess number ($N = N_\uparrow - N_\downarrow$) of dipoles in one orientation state determines the dipole potential $\Delta\chi$ which should play a part in the rate at which charges cross the interface. Secondly, when $N = 0 = N_\uparrow - N_\downarrow$, water is held least strongly to the electrode and can be most easily displaced by organic molecules adsorbing on an electrode; this is the water-competition model of organic adsorption. For the adsorption of uncharged species on the electrode, this model provides a good rationalization of the parabolic shape of the coverage–potential curve.

The excess charge on the metal not only determines the $\Delta\chi$ potential and the extent of organic adsorption; it also governs, along with the solution composition, the arrangement of the excess charges in the solution.

There are two types of interphase structure in *concentrated* solutions. Firstly, there can be a layer of hydrated ions touching a charged hydrated electrode; this is strictly a *double* layer to which a simple parallel-plate condenser model can be applied. Secondly, some ions with little or no primary hydration can displace the primary water molecules next to the electrode and populate the IHP by contact adsorption; this is a *triple* layer. There are thus layers of charge on the metal and on the IHP and OHP, but the sum of the excess-charge densities in the two solution layers is equal in magnitude to the excess charge on the metal.

The triple-layer model can be mathematically treated by taking into account the lateral repulsion of the contact-adsorbed ions occupying the IHP. The model is able to account satisfactorily for much of the complicated fine structure of the experimental differential capacity *versus* potential curve obtained from solutions containing contact-adsorbable ions.

In addition to these two or three *layers* of charge, there can also be a *region* of diffuse charge in the solution. This region spreads out more and more as the solution is diluted. It constitutes the ionic cloud of the electrode, very similar in nature to the atmosphere of ionic charge round an ion in solution (Chapter 3).

Essential for the existence of a region of diffuse charge is a low concentration of charged particles (ions) in solution. The same argument applies to charge carriers inside the solid electrode. In the case of a metal, the high concentration of charge carriers (electrons) does not permit the existence of a diffuse-charge region—all the charge q_M is within about an atom diameter of the metal surface. In semiconductors, however, the concentration of charge carriers (electrons and holes) is far, far lower than for a metal. Hence, there can be a diffuse or space-charge region and a potential drop inside the semiconductor at a semiconductor–electrolyte interface.

The possibility of a space charge within the solid electrode, on the one hand, distinguishes semiconductor electrodes from metal electrodes but, on the other hand, makes the space charge inside the semiconductor analogous to the diffuse charge in the solution. The similarity between the situations inside the semiconductor and inside the solution is further exemplified by the fact that, corresponding to the phenomenon of contact adsorption of ions from solution, there can be "contact adsorption" of electrons onto surface states on the semiconductor surface.

7.9. DOUBLE LAYERS BETWEEN PHASES MOVING RELATIVE TO EACH OTHER

7.9.1. The Phenomenology of Mobile Electrified Interfaces: Electrokinetic Properties

The double layer has hitherto been considered still, or static, in the sense that the bulk phases which meet at the interface are at rest relative to each other. When, however, one of the phases moves relative to the other, interesting electrical phenomena arise—called *electrokinetic phenomena*. These phenomena will first be described in a gross, nonmechanistic way, and then it will be shown that the underlying mechanisms are based on mobile electrified interfaces.

Consider that a potential difference is applied across a glass capillary tube filled with an electrolytic solution (Fig. 7.118). What would one expect? Of course, one would expect a current to flow through the capillary according to Ohm's law. In practice, however, a remarkable and unexpected phenomenon is observed. In addition to the current, the solution itself begins to flow—the phenomenon of electro-osmosis. This is strange, because liquid flow is generally associated with the application of a pressure gradient. It appears, therefore, that a potential difference is doing the job normally achieved by a pressure difference.

This phenomenon of electro-osmosis can be treated in mathematical form. The fact is that the velocity of flow of electrolyte, V, depends not only

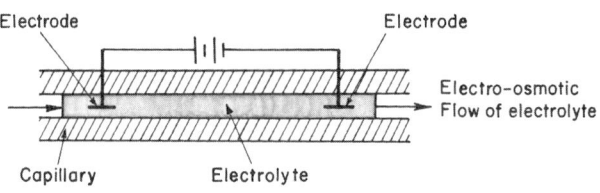

Fig. 7.118. The phenomenon of electro-osmosis.

on its usual driving force, i.e., a pressure gradient ΔP but also on the electric field X. When the driving forces are small, it can always be assumed that there are linear relations between driving forces and the resulting flows (cf. Section 4.5.6). Hence,

$$v = a_1 \Delta P + a_2 X \qquad (7.235)$$

Even when the usual driving force ΔP is absent ($\Delta P = 0$), one still has

$$v = a_2 X$$

or

$$\frac{v}{X} = a_2 \qquad (7.236)$$

Thus, the coefficient a_2 describes the electro-osmotic flow velocity per unit of potential gradient, i.e., the electro-osmotic mobility.

An obvious idea arises now. If an electric field X can achieve what a pressure difference ΔP normally does, namely, produce a liquid flow, then perhaps a pressure difference will produce an electric current which is normally the result of an electric field. Experiment (Fig. 7.119) once again yields an interesting answer; an electric current known as a *streaming current* is in fact produced by a pressure difference.

One can transcribe the phenomenon in the form of an equation following the same thinking as for electro-osmosis. One says: A current density i results not only from an electric field but also from a pressure difference ΔP, and, for small X and ΔP,

$$i = a_3 \Delta P + a_4 X \qquad (7.237)$$

Even when the usual driving force X is absent ($X = 0$), one still has

$$i = a_3 \Delta P \quad \text{or} \quad \frac{i}{\Delta P} = a_3 \qquad (7.238)$$

The streaming-current constant a_3 is the streaming-current density produced by a unit pressure difference.

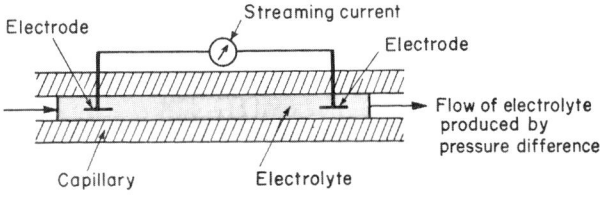

Fig. 7.119. The phenomenon of streaming current.

If both sides of Eq. (7.238) are divided by the specific conductivity σ of the electrolyte, it is clear that

$$X = \frac{i}{\sigma} = \frac{a_3}{\sigma} \Delta P \tag{7.239}$$

In words, the application of a pressure difference in an electrolyte should produce a potential difference and a corresponding electric field. This is the phenomenon of *streaming potential.*

What has been done so far is to take experimental laws and express them in the form of phenomenological equations.

$$v = a_1 \Delta P + a_2 X \tag{7.235}$$

$$i = a_3 \Delta P + a_4 X \tag{7.237}$$

Just as the phenomenological equations describing the equilibrium properties of material systems constitute the subject matter of equilibrium thermodynamics, the above phenomenological equations describing the flow properties fall within the purview of nonequilibrium thermodynamics. In this latter subject, the Onsager reciprocity relation occupies a fundamental place (*cf.* Section 4.5.6).

Now, the Onsager reciprocity relation when applied to the present context predicts that the *cross coefficients* a_2 and a_3, which determine the rate of flow of liquid due to the applied electric field and the current passing due to a hydrostatic pressure difference, respectively, are equal, i.e.,

$$\left(\frac{i}{\Delta P}\right)_{X=0} = a_3 = a_2 = \left(\frac{v}{X}\right)_{P=0} \tag{7.240}$$

In words, the prediction is that the current per unit of pressure gradient should be equal to the fluid flow velocity per unit of electric field. Experiments prove that this is indeed the case.

As long as one remains within the framework of thermodynamics (whether the equilibrium or nonequilibrium variety), one has always to appeal to experiment for the values of coefficients. To calculate them, one must leave phenomenology and turn to models so that the atomistic mechanisms underlying the phenomenological laws are revealed. This will be done now, and, interestingly enough, it will be seen that the electrokinetic phenomena—electro-osmosis, streaming current, and streaming potential—depend on the electrification of the interface between the two phases.

7.9.2. The Relative Motion of One of the Phases Constituting an Electrified Interface Produces a Streaming Current

To give an atomistic interpretation of electrokinetic phenomena, one must consider questions such as: What happens when one of the phases moves relative to the other? For example, what happens when the electrolyte is made to flow past an electrode at rest?

Consider a plane electrode in an electrolytic solution dilute enough for there to be a "thick" diffuse layer, i.e., one which is hundreds or thousands of angstroms thick. It has been shown (Section 7.4.5) that the thickness of the diffuse-charge region is given by $\varkappa^{-1}$, where $\varkappa$ is the Debye–Hückel parameter. This diffuse layer can be considered equivalent to the Gouy charge density q_d placed at a distance $\varkappa^{-1}$ from the electrode.

Suppose now that a pressure difference is applied on the electrolytic solution in a direction parallel to the electrode. The electrolyte will begin to flow. This flow will be opposed by the viscous force, which is given by the viscosity times the velocity gradient. When the liquid attains a steady velocity (i.e., zero acceleration), the pressure difference is equal to the viscous force

$$\Delta P = \eta \frac{dv}{dx} \qquad (7.241)$$

When a viscous fluid flows past a solid, the velocity of any volume element of the fluid depends on its distance x from the solid surface (Fig. 7.120). At $x = 0$, i.e., at the solid surface, the velocity is zero because the solid exerts forces on the fluid particles and does not allow them to slip past. This hydrodynamic fact has been frequently expressed in another way, namely, the charges on the IHP and OHP plane have been considered fixed and immobile. As one goes away from the OHP, the fluid velocity increases (it will be assumed that this increase is linear with distance).

Let the fluid velocity be v at the distance $\varkappa^{-1}$. Then the velocity gradient can be written as

$$\frac{dv}{dx} = \frac{v}{\varkappa^{-1}} \qquad (7.242)$$

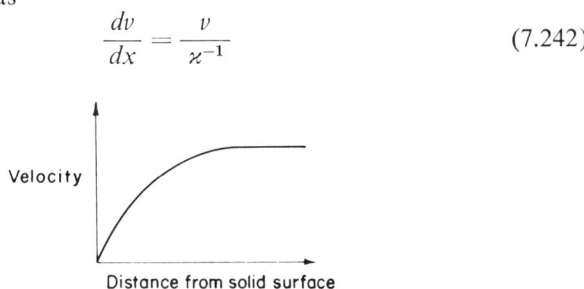

Fig. 7.120. The variation of fluid velocity with distance from the solid surface.

Now the situation is similar to that shown in Fig. 7.66 (*cf.* Section 7.4.5), where the diffuse charge q_d is assumed to be situated at a distance $\varkappa^{-1}$ from the OHP. This gives rise to a parallel-plate condenser model. The potential at one plate (in the solution side) is taken as zero, while the potential at the other plate (which coincides with the OHP) is ψ_0. This latter potential is often referred to in the study of electrokinetic phenomena as the zeta (ζ) potential. It is given, in the framework of the present simplified model, by the application of the expression for the potential difference, ψ_0, as a function of distance $(1/\varkappa)$ inside a parallel-plate condenser. Thus,

$$\psi_0 = \frac{4\pi q_d \varkappa^{-1}}{\varepsilon} \tag{7.243}$$

or

$$\frac{1}{\varkappa^{-1}} = \frac{4\pi q_d}{\psi_0 \varepsilon} \tag{7.244}$$

This expression (7.244) for $1/\varkappa^{-1}$ can be inserted into Eq. (7.242), and the result, into Eq. (7.241) to obtain

$$\Delta P = \left(\frac{4\pi \eta}{\psi_0 \varepsilon}\right) q_d v \tag{7.245}$$

According to the model (Fig. 7.121), a charge of q_d coulombs is in a unit area (1 cm²) of a plane parallel to the immobile, solid phase. When this charge moves through a transit plane of unit area (1 cm²) normal to the solid phase, it means that a charge of q_d coulombs cm⁻³ moves with a velocity of v cm sec⁻¹ and gives rise to a current density $i = q_d v$. Hence, substituting this equation into Eq. (7.245), and transforming,

$$\frac{i}{\Delta P} = \frac{\psi_0 \varepsilon}{4\pi \eta} = a_3 \tag{7.246}$$

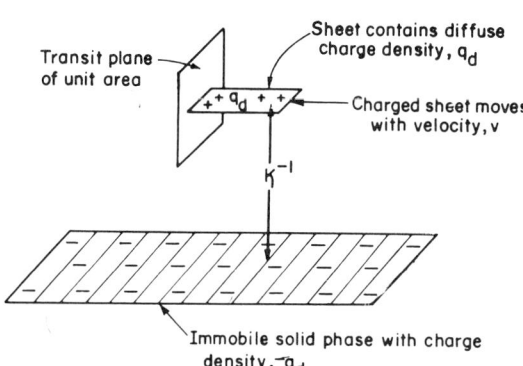

Fig. 7.121. A schematic diagram for the computation of the streaming-current density.

Thus, one has obtained an atomistic picture for the streaming-current density, i.e., the current density produced per unit of pressure gradient.

If one divides both sides by the specific conductivity of the electrolyte, one gets an atomistic expression for the streaming potential [*cf.* Eq. (7.238)],

$$\frac{X}{\Delta P} = \frac{\psi_0 \varepsilon}{4\pi\eta\sigma} = \frac{a_3}{\sigma} \qquad (7.247)$$

where X is the electric field, or the gradient of the streaming potential in the direction of movement of the liquid, caused by the pressure difference ΔP.

The phenomenon turns out to depend on the fact that there *is* a diffuse-charge region near the electrode. If there is no diffuse charge, i.e., if all the diffuse charge is "fixed" on the OHP (as in concentrated solutions), there are no excess charges in any volume element in the solution and the phenomenon of a streaming current and streaming potential disappears. Any factor (e.g., contact adsorption of ions or concentration of electrolyte) which affects ψ_0 also affects the streaming current.

7.9.3. A Potential Difference Applied Parallel to an Electrified Interface Produces an Electro-osmotic Motion of One of the Phases Relative to the Other

Instead of applying a pressure difference, one can apply a potential difference to the electrolyte in a direction parallel to the interface (Fig. 7.122). Once again a layer of charge q_d at a distance $\varkappa^{-1}$ from the solid, immobile phase will be assumed to *represent* the diffuse-charge region of the interface.

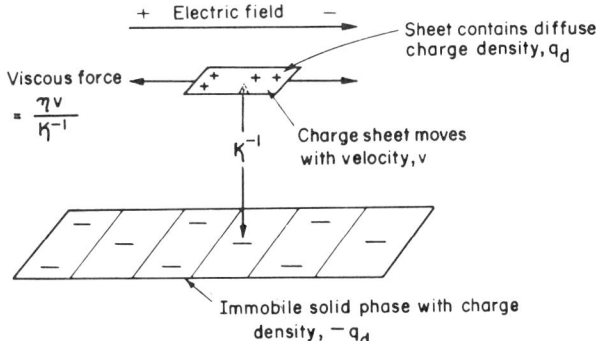

Fig. 7.122. A schematic diagram for the computation of the electro-osmotic mobility.

This layer of diffuse charge q_d will experience an electric force $q_d X$ (where X is the electric field). The charged layer on the solution side will begin to move. But the motion of the charged fluid is opposed by a viscous force which is once again given [cf. Eq. (7.245)] by $\eta(v/\varkappa^{-1})$ or $(4\pi\eta/\psi_0\varepsilon)q_d v$. When the electrolyte attains a steady velocity, the electric and viscous forces are exactly equal. Hence,

$$Xq_d = \frac{4\pi\eta}{\psi_0\varepsilon} q_d v \qquad (7.248)$$

or [from (7.240)],

$$\frac{v}{X} = \frac{\psi_0\varepsilon}{4\pi\eta} = a_2 \qquad (7.249)$$

But v/X is the electro-osmotic velocity of the fluid per unit of electric field, i.e., the electro-osmotic mobility. It is interesting to note that both the electro-osmotic mobility $v/X = a_2$ and the streaming-current coefficient $i/\Delta P = a_3$ have been proved to be equal to each other and to $\psi_0\varepsilon/4\pi\eta$. This only means that the Onsager reciprocity relation has been shown to be consistent with a simple model of some electrokinetic phenomena.

7.9.4. Electrophoresis: Moving Solid Particles in a Stationary Electrolyte

Electrokinetic phenomena depend on the relative motion of the phases constituting the double layer. In the treatment of electro-osmotic mobility, the electrolyte was considered to move within a stationary capillary—a moving cylinder of liquids within a static cylinder of solid. But the arguments only need *relative* motion; the arguments would be equally valid if one considered a moving cylindrical solid within a stationary liquid.

The solid phase must be large enough to have at its interface a double layer. Then, equating the viscous force created by the movement of the particle to the electrical force due to the interaction of the field with the charge on the particle, one obtains

$$\frac{v}{X} = \frac{\psi_0\varepsilon}{4\pi\eta} \qquad (7.249)$$

This equation has been derived above for the movement of a liquid through a stationary solid phase. Its application here to the movement of colloidal particles under experimental conditions which render the liquid medium immobile implies that the solid particle is large compared with the dimensions of the diffuse double layer $\varkappa^{-1}$. It is customary to term this movement of the solid phase *electrophoresis*. The phenomenon is observed with particles suspended in a liquid (Fig. 7.123).

The charge q_d in the diffuse double layer around a colloidal particle moving in the electric field has the same problem as the ionic atmosphere around a moving ion (*cf.* Section 4.6.4). On the one hand, it tries to move with the particle to which it is attached. On the other hand, it is influenced by the electric field which pulls it in the opposite direction. The particle apparently wins. It moves through the liquid with a diffuse double layer surrounding it, although it is not actually carrying the oppositely charged ionic atmosphere with it but rather leaving part of it behind and rebuilding it in front as it moves along. The tendency of the ions in the diffuse double layer to move in the direction opposite to the movement of the particles has its effect on the particle. It produces a "drag" which slows down the movement of the particle much as the ionic cloud slows down the movement of ions and affects their mobility (*cf.* Section 4.6.4).

Equation (7.249) indicates that the electrophoretic mobility is independent of the shape of the particles. Suppose, however, that the particle is spherical. Then one could arrive at the electrophoretic mobility in a completely different way and in a manner used to calculate the electrophoretic effect in conduction (*cf.* Section 4.6.4). One starts with Stokes' law (*cf.* Section 4.4.8)

$$\text{Viscous force} = 6\pi\varkappa^{-1}\eta v \qquad (7.250)$$

and say that (Fig. 7.123) the moving entity has a radius $\varkappa^{-1}$ (i.e., a particle of radius a contained in its thick ionic cloud of radius $\varkappa^{-1}$). The electric force on the particle plus cloud is given by the field times the charge on the cloud, which is q_d, i.e.,

$$\text{Electric force} = Xq_d$$

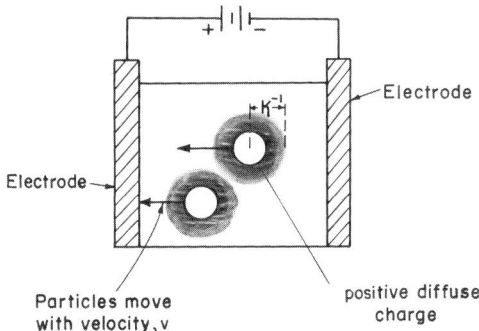

Fig. 7.123. The phenomenon of electrophoresis.

Now, q_d can be related to the potential ψ_0 (at the surface of the particle) in a manner characteristic of the decay of potential from a charged sphere, i.e.,

$$\psi_0 = \frac{q_d}{\varepsilon \varkappa^{-1}} \tag{7.251}$$

(This assumes that the radius a of the sphere $\ll \varkappa^{-1}$.) Hence, the electric force is $X\psi_0\varepsilon\varkappa^{-1}$.

When the particle attains a steady-state velocity, the electric and viscous forces are exactly equal and, hence, by utilizing Stokes' law with $\varkappa^{-1}$ as the radius,

$$\frac{v}{X} = \frac{\psi_0 \varepsilon}{6\pi\eta} \tag{7.252a}$$

If one compares Eqs. (7.252a) and (7.249), everything is fine, except that this Stokes' law approach gives a numerical factor $f = \frac{1}{6}$, whereas the electroosmotic approach gives $f = \frac{1}{4}$. It turns out that each is right for a particular set of conditions. This conclusion comes out of an accurate mathematical treatment which results in the following expression for the

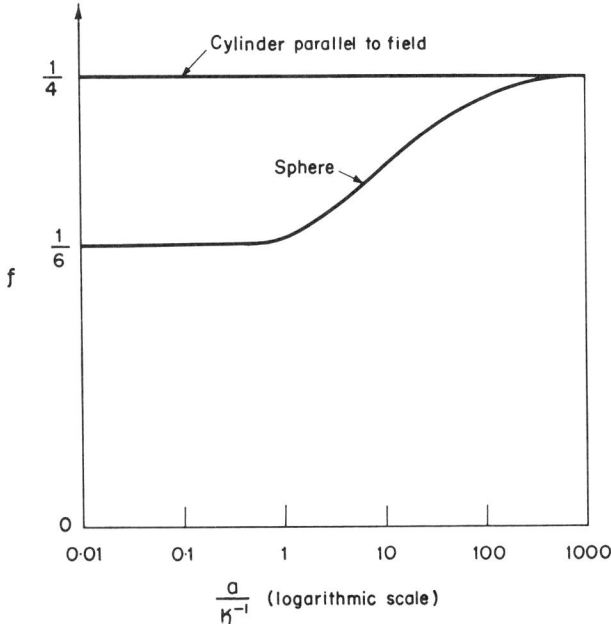

Fig. 7.124. The variation of the numerical factor f with the ratio of the radius a of the particle to the effective thickness $\varkappa^{-1}$ of the diffuse-charge region.

electrophoretic velocity

$$\frac{v}{X} = f \frac{\varepsilon \psi_0}{\pi \eta} \qquad (7.252b)$$

The quantity f is a numerical factor which depends on the ratio $a/\varkappa^{-1}$, where a is the radius of the spherical or cylindrical particle. In other words, f depends on the ratio of the radius of the particle to the effective thickness $\varkappa^{-1}$ of the diffuse layer. When $a/\varkappa^{-1}$ is large (the particle large in comparison with the diffuse-charge thickness), the numerical factor is always equal to $\frac{1}{4}$ irrespective of the shape of the particle. But, when the particle is small compared with the thickness of the double layer, f is $\frac{1}{4}$ for cylindrical particles parallel to the field and $\frac{1}{6}$ for spherical particles (Fig. 7.124).

Further Reading

1. H. L. von Helmholtz, *Ann. Physik*, **7**: 337 (1879); the English translation is published as Engineering Research Bull. No. 33 (1951) of the University of Michigan.
2. F. Booth, *Nature* **161**: 83 (1948).
3. H. R. Kruyt, ed., *Colloid Science*, Vols. I and II, Elsevier Publishing Co., New York, 1952.
4. J. J. Davies and E. K. Rideal, *Electrokinetic Phenomena in Interfacial Phenomena*, Academic Press, Inc., New York, 1963.
5. P. Sennett and J. P. Olivier, in: S. Ross, ed., *Chemistry and Physics of Interfaces*, p. 75, A.C.S. Publications, Washington, D.C., 1964–1965.
6. P. N. Sawyer, *Biophys. J.*, **6**: 8641 (1966).

7.10. COLLOID CHEMISTRY

7.10.1. Colloids: The Thickness of the Double Layer and the Bulk Dimensions Are of the Same Order

The *sizes* of the phases forming the electrified interface have not quantitatively entered the picture so far. There has been a certain extravagance with dimensions. If, for instance, the metal in contact with the electrolyte was a sphere (e.g., a mercury drop), its radius was assumed to be infinitely large compared with any dimensions characteristic of the double layer, e.g., the thickness $\varkappa^{-1}$ of the Gouy region. Such large metal spheres, dropped into a solution, sink to the bottom of the vessel and lie there stable and immobile.

What would happen if the radii of the spheres were taken smaller and smaller? In general, changes in the magnitude of a parameter (size, temperature, time, velocity, field, etc.) ultimately lead to new phenomena. Thus, the engineer knows well that "scaling up" or "scaling down" generally results in new modes of behavior.

In the case of the "shrinking" metal spheres, too, important new aspects of behavior arise when such spheres attain submicroscopic dimensions (10 to 10,000 Å), i.e., dimensions of the same order of magnitude as, and smaller than, the wavelength of the light used by microscopes. The little metal spheres begin to show the behavior of what is called the *colloidal state of matter*—an in-between world where the particles are too gross to display the fine behavior of atoms and too minute to reveal the bulk properties of macroscopic matter. The key to the understanding of the colloidal state lies in knowledge of the structure of electrified interfaces: colloid chemistry is electrochemistry. How double layers assume such significant roles will now be briefly sketched.

Referring again to the metal spheres of submicroscopic dimensions, one point becomes clear. The smaller they are ($\sim$ microns), the more they react to the thermal "kicks" from the ions and water molecules of the electrolyte; they take off on a random walk through the solution. Large ($\sim$ centimeters) spheres, too, exchange momentum with the particles of the solution, but their masses are huge compared with those of ions or molecules, so that the velocities resulting (to the spheres) from such collisions are essentially zero.

Once the microspheres begin to jump about in Brownian movement in the solution, some of them collide with each other. What should happen when two approximately 10^{-5}-cm metal spheres collide? Many aspects of colloidal chemistry—and hence of molecular biology, including the electrochemical basis of the stability of blood and the forming of clots—are illuminated by a consideration of this subject.

7.10.2. The Interaction of Double Layers and the Stability of Colloids

The first thing to remember is that each metal sphere sees its environment through its charged interface; each sphere is enveloped in a double layer. All the concepts and pictures of the electrified interface that have been developed in this chapter are of immediate relevance[†] to the microspheres rushing toward a collision.

[†] The surfaces being considered are not planar, and, therefore, instead of Helmholtz–

Considering dilute solutions and no contact-adsorbing ions, one can picture each metal sphere surrounded by a (Gouy–Chapman) region of diffuse charge. Note, however, that the Gouy–Chapman layers of both colliding spheres contain charges of the *same sign*. Thus, *there is coulombic repulsion as the two spheres come close*. This repulsion energy depends on the distance apart r of the spheres and varies with distance in the same way as the Gouy–Chapman potential. This dependence on distance is approximately given by $\psi_0 e^{-\varkappa r}$ (see Section 7.4.5). One is talking here about the *interaction of two double layers*.

But not only do double layers interact with double layers, the metal of one sphere also interacts with the metal of the second sphere. There is what is called the Van der Waals *attraction*, essentially a dispersion interaction, which depends on r^{-6}, and the electron overlap *repulsion*, which varies as r^{-12}. These interactions between the bulk of the two colloidal metal spheres shall be represented together by a term $-Ar^{-6} + Br^{-12}$, where A and B depend essentially on the chemical composition of the phase which is dispersed in the solution.

The total interaction between the two metal spheres can therefore be classified into two parts, (a) the *surface*, or double-layer, interaction determined by the Gouy–Chapman potential $\psi_0 e^{-\varkappa r}$ and (b) the *volume*, or bulk, interaction $-Ar^{-6} + Br^{-12}$. The interaction between double layers ranges from *indifference* at large distance to *increasing repulsion* as the particles approach. The *bulk* interaction leads to an attraction unless the spheres get too close, when there is a sharp repulsion (Fig. 7.125). The total interaction energy depends on the interplay of the surface (double-layer) and volume (bulk) effects and may be represented thus

$$U_{\text{total}} = \psi_0 e^{-\varkappa r} + (-Ar^{-6} + Br^{-12}) \qquad (7.253)$$

This approximate formula contains information concerning what happens when two colloidal particles (the two metal spheres) collide. One has to plot this total interaction energy U_{total} against the distance apart of the particles.

Consider one type of energy–distance diagram (Fig. 7.125). It is seen that, for the first type of behavior where the electrostatic repulsion predominates, the net energy U_{total} is always positive; this means that two metal

---

Perrin parallel-plate condensers, one has concentric-sphere capacitors; Gouy–Chapman regions show radial, instead of planar, symmetry. All such points sophisticate the mathematics, but lead to few new truths. Hence, such details will be ignored in this very simple account of the dominating role of double layers in colloid chemistry.

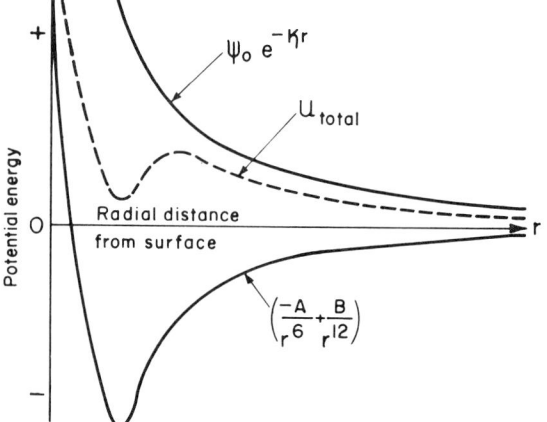

Fig. 7.125. The energy of interaction between two colloidal particles as a function of their distance apart, when the conditions favor stability of the colloid.

spheres under this condition cannot stick together stably. Note that (Fig. 7.125), if the spheres did not wrap themselves in double layers, the interaction between the particles themselves, neglecting the double-layer repulsion, would predominate and have a minimum in a negative potential energy region corresponding, therefore, to a favoring of the *aggregation* of colloidal particles.

Thus, particles of colloidal dimensions survive aggregation into macroscopic phases *only* because their boundaries are guarded by electrified interfaces. The repulsion between double layers is the key to the stability of colloids.

It has been shown, however, in some detail (Section 7.4.5), that the structure of an electrified interface and therefore the potential drop across it markedly depend on the composition of the electrolyte. Make the solution concentrated by adding some electrolyte, and the Gouy–Chapman region starts being reduced in thickness and the potential falls sharply. Put in contact-adsorbing ions, and they start populating the IHP which gives a region of linear potential drop. All this means that one has by variation of the solution composition an indirect control over the double-layer contribution and therefore the total interaction energy for two colloidal particles. One can control the stability of colloids.

How can colloidal particles be made to aggregate? The bulk-interaction curve is given by nature for a given material; it cannot be altered. Hence, what one has to do is to get lower Gouy–Chapman potentials at the r_{min}

distance. This is easy; one adds more electrolyte to the solution. The n^0 of Eq. (3.43) increases, $\varkappa$ increases, and, since $\psi = \psi_0 e^{-\varkappa x}$, ψ falls more sharply with distance. In other words, the Gouy–Chapman region is compressed, and the total interaction curve becomes negative and shows a minimum at r_{min} (Fig. 7.126). Two metal spheres approaching each other get irreversibly stuck together at this distance. The colloid has lost its stability. This is known as *coagulation*, or *flocculation*.

There is another way of bringing about this irreversible flocculation. Recall that, by contact adsorption of ions, the bulk of the potential drop across the interface can be made to occur between the metal and the IHP. Thus, by the addition of contact-adsorbing ions, the value of ψ_0 can be reduced without significantly changing the concentration of the bulk electrolyte. The effect of this will be qualitatively similar to that shown in Fig. 7.126 and is shown in Fig. 7.127. The value of U_{total} again comes into the negative potential-energy region, i.e., a *stable* configuration of particles in contact may exist, and a flocculation thus again occurs.

7.10.3. Sols and Gels

The essence of the behavior characteristic of the colloidal state is that double-layer interactions are as significant as bulk interactions. In other words, surface interactions are on a par with volume interactions. This

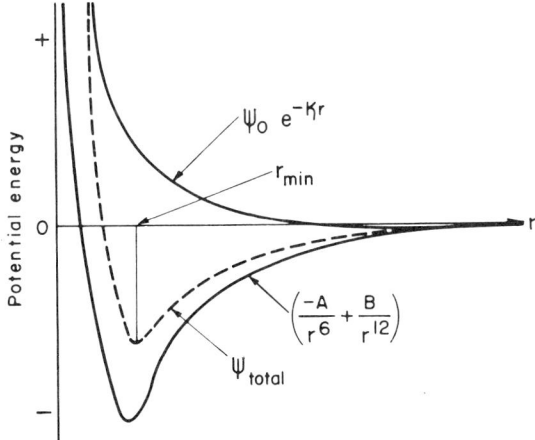

Fig. 7.126. The energy of interaction between two colloidal particles as a function of their distance apart, when the conditions favor coagulation of the colloid.

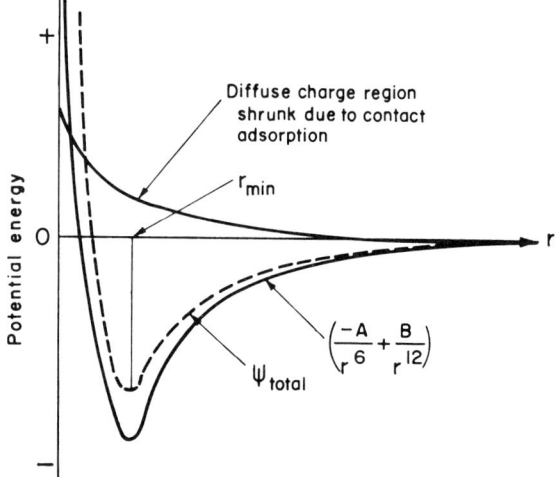

Fig. 7.127. The effect of the contact adsorption of ions on the condition of the stability of a colloid.

condition can therefore be realized in all systems where the surface to volume ratios are high, i.e., at submicroscopic dimensions.

One type of colloidal system has been chosen above for discussion, a system in which the solid metal phase has been shrunk in three dimensions to give small solid particles in Brownian motion in a solution. Such a colloidal suspension consisting of discrete, separate particles immersed in a continuous phase is known as a *sol*. One can also have a case where only *two* dimensions, e.g., the height z and breadth y of a cube, are shrunk to colloidal dimensions. The result is long spaghettilike particles dispersed in solution—macromolecular solutions.

Instead of having one phase discontinuous and in the form of separate particles, it is possible to have the phase as a continuous matrix with pores of very fine dimensions running through it. This is a porous mass, or membrane, also known as a *gel*. In such membranes, interactions inside the pores become highly dependent on double-layer interactions.

Sols and gels are frequently partakers in biological processes. A living cell is separated from the outside by a membrane (a gel), and inside is a collection of colloidal particles held in suspension by interacting Gouy layers. A vivid example of this is given by the electrochemical mechanism of the clotting of blood. Thus, Sawyer and Srinivasan have shown that blood clots at a metal–solution interface when the potential difference across it exceeds a critical value. A basis for the control of thrombosis is hence provided. Electrified interfaces are therefore indeed essential to life, but there is something else which is essential, too. Charge transfer must occur for life to go on. In other words, charges must leak across electrified interfaces, as for example in the consumption of oxygen at the interfaces of biological cells (Del Ducca and Fucsoe). This transfer of charge across electrical double layers, which constitutes the very extensive field of *electrodics*, will now be examined.

Further Reading

1. T. M. Riddick, *Control of Colloid Stability Through Zeta Potential*, Livingston Publishing Co., New York, 1968.

Appendix 7.1. Measurement of the Electrode–Solution Volta Potential Difference

The value of the Volta potential difference between a metal α and a solution can be obtained by performing two measurements, in one of which the electrode α is in contact with the solution (Fig. A7.1), whereas, in the second, it is separated from the latter by a space filled with ionized unreactive gas at a low pressure (Fig. A7.2). Electrode α and another electrode of a metal γ, reversible to ions M^{z+} in solution, are connected to a potentiometer P by wires β and β' made of the same metal.

Consider first the arrangement shown in Fig. A7.1. By adjusting the potentiometer to a reading V such that the galvanometer G indicates no current flowing, conditions of electronic equilibrium are created between α and β, i.e.,

$$\bar{\mu}_{el}^{\alpha} = \bar{\mu}_{el}^{\beta} \tag{A7.1}$$

and between β' and γ, i.e.,

$$\bar{\mu}_{el}^{\beta'} = \bar{\mu}_{el}^{\gamma} \tag{A7.2}$$

The potentiometer reading V is equal to the Galvani potential difference between β and β', i.e.,

$$V = \phi^{\beta} - \phi^{\beta'} \tag{A7.3}$$

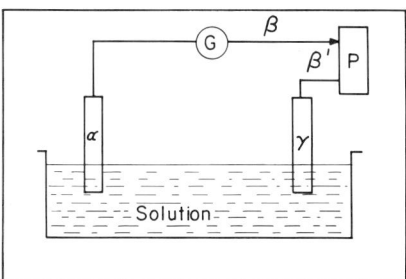

Fig. A7.1. The measurement of the Volta potential.

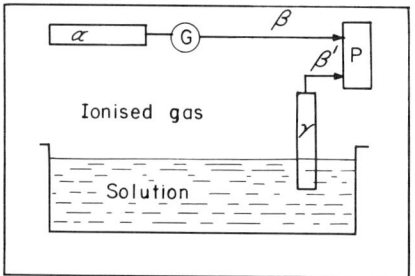

Fig. A7.2. Principle of measurement of the Volta potential difference.

By multiplying Eq. (A7.3) by F and adding and subtracting μ_{el}^{β} (chemical potential of electrons in phase β),

$$FV = \bar{\mu}_{el}^{\beta} - \bar{\mu}_{el}^{\beta'} \tag{A7.4}$$

Substituting Eq. (A7.1) and (A7.2) in Eq. (A7.4), one obtains

$$FV = \bar{\mu}_{el}^{\gamma} - \bar{\mu}_{el}^{\alpha} \tag{A7.5}$$

At the γ-solution interface, an equilibrium exists

$$M \rightleftharpoons M^+ + \bar{e}$$

(for simplicity univalent ions are considered), and thus

$$\mu^{\gamma} = \bar{\mu}_{M^+}^{sol} + \bar{\mu}_{el}^{\gamma} \tag{A7.6}$$

Eliminating $\bar{\mu}_{el}^{\gamma}$ between Eqs. (A7.5) and (A7.6), one obtains

$$FV = \mu^{\gamma} - \bar{\mu}_{M^+}^{sol} - \bar{\mu}_{el}^{\alpha} \tag{A7.7}$$

Using Eq. (7.31),

$$\bar{\mu}_{M^+}^{sol} = \mu_{M^+}^{sol} + F\psi^{sol} + F\chi^{sol} \tag{7.31}$$

and

$$\bar{\mu}_{el}^{\alpha} = \mu_{el}^{\alpha} - F\psi^{\alpha} - F\chi^{\alpha} \tag{7.31}$$

and denoting

$$\mu_{M^+}^{sol} + F\chi^{sol} = \alpha_{M^+}^{sol}$$

and

$$\mu_{el}^{\alpha} - F\chi^{\alpha} = \alpha_{el}^{\alpha}$$

Eq. (A7.7) may be transformed into

$$FV = \mu^{\gamma} - \alpha_{M^+}^{sol} - \alpha_{el}^{\alpha} - F(\psi^{sol} - \psi^{\alpha}) \tag{A7.8}$$

We shall return to this equation after consideration of the arrangement shown in Fig. A7.2.

By adjusting the potentiometer to a reading V' such that the galvanometer indicates no current flowing when α is moved relative to the solution, conditions are created such that there is no *field* between α and the solution and thus

$$\psi^\alpha = \psi^{\text{sol}} \tag{A7.9}$$

Equations describing equilibriums between phases α and β, β' and γ, γ and solution obtain as previously, and by the same reasoning one arrives at an equation analogous to (A7.8):

$$FV' = \mu^\gamma - \alpha^{\text{sol}}_{\text{M}+} - \alpha^\alpha_{el} - F(\psi^{\text{sol}} - \psi^\alpha) \tag{A7.8a}$$

Now, however, $\psi^\alpha = \psi^{\text{sol}}$ [Eq. (A7.9) and thus (A7.8a)] reduces to

$$FV' = \mu^\alpha - \alpha^{\text{sol}} - \alpha^\alpha \tag{A7.10}$$

Subtracting Eq. (A7.10) from equation (A7.8), one obtains

$$\psi^\alpha - \psi^{\text{sol}} = V' - V \tag{A7.12}$$

in which the left-hand side is the single-electrode Volta potential between phase α and the solution.

It is obvious from the above reasoning that introduction of phases β and β' is immaterial; the same argument could be carried out if potentiometer leads were made of metal α or γ or if any amount of various metallic phases was introduced between α and γ.

CHAPTER 8
ELECTRODICS

8.1. INTRODUCTION

8.1.1. The Situation Thus Far

The present stage of understanding has been reached from a starting point which was the state of the ions and solvent molecules in the bulk of the solution. What emerged was a picture of intense activity with the particles in the solution engaged in diverse forms of interaction (Chapters 2, 3, and 6) and motion (Chapters 4 and 5).

Ions enter into interactions with solvent molecules and become wrapped in solvent sheaths. The ions also interact with other ions and surround themselves, in dilute solutions, with ionic clouds.

But there is no rest for the ions in the solution. They are engaged in a ceaseless random walk, and their darting hither and thither is associated with the deformation of ionic clouds and with a layer of water molecules—the primary hydration sheath—accompanying the ion in its motions. A pure random walk gets the ions nowhere, i.e., their time average displacement is zero, and there is no net transport of ions. When, however, the ions feel a driving force (an externally applied electric field or an internally produced concentration gradient), a net *drift* is superimposed on the random walk in solution.

Thus, ions arrive at the interphase regions (Chapter 7), where the anisotropy of the forces causes ions to adopt arrangements unknown in the bulk of the solution. There is an electric atmosphere at the interface:

charge separation, a potential gradient, orientation of water dipoles, adsorption of organic molecules, etc.

What next? That depends on the type of interface. If the interface is ideally polarizable, a specific restriction is placed on the charge populating the interface: they must not cross it. If these are the rules of the game, then all there is to do is to describe the *structure* of the interface, i.e., how, statistically speaking, the ions distribute themselves, how dipoles are aligned in the interphase region, etc. There can be nothing to say about how charges cross the interface.

But the majority of interfaces do not observe the restriction that there should be no leakage of charge across the interface. Once the transfer of charges *across* the interface is considered, the perspective changes. The most exciting aspects of electrochemistry remain to be described. The whole subject of electrode kinetics, or electrodics, lies ahead. The transfer of charge across electrically charged interfaces is a very significant part of nature. It compels electrochemistry to evolve new concepts and link itself with the everyday world, with batteries and brain cells, with energy producers and the prevention of corrosion, with the avoidance of thrombosis and the synthesis of nylon.

In a sense, therefore, the book so far has been a preparation for the understanding of interfacial charge transfer and the treatment of *electrodics*. Modern electrochemistry is the study of charge transfer across interfaces, and *ionics*, the physical chemistry of the ionic side of the double layer, a prerequisite.

8.1.2. Charge Transfer: Its Chemical and Electrical Implications

The transfer of charge across the electrified interface consists essentially of the exchange of electrons between the electrode and particles on the solution side of the interface. The particle involved in the electron transfer may be a neutral molecule (water, methanol, etc.), a complex ion (e.g., cuprammonium ion), or a simple ion (e.g., cupric). What are the implications of charge transfer?

Suppose that the particle is an ion positioned on the OHP and it accepts from or donates an electron to the electrode. Its valence (or oxidation) state necessarily changes when such a transfer occurs. If, for instance, the ion donates electrons to the electrode, i.e., the electron-transfer reaction is given by

$$M^+ \xrightarrow[\text{donation to electrode}]{\text{electron}} M^{++} \tag{8.1}$$

the valence state of the ion increases by 1, i.e., the ion is de-electronated, or oxidized. An example of de-electronation, or oxidation, is the donation of an electron from a ferrous ion to the electrode, in which case, the divalent ferrous ion undergoes oxidation to a trivalent ferric ion. One may write

$$Fe^{++} \rightarrow Fe^{+++} + e \qquad (8.2)$$

a description which clearly reveals the change of valence state, or oxidation state, following electron transfer but conceals the active role of the electrode as an electron sink.

A simple conclusion arises from this introductory material so far: *Charge transfer across an interface implies chemical transformations*, i.e., the transformation of substances into other substances. *By controlling the direction, extent, and rate of electron transfer across an interface, one can control chemical reactions.*

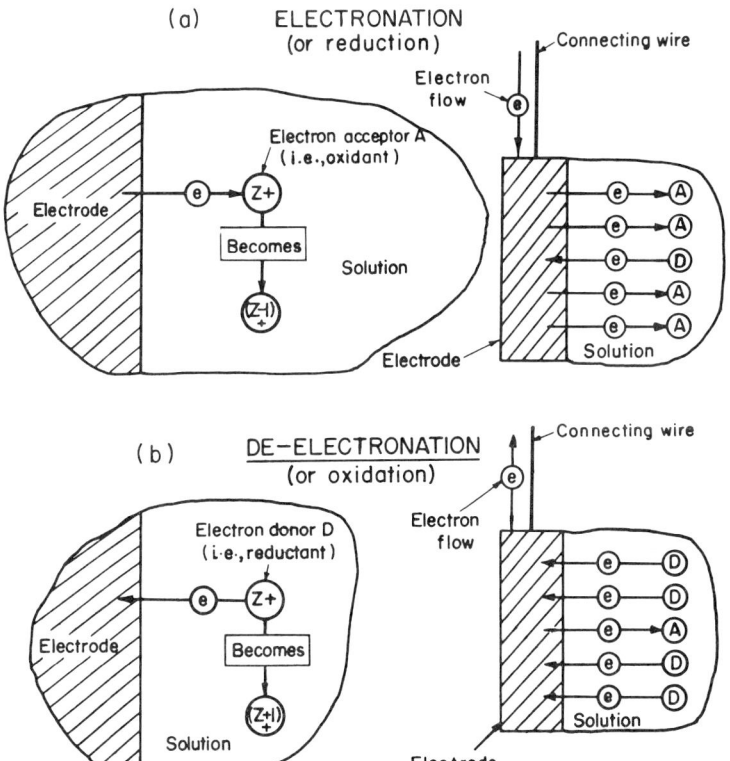

Fig. 8.1. The electrons in electionation go from the electrode into the solution (a), while in the de-electronation they go in the opposite direction (b).

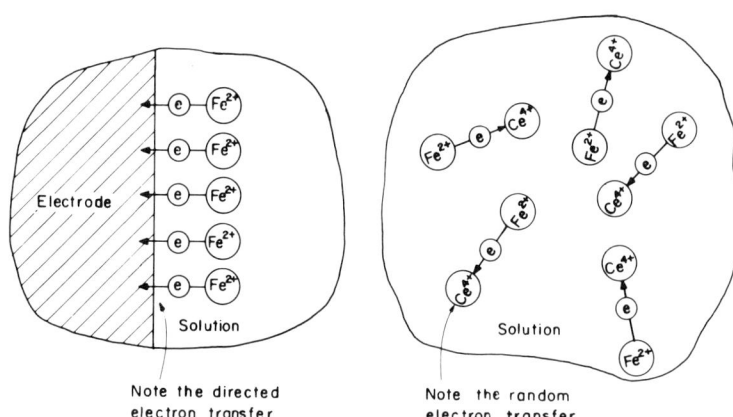

Fig. 8.2. In solution, the direction of the electron transfer is random and there can be no net current. At the interface, however, the net unidirectional flow of electrons results in current.

There is, however, another aspect to charge transfer. The transfer of electrons between the electrode and particles at the IHP or OHP can be considered to be directed normal to the place of the interface. What is a directed electron transfer? It is the transport of charge, i.e., a *current* of electricity. *Charge transfer implies an electric current across the interface.*

Two interesting points arise. If the electron transfer is from electrode to particle (electronation, or reduction) and leads to the diminution of the positive charge on the particle, the current goes one way (Fig. 8.1a); if, however, the electron transfer is in the other direction, i.e., de-electronation, the current goes the other way (Fig. 8.1b). Whether there is a *net* current flowing across the interface or not, depends upon whether or not there are more electron-transfer acts in one direction than the other. This question of net current and how it is composed of electron currents both from and to the electrode will be treated in detail later.

The second point concerns a comparison between charge transfer at the interface (Fig. 8.2a) and in the bulk of the solution (Fig. 8.2b). The de-electronation (oxidation) of ferrous to ferric ions can take place either at the electrode–electrolyte interface or in the bulk of the solution by ceric ions, for example. In both cases, the reaction occurs by ferrous ions surrendering electrons: either to the electrode at the interface or to ceric ions in the

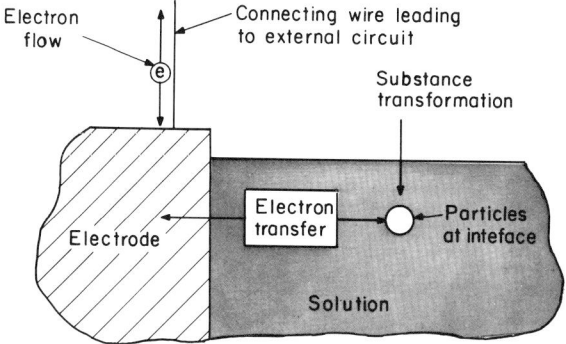

Fig. 8.3. Electron transfer across interfaces has a chemical implication, the transformation of substance, and an electrical implication, electron flow.

bulk of the solution. Each act of electron transfer can be represented by a vector, showing the direction of the electron current (Fig. 8.2). In the bulk of the solution, the vectors are all random in direction; hence, there is no current in solution (Fig. 8.2b). At the interface, however, one can consider all the vectors' pointing toward the electrode (Fig. 8.2a). The bulk situation is like the random walk of electrons; it can never yield electron transport. The interface situation is like the drift, or flow, of electrons; there *is* net transport and a current.

In conclusion, electron transfer across charged interfaces has two implications which have far-reaching importance in nature (Fig. 8.3), a chemical implication and an electrical implication. Substances can be produced (chemistry), and currents can be generated (electricity). To realize these possibilities, however, one must understand more about the mechanics of charge transfer at interfaces.

8.1.3. Can an Isolated Electrode–Solution Interface Be Used as a Device?

An interesting possibility has emerged from the qualitative discussion of electron transfer across an electrode–electrolyte interface. Perhaps, a charged interface can be adapted for the purpose of producing new chemical substances or electrical power. In other words, can an interface be made into a device?[†]

The answer to this question shall be approached through a comparison

---

[†] A device is taken to mean a thing which has been fabricated *to serve a certain purpose*.

between a semiconductor n–p junction and a metal–solution junction, or interface (Fig. 8.4).

There is an aspect of similarity between these two systems in that the conduction mechanism differs in the two phases forming the junction. Thus, in the case of semiconductor n–p junctions, conduction is largely by electrons in the n type of semiconductor and largely by holes in the p type of semiconductor (*cf.* Section 7.7.7). In the case of the metal–solution junction, electrons are the charge carriers in the metal, whereas ions carry the current through the electrolytic solution.

There is more to the similarity than the fact that there is a change of conduction mechanism at the two types of junctions, n–p (solid state) and e–i (electron–ion, or electrochemical). The n–p junction passes current more easily in one direction than the other (*cf.* Section 8.4.1). It can therefore be made into a rectifying device by attaching metal contacts to the n and p types of material.

The metal–solution e–i junction, too, usually permits easier current flow in one direction than the other (more about this later). But, to use the e–i interface as a rectifying *device*, one has to connect up the metal and solution phases with desired circuits. Here comes the problem that proved awkward in the measurement of the potential difference across the interface (*cf.* Section 7.2.5*a*). The metal phase can be contacted with a metallic wire, but the introduction of a metallic wire into the electrolytic

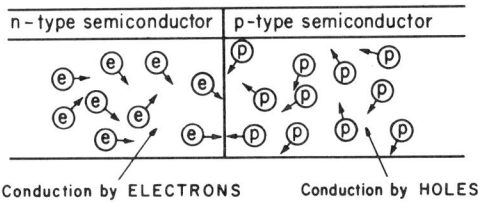

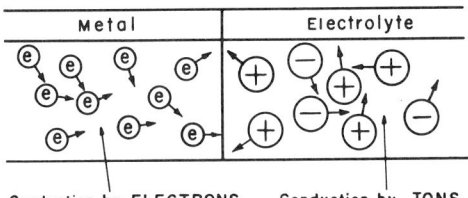

Fig. 8.4. Metal–solution junctions and semiconductor junctions are similar in that, at both, there is a change in the type of conduction of electricity.

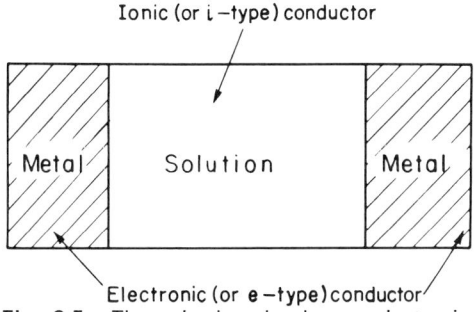

Fig. 8.5. The e–i–e junction has an electronic conductor at both ends.

solution produces a *second* metal–solution interface, i.e., a second e–i junction.

The potential difference across a single e–i junction cannot therefore be measured. Correspondingly, the *single* metal–solution interface cannot be used as a device.

To summarize: The isolated *single* interface holds out hypothetical possibilities of producing power and doing chemistry, but these possibilities cannot be tapped for the only way of tapping them is with metallic contacts which, when dipped into the i type of conductor (i.e., the solution), generate another interface.

8.1.4. Electrochemical Systems Can Be Used as Devices

It has been seen, however, that by assembling *two* metal–solution interfaces, i.e., forming an e–i–e junction, one has an electronic conductor at both ends of the assembly (Fig. 8.5). The potential difference across the whole assembly, i.e., across the e–i–e junction, can be measured.

A simple procedure for this measurement is to connect up the two ends of the two-junction assembly (e–i–e) to the metallic terminals of an electronic voltmeter. Thus, a two-interface metal–solution–metal electrochemical system (or *cell*, as it is generally called) can be incorporated into electrical circuits. *It has device possibilities.* Such possibilities will now be described.

8.1.5. An Electrochemical Device: The Substance Producer

Let the two-interface assembly (the electrochemical system, or cell) be connected to a source of direct current or a constant unidirectional flow of electrons (Fig. 8.6). The dc source may be an electronic power supply,

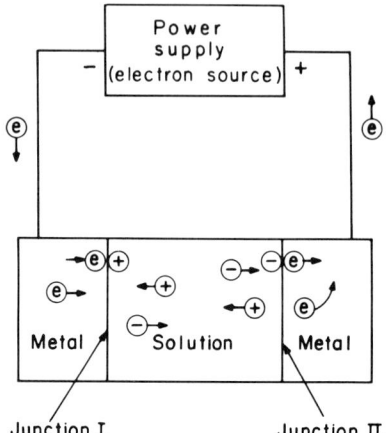

Fig. 8.6. The situation in an e–i–e system subjected to a unidirectional flow of electrons (direct current).

for instance. A continuous path or circuit has been provided for the uninterrupted flow of charge. There will be a flow of electrons in the metallic part of the circuit and an ionic flow in the electrolytic solution.

Follow the path of the drifting charge. Electrons flow out of the dc source from its negative terminal, through the metallic leads, and then up to the metal–solution interface I (the *e–i* junction). At the junction, there is a change of charge carriers from electrons to ions. The positive and negative ions carry the charge through the solution (according to their transport numbers, *cf.* Section 4.5.2); and, at the second junction (II), there is another change of charge carriers from ions to electrons.

What this means is that electrons flow into the electrochemical system *e–i–e* from one side and flow out through the other side. Thus (Fig. 8.7), one electrode–electrolyte interface (or *e–i* junction) serves as an *entrance* for the electron flow; and the other interface, as an *exit*. At the entry junction, electrons leave the electrode and are transferred to the particles of the solution, which are therefore electronated, or reduced; at the exit interface, particles get de-electronated, or oxidized, by donating electrons to the electrode.[†]

[†] At an electronation electrode (electrons *leave* the metal and go to the solution), the statement that electrons *leave* refers only to the interface of the electronic conductor with the *solution*. Electrons must *enter* an electronation electrode if one took the interface to which one refers as that of the electrode material with the connector. How-

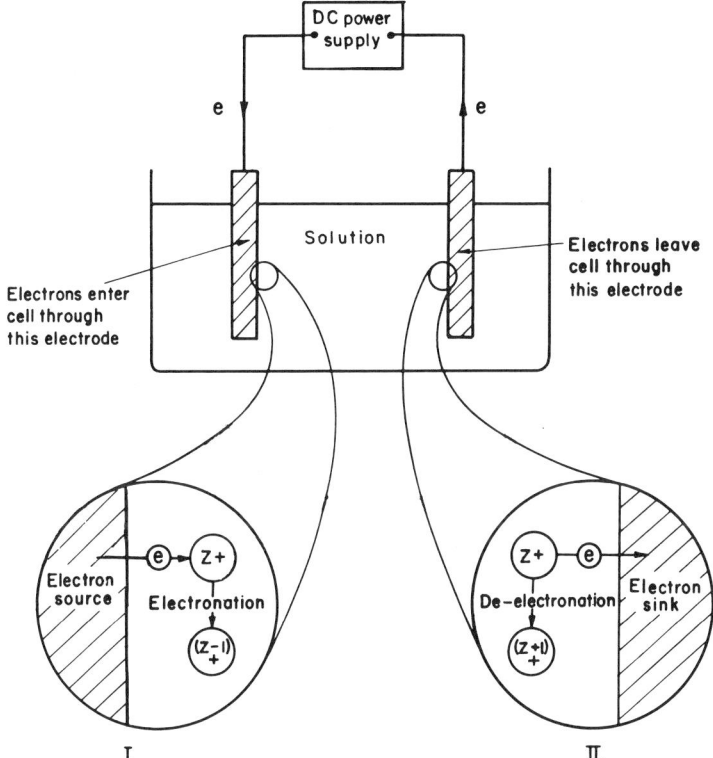

Fig. 8.7. An *e–i–e* system can be considered as composed of an electron source (I) and an electron sink (II).

The electron-entry electrode (Faraday called such electrodes *cathodes*) is the *electron source*; it *gives* electrons to particles in the solution. The electronation of electron-accepting particles occurs at such electron sources (Fig. 8.7). The electron-exit electrode (Faraday's term was *anode*) is the electron *sink*; it *takes* electrons from particles in the solution. The de-electronation of electron-donating particles occurs at such electron sinks (Fig. 8.7).

All this can be exemplified (Fig. 8.8) by the following electrochemical system: Two electronic conductors (e.g., two pieces of platinum) are dipped

---

ever, in electrochemistry, it is the interface between the electron conduction and the ionic conductor which calls the tune, and hence it is whether electrons leave, or arrive at, that interface which is the reference situation when making out a terminology that describes the situation without resource to Greek words.

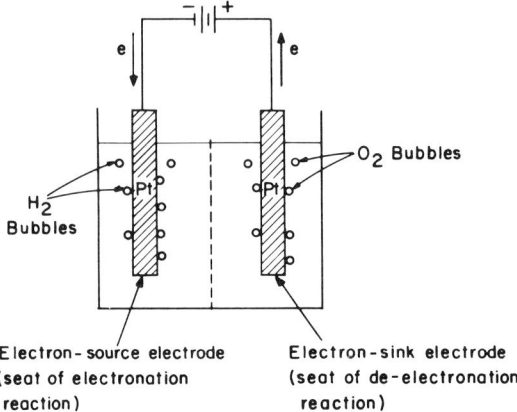

Fig. 8.8. An electrochemical system composed of two pieces of platinum immersed in water becomes a substance producer upon passage of current.

into water (made ionically conducting by the addition of some electrolyte) and then connected to an external electron source. At that conductor, which serves as an electron-source, hydrogen ions, which may be present if the electrolyte is an acid, are electronated and thus *transformed* into hydrogen atoms, which combine to form hydrogen molecules. The transformation can be represented thus

$$\text{Hydrogen ions} \xrightarrow[\text{from source}]{+ \text{ electrons}} \text{Hydrogen molecules} \qquad (8.3)$$

At the electron-sink conductor, water molecules are de-electronated to produce oxygen molecules

$$\text{Water molecules} \xrightarrow[\text{to sink}]{- \text{ electrons}} \text{Oxygen molecules} \qquad (8.4)$$

If the external power source can keep pumping electrons through the system, or cell, it can sustain the production of a gas of hydrogen molecules at the electron-source electrode and a gas of oxygen molecules at the electron-sink electrode. An electrochemical device has been mentally fabricated, a device which serves to produce the substances hydrogen and oxygen.

The hydrogen–oxygen generator has only been mentioned as an example, as it will be seen that, by assembling the two appropriate metal–solution interfaces, i.e., by immersing the right electrodes in the right electrolytic solutions, a host of substances can be produced. Thus, *an externally driven electrochemical system is a device which can be used as an electrochemical substance producer.*[†] This is electrochemistry. It deals with electron transfer at electrified interfaces from electronic conductors to particles in an ionic conductor and vice versa. It is a specific way of commanding the course of chemical reactions.

What does it take to produce hydrogen and oxygen in a chemical (i.e., nonelectrochemical) way? Firstly, two entirely separate reactions have to be used. Secondly, one has only a coarse control (the variation of the temperature) over the rates of the reactions. But, in electrochemical substance producers, the rate and type of substance synthesized can be controlled by controlling the potential difference across the two metal–solution interfaces. Usually, no high temperatures are required. Electrochemical substance producers run cool. It will be seen that their special property lies in the effects upon the rates of reaction of the very intense electric fields ($\sim 10^7$ V cm^{-1}) which exist (*cf.* Chapter 7) at electrode–electrolyte interfaces.

8.1.6. Another Electrochemical Device: The Energy Producer

The electrochemical substance producer did not work spontaneously. Energy (contained in the stream of electrons from an outside power source) had to be pumped in to *make* the reaction go and produce the desired substances. The system had to be driven.

In chemistry, reactions can be classified into two great divisions, (1) those for which the overall free-energy change is positive and (2) those for which the free-energy change is negative. This classification, one might say in homely terms, corresponds to reactions which have to be *made* to go ($\Delta G > 0$) and those which can go by themselves ($\Delta G < 0$).

In electrochemistry, the analogue to the $\Delta G > 0$ situation is the electrochemical substance producer; it has to be driven or forced to do the required chemistry (Fig. 8.9). What is the analogue to the $\Delta G < 0$ situation? It is the *electrochemical energy producer*.[‡] In it, the charge transfer reactions run spontaneously. But, whenever charge transfer takes place at an interface, substances are transformed and currents are generated. So the spontaneous reactions produce a flow of electrons; *the electrochemical energy producer is a device which can be used to generate electrical energy.*

[†] An electrochemical substance producer is also known as an electrolytic cell, electrolyser, or *electrochemical reactor*.
[‡] An electrochemical energy producer is also known as a *galvanic cell*. Fuel cells (11.2.1) and batteries (11.3.1) are examples of such energy producers.

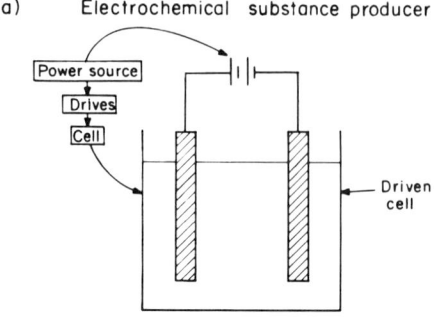

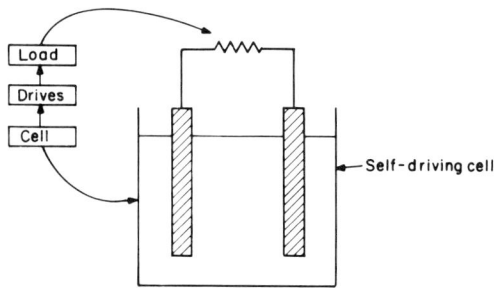

Fig. 8.9. Electrochemical devices can be divided according to the direction of the free-energy change into (a) substance producers $\Delta G > 0$ and (b) energy producers $\Delta G < 0$. In the first case, they are consumers of energy from outside, while, in the second, they are consumers of the substance fed into the cell.

The terms, electrochemical substance producer and electrochemical energy producer, are intended to emphasize the device aspects of two types of electrochemical systems. The question is: What function is the system designed to fulfill, produce substances or electrical energy? Chemical transformations do occur and substances are produced in an energy producer, but the point is that one does not usually fabricate an energy producer to produce substances. One could not directly control the production rate, e.g., for there is no outside source to control the available electron flow, and, often, the substances one wants to produce are not produced by a spontaneous reaction, e.g., H_2 does not come from water spontaneously but has to be forced to come by use of an external electrical source.

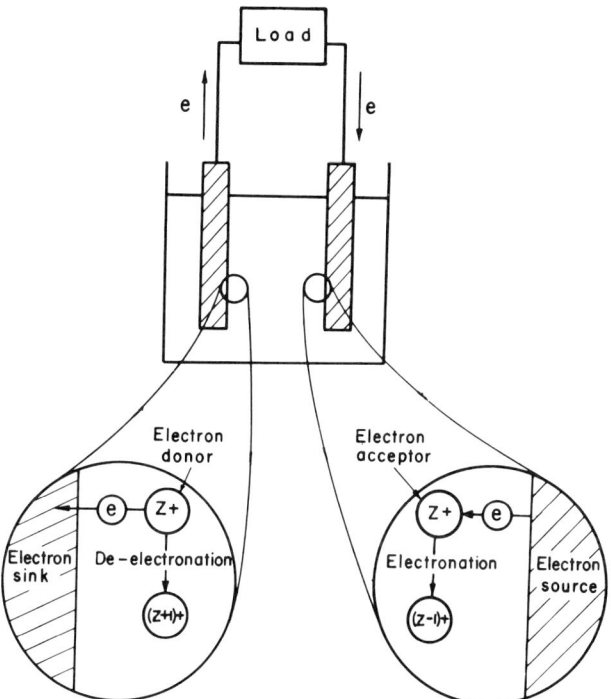

Fig. 8.10. A schematic representation of the processes occurring in an electrochemical energy producer.

How do such electrochemical power sources work? An elementary first description would run as follows (Fig. 8.10): Suppose that, at one interface, particles of the solution (electron donors) are spontaneously de-electronated by the electron-sink electrode. Further, consider that, at the other interface, spontaneous electron transfer occurs from the source electrode to electron acceptors in solution. Of course, for the flow of charge to be maintained around the circuit, the two metal electrodes must be connected through a circuit element (some external "load," e.g., an electric motor) which consumes power by utilizing the electron flow.

An example of an electrochemical energy producer is the electrochemical system (Fig. 8.11) consisting of a lithium electrode, a LiCl liquid electrolyte, chlorine gas, and a graphite electrode. At the lithium electrode, the lithium from the metallic lattice undergoes transfer to the solution; this is the de-electronation of lithium

$$\text{Li} \xrightarrow[\text{to sink}]{-\text{ electrons}} \text{Li}^+ \qquad (8.5)$$

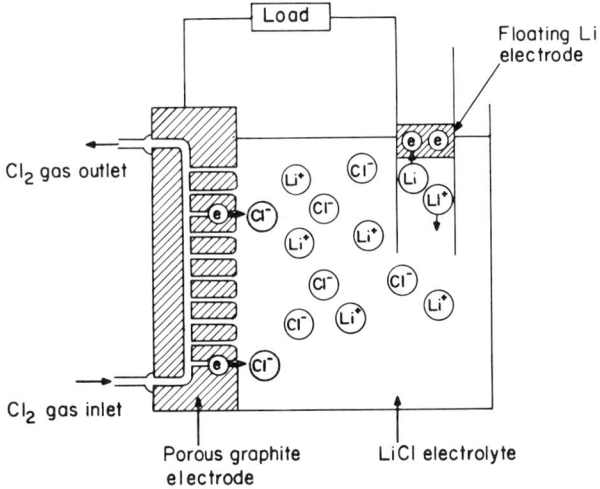

Fig. 8.11. The lithium–chlorine cell as an electrochemical energy producer (schematic).

Thus, the lithium electrode is the electron sink. At the graphite electrode, chlorine molecules (from the gas) are electronated to form chloride ions, which wander off into solution,

$$Cl_2 \xrightarrow[\text{from source}]{+ \text{ electrons}} 2Cl^- \tag{8.6}$$

The graphite is, therefore, the electron-source electrode.

Another example of an energy producer is the *Daniell cell*, as it is called, zinc and copper electrodes in a zinc-sulfate–copper-sulfate solution (Fig. 8.12). At the zinc electron-sink electrode, zinc is spontaneously de-electronated and dissolves

$$Zn \xrightarrow[\text{to sink}]{- \text{ electrons}} Zn^{++} \tag{8.7}$$

and at the copper electron-source electrode, copper ions from the solution are electronated and deposit

$$Cu^{2+} \xrightarrow[\text{from source}]{+ \text{ electrons}} Cu \tag{8.8}$$

What is happening inside such electrochemical energy producers? The chemical transformations (the de-electronation and electronation reactions) at the two interfaces provide the stream of electrons available for external use. Thus, the energy stored in the chemical constituents at the two interfaces is being converted (noiselessly, without moving parts, directly, and without the production of fumes) into useful electrical energy. Such a happening is the basis of the large effort being increasingly made throughout the world in the direction of developing practical electrochemical energy producers which are of sufficient power density (power per unit of weight) to take the place of the common heat engines, effecting the same conversion but with heat energy as an intermediate (11.2.5).

8.1.7. The Electrochemical Undevice: The Substance Destroyer and Energy Waster

Two types of electrochemical systems have been described. One of them was driven from an external source of potential and could be used to synthesize desired substances; the other was self-driving and could be used to generate electric power. In both of them, the electronic conductors (the metals) involved in the two interfaces were spatially isolated and separate pieces of material.

A third possibility may now be considered (Fig. 8.13). Let the two electronic conductors be two *areas* of the same chunk of material. The

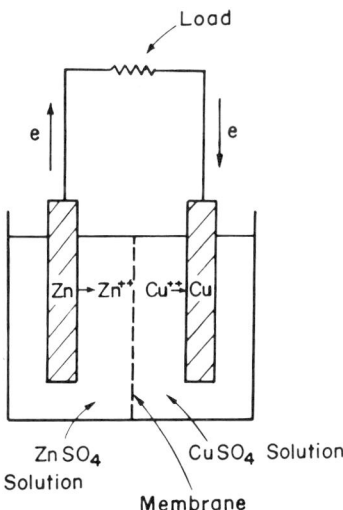

Fig. 8.12. The classical Daniell cell consists of zinc and copper electrodes in corresponding sulfate solutions.

860   CHAPTER 8

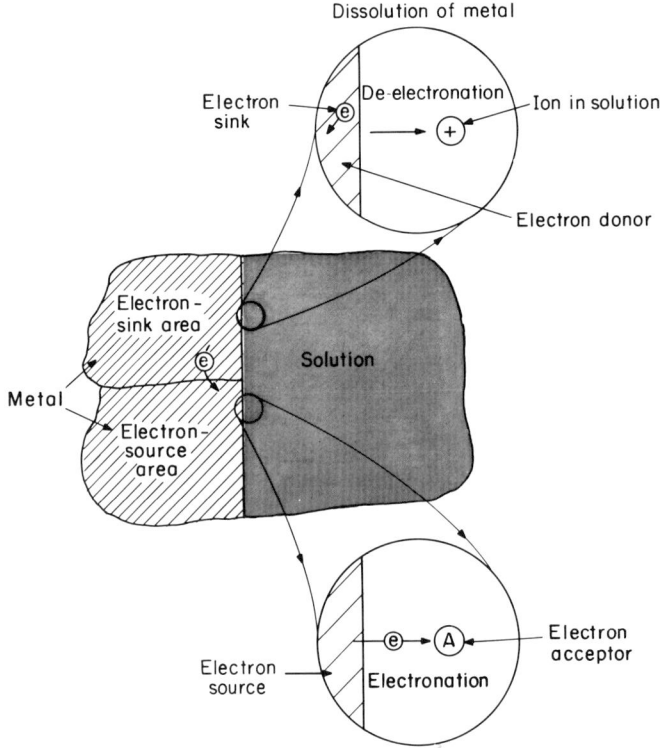

Fig. 8.13. Processes occurring in an electrochemical undevice, the electrochemical corrosion of metals.

relative size of these areas is irrelevant at this moment. Let de-electronation occur spontaneously at one local area of the electronic conductor; and electronation, at another local area. In other words, certain areas of the same chunk of metal are electron sinks, and other areas are electron sources. The system can pump electrons. Further, there is a continuous path for the flow of ions through the electrolyte, electron-transfer reactions occur in opposite directions at the two local areas, and electron flow occurs in the electronic conductor (Fig. 8.13).

An external electron pump is not needed to keep these spontaneous electronation–de-electronation reactions going. The system is self-driven and produces power. But, the corresponding electrical energy is unavailable because the two interfaces are short circuited. This curious assembly of two interfaces produces its own power, but wastes it because it is not possible to apply the power to an external system. The system has no device possibilities as far as energy production is concerned.

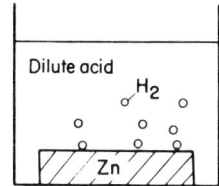

Fig. 8.14. Acid corrosion of zinc is an example of substance wastage.

Is there also a wastage of substances in such a short-circuited electrochemical system? Consider an example, a piece of zinc immersed in dilute acid (Fig. 8.14). Hydrogen bubbles off the surface nearly everywhere. At these areas, the electronation of the hydrogen ions of the electrolyte is occurring. But, at the areas at which hydrogen does not evolve, the de-electronation of zinc (i.e., the electrochemical dissolution of zinc) is taking place. The electron flow through the zinc keeps the system running.

Dissolution of zinc occurs in the case of the Daniell cell, too (Fig. 8.12). The difference lies in the fact that, in the case of the zinc–copper system, the electrons could be made to flow in an external circuit and do work. The system could act as a power producer. The zinc–acid system, however, presents no such device possibilities; *the electron flow cannot be utilized:* it is an undevice.

Further, a permanent substance transformation is occurring in the system due to the dissolution of the zinc metal at the de-electronation areas. This *metal decay* leads to cavities in, and a general disintegration of, the piece of zinc. The electrochemical system has driven itself to destruction while wasting a lot of energy on the way. This spontaneous, wasteful metal breakdown is known as *corrosion*, a chief limiting factor in the life of metals. Metals are won from ores with great difficulty; by the spontaneous electrochemical dissolution of metals they revert to their natural state.

8.1.8. Some Basic Questions

It has become clear that systems of practical interest are systems with teams or pairs of interfaces with one electron-source electrode and one electron-sink electrode. Such systems can be set up to synthesize substances and produce power, but they can also drive themselves to decay and destruction. The basic possibilities have been listed.

What controls these happenings? What controls the rate of charge flow through these two-interface systems? This rate must be understood;

it decides the rate at which substances are synthesized and the rate of output of electrical energy. It determines how fast a metal will disintegrate.

In Chapter 4, the basic factors controlling the drift of ions through the solution have been discussed. That leaves two regions of the electrochemical system in which the rate of charge transport has yet to be understood. These regions are the two metal–solution interfaces.

This is the basic problem: One must understand electron transfer across interfaces, whether it can occur spontaneously or not, what factors control its rate, the role of the metal, the role of the solution. There must be some electrical factors controlling this electron-transfer rate; how else could one use external electron sources to make the interfaces perform the required chemistry at a desired rate?

Why tackle the whole system of two interfaces at once? One can conceptually dismantle the assembly, isolate a single metal–solution interface, and analyze it. This then is the problem stripped down to essentials: What determines the rate of electron transfer at a *single* metal–solution interface?

8.2. THE BASIC ELECTRODIC EQUATION: THE BUTLER–VOLMER EQUATION

8.2.1. The Instant of Immersion of a Metal in an Electrolytic Solution

One could take off on this discussion from a picture of a fully electrified metal–solution interface with charges positioned on the metal surface and on the solution side of the interface and with a field operating across the two phases, etc. But the basic factors reveal themselves more clearly if one follows through the events which occur when a metal electrode M is dipped into an electrolytic solution containing M^+ ions, e.g., a silver electrode immersed in a $AgNO_3$ solution. The instant of immersion, $t = 0$, serves therefore as the starting point of the discussion.

This evolutionary approach to a problem has been used before. Thus, it was applied to elucidate the concept of surface excess (*cf.* Section 7.2.6*b*); the concentration–distance profiles were analyzed at $t = 0$ and at $t \to \infty$ after the interface has acquired a steady-state electrification and structure.

The initial condition shall first be sketched in greater detail. At the instant of immersion, the metal is electroneutral, or uncharged, $q_M = 0$. Since the interface region as a whole must then be electroneutral, there must be zero excess charge on the solution side of the interface, i.e., $|q_M| = |q_S|$

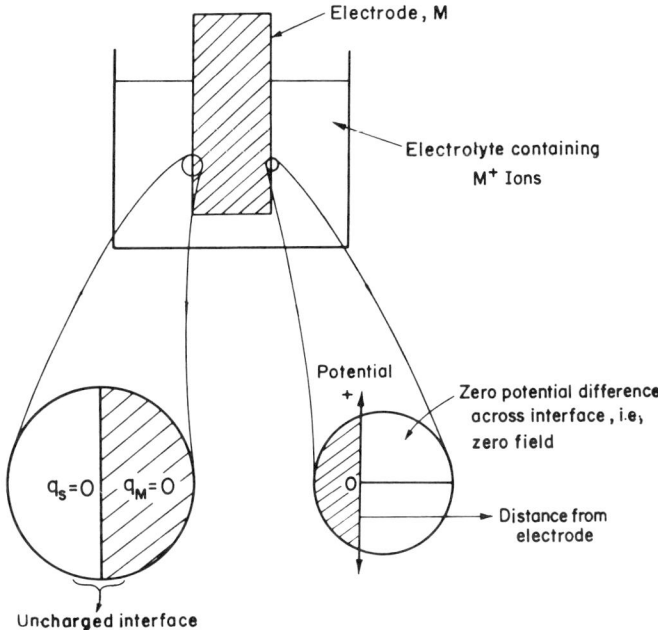

Fig. 8.15. At the instant of immersion there is electroneutrality on both sides of the interface and consequently a zero potential difference and zero field across it.

$= 0$. Hence, there is a zero potential difference[†] and a zero field operating in the interphase region (Fig. 8.15). This is, of course, a too simple picture. Thus, one has ignored the fact that, even when the metal charge is zero, there is a small net orientation of water molecules and, hence, water will immediately start orienting itself in respect to the surface, which will create a dipole field (*cf.* the asymmetric water molecule, Section 7.5.5). Such aspects can always be brought into a more refined picture. As a first step, simplicity shall have the priority.

Under these conditions of zero field, there are no electrical effects and one has pure chemistry, no electrochemistry.

The simplest possible reaction is initially considered, an elementary one-step electrodic reaction: The electrode donates an electron to the

---

[†] If measured (*cf.* 8.2.14) with respect to a common reference electrode at different metals, a stable zero-charge situation can be made to arise at different potentials, the pzc, as shown in Table 8.1 (*cf.* also Section 7.2.5*m*).

TABLE 8.1
Some Potentials of Zero Charge

Metal	Potential of zero charge, V (normal hydrogen scale)
Mercury	-0.19 ± 0.01
Chromium	-0.45 ± 0.05
Copper	-0.16 ± 0.05
Gold	$+0.15 \pm 0.05$
Iron	-0.37 ± 0.03
Lead	-0.60 ± 0.05
Platinum*	$+0.56 \pm 0.03$
Silver	-0.44 ± 0.02
Zink	-0.63 ± 0.05
Nickel	-0.28 ± 0.03

* Values of the pzc for Pt are controversial. That given concerns an electrode free from hydrogen at pH = 0.

electron-acceptor ion A^+. After the receipt of the electron, the electron acceptor is transformed into a new substance D.

$$A^+ + e \rightarrow D \tag{8.9}$$

The reaction might be, for instance, the conversion of silver ions into metallic silver or the electronation of ferric ions into ferrous ions.[†]

Now, the question is: Will the electron-transfer reaction occur of its own free will, or must it be driven? Will it occur spontaneously or not? Such questions are answered by thermodynamics. One looks up whether the metal–solution system is in a state of minimum free energy, i.e., whether the system is in equilibrium. The precise criterion (cf. Section 7.3.3c) for equilibrium across an interface is the equality of the electrochemical potentials of the species which can leak across the interface. In the present context, therefore, the criterion should be stated thus: Is $(\bar{\mu}_{A^+})_{\text{solution}}$ equal to $(\bar{\mu}_{A^+})_{\text{electrode}}$?[‡] In other words, is $^M\varDelta^S\bar{\mu}_{A^+} = 0$? But this criterion can

---

[†] Note that, in general, the electron acceptor need not be an ionic species. It was given a positive sign in this example so as to stress that it differs from the donor by one positive charge.

[‡] There is sometimes difficulty in accepting the equality of the electrochemical potentials of the ions in the solution and in the metal. However, consider $M \rightleftharpoons M^+ + e$. Within the metal itself, $\mu_{\text{metal}} = \bar{\mu}_{M^+,\text{metal}} + \bar{\mu}_{e,\text{metal}}$. In respect to the equilibrium between metal and solution, $\mu_{\text{metal}} = \bar{\mu}_{M^+,\text{soln}} + \bar{\mu}_{e,\text{metal}}$. Hence, $\bar{\mu}_{M^+,\text{soln}} = \bar{\mu}_{M^+,\text{metal}}$.

be written in terms of the chemical and inner potentials (*cf.* Section 7.3.3*a*) thus

$$^M\Delta^S\bar{\mu}_{A^+} = {}^M\Delta^S\mu_{A^+} + F{}^M\Delta^S\phi \qquad (8.10)$$

Since one has stipulated that there is zero field, the term $^M\Delta^S\phi$ can be set equal to zero. Hence, to know whether the interface is at equilibrium, one must check whether the chemical potentials of A^+ are the same on both sides of the interface.

It is as in the process of diffusion. If there is a difference of the chemical potential of a species in two regions, then the gradient of chemical potential acts as the driving force for diffusion. If, therefore, at $t = 0$, the chemical potentials of A^+ are *not* equal on both sides of the interface, there is no equilibrium across the interface. Thermodynamics allows the electrode reaction to proceed spontaneously. What happens when it does so proceed?

8.2.2. The Rate of Charge-Transfer Reactions under Zero Field: The Chemical Rate Constant

The essence of an electrodic reaction is interfacial electron transfer.

At this stage of the treatment, the detailed mechanics of electron transfer will be ignored. Thus, no attention will be given to such questions as: Does the electron jump between electrode and particle A^+ in solution, or does the electron wait for particle A^+ to move across the electrode–solution interface before it jumps? The key point is that a *charge* must move across the interface.

Incidentally, it may be noted (Fig. 8.16) that, whether negatively charged electrons move from electrode to solution or positively charged ions move from solution to electrode, the current is in the same direction. So, from the *formal* point of view, it is all the same whether electron transfers to the ion or ion transfers to the electrode are considered.

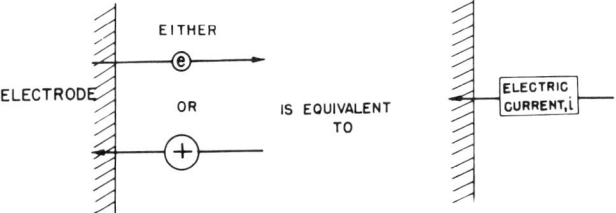

Fig. 8.16. In the conventional definition of current, the current from the solution to the electrode means either the flow of positive ions across the interface in the same direction or the flow of electrons in the opposite direction. These two flows are formally equivalent.

866 CHAPTER 8

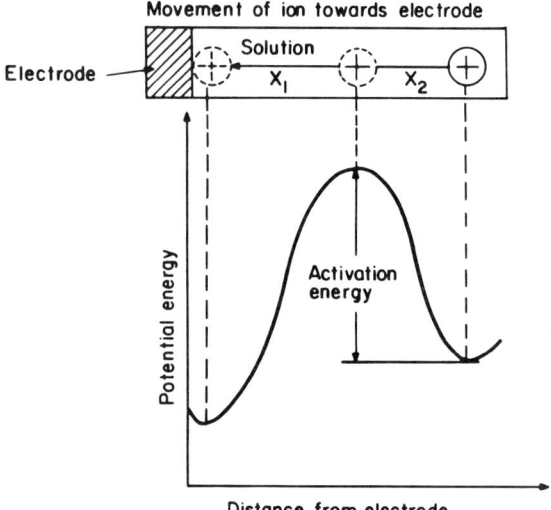

Fig. 8.17. Construction of a potential-energy–distance profile by consideration of the potential-energy changes produced by varying x_1 and x_2.

The approach of the rate-process theory (*cf.* Section 4.2.18) can be used at this point. Consider the movement of the positive ion A^+ from the solution side of the interface, across the few angstrom units of the double layer, to the metal surface. Somewhere along the way the electron transfer occurs from electrode to ion; precisely where, it does not matter at present. The progress of the moving charge can be charted by specifying the values assumed by x_1 and x_2 in the movement (*cf.* Fig. 8.17)

Site on solution side of double layer → Site on metal surface

As the ion moves (i.e., as x_1 and x_2 of Fig. 8.17 vary), its potential energy changes. By plotting the potential energy against the distance coordinates x_1 and x_2, the potential-energy diagram of the quoted figure is obtained.

Each point on this diagram represents the energy corresponding to a certain location of the moving ion. The occurrence of the charge-transfer reaction is represented by the motion of the point from a position on the diagram corresponding to the initial state to one corresponding to the final state.[†] Hence, the positive ion has to have a certain activation energy before

---

[†] This picture represents an oversimplification which will be corrected later (*cf.* Section 8.5).

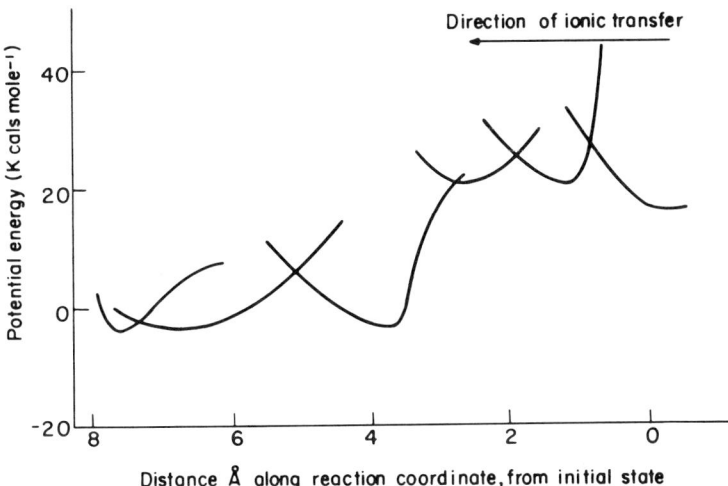

Fig. 8.18. The potential-energy–distance profile for the various successive steps of a metal ion undergoing electronation. (For elucidation of such steps, see Section 10.2.2.)

the charge-transfer reaction is accomplished. (An example of an actual diagram for such a situation is shown in Fig. 8.18).

This process of an ion's jumping from a solution site to the metal is similar to the elementary act of diffusion, namely, the jump of an ion from one site to another (cf. Fig. 4.39). In the case of diffusion, too, there is a potential-energy barrier. The frequency with which an ion successfully jumped the energy barrier for diffusion (i.e., the jump frequency) has been shown (Section 4.2.17) to be[†]

$$\vec{k} = \frac{kT}{h} e^{-\vec{\Delta G}^{0\ddag}/RT} \tag{4.111}$$

where $\vec{\Delta G}^{0\ddag}$ is the standard free energy of activation, the change in free energy required to climb to the top of the barrier (cf. Section 4.2.18) when there is zero electric field acting on the ion in its motion.

A similar expression can be used for the frequency with which an ion can climb the activation-energy barrier in the transfer from solution to metal (Figs. 8.17 and 8.18) and accomplish the charge-transfer reaction. When this jump frequency is multiplied by the concentration c_{A^+} of electron-acceptor ions A^+ on the solution side of the interface, one obtains

---

[†] It will be recalled that the k in the term kT/h is the Boltzmann constant and not the jump frequency.

the rate of the electronation reaction under zero electric field

$$\vec{v}_c = \frac{kT}{h} c_{A^+} e^{-\Delta G_c^{0\ddagger}/RT} \qquad (8.11)$$

(The arrow over the v_c indicates that it is the electronation reaction that is being considered, and the subscript c indicates that it is a chemical, or zero-field, rate). This expression can be separated into a concentration-independent portion $\vec{k}_c$ and a concentration term c_{A^+}

$$\vec{v}_c = \vec{k}_c c_{A^+} \qquad (8.12)$$

where

$$\vec{k}_c = \frac{kT}{h} e^{-\Delta G_c^{0\ddagger}/RT} \qquad (8.13)$$

The concentration-independent portion $\vec{k}_c$ is the *rate constant*. It is the frequency of *successful* jumps, i.e., those in which the particle succeeds in passing over the barrier.

The dimensions of $\vec{k}_c$ can be obtained from Eq. (8.12). The velocity $\vec{v}_c$ of the reaction is the number of moles transformed per unit area of the electrode surface per second; and the concentration is in moles per square centimeter. Hence, $\vec{k}_c$ has the dimension second^{-1}.

This first description of the state of affairs at the instant of immersing a metal in a solution can be summarized. The metal and solution sides of the interface are at first uncharged. There is no potential difference and no electrical field across the interface. Nevertheless, the metal–solution interface may not be at equilibrium, in which case thermodynamics indicates that an electron-transfer reaction will occur. The rate of the electron-transfer reaction under zero field is given by purely chemical-kinetic considerations.

8.2.3. Some Consequences of Electron Transfer at an Interface

Once the electron-transfer reaction occurs, a train of consequences ensues (Fig. 8.19).

The emigration of the electron from the electrode to the electron acceptor A^+ leaves the metal poorer by one negative charge. The metal has become charged positive. Its electroneutrality has been upset.

A similar argument can be applied to the solution side of the interface. Prior to the electronation reaction, this solution side was electroneutral. The electronation reaction involving the transfer of a positive ion toward the metal has the effect of reducing the positive charge on the solution side of the interface which thus acquires a net negative charge. Thus, both sides of the metal–solution interfaces have become charged. This is the embryo

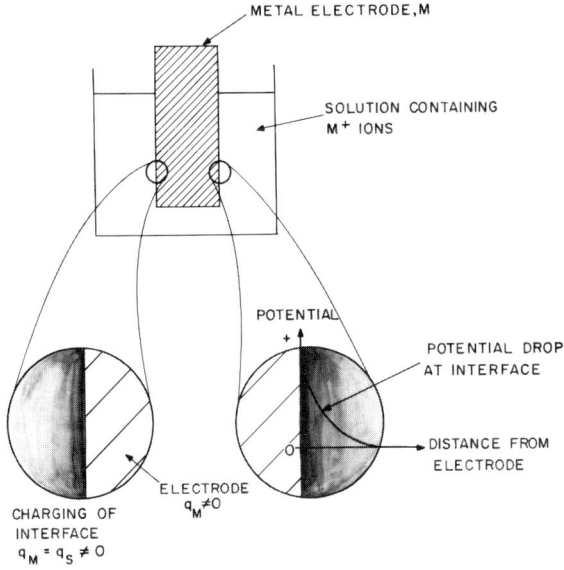

Fig. 8.19. The consequences of an electron-transfer reaction are numerous; the charging of the interface and the establishing of a potential drop are the most direct ones.

of the electrified interface. Further, the charge separation at the interface implies a potential difference across the interface. An electrical field has been created.

Now, all electrodic reactions (electronation and deelectronation) involve movements of charge across the interface. Since electric fields affect the rate of movement of charges, the rates of electrodic reactions must be affected by the embryonic field at the interface.

Thus, the unelectrical state of affairs existing at the instant of immersion is radically changed by the occurrence of the electron-transfer reaction. The metal charges up, an electrical field develops, the zero-field rate of the reaction becomes altered.[†] Chemistry becomes electrochemistry.

8.2.4. What Is the Rate of an Electron-Transfer Reaction under the Influence of an Electric Field?

The electric field at the interface is a vector. It is a quantity directed normally to the interface. If the metal is positively charged and the solution

---

[†] All these events, sketched here in some detail, occur over a small time, which is in the microsecond range for some metal–solution interfaces.

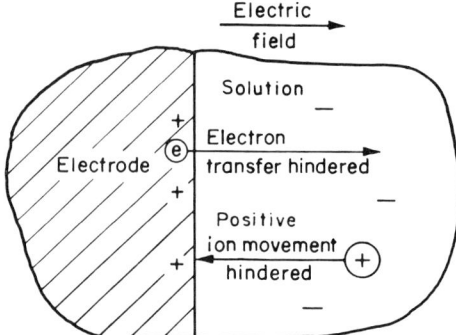

Fig. 8.20. The electric field developing across the interface hinders further charge transfer.

negatively charged, further electron transfer from the electrode to an electron acceptor in the solution (or positive-ion movement in the opposite direction) is opposed by the field (Fig. 8.20). How is the rate of the electron-transfer reaction affected by the interaction between the moving charge and the directed field?

When there is a field across the interface, the work done by the positive ion in climbing the potential-energy barrier (Fig. 8.17) has to include the electrical work.

How can one compute this electrical work? To do the job rigorously, one has to consider the effect of the electrical field on the potential barrier. Each point on the potential-energy curve represents the energy of the particles of the system when they are in a certain configuration. And, in the presence of the field, the energies of all the charged particles will be altered. So, the points and thus the curve may shift up or down.

A highly simplified and approximate approach will be adopted. The electron will be forgotten, and it will be assumed that the electronation reaction consists in moving the ion from its initial state *right across* the interface to its final position on the metal.

On this basis, the electrical work of *activating* the ion so that the energy representing it "passes over the top of the barrier" is given by the charge e_0 on the ion times the potential difference through which the ion is moved to reach the top (Fig. 8.21).

The electrical contribution to the standard free energy of activation can be estimated by reference to Fig. 8.21. As the positive ion begins to move across the double layer, it has to do electrostatic work against the field in the double layer, i.e., the field does work on the ion. Let the total potential difference through which the ion passes be, say, $\Delta\phi$. However,

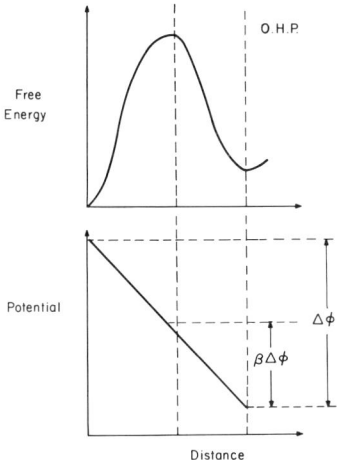

Fig. 8.21. The electrical work of activating the ion is determined by the potential difference across which the ion has to be moved to reach the top of the free-energy–distance relation.

with respect to the contribution of this electrostatic work to the standard free energy of activation for the *forward* reaction (the ion *from* the solution *to* the electrode), only a *part* of the total $\Delta\phi$ is of importance, namely, that part through which the ion passes during passage to a point (perhaps somewhat halfway across the double layer) when the energy of the ion passes the summit of the energy barrier of Fig. 8.21. The summit has been lowered by the electrical work done, i.e., the potential difference passed through multiplied by the charge (e_0 on a unicharged ion but F, the Faraday, for a gram-ion of ions). Suppose that, instead of writing the important part of the potential difference through which the ion moves as $1/2\Delta\phi$, one writes it as $\beta\Delta\phi$, where β is a factor greater than zero but less than unity. The title for β is unusually logical; it is called *the symmetry factor*

$$\beta = \frac{\text{Distance across double layer to summit}}{\text{Distance across whole double layer}}$$

Then, $\beta\Delta\phi F$ is the amount by which the energy barrier for the ion-to-electrode transfer is *lowered*, and, hence, $(1 - \beta)\Delta\phi F$ is the amount it is *raised* for the metal-to-solution reaction (see Fig. 8.21). Thus, once more,[†]

[†] The argument given is a very crude one and will be transformed later in this chapter (8.5) to a quantum-mechanical one.

for the forward reaction,

Electrical contribution to the free energy of activation $= +\beta F \Delta\phi$ (8.14)

In the presence of the field, therefore, the total free energy of activation for the electronation reaction is equal to the chemical free energy of activation $\Delta \vec{G}_c^{0\ddagger}$ plus the electrical contribution $\beta F \Delta\phi$[†]

$$\Delta \vec{G}^{0\ddagger} = \Delta \vec{G}_c^{0\ddagger} + \beta F \Delta\phi \qquad (8.15)$$

Thus, the rate[‡] $\vec{v}_e$ of the electronation reaction under the influence of the electrical field can be written in any one of the following forms

$$\vec{v}_e = \frac{kT}{h} c_{A^+} e^{-\Delta \vec{G}^{0\ddagger}/RT} \qquad (8.16)$$

$$= \frac{kT}{h} c_{A^+} e^{-\Delta \vec{G}_c^{0\ddagger}/RT} e^{-\beta F \Delta\phi/RT} \qquad (8.17)$$

$$= \vec{v}_c e^{-\beta F \Delta\phi/RT} \qquad (8.18)$$

$$= \vec{k}_c c_{A^+} e^{-\beta F \Delta\phi/RT} \qquad (8.19)$$

This rate $\vec{v}_e$ is the number of moles of positive ions reacting per second by crossing unit area of the interface. When this is multiplied by the charge per mole of positive charges, one obtains the electronation-current density $\vec{i}$

$$\vec{i} = F\vec{v}_e \qquad (8.20)$$

$$= F\vec{k}_c c_{A^+} e^{-\beta F \Delta\phi/RT} \qquad (8.21)$$

(Since $\vec{v}_e$ and F have the dimensions of moles per square centimeter per second and coulombs per mole (or amperes × second per mole), respectively, $\vec{i}$ has the dimensions of amperes per square centimeter.) This exponential relationship, first established by Volmer and Erdey-Gruz but indicated earlier in a rough and inaccurate way by Butler, symbolizes the link between the electric field and the rate of electron transfer across the interface. It shows (Fig. 8.22) that small changes in the field at the electrified interface produce large changes in the current density (e.g., if $\beta \simeq \frac{1}{2}$, it can be shown that a 120-mV change in $\Delta\phi$ produces a tenfold change in $\vec{i}$). The exponential

---

[†] Thus, for an electronation reaction, $\Delta\phi$ is negative and the electrochemical standard free energy of activation $\Delta \vec{G}^{0\ne}$ is *lowered* by $\Delta\phi$, i.e., the rate is increased.

[‡] The subscript e is to emphasize that the velocity under discussion is that in the presence of an electrical field in contrast to v_c, the velocity under zero-field conditions.

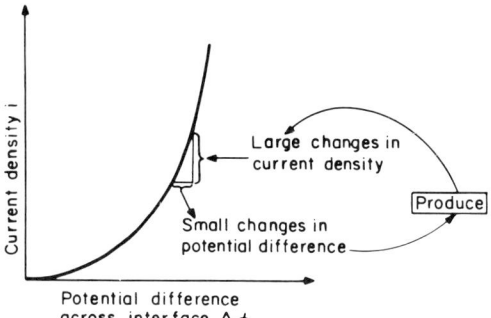

Fig. 8.22. The exponential nature of the current–potential dependence results in large changes in current for small changes in potential.

relation (8.18) is the link between chemistry $\vec{v}_c$ and electrochemistry $\vec{v}_e$. It may be recalled (Section 4.4.6) that, in the transport of ions, too, the drift in the presence of a field (i.e., conduction) was related to the zero-field drift (i.e., diffusion) by a similar type of expression.

For each occurrence of the electron-transfer reaction, the argument can be repeated. If the metal is not connected to any other source of charge, every electronation of A^+ ions charges the metal less negatively and the solution less positively, decreases the potential difference and field across the interface, increases the electrical work of activating the ion to the top of the barrier, decreases the electrical factor $e^{-\beta F \Delta\phi/RT}$, and reduces the rate of the electronation reaction. There is a cumulative tendency here.[†] The larger the number of electrons transferred, the smaller is the attracting electric field and the smaller is the rate of the reaction. It looks as if, after a sufficient number of electron transfers, the electronation reaction should slow down or stop, $\vec{v}_e \to 0$.

8.2.5. The Two-Way Electron Traffic across the Interface

The description given suggests that every interface should eventually stop letting charge leak through because the transfer of positive charge from solution to the electrode makes the latter more positive and hence hinders further transfers. According to this picture, every interface should

---

[†] The extremely simple picture being built up here is that of the approach to equilibrium after a metal has been introduced into a solution. It is, of course, an artificial case in many ways, one of which is explained below. In an actual electrode connected to an external current source, there is a supply of electrons to or from the electrode from the outer circuit.

behave after a short time as an ideally polarizable interface (Section 7.2.5*d*). But most interfaces *do sustain* continuous electron leakage. Something vital has been left out of the discussion.

The missing aspect is based on the principle of microscopic reversibility—a formal way of saying that, *among atoms*, there are no one-way streets. What can happen in one direction can also happen in the opposite direction. If positive ions can move from the solution to the electrode, they can jump back in the opposite direction. Not only is there an electronation reaction

$$A^+ + e \to D$$

but also a de-electronation reaction

$$D \to A^+ + e$$

In stating that the electronation reaction encounters a field which makes ion movements (from the solution toward the electrode) more and more difficult and that the reaction would stop, the possibility of reverse *de*-electronation jumps was not considered.

Let this reverse reaction be considered now. Its rate under the influence of the field is given by arguments almost identical with those applying to an electronation reaction. There is, however, an extremely important difference. The charges are sensitive to whether they are moving with the field or against the field. If they are moving with the field, the greater the field is, the less is the work required to move them. Hence, if the directed field *hinders* the ion transfer from solution to electrode, it *helps* along the jumps in the opposite direction.

Thus, if, in the electronation reaction, the positive ion moves *against* the directed field, it moves *with* the field in the de-electronation (Fig. 8.23). Further, if the positive ion has to be activated through a potential difference

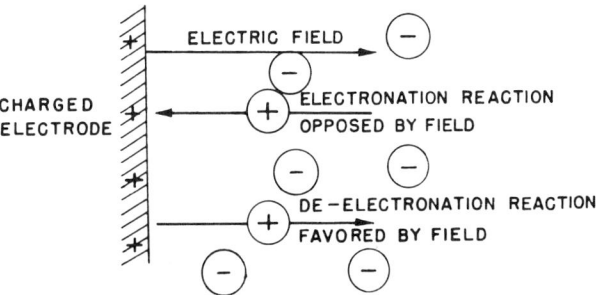

Fig. 8.23. If the field hinders the electronation reaction it helps the de-electronation.

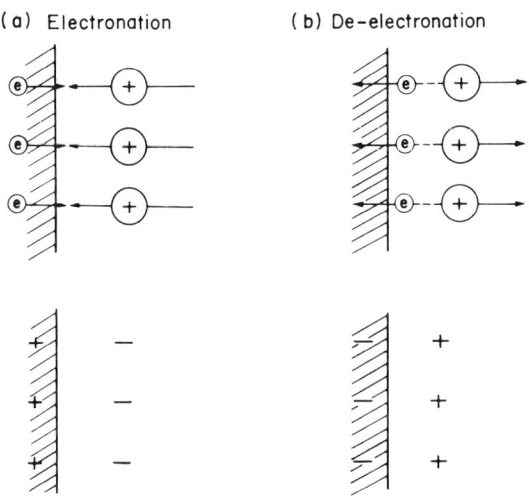

Fig. 8.24. The de-electronation reaction (b) opposes the achievement of the electronation reaction (a) in charging the interface.

of $\beta\Delta\phi$ in the forward direction (i.e., from solution to metal), it has to be activated through the remainder $(1 - \beta) \Delta\phi$ in the de-electronation reaction. Hence, the electrical work of activation for the reverse reaction is $+F[(1 - \beta) \Delta\phi]$, the plus sign is because the field *assists* ion transfer. The rate of the de-electronation reaction becomes therefore[†]

$$\vec{v}_e = \vec{k}_c c_D e^{(1-\beta)F\Delta\phi/RT} \tag{8.22}$$

and the de-electronation current density becomes

$$\vec{i} = F\vec{k}_c c_D e^{(1-\beta)F\Delta\phi/RT} \tag{8.23}$$

The discussion proceeded on the basis that, starting from the instant of immersion (zero field), the electronation reaction built up the field as it

---

[†] It may be noted that, when the donor is a metal atom (e.g., for the reaction Ag → Ag$^+$ + e), the concentration term in (8.22) vanishes. This is because the reactant in that case is pure metal M and the activity of a pure substance is unity. But this is a formal answer; for a physical picture, one may think in terms of a collision theory. Thus, for the M$^+$ + e → M reaction, one would expect that the number of successful encounters with the electrode (and, hence, the rate) would depend on the number per unit volume (or concentration) of M$^+$ particles; the fewer is the number of M$^+$ ions, the fewer are the collisions and the smaller is the rate. For the M → M$^+$ + e reaction, however, the number of M atoms facing the solution does not change when the solution concentration changes, and hence can be thought of as included in the rate constant $\vec{k}_c$.

occurred. The reaction charged the metal less negative, or more positive. Notice (Fig. 8.24), however, that, at the same time, the de-electronation reaction is splitting a neutral atom into an ion and an electron and thus putting electrons into the metal and opposing the increase of positive charge on the metal (i.e., opposing the buildup of the field).

This, then, is the explanation of why the electronation reaction does not come to a halt. Just as the excess positive charge on the metal builds up and decreases the electronation rate, so the reverse reaction *increases* in rate by pumping electrons into the metal, decreasing the excess positive charge on the metal, and negating the tendency to stop the electronation reaction.

What is more, it is obvious that, if no *external* electron source (power supply, e.g.) or sink (a load) is connected up to the electrode, a stalemate must be reached between the electronation and de-electronation reactions which are trying to change the field, electrode charge, etc., in opposite ways. There must be some value of the field or potential difference $\Delta\phi$ at which the electrode's rate of loss of electrons (the electronation reaction) and its gain of electrons (the de-electronation reaction) must become *equal*, i.e., the electronation and de-electronation currents become equal

$$\vec{i} = \overleftarrow{i} \tag{8.24}$$

The charge on the metal then becomes constant, and so does the charge on the solution. The field across the interface becomes constant.

The interface has attained equilibrium, and there is a characteristic equilibrium potential difference $\Delta\phi_e$ across the interface. The value of $\Delta\phi_e$ is a characteristic of the reaction $A^+ + e \to D$.

Books were written about this particular potential $\Delta\phi_e$, and, for about fifty years, electrochemists remained preoccupied with it. Being strictly an equilibrium concept pertaining to the special case $\vec{i} = \overleftarrow{i}$, they described a situation at which nothing *net* happened. Meanwhile, electrochemical devices were being used, e.g., to synthesize substances and generate power, happenings in principle beyond the scope of an equilibrium viewpoint. Had no new concepts been evolved, concepts which did not concentrate on $\Delta\phi_e$, progress in electrochemistry would have been in danger of becoming like the net current of Eq. (8.24).

8.2.6. The Interface at Equilibrium: The Equilibrium Exchange-Current Density i_0

The metal–solution interface in its equilibrium state seems at first to present a tranquil scene. There is no *net* current, no net electronation and deelectronation, no substances produced, and no change in potential difference or field across the interface. But, beneath the apparent lack of motion, charges pass constantly to and fro across the interface. The electronation and deelectronation reactions continue to occur—*but at the same rate*. The currents corresponding to these reactions are equal in magnitude and opposite in direction

$$\vec{i} = F\vec{k}_c c_{A^+} e^{-\beta F \Delta\phi_e/RT} = \overleftarrow{i} = F\overleftarrow{k}_c c_D e^{(1-\beta)F\Delta\phi_e/RT} \qquad (8.25)$$

where $\Delta\phi_e$ is the absolute potential difference across the interface at equilibrium.

As shown first by Butler, an equality of this type can lead to a kinetic formulation of the equilibrium potential $\Delta\phi_e$.

What is the significance of these equilibrium currents? They are a quantitative measure of the rate of the reaction which occurs (in opposite directions at equal rates) at an interface at equilibrium. They express, in terms of numerical magnitudes, the rate of the two-way electron traffic between the electrode and particles in the electrolyte when there is no *net* charge transport from one phase to the other. They characterize the rate, in terms of current, of the electron exchange between the metal and the solution under equilibium conditions.

At the characteristic potential difference $\Delta\phi_e$ corresponding to the equilibrium, Eq. (8.24) indicates that $\vec{i} = \overleftarrow{i}$. Hence, as suggested by Butler, the individual electronation and deelectronation currents underlying the state of equilibrium can be designated by the same term. Such a magnitude is designated the *equilibrium exchange-current density* i_0. It is *one* particular value of the field-dependent current densities, the value at equilibrium. One has

$$i_0 = \vec{i} = F\vec{k}_c c_{A^+} e^{-\beta F \Delta\phi_e/RT} = \overleftarrow{i} = F\overleftarrow{k}_c c_D e^{(1-\beta)F\Delta\phi_e/RT} \qquad (8.26)$$

Exchange-current densities reflect the kinetic properties of the particular interfacial systems concerned and thus can vary from one reaction to another and from one electrode material to another by many orders of magnitude (*cf.* Table 8.2).

Can the equilibrium exchange-current density i_0 be measured directly by an instrument? It cannot, for all current-measuring instruments are based on the effects (magnetic field, voltages developed across resistors, etc.), produced by a *drift* of electrons, i.e., by a net transport of electrons. A pure random walk of electrons, not leading to a net flow, cannot be thus sensed. Now, the equilibrium exchange-current density is like a one-dimensional

TABLE 8.2
Exchange-Current Densities i_0 at 25 °C for Some Electrode Reactions

Metal	System	Medium	$\log i_0$, A cm^{-2}
Mercury	Cr^{3+}/Cr^{2+}	KCl	-6.0
Platinum	Ce^{4+}/Ce^{3+}	H_2SO_4	-4.4
Platinum	Fe^{3+}/Fe^{2+}	H_2SO_4	-2.6
Rhodium	Fe^{3+}/Fe^{2+}	H_2SO_4	-2.76
Iridium	Fe^{3+}/Fe^{2+}	H_2SO_4	-2.8
Palladium	Fe^{3+}/Fe^{2+}	H_2SO_4	-2.2
Gold	H^+/H_2	H_2SO_4	-3.6
Platinum	H^+/H_2	H_2SO_4	-3.1
Mercury	H^+/H_2	H_2SO_4	-12.1
Nickel	H^+/H_2	H_2SO_4	-5.2
Tungsten	H^+/H_2	H_2SO_4	-5.9
Lead	H^+/H_2	H_2SO_4	-11.3

random walk of electrons across the interface with equal numbers of electrons walking in both directions (Fig. 8.25). Such a situation was analyzed in taking a fundamental look at self-diffusion (*cf.* Section 6.4.2*a*). It was realized that, when equal numbers of *identical* particles moved in opposite directions between two parallel planes, their motions could not be detected. The equilibrium exchange current cannot therefore be directly and simply measured because there is no net current at equilibrium. However, it is possible to determine the i_0, and some methods of doing this will be mentioned later.

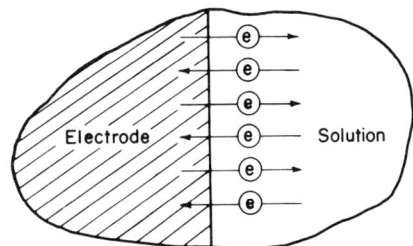

Fig. 8.25. The equilibrium exchange current implies an equal number of electrons walking in both directions across the interface.

8.2.7. The Interface Departs from Equilibrium: The Nonequilibrium Drift-Current Density i

How can the individual electronation and de-electronation currents produce, or be made to produce, a *net* flow of electrons? This must occur directly they are made unequal to each other. The electron drift is then equal to the difference between the charge transfers due to the electronation reaction and those due to the de-electronation reaction. Electron drift, however, implies a net current and net chemical transformations, and this means that the interface is no longer at equilibrium. The nonequilibrium drift-current density i (or simply the current density) is given by the difference between the de-electronation $\overleftarrow{i}$ and the electronation $\overrightarrow{i}$ current[†]

$$i = \overleftarrow{i} - \overrightarrow{i} \tag{8.27}$$

On the basis of this understanding, it is easy to see that the expression for the nonequilibrium drift-current density is

$$i = \overleftarrow{i} - \overrightarrow{i} = F\overleftarrow{k}_c c_D e^{(1-\beta)\Delta\phi F/RT} - F\overrightarrow{k}_c c_{A^+} e^{-\beta\Delta\phi F/RT} \tag{8.28}$$

where $\Delta\phi$ is the nonequilibrium potential difference across the interface ($\Delta\phi \neq \Delta\phi_e$) corresponding to the current density i. One can split this nonequilibrium $\Delta\phi$ into the equilibrium potential difference $\Delta\phi_e$ and another portion, namely, the extra part η by which the potential of the electrode departs from that at equilibrium, i.e., $\Delta\phi - \Delta\phi_e = \eta$ and write

$$\Delta\phi = \Delta\phi_e + (\Delta\phi - \Delta\phi_e) = \Delta\phi_e + \eta \tag{8.29}$$

The term $\eta = \Delta\phi - \Delta\phi_e$ measures how much the potential[‡] has departed from the equilibrium value $\Delta\phi_e$. One can now write a net current density [*cf.* Eq. (8.28)]

$$i = \{F\overleftarrow{k}_c c_D e^{(1-\beta)F\Delta\phi_e/RT}\}e^{(1-\beta)F\eta/RT} - \{F\overrightarrow{k}_c c_{A^+} e^{-\beta F\Delta\phi_e/RT}\}e^{-\beta F\eta/RT} \tag{8.30}$$

However, the two terms inside the brackets are simply the expressions for

---

[†] Current densities have directions (they are vectors): Unfortunately, therefore, an arbitrary decision must be made regarding the *sign* of the net current density i. Here, putting the de-electronation current density $\overleftarrow{i}$ first is meant to imply that, when the magnitude of $\overleftarrow{i}$ is greater than the magnitude of $\overrightarrow{i}$, the net current i is taken as positive. Hence, when there is *net* flow of electrons from solution to metal (i.e., net de-electronation), the net current is taken as positive. Of course, other conventions are possible.

[‡] In electrochemistry, it is customary to avoid repetition of the term *potential difference* and to simply say *potential* even though a potential difference is implied.

the equilibrium exchange-current density, i_0 [see Eq. (8.26)]. Hence, a convenient way of writing (8.30) is

$$i = i_0[e^{(1-\beta)\eta F/RT} - e^{-\beta\eta F/RT}] \tag{8.31}$$

This is a rather fundamental equation in electrodics. It may be termed the *Butler–Volmer equation* after the workers from whose work it was evolved. It shows how the current density across a metal–solution interface depends on the difference η between the actual nonequilibrium and equilibrium potential differences. Small changes in η produce large changes in i. This achievement of electrical control over reaction rates is the point which distinguishes electrochemical from chemical kinetics.

8.2.8. The Current-Producing (or Current-Produced) Potential Difference: The Overpotential η

The quantity η [*cf.* Eq. (8.31)] is apparently of crucial importance to electrodics. It shall therefore be given closer examination. A potential difference in the case of a linear potential drop is equal to an electric field times distance. Thus, one can write (Fig. 8.26) for the potential difference $\Delta\phi$ existing at an interface across which a current density i is flowing

$$\Delta\phi = lX \tag{8.32}$$

where X is the electric field, and l the distance between the metal surface and the *locus* of centers of the particles positioned for reaction on the solution side of the interface (this distance will be considered in greater detail later). Similarly, for the equilibrium potential,

$$\Delta\phi_e = lX_e \tag{8.33}$$

Thus,

$$\eta = \Delta\phi - \Delta\phi_e = l(X - X_e) = l\,\delta X \tag{8.34}$$

and

$$i = i_0[e^{(1-\beta)Fl\delta X/RT} - e^{-\beta Fl\delta X/RT}] \tag{8.35}$$

where $\delta X = X - X_e$ is the difference between the nonequilibrium field X corresponding to a potential $\Delta\phi$ and the equilibrium field X_e corresponding to the equilibrium potential $\Delta\phi_e$. Note (Fig. 8.26) that, as the current I flows through the solution, there appears an additional potential difference IR, between the two electrodes of an actual cell, owing to the passage of

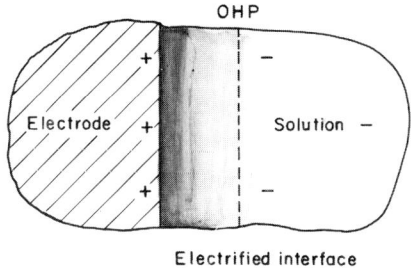

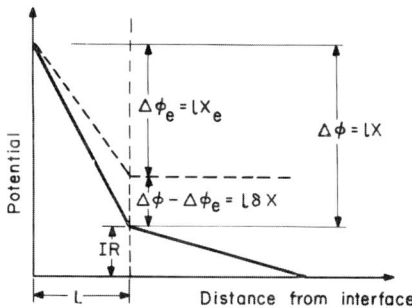

Fig. 8.26. The potential difference across the interface is equal to the product of the field and the distance between the metal surface and the centers of the ionic particles on the solution side at the distance of closest approach.

current through the resistance of the solution. This acts as a driving force for the ionic motion required in the process, i.e., that force required to give the ions in solution a preferential drift between electrodes.

What does Eq. (8.35) teach? It shows that the equilibrium field (corresponding to which, $\delta X = 0$) cannot produce a *net* current; only an *excess* field ($\delta X \neq 0$) can drive a current i. One has here a situation which is common in the phenomena of transport (*cf.* Section 4.2.1); corresponding to every flow or flux, there must be a driving force. The flux is the current density (the flow or charge across the interface); the driving force is the excess field $\delta X \neq 0$ acting on the charges concerned.

Why should the driving force be an *excess* field and not just a field? In the expression for the conduction current through an electrolytic solution [*cf.* Section 4.4.12 and Eq. (4.205) in Section 4.4.13],

$$i = Fk_D lc(e^{-pX} - e^{+pX})$$

one found that the current density i was driven by the field, not by the excess field. The reason is simple. In conduction by ionic migration, this *entire* field inside the electrolyte arises from the externally applied field. Switch off the externally applied field, and the net potential drop inside the electrolyte collapses to zero, and so does the ion migration, or conduction current. This, however, is not the case at the interface. If one switches off the externally applied field, the *excess* field δX and therefore the current drops to zero but the field at the interface does not vanish. The equilibrium field still remains (Fig. 8.27). It drives the electronation and de-electronation current densities at an equal and opposite rate, i.e., gives no net current.

In the case of interfaces, therefore, the net current is driven only by the excess field. This excess field δX is best termed the *current-producing field*.

So far, the discussion of Eq. (8.35) has been presented in terms of a substance-producing electrochemical device. What is the situation in an

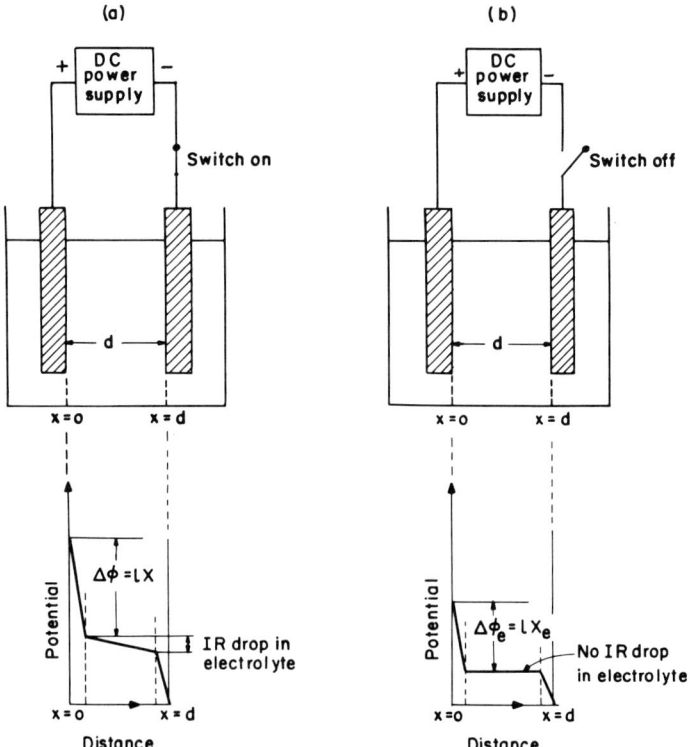

Fig. 8.27. It takes an excess field to drive current (a). But, when the externally applied field is switched off, the field at the interface does not vanish. The equilibrium field remains (b).

energy-producing electrochemical device? It will be recalled that this is a self-driving rather than an externally driven electrochemical system, or cell. The substance within the system undergoes spontaneous charge-transfer reactions which drive a current through the load in the external circuit and, in the process, create an excess field on it; this is a *current-produced field*.

It is customary, however, to talk in terms of excess potential differences (or, simply, overpotentials) rather than in terms of excess fields. Thus, the expression for the net current density i is generally written with an $\eta = \Delta\phi - \Delta\phi_e$

$$i = i_0[e^{(1-\beta)\eta F/RT} - e^{-\beta\eta F/RT}] \tag{8.31}$$

rather than with a $\delta X = X - X_e$ and

$$i = i_0[e^{(1-\beta)Fl\delta X/RT} - e^{-\beta Fl\delta X/RT}] \tag{8.35}$$

When an externally driven electrochemical system, or cell, is considered, the excess potential η (excess with reference to the equilibrium potential difference $\Delta\phi_e$) is the potential difference that drives the current; it is the *current-producing potential*. On the other hand, if the system is a self-driving electrochemical system, or cell, then the current driven through the external load generates an excess potential η; this is a *current-produced potential*. The term *overpotential* is used to refer both to the current-producing potential η in a driven system and to the current-produced potential η in a self-driving cell.

8.2.9. The Basic Electrodic (Butler–Volmer) Equation: Some General and Special Cases

Before engaging in discussion, one should issue a reminder here. All relations derived so far cover only one, the simplest, of all possible cases: a single-step single-electron-exchange reaction. In more complex cases, important changes appear, but these will be discussed later (particularly in Section 9.1).

Even from this simple case, much can still be learned.

A better feel for the indications of the Butler–Volmer relation [Eq. (8.31)] is obtained by plotting i against η. The i versus η curve so obtained (Fig. 8.28a) looks much like the plot of a hyperbolic sine function. There is in fact a basis for this resemblance to a sinh function. It has been said that the symmetry factor is about $\frac{1}{2}$. Let it be assumed to be $\frac{1}{2}$. Then Eq. (8.31) becomes

$$i = i_0(e^{+F\eta/2RT} - e^{-F\eta/2RT}) \tag{8.36}$$

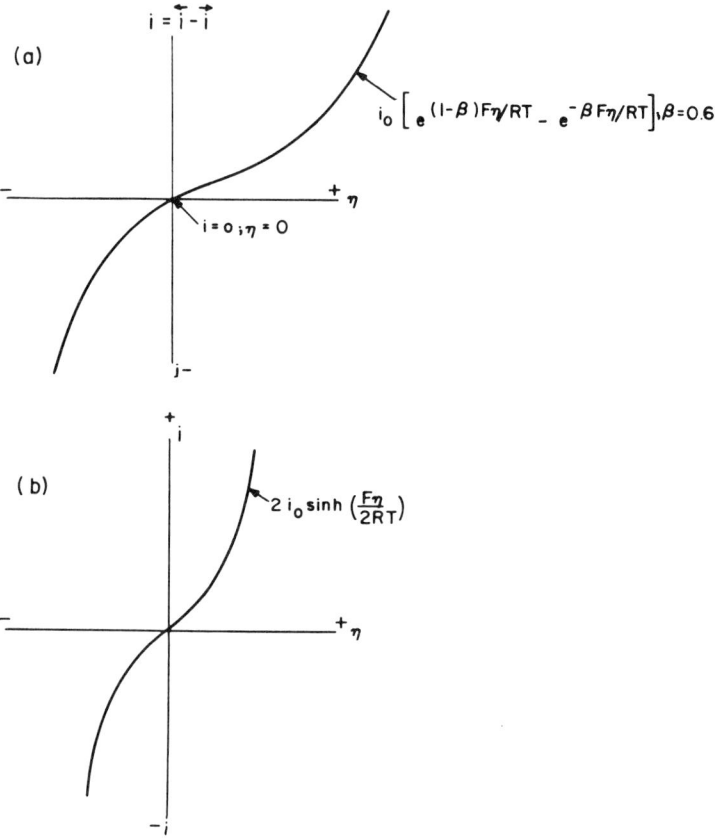

Fig. 8.28. The dependence of current density on overpotential (a) is of the shape of a hyperbolic sinh function (b).

and, since

$$\frac{e^x - e^{-x}}{2} = \sinh x \tag{8.37}$$

$$i = 2i_0 \sinh \frac{F\eta}{2RT} \tag{8.38}$$

The i versus sinh η curve, however, is symmetrical (cf. Fig. 8.28b). A symmetry factor of $\frac{1}{2}$ corresponding to a symmetrical barrier yields a symmetrical i versus η curve. Hence, equal magnitudes of η on either side of the zero produce equal currents; and, conversely, equal de-electronation and electronation currents should produce equal overpotentials, or current-produced potentials, η. This means that the interface cannot rectify a

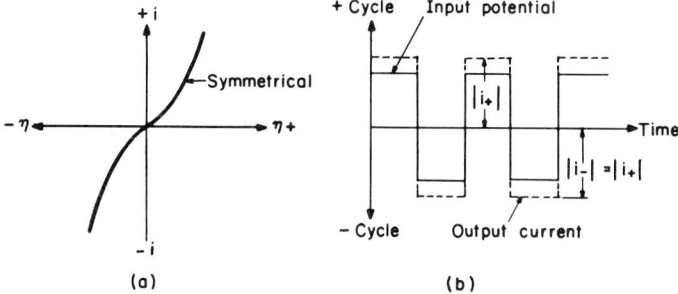

Fig. 8.29. When the i versus η relation is perfectly symmetrical (a), the interface cannot rectify the current responding to a periodically varying potential.

periodically varying potential or current (Fig. 8.29). On the other hand, if $\beta \neq \frac{1}{2}$, then the i versus η curve would not be symmetrical and the interface would have rectifying properties (Fig. 8.30). The effect, known as *faradaic rectification*, was discovered by Doss.

The hyperbolic sine function has two interesting limiting cases (Fig. 8.31). The first limiting case is when the overpotential η [Eq. (8.31)] or the excess field δX [Eq. (8.35)] is numerically large. *This is the high-overpotential or high-field approximation.* Under these conditions of large η (if the example of a net de-electronation reaction is taken),

$$e^{F\eta/2RT} \gg e^{-F\eta/2RT} \tag{8.39}$$

and, since the $e^{-F\eta/2RT}$ term tends to zero,

$$2\sinh \frac{F\eta}{2RT} \simeq e^{F\eta/2RT} \tag{8.40}$$

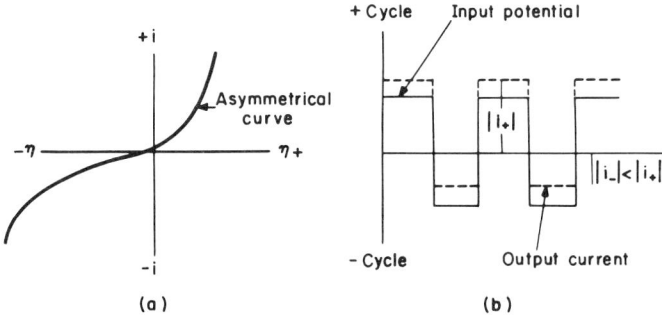

Fig. 8.30. If the symmetry factor is different from $\frac{1}{2}$, the i versus η curve is asymmetrical (a) and there is a faradaic rectification effect or a periodically varying potential.

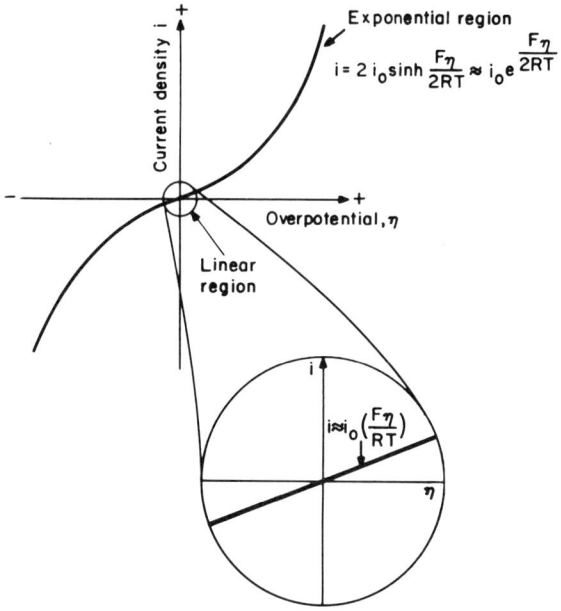

Fig. 8.31. In the narrow region of small overpotentials, the *i* versus η relationship is linear, and, at sufficiently high positive or negative overpotentials, the *i* versus η relationship becomes exponential.

Hence, under high fields, the Butler–Volmer equation reduces [from (8.38)] to

$$i \simeq i_0 e^{F\eta/2RT} \tag{8.41}$$

i.e., the current density increases exponentially with the overpotential η or with the driving force of the excess electric field across the double layer.

The second limiting case is when the overpotential η or the excess field δX is small. *This is the low-overpotential or low-field approximation.* Under these conditions [see Eq. (3.12)] of small η, one can consider that $(F\eta/2RT) \ll 1$ and use the approximation

$$\sinh \frac{F\eta}{2RT} \simeq \frac{F\eta}{2RT} \tag{8.42}$$

The low-field approximation thus reduces the Butler–Volmer equation to the special case

$$i = \frac{i_0 F\eta}{RT} \tag{8.43}$$

a linear relationship between the current density i and the overpotential η or driving force δX.

What excess fields or overpotentials η are low, and what fields are high? What are the quantitative criteria of low and high fields (or low and high overpotentials)? Consider the high-field approximation. It is based on

$$e^{F\eta/2RT} \gg e^{-F\eta/2RT} \tag{8.39}$$

Let it be assumed that the right-hand side of (8.39) should be less than 1% of the left-hand side. Then, the condition for the high-field approximation is[†]

$$\frac{F\eta}{RT} > 2 \ln_e 10 \tag{8.44}$$

or

$$\eta > 0.12 \text{ V} \tag{8.45}$$

where $2.303RT/F$ is equal to 0.058 V at 298°K. Hence, when the interfacial potential difference $\Delta\phi$ exceeds the equilibrium potential $\Delta\phi_e$ by about 0.120 V for a one-electron transfer process, one can use the exponential i versus η high-field law with an applicability of about 99%.

The condition for the low-field approximation is $(F\eta/2RT) \ll 1$. Let this be taken to be

$$\frac{F\eta}{RT} < \frac{1}{5} \tag{8.46}$$

then,

$$\eta < 0.01 \text{ V}$$

Hence, when the overpotential η is about 0.01 V or less for a one-electron transfer reaction, the linear i versus η law can be used with good justification.

It will be recalled [Eq. (8.34)] that the overpotential η is simply related to the driving force for the electrodic reaction and is thus related to the electric field in excess of that present at equilibrium

$$\eta = l \, \delta X \tag{8.34}$$

Thus, the magnitude of the driving force is a measure of how far the system has departed from equilibrium. Hence, the low-field approximation (a small η or δX) implies that the interface is in a near-equilibrium condition, and the high-field (a large η or δX) approximation implies that the interface has been pushed far away from equilibrium.

[†] The condition is for a one-electron transfer reaction.

8.2.10. The High-Field Approximation: The Exponential i versus η Law

One does not, of course, necessarily have to go through the sinh-containing version of the Butler–Volmer equation to obtain the high-field approximation for the current–overpotential law in electrodics.

The Butler–Volmer equation [Eq. (8.31)] contains two terms, one of them representing the deelectronation-current density $\overleftarrow{i}$ and the other, the electronation-current density $\overrightarrow{i}$

$$i = \overleftarrow{i} - \overrightarrow{i} \tag{8.27}$$

where [cf. Eq. (8.31)]

$$\overleftarrow{i} = i_0 e^{(1-\beta)F\eta/RT} \quad \text{and} \quad \overrightarrow{i} = i_0 e^{-\beta F\eta/RT}$$

What happens (Fig. 8.32) when η is increased? The electronation-current

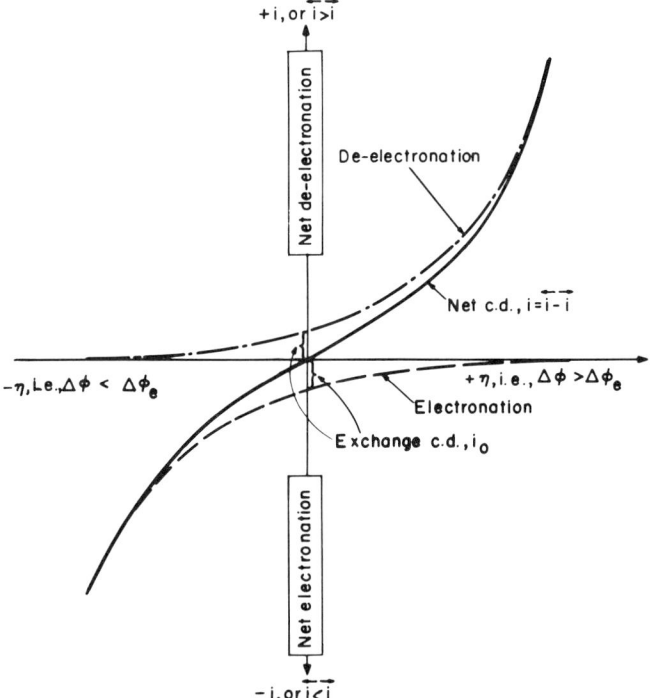

Fig. 8.32. The graphical representation of the electronation- and de-electronation-current densities and the approach of the Butler–Volmer equation to the high-field approximation as $\overrightarrow{i}$ becomes small.

TABLE 8.3

High-Field Approximation of the Butler–Volmer Equation
$\beta = 0.5$, $i_0 = 1$ mA cm^{-2}, and T = 25 °C

η, V	$\dfrac{F\eta}{RT}$	$i_0 e^{(1-\beta)F\eta/RT}$, mA cm^{-2}	$i_0[e^{(1-\beta)F\eta/RT} - e^{-\beta F\eta/RT}]$, mA cm^{-2}	Error % in the high-field approximation
0	0	1.00	0	∞
0.001	0.039	1.02	0.04	+2450
0.005	0.195	1.10	0.195	+464
0.010	0.390	1.21	0.39	+210
0.020	0.780	1.48	0.80	+85
0.030	1.17	1.79	1.23	+45.5
0.050	1.95	2.65	2.27	+16.7
0.100	3.90	7.03	6.89	+2.0
0.200	7.80	49.4	49.38	<+0.1

density $\vec{i}$ decreases and deelectronation-current density $\overleftarrow{i}$ increases. When η is large enough, $\overleftarrow{i} \gg \vec{i}$ and the $\vec{i}$ becomes so small that it can be dropped out of the expression. Thus, the *high*-field approximation of the Butler–Volmer equation (valid at η's greater than about 0.10 V)† yields

$$i \simeq i_0 e^{(1-\beta)F\eta/RT} \tag{8.47}$$

The error involved in replacing the Butler–Volmer equation with (8.47) as a function of potential is shown in Table 8.3.

For convenience of plotting, it is useful to put this Eq. (8.47) into a logarithmic form by taking logarithms

$$\eta = -\frac{RT}{(1-\beta)F}\ln i_0 + \frac{RT}{(1-\beta)F}\ln i \tag{8.48}$$

or

$$\eta = -\frac{2.303RT}{(1-\beta)F}\log i_0 + \frac{2.303RT}{(1-\beta)F}\log i \tag{8.49}$$

---

† The corresponding equation for the electronation-current density at overpotentials sufficiently negative with respect to the equilibrium region is hence

$$i = i_0 e^{-\beta \eta F/RT} \tag{8.47a}$$

[*cf.* (8.41), the corresponding equation for $\beta = \frac{1}{2}$].

Thus, when the electronation current $\vec{i}$ becomes too small for consideration, *the current-producing or the current-produced potential (the overpotential) is a linear function of log i*. Since $\eta = \Delta\phi - \Delta\phi_e$ and $\Delta\phi_e$ (the equilibrium potential) is a constant, the variation in η with log i is the same as the variation in $\Delta\phi$ and, if the potential difference across the interface is plotted against the logarithm of the net current density, a straight-line plot is obtained.

It has often been stressed that the absolute value of the potential difference, $\Delta\phi$, across an interface cannot be measured. How then is it possible to determine the overpotential, $\eta = \Delta\phi - \Delta\phi_e$, which is the difference between two absolute potentials, one (*viz.*, $\Delta\phi$) corresponding to a current density i, and the other (*viz.*, $\Delta\phi_e$) to equilibrium?

In principle, the approach is similar to that of Section 7.2.5. There it was shown that by setting up a two-electrode system, i.e., by coupling the interface under study (the so-called *test or working electrode*) with a nonpolarizable interface (the *reference electrode*), two quantities can be measured: (1) the changes in the potential of the test electrode, and (2) the potential of the test electrode *relative* to the reference electrode. Such a two-electrode system is quite adequate for the measurement of equilibrium relative electrode potentials.

To determine the overpotential, however, it is necessary to alter the above two-electrode system by introducing an extra auxiliary electrode, which is termed the *auxiliary or counter electrode*. Thus a three-electrode arrangement is set up as in Fig. 8.33. In such a setup, the counter electrode is connected to the test electrode via a *polarizing* circuit (e.g., a power source) through which a controllable current is made to pass and produce alterations of the potential of the test electrode. Between the nonpolarizable reference electrode and test electrode is connected an instrument which is capable of measuring the potential difference between these electrodes. (This instrument must have a high input impedance and therefore draw negligible current for the reason discussed in Section 8.2.12.)

When no current flows through the polarizing circuit and there is equilibrium at the test-electrode–electrolyte interface, the potential difference, E_e, between the test and reference electrodes is given by (*cf.* Section 7.2.5)

$$E_e = \Delta\phi_e + \Delta\phi_{\text{contact}} + \Delta\phi_{\text{ref},e} \tag{8.50}$$

where $\Delta\phi_{\text{contact}}$ and $\Delta\phi_{\text{ref},e}$ are the potential differences across the metal–metal contact and reference interfaces.

When a current I is passed through the polarizing circuit (i.e., between

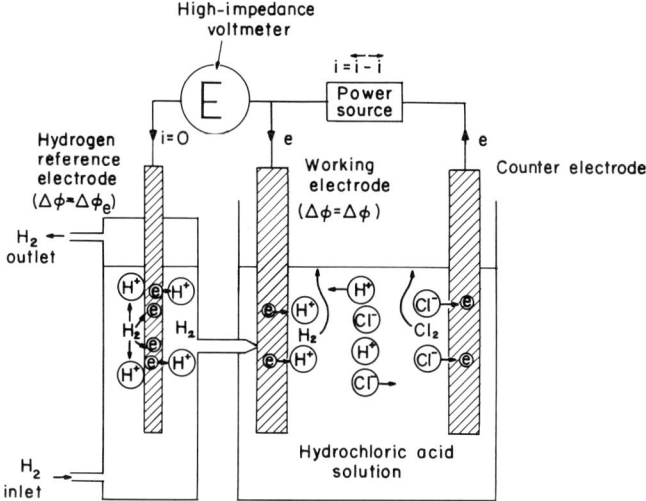

Fig. 8.33. The three-electrode system required to measure electrode overpotentials, i.e., $\Delta\phi - \Delta\phi_e$. The potential between the working electrode and the reference electrode when both $\Delta\phi$ and $\Delta\phi_e$ correspond to the same reaction is equal to the overpotential η. The tube joining reference electrode and working electrode is called a Luggin capillary. It helps diminish the inclusion of illicit IR drop in the measurement.

the test and counter electrodes), (1) the potential of the test electrode changes from $\Delta\phi_e$ to $\Delta\phi$; (2) the potential difference across the metal–metal contact can be considered to be unchanged; (3) the potential of the reference electrode remains at the equilibrium value, $\Delta\phi_{\text{ref},e}$, because no current flows through the measuring circuit (i.e., between the reference and text electrodes); and (4) a potential drop arises in the electrolyte through which the polarizing current flows. Thus, under these conditions, the measured potential difference between the test and reference electrodes is

$$E = \Delta\phi + \Delta\phi_{\text{contact}} + \Delta\phi_{\text{ref},e} - IR \qquad (8.51)$$

In this equation IR is the potential drop developed when the polarizing current I overcomes the resistance $R = l/\sigma A$ of the electrolyte between the test electrode and the so-called *Luggin tip* or probe by means of which the reference electrode makes ionic or electrolytic contact with the test electrode.

From Eqs. (8.50) and (8.51), it follows that

$$\eta = \Delta\phi - \Delta\phi_e = (E - E_e) + IR \qquad (8.52)$$

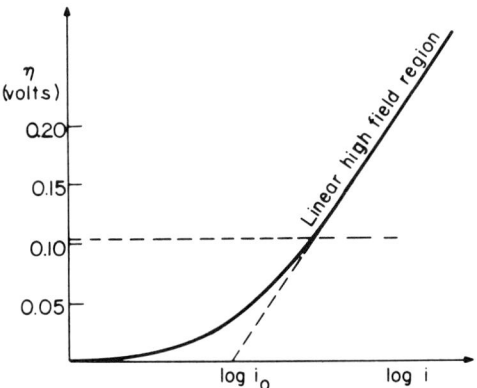

Fig. 8.34. A typical Tafel line for a one-electron-transfer electrode reaction, showing the exponential relationship at high overpotentials, which makes the relation between η and $\log i$ linear.

The IR error can be virtually eliminated by minimizing R, i.e., by choosing high-conductivity electrolytes and using small distances between the test electrode and Luggin tip.

When η values are obtained at various currents, it is possible to obtain η vs. $\log i$ plots.

Such η versus $\log i$ plots are known as *Tafel* lines, in recognition of Tafel who first published measurements which showed the behavior of Eq. (8.47). An example is shown in Fig. 8.34. It will be noticed that the intercept a in the Tafel plot permits a determination of the equilibrium exchange-current density i_0. The slope b has a meaning given by Eq. (8.49) only for the particular case of simple one-step electron-transfer reactions [such as that of Eq. (8.9)]. For other electrode reactions, the relation is more complex and depends on precisely what path and mechanism applies. This aspect of electrodics will be dealt with later (Section 9.5.5*b*).

8.2.11. The Low-Field Approximation: The Linear *i* versus η Law

The exponential high-field law has been derived (Section 8.2.10) by neglecting one of the currents constituting the net current density *i*. The linear low-field law is obtained (*cf*. Section 8.2.9) by expanding the exponentials and, since η is by definition small in this approximation, by retaining only the first two terms of the expansion of each exponential term.

$$i = i_0[e^{(1-\beta)F\eta/RT} - e^{-\beta F\eta/RT}]$$

$$\simeq i_0\left[1 + \frac{(1-\beta)F\eta}{RT} - 1 + \frac{\beta F\eta}{RT}\right]$$

$$\simeq \frac{i_0 F\eta}{RT} \tag{8.53}$$

The error involved in replacing the Butler–Volmer equation by (8.53) is shown in Table 8.4 as a function of overpotential.

This special case of the electrodic reaction slightly off, but near, equilibrium (i.e., small η) shows that electrodic reactions across interfaces exhibit ohmic behavior under low-field conditions. The current density is proportional to the current-producing or current-produced potential difference (the overpotential η). One could write

$$i = \sigma_{M/S}(\eta) = \sigma_{M/S} l \delta X \tag{8.54}$$

where $\sigma_{M/S}$ is the *conductivity* of the metal–solution interface. Equation (8.54) is another example of the fact that, near equilibrium, all flows can be taken to be proportional to their corresponding driving forces (Table 8.5). This linear expression serves to emphasize that the potential producing the current in electrodic reactions driven from an outside source is the *excess* potential difference η.

TABLE 8.4

Linear Approximation of the Butler–Volmer Equation
$\beta = 0.5$, $i_0 = 1$ mA cm^{-2}, and T = 25 °C

η, V	$\dfrac{\eta F}{2RT}$	$i_0 \dfrac{\eta F}{RT}$, mA cm^{-2}	$i_0[e^{(1-\beta)\eta F/RT} - e^{-\beta\eta F/RT}]$, mA cm^{-2}	Error % in linear law
0	0	0	0	0
0.001	0.02	0.039	0.039	<-0.1
0.002	0.039	0.078	0.078	<-0.1
0.005	0.10	0.195	0.195	<-0.5
0.010	0.195	0.390	0.393	-0.7
0.020	0.390	0.780	0.80	-2.5
0.030	0.585	1.17	1.23	-4.9
0.050	0.975	1.95	2.27	-14.1
0.100	1.95	3.90	6.89	-43.4

TABLE 8.5
Forces, Fluxes, and Laws for Different Phenomena Caused by Small Perturbations of Equilibrium

Phenomenon	Force	Flux	Law Form	Law Name
Diffusion	Concentration gradient $\partial c/\partial x$	Molecular flux $\partial n/\partial t$	$\dfrac{\partial n}{\partial t} = D\dfrac{\partial c}{\partial x}$	Fick's law
Electrical conduction	Potential gradient $\partial V/\partial x$	Electrical current I	$I = \dfrac{V}{R}$	Ohm's law
Heat conduction	Temperature gradient $\partial T/\partial x$	Heat flow $\partial Q/\partial t$	$\dfrac{\partial Q}{\partial t} = \varkappa \dfrac{\partial T}{\partial x}$	Fourier's law
Viscous flow	Velocity gradient $\partial v/\partial x$	Shear stress F/A	$\dfrac{F}{A} = \eta \dfrac{\partial v}{\partial x}$	Newton's law
Occurrence of a chemical process	Chemical potential difference ΔG^0	Chemical reaction rate $\partial n/\partial t$	$\dfrac{\partial n}{\partial t} = \vec{k}\Delta c$	A Form of the Mass Action Law ($\Delta c \equiv$ departure of concentration value from that of equilibrium)
Occurrence of an electrochemical reaction	Overpotential η	Electrical current density i	$i = i_0 \dfrac{F}{RT} \eta$	Linearized Butler–Volmer equation

8.2.12. Nonpolarizable and Polarizable Interfaces

The linear low-field law provides a simple way of understanding nonpolarizable and polarizable interfaces. It will be recalled that a nonpolarizable interface is one at which the potential difference does not change easily with the passage of current. It does not *polarize*. As long as one avoided (Chapter 7) considering the charge-transfer reactions, one could not really understand how a nonpolarizable interface works. Now the situation can be made clearer.

Consider the linear law (8.53) written thus

$$\eta = \frac{RT}{F}\frac{i}{i_0}$$

By rearrangement, one obtains

$$\frac{\eta}{i} = \frac{RT}{Fi_0} = \varrho_{M/S} \qquad (8.55)$$

What is the significance of this η/i ratio? The linear law is that analogue of Ohm's law which applies to the interface. The term η/i corresponds to the resistance $\varrho_{M/S}$ of the interface to the charge-transfer reaction. The reaction resistance, which mainly depends upon the exchange-current density i_0, determines what may be termed the *polarizability*, i.e., what overpotential a particular current density needs (for a driven cell) or produces (for a spontaneously performing cell). One can improve the treatment by assuming that perhaps i_0 or, rather, the concentration inside i_0 may not be a constant with current, whereupon it is better to use the differential resistance, which, for η in the linear region, is

$$\left(\frac{\partial \eta}{\partial i}\right)_{c_A,c_D,T} = \frac{RT}{Fi_0} = \varrho_{M/S}$$

Now observe what happens if the equilibrium exchange-current density i_0 tends to very high values, i.e., toward infinity. As $i_0 \to \infty$, $\varrho_{M/S} = (\partial \eta/\partial i)_{c_A,c_D,T} \to 0$. Then, the slope of the η versus i curve is zero, i.e., despite the passage of a current density i across the interface, the overpotential tends to be zero. The interface remains virtually at its equilibrium potential difference. This is precisely the behavior of an ideally nonpolarizable interface.

Earlier (Section 7.2.5c), the nonpolarizable interface has been described as maintaining its potential by leaking charge easily. This picture is correct. But, now, one has a quantitative criterion of how polarizable an interface is; the criterion is the equilibrium exchange-current density i_0. The higher the value of i_0 is, the less does the potential difference across an interface depart from the equilibrium value on the passage of a current. The (hypothetical) ideally nonpolarizable interface (with $i_0 \to \infty$), therefore, is always at the equilibrium potential (Section 7.2.5c).

An exchange-current density of infinity is of course an idealized case. All values of i_0 must be finite, which means that all interfaces show some degree of polarizability. As shown in Table 8.6, the larger the values of i_0 are, the greater is the current density i required to produce a given change

TABLE 8.6

The Influence of i_0 upon the Current Density Required to Attain a Given Overpotential $\Delta\phi - \Delta\phi_e$
$\beta = 0.5$ and $T = 25\,°C$

i_0, A cm^{-2}	Current density [A cm^{-2}] required for the following overpotentials [V]			
	0.001	0.010	0.100	0.200
10^{-6}	4×10^{-8}	3.9×10^{-7}	6.9×10^{-6}	4.9×10^{-5}
10^{-3}	4×10^{-5}	3.9×10^{-4}	6.9×10^{-3}	4.9×10^{-2}
1	4×10^{-2}	0.39	6.9	49.4

of potential from the equilibrium value $\Delta\phi_e$ characteristic of the given reaction.

Similarly, one can conceive of the other extreme, the case of $i_0 \to 0$ (Fig. 8.35). Here, $d\eta/di$, the polarizability, and the reaction resistance $\varrho_{M/S}$ become infinite. The potential departs from the equilibrium values even with a very small current density leaking across the interface. This, however, is precisely what could be expected for a highly polarizable interface; its potential is easily changeable, it can be varied at will by an external power source without passing significant currents.

Thus, the concepts of polarizable and nonpolarizable interfaces are quantified. The value $i_0 \to 0$ is the idealized extreme of a polarizable interface; $i_0 \to \infty$ is the idealized extreme of a nonpolarizable interface.

The concept of the polarizability $d\eta/di$ also shows why instruments with *high input impedance* must be used by those making measurements of the potential differences across electrochemical systems or cells.

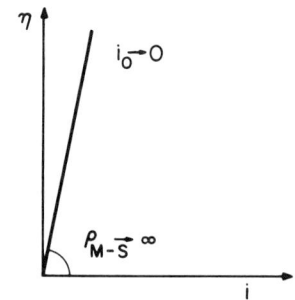

Fig. 8.35. At low i_0 values, the interface has high resistance or polarizability.

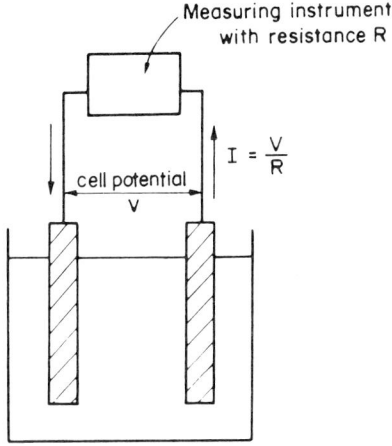

Fig. 8.36. When the measuring instrument draws appreciable current, the measured cell potential V is smaller than that which the cell has in the absence of the current flow drawn by the measuring instrument. Hence, the latter should have a very high impedance.

Thus, an instrument which measures potential differences has itself a particular resistance R. When connected across the cell (Fig. 8.36), the instrument draws a current given by Ohm's law

$$I = \frac{\text{Cell potential}}{R} \tag{8.56}$$

The passage of such a current across an interface would produce an excess —artifactual—potential difference $\Delta \eta$, which would mean that the potential of the interface during measurement had changed from the potential before measurement. This would be poor experimentation; one aims at measurement of the interfacial potential difference without altering it by the process of measurement. It follows from (8.56) that one must use instruments with very high input impedance (or resistances) R. Then, the currents flowing across the electrode during measurement and therefore the disturbances to the potential difference under measurement become negligible.

8.2.13. Zero Net Current and the Classical Law of Nernst

Two *regions* of the general interfacial current-density–overpotential curve have been discussed, the exponential region far from equilibrium and the linear region near equilibrium. Now attention will be focused on one single *point* on the i versus η curve (Fig. 8.37). This is the unique point corresponding to *equilibrium* where the overpotential η and the driving force $I\,\delta X$, are equal to zero; as a consequence, the net current density i is also zero and the de-electronation $\overleftarrow{i}$ and electronation $\overrightarrow{i}$ current densities are equal to each other, $\overleftarrow{i} = \overrightarrow{i}$.

Since, under equilibrium conditions, the individual electronation- and de-electronation-current densities are equal to the equilibrium exchange-current density, i.e. (*cf.* Section 8.2.6),

$$i_0 = \overleftarrow{i} = F\overleftarrow{k}_c c_D e^{(1-\beta)F\Delta\phi_e/RT} = \overrightarrow{i} = F\overrightarrow{k}_c c_{A^+} e^{-\beta F\Delta\phi_e/RT} \tag{8.26}$$

it follows that

$$e^{F\Delta\phi_e/RT} = \frac{\overrightarrow{k}_c}{\overleftarrow{k}_c} \frac{c_{A^+}}{c_D} \tag{8.57}$$

Upon taking logarithms,

$$\Delta\phi_e = \frac{RT}{F} \ln \frac{\overrightarrow{k}_c}{\overleftarrow{k}_c} + \frac{RT}{F} \ln \frac{c_{A^+}}{c_D} \tag{8.58}$$

Now it will be recalled that the chemical rate constants pertain to the condition of zero field at the interface. Can they be measured? No, because, to measure them, one must know when there is zero field, i.e., one must be able to measure the field or the absolute potential difference across a *single* metal–solution interface. This, however, is impossible because, as explained earlier (*cf.* Section 7.2.5*b*), the moment such an attempt is made, one inevitably generates at least one extra phase boundary and therefore one extra electrified interface. So, one always has two or more unknowns,

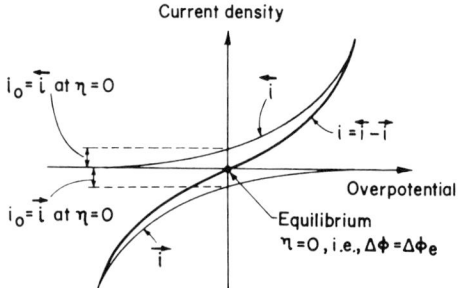

Fig. 8.37. The i versus η relation has one point of particular interest, the point of zero overpotential and zero current. It corresponds to equilibrium at the electrode surface.

the $\Delta\phi$'s across the two or more interfaces, and only one equation, namely, Eq. (7.8), from which to obtain the $\Delta\phi$ across the interface under study.

The conclusion is that one cannot experimentally determine the term $(RT/F) \ln (\vec{k}_c/\overleftarrow{k}_c)$ in Eq. (8.58) because of this lack of knowledge of the absolute value of $\Delta\phi_e$.

Think for a moment about the term $(RT/F) \ln (\vec{k}_c/\overleftarrow{k}_c)$. It is obvious that its value is given by the value assumed by $\Delta\phi_e$ when the concentration ratio c_{A+}/c_D is unity and the term $(RT/F) \ln (c_{A+}/c_D)$ becomes zero. It appears therefore that the term $(RT/F) \ln (\vec{k}_c/\overleftarrow{k}_c)$ represents a particular value of $\Delta\phi_e$, namely, *that value when the concentration ratio of electron acceptor (oxidant) to electron donor (reductant) is standardized or normalized to unity.* For convenience, this standardized or, simply, standard value of $\Delta\phi_e$ is given the symbol $\Delta\phi_e^0$. [Remember that, like $\Delta\phi_e$, its absolute value is not experimentally measurable. It is only possible to relate it to the total potential in a cell (*cf.* Section 8.2.14).] Then

$$\Delta\phi_e^0 = \frac{RT}{F} \ln \frac{\vec{k}_c}{\overleftarrow{k}_c} \qquad (8.59)$$

and

$$\Delta\phi_e = \Delta\phi_e^0 + \frac{RT}{F} \ln \frac{c_{A+}}{c_D} \qquad (8.60)$$

an equation which shows that the equilibrium potential difference varies linearly with the logarithm of the concentration ratio of electron acceptor to electron donor (Fig. 8.38).

Had one started out with the rigorous form of Eq. (8.38), containing the activities a_A and a_D rather than the concentrations c_A and c_D, one would have ended up with the following equation[†]

$$\Delta\phi_e = \Delta\phi_e^0 + \frac{RT}{F} \ln \frac{a_A}{a_D} \qquad (8.61)$$

An equation of this form for the equilibrium or zero-current value of the potential difference $\Delta\phi_e$ across an interface is without doubt the most well-known equation of classical electrochemistry. Here, it has been obtained as the equation for a particular point, namely, equilibrium, on the i versus η curve. But it was first derived with the use of thermodynamic

---

[†] This is why, in Fig. 8.38, there is a deviation from linearity at high concentration ratios; the plot is in terms of molalities (gram-ions of solute per thousand grams of solvent), and the activity coefficient terms have been neglected.

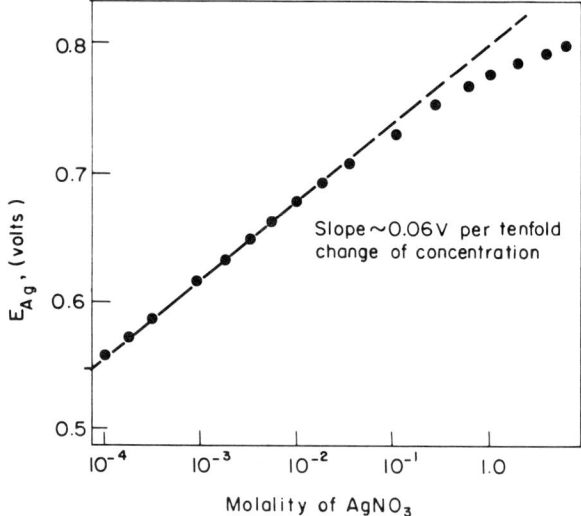

Fig. 8.38. In the case of a silver electrode in a solution of silver ions, the linear relationship between the equilibrium potential and log of concentration of the potential-determining species is maintained over a considerable region of concentrations.

cycles, by the great nineteenth-century physical chemist Walter Nernst, whose name it bears.

It is easy to derive the Nernst equation (8.60) by means of equilibrium, thermodynamic reasoning. One simply applies the condition for equilibrium [cf. Eq. (7.32) of Section 7.3.3c] to an interface. Thus, in a general case of a z-electron exchange in a reaction $M^{z+} + ze \rightleftharpoons M$ at M^{z+}/M interface, the equality of electrochemical potentials of the final and initial states reads

$$\mu_M = \bar{\mu}_{M^{z+}} + z\bar{\mu}_e \tag{A.7.6}$$

since the electrochemical potential of the neutral metal atoms is equal to the chemical potential μ_M. Now,

$$\bar{\mu}_{M^{z+}} = \mu_{M^{z+}} + zF\phi_S$$

and

$$\bar{\mu}_e = \mu_e - F\phi_M$$

the species being on the solution side and on the metal side of the interface, respectively. The ϕ_M and ϕ_S are the inner potentials of those two phases, and the negative signs on $F\phi_M$ reflect the fact that the charge on an electron is -1.

Introducing these into (A7.6), one obtains

$$\mu_M = \mu_{M^{z+}} + \mu_e + zF(\phi_S - \phi_M)$$

Thus, according to the convention that the potential difference across the interface is the inner potential of the metal minus the inner potential of the solution, one has

$$\Delta\phi_e = \phi_M - \phi_S = \frac{1}{zF}(\mu_{M^{z+}} + \mu_e - \mu_M) = -\frac{\Delta G}{zF} \quad (8.62)$$

But it will be recalled that chemical potentials are related to activities by the familiar expression

$$\mu_i = \mu_i^0 + RT \ln a_i \quad (3.54)$$

Since the metal atoms in the metal are of a metal in the pure form, they are in its standard state, so that their activity is considered to be 1, $a_M = 1$. The electrons are also in their standard state, $a_e = 1$, and the equation becomes

$$\Delta\phi_e = \frac{\mu_{M^{z+}}^0 + \mu_e^0 - \mu_M^0}{zF} + \frac{RT}{zF} \ln a_{M^{z+}}$$

and, when $a_{M^{z+}} = 1$,

$$\Delta\phi_e^0 = \frac{\mu_{M^{z+}}^0 + \mu_e^0 - \mu_M^0}{zF} = -\frac{\Delta G^0}{zF}$$

which results in

$$\Delta\phi_e = \Delta\phi_e^0 + \frac{RT}{zF} \ln a_{M^{z+'}} \quad (8.61)$$

the classical law of Nernst.

Correspondingly, it can be shown that, for a more general case of an acceptor A^{z+} which accepts z electrons in its equilibrium with a donor D,

$$\Delta\phi_e = \Delta\phi_e^0 + \frac{RT}{zF} \ln \frac{a_{A^{z+}}}{a_D} \quad (8.61)$$

8.2.14. The Nernst Equation

When the Nernst equation for an interface at equilibrium is written in the form of (8.61) it has an awkward feature. One cannot experimentally measure the $\Delta\phi$'s (cf. Section 7.2.5b). The equation needs some modification before it can be used.

In discussing this modification, it is helpful to consider a concrete example. Suppose that the elctron-transfer reaction at the interface under study is

$$Cu^{++} + 2e \rightleftharpoons Cu$$

i.e., the electron acceptor A is Cu^{++} and the electron donor D is Cu. The Nernst equation for the interface reads (on setting $a_{Cu} = 1$)

$$^{Cu}\Delta^{S}\phi_e = {}^{Cu}\Delta^{S}\phi_e^{0} + \frac{RT}{2F} \ln a_{Cu^{++}} \qquad (8.63)$$

At this stage, two facts are recalled. Firstly, the potential difference across an electrochemical cell, or system, is measurable. Thus, if the Cu^{++}/Cu interface is incorporated in a cell along with a second metal–solution interface, the potential difference across the whole cell is measurable. Secondly, if the second interface is nonpolarizable (i.e., its potential does not depart significantly from the equilibrium value on the passage across it of a small current), it contributes a constant value to the potential difference across the cell. Thus, by choosing a standard hydrogen electrode as the nonpolarizable interface, the following system has been built (Fig. 8.39)

$$\text{Pt, } H_2(P_{H_2} = 1)/H^+(a_{H^+} = 1)//Cu^{++}/Cu/Pt'$$

In this representation, Pt' has been written in at the right to show that a contact potential difference will arise where the platinum wire from the high-input-impedance voltmeter (Fig. 8.39) contacts the copper electrode. The symbol // is used to indicate that the potential due to the junction between the solutions containing the H^+ and Cu^{++} has been minimized.

The potential difference E_e across this cell (i.e., the reading on the voltmeter) consists of the following potential differences [cf. Eq. (7.3), Section 7.2.5c]

$$E_e = {}^{Pt}\Delta^{Pt'}\phi = {}^{Cu}\Delta^{S}\phi_e + {}^{S}\Delta^{Pt}\phi_e^{0} + {}^{Pt'}\Delta^{Cu}\phi \qquad (8.64)$$

Similarly, if the copper ions are at unit activity, $a_{Cu^{++}} = 1$, one has

$$E^0 = {}^{Pt}\Delta^{Pt'}\phi^0 = {}^{Cu}\Delta^{S}\phi_e^{0} + {}^{S}\Delta^{Pt}\phi_e^{0} + {}^{Pt'}\Delta^{Cu}\phi \qquad (8.65)$$

Note that both ${}^{S}\Delta^{Pt}\phi_e^{0}$ and ${}^{Pt'}\Delta^{Cu}\phi$ are constant quantities, the former because the hydrogen electrode is nonpolarizable and the latter because the inner potential difference of two metals *in contact*, ${}^{Pt}\Delta^{Cu}\phi$, depends purely on the two metals and not on what is happening inside the solution.

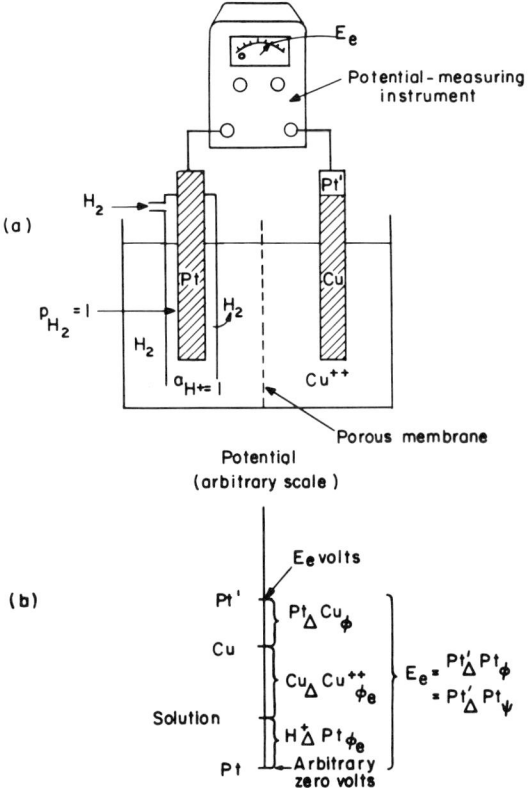

Fig. 8.39. A galvanic cell composed of a copper electrode in cupric ion solution and a standard hydrogen electrode (a) gives a measurable potential difference E_e (b).

Now, if the Nernst equation were of the form

$$E_e = E^0 + \frac{RT}{2F} \ln a_{\text{Cu}^{++}} \tag{8.66}$$

a very convenient expression would result. One could determine E_e^0, the *cell* potential, when the solution has $a_{\text{Cu}^{++}} = 1$, and, once E_e^0 was known, one could predict the cell potential for any other activity of copper ions, from (8.66).

The question is: How can one get an equation of the form of (8.66) from Eq. (8.63)? One is at liberty to add constants to both sides of an equation. So, one can add $^S\!\Delta^{\text{Pt}}\phi_e^0 + {^{\text{Pt}'}}\!\Delta^{\text{Cu}}\phi$ to both sides of Eq. (8.63).

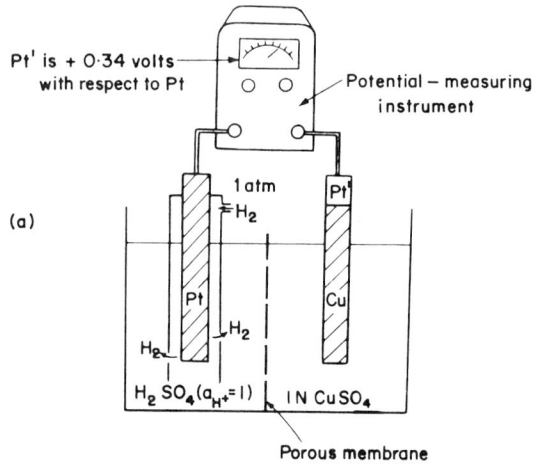

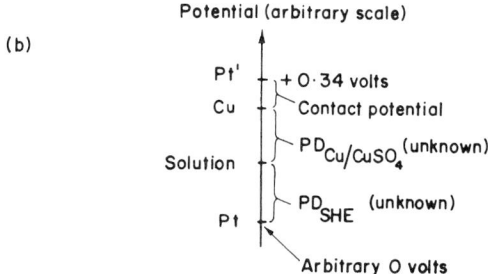

Fig. 8.40. When the cell is so constructed that the activity of cupric ions is unity, the voltmeter should show 0.34 V at 25°C, which is then the standard potential of copper on the standard hydrogen scale.

One obtains

$$[^{Cu}\Delta^S\phi_e + {}^S\Delta^{Pt}\phi_e^0 + {}^{Pt'}\Delta^{Cu}\phi] = [^{Cu}\Delta^S\phi_e^0 + {}^S\Delta^{Pt}\phi_e^0 + {}^{Pt'}\Delta^{Cu}\phi]$$
$$+ \frac{RT}{2F} \ln a_{Cu^{++}}$$

or [cf. Eqs. (8.64) and (8.65)]

$$E_e = E^0 + \frac{RT}{2F} \ln a_{Cu^{++}} \qquad (8.66)$$

What is E^0? It is the potential difference $^{Pt}\Delta^{Pt'}\phi$ across the cell (shown in Fig. 8.39) containing the standard hydrogen electrode.

The measured reading on the voltmeter can be used to define the *equilibrium potential* for the reaction $Cu^{++} + 2e \rightleftharpoons Cu$. For example, it is said that the *standard electrode potential* E^0 of the $Cu^{++} + 2e \rightleftharpoons Cu$ reaction is $+0.337$ V at 25°C. This means (Fig. 8.40) that, when a cell is constructed thus

$$Pt, H_2[p_{H_2} = 1]/H^+[a_{H^+} = 1]//Cu^{++}[a_{Cu^{++}} = 1]/Cu$$

the voltmeter shows that the cell potential is 0.337 V at 25°C with the copper electrode positive. From Eq. (8.64), it is obvious that measuring the standard electrode potential on the standard hydrogen scale (i.e., the special kind of cell potential which arises when one has the electrode concerned made up in a cell with a standard H_2 electrode) is a far cry from measuring the absolute potential difference $\Delta\phi$ across the Cu/Cu^{++} interface; the value E^0 contains the $\Delta\phi$ across the Pt/Cu interface, the unmeasurable $\Delta\phi$ across the Pt/H^+ interface, and, if not reduced to a negligible quantity (see Section 4.54), a small liquid junction potential at the junction between the H^+/Cu^{++} solutions.

Once the standard electrode potential is known, the Nernst equation permits a calculation of what the electrode potential will be when the solution has any other $a_{Cu^{++}}$ in Eq. (8.66). For example, the electrode potential of a copper electrode immersed in a Cu^{++} solution of $a_{Cu^{++}} = 2$

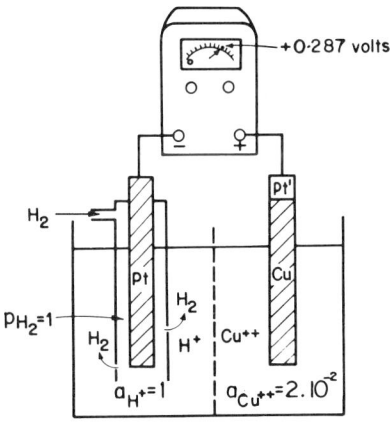

Fig. 8.41. If the activity of cupric ions is changed, e.g., to 2×10^{-2} mole/liter, the voltage of the cell should decrease to 0.287 V, as can be calculated from the Nernst equation.

× 10^{-2} would be (Fig. 8.41)

$$E = 0.337 + \frac{0.059}{2} \log (2 \times 10^{-2}) = 0.287 \text{ V}$$

8.2.15. The Nernst Equation: Its Sphere of Relevance

It is clear that the Nernst equation is the law describing a unique point on the i versus η curve, the law which represents the potential difference across an interface at equilibrium ($i = 0$), and the special case of the general Butler–Volmer equation (8.31) for $\eta = 0$. The heyday of the uses of this equilibrium equation is over. But there was a time when it was pushed into almost every electrochemical situation (including the nonequilibrium ones); and it lives on in an overactive form in University classrooms in 1969.

Perhaps a reason for a certain overuse up to the early 1950's of this equilibrium equation in attempts to unravel electrodic phenomena arose in the following way: One may rewrite the Nernst equation in the exponential form thus

$$\frac{c_A}{c_D} = e^{-zF(\Delta\phi_e - \Delta\phi_e^0)/RT} \tag{8.68}$$

The equation now bears a certain superficial resemblance to the exponential i versus η law [cf. Eq. (8.47) with $\eta = (\Delta\phi - \Delta\phi_e)$]

$$i = i_0 e^{[(1-\beta)F/RT](\Delta\phi - \Delta\phi_e)} \tag{8.47}$$

This i versus η law says that changes in the potential difference across the interface are current producing. The exponential form of the Nernst equation [Eq. (8.68)] ascribes changes in potential to the concentration ratio c_A/c_D. It was previously thought, before the development of electrodics, that the rate of electron exchange across interfaces was very high ($i_0 \to \infty$) so that interfaces were always *at equilibrium* in respect to the electron exchange, and one could then use (8.68) to relate c_A/c_D to the rate of reaction and finally to current density by connecting the values of the concentration of reactant near the interface to those in the bulk of the solution by diffusion equations (Section 4.2.11). Such an attempt to apply the Nernst approach—thermodynamical—to kinetic processes *far* from equilibrium was of course bound to fail. It can lead to physically meaningless conclusions.

Consider, e.g., the hydrogen-evolution reaction, $2H^+ + 2e = H_2$, and suppose that one tried to interpret the i versus η exponential law by means of the Nernst equation. One could say that the hydrogen-evolution reaction requires an initial step in which hydrogen ions discharge (are electronated) to form atomic hydrogen H

$$H^+ + e \to H$$

If this reaction is at equilibrium, the Nernst equation would read

$$E = E^0 - \frac{RT}{F} \ln p_H + \frac{RT}{F} \ln c_{H^+}$$

where E^0 is the standard potential for the reaction, and p_H is the partial pressure of atomic hydrogen. The value of E^0, calculated with the standard free energy of the reaction $2H \to H_2$, is -1.98 V.

To push the Nernst view further, one can attribute the entire overpotential η produced by a current density i to changes in partial pressure of atomic hydrogen (instead of delay and nonequilibrium in electronation). Thus, one would say that any given potential E observed under the passage of current was an equilibrium potential determined by the corresponding pressure of atomic hydrogen which accumulated or became depleted while the current was passing. The overpotential of the hydrogen electrode, η, could then be expressed by

$$\eta = E - E_{e[\text{hydrogen electrode}]} = \left[-1.98 - \frac{RT}{F} \ln \left(\frac{p_H}{c_{H^+}}\right)\right] - \frac{RT}{F} \ln \frac{c_{H^+}}{p_{H_2}}$$

and, for $p_{H_2} = 1$ atm,

$$\eta = -1.98 - \frac{RT}{F} \ln p_H \tag{8.69}$$

It is even possible to connect up p_H to the current density i and have a linear Tafel relation between η and $\ln i$.

Things appear to go well until one asks: What values of partial pressure are required to produce, say, 1 V of cathodic overpotential? In other words, what pressure of atomic hydrogen corresponds to the potential of -1 V? Substituting this value in Eq. (8.69),

$$-\log p_H = -\frac{-1 + 1.98}{0.058}$$

or

$$p_H = 10^{-16.9} \text{ atm}$$

This pressure corresponds to about 150 atoms per cm³, or to less than one hydrogen atom per 100,000 cm² of the surface:[†] That means that on an electrode of, say, 10 or 100 cm² surface, there is hardly any possibility of finding a single adsorbed hydrogen atom.

Obviously, then, one cannot attribute physical significance to such a Nernstian interpretation of the rate of a reaction. Any equation, and the approach on which it is based, has a sphere of relevance. The Nernst equation is properly derived for an equilibrium situation. It should not be forced into interpreting a situation which is far from equilibrium (as electron transfer reactions often are). But that is what was done all too frequently in an attempt to treat electrodic phenomena before about 1950, with the frustrating failures which impeded progress in the electrochemical field for many decades.

8.2.16. Looking Back

Once the device possibilities of electrochemical systems were glimpsed, the basic problem of electron transfer across electrode–electrolyte interfaces demanded understanding. It was to consider this problem that one began to look at what happens at the interface.

Things are simple at the instant of immersion of a metal in an electrolytic solution. There is no field and no potential difference across the interface. Reactions (e.g., $M^+ + e \rightarrow M$) run for a very short while *chemically*. However, the very occurrence of a charge-transfer reaction across the interface in one direction creates an electric field, a fraction of which puts a brake on the reaction $M^+ + e \rightarrow M$. The same field, however, has an accelerating effect on the charge-transfer reaction in the opposite direction, $M \rightarrow M^+ + e$.

Thus, the electronation and de-electronation reactions modify the field, and the field, in feed-back style, alters the rates until the two rates of $M^+ + e \rightarrow M$ and $M \rightarrow M^+ + e$ become equal. This is equilibrium. Underlying the condition of zero net currents, an equilibrium exchange-current density i_0 flows across the interface in both directions. The potential difference across the interface at equilibrium depends upon the activity ratio of electron acceptor to electron donor. Alter the ratio, and the equilibrium potential changes.

If all interfaces remained at equilibrium, electrochemical devices would not work. Substances could not be produced electrochemically; neither would power production be possible. Net currents must flow across inter-

---

[†] Assuming the surface layer to be about 5 Å thick, there will be $150.5 \times 10^{-8} = 7.5 \times 10^{-6}$ atoms cm⁻², i.e., about 1 atom in 100,000 cm².

faces for devices to work. There must be net electronation or net de-electronation. Interfaces need to move away from equilibrium and its characteristic potential $\Delta\phi_e$.

It turned out that the current density i across an interface is linked to the *overpotential*, or excess potential, $\eta = \Delta\phi - \Delta\phi_e$. In a driven electrochemical system, it is the excess-potential difference η (or, rather, the excess field δX) which drives the current density. Make $\eta = \Delta\phi - \Delta\phi_e$ more positive, and the net de-electronation-current density $i = \overleftarrow{i} - \overrightarrow{i}$ increases. In a self-driving cell, it is the current density i which sets up an excess field δX and an overpotential η. Increase the net de-electronation-current density, and the excess potential at the electron-sink electrode (anode) becomes more positive.

The quantitative description of these features of an interface during charge transfer is crystallized in the most fundamental equation of electrodics, the Butler–Volmer equation, which is essentially an i versus η relation.

This general equation covers charge transfer at electrified interfaces under conditions both of zero excess field, low excess fields, and high excess fields and of low and high overpotentials. Thus, the Butler–Volmer equation spans a whole range of conditions. At equilibrium, it settles down into the Nernst equation, near equilibrium it elegantly reduces to a linear i versus η Ohm's law for interfaces, and far away from equilibrium it becomes an exponential i versus η law.

The Butler–Volmer equation has therefore yielded much which is essential to the first appreciation of electrode kinetics. It has not however been mined out. One has to dig deeper, and, after electron transfer at one interface has been understood, in a more general way, electrochemical systems, or cells, with two electrode–electrolyte interfaces must be tackled. It is the theoretical descriptions of these systems which give the bases to electrochemical-energy production, electrochemical synthesis, the stability of metals, and the functioning of some, perhaps many, biological systems.

Further Reading

1. J. Tafel, *Z. Physik. Chem.* (*Leipzig*), **50**: 641 (1905).
2. J. A. V. Butler, *Trans. Faraday Soc.*, **19**: 729 (1924).
3. T. Erdey-Gruz and M. Volmer, *Z. Physik. Chem.* (*Leipzig*), **150**: 203 (1930).
4. H. Eyring, S. Glasstone, K. J. Laidler, *J. Chem. Phys.*, **7**: 1053 (1939).
5. P. Dolin and B. Erschler, *Acta Physicochim. USSR* **13**: 747 (1940).
6. J. O'M. Bockris, "Electrode Kinetics," in: J. O'M. Bockris, ed., *Modern Aspects of Electrochemistry*, Vol. I, Butterworth's Publications, Ltd., London, 1954.

7. B. E. Conway, "Electrochemical Kinetic Principles," in: *Theory and Principles of Electrode Processes*, (*Modern Concepts in Chemistry*), Chapter 6, The Ronald Press Co., New York, 1965.
8. J. O'M. Bockris and S. Srinivasan, "Electrode Kinetics," in: *Fuell Cells: Their Electrochemistry*, Chapter 2, McGraw-Hill Book Company, New York, 1969.

8.3. THE BUTLER–VOLMER EQUATION: FURTHER DETAILS

8.3.1. The Need for a Careful Look at Some Quantities in the Butler–Volmer Equation

The interphase regions constitute the essential parts of an electrochemical system. It is there that charge is pumped into and out of the system. The behavior of the system depends therefore on the charge-transfer reactions which occur at the interfaces. The basic law of charge-transfer reactions has been expressed through the Butler–Volmer electrodic equation (8.31)

$$i = i_0[e^{(1-\beta)F\eta/RT} - e^{-\beta F\eta/RT}] \tag{8.31}$$

with[†]

$$i_0 = F\vec{k}c_A^0 e^{-\beta F \Delta\phi_e/RT} = F\overleftarrow{k}c_D^0 e^{(1-\beta)F\Delta\phi_e/RT} \tag{8.70}$$

There are, however, several questions which can be raised. For example, is the simple and distinctly naïve interpretation (Section 8.2.4) of β (the symmetry factor) adequate? It has been brought in as a factor expressing the fraction of the potential difference across the double layer which affects the rates of the electronation and de-electronation reactions. Should one not get a better comprehension of β? Can one relate it to molecular quantities?

The basic electrodic equation also conceals a geographic problem. The whole analysis has proceeded from the statement that the electron acceptors are positioned near the electrode before being involved in the charge-transfer reaction. Where? Does it matter? It would surely be expected to, and very much. It will be recalled (Chapter 7) that both the potential and concentrations of various species can vary near the interface. As the location of the initial state of the reaction is altered, the potential differences and concentrations appearing in the basic equation also vary.

What, therefore, is the potential difference to be used? $^M\Delta^{\mathrm{IHP}}\phi_e$ the potential difference from the metal to the contact adsorption plane, or IHP

[†] To emphasize the point, $c_A{}^0$ and $c_D{}^0$ have been written with superscripts to indicate that they refer to *bulk* concentrations.

(inner Helmholtz plane, cf. Section 7.1.8)? ${}^M\varDelta^{\text{OHP}}\phi_e$, the potential difference from the metal to the OHP (outer Helmholtz plane cf. Section 7.1.8), or ${}^M\varDelta^S\phi_e = \varDelta\phi_e$, the potential difference from the metal to the bulk of the electrolytic solution? Does one consider the fraction β of the whole potential difference across the interface or only a fraction β of a part of this potential difference?

Similarly, what concentrations of electron acceptors and donors must be fed into the basic equation? Bulk values or the values at the OHP or the values of the contact-adsorbed species?

It is clear that these questions cluster around some basic quantities which appear in the Butler–Volmer electrodic equation, such as β, the interfacial concentrations of electron acceptors and donors, and the potential difference which affects the reaction rate. Some attempt must be made to tackle these questions.

8.3.2. The Relation between Structure at the Electrified Interface and the Rate of Charge-Transfer Reactions

The question of what potential difference should be fed into the basic electrodic equation (8.70) will be considered now. To answer the question, one has to ask: Where must the reactant particles be positioned in order to undergo charge transfer? Suppose that, in the electronation reaction,

$$A^{z+} + e = D^{(z-1)+}$$

the z-valent A^{z+} ions start the charge-transfer reaction from the OHP (Fig. 8.42), the plane defined by the *locus* of centers of the hydrated A^{z+} ions in contact with the hydrated electrode. It is actually fair to consider as electron acceptors only those ions which are in the first layer near the electrode surface, as will be shown shortly (*cf.* Section 8.5).

On assuming for simplicity no contact adsorption, there is an (approximately) linear potential drop from the metal to the OHP and the potential difference is ${}^M\varDelta^{\text{OHP}}\phi$ (Fig. 8.42). It is this potential difference which must be inserted into the expression for the equilibrium exchange-charge density i_0 because the electron transfer from the metal to the acceptor ions in solution (and conversely from donors to the electrode) is likely to involve predominantly those ions which are closest to the electrode, and when there is no contact adsorption, this closest layer of ions will be located at the OHP.

Thus, one must replace $\varDelta\phi_e = {}^M\varDelta^S\phi_e$ in equation (8.70) with ${}^M\varDelta^{\text{OHP}}\phi_e$.

Does ${}^M\varDelta^{\text{OHP}}\phi_e$ constitute the whole potential difference ${}^M\varDelta^S\phi_e = \varDelta\phi_e$ between the metal and the bulk of the solution? In general, it does not

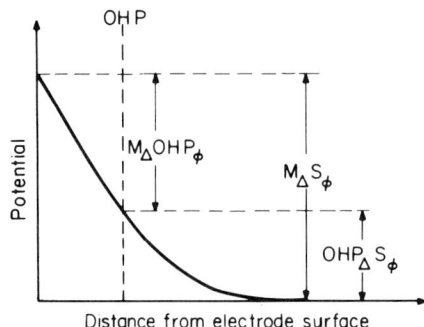

Fig. 8.42. The potential difference across the interface can be divided into the linear portion of the layer extending to the OHP, at which the ions ready to discharge are located, and a portion in the diffuse part of the double layer, which is called the *electrokinetic* or ζ *potential*.

(Fig. 8.42), i.e.,

$$^{M}\Delta^{\mathrm{OHP}}\phi_e \neq {}^{M}\Delta^{S}\phi_e$$

because, if there is a Gouy–Chapman diffuse-charge region, there will be a further potential drop $^{\mathrm{OHP}}\Delta^{S}\phi_e$ between the OHP and the bulk of the solution (*cf.* Section 7.4.8).

Having considered the potential difference to be used in the expression for the exchange current density, the concentration of electron acceptors A^{z+} will now be examined. If there is a potential drop between the solution *bulk* and the OHP, the electron-acceptor ions A^{z+} have to do work to climb this potential hill $^{\mathrm{OHP}}\Delta^{S}\phi_e$. Consequently, at equilibrium, the concentration $c_{A^{z+},\mathrm{OHP}}$ of A^{z+} particles at the OHP will not be equal to their bulk concentration $c^{0}_{A^{z+}}$.

Assuming that there is an equilibrium distribution of A^{z+} ions, one can use the Boltzmann distribution law to relate the two concentrations

$$c_{A^{z+},\mathrm{OHP}} = c^{0}_{A^{z+}} e^{-W/kT} \qquad (8.71)$$

where W can be taken to be the electrical work to carry the A^{z+} ions through the potential difference from the solution bulk to the OHP. This electrical work is $zF^{\mathrm{OHP}}\Delta^{S}\phi_e$.

Hence, instead of assuming that the concentration term which will go into equations relating current to potential is simply that of the bulk (or the equivalent of it expressed in moles per square centimeter), one should take the relevant concentration as that at the OHP, i.e., c^{0}_{A} or $c^{0}_{A^{z+}}$

in equation (8.70) must be replaced with $c_{A^{z+},OHP}$. Or, from (8.71), and generalizing the symbols

$$c_{OHP} = c_{bulk}e^{-z_i(^{OHP}\Delta^S\phi F/RT)} \quad (8.72)$$

Let these two corrections for the presence of the double layer with structure be worked through in a very simple case. One has, as before,

$$i = i_0[e^{(1-\beta)\eta F/RT} - e^{-\beta\eta F/RT}] \quad (8.31)$$

Suppose one takes i_0 as pertaining to an electron-accepting solute, e.g., an M^{z+} cation, while the concentration of donor species (which might, for simplicity, be looked on as the corresponding metal M) is kept constant.

The corresponding i_0 value under conditions such that the double-layer structure is neglected (electrons exchange with layers of ions at bulk concentration) would be

$$(i_0)_{ideal} = F\vec{k}_c c_{A^{z+}}e^{-\beta\Delta\phi_e F/RT} \quad (8.73)$$

If (8.73) is now rewritten, taking into account a simple double-layer structure (e.g., that described in 7.4.8), one has (substituting for the bulk concentration $c_{A^{z+}}$ of 8.73 with the concentration in the OHP; and changing $\Delta\phi_e$ of 8.73 into the potential difference between the metal and the OHP):

$$(i_0)_{real} = F\vec{k}c_{A^{z+}}e^{-z_{A^{z+}}(F^{OHP}\Delta^S\phi/RT)}e^{-\beta F(\Delta\phi_e - ^{OHP}\Delta^S\phi)} \quad (8.74)$$

Thus,

$$(i_0)_{real} = (i_0)_{ideal}e^{-(F^{OHP}\Delta^S\phi/RT)(z_{A^+}-\beta)} \quad (8.75)$$

Such a change in the value of i_0, neglecting the oversimplifications made in view of the discussion below, makes a significant difference to the expected dependence of i_0 on concentration because the value of the potential of the OHP is concentration dependent. The dependence need not follow Nernst's law for $^{OHP}\Delta^S\phi$ is not a reversible potential of the type considered there. Nevertheless, under some simple limiting conditions one can show from the theory of the electrical double layer that[†]

$$^{OHP}\Delta^S\phi = ^{OHP}\Delta^S\phi^0 + \frac{RT}{z_i F}\ln c_i \quad (8.78)$$

[†] Thus, (7.102), written fully, is

$$C_{D.L.}(\Delta\phi - ^{OHP}\Delta^S\phi) = 2\left(\frac{\varepsilon nkT}{2\pi}\right)^{1/2}\sinh\frac{ze_0\,^{OHP}\Delta^S\phi}{2kT} \quad (8.76)$$

For $\Delta\phi > ^{OHP}\Delta^S\phi$, $^{OHP}\Delta^S\phi$ (say) sufficiently negative, and $\Delta\phi$ negative,

$$-C_{D.L.}\Delta\phi = 2\left(\frac{\varepsilon kT}{2\pi}\right)^{1/2}n^{1/2}(-e^{-ze_0\,^{OHP}\Delta^S\phi/2kT}) \quad (8.77)$$

(Footnote continued on p. 914)

Consider, thus, i_0 as a function of concentration for a double layer in which the existence of a significant difference between the potential of the OHP and the solution bulk is neglected, i.e., the primitive case of Eq. (8.73): The $\Delta\phi_e$ is given by Nernst's law, so that, substituting this Eq. (8.60) in (8.73), taking logarithms, and differentiating with respect to log $c_{A^{z+}}$, one obtains

$$\frac{\partial \log i_0}{\partial \log c_{A^{z+}}} = 1 - \frac{\beta}{z_{A^{z+}}} \tag{8.79}$$

However, if one uses the double-layer structure-corrected version of i_0 [Eq. (8.74)] and takes into account the variation due to (8.72), then a similar substitution of Nernst's law for $\Delta\phi_e$ and of Eq. (8.78) for $^{\text{OHP}}\Delta^s\phi$ in (8.76) gives the very different result

$$\frac{\partial \ln i_0}{\partial \ln a_{A^{z+}}} = 0 \tag{8.80}$$

The effect of double-layer structure on electrode kinetics was first introduced by the great Russian electrochemist Frumkin, who thus opened an entire area of electrode kinetics, namely, that of the relation of the interfacial structure to the velocity of reactions across it. Frumkin thereby produced the broadening and the clarification which always comes when there is a coupling of two apparently separate but related fields.

In this simple account only the effect of double-layer structure on exchange current density has been worked out. However, other important and central consequences exist, for example, the effect of introducing the modifications of (8.75) into (8.31). What will happen when one thinks of the dependence of i upon η, i.e., upon the evaluation of the Tafel slope, $\partial \ln i/\partial \eta$? This will no longer be equal simply to $RT/(1 - \beta)F$, for now, the dependence of $^{\text{OHP}}\Delta^s\phi$ upon potential must be accounted for. Luckily for the simplicity of interpretations of data in electrode kinetics, it is easy to show that such an influence is not numerically significant except when

---

If $c_{\text{D.L.}}$ and $\Delta\phi$ are not dependent on n, the number of ions per cubic centimeter in the solution,

$$^{\text{OHP}}\Delta^s\phi = {}^{\text{OHP}}\Delta^s\phi^0 + \frac{RT}{zF} \ln n$$

where

$$-\ln \frac{2}{C_{\text{D.L.}}\Delta\phi} \left(\frac{\varepsilon kT}{2\pi}\right)^{1/2} = {}^{\text{OHP}}\Delta^s\phi^0$$

Conversion of n to c, the corresponding gram-ionic quantity, gives (8.78).

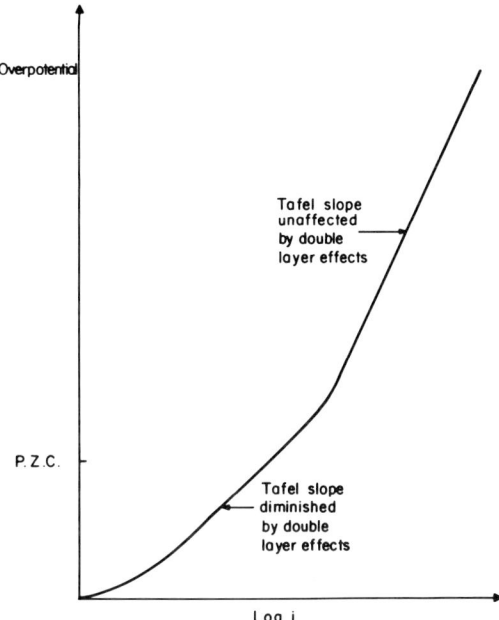

Fig. 8.43. The existence of the region of diffuse charge in the electrified interface has an effect on the Tafel relation, making it change the slope and deviate from linearity.

the electrode potential is in the vicinity of the potential of zero charge. Then, the value of $\partial \ln i/\partial \eta$ increases compared with that under conditions far from the pzc (Fig. 8.43).

The present discussion involves almost an orgy of oversimplification with the justification that simple and easily understood results could be obtained. Such results will hold their validity only if a large number of implicit assumptions are granted. Having got the reader to see the entrance to the field of the effects of double-layer structure on electrode kinetics, let the tacit assumptions be clearly stated.

The first is an almost universal one in electrodics, that it is only the layer of ions next to the electronic conductor which provides it with electrons or takes electrons from it. There will be something to say about this when the dependence of the tunneling of electrons through energy barriers is discussed in Section 8.5.

The second is that the assumed structure of the double layer, though no longer absolutely primitive, is a very simple one. It is not desirable (as has been done in the above deductions) to assume the absence of contact-adsorbed ions since, particularly at solid electrodes, significant concentrations of these are present (*cf.* Section 7.4).

The third assumption made here is that the dependence of $^{\text{OHP}}\Delta^s\phi$ on concentration is given by Eq. (8.78); this is only true as an approximation and under the conditions already mentioned. A more subtle point is that $^{\text{OHP}}\Delta^s\phi$ has been assumed to be independent of the value of $\Delta\phi_e$ and more detailed examination again shows that the approximation is only valid far from the pzc.

One helpful point is that, although these double-layer effects are still present to a small extent at concentrations in the realm of $1N$, they tend to become small in more concentrated (and hence many practical) solutions. For this reason and because of the complexity of their more realistic treatment, they will often be suppressed in equations written below.

One last point—another door to the same general area—should be mentioned. It is that there are many double-layer structure effects which arise during the adsorption of organic reactants on electrodes. These are usually uncharged. For this reason, they can adsorb on the electrode to a greater extent than anions can (where the ionic charge causes lateral repulsion, *cf.* 7.4.26). The organic molecules displace water and change $\Delta\chi$, the surface potential (7.4.30). The area concerned touches on many effects, e.g., in electrodeposition theory (the effect of additives) and the effect of corrosion inhibitors (11.1.14).

8.3.3. The Interfacial Concentrations May Depend on Ionic Transport in the Electrolyte

Observe (Fig. 8.43) what happens at the OHP (referred to as the $x = 0$ plane) when a constant current is driven across the interface. The electron acceptors at the $x = 0$ plane are *consumed* at a constant rate by the reaction

$$A^{z+} + e \to D^{(z-1)+} \tag{8.81}$$

To keep the reaction going and the electronation-current constant, a steady supply of electron-acceptor ions must be maintained by transport from the electrolyte bulk. This transport may be by diffusion (random walk) or migration under an electric field (drift).

In considering the effect of double-layer structure on electrode kinetics, it was pointed out that the existence of a diffuse charge region causes the concentration at the outer Helmholtz plane to differ from the bulk concentration. The consumption of electron acceptors by the electronation reaction, and the need for transport processes to maintain the supply may

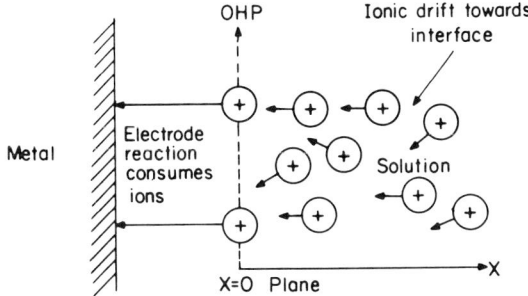

Fig. 8.44. As the electrode reaction consumes ions from the OHP, their concentration is bound to decrease for the ionic drift diffusion from the bulk of the solution can start only after a certain concentration gradient is established.

be yet another reason for the interfacial concentration of electron acceptors to deviate from the bulk concentrations. Thus, arises an effect of transport processes upon the rates of electrode reactions (Fig. 8.44). The possibility of this effect is only mentioned here; this influence of ionics upon electrodics is elaborated upon in Chapter 9.

8.3.4. What is the Physical Meaning of the Symmetry Factor β?

8.3.4a. The Factor β Is at the Center of Electrode Kinetics. The central equation of electrode kinetics is the Butler–Volmer equation

$$i = i_0[e^{(1-\beta)F\eta/RT} - e^{-\beta F\eta/RT}] \tag{8.31}$$

The paramount importance of this repeatedly mentioned equation derives from the fact that it connects two basic aspects of charge transfer at an electrified interface, namely, the equilibrium current, which represents the rate of the transformation of substances at the interface without the accelerating effect of overpotential, and the electrical effects that result from the application of overpotential. Thus, the equation embraces the chemical and electrical aspects of charge transfer.

Now examine the exponential terms in the Butler Volmer equation. The two quantities of interest in the exponents are β and η. It is these two quantities which determine the interrelationship between the potential difference across the interface and the rate of the charge-transfer reaction; they determine how the electrode potential affects or is affected by the rate.

The quantity $\eta = \Delta\phi - \Delta\phi_e$, which determines how far the interfacial

TABLE 8.7

Values of β for Some Reactions

Metal	System	β
Platinum	$Fe^{3+} + e \to Fe^{2+}$	0.58
Platinum	$Ce^{4+} + e \to Ce^{3+}$	0.75
Mercury	$Ti^{4+} + e \to Ti^{3+}$	0.42
Mercury	$2H^+ + 2e \to H_2$	0.50
Nickel	$2H^+ + 2e \to H_2$	0.58
Silver	$Ag^+ + e \to Ag$	0.55

potential difference has departed from equilibrium, can be externally controlled. Thus, in a *driven* electrochemical system (e.g., a substance-producing device), it can be chosen at will by turning a knob on the external power supply, and, in a *self-driving* cell (e.g., an energy-producing device), it depends on the current drawn from the cell and this current in turn depends on the resistance of the external load. Thus, η is at the command of the experimenter.

The symmetry factor β, however, is an intrinsic characteristic of the given charge-transfer reaction at the given interface. It determines what fraction of the electrical energy resulting from the displacement of the potential from the equilibrium value affects the rate of electrochemical transformation. In short, β depends upon the interfacial reaction, not upon an experimenter's manipulation. It is one of the most important parameters in electrodics. Experimentally obtained β values for *simple*, *single* electron-exchange reactions gather around 0.5 (Table 8.7).

8.3.4b. A Preliminary to a Second Theory of β: Potential-Energy–Distance Relations of Particles Undergoing Charge Transfer. A simple transfer reaction will be considered, e.g., the electronation of a hydronium ion, i.e., a proton hydrated by one water molecule, H^+—H_2O,

$$M(e) + H^+\text{---}H_2O \to M\text{---}H + H_2O$$

The symbol $M(e)$ represents an electron in the metal, and the symbol M—H represents a chemisorbed hydrogen atom.

It can be seen that the progress of the charge-transfer reaction involves the *breaking* of the bond between H^+ and H_2O and the *making* of a bond between the metal M and the electronated electron acceptor H.

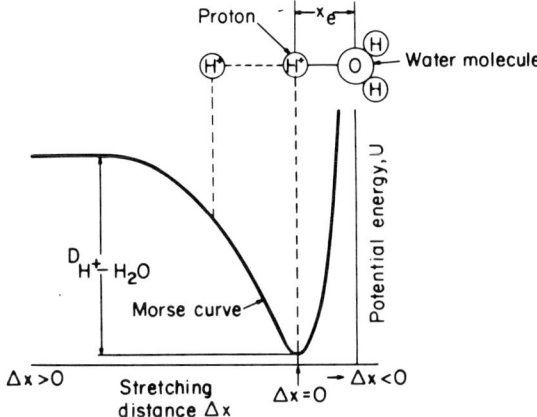

Fig. 8.45. A plot of potential energy with distance between atoms reflects the increase in energy as the bond between the hydrogen ion and the water molecule is stretched.

An interesting way of looking at such a bond-break–bond-make reaction suggests itself. At first, one can forget the bond-make part of the reaction and consider only the bond-break portion. As this bond stretches, the energy of the system increases, and, finally, when the stretching is sufficient, the bond is broken; the molecule dissociates. One can plot the energy as a function of stretching distance, i.e., the distance between the particles, as in Fig. 8.45.

A precise potential energy *versus* distance relation for the systems of particles, H^+—OH_2, *in solution* is relatively complicated. But a simple relation for the analogous process of stretching and rupture of a *diatomic* molecule is known from spectroscopy. This is called the *Morse equation*

$$U_x = D(1 - e^{-a\Delta x})^2 \qquad (8.82)$$

where U_x is the potential energy of the system; Δx is the stretching distance which goes to infinity when the bond is broken, in which case $e^{-a\Delta x} \to 0$; and $U = D$, the dissociation energy (Fig. 8.45). What is being done here is to use the Morse equation, which correctly applies to diatomic molecules in the gas phase, for a pseudo-diatomic system H^+—(OH_2), i.e., the OH_2 is treated as if it were a single particle when the dissociation of H^+ from OH_2 is under discussion.

Now one can ignore the bond-break portion and consider only the bond-make portion. In short, one can consider the energy change as the

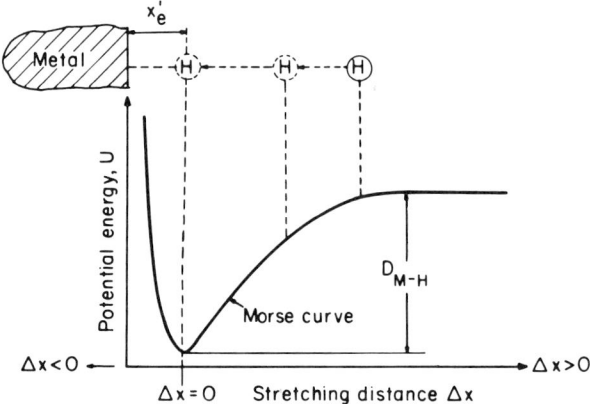

Fig. 8.46. A similar change in potential energy but of opposite direction to that in Fig. 8.45 is obtained as a hydrogen atom is approaching a metal surface. The Morse curve reflects the energy of adsorption of the atom.

hydrogen atom approaches the metal. Again one has a Morse type of curve (Fig. 8.46).

The two Morse curves—one for the H^+—OH_2 stretching and the other for the M—H stretching—can now be associated so that one obtains a single diagram. To do this, the minima of the two curves must be positioned properly relative to each other.

This positioning can be done on the basis of a reasonable (but assumed) model for the way in which the hydrogen atom (the product of the reaction $M(e) + H^+$—$OH_2 = M$—$H + H_2O$) and H^+—OH_2 (the reactant) are located with respect to the metal electrode. For the distance between the electron (in M) and H^+—OH_2, one has to remember the picture of the electrified interface and think of how close *hydrated* ions can approach the electrode (*cf.* Section 7.1.8). It will be recalled that hydrated ions can approach as close as the OHP. One must also have some knowledge of internuclear distances in metal hydrides and, in this way, try to get at the M—H distance. This allows an approximate estimate to be made of the relative positions of the minima of the two curves.

When the two Morse curves are combined into a single diagram, an interesting result emerges (Fig. 8.47): The middle section of the potential-energy–distance curve has the shape of a potential-energy barrier. This barrier represents the energy of the system as the particle H^+ jumps from the OHP to the metal.

Of course, a barrier synthesized in this way is quite a poor approxi-

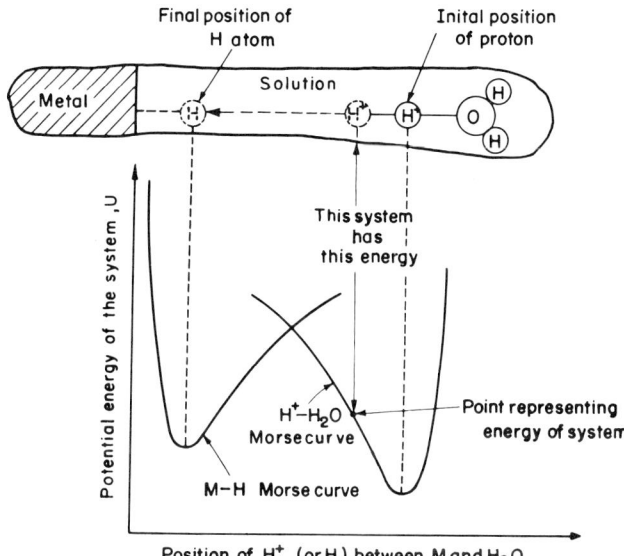

Fig. 8.47. When the minima of the two potential-energy curves are positioned at a proper place, the curves intersect, thus forming the potential-energy barrier.

mation to the actual barrier. When the H^+—H_2O bond is stretching, the influence of the metal on the energy of the system has been ignored. Similarly, one has ignored the influence of the water molecule on the energy as the M—H bond is stretched. Thus, the superposition of Morse curves assumes that the energy of a three-body system (M, H^+, and H_2O) with three-body interactions can be obtained by properly superposing the energies from two two-body systems each involving two-body interactions.[†] If one takes into account the effects of the third atom on the energy of the system, one result is a lowering of the energy in all the configurations, particularly the energy near the intersection of the Morse curves. This contributes (Fig. 8.48) to the *rounded top* of an actual barrier obtained by sectioning the corresponding potential-energy surface. Nevertheless, it is useful to analyze the approximate barrier obtained by the combination of two Morse curves. It enables one to clutch some aspects of reality out of a situation, the more realistic treatment of which would be too complex to treat as yet.

[†] This is a procedure for treating three-body systems approximately. For example, the sun–moon–earth problem is handled this way.

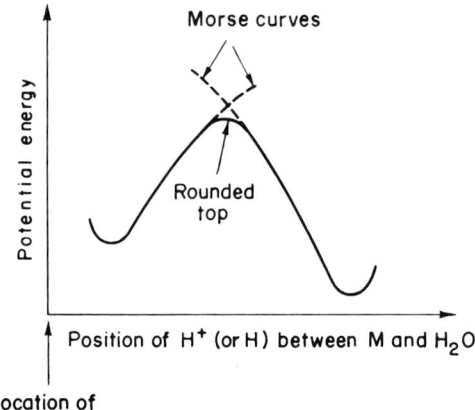

Fig. 8.48. Since, in reality, the potential energy of any position is a result of three-body interaction, its top is somewhat rounded compared with the result of the intersection of the Morse curves.

8.3.4c. A Simple Picture of the Symmetry Factor. In order to employ simple geometry, one now ignores the curvature of the Morse curves and considers that the potential-energy barrier near the intersection point is made up of straight lines (Fig. 8.49). This simplifying analogue of the barrier is useful for a first-base discussion of the symmetry factor β.

At the outset, recall how the symmetry factor was introduced (Section 8.2.4). The charge-transfer reaction was roughly pictured as the jump of an electron acceptor toward the electrode during which, somewhere on route, an electron jumped to the particle and completed its job of electro-

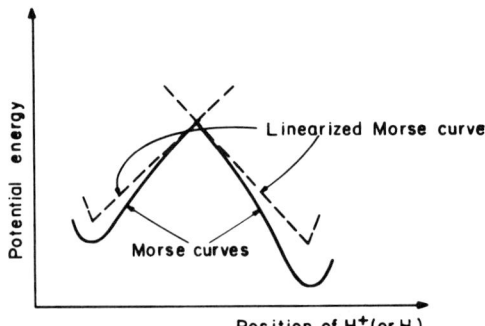

Fig. 8.49. For the sake of simplicity, the Morse curves can be linearized, and the model thus obtained can be considered sufficiently real for general (if approximate) considerations.

nation. Representing the energy of the system by a point on an energy–distance curve, the progress of the jump could be charted by the movement of the representative point across a potential-energy barrier. Once the point climbed to the top of the barrier (the activated state), the rest of the jump was assured (automatic). But this climb to the peak (or activation) requires some work to be done. The chemical activation work is altered by the presence of an electric field. More electrical work is done by or on the ion in climbing the barrier in the presence of the electric field at the interface than without it. The question is: How is the extra electrical work of activation to be computed?

The first approach (Section 8.2.4) at this computation ran along the following lines: The electrical work of activation arises because, in the activation process, charges have to be moved through the difference of potential between the initial and activated states, i.e., from $x_1 + x_2$ to x_1 in Fig. 8.17. It was necessary, therefore, to know what fraction of the total jump distance is the distance between the initial state and the barrier peak. This distance ratio was defined as the symmetry factor β, i.e.,

$$\beta = \frac{\text{Distance along reaction coordinate between initial and activated states}}{\text{Distance along reaction coordinate between initial and final states}}$$

The essential point which emerged from this first discussion of β is that only a fraction of the potential difference across the double layer, not the whole potential difference, is operative on the reaction. That there is a fraction β becomes clear; what the fraction is remains a problem as long as the barrier shape is not known. This point of view must only be considered as the *first murmuring* of a theory of β, the symmetry factor.

A different (second) approach may be adopted. The main point in this new approach is that the value of β will be shown to depend on the *relative slopes of the potential-energy–distance curves* representing the energies of the particles (rather than the position of the summit of the potential energy barrier within the path of the ion in its jump from double layer to electrode [Section 8.2.4]).

Suppose that a potential difference $\Delta\phi$ is applied across the interface. How does this affect the barrier obtained if one linearizes the Morse curves for the electrodic reaction

$$M(e) + H^+ - H_2O \rightarrow M - H + H_2O? \tag{8.83}$$

The curve (or rather its linearized version) for the stretching of the M—H bond in the system M—H + H_2O will not be influenced by the field because the particles M—H and H_2O are not charged.

924 CHAPTER 8

The effect of the electric field on the linear curve for the stretching of the H—OH bond in the system $M(e) + H^+$—OH_2 has to be thought about now. When the potential difference across the interface is changed (from 0 to ϕ), the energy of the H^+—OH_2 part of the system suffers little change, but the energy of the electron in the metal [and thus the energy of the left-hand side of (8.83)] is altered by an amount which is easily calculable. The change in electron energy is equal to the change in potential times the electronic charge. Hence, the total change of energy of the initial state (the electron in the metal and H^+—OH_2 at the OHP) is $e_0 \Delta\phi$ per system, or $F\Delta\phi$ per mole of systems $M(e) + H^+$—H_2O. What this implies is that the linear version of the Morse curve for H^+—OH_2 stretching in the system $M(e) + H^+$—H_2O is *shifted vertically* through an energy $F\Delta\phi$ (Fig. 8.50).

The vertical shift has arisen from the application of an absolute potential difference of $\Delta\phi$ to a hypothetical interface, initially with zero potential difference across it, i.e., $\Delta\phi = 0$. But the argument is valid for any change of potential across the interface. Thus, if the double-layer potential is initially $\Delta\phi_e$ (i.e., the interface is at equilibrium) and then the potential is changed to $\Delta\phi$, the Morse curve for the initial state is shifted vertically through an energy $F(\Delta\phi - \Delta\phi_e)$, or $F\eta$.

As a consequence of the vertical shift of one linear curve, the critical activation energy for the reaction (the main factor upon which its rate depends) is altered from $E_e^{\neq}$ at equilibrium (i.e., AF in Fig. 8.51) to $E_\eta^{\neq}$ at the overpotential η (i.e., HD of Fig. 8.51). *The difference $\Delta E^{\neq}$ between*

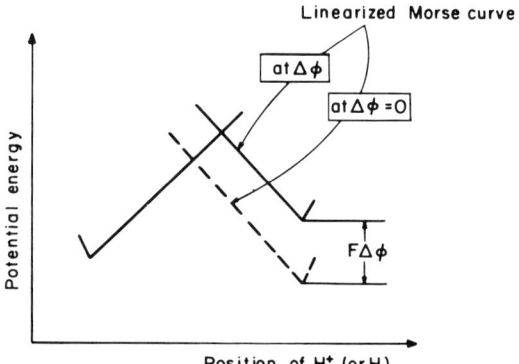

Fig. 8.50. When the potential difference across the interface is changed from zero to a value $\Delta\phi$, the Morse curve of the initial state is shifted vertically by the amount of electrical energy $F\Delta\phi$.

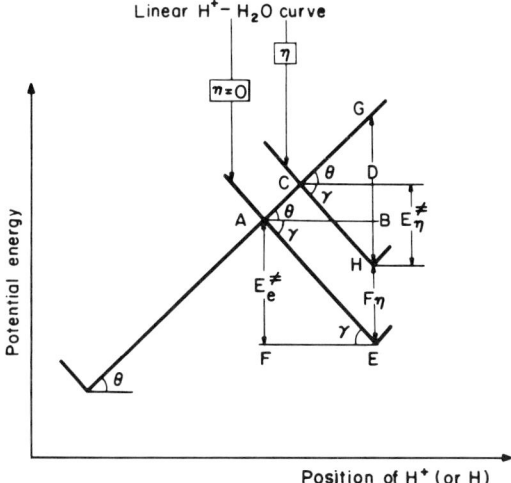

Fig. 8.51. As a consequence of the vertical shift of one linear curve, the critical activation energy is altered.

the two activation energies has resulted from the electrical energy $F\eta$ that has been introduced into the reaction. What is the relationship between $\Delta E^{\neq}$ and $F\eta$? The change $\Delta E^{\neq}$ in activation energy decides the net current output; the $F\eta$ is the input electrical energy channeled into the interface. One seeks to know: How much did the activation energy decrease for the given energy input $F\eta$?

In terms of the linear analogue, the question is answered by a trivial exercise in geometry (see Fig. 8.51). One has

$$AB = FE = \frac{E_e^{\neq}}{\tan \gamma} \text{[from } \triangle AEF\text{]} \quad \text{and} \quad AB = \frac{GE - E_e^{\neq}}{\tan \theta} \text{[from } \triangle ABG\text{]}$$

and, therefore,

$$E_e^{\neq} = \frac{\tan \gamma}{\tan \theta} (GE - E_e^{\neq}) \tag{8.84}$$

Further,

$$CD = \frac{E_\eta^{\neq}}{\tan \gamma} \text{[from } \triangle CDH\text{]} \quad \text{and} \quad CD = \frac{GH - E_\eta^{\neq}}{\tan \theta} \text{[from } \triangle CDG\text{]}$$

Hence,

$$E_\eta^{\neq} = \frac{\tan \gamma}{\tan \theta} (GH - E_\eta^{\neq}) \tag{8.85}$$

By making use of Eqs. (8.84) and (8.85), it follows that change in activation energy

$$\Delta E^{\neq} = E_e^{\neq} - E_n^{\neq}$$
$$= \frac{\tan \gamma}{\tan \theta}(GE - GH) - (E_e^{\neq} - E_n^{\neq})$$
$$= \frac{\tan \gamma}{\tan \theta}(F\eta - \Delta E^{\neq})$$

or

$$\Delta E^{\neq} = \left(\frac{\tan \gamma}{\tan \gamma + \tan \theta}\right) F\eta \tag{8.86}$$

This is a basic result. The change in activation energy due to a change in the electric field in the double layer has been computed. It depends on the input electric energy $F\eta$ and a trigonometric function which cannot exceed unity. This fraction determines how much of the input electric energy fed into the interface goes toward affecting the activation energy and therefore the net rate of the reaction. The fraction has the basic characteristics of the symmetry factor, with which it will be identified.

Thus, it has been shown, by linearizing Morse curves, that (*cf.* Fig. 8.52):

$$\beta = \frac{\text{Change of activation energy, } \Delta E^{\neq}}{\text{Change of electrical energy, } F\eta} = \frac{\tan \gamma}{\tan \theta + \tan \gamma} \tag{8.87}$$

One had proceeded previously (in the derivation of the Butler–Volmer equation, *cf.* Section 8.2.4) on the basis that $\Delta E^{\neq} = \beta F\eta$, and it is fair to ask: What new knowledge has emerged? The symmetry fractor has now been given in terms of the slopes (tan γ and tan θ) of the linearized Morse curves, and these slopes are related to those *molecular* quantities (e.g., force constants of the molecular bonds involved) which determine the shape and slopes of potential-energy–distance relations (linearized for simplicity).

The symmetry factor β is obviously a central entity in electrodics and a fundamental quantity in the theoretical treatment of charge transfer at interfaces, particularly in the relating of electrode kinetics to solid state physics.

8.3.4d. Is the β in the Butler–Volmer Equation Independent of Overpotential? In order to consider the influence of the current-producing (or current-produced) overpotential η on the activation energy, the Morse curves used to synthesize the potential-energy barrier were linearized, and then one linear curve was shifted vertically through an energy $F\eta$. During this shift brought about by a change of interfacial potential difference, the

(a)

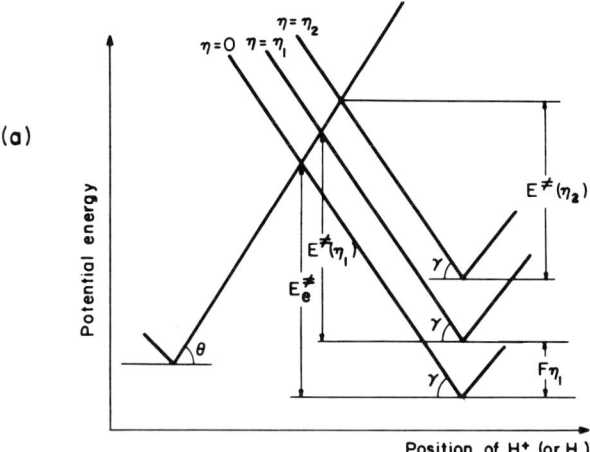

(b)

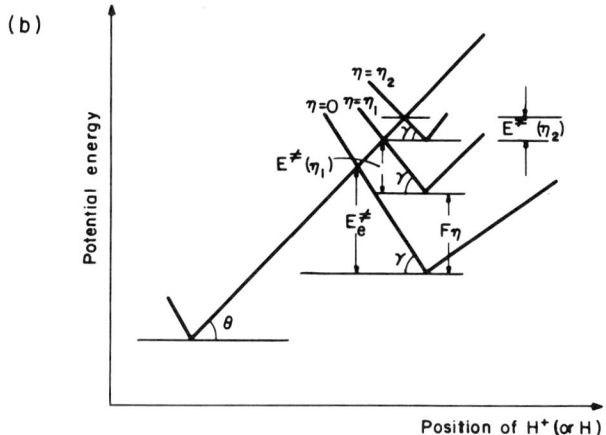

Fig. 8.52. The influence of overpotential on the symmetry factor β: (a) reaction of low i_0: β independent of η; (b) reaction of high i_0: β dependent on η, tending to zero in the limit of high η.

slopes $\tan \gamma$ and $\tan \theta$ of the linear curves were maintained constant (Fig. 8.51). On this approximate basis, the symmetry factor, which is a function only of the slopes γ and θ of the linear curves, appears to remain constant during a change of potential.

Is this result a feature of barriers at interfaces or merely a consequence of shifting a *linear* curve? It is clear that, once a linear curve is displaced vertically, it cannot but yield a *parallel* shift of the curve and therefore a constant β. The apparent constancy of β with potential is a result of the linearization of the potential-energy–distance curves.

A glance at Fig. 8.52 shows, however, that, when the activation energy at equilibrium, $E_e^{\pm}$, is large, i.e., for electrode reactions of low exchange-current density i_0 [*cf.* Eq. (8.70)], the slopes of the linear curves and, hence, β do not change significantly with overpotential. Such changes in β become likely only for reactions which have very low equilibrium activation energies (i.e., very high i_0's). One can take it, therefore, that, for all but very fast electrode reactions, the symmetry factor β will be independent of overpotential η over a reasonably large (e.g., hundreds of millivolts) range of potentials. At sufficiently high overpotentials, the curves will be changed in relative position sufficiently that they will begin to intersect at positions of differing curvature (real, rather than idealized, potential energy curves tend to decrease in slope near their minima). Directly the curvature of one of the potential-energy–distance curves at the point of intersection begins to change compared with that of the other curve, β begins to change [*cf.* Eq. (8.86)].

8.3.5. Summing-up of Further Details on the Butler–Volmer Equation

The basic task of electrodics is to understand the relation between the current density and the potential difference across an interface.

Simple considerations of potential-energy barriers for the occurrence of electrodic reaction led to the development of the fundamental equation of electrodics, the Butler–Volmer equation

$$i = i_0[e^{(1-\beta)F\eta/RT} - e^{-\beta F\eta/RT}] \quad (8.31)$$

with

$$i_0 = Fc_A \frac{kT}{h} e^{-\Delta \vec{G}_c^{0\pm}/RT} e^{-\beta F \Delta \phi_e/RT}$$

$$= Fc_D \frac{kT}{h} e^{-\Delta \vec{G}^0/RT} e^{(1-\beta)F\Delta\phi_e/RT} \quad (8.70)$$

This section (8.3) has been embarked upon to throw light on three questions: (1) What is the plane on the solution side up to which the metal–solution potential drop is considered? (2) What concentrations c_A and c_D must be used in the Butler–Volmer equation, bulk or interfacial concentrations? (3) What is the physical significance of the symmetry factor β?

Analysis revealed that the answers to the first two questions are straightforward. If the electron acceptor starts off the reaction (the climb of the barrier) positioned at the OHP, the potential difference $\Delta\phi$, in the Butler–Volmer equation is that between the metal and the OHP. Further, the concentrations also pertain to this plane.

It is possible, however, to write the exchange-current density i_0 so that it always refers to the *bulk* concentrations and to the *total* potential difference across the whole interphase region. In this case, double-layer and concentration corrections enter into the Butler–Volmer equation as modified to take into account the fact that $\phi_{\text{OHP}} \neq \phi_{\text{bulk}}$ and $c_{\text{A,D}}$ [at the OHP] $\neq c_{\text{A,D}}^0$ [in the bulk].

A consideration of the physical significance of the symmetry factor was made in an approximate fashion by synthesizing a potential-energy barrier from two Morse curves. Each of these represents the bond-make and bond-break part of a simple electron-transfer reaction. A simplifying linearization of the Morse curves (themselves simplifications of the actual energy–distance plots) near the barrier peak permitted a simple approximate geometric interpretation of β in terms of the slopes of the linearized curves. The symmetry factor was shown to represent the fraction of the input electrical energy by which the electrical part of the activation energy decreases and thus the net rate of the reaction increases. An important practical function of electrodics is to form the rational basis of the application both of electrical energy for the direct synthesis of compounds and of conversion of chemical to electrical energy. It can now be realized what a central role β plays in these practical subjects, for its value determines how much change of rate of production of a substance one gets when one changes the potential on the electrodes by a given amount or how much change in power of an energy converter will result from a change of the external load. The acceleration of an electric car would thus, for a given source of stored electrical energy, be a function of β for the electrode reactions concerned (*cf.* 11.3).

Further Reading

1. W. Nernst, *Z. Physik. Chem.*, **47**: 52 (1904).
2. T. Erdey-Gruz and M. Volmer, *Z. Physik. Chem.*, **150A**: 203 (1930).
3. A. N. Frumkin, *Z. Physik. Chem.*, **A164**: 121 (1933).
4. J. Horiuti and M. Polanyi, *Acta Physicochim. USSR*, **2**: 505 (1935).
5. J. A. V. Butler, *Proc. Roy. Soc. (London)*, **A157**: 423 (1936).
6. P. Delahay, *Double Layer and Electrode Kinetics*, Interscience Publishers, Inc., New York, London, Sydney, 1965.
7. K. J. Vetter, *Electrochemical Kinetics 2, Theory of Overvoltage*, Academic Press, Inc., New York, London, 1967.

8.4. THE CURRENT–POTENTIAL LAWS AT OTHER TYPES OF CHARGED INTERFACES

8.4.1. Semiconductor n–p Junctions

One important method of sharpening one's understanding of a situation is to compare and contrast it with other situations. The listing of the samenesses and differences generates clearer vision and fresh insight.

The purpose here is to have a look at the current–potential relations for some other types of interfaces while keeping in mind charge transfer at an electrode–electrolyte interface and its basic law, the Butler–Volmer equation

$$i = i_0[e^{(1-\beta)F\eta/RT} - e^{-\beta F\eta/RT}] \qquad (8.31)$$

Consider an interface formed by joining together an n type of semiconductor (e.g., germanium "doped" with arsenic atoms which donate electrons, *cf.* Table 8.8) and a p type of semiconductor (e.g., germanium doped with indium atoms which accept electrons). The charge carriers in the n type of material will be mostly electrons, and the p type of material will have a majority of holes (*cf.* Section 7.7.7).

Before the junction is formed, there is overall electroneutrality in both types of material. Thus, the positive charge on the immobile arsenic donors in the n type of material exactly balances the negative charge of the free

TABLE 8.8

Summary of Features of Impurity Conduction in Silicon and Germanium

Conductivity type	n type or excess	p type or defect
Conduction by	(Excess) electrons	Holes
Energy band in which carrier moves	Conduction	Valence bond
Sign of carrier	Negative	Positive
Valence of impurity atom	5	3
Name for impurity atom	Donor	Acceptor
Typical impurities	Elements of Group V: Phosphorus, P Arsenic, As Antimony, Sb	Elements of Group III: Boron, B Aluminum, Al Gallium, Ga Indium, In

electrons; and the negative charge on the fixed indium acceptors in the p type of material balances out the positive charge on the mobile holes.

Once the two types of material are brought face to face to form a junction, electrical contact is established and a path is provided for electrons and holes to move from one side of the junction to the other (Fig. 8.53a).

Consider the hole movement first. In the n type of material, holes are generated when electrons from the valence band jump to acceptor atoms (*cf.* Section 7.7). These holes can random-walk across the junction into the p type of material. Conversely, holes from the p side can random-walk into the n type of material where they are consumed in a hole–electron recombination process (the reverse of a hole-generation process). Both electrons and holes have considerable mobility (Table 8.9).

Since one starts off with a far larger hole concentration in the p type of material than the n type of material, there will initially be more holes taking the $p \to n$ random walk than the $n \to p$ random walk. One has stated in microscopic language that there will be diffusion of holes in the $p \to n$ direction (Fig. 8.53b).

TABLE 8.9

Room-Temperature Energy Gap and Electron and Hole Mobilities for Some Semiconductors

Semiconductor	Energy gap, eV	Electron mobility v_e, cm^2 V^{-1} sec^{-1}	Hole mobility v_p, cm^2 V^{-1} sec^{-1}
Carbon (diamond)	5.6	1,800	1200
Germanium	0.66	3,900	1700
Silicon	1.09	1,420	250
Tellurium	0.34	300	200
PbS	0.37	500	150
ZnO	3.1–3.2	85	—
AlSb	1.52	50	150
GaAs	1.35	4,000	400
GaSb	0.7	5,000	1000
InP	1.3	3,500	700
InSb	0.17	80,000	4000
Mg$_2$Si	0.7–0.8	400	70
Mg$_2$Ge	0.6–0.7	500	100
Mg$_2$Sn	0.2	300	250

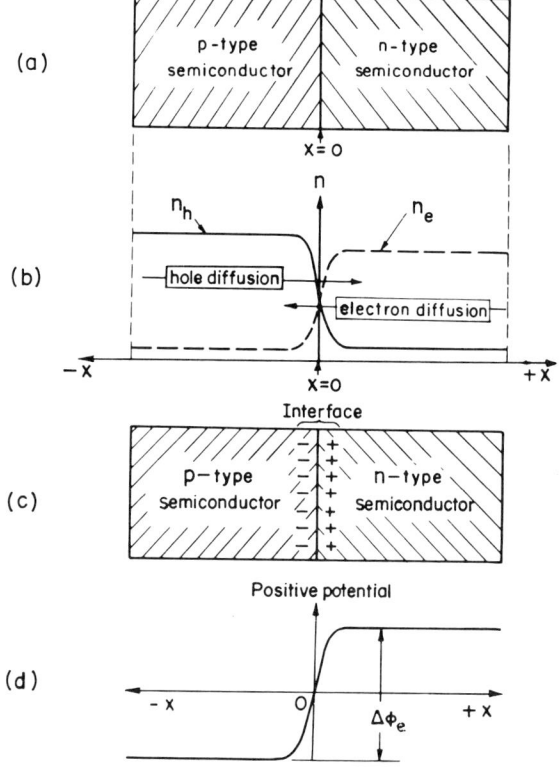

Fig. 8.53. When a junction between a *p* type and an *n* type of semiconductor is established (a), a diffusion of holes and electrons in opposite directions takes place (b). This results in a separation of charge (c) and a formation of electrical potential difference across the interface (d).

But what is the result of this $p \rightarrow n$ hole diffusion? The net $p \rightarrow n$ transport of holes leaves a negative charge on the *p* material and confers a positive charge on the *n* material (Fig. 8.53c). A potential difference develops (Fig. 8.53d). Further, this charging of the two sides of the interface and the resultant potential difference acts precisely in such a manner as to oppose further $p \rightarrow n$ hole diffusion (Fig. 8.54).

Thus, equilibrium is reached when the driving *force* for the diffusion (the concentration gradient) is just compensated for by the electric field (the potential gradient). Under these equilibrium conditions, there is an equilibrium net charge on each side of the junction and an equilibrium

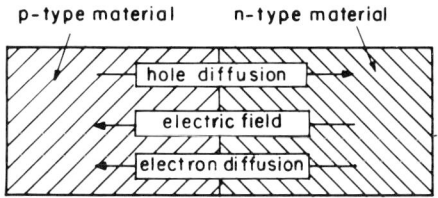

Fig. 8.54. The established electric field opposes further diffusion of holes across the interface.

potential difference $\Delta\phi_e$. This whole process is analogous to the way charge transfer across a nonpolarizable electrode–solution interface results in the establishment of an equilibrium potential difference $\Delta\phi_e$ across the interface.

Since there is no net diffusion under equilibrium conditions, the $n \to p$ hole current is equal to the $p \to n$ hole current. These equilibrium currents are analogous to the equilibrium exchange currents at an electrode–solution interface. They represent the exchange of holes across the junction between the n and p types of material and shall be designated by the symbol $i_{0,h}$. This i_0 shall now be examined more carefully.

Consider the holes which are making the $p \to n$ crossing. The number of holes approaching a unit of area of the junction per second is proportional to the number per unit volume of holes on the p side, i.e., $n_{h,p}$. Will they cross the junction? Each hole approaching the interface finds that it has to surmount the potential difference $\Delta\phi_e$, and the probability that it will climb the barrier is given by the Boltzmann term[†] $e^{-e_0 \Delta\phi_e / kT}$. Hence, the $p \to n$ hole current density at equilibrium is

$$\vec{i}_{h, \Delta\phi_e} \propto n_{h,p} e^{-e_0 \Delta\phi_e / kT}$$
$$= \vec{k} n_{h,p} e^{-e_0 \Delta\phi_e / kT} \tag{8.88}$$

where the arrow $\to$ over the i represents $p \to n$ crossings and $\vec{k}$ is the proportionality constant.

Now think of the $n \to p$ hole current. When the holes from the n type of medium reach the junction (cf. Fig. 8.53c), they do not at all see any barrier due to an electrical potential difference, so they simply tumble over the potential drop. Hence, the $n \to p$ hole current density at equilibrium

---

[†] The absence of β in this expression will be commented on below.

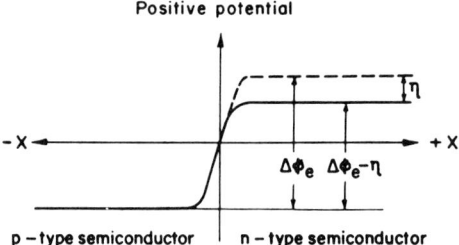

Fig. 8.55. As the potential difference across the interface is lowered by superimposing an external field, the current starts flowing.

is controlled only by diffusion and is simply proportional[†] to the number $n_{h,n}$ [n side] of holes in the n type of material

$$\overleftarrow{i}_{h,\Delta\phi_e} = \overleftarrow{k} n_{h,n} \tag{8.89}$$

where the arrow $\leftarrow$ over the i indicates $p \leftarrow n$ crossings.

Hence, at equilibrium,

$$\vec{i}_{0,h} = \vec{i}_{h,\Delta\phi_e} = \overleftarrow{i}_{h,\Delta\phi_e} \tag{8.90}$$

Now, what happens if the potential difference across the junction is lowered by an amount η (Fig. 8.55)? The holes making the $p \rightarrow n$ crossing find a smaller barrier to climb, and, hence, the $p \rightarrow n$ hole current density becomes

$$\vec{i}_{h,(\Delta\phi_e-\eta)} = \vec{k} n_{h,p} e^{-e_0(\Delta\phi_e-\eta)/kT}$$
$$= i_{0,h} e^{e_0\eta/kT} \tag{8.91}$$

But the holes crossing from the n to p type of material still have no barrier to climb at all. Hence, the $n \rightarrow p$ hole current density still depends only on the number of holes in the n type of material and not on the potential difference across the junction, i.e., i_h [$n \rightarrow p$] is unaffected by the field

$$\overleftarrow{i}_{h,(\Delta\phi_e-\eta)} = i_{0,h} \tag{8.92}$$

The net hole current density is given by the difference in the hole

---

[†] The proportionality constant depends on the fraction moving normal to, and colliding with, the interface per second. Hence, if the velocities of holes in both types of material are the same, the $\vec{k}$'s in Eqs. (8.88) and (8.89) can be assumed to be equal.

current densities for the two directions

$$\begin{aligned} i_h &= \vec{i}_{h,(\Delta\phi_e-\eta)} - \overleftarrow{i}_{h,(\Delta\phi_e-\eta)} \\ &= i_{0,h} e^{+e_0\eta/kT} - i_{0,h} \\ &= i_{0,h}(e^{+e_0\eta/kT} - 1) \end{aligned} \qquad (8.93)$$

All these arguments can be applied to the *electrons* making $n \to p$ and $p \to n$ crossings and giving rise to electron current densities. The net electron current density is given by an expression similar to (8.93), i.e.,

$$i_e = i_{0,e}(e^{e_0\eta/kT} - 1) \qquad (8.94)$$

The total current density across the junction is therefore equal to the sum of the electron and hole current densities, just as the total ionic-migration current density in an electrolyte is equal to the sum of the current densities due to positive and negative ions (*cf.* Section 4.3).

Hence, the current–potential law for an *n–p* junction is (Fig. 8.56)

$$i = i_0'(e^{e_0\eta/kT} - 1) \qquad (8.95)$$

where

$$i_0' = i_{0,h} + i_{0,e} \qquad (8.96)$$

Notice that, as η (the departure from the equilibrium potential) increases, $e^{e_0\eta/kT}$ increases in comparison to unity until, when $e^{e_0\eta/kT} \gg 1$,

$$i = i_0' e^{e_0\eta/kT} \qquad (8.97)$$

the exponential law for *n–p* junctions.

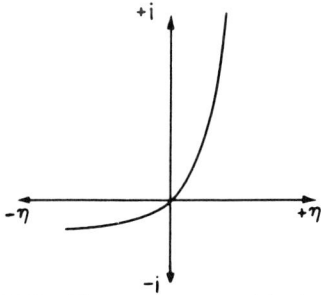

Fig. 8.56. The current–potential relation at a *p–n* semiconductor junction differs from that at an electrode–solution interface by being totally asymmetrical.

One is now in a position to compare the current–potential law for an electrode–electrolyte interface[†] (which has been referred to as an *e–i* junction) with that for any *n–p* junction

$$i = i_0[e^{(1-\beta)e_0\eta/kT} - e^{-\beta e_0\eta/kT}] \quad [e\text{–}i \text{ junction}] \quad (8.98)$$

$$i = i_0'[e^{e_0\eta/kT} - 1] \quad [n\text{–}p \text{ junction}] \quad (8.95)$$

For large departures from equilibrium, i.e., large η, both types of interfaces tend to give an exponential *i* versus η law. Thus,

$$i = i_0 e^{(1-\beta)e_0\eta/kT} \quad [e\text{–}i \text{ junction}] \quad (8.99)$$

$$i = i_0' e^{e_0\eta/kT} \quad [n\text{–}p \text{ junction}] \quad (8.97)$$

For small departures from equilibrium, i.e., small η, a linear *i* versus η law is obtained for both *e–i* and *n–p* junctions.

$$i = i_0 e_0 \eta / kT \quad [e\text{–}i \text{ junction}] \quad (8.100)$$

$$i = i_0' e_0 \eta / kT \quad [n\text{–}p \text{ junction}] \quad (8.101)$$

It is seen, therefore, that there are basic similarities in the *i* versus η laws for both types of interfaces, but there is an important difference. There is no symmetry factor β in the exponential *i* versus η law for semiconductor *n–p* junctions. Why?

Think back to the origin of the symmetry factor in *electrodic* expressions. The main point to be noted is that there is a *hill-shaped* potential-energy barrier *even in the absence of an electric field*. This barrier has to do with the atomic movements in bond stretching (*cf.* Section 8.3.4) which are a prerequisite of processes such as chemical reactions and diffusion of atoms and ions. What the electric field does in the case of charged particles is to modify the *already existing* potential barriers. The modification is such that only a fraction $(1 - \beta)$ of the input electric energy $e_0\eta$ turns up in the change of activation energy and hence in the rate expression. This is because the atom movements necessary for the system to reach the barrier peak are only a *fraction* of the total distance over which the potential difference extends.

The situation in the case of the transfer of holes (or electrons) across *n–p* junctions is different. Firstly, the only difficulty which the electron has

---

[†] To facilitate the comparison, e_0/k is written instead of $Ne_0/Nk = F/R$.

to overcome is that due to the electric field. When there is no field, there is no barrier. This is because the barrier is not an expression of the energies involved in atomic movements; there are no atomic movements as prerequisites to the movements of holes or electrons. Secondly, whereas potential-energy barriers for atom movements and reactions are like hills, the barrier for hole and electron movements is like a cliff with its attendant implication that "falling over the cliff" does not involve an activation energy (Fig. 8.55). Finally, since the holes and electrons reach the barrier top only after traversing the *whole* distance over which the field extends, the entire $e_0\eta$—not a fraction $(1 - \beta)e_0\eta$—affects the hole and electron movements.

There is therefore one essential conclusion from the comparison of electrodic e–i junctions and semiconductor n–p junctions: *The symmetry factor β originates in the atomic movements which are a necessary condition for the charge-transfer reactions at electrode–electrolyte interfaces. Interfacial charge-transfer processes which do not involve such movements do not involve this factor.*

8.4.2. The Current across Biological Membranes

The exponential current–potential law has been shown to be obtained for charge transfer across solid–liquid (e.g., electrode–electrolyte) and solid–solid (e.g., semiconductor-junction) interfaces. What fundamental condi-

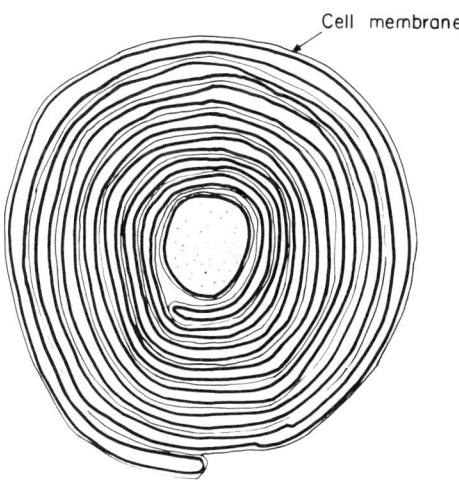

Fig. 8.57. The inside of a nerve cell is wrapped by folded membranes which fuse into a tough, compact wrapping.

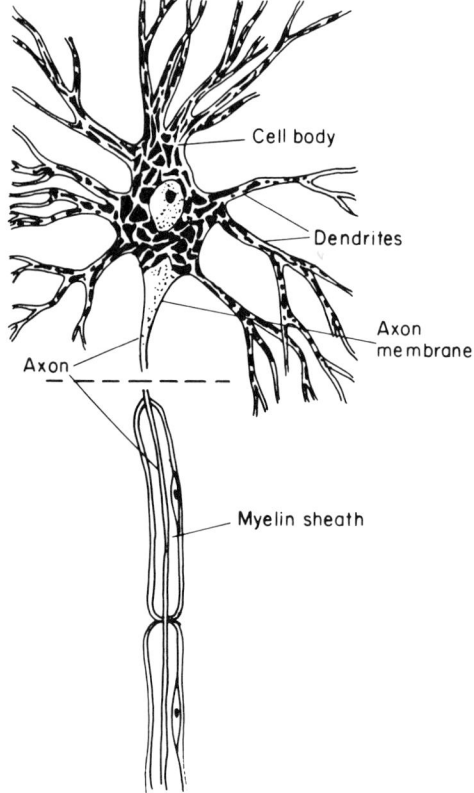

Fig. 8.58. The long nerve cell, the axon, is covered by the membrane 50 to 100 Å thick, which separates the biological fluids inside and outside the cell.

tions are sufficient to provoke the existence of an exponential dependence of current on the potential difference across an interface? The principal requirement is that there must be a potential-energy barrier (at least in one direction) which hinders the movement of charged particles across the interface and which can be modified in height by the interfacial potential difference. Could liquid–liquid or, perhaps, gel–liquid interfaces display an exponential current–potential law for charge transfer across their interfaces? Such gel–liquid interfaces are common in biological systems. The inside of a living cell is separated from the outside by a membrane which is a gel (Fig. 8.57).

Consider the *axon* of a squid (~1 mm in diameter).

The axon membrane (~50 to 100 Å thick) separates the biological fluids inside and outside the cell (Fig. 8.58). These fluids are aqueous electrolytic solutions of almost equal conductivity, but the chemical composition of the two solutions is different. Thus, Na^+ and Cl^- ions constitute more than 90% of the charged species *outside* the cell. These ions, however, together account for less than 10% of the charged particles inside the cell. Here, the charged constituents are principally K^+ ions and a variety of negatively charged organic ions that are too large to move across the membrane, which is permeable only to Na^+, K^+, and Cl^- ions. Measurement shows that the Na^+ concentration is normally 10 times higher outside the axon than inside it, whereas the K^+ concentration is 30 times higher inside the axon than outside it.

Further, the mobilities of these different ionic species in their passage through the membrane are not the same; K^+ and Cl^- ions can permeate the membrane more easily than Na^+, and the inorganic ions, much more easily than the large organic ions. These differential mobilities or permeabilities (or currents) across the membrane result (*cf.* Section 4.5.4) in the development of a potential difference (~0.08 V) across the membrane such that the inside is *negative* with respect to the outside (Fig. 8.59). Choosing the outside potential as an arbitrary zero, the potential difference between the inside and outside of the membrane is the membrane potential.

Now suppose that a pulse of current is applied across the membrane with a polarity such that the net membrane potential is increased in the positive direction to a value above a certain threshold potential. Then, by some as yet uncertain mechanism, the permeability of the membrane to the different ions is found to alter, and a flow of Na^+ ions occurs from the outside solution through the membrane to the solution inside it. It is said that the membrane "opens up the sodium gates." As a result of this transfer of positive charge into the cell, the inside of the axon becomes charged

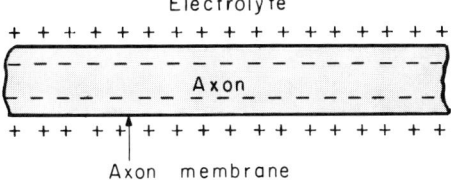

Fig. 8.59. Different mobilities of ions across the membrane result in the separation of charge on the two sides of the membrane, which makes the inside negative in relation to the outside.

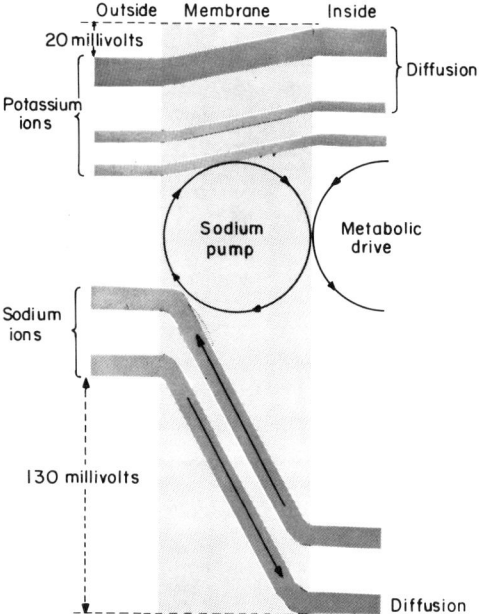

Fig. 8.60. An unknown mechanism makes the membrane suddendly change in permeability to sodium ions; it opens the sodium gate. As a result of the transfer of positive ions into the cell, the potential locally reverses the sign and develops into the action potential. Then another mechanism steps in, the sodium pump, which pumps both sodium and potassium ions against the concentration gradient to reestablish the original state.

positively with respect to the outside, and the potential difference across the nerve cell changes from about -0.08 V (the outside potential taken as zero) to about $+0.04$ V (Fig. 8.60). This sudden inflow (or current) of Na^+ ions and the resulting change of potential difference across the membrane is termed the *action potential* or nerve impulse. After this potential change has occurred, the original Na^+ and K^+ concentrations start recovering by some other yet to be established mechanism which makes the Na^+ and K^+ ions go back through the membrane against the concentration gradient until the relative concentrations inside and outside the cell assume the values they had before the potential difference was applied by the external source.

The action potential developed at a given point of the axon changes

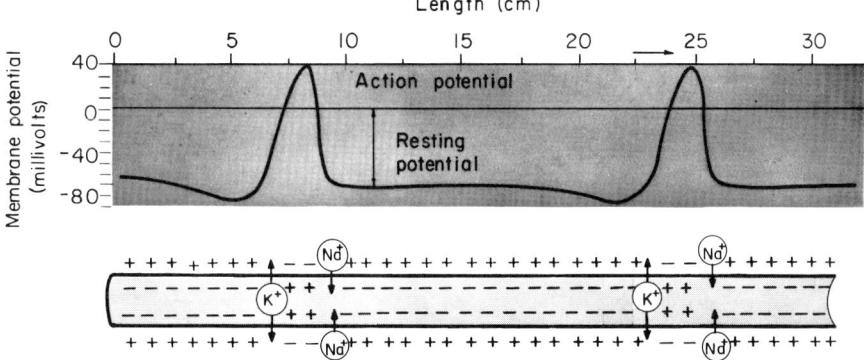

Fig. 8.61. The action potential, developed at one point, sets up the conditions for opening the sodium gates at the adjacent point, and thus, the nerve impulse propagates along the axon.

the membrane characteristics at an adjacent point and sets up the conditions for opening the Na⁺ gates at this adjacent point; thus, a nerve impulse propagates along the axon (Fig. 8.61). It is the propagation of such impulses which provides the basis for the communication between the various parts of the body and the brain.

Now, the interesting point is that the transmembrane Na⁺ current depends upon the potential difference across the membrane. What is the nature of the variation of the Na⁺ ion current with the membrane potential?

Experiment yields an interesting answer (Cope, Mandell). The plot of $\Delta\phi$, the membrane potential, against log i_{ion} is a straight line (see Fig. 8.62). In other words, the ion current varies with the membrane potential according to an exponential law

$$i_{\text{ion}} = Ae^{B\Delta\phi} \qquad (8.102)$$

where A and B are constants.

How this exponential current–potential relationship comes about will constitute exciting research for the immediate future. That the interpretations will be along electrochemical lines seems rather probable, for example, a number of examples show that in (8.102) $B = \frac{1}{2}F/RT$, as expected from the Butler–Volmer equation. That the research will have a far-reaching impact is quite certain because membrane–liquid interfaces occur in all biological systems, and the mechanism of their function is one of the most important aspects of molecular biology; in fact, it is the basis of neurophysiology.

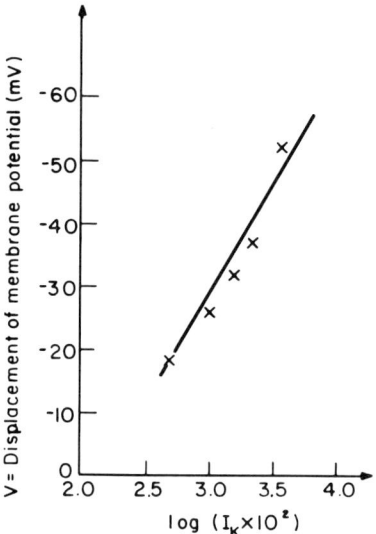

Fig. 8.62. The plot of the change in membrane potential with log of the ionic flux (current) is linear.

8.4.3. The Hot Emission of Electrons from a Metal into Vacuum

Another interesting interface and one that has been studied for some time is that between a metal and vacuum. By raising the temperature sufficiently, it is possible to *evaporate* the electrons from the metal. But for this phenomenon, a television screen would not receive its supply of fluorescence-causing electrons, and a vacuum tube would not be able to maintain electron currents. What is happening in this *hot* emission of electrons?

As a conduction electron from the metal tries to jump into vacuum, it feels a force pulling it back inside. This force arises from the image charge induced in the metal as the electron leaves the surface (*cf.* Section 7.2.5*h*). One can say, therefore, that, as an electron inside the metal approaches the interface, it encounters a steep potential cliff (see Figure 8.63). Hence, the emission current, i.e., the number of electrons jumping out of 1 cm² of the surface per second will be the electronic charge e_0 times the number N_e striking the surface times a probability factor $e^{-\Phi/kT}$

$$i = e_0 N_e e^{-\Phi/kT} \tag{8.103}$$

where Φ is the critical energy required to lift an electron (of energy E) out from the metal into vacuum. The quantity Φ is known as the *work function*. The values of work functions for different metals are shown in Table 8.10. The N_e must depend on the temperature because it reflects how fast the electrons are random walking inside the metal, etc. To emphasize this point, one can write $N_e(T)$, which indicates formally a functional dependence on temperature. Thus,

$$i = e_0 N_e(T) e^{-\Phi/kT} \tag{8.104}$$

An equation of this type governs the emission current from the electron-source electrode in a vacuum tube. What, then, happens if the majority of electrons inside the metal cannot climb the work-function cliff? Then, $i \simeq 0$, i.e., hardly any emission. By raising the temperature of the metal, Φ/kT decreases, and i increases. In such a situation, one has a temperature-produced (thermionic) emission current.

From this point of view, the exponential law for the charge-transfer current across a metal–vacuum interface differs from the exponential law for the current at an electrode–electrolyte interface. In thermionic emission, one has to increase the temperature in order to increase the current. At a metal–solution interface, $i = Ae^{B\Delta\phi}$, and one can achieve the same result in

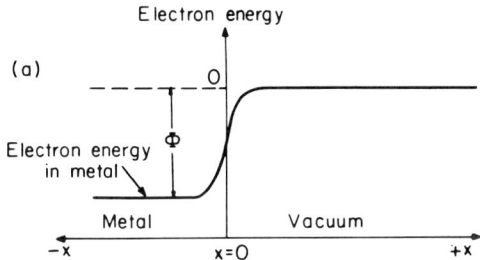

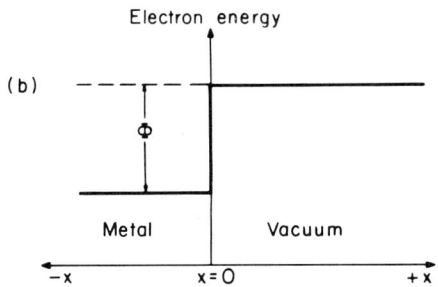

Fig. 8.63. As an electron inside the metal approaches the surface, it encounters a steep potential cliff (a) which can be represented with sufficient accuracy as a potential step (b).

TABLE 8.10

Some Values for the Work Function Φ of Metals

Metal	Φ, eV
Mercury	4.5
Chromium	4.4
Copper	4.4
Gold	4.8
Iron	4.5
Lead	4.0
Platinum	5.0
Silver	4.3
Zink	4.0
Nickel	4.8

a much easier way by changing the potential difference across the interface by turning a knob on an external power supply. This easy ability to change the current by many orders of magnitude is a convenience in experimentation with electrochemical, as opposed to thermionic, emission of electrons, just as is the ease of control of the velocity of electrochemical rather than that of chemical reactions.

8.4.4. The Cold Emission of Electrons from a Metal into Vacuum

Instead of boiling off the electrons from a metal into vacuum, suppose that a strong electric field ($\sim 10^6$ V cm^{-1}) is turned on to help the electrons jump out.

The potential barrier at the interface will be altered. The barrier in the presence of the field can be synthesized in a way similar to that for a charge-transfer reaction can be approximated from two potential-energy–distance curves. In this case, however, one would combine into a single diagram the following curves: (1) an image energy–distance curve and (2) an electric potential–distance curve, which will be a straight line if the applied electric field is a constant. The calculated values of the image interaction energy as a function of distance are given in Table 8.11. The resulting barrier is shown in Fig. 8.64.

Common sense tells one that, if an electron had an energy less than the barrier peak, it would be imprisoned for life inside the metal. Electrons, however, live by different laws, and these are not consistent with common-

TABLE 8.11

Image Energy of an Electron Leaving an Infinite Plane Metal Surface in a Vacuum, as a Function of the Distance x from the Metal Surface

x, cm	Image energy, eV
10^{-8}	-3.6
10^{-7}	-0.36
10^{-6}	-0.036
10^{-5}	-0.004

sense notions because *common* sense has been built up from senses observing macroscopic objects very much larger than electrons. For example, electrons have the property of being able to penetrate into or leak *through* a barrier (Fig. 8.65). They come out with the same energy as they had inside

$$E_{\text{inside}} = E_{\text{outside}} \tag{8.105}$$

Thus, there can be a field-induced *cold* emission of electrons and therefore a tunneling current I, which depends exponentially on the electric field at the interface—the larger the field, the higher the tunneling current.

What has all this to do with charge transfer at an electrode–electrolyte interface? The relevance of cold emission to electrodics shall not be anticipated. Note, however, that the picture of charge transfer (at electrode–solution interfaces) sketched so far has been conspicuously silent on the question of what the electron is doing. The picture has been completely

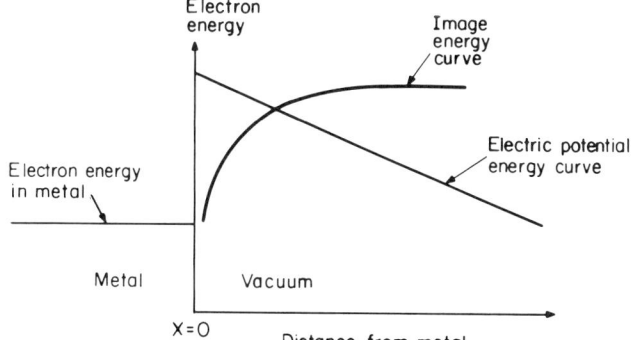

Fig. 8.64. When an external electric field is present, the energy barrier will be represented by a combination of the electric-potential and the image-energy curves.

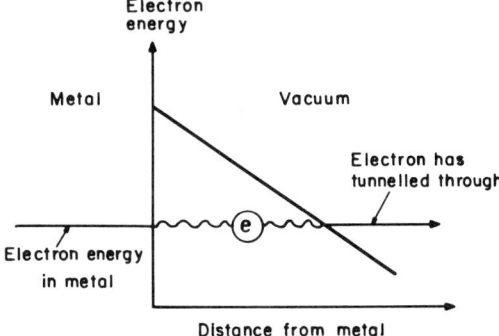

Fig. 8.65. An electron can leak through a sawtooth barrier and come out with the same energy. This is electron tunneling.

ion-centric, and the role of the electron has been dismissed with the statement that somewhere along an ion's movements to and from an electrode, electron transfer between the metal and the electrode occurs. What are the details of this electron transfer? Is there an analogy to cold emission? There is an electric field at a metal–solution interface also, and, in place of a vacuum into which electrons can penetrate, there are electron-acceptor ions or molecules to which electrons can tunnel. Is there electron tunneling between an electrode and electron-acceptor ions?

One must explore an *electron-centric* point of view toward charge transfer at an electrodic interface, i.e., investigate the mechanics of electron jumps at metal–solution boundaries.

Further Reading

1. W. F. Floyd, "Electrochemical Properties of Nerve and Muscle," in: J. O'M. Bockris, ed., *Modern Aspects of Electrochemistry*, Vol. I, Butterworth's Publications, Ltd., London, 1954.
2. J. F. Dewald, "Semiconductor Electrochemistry," in: N. B. Hannay, ed., *Semiconductors*, Chapter 17, Reinhold Publishing Corp., New York, 1959.
3. L. Azaroff, Chapter 12, "Properties of Semiconductors," in: McGraw-Hill ed., *Introduction to Solids*, Book Company, New York, 1960.
4. H. Gerischer, "Semiconductor-Electrode Reactions," in: P. Delahay, ed., *Recent Advances in Electrochemistry*, Interscience Publishers, Inc., New York, 1961.
5. M. Del Ducca and P. Fucsoe, "An Electrochemical Mechanism for the Action of Biological Fuel Cells," *International Science and Technology*, March, 1965.

6. L. Y. Wei, "A New Theory of Nerve Conduction," *IEEE Spectrum*, **3**: 123 (1966).
7. L. F. Stevens, *Neurophysiology, a Primer*, John Wiley & Sons, Inc., New York, 1967.
8. D. Krech, "The Chemistry of Learning," *Saturday Review*, Jan. 20, (1968).
9. L. Bass and W. J. Moore, "A Model of Nervous Excitation Based on the Wien Dissociation Effect," in: *Structural Chemistry and Molecular Biology*, A. Rich and N. Davidson, eds., Freeman, San Francisco, 1968.

8.5. THE QUANTUM ASPECTS OF CHARGE-TRANSFER FRACTIONS AT ELECTRODE–SOLUTION INTERFACES

8.5.1. A Few Words on the Mechanics of Electrons

It has hardly been a secret for half a century that the quantum laws governing the behavior of electrons and other small particles are entirely different from the Newtonian laws describing the motion of missiles and other macroscopic objects. Since electron transfer is the essential part of an electrodic reaction, as Gurney (1931) was the first to have stressed, it is vital to have an idea of how these laws determine the movement of electrons which leap out from a metal to electron-acceptor ions in solution in the presence of an electric field. In other words, one seeks the *electrodic* version of cold emission.

The essential feature of events in the world of electrons is the "uncertainty principle." According to this principle, the knowledge of the momentum p of an electron is necessarily accompanied by an ignorance regarding its exact position x. One can only hope, therefore, to discuss the *probability* of finding the particle at a particular position x at a particular time t.

This probability is given by the square of the absolute value of a quantity known as the *probability amplitude* ψ. The central problem, therefore, is to calculate ψ. Once ψ is known, $|\psi|^2$ reveals the probability of finding the electron at a given place at a given time.

Now the probability amplitude ψ of finding an electron, e.g., in different places at different times, varies with space and time according to the Schrödinger equation. When this equation is solved for an electron in a uniform force field, it is found that the probability amplitude varies as $e^{-i\omega t}$. Here ω is the angular frequency given in terms of the conventional frequency ν by the expression

$$\omega = 2\pi\nu \tag{8.106}$$

One has, however, from De Moivre's theorem,

$$e^{-i\omega t} = \cos \omega t - i \sin \omega t \tag{8.107}$$

Hence, in a uniform force field, the probability amplitude varies as $\cos \omega t$, i.e., the probability shows periodicity. One can refer to ψ waves.

If the probability amplitude ψ is known at a certain point at a certain time t, then the probability amplitude at a distance x from that point is given by $e^{-i\omega(t-\delta)}$, where δ, the phase factor, is the time taken by the wave to travel through the distance x. It is easy to see that $\delta = x/v$, where v is the velocity of the ψ wave and related to its wavelength λ and the frequency v by the relation

$$v = \lambda v \tag{8.108}$$

Hence,

$$e^{-i\omega(t-\delta)} = e^{-i\omega t + (i\omega x/v)} = e^{-i[\omega t + (2\pi x/\lambda)]}$$
$$= e^{-i(\omega t - \dot{k} x)} \tag{8.109}$$

where $\dot{k}$ is known as the *wave number, wavelength constant,* or *propagation constant* and is related thus to the wavelength

$$\dot{k} = 2\pi/\lambda \tag{8.110}$$

Thus, in a uniform force field, the probability amplitude ψ for the electron varies with time and space as $e^{-i(\omega t - \dot{k} x)}$.

According to the quantum laws, therefore, electrons display particle characteristics (energy E and momentum p) as well as wave characteristics (angular frequency ω and wave number $\dot{k}$). There are, however, two fundamental relationships between the particle and wave aspects. These are

$$E = h v \qquad \text{[Planck's relation]} \tag{8.111}$$

and

$$p = \frac{h}{\lambda} \qquad \text{[de Broglie's relation]} \tag{8.112}$$

which become

$$E = \hbar \omega \tag{8.113}$$

and

$$p = \hbar \dot{k} \tag{8.114}$$

in terms of

$$\hbar = \frac{h}{2\pi} \tag{8.115}$$

TABLE 8.12
De Broglie Wavelengths of Various Particles at a Velocity of 10^5 cm sec^{-1}

Particle	Particle mass, g	λ, Å
Electron	9.11×10^{-28}	7000
Proton	1.67×10^{-24}	4
Deuteron	3.34×10^{-24}	2
Helium	6.64×10^{-24}	1
Lithium atom	11.6×10^{-24}	0.6
Sodium atom	38.1×10^{-24}	0.2
Potassium atom	64.9×10^{-24}	0.1

The de Broglie relation (8.112) makes it possible to understand why only the lighter particles, and above all electrons, exhibit wavelike character. Table 8.12 shows the calculated wavelength of different material particles (for a common velocity of 10^5 cm sec^{-1}.) It is seen that the wavelengths attaching to the large particles are smaller than the dimensions of the particle, thus the wave character of large particles can be ignored.

By using the relations (8.111) to (8.115), the ψ, which determines the probability of finding the electron at a given place and at a given time, varies [from (8.109)] with time and space as

$$e^{-(i/\hbar)[Et-px]} \tag{8.116}$$

Note that the energy E in the exponent is the *total energy*, i.e., the sum of the kinetic energy E_k and the potential energy E_p

$$\begin{aligned} E &= E_k + E_p \\ &= \frac{mv^2}{2} + E_p \\ &= \frac{p^2}{2m} + E_p \end{aligned} \tag{8.117}$$

Thus, by substituting for E, the probability amplitude ψ turns out to vary as

$$\psi \sim e^{-\frac{i}{\hbar}[(\frac{p^2}{2m}+E_p)t-px]} \tag{8.118}$$

in a region where the potential energy of the electron is E_p.

8.5.2. The Penetration of Electrons into Classically Forbidden Regions

The purpose of going into this description of the behavior of electrons is to understand how the essential act of an electrodic reaction occurs, i.e., how an electron jumps from the metal to electron-acceptor ions in the solution side of the double layer. It is easy to show (though it was not explicitly pointed out until 1966) that there is no possibility of the electrons going *over* the barrier between the metal in which they exist and the surrounding environment (as do particles in thermionic emission). Currents observed at electrodes at room temperatures are much too high for that. Electrons go *over* barriers at appreciable rates only at high temperatures (*cf.* thermionic emission).

Thus, the first question which must be asked is: How *do* electrons get through the barriers over which they cannot mount? A way of approaching this problem is to consider the following situation, a metal–vacuum interface. Suppose that, at $x < 0$, i.e., inside the metal, the potential energy of the electron is V_1; outside the metal (at $x > 0$), there is another region (a vacuum) where the potential energy is V_2 (Fig. 8.66). Then there is a step-like potential-energy barrier at $x = 0$. How do electrons negotiate this barrier?

The total energy E of the electron in the metal is

$$E_1 = \frac{p_1^2}{2m} + V_1 \tag{8.119}$$

and the total energy of the electron in vacuum is

$$E_2 = \frac{p_2^2}{2m} + V_2 \tag{8.120}$$

The question is: Can an electron in the metal with an energy E_1 leap into vacuum?

The total energy of the electron must be the same on both sides of the barrier so long as there is no emission or absorption of radiation, i.e., the electron transfer is a radiationless transition; this is the law of conservation of energy. Hence,

$$E_1 = E_2 \tag{8.121}$$

or

$$\frac{p_2^2}{2m} = \frac{p_1^2}{2m} - (V_2 - V_1) \tag{8.122}$$

Suppose now that the kinetic energy of the electron in the metal, i.e.,

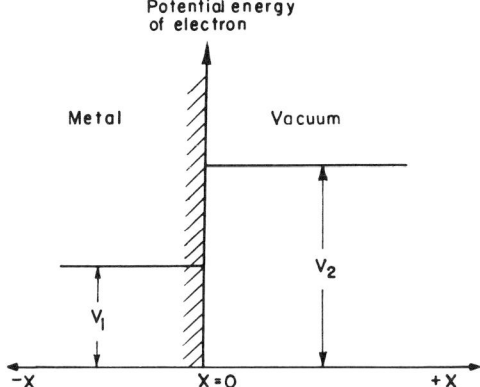

Fig. 8.66. Potential energies of the electron in the metal and in vacuum are different—an impossible jump.

$p_1^2/2m$, is less than the height of the potential-energy barrier, i.e., $V_2 - V_1$. Classical notions indicate that such a particle with an energy too low to surmount the barrier would not be able to jump into vacuum. Since

$$\frac{p_1^2}{2m} < V_2 - V_1$$

it follows from (8.122) that $p_2^2/2m < 0$, i.e., $p_2^2/2m$ would be a negative quantity, or the kinetic energy of the electron if it leaped into vacuum would be *negative*. Now, in the expression for the kinetic energy, as p_2^2 cannot be negative, the mass m of the electron ought to be negative—a classically impossible result. Thus, by following the laws of macroscopic objects (the laws of Newton) an electron can never jump out of the metal into vacuum for the situation shown in Fig. 8.66.

According to the quantum laws for electrons, however, one must analyze the situation in terms of probability amplitudes. The chance of finding the electron in vacuum is the square of the absolute value of the probability amplitude ψ_2 in vacuum. (Remember that it is the probability amplitude ψ which decides where electrons are likely to be found). But what this ψ_2 depends on is given by (8.118), in which one writes V_2 for E_p

$$\psi_2 \sim e^{-\frac{i}{\hbar}[(\frac{p_2^2}{2m} + V_2)t - p_2 x]} \qquad (8.123)$$

To evaluate the exponent, one must get at p_2.

By the law of conservation of energy, it has been shown that $p_2^2/2m$

is negative when $p_1^2/2m < V_2 - V_1$. Since, however, 2 and m are positive, p_2^2 (the square of the electron momentum in vacuum) is negative. What is the mathematical meaning of the square of a quantity being negative? It must mean that the quantity is imaginary and equal to a real quantity times i, where $i = \sqrt{-1}$. Using the symbol p_2' for the real part of the electron momentum, one can write

$$p_2 = ip_2' \tag{8.124}$$

This expression for p_2 can be inserted in Eq. (8.123) for ψ_2 to give

$$\psi_2 \sim e^{-\frac{i}{\hbar}\left\{\left[\frac{(ip_2')^2}{2m} + V_2\right]t - ip_2'x\right\}} \tag{8.125}$$

By restricting oneself to the space variation of ψ_2, i.e., how ψ_2 depends on the distance x from the metal–vacuum interface, the result is

$$\psi_2 \sim e^{(i/\hbar)ip_2'x} = e^{-p_2'x/\hbar} \tag{8.126}$$

Hence, in going to the wave aspects of the particle, a wave-mechanical formula has been evolved which does have the imaginary term (8.125). All the quantities p_2', x, and $\hbar$ in the exponent of Eq. (8.126) are real quantities, and, hence, the space dependence of ψ_2 becomes a real exponential.

There is a great deal of significance in the result that ψ_2 *in vacuum* on the *other side of the barrier* has *real* space dependence. Since the probability of finding the electron in vacuum is the square of the electron's probability amplitude ψ_2 in that region, it turns out that there is a finite probability that the electron will penetrate into the barrier, as was impossible in classical theory. This means that, even though the kinetic energy of the electron in the metal is inadequate for the electron to climb over the potential barrier, its quantum behavior, which is described by its probability amplitude, permits it to penetrate or *leak* or *tunnel* into the apparently impossible barrier at a metal–vacuum interface. This is the crux of why it is possible for electrons to be emitted from an electrode in the cold.[†]

[†] This is not quite the whole story. In the cold emission of electrons through a vacuum, there is an electric field applied at the surface, too, and these electrons which are penetrating *into* the barrier in the above account have somewhere to go, i.e., energy states equal to those in the metal. Such electrons not only penetrate into the barrier at the metal–vacuum interface, they tunnel through it and can be collected on some other electrode. Tunneling *through* barriers, not into them, is considered in Section 8.5.3.

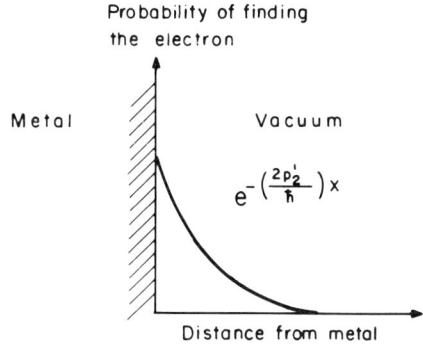

Fig. 8.67. The probability of finding an electron in vacuum outside the metal decreases exponentially with distance.

Now, the probability of finding the electron in vacuum, i.e., its tunneling probability P_T, is given by

$$P_T = |\psi|^2 = e^{-2p_2'x/\hbar} \tag{8.127}$$

Hence, though there is a finite chance of finding the electron in vacuum, this probability decreases as $e^{-(p_2'x/\hbar)x}$; it falls off exponentially with distance x from the interface (Fig. 8.67). The farther out one goes from the electrode, the smaller the chance is of finding the electron.

8.5.3. The Probability of Electron Tunneling through Barriers

What can one do to increase the emission of electrons from a metal into vacuum? There is, of course, the classical solution to this problem. In this classical method, one avoids relying on any tunneling of electrons. Instead, the kinetic energy of the electrons inside the metal can be increased (by increasing the temperature) until they can climb *over* the barrier, i.e., classical (non-quantal) *thermionic* emission has been achieved.

The quantum solution to the problem of increasing electron emission is on different lines. The tunneling probability depends on $e^{-(2p_2')x/\hbar}$, so why not[†] decrease x? In tunneling into a vacuum, there are essentially no available electronic states of energy equal to those in the metal. Penetration can occur, but no passage of electrons *through* a barrier to remain on the

---

[†] Of course, one can reduce p_2', which depends on the magnitude of the negative number p_2, and this in turn is controlled by the height of the barrier $V_2 - V_1$ [*cf*. Eq. (8.127)]. That is, the lower the barrier, the more the tunneling.

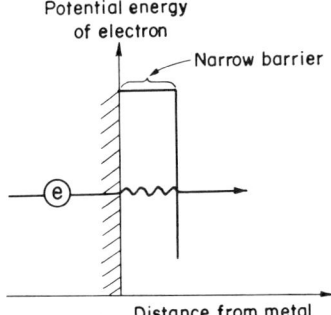

Fig. 8.68. Tunneling probability depends on how narrow the energy barrier is.

far side was possible; there was no far side within sufficiently small x. Suppose now one makes x finite and quite small (tens of angstroms, or less) by having some molecular entities having possible energy states for the electron equal to those which the electrons have in the metal (for this, see next section). Then, the condition for radiationless tunneling (8.121) can be fulfilled at a value of x small enough to let $|\psi|^2$ have a finite value. The barrier has been narrowed and becomes penetrable (Fig. 8.68).

In the case of a rectangular barrier of width l, the tunneling probability P_T is given [from (8.127)] by

$$P_T = e^{-(2p_2')l/\hbar} = e^{-4\pi l p_2'/h} \tag{8.128}$$

Now, from Eqs. (8.122) and (8.124),

$$-(p_2')^2 = 2m \left(\frac{p_1^2}{2m} + V_1 - V_2 \right) \tag{8.129}$$

and, since $(p_1^2/2m) + V_1 =$ the total energy E of the electron in the metal, one has

$$p_2' = \sqrt{2m(V_2 - E)} \tag{8.130}$$

and, therefore, by substituting for p_2' in Eq. (8.128), the tunneling probability becomes

$$P_T = e^{-(4\pi l/h)\sqrt{2m(V_2-E)}} \tag{8.131}$$

Of course, it is not essential that the barrier be rectangular; one can work with all sorts of barrier shapes. The only change in the analysis would be to treat V_2, not as a constant, but as a function of x. This is equivalent

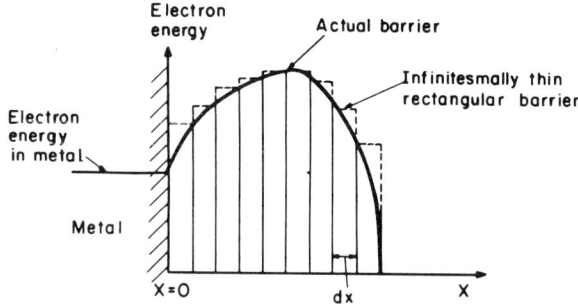

Fig. 8.69. Any nonrectangular energy barrier can be represented as a sum of a number of thin, rectangular barriers.

to looking upon the nonrectangular-shaped barrier as a number of rectangular barriers, each of infinitesimal thickness (Fig. 8.69).

The total probability is given, therefore, by multiplying together the probabilities for the various dx-thick rectangular barriers

$$P_T = \prod (P_T)_i = \prod e^{-(4\pi/h)\sqrt{2m[(V_2)_i - E]}\,dx} \qquad (8.132)$$

where i stands for the ith rectangular barrier, and $\prod$ is the symbol for the product of terms $(P_T)_i$, i.e., $(P_T)_1, (P_T)_2, \ldots, (P_T)_i$. Instead, however, of multiplying the exponential terms, one can add the exponents, i.e., one can integrate the exponent over dx, if the function $V_2 = f(x)$ is known

$$P_T = \exp\left[-\frac{4\pi}{h}\int_{x=0}^{x}\sqrt{2m[f(x) - E]}\,dx\right] \qquad (8.133)$$

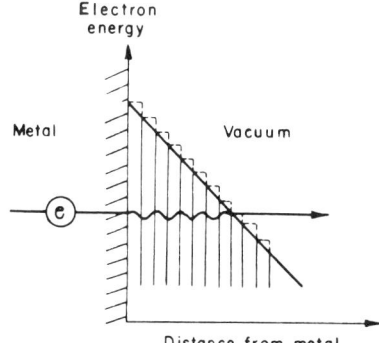

Fig. 8.70. In cold emission, a right-angled triangle barrier can be assumed in calculating tunneling probability.

It is by such an integration that one can evaluate the tunneling probability for the right-angled triangle barrier (Fig. 8.70) corresponding to cold emission.

The tunneling of electrons, which has been treated here in an elementary way, is of fundamental importance in diverse phenomena apart from the cold emission of electrons at a metal–vacuum interface. It is, for instance, the basis of the functioning of some types of semiconductor devices, e.g., the Esaki tunnel diode. Electron tunneling will prove to be of great significance to electron-transfer reactions at a metal–solution interface. Indeed, it is the central act in the theory of electrochemical reactions.

8.5.4. The Distribution of Electrons among the Energy Levels in a Metal

The tunneling of electrons through barriers has an important feature: For a radiationless transition, *the electron has to have the same energy on both sides of the barrier.* In other words, if electrons tunnel through the barrier at the surface of a metal, they come out with the same energy that they had in the metal. Then, what is the energy or distribution of energies of electrons inside a metal? A simple way of deriving the distribution of energies among the electrons in the metal is presented here.

Suppose that, in the lattice of the metal, there are impurity atoms which can exist either in their ground states with energy zero or in excited states with energy ε. Electrons which collide with these impurity atoms can exchange quanta of energy ε. Let the initial energy of an electron be E, and, after collision with the impurity atom in its excited state, let the electron energy be raised to $E + \varepsilon$ (Fig. 8.71).

Then, the rate $r_{E \to E+\varepsilon}$ of the $E \to E + \varepsilon$ electronic transitions is pro-

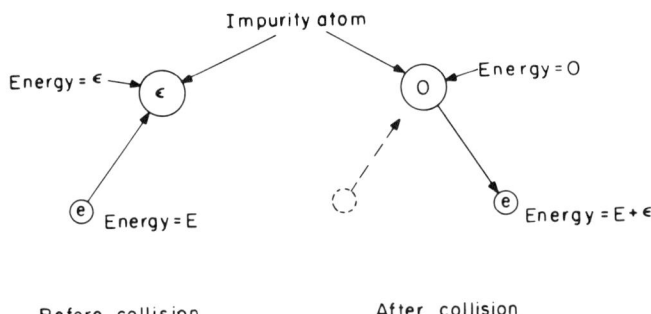

Fig. 8.71. After collision with excited impurity atoms, the energy of electrons is increased by quanta released by the atoms.

portional to the product of three probabilities: (1) P_E, the probability of finding the electron before collision in the energy state E; (2) times P_ε, the probability of the impurity atom's being in the excited state of energy ε; and (3) times the probability that a state of electron energy $E + \varepsilon$ is *vacant* in the metal. This third probability is $1 - P_{E+\varepsilon}$, where $P_{E+\varepsilon}$ is the probability that the $E + \varepsilon$ acceptor-energy state is occupied. Herein enters the essential difference between the distribution law for electrons and that for ions and other particles in solutions. The ions and other larger particles are treated with good approximation by the classical Boltzmann distribution law. Electrons, however, obey the Pauli exclusion principle: One cannot push more than two electrons into the same energy state.(This is why one has to consider the probability of the acceptor-energy state's being *vacant*).

Hence,

$$r_{E \to E+\varepsilon} \propto P_E P_\varepsilon (1 - P_{E+\varepsilon}) \tag{8.134}$$

By a similar argument, the reverse rate, or the rate of $E + \varepsilon \to E$ electronic transitions is

$$r_{E+\varepsilon \to E} \propto P_{E+\varepsilon} P_0 (1 - P_E) \tag{8.135}$$

where P_0 is the probability of the impurity atom's being in the ground state.

When thermal equilibrium is attained, the two rates are equal

$$r_{E \to E+\varepsilon} = r_{E+\varepsilon \to E} \tag{8.136}$$

and, ignoring the proportionality constants, which can be shown to be equal, one has

$$P_E P_\varepsilon (1 - P_{E+\varepsilon}) = P_{E+\varepsilon} P_0 (1 - P_E)$$

$$\frac{(1 - P_E)/P_E}{(1 - P_{E+\varepsilon})/P_{E+\varepsilon}} = \frac{P_\varepsilon}{P_0} \tag{8.137}$$

But the ratio P_ε/P_0 of the probabilities of the impurity *atoms'* being in the excited and ground states is given by the classical *Boltzmann* probability factor because atoms and not electrons are being dealt with. Thus,

$$\frac{P_\varepsilon}{P_0} = e^{-\varepsilon/kT} = e^{-[(E+\varepsilon)-E]/kT} \tag{8.138}$$

By substituting this expression (8.138) in Eq. (8.137), the result is

$$\frac{(1 - P_E)/P_E}{(1 - P_{E+\varepsilon})/P_{E+\varepsilon}} = \frac{e^{E/kT}}{e^{E+\varepsilon/kT}}$$

or

$$\frac{1-P_E}{P_E}e^{-E/kT} = \frac{(1-P_{E+\varepsilon})}{P_{E+\varepsilon}}e^{-E+\varepsilon/kT} \qquad (8.139)$$

The left-hand side of Eq. (8.139) depends only on the electron-energy state E (apart from the temperature); similarly, the right-hand side depends only on the state $E + \varepsilon$ (apart from the temperature). Hence, Eq. (8.139) must remain true for all values of E and $E + \varepsilon$, i.e., E and ε. The only way this can happen is if each side of Eq. (8.139) is independent of the energy states E and ε. That is,

$$\frac{1-P_E}{P_E}e^{-E/kT} = K \qquad (8.140)$$

where K is a function of temperature but independent of E and ε. By rearrangement, one has

$$\begin{aligned}P_E &= \frac{e^{-E/kT}}{K + e^{-E/kT}} \\ &= \frac{1}{Ke^{E/kT}+1}\end{aligned} \qquad (8.141)$$

To stress the exponential nature of the temperature dependence of K, it is customary, however, to write K in the equivalent form

$$K = e^{-E_F/kT} \qquad (8.142)$$

where E_F is independent of E and ε. Thus, combining (8.142) and (8.141),

$$P_E = \frac{1}{e^{(E-E_F)/RT}+1} \qquad (8.143)$$

This is the Fermi–Dirac distribution function for the probability that an electron in a metal will have an energy E. If there is a population of electrons in a metal, the probability of an electron's having an energy E, i.e., the fraction having an energy E, is given by the expression (8.143) for P_E.

A better feel for P_E is obtained by considering the fate of the expression at 0°K. There are two cases here: (1) If $E < E_F$, then $E - E_F$ is negative and $e^{E-E_F/kT} \to 0$, i.e., $P_E \to 1$. (2) If $E > E_F$, then $E - E_F$ is positive, $e^{E-E_F/kT} \to \infty$, and $P_E \to 0$. Hence, at 0°K, all energy states with energy less than E_F are *completely populated*, and all states with energy greater than E_F are empty. At the absolute zero, the energy E_F is like a cutoff energy, and it is known as *the Fermi energy*. The distribution of the number of electrons with an energy E at 0°K is shown in Fig. 8.72.

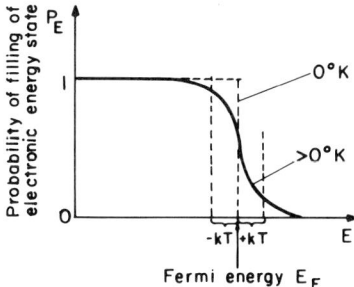

Fig. 8.72. There is a rather sharp transition in the probability of finding an electron in an energy level at 0°K from 1 to 0 around a value of energy known as Fermi energy.

Another way of looking at the Fermi energy is to consider a situation of $T > 0°\text{K}$ and to set $E = E_F$ in Eq. (8.143). Then, $e^{E-\eta_F/kT} = 1$, and $P_{E=E_F} = \tfrac{1}{2}$. In other words, the Fermi energy is the energy level which is half populated by electrons at $T > 0°\text{K}$.

The distribution function at $T > 0°\text{K}$ is shown (the full line) in Fig. 8.72. It will be seen that the distribution function takes values in between zero and unity only in a relatively narrow energy range kT around the Fermi energy. *It is only electrons occupying these energy states that need to be reckoned with as free electrons.* For, when $P_E = 1$, all the energy states in the metal are filled, the electrons have no states to move into in the metal, and hence they cannot be taken as free. When $P_E = 0$, the energy state E will have a probability of occupation of zero and hence can be ignored. One can see, therefore, that the *average* energy of the free electrons can be taken to be approximately the Fermi energy since it is situated symmetrically between $P_E = 0$ and $P_E = 1$. Hence, in discussions on electron tunneling from the metal to ions in solution, the energy of the mobile electrons in the metal will be taken to be the Fermi energy. At 0°K, this view is exact; at temperatures above 0°K, e.g., at room temperatures, it is easy to show that nearly all electrons which tunnel have an energy within kT of the Fermi energy, and, as kT is usually very small compared with the Fermi energy, there is a small loss in accuracy by taking the electrons which tunnel to molecules in solution as all having the Fermi energy.

8.5.5. Under What Conditions Do Electrons Tunnel between the Electrode and Ions in Solution?

Electrons can tunnel through a barrier only if they can emerge with the *same* total energy as they had when they went in; this is a fundamental basis of the quantum-mechanical theory of radiationless tunneling (see Section 8.5.2). In other words, the energy of the tunneling particles must be the same on both sides of the barrier (Fig. 8.73).

This tunneling rule will now be applied to the special case of electrons tunneling from the metal to hydrated protons in the electronation reaction

$$M(e) + H^+\!\!-\!\!H_2O \rightarrow M\!\!-\!\!H + H_2O$$

But, first, one must imagine the shape of the barrier. This will be done by using the same type of approach as was used for a hydrogen nucleus' moving from a water molecule to the metal (*cf*. Section 8.3.4*b*). The barrier was synthesized from two curves. One curve represented the energy–distance relationship as a hydrogen ion moved away from a water molecule, and the other represented the energy–distance relationship as a hydrogen atom moved away from the metal (Fig. 8.48).

By analogy, the barrier for electron tunneling can be synthesized in the following manner: Two questions are asked. Firstly, how does the energy change as an electron moves away from a metal? Secondly, how does the energy vary as an electron moves away from a singly charged hydrogen ion?

The answers to these questions are straightforward. As the electron–metal distance increases, the attractive image force $-e_0^2/4x^2$ decreases and, hence, the image energy $e_0^2/4x$ increases (in the sense of becoming more positive, which means less attraction between electron and metal). This variation of image energy with distance yields one energy–distance curve (Fig. 8.74). The interaction between an electron and a hydrogen ion can

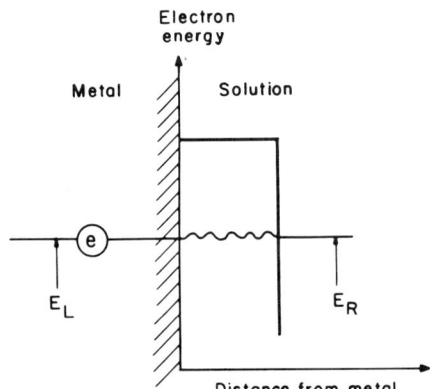

Fig. 8.73. In radiationless transitions, the energy of the tunneling particles must be the same on both sides of the barrier.

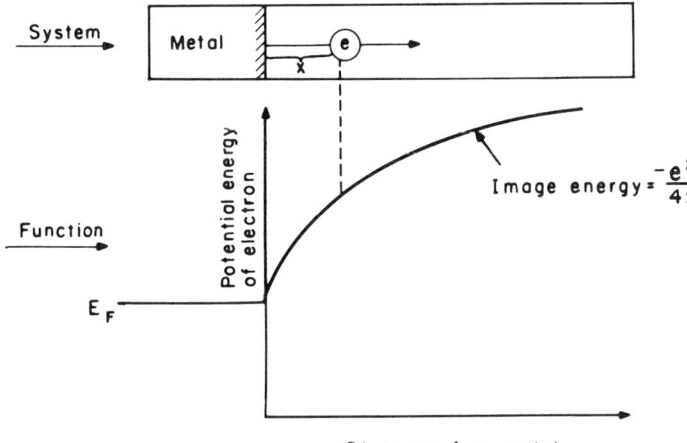

Fig. 8.74. Image energy increases (becoming more positive) with distance between the electrode and the escaping electron.

be assumed, for expository purposes, to arise only from an attractive coulombic force $-e_0^2/x^2$. Hence, as the electron moves away from the hydrogen ion, the coulombic energy e_0^2/x increases, which gives rise to a second energy–distance curve (Fig. 8.75). When these two energy–distance curves are combined into a single diagram, with the metal–hydrogen-ion distance fixed by the same arguments as in Section 8.3.4b, one obtains a

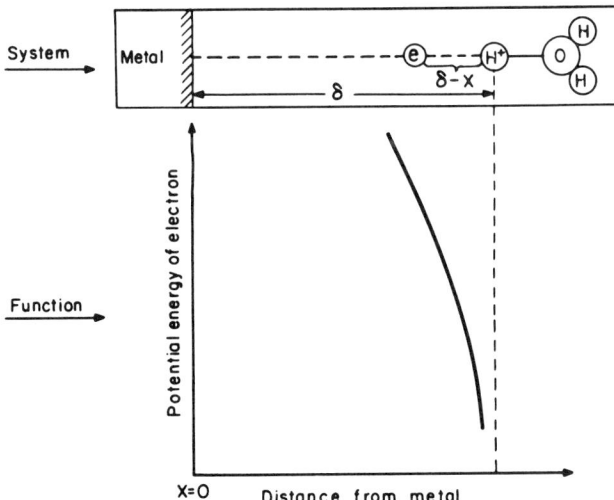

Fig. 8.75. As the electron leaves for the electrode, the energy increases with distance from the hydrated proton.

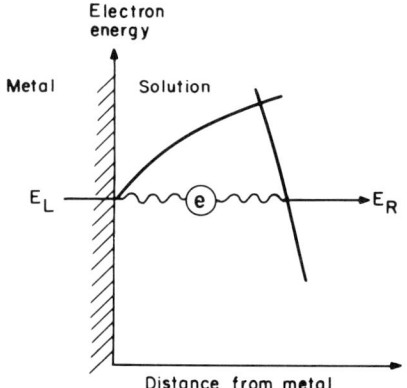

Fig. 8.76. The two potential-energy–distance functions define the barrier through which electrons are tunneling from the metal to hydrated hydrogen ions.

barrier (Fig. 8.76) for the tunneling of electrons from the metal to hydrated hydrogen ions.

The tunneling rule says that the electron energy before entering the barrier must be equal to the electron energy after leaving the barrier. Let the zero of energy for the electron be taken at the energy of an isolated electron in vacuum far away (i.e., at an infinite distance) from the metal and the hydrogen ion. Then, as stated before, the tunneling rule is that the electron energy E_L to the left of the barrier must be equal to the energy E_R to the right of the barrier (Fig. 8.76). But what are E_L and E_R?

The average energy of an electron in the conduction band of the metal electrode has been shown (Section 8.5.6) to be its Fermi energy E_F. To pull the electron out (no tunneling) from this Fermi energy to a final resting place in vacuum requires that an amount of work (equal to the work function Φ, cf. Section 8.4.3) be done on the electron or by the electron. Hence, if an electron is brought in from vacuum to the Fermi energy, it does an amount of work Φ (Fig. 8.77). Hence, the mean electron energy in the metal (referred to zero in vacuum) is Φ, i.e.,

$$E_L = \Phi \tag{8.144}$$

Since, however, the tunneling rule says that

$$E_L = E_R \tag{8.145}$$

it means that the tunneling electrons would like to emerge from the barrier with an energy $E_R = \Phi$.

But will these electrons (with an energy Φ) be accepted by the hydrated ions? Pauli's exclusion principle rules the electrons by stipulating that every energy state is permitted a fixed quota of electrons, namely, two electrons. Once this quota is filled, no more electrons are accepted into that energy state. The hydrated protons hence cannot receive tunneling electrons into states which already have two electrons.

Suppose that E_R is the electron energy level (referred to an energy of zero vacuum) down to which there are vacant acceptor states in the hydrated proton. Then, electron tunneling is impossible (Fig. 8.78) if $\Phi > E_R$ because the electrons will be trying to jump into filled states and the *exclusion principle* blocks this attempt. The condition for tunneling (Gurney, 1932) is illustrated in Fig. 8.79, and given by

$$\Phi \leq E_R \qquad (8.146)$$

What is the value of E_R (Fig. 8.80), the energy of the electron level in the particle in solution which must have an energy equal to that of electrons in the metal? To get the value of this energy, a calculation can be made which is in principle the same as the one made above for the electron in the metal. There, one found that to take an electron from infinity in

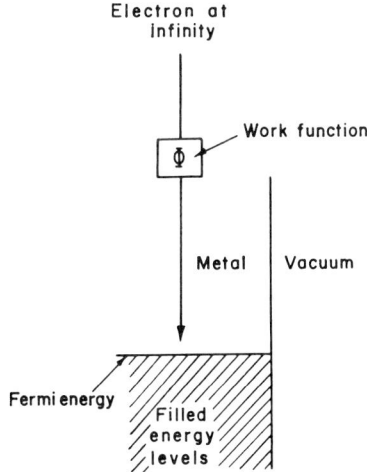

Fig. 8.77. When brought from vacuum into metal, an electron does work numerically equal to the work function.

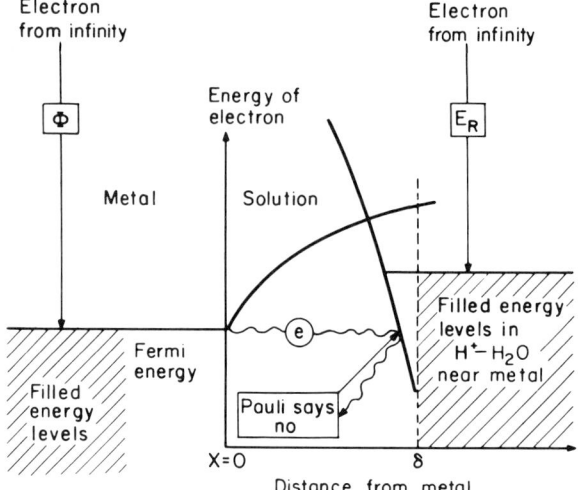

Fig. 8.78. Tunneling to hydrated protons is impossible when these protons do not have vacant states of the same or greater energy than that of the Fermi level.

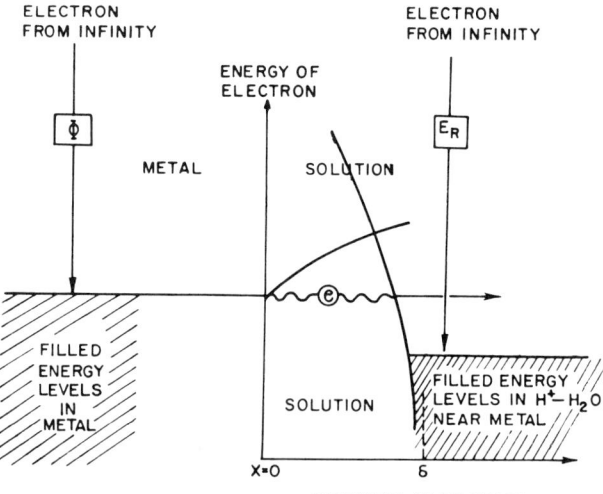

Fig. 8.79. The condition for tunneling is that E_R be equal to or larger than the Fermi energy, for then there are vacant states to which an electron may jump.

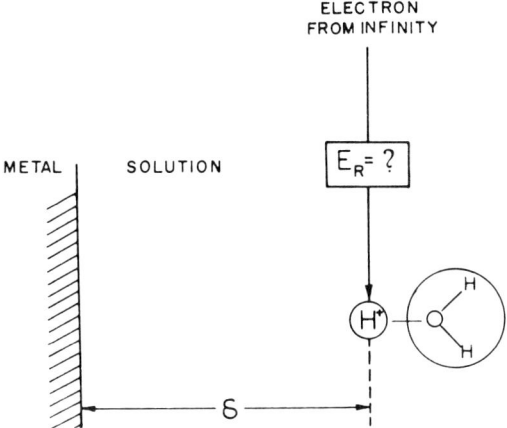

Fig. 8.80. The energy level E_R is determined by the amount of energy which is released when an electron drops from infinity into a hydrated proton, the H atom from which bonds with the metal.

vacuum to the Fermi level needed a work Φ. To obtain the corresponding work necessary to take an electron from the vacuum to an energy level in a hydrated proton near to a metal is a little more complex.

Thus, consider firstly the work which must be done to remove the atom from its interaction with the metal. Let this work be the heat of adsorption A (Fig. 8.81). The hydrogen atoms interact repulsively, at sufficiently short distances, with the neighboring oxygen of water. Let the overcoming of the associated energy be represented by $-R$ (Fig. 8.82). Thus, $-A - R$ gives us a free atom. One must now ionize it to get an electron at infinity, energy I

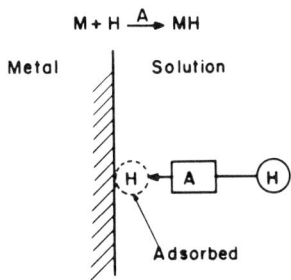

Fig. 8.81. The energy of adsorption is equal to the energy released when a neutral atom establishes a bond with the metal substrate.

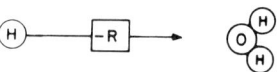

Fig. 8.82. As the neutral atom is returned to the water, the interaction of H with the associated OH_2 must be overcome.

(Fig. 8.83), and replace the ion in solution (Fig. 8.84), energy L (the heat of hydration). Hence,

$$E_R = -A - R + I + L$$
$$= (I + L) - (R + A) \tag{8.147}$$

So the condition for radiationless tunneling of an electron from the metal to a hydrated proton is (see Fig. 8.85)

$$\Phi \lesssim E_R$$

or

$$R + A \lesssim I + L - \Phi \tag{8.148}$$

8.5.6. The Tunneling Condition and the Proton-Transfer Curve

Consider the initial state of the reaction leading to the electronation of a hydrated proton. The stable hydrated proton is positioned at the OHP, and the electron is in the metal. Will the electron tunnel through to the proton?

The answer is simply found. One has to check on the comparative magnitudes of $R + A$ and $I + L - \Phi$. If $R + A \lesssim I + L - \Phi$, electron tunneling can occur [cf. Eq. (8.148)]. If $R + A \gtrsim I + L - \Phi$, tunneling is impossible. When the numbers are put in, it usually turns out that (for the same value of the distance of the H nucleus from the electrode) $R + A \ll I + L - \Phi$. This means that the electron in the metal is forbidden to tunnel across the interface. This implies of course that the electronation reaction $M(e) + H_3O^+ \rightarrow M-H + H_2O$ cannot occur to an H_3O^+ ion in its ground state. Then, how *does* it occur?

The mechanism is interesting. Examine the quantities in the condition for electron tunneling, namely,

$$R + A \leq I + L - \Phi \tag{8.148}$$

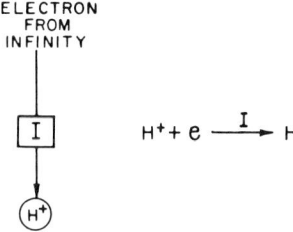

Fig. 8.83. The ionization potential is numerically equal to the energy released upon electronation of a proton by an electron coming from infinity.

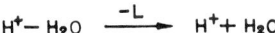

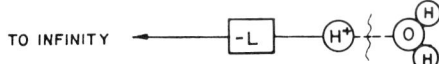

Fig. 8.84. The removal of a proton from its hydration sheath involves the loss of the energy of hydration of the proton, which is equal to the negative value of the hydration energy.

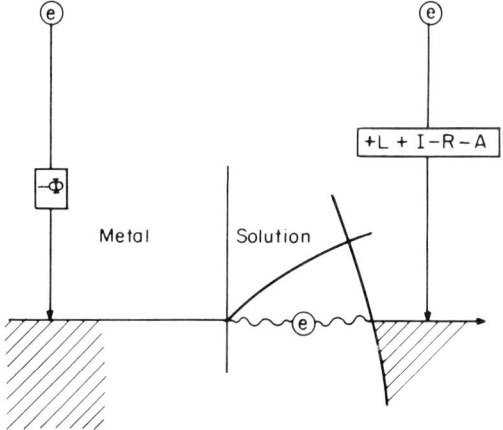

Fig. 8.85. The condition for electron tunneling is that the energy released upon bringing an electron to the hydrated ion in solution be equal to, or larger than, that of bringing it into the metal.

TABLE 8.13
Energy Terms in Eq. (8.148) for the Hydrogen Electronation Reaction on Platinum

Term	Energy, kcal mole^{-1}
R	23
A	-50
I	313
L	-269
Φ	120
$R + A$	-27
$I + L - \Phi$	-60

The quantities R, A, and L are not constants; they vary as the interparticle distances change (the values in the table are the equilibrium ones). Thus, the H—H$_2$O repulsion energy R depends on the distance between the hydrogen atom and the water molecule; the M—H adsorption energy A depends on the metal–hydrogen-atom distance, and, finally, the H$^+$—H$_2$O hydration energy L varies with the proton–water distance. (By viewing the H$^+$—H$_2$O and M—H as diatomic systems undergoing vibrations like a harmonic oscillator, it will be realized that the hydration energy L of the proton and the adsorption energy A of the hydrogen atom vary periodically with the vibratory movements of the H$^+$—H$_2$O and M—H interactions.) Obviously, therefore, the magnitudes of $R + A$ and $I + L - \Phi$ can be varied by altering the distances mentioned. One glimpses a possibility here of satisfying the tunneling condition $R + A \leq I + L - \Phi$ by moving the proton and thus changing the energy of the electronic state in the hydrated proton available for the electron tunneling from the metal. This possibility will now be explored.

Consider the right-hand side of the tunneling condition, i.e., the quantity $I + L - \Phi$. The energy $I - \Phi$ represents the amount of energy released when the electron is transferred from the metal to the isolated proton ($-\Phi$ is released when the electron is taken from the metal into vacuum; and $+I$, when it is put into the proton). The energy $+L$ is the hydration energy of the hydrated proton H$^+$—H$_2$O. Hence, $I + L - \Phi$ is the energy of an electron in the metal, M(e), *and* a hydrated proton H$^+$—H$_2$O i.e., $I + L - \Phi$ is the energy of M(e) + H$^+$—H$_2$O (with respect to the energy of hydrogen atom).

It has been pointed out, however, that L does not remain constant as the H$^+$—H$_2$O bond vibrates. If the water molecule attached to the hydrogen ion is considered at a fixed distance from the metal, vibrations of the H^+—H_2O bond correspond to movements of the proton H^+ toward and away from the metal. As the proton oscillates between the metal and water molecules, the H^+—H_2O hydration energy L undergoes periodic changes.

Consequently, the energy of the system $M(e) + H^+$—H_2O also undergoes changes depending upon the position of the proton between the metal and the water molecule. One can therefore plot the energy $I + L - \Phi$, i.e., of $M(e) + H^+$—H_2O as a function of the distance between the H^+ and H_2O, which means as a function of the extent to which the H^+—H_2O bond stretches (Fig. 8.86).

In the same way, one can analyze the left-hand side of the tunneling condition, i.e., $R + A$. Since the energy A is the adsorption energy of the hydrogen atom adsorbed on the metal and R is the H—H_2O repulsion energy, the term $R + A$ represents the energy of M—H + H_2O (with respect to the energy of an hydrogen atom). Further, the M—H bond can be considered to stretch and contract with corresponding periodic changes in the adsorption energy A. Now consider that the metal and water molecule are fixed at the same distance apart as in the treatment of the $M(e) + H^+$—H_2O. Then, as the M—H bond vibrates, the hydrogen atom moves periodically closer and farther with respect to the fixed water molecule, which produces corresponding changes in the H—H_2O repulsion energy R.

Hence, as there is a variation in the position of the hydrogen atom between the metal and water molecule, there are changes in $R + A$, i.e.,

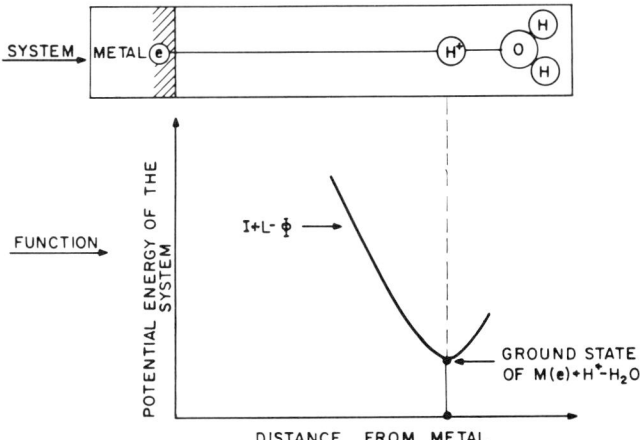

Fig. 8.86. The energy $I + L - \Phi$ can be plotted as a function of the extent to which the H^+—H_2O bond stretches.

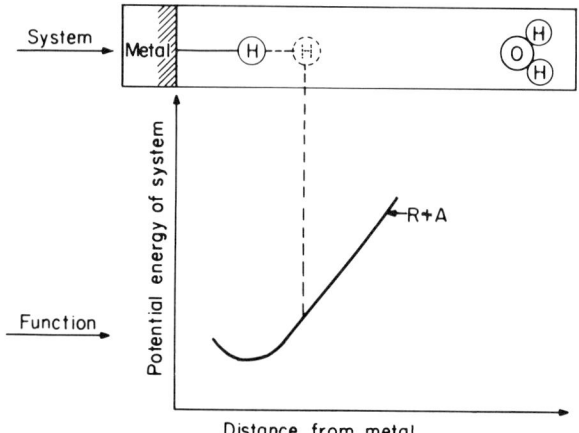

Fig. 8.87. The energy of the system is increased as it moves against the repulsive forces of the water molecules.

changes in the energy of M—H + H_2O. This means that one can plot the energy $R + A$ of M—H + H_2O as a function of the position of the hydrogen atom between the metal and the water molecule (Fig. 8.87).

At this stage, an elementary step will be taken. The curves for the energies $I + L - \Phi$ of $M(e) + H^+$—H_2O and $R + A$ of M—H + H_2O as a function of the position between the metal and water molecule of H^+ and H, respectively, will be represented on the same diagram (Fig. 8.88).

Usually, one plots two functions in the same diagram only if they vary with the same quantity represented on the abscissa. But, here, for the

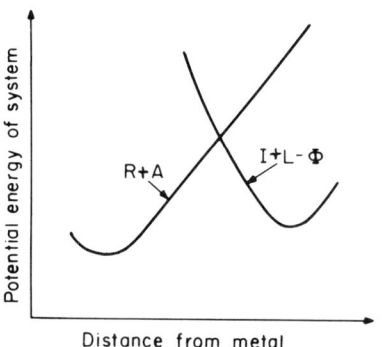

Fig. 8.88. When Figs. 8.86 and 8.87 are plotted on the same diagram, the potential-energy curves intersect.

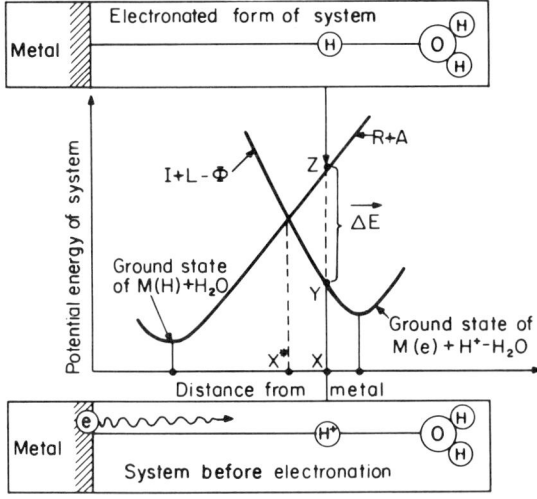

Fig. 8.89. In the electronation reaction, the H^+—H_2O bond has to stretch in order that the energy $\vec{\Delta E}$ be squeezed to zero.

$I + L - \Phi$ curve, the movement of the proton is represented on the x axis; and, for the $R + A$ curve, the movement of the hydrogen atom is represented on the same axis. This means that the $I + L - \Phi$ curve represents the situation *before* electronation, i.e., the $M(e) + H^+$—H_2O state, and the $R + A$ curve represents the situation *after* electronation, i.e., the M—H $+ H_2O$ state (Fig. 8.89). As long as electronation has not occurred, the energy of the system $M(e) + H^+$—H_2O is given by a point on the $I + L - \Phi$ curve.

Now, the threads of the argument can be gathered. Suppose that the hydrogen ion is at a particular position X between the metal and the water molecule (Fig. 8.89). The energy of the $M(e) + H^+$—H_2O system in the state will be given by the point Y on the $I + L - \Phi$ curve corresponding to the position X. If electron transfer could take place from the metal to the H^+ at this point, the transformation

$$M(e) + H^+\text{—}H_2O \rightarrow M\text{—}H + H_2O$$

would have occurred, the system would be in the electronated state, M—H $+ H_2O$, and its energy would be given by the point Z on the $R + A$ curve corresponding to X.

But can electron transfer occur to the H^+ in whatever position it is?

How can this question be answered? Simply, by referring to the tunneling condition, namely,

$$R + A \leq I + L - \Phi \tag{8.148}$$

But we notice that, when the H$^+$ is in the position X, (Fig. 8.89)

$$R + A > I + L - \Phi$$

or,[†] writing

$$(R + A) - (I + L - \Phi) = \vec{\Delta E} \tag{8.149}$$

one has $\vec{\Delta E} > 0$. Hence, as long as $\vec{\Delta E} > 0$, tunneling is forbidden.

Looking at the curves (Fig. 8.89) representing the energy of the system as a function of the position of the hydrogen nucleus, it becomes clear that, as long as the H$^+$ is farther away from the metal than a critical distance X^*, tunneling cannot occur.

Thus, when the hydrated hydrogen ion is in its ground state, the H$^+$ is at a maximum distance from the metal (for the fixed M—H$_2$O distance), $\vec{\Delta E}$ is a maximum, and the electron is forbidden a crossing over to the H$^+$. As the proton–water bond stretches, $\vec{\Delta E}$ decreases, but, as long as it is positive, there are no empty electronic states in the H$^+$—H$_2$O into which the electron in the metal can tunnel. When the H$^+$—H$_2$O bond stretches through a distance such that the H$^+$ is at a distance X^* from the metal, $\vec{\Delta E} = 0$, or $R + A = I + L - \Phi$, the electron tunneling condition is satisfied, and electron-energy state is available in the acceptor molecule, and the electron may tunnel to the proton. The moment the proton becomes a hydrogen atom, the relevant curve becomes the $R + A$ curve and the repulsion R of the water molecule and the attraction A of the metal set in and cooperate to make the hydrogen atom adsorb on the metal.

Reflect on this analysis. The condition for electron tunneling was initially developed (Section 8.5.5) in terms of a barrier for electron transfer. However, by thinking carefully about the quantities $R + A$ and $I + L - \Phi$, one was able to express the tunneling condition in terms of proton movements. But look at the diagram representing the energy of the system as the proton moves from the water molecule toward the metal. The proton has to climb a barrier. In fact, this barrier for the proton movement is similar to barriers for ion movement and hence for electronation reactions which have been drawn earlier (cf. Section 8.2.2).

[†] The arrow over $\vec{\Delta E}$ is intended to specify the direction of electron tunneling, i.e., from electrode to ion (the electronation reaction).

The new view has linked itself with the old, but there is an advance. In the old picture for an electronation reaction, when the ion had moved till the energy was at the peak of the barrier (the activated state), the reaction occurred rather magically. The magic lay in the facile and tacit implication that, when the activated state was reached for a bond breaking of H^+ from OH_2 and the formation of H—M, an electron was automatically available and present at the transition state. In the present model, the significance of the intersection point, in terms of the point at which the tunneling condition is fulfilled, is made clear, and the dependence of electron transfer upon bond stretching is defined.

The mechanism of electron transfer at interfaces has now become clearer. The proton-transfer barrier contains all the quantities necessary to define the condition for electron tunneling. It turns out that electron tunneling to hydrated protons cannot occur till the proton moves to a position corresponding to the barrier peak. This quantum-mechanical view clarifies the simpler one given earlier (Section 8.3.4b) and reveals the significance of the barrier peak in diagrams such as shown in Fig. 8.49.

8.5.7. Electron Tunneling and the De-electronation Reaction

Can the electron-tunneling mechanism for charge transfer explain the reverse de-electronation reaction? There is no problem here. Electrons can tunnel from either side of a barrier to the other, either from the metal to

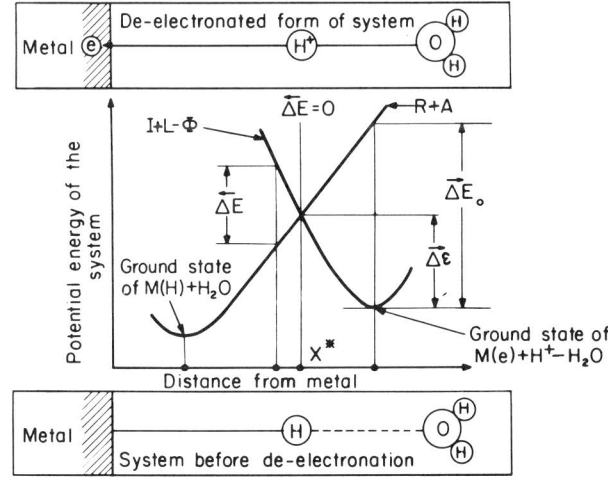

Fig. 8.90. In the de-electronation reaction, the M—H bond has to stretch in order that the energy $\overleftarrow{\Delta E}$ be squeezed to zero.

an electron acceptor (e.g., H^+) or from an electron donor (e.g., hydrogen) to the metal. The tunneling condition, however, has to be altered. Electrons from hydrogen atoms cannot tunnel into filled electron states in the metal. Hence, the tunneling condition becomes

$$R + A \geq I + L - \Phi \qquad (8.150)$$

Switching over to the proton-transfer curve, it is clear that, when the M—H bond is in its stable, unstretched condition, an electron cannot tunnel from the hydrogen atom to the metal, i.e., the hydrogen atom cannot be de-electronated. The M—H bond has to stretch to the critical distance X^* so that the energy difference $\vec{\Delta E}$ can be squeezed to zero (Fig. 8.90), just as, in the electronation reaction, the H^+—H_2O bond had to stretch sufficiently far for $\vec{\Delta E}$ to become zero.

8.5.8. A Perspective View of Charge-Transfer Reactions at an Electrode

The charge-transfer reaction has turned out to be something of a drama. The scene is the electrified interface. The actors are the metal teeming with free electrons in a hierarchy of levels, the proton whose conversion into a hydrogen atom is the theme of the play, and the water molecules whose association with the proton earns for it an important role. Electrons are too weak in energy to climb the potential-energy hill at the metal–solution interface. Nevertheless, their quantum qualities permit the electrons to sneak through the barrier in ghostlike fashion provided there are hospitable acceptor states for the electron, i.e., welcoming vacant energy levels in the hydrated hydrogen ions.

In a stable hydrated proton the levels into which a tunneling electron would like to be received are all occupied. Tunneling becomes possible, therefore, only if the proton–water bond stretches until the electron can be welcomed into a vacant electronic state. Thus, charge transfer depends on the cooperative movements of both the proton and the electron with the metal and the water molecule influencing the energetics of these electron and proton motions.

The model of charge transfer at interfaces presented here is essentially due to Gurney, 1932. It is little realized that one of the first applications of quantum mechanics to chemistry was to the basic model for electrodic kinetics. Gurney's theory became criticized for much that later came to be recognized as fairly trivial objections. It was reexpressed and elaborated upon in the 1960's by Gerischer. The delay in the acceptance of the quantum-mechanical model for charge transfer at interfaces did not help the progress of electrochemistry and the associated fields involving interfacial charge transfer.

8.5.9. The Symmetry Factor β: A Better View

An important point has come out of the picture of charge transfer. Bonds have to be stretched as a *prerequisite* to electron tunneling. Whether tunneling can occur or not can be expressed, for the electronation reaction, by a parameter[†] $\vec{\Delta E_0}$, which is the difference $(R + A) - (I + L - \Phi)$. To make tunneling possible, $\vec{\Delta E_0}$ must be "zeroed."

The way the original $\vec{\Delta E_0}$ gets squeezed to zero is by stretching the H^+—H_2O bond through a critical distance such that the proton is at a distance X^* from the metal. That is, the energy of the state $M(e) + H^+$—H_2O must be raised to a value at which electron tunneling begins to occur, i.e., at the intersection point of the potential energy curve of (Fig. 8.91). Let the critical stretching energy be $\vec{\Delta \varepsilon}$. How much stretching energy $\vec{\Delta \varepsilon}$ must be put in to zero $\vec{\Delta E_0}$ and thus make tunneling possible? *It appears that the ratio $\vec{\Delta \varepsilon}/\vec{\Delta E_0}$ (cf. Fig. 8.90) is of fundamental importance in charge-transfer theory.* Is it a new quantity or a familiar quantity in a different garb?

Some geometry is indicated. Consider the linear analog of Fig. 8.90. This figure is an extreme simplification[‡] of the potential-energy–distance relations when there is a vibrational stretching of the H^+—O bond in the system $M(e) + H^+$—OH_2 or the M—H bond in the system M—H + H_2O. It is concerned with proton stretching as a precondition for electron tunneling. It is obvious (Fig. 8.91) that

$$\frac{\vec{\Delta \varepsilon}}{\vec{\Delta E_0}} = \frac{AB \tan \gamma}{AB(\tan \gamma + \tan \theta)} = \frac{\tan \gamma}{\tan \gamma + \tan \theta} \qquad (8.151)$$

But $\tan \gamma / (\tan \gamma + \tan \theta)$ is none other than the symmetry factor β of Eq. (8.87). This is perhaps a surprising realization. The ratio $\vec{\Delta \varepsilon}/\vec{\Delta E_0}$ which is so crucial to the fundamental picture of charge transfer according to

---

[†] To emphasize that the discussion pertains to zero-field conditions, a subscript 0 is used thus: $\vec{\Delta E_0}$.

[‡] It will be noted that if the initial state involves a particular electronic state (e.g., the ground state), then that state is considered to be maintained throughout the passage of the electron. In other words, it is assumed that there are no transitions to a higher electronic state during the reaction. Such reactions are termed "adiabatic."

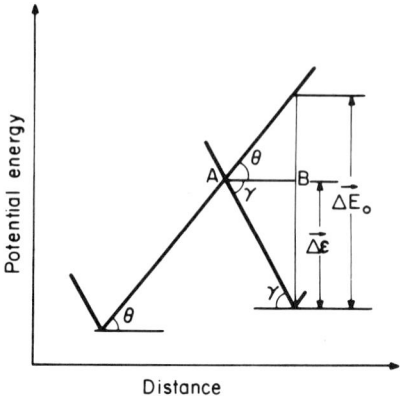

Fig. 8.91. A linear analogue to the potential-energy barrier for electron tunneling shows that the ratio of tan γ to tan γ + tan θ is equal to the ratio of energies $\vec{\Delta\varepsilon}$ to $\vec{\Delta E_0}$, the former being equal to the symmetry factor β. This gives the symmetry factor new physical meaning.

the Gurney model involving a critical bond stretching as a precondition for tunneling (Section 8.3.4) is, in fact, the symmetry factor β. Since, however, β has already been shown to be equal to the change in the activation energy of the reaction produced by a change $\Delta\phi$ in the potential difference from the metal to the OHP, one can write

$$\frac{\vec{\Delta\varepsilon}}{\vec{\Delta E_0}} = \beta = \frac{\text{Change in activation energy}}{\text{Change in interfacial potential}} \quad (8.152)$$

The picture begins to come somewhat into focus. Starting off with some basic mechanics of electrons, one was able to define the quantum-mechanical condition for the tunneling of electrons from a metallic donor to electron acceptors through an electron-energy barrier. The tunneling condition could be expressed in terms of an energy barrier for ion movement, e.g., the movement of protons toward the metal in the reaction

$$M(e) + H^+\text{—}H_2O \rightarrow M\text{—}H + H_2O$$

Tunneling becomes possible only when the proton has reached a position in between the metal surface and the H_2O corresponding to the barrier peak in, e.g., Fig. 8.89. The former assumption that a reaction step occurs when the energy of the system climbs to the peak is correct but now a rational quantum-mechanical basis has been given, and the electrons rather rightfully have been placed in the central rôle in charge transfer, instead of the ions and bond stretching as in the earlier model (Section 8.2.2). The ion stretching is very important, but this is so *only because of the need for the tunneling electrons to find a state in the acceptor particle equal in energy to the one (in the metal) from which they came, the condition for radiationless tunneling.*

The energy $\vec{\Delta\varepsilon}$ required to stretch the H^+—H_2O bond to the critical condition for electron tunneling is *a fraction* of the energy gap $\vec{\Delta E_0}$ which must be closed to make tunneling possible. The fraction $\vec{\Delta\varepsilon}/\vec{\Delta E_0}$ of the earlier discussion turns out to be none other than the symmetry factor β of the former treatment (*cf.* Section 8.3.7c). This is the fundamental (third) theory of the symmetry factor in terms of electron and proton mechanics. It bears out the following intuition which arose from the comparison of current–potential laws for electrode–electrolyte and semiconductor *n–p* junctions (*cf.* Section 8.4.1): a symmetry factor arises in the relation of current to potential when atom movements are a prerequisite for charge transfer across an electrified interface, but is absent when this is not so.

8.5.10. Quantifying the Charge-Transfer Picture

The object now is to develop an expression for the electronation current density for the reaction

$$M(e) + H_3O^+ \rightarrow MH + H_2O$$

The basis of the expression will be the charge-transfer mechanism just presented, a mechanism which involves H^+—H_2O bond stretching as a precondition for electron tunneling. The electronation-current density is given by

$$\vec{i} = \text{(Charge on one electron)(Number of successful electron transfers cm}^{-2} \text{ sec}^{-1}\text{)} \quad (8.153)$$

What does the number of successful electron transfers depend on? Firstly, on the number $n(E_F)$ of electrons with Fermi energy E_F that collide per second from inside the metal with a unit area of the metal–solution interface; secondly, on the probability P_T of their being able to tunnel to the ions; thirdly, on the probability $P(\vec{\Delta\varepsilon})$ that the H^+—H_2O bond is suitably stretched to welcome tunneling electrons, and, finally, on the number $n(H_3O^+)$ of H_3O^+ ions populating a unit area of the OHP, i.e.,

$$\vec{i} = e_0 n(E_F) P_T P(\vec{\Delta}\varepsilon) n(H_3O^+) \tag{8.154}$$

The evaluation of the number $n(E_F)$ of electrons of energy E_F striking 1 cm^2 of the surface of the metal from inside is rather cumbersome. It turns out that (*cf.* Appendix 8.1)

$$n(E_F) = \frac{4\pi m(kT)^2}{h^3} \tag{8.155}$$

The tunneling probability P_T has been shown (see Section 8.5.3) to be given by

$$P_T = \exp\left\{-\frac{4\pi l}{h}[2m(E_x - E_F)]^{\frac{1}{2}}\right\} \tag{8.131a}$$

where V_2 is replaced by E_x, the energy corresponding to the peak of the barrier (Fig. 8.76), and E is replaced by E_F, the Fermi level energy.

The next question concerns the probability $P(\vec{\Delta}\varepsilon)$ that the H$^+$—H$_2$O bond is adequately stretched to permit electron tunneling. In other words, what is the fraction of H_3O^+ ions which are suitably stretched? One may use a Boltzmann expression for this probability, i.e.,

$$P(\vec{\Delta}\varepsilon) = e^{-\vec{\Delta}\varepsilon/kT} \tag{8.156}$$

Finally, one can evaluate the number $n(H_3O^+)$ of hydrated protons at the OHP by the same argument that one used in the theory of contact adsorption [see Eq. (7.133)].[†] Thus,

$$n_{(H_3O^+)} = 2rNc_{H_3O^+} \tag{8.157}$$

Inserting Eqs. (8.157), (8.131a), and (8.155) into the expression (8.154) for the electronation-current density, one obtains

$$\vec{i}_{\Delta\phi=0} = e_0\left[\frac{4\pi m(kT)^2}{h^3}\right] e^{(-4\pi l/h)[2m(E_x-E_F)]^{1/2}} P(\vec{\Delta}\varepsilon_0) 2rNc_{H_3O^+} \tag{8.158}$$

---

† Of course, one assumes here the very simplest model of the interface to illustrate bare principles. It is quite improbable in practice. The substitution of $2rNc_{H_3O^+}$ as the total possible *number* of receptor particles for electrons is tantamount to the assumption that there is no contact adsorption and that the solution is concentrated enough so that $^{OHP}\Delta^s\phi$ is negligible (*cf.* Sections 7.7.10 and 8.3.2).

where the subscript $\Delta\phi = 0$ has been inserted to emphasize that this expression is valid only when the absolute potential difference across the interface is zero.

Since, however, it [cf. Eq. (8.152)] has been shown that $\vec{\Delta\varepsilon_0}$ is a fraction β of $\vec{\Delta E_0}$, the energy difference which has to be reduced to zero in order to satisfy the tunneling condition, one can write [cf. Eq. (8.156)]

$$P(\vec{\Delta\varepsilon_0}) = e^{-\vec{\Delta\varepsilon_0}/kT} = e^{-(\beta/kT)\vec{\Delta E_0}} \tag{8.159}$$

and, therefore,

$$i_{\Delta\phi=0} = 2e_0 \left[\frac{4\pi m(kT)^2}{h^3}\right] e^{(-4\pi l/h)[2m(E_x - E_F)]^{1/2}} e^{-\beta\vec{\Delta E_0}/kT} N_{H_3O^+} rc_{H_3O^+} \tag{8.160}$$

The effect of a potential difference $\Delta\phi$ on the electronation-current density must now be considered.

This will be done by recalling that the first direct effect of the application of potential is to change the energy of the electrons in the metal. Instead of its being Φ, it will be changed by $-e_0\Delta\phi$ to some other value, say, $\Phi - e_0\Delta\phi$. This change in the electron energy alters the energy of $M(e) + H^+ - H_2O$, and its energy–distance curve is raised or lowered depending on the sign of $\Delta\phi$. Thus (Fig. 8.92), the energy gap is altered from the value $\vec{\Delta\varepsilon_0}$ at $\Delta\phi = 0$, i.e., at a zero Galvani potential difference at the electrode, to a value of $\vec{\Delta\varepsilon}_{\Delta\phi}$ at the potential difference $\Delta\phi$.

Since the energy gap has changed, there will also be a change in the

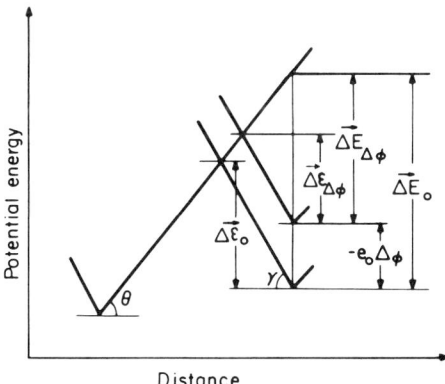

Fig. 8.92. A change in electron energy alters the energy of the initial state and produces a vertical shift in the potential-energy curve of $e_0\Delta\phi$.

probability of the H^+—H_2O being stretched sufficiently to satisfy the tunneling condition. The fraction of H_3O^+ ions which are suitably stretched at a potential $\Delta\phi$ is given by

$$P(\vec{\Delta\varepsilon}_{\Delta\phi}) = e^{-\vec{\Delta\varepsilon}_{\Delta\phi}/kT} = e^{-(\beta/kT)\vec{\Delta}E_{\Delta\phi}}$$
$$= e^{-\beta\vec{\Delta}E_0/kT} e^{-\beta e_0 \Delta\phi/kT} \quad (8.161)$$

since (cf. Fig. 8.92)

$$\vec{\Delta}E_{\Delta\phi} = \vec{\Delta}E_0 + e_0\Delta\phi \quad (8.162)$$

Hence, from Eqs. (8.161) and (8.159),

$$P(\vec{\Delta\varepsilon}_{\Delta\phi}) = P(\vec{\Delta\varepsilon}_0) e^{-\beta e_0 \Delta\phi/kT} \quad (8.163)$$

The electronation-current density corresponding to the potential $\Delta\phi$ is therefore given by an equation similar to (8.158), except that the $P(\vec{\Delta\varepsilon}_0)$ of Eq. (8.158) is replaced by $P(\vec{\Delta\varepsilon}_{\Delta\phi})$ of Eq. (8.163), i.e.,

$$i_{\Delta\phi} = 2e_0 \frac{4\pi m (kT)^2}{h^3} e^{-(4\pi l/h)[2m(E_x - E_F)]^{1/2}} e^{-\beta\vec{\Delta}E_0/kT} N_{H_3O^+} rc_{H_3O^+} e^{-\beta e_0 \Delta\phi/kT}$$
$$\simeq i_{\Delta\phi=0} e^{-\beta e_0 \Delta\phi/kT} \quad (8.164)$$

But

$$e^{-\beta e_0 \Delta\phi/kT} = e^{-\beta e_0 \Delta\phi_e/kT} e^{-\beta e_0 \eta/kT} = e^{-\beta F \Delta\phi_e/RT} e^{-\beta F\eta/RT} \quad (8.165)$$

where $\Delta\phi - \Delta\phi_e$ is replaced by the overpotential η, and Ne_0/Nk by F/R. Thus,

$$\vec{i}_{\Delta\phi} = (\vec{i}_{\Delta\phi=0} e^{-\beta F \Delta\phi_e/RT}) e^{-\beta F\eta/RT} \quad (8.166)$$

and, since the current density at equilibrium is the exchange-current density i_0, Eq. (8.166) can be rewritten in the form

$$\vec{i}_{\Delta\phi} = i_0 e^{-\beta F\eta/RT} \quad (8.167)$$

or, by dropping the subscript for the i,

$$i = i_0 e^{-\beta F\eta/RT} \quad (8.168)$$

This is the familiar exponential current–potential law already derived in a different way earlier (cf. Section 8.2.9). But it now has a quantum-mechanical basis which will allow us to relate currents at electrodes to the properties of the solid substrate and to solid-state physics. Hence, one obtains the beginnings of an ability to relate, e.g., the velocities of the corrosion of substances, the catalytic properties of substrates, and *perhaps* the functioning of some biological charge-transfer processes,[†] to the properties of the solid phase, through the bonding energy and the quantum-mechanical conditions for electron tunneling through the interfacial energy barriers.

8.5.11. Some Desirable Refinements and Generalizations

Of course, there are many refinements that can be made in the present statement, at a quite rudimentary level, of the bare principles of interfacial charge transfer, in quantum-mechanical terms. What these refinements are becomes clear if one thinks of the many approximations and considerable degree of simplifications contained in the picture presented above.

For example, the only electrons considered in the tunneling process are those with the Fermi energy E_F. However, it is only at $0°K$ that there are no electrons in the energy levels above the Fermi level. But the modifications necessary to take into account the electrons of energies higher than E_F involve no new principle. It is only a matter of using an expression similar to (8.158) for each value of the electron energy and integrating over all energies of electrons which can tunnel. Further, the electrons were considered to tunnel to a fixed and arbitrary distance from the electrode, whereas one should consider electrons tunneling through to ions at various distances out in the solution. Another gross approximation is the use of oversimple equations, such as Morse's equation, to represent the energy–distance relations for the stretching of bonds. This defect could be remedied by using more appropriate (but much more complicated) force–distance laws for the stretching and then devising computer programs to handle the lengthy complicated calculations corresponding to the more accurate models.

Hence, what has been sketched here is a very simple picture of fundamental charge-transfer theory for the H_3O^+ electronation reaction. The picture involves both proton movements with respect to the water molecules and the quantum mechanics of electron behavior in *tunneling through* energy barriers at surfaces.[‡]

[†] Quantum-mechanical tunneling of electrons is also the probable mechanism of electric-charge transport in biological membranes (e.g., electrons tunnel over 30 to 40 Å gaps in the jumps from cytochromes to proteins).

[‡] The reader *must* be clear that the barrier which is being tunneled through is a barrier such as that of Fig. 8.76. It is not the barriers such as those set up by the stretching H^+—O, or H—M bonds. These latter energy barriers (e.g., Fig. 8.88) are those which determine the number of times per H_3O^+ per second the condition is achieved that makes tunneling *possible* (the establishment of vibrational-energy levels in the O–H^+ bond

The basic picture is capable of generalization. It requires little imagination to develop the expression for the corresponding de-electronation current density for the reaction: M—H + $H_2O \rightarrow$ M(e) + H^+—H_2O. One works in terms of the probability of the M—H bond being suitably stretched, the population density of adsorbed hydrogen atoms having various bond energies in their attachment to the surface, etc. Once an expression for the de-electronation current density is obtained, one can easily get the full Butler–Volmer equation by considering the difference in the electronation and de-electronation current densities.

Further, there is no reason why the model cannot, in principle, be used in the case of other electrodic reactions. Of course, the systems become more complicated. For example, consider a reaction involving the electronation of a hydrated metal ion leading to deposition on to the electrode, e.g.,

$$[Ag(H_2O)_n]^+ + M(e) \rightarrow M\text{—}Ag + nH_2O$$

A complication here is that the ion, e.g., a Ag^+ ion, is enveloped in a hydration shell in which there are n water molecules. In this case, too, the condition for the tunneling of an electron from the electrode to the ion involves characteristic displacement of the ion with respect to the surrounding water molecules until the tunneling condition is satisfied, i.e., the electron energy in $[Ag(H_2O)_n]^+$ is equal to the Fermi energy of electrons in the metal.

The same sort of model applies to the *redox* reactions in which both electron acceptor and electron donor are ions in solution. It is customary to write such reactions as simple electron-transfer reactions, e.g.,

$$Fe^{+++} + e \rightarrow Fe^{++}$$

for the electronation of ferric ions. In actual fact, both ions are solvated, so one must write

$$[Fe(H_2O)_n]^{+++} + M(e) \rightarrow [Fe(H_2O)_m]^{++} + H_2O_{n-m} + M$$

Here, too, appropriate displacement of the ion with respect to the hydration shell gives rise to an electron-energy level in the receptor ion equal to that in the electron donor ion, and tunneling of an electron may then occur.

---

equal in energy to that of the Fermi electronic level in the metal). Therefore, the electron can tunnel through the barrier to the (equal-energied) state available in the solution (Fig. 8.79) without radiation.

An approach to the electron-transfer kinetics of homogeneous redox reactions was developed in the 1950's, apparently independent of the earlier tunneling equations developed by Gurney for the electrode–solution interface. Important contributions in this area have been made by Libby, by Marcus, and by Levich. These workers have stressed the purely electrostatic changes which occur during rearrangements of the solvation sheath—the changes in solvation energy—which would be associated with the receipt of an electron tunneling from Fe^{3+} to Fe^{2+}. They have also developed such approaches to charge-transfer theory for redox reactions which occur at interfaces. Here, important contributions have been made also by Dogonadze and Chimadzev. Electrons may tunnel through barriers at interfaces only if electronic levels are available to them on *both* sides. In practice, the difference in approach is only in which way the energy of rearrangement of the ion in solution is expressed. In the earlier, Gurney approach, the stretching of bonds and the potential-energy surface was stressed. In the later approach, the Born solvation energy is the aspect of the ion–solvent interactions which is stressed. The bond-stretching approach is of course more general, although also more difficult in a numerical sense. There is some disadvantage in approaches which assume that a continuous dielectric structure is present up to the electrode surface.

These, however, are long stories—largely unfinished and some hardly started—and will not be presented in this treatment, which has concerned itself with the basic and bare principles of interfacial charge-transfer theory from the quantum-mechanical point of view.

8.5.12. Surveying the Progress

The picture of the electrified interface which was sketched in the previous chapter (Chapter 7) permitted a broad understanding of the arrangement of the various particles in the electrified interface between an electrode and a solution. One also came out with an idea of the various potential drops in the interphase region. That sketch of the situation at the electrified interface was done, however, without considering the multifold possibilities of charge-transfer reactions.

The present chapter commenced with a consideration of the consequences of electron-transfer reactions. Device possibilities—substance producers and energy producers—were glimpsed. It was quickly realized, however, that only systems of *pairs* of electrode–electrolyte interfaces could be fabricated into devices. Nevertheless, attention was turned toward the understanding of single electrode–electrolyte interfaces as a preliminary step before tackling electrochemical devices.

In any charge-transfer reaction at an electrode–solution interface, at least the electron acceptor or donor must be an ion; otherwise, the reaction would be a purely chemical one. Further, a *necessary* condition for the successful transfer of charge is that there should be ion movement across at least part of the interface.

Based on the picture of *ion* movement, an analysis of the current due to the passage of electron-acceptor ions toward the electrode led to the derivation of the Butler–Volmer equation, a basic equation in electrodics. This equation displayed two special cases, firstly, a linear current–potential relation and, secondly, an exponential current–potential relation.

The Butler–Volmer equation also included two interesting quantities. One was the exchange-current density i_0, which represented the ceaseless two-way flow of charge across an interface at which there is equilibrium and a zero net current across the metal–solution interface. The i_0 was the quantity characterizing the dynamics of the interface. The second quantity in the equation was the symmetry factor β, which was formally given by geometrical arguments regarding the shape of the potential-energy–distance relations for the ion movement from the OHP toward the electrode. Physically speaking, however, the symmetry factor decided how much of the electrical energy $F\Delta\phi$ at the interface was available for the rate of the charge-transfer reaction. The Butler–Volmer equation being the theoretical expression of the velocity–potential behavior of reactions at an electrified interface, one had attained, at that stage, a *formal* understanding of electronation and de-electronation reactions.

But looming in the background was the feeling that one had turned a shaded eye to the act of electron transfer without which charge transfer is not charge transfer. The point is that ion movement is a necessary but not a sufficient condition for charge transfer. There must be electron transfer, too. But, at what stage in the motion of an ion toward the electrode does electron transfer occur? Does *electron* transfer occur *per se*, independent of the ion stretching? Has it a potential-energy barrier of its own? Are the energies of electrons in the metal sufficient to climb the barrier? The act of electron transfer to an ion needed a somewhat more discerning look.

In a sense, one had to begin at the beginning again and say something about the laws which govern how electrons move. These quantum laws permitted the electrons to behave in nonclassical ways. For instance, if the potential-energy hill was too high to climb over, the electrons could tunnel through the barrier. This tunneling property proved of vital significance to charge-transfer reactions.[†] A simple tunneling condition could be written down, and, from this, it turned out that electron tunneling became possible precisely at the point at which there had been enough bond stretching so that the electronic states available to the electron when it tunneled to the electron-acceptor particle were equal to those it had had in the metal.

It turned out that this condition for electron tunneling was attained when the proton's potential-energy–distance curve intersected the curve of the potential energy of the hydrogen atom in its vibrations on the substrate surface. The distance of the point of intersection of these two bond-stretching curves from the metal surface established the distance to which electrons have to tunnel after they have left the Fermi level in the metal. It gave the basis for the construction of the potential-energy barrier for electron tunneling. Such tunneling proved potential dependent because of the effect of the Galvani potential of the metal on the frequency of attainment of the condition for tunneling to occur.

Thus, a fairly comprehensive (though extremely oversimplified) picture of charge transfer at metal–solution interfaces emerged. For the electronation of ions, it is a picture of a hydrated ion reaching the electrified interface by one of the conventional mechanisms of transport, diffusion and conduction. Then the bond between the ion and a water molecule must stretch, and, the moment the stretching is adequate to satisfy the electron-tunneling condition, an electron leaps from the electrode to the ion, electron transfer has occurred, and the particle moves on to its final state. The picture is complete in principle, although approximate and crude in execution, and one goes on to consider other problems, armed with a sound basis for the Butler–Volmer equation, which furnishes the current–potential relation for a single-step, single-electron transfer reaction at a single metal–solution interface.

[†] In fact, one might say, "Electrochemical reactions always are quantal; chemical reactions not involving electron transfer seldom are," in the sense that (although, of course, the behavior and distribution of energy in the bonds of all molecules are quantized) the basic calculations and theory of activated collisions in thermal reactions are essentially the same when approached quantally or classically so long as the atoms are sufficiently heavy. In such cases, the energy level of the particles concerned in the reactions must exceed the *height* of the energy barrier. The passage *through* the barrier is negligible. A few chemical reactions are quantal in this (tunneling) sense, and these are exclusively reactions involving hydrogen atoms or H^+.

Further Reading

1. R. W. Gurney, *Proc. Roy. Soc. (London)*, **134A**: 137 (1932).
2. M. Green, "Electrochemistry of the Semiconductor–Electrolyte Interface," in: J. O'M. Bockris, ed., *Modern Aspects of Electrochemistry*, Vol. II, Chapter 5, Butterworth's Publications, Ltd., London, 1959.
3. H. Gerischer, in: P. Delahay, ed., *Recent Advances in Electrochemistry*, Vol. I, Chapter 2, Interscience Publishers, Inc., New York, 1962.
4. R. A. Marcus, *J. Chem. Phys.*, **43**: 679 (1965).
5. Y. Levich, in: P. Delahay, ed., *Recent Advances in Electrochemistry*, Chapter 4, Interscience Publishers, Inc., New York, 1965.
6. D. B. Matthews, *Proc. Roy. Soc. (London)*, **292A**: 479 (1966).
7. D. B. Matthews and J. O'M. Bockris, "Quantum Mechanics of Charge Transfer at Interfaces," in: J. O'M. Bockris and B. E. Conway, eds., *Modern Aspects of Electrochemistry*, Vol. VI, Chapter 1, Plenum Press, New York, 1969.

8.6. ELECTRODIC REACTIONS AND CHEMICAL REACTIONS

One way of increasing one's understanding of a thing is to see it in perspective. The attempt to set anything in perspective compels one to establish connections with related things and to pick out the samenesses and differences.

The thing of interest here is an electrodic reaction—an electronation or a de-electronation charge-transfer reaction at an electrode–solution interface. Very early in the treatment, it was stressed that, when one is electronating and de-electronating substances, one is doing chemistry at a surface of an electronic conductor in contact with an ionic solution.

The question naturally arises: How do electrodic reactions compare and contrast with chemical reactions occurring on a surface, i.e., heterogeneous chemical reactions.

At first, one can examine how the rates of the two types of reactions are expressed. The rate v of a heterogeneous chemical reaction is expressed in moles of substance transformed per second per square centimeter of the surface. The electrodic reaction rate, however, is expressed in terms of a current density i, i.e., in amperes per square centimeter.

The apparent dissimilarity between the electrodic and chemical rates can be removed if the current density is divided by the charge F transferred per mole of reactant. One obtains the number of moles of electron acceptor (or donor) electronated (or deelectronated) per second per square centimeter of the interfaces. That is, i/F has the same units (moles per square centimeter per second) as the rate of a surface chemical reaction. Hence, the *current density* across an electrode–electrolyte interface is to an electrodic

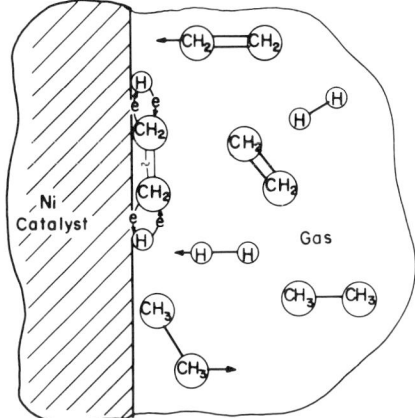

Fig. 8.93. In heterogeneous catalysis, the species must come to the surface and get adsorbed there before the reaction takes place. The products must move away to make room for other molecules of reactants.

reaction what *rate* is to a heterogeneous chemical reaction at a solid–gas or solid–solution interface.

Now look at the overall process of a heterogeneous chemical reaction, catalysis (Fig. 8.93). The reactants must travel up to the catalyst surface and get adsorbed there, then the reaction must occur, and, finally, the products must move away from the surface. For example, think of the gas-phase hydrogenation of ethylene at a nickel surface. The ethylene and hydrogen molecules adsorb on the nickel surface, the reaction occurs, and then the ethane quits the surface.

All this is quite similar to what happens in some electrodic reactions (Fig. 8.94). The ions have to reach the interphase region, get positioned, and undergo the charge-transfer reaction, and the electronated particle (in a one-step reaction) then moves off into the solution. The electrode is acting as a catalyst—an *electrocatalyst*.

What about the mathematical expressions for the rates of chemical and electrodic reactions? Chemical reactions, too, involve the movements of atoms to break old bonds and make new bonds. Consequently, they involve potential-energy–distance relations and barriers and activated states. In fact, the expression for the rate of a heterogeneous reaction is identical to the expression for the *zero-field* rate of an electrodic reaction, i.e.,

$$v = \frac{kT}{h} \prod_i c_i \, e^{-\Delta G^{0\ddagger}/RT} \qquad (8.169)$$

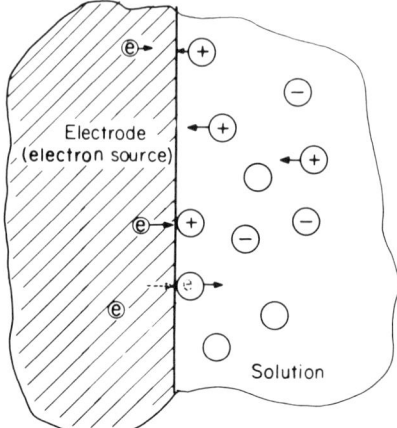

Fig. 8.94. In electrodic reactions, a set of circumstances has to take place similar to that of any heterogeneous reaction. The ions have to move to the surface, get adsorbed, undergo the reaction, and move away.

where Πc_i is the symbol used to indicate the *product* of the concentrations of the species involved in the reaction, and $\Delta G^{0\neq}$ is the standard free energy of activation for the chemical reaction (see Section A.4.3).

The point is, however, that the account of a heterogeneous chemical reaction ends essentially with the rate expression [Eq. (8.169)] just restated. The heterogeneous chemical reaction is distinguished from the electrodic reaction in that it does not involve a *net* charge transfer. The electrodic reaction, since it does involve a net charge, is therefore potential dependent in its rate, namely,

$$v_{\text{electrodic}} = \frac{i}{F} = \frac{i_0}{F} e^{-\beta F \eta / RT} \tag{8.170}$$

This is the crux of the difference between the electrodic reaction and the heterogeneous chemical reaction; the latter has a *fixed* activation energy and for fixed concentrations of reactants, the only way one can speed up the reaction is to increase the temperature, a drastic procedure. The relevant exponential factors in the electrodic or chemical reaction rates are $e^{-\beta F \eta / RT}$ and $e^{-\Delta E / RT}$. Suppose one assumes a typical value of 10,000 cal mole^{-1} for ΔE, $\beta = \frac{1}{2}$, and the starting $T = 300°K$. If one increases the overpotential by only -0.5 V, the electrodic rate increases about 10^5 times. A similar increase in the chemical rate by temperature alone would need the usually impractical temperature increase of about 1000°C.

In an electrodic reaction, the activation energy can be varied by the field across the interface to make the electrodic reaction go at speeds which differ by many orders of magnitude. Hence, if one prefers, one can say that electrodic reactions are heterogeneous chemical reactions with variable activation energies. Electrodic reactions are clearly also heterogeneous charge-transfer reactions. These are semantic matters; the point is, electrodic reactions have a potential-dependent activation energy.

What about homogeneous charge-transfer reactions, say, the electronation of a ferrous ion by a ceric ion,

$$Ce^{++++} + Fe^{++} \rightarrow Ce^{+++} + Fe^{+++}$$

One could compare the ceric ion with a microelectrode which contains the electron required for electronation. It turns out (8.5.12) that the picture of this homogeneous charge-transfer reaction is similar to that presented in Section 8.4. The bonds between the water molecules and the ferrous ion have to stretch, and, when the tunneling condition is thus satisfied, the electron tunnels through to complete the electronation of the cerium ion (when *it* is stretched in respect to its surroundings). But, for homogeneous charge-transfer reactions, one cannot turn a knob and change the reaction rate.

One can conclude from this discussion that electrodic reactions are basically similar to heterogeneous chemical reactions with one essential operational difference—the reaction rate depends on the interfacial field and this can be controlled by external means—and with one fundamental difference—the electrochemical reactions are always quantal in character. Whether, now, electrodics (i.e., electrode kinetics, or interfacial charge-transfer kinetics) should be considered a special case of heterogeneous-reaction kinetics or the latter should be considered the zero-field special case of electrodics is a matter of definition.

Further Reading

1. H. Eyring and E. N. Eyring, *Modern Chemical Kinetics*, Reinhold Publishing Corp., New York, 1965.
2. S. Srinivasan and H. Wroblowa, "Electrocatalysis," in: *Recent Advances in Catalysis*, ed. P. Emmett Vol. XVII, Chapter 6, Academic Press, Inc., New York, 1967.
3. J. O'M Bockris and S. Srinivasan, *Fuel Cells: Their Electrochemistry*, McGraw-Hill Book Company, New York, 1969.

Appendix 8.1. The Number of Electrons Having Energy E_F Striking the Surface of a Metal from the Inside

To avoid a more complete development of the Fermi–Dirac statistics, let the derivation be carried out by consideration of the equilibrium between a (classical) gas of electrons outside the metal, and those electrons inside. Thus, the situation considered is that of thermionic emission when the net emission rate is limitingly small.

One may write

$$\mu_{\text{(electrons in gas)}} = \mu_{\text{(electrons in metal)}} \qquad (A8.1)$$

But

$$\mu_{\text{elec gas}} = -kT \ln f_{\text{trans}} + u_g \qquad (A8.2)$$

where f_{trans} is the partition function for translational motion in which a given particle is confined to a volume v, and u is the potential energy; also,

$$\mu_{\text{elec met}} = E_F + u_m \qquad (A8.3)$$

But

$$f_{\text{trans}} = \frac{g(2\pi m_e kT)^{\frac{3}{2}} v}{h^3} \qquad (A8.4)$$

where g is the degeneracy in the gas phase (2 for electrons) and v is the space available per electron. One may take $v = 1/n$, the number of electrons per cubic centimeter. Hence,

$$-kT \ln \left[\frac{g(2\pi m_e kT)^{\frac{3}{2}}}{h^3 n} \right] + u_g = E_F + u_m \qquad (A8.5)$$

$$n = \frac{g(2\pi m_e kT)^{\frac{3}{2}}}{h^3} e^{-[(u_g - u_m) - E_F]/kT} \qquad (A8.6)$$

where n is the number of electrons cm^{-3} *in the gas*.

Now, the velocity in one direction for a gas is

$$u = \left(\frac{kT}{2\pi m} \right)^{\frac{1}{2}} \qquad (A8.7)$$

Thus, at equilibrium, the number of electrons reentering the metal is nu. For $g = 2$, this gives

$$\frac{4\pi m_e (kT)^2}{h^3} e^{-(u_g - u_m - E_F)/kT} \qquad (A8.8)$$

This must also be the number coming out. But the number coming out is

$$(\text{Number striking surface from inside}) \; e^{-\Phi/RT} \qquad (A8.9)$$

From (8.8) and (8.9), it follows that the number of electrons striking the surface from inside is

$$\frac{4\pi m_e (kT)^2}{h^3} \qquad (8.155)$$

CHAPTER 9
ELECTRODICS: MORE FUNDAMENTALS

9.1. MULTISTEP REACTIONS

9.1.1. The Question of Multistep Reactions

The analysis of electrodic reactions has been restricted so far to *one-step*, one-electron charge-transfer reactions. In practice, however, few reactions consist of only one step. For example, after the electronation of H_3O^+ ions,

$$M(e) + H_3O^+ \rightarrow MH + H_2O$$

one has a hydrogen atom adsorbed on the metal. The reaction does not terminate at this point. The adsorbed hydrogen atoms go on to combine and form molecular hydrogen, and this hydrogen is evolved thus:

$$2MH \rightarrow 2M + H_2\uparrow$$

In fact, therefore, the hydrogen-evolution charge-transfer reaction consists of several steps:

Step 1. H_3O^+ [in bulk of solution] $\xrightarrow{\text{diffusion}}$ H_3O^+ [at OHP]

Step 2. $M(e) + H_3O^+$ [at OHP] $\xrightarrow{\text{charge transfer}}$ $MH + H_2O$

Step 3. $2MH \xrightarrow{\text{combination}} 2M + H_2$ [at electrode]

Step 4. H_2 [at electrode] $\xrightarrow{\text{bubble formation}}$ H_2 [in atmosphere]

992 CHAPTER 9

It has been assumed so far that the rate of the reaction is determined by the charge transfer, step 2. Is this always true? Should not one conceive of the possibility of the rate's being controlled by steps other than the charge transfer, step 2? In fact, how does one analyze the rate of a series of consecutive reactions, i.e., multistep reactions, which occur in a sequence of several steps?

9.1.2. Some Ideas on Queues, or Waiting Lines

Before examining multistep reactions, it is worthwhile giving qualitative consideration to the general problem of the formation of queues, or waiting lines. This problem can be posed in the following familiar terms (Fig. 9.1).

Suppose that passengers (undertaking a train journey) arrive at the ticket counter. At the counter, the passengers have to be serviced, i.e., issued tickets, information, etc. It is obvious that, if the arrival rate of passengers at the ticket counter (to be referred to as a *servicing center*) is greater than the servicing rate, then a queue, or waiting line, of passengers builds up.

This is not an isolated example. One can speak of queues, or waiting lines, of customers at a supermarket counter, of automobiles at traffic lights, of airplanes at an airport, of patients at the receiving ward of a hospital, of telephone calls at a switchboard or exchange, of components on a factory assembly line, of fluids flowing through narrow constrictions, etc.

In all these cases, something arrives at a servicing center, and any servicing delay leads to the buildup of a queue. Since the basic pattern in all these examples is the same, a general theory, known as *queueing* or *waiting-time theory* has been developed. Its concern is to relate the magnitude of the queue to the arrival and servicing rates. Its utilitarian purpose is of course to understand how to minimize and possibly eliminate the queue.

Now, servicing centers are generally quite complex. They invariably consist of subcenters. An intermediate-stop airport, e.g., is a servicing center for airplanes, but there are several subcenters—landing, taxiing, unloading passengers and freight, refueling, loading passengers, etc. Similarly,

Fig. 9.1. If the rate of arrival is high enough, the throughput of passengers will depend entirely on the rate of servicing and a waiting line will build up at the servicing center.

a town through which automobiles pass is a complex servicing center consisting of several traffic lights.

Whenever there is more than one subcenter in the servicing center, a central question emerges in queueing theory: Which subcenter of the complex servicing center is mainly responsible for the queue? Or, alternatively, which subcenter determines the overall servicing rate? For example, which particular traffic light is mainly responsible for the traffic jam? Or, in an automobile factory, what controls the overall rate of production, component manufacture, assembling the parts, or the final finishing?

9.1.3. The Overpotential η Is Related to the Electron Queue at an Interface

What have these thoughts on queues to do with electrodic reactions? Think of the electrified metal–solution interface as a servicing center for the electrons which flow into it from the metal to participate in the electronation reaction. The electrodic reaction represents the servicing of the electrons.

Any *servicing difficulties* and delays, such as preconditions which must be satisfied before electron tunneling occurs, lead to a queue of electrons on the electrode. In other words, the excess charge q_M on the electrode becomes more negative, and, thus, the potential difference across the interface departs from the equilibrium value. The overpotential, therefore, is determined by the electron queue.

As in other complex servicing centers, the electrodic reaction may consist of a number of steps. For example, it has just been indicated that the overall electrodic reaction

$$2H_3O^+ + 2M(e) \rightarrow 2M + 2H_2O + H_2 \tag{9.1}$$

includes the following steps

$$H_3O^+ + M(e) \rightarrow MH + H_2O$$

and

$$2MH \rightarrow 2M + H_2$$

Or, as another example, the discharge of silver ions may consist of the transport of ions from the bulk of the solution

$$Ag^+ \text{[solution]} \xrightarrow{\text{diffusion}} Ag^+ \text{[OHP]}$$

and charge transfer

$$Ag^+ \text{[OHP]} + e \rightarrow Ag \text{[metal]}$$

Thus, the basic queueing problem arises in the case of electrodic reactions, too. Which particular subcenter (i.e., step) of the overall servicing center (i.e., electrode reaction) is the cause of the waiting line, or queue, of electrons (and, therefore, of the overpotential η)? Correspondingly, which particular subcenter (or step) controls the overall servicing rate?

It can easily be seen that, whichever of the unit steps in the above reactions causes the hold up in the servicing center, there will be a waiting line of electrons forming. If, e.g., the recombination reaction of hydrogen evolution (succeeding the charge transfer) is slow, the electrons will accumulate waiting for the product of the transfer to pass through this bottleneck. If, on the other side, the transport of Ag^+ ions (as a step preceding electron transfer) is slow, the electrons will accumulate, waiting for their partners like unfinished products on an assembly line waiting for parts to come to a particular place.

The above questions are obviously of crucial importance. Once there is an understanding of electron waiting lines, i.e., of the origin of the current-produced potential, η, then one can consider how to control the factor that causes the electron waiting line and, therefore, how to control η and perhaps significantly reduce it.

9.1.4. A Near-Equilibrium Relation between the Current Density and Overpotential for a Multistep Reaction

It has been stated that, though an overall electrodic reaction may consist of several steps, it is usually possible to single out one step and regard it as the essential cause of the overall electron queue and hence the overpotential η. What is the justification for this discriminatory attitude toward the one step?

Consider an overall electrodic reaction that takes place in n steps (Fig. 9.2). Let it be assumed, for convenience of exposition, that each step is a charge-transfer reaction with an electron acceptor's receiving an electron.

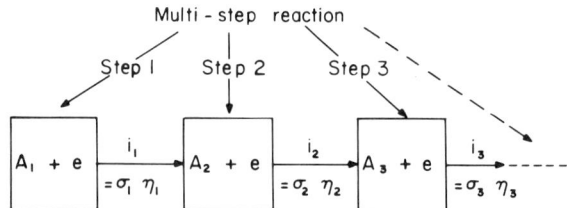

Fig. 9.2. In a multistep electron-exchange reaction, each step produces its individual current density. At a steady state, all these currents must be equal.

To simplify the treatment, let it also be assumed that the n individual electronation reactions are only slightly off equilibrium and that, therefore, for each reaction, one can use the *linear* current-density–overpotential law [Eq. (8.53), Section 8.2.11]. The rate of any one step in such a case is proportional to its overpotential $\eta_j{}^\dagger$ (*cf.* Section 8.2.9)

$$i_j = \sigma_j \eta_j \tag{9.2}$$

where, in analogy to Ohm's law, σ_j is the reciprocal *resistance*, or the *conductivity*, of the reaction step j.

Each of the n electronation steps has associated with it an individual current density which is produced by a corresponding overpotential at the interface. Thus, one can write

$$\begin{aligned} i_1 &= \sigma_1 \eta_1 \\ i_2 &= \sigma_2 \eta_2 \\ &\vdots \\ i_n &= \sigma_n \eta_n \end{aligned} \tag{9.3}$$

Now consider the situation where the overall reaction settles down and the intermediates do not change with time, i.e., when steady-state conditions are reached. Since consecutive currents are being considered, the current density from one reaction must be equal to the current density for the following reaction. Thus, the current densities of all the steps are equal to each other, i.e.,

$$\begin{aligned} i_1 = i_2 = \cdots &= i_n = i_j \quad [j = 1, 2, 3, \ldots, n] \\ &= \sigma_1 \eta_1 = \sigma_2 \eta_2 = \cdots = \sigma_n \eta_n \\ &= \frac{\eta_1}{1/\sigma_1} = \frac{\eta_2}{1/\sigma_2} = \cdots = \frac{\eta_n}{1/\sigma_n} \end{aligned} \tag{9.4}$$

---

† One can ask at this point how it is possible to have different overpotentials for different steps of a reaction at one and the same metal–solution interface. Here it is necessary to remember that overpotential as a driving force is the difference between the reversible potential and the actual electrode potential. Hence, different overpotentials at one and the same actual potential attained by the electrode mean that, in the process of establishing the steady state, each unit step has established a different reversible potential. This is achieved by changing the concentrations of various intermediate species or participants in the steps.

The equalities (9.4) can also be written as

$$i_j \frac{1}{\sigma_1} = \eta_1$$

$$i_j \frac{1}{\sigma_2} = \eta_2 \qquad (9.5)$$

$$i_j \frac{1}{\sigma_n} = \eta_n$$

and, summing all the equations, one obtains

$$i_j \left(\frac{1}{\sigma_1} + \frac{1}{\sigma_2} + \cdots + \frac{1}{\sigma_n} \right) = \eta_1 + \eta_2 + \cdots + \eta_n \qquad (9.6)$$

or

$$i_j = \frac{\sum_{j=1}^{n} \eta_j}{\sum_{j=1}^{n} (1/\sigma_j)} \qquad (9.7)$$

But the steps behave as parallel to each other as far as the electron flow through the interface is concerned. Hence, the total current must also be equal to the sum of those individual currents, i.e.,

$$i = i_1 + i_2 + \cdots + i_n = n i_j \qquad (9.8)$$

Hence, the total current flowing through the interface is from Eqs. (9.7) and (9.8)

$$i = \frac{\sum_{j=1}^{n} \eta_j}{(1/n) \sum_{j=1}^{n} (1/\sigma_j)} \qquad (9.9)$$

The inverse of the conductivity of each reaction is its resistivity, and the sum of all the resistivities divided by their number gives the average resistivity, R_F, of the reaction,

$$\frac{1}{n} \sum_{j=1}^{n} \frac{1}{\sigma_j} = R_F \qquad (9.10)$$

This average resistivity, R_F, shall be called the faradaic resistance of the interface.

The argument can be generalized without restricting it to near the equilibrium. The only difference is that, far from equilibrium, exponential current-density–potential relations are operative and the resistances of individual reactions as well as the faradaic resistance are not constant any more but dependent on overpotential. This is the basic operational difference between the faradaic and ohmic resistances.

9.1.5. The Concept of a Rate-Determining Step

Observe what happens in expressions (9.7) and (9.9) if the conductivity σ for one step r is much smaller than that for any other step, $j \neq r$, i.e.,

$$\sigma_r \ll \sigma_j \quad [j \neq r] \tag{9.11}$$

In that case,

$$\sum_{j=1}^{n} \frac{1}{\sigma_j} = \frac{1}{\sigma_1} + \frac{1}{\sigma_2} + \cdots + \frac{1}{\sigma_r} + \cdots + \frac{1}{\sigma_n} \approx \frac{1}{\sigma_r} \tag{9.12}$$

because all the terms $1/\sigma_j \, [j \neq r]$ become insignificant in comparison with $1/\sigma_r$.

Also, because of the equalities expressed in (9.4), the same must apply to the overpotentials η. They must all become insignificant compared with η_r

$$\eta_r \gg \eta_j \quad [j \neq r] \quad \text{or} \quad \eta \approx \eta_r \tag{9.13}$$

Thus Eq. (9.9) can be rewritten as

$$i \simeq \frac{\eta}{(1/n)(1/\sigma_r)} \tag{9.9a}$$

Hence, a single step will control the overall rate if its conductivity is much smaller (or its resistivity is much larger) than that of any other step.

It is seen, however, from Eqs. (8.43) and (8.55), that the conductivity σ_j of any step is determined largely by its equilibrium exchange-current density $i_{0,j}$. The smaller the $i_{0,j}$ is for the step, the lower is its conductivity. Thus, one can say that the step with the smallest $i_{0,j}$ generally determines the overall current.[†]

In fact, one can imagine (Fig. 9.3) that the electrodic reaction is like a resistor and the faradaic resistance of the overall reaction is a series com-

---

[†] The relation of σ_r to the exchange-current density will be discussed in more detail at the end of this section.

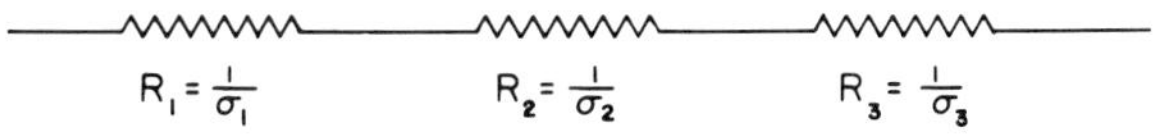

$$R_1 = \frac{1}{\sigma_1} \qquad R_2 = \frac{1}{\sigma_2} \qquad R_3 = \frac{1}{\sigma_3}$$

Fig. 9.3. The total electrical resistance of an electrode reaction consisting of a series of consecutive steps is obtained as a series combination of individual resistors.

bination of resistors in an electrical circuit. Then, the overall conductance of the circuit is approximately given by the smallest conductance or largest resistance so long as one of the resistors is significantly—say, 10 times— larger than any of the other resistors.

One should note that $i_{0,j}$ consists of two factors, the rate constant and the concentration of the substrate in the given step. Hence, either of those being small can be the cause of a slow rate-determining step (rds).

There is another interesting result of the concept of an rds. If all the exchange-current densities except that for the rds are very large, it means that the overpotentials due to all other steps are negligibly small [cf. Eq. (9.13)]. Since the magnitude of the overpotential for a step is a measure of how far the step is away from equilibrium, then if $\eta_j \to 0$ $[j \neq r]$, one concludes that the jth step is almost in equilibrium, i.e., it is in *quasi equilibrium*. Hence, the existence of a unique rds usually implies that other steps are virtually in equilibrium.

The electron waiting-line problem is hence clear. In a particular multistep electron-transfer reaction, the step with the lowest servicing rate or conductivity produces the largest queue and, indeed, the total queue is virtually a simple multiple of the queue at the rds. In other words, *in the steady state, all n steps proceed at the rate of the rate-determining step i_r* [cf. Eq. (9.4)], and the total net current is

$$i = n i_r \tag{9.14}$$

where n is the number of the single-electron transfer steps in the overall reaction.

Since

$$i_r = \vec{i}_r - \overleftarrow{i}_r$$

then

$$i = n(\vec{i}_r - \overleftarrow{i}_r) \tag{9.14a}$$

In order to develop the Butler–Volmer equation for a multistep reaction, expressions for $\vec{i}_r$ and $\overleftarrow{i}_r$ must be found for this case. Consider a multistep reaction

$$\begin{array}{ll} A + e \rightleftharpoons B & [\text{Step 1}] \\ B + e \rightleftharpoons C & [\text{Step 2}] \\ \vdots & \vdots \\ P + e \rightleftharpoons R & [\text{Step } \vec{\gamma}] \\ \underline{R + e \rightarrow S} & [\text{rds}] \\ S + e \rightleftharpoons T & [\text{Step } \overleftarrow{\gamma} \equiv n - \vec{\gamma} - 1] \\ \vdots & \vdots \\ Y + e \rightleftharpoons Z & [\text{Step } n] \end{array} \qquad (9.15)$$

in which $R + e \rightarrow S$ is the single-electron transfer rds preceded by $\vec{\gamma}$ other single-electron transfer steps and followed by $\overleftarrow{\gamma}$ †such steps.

The current $\vec{i}_r$ of the forward (electronation) reaction in the rds is equal to

$$\vec{i}_r = F\vec{k}_R c_R e^{-\beta F \Delta\phi/RT} \qquad (8.21)$$

Equation (8.21), in which c_R is the concentration of an intermediate, may give an erroneous impression that the current–potential relation is completely determined by the exponential term in $\Delta\phi$. However, species R was the result of a series of charge-transfer mechanisms, and, thus, its concentration, as shown below, is also potential dependent. To unravel this dependence, it will be recalled that all steps preceding and following the rds can often be assumed to be at equilibrium. Then, one can equate their forward and backward rates, e.g., for the first step $A + e^- \rightleftharpoons B$,

$$\vec{i}_1 \simeq \overleftarrow{i}_1$$

or, using Eqs. (8.21) and (8.23),

$$F\vec{k}_1 c_A e^{-\beta F \Delta\phi/RT} \simeq F\overleftarrow{k}_1 c_B e^{(1-\beta)F \Delta\phi/RT}$$

Therefrom,

$$c_B = K_1 c_A e^{-F \Delta\phi/RT} \qquad (9.16)$$

where

$$K_1 = \frac{\vec{k}_1}{\overleftarrow{k}_1}$$

---

† The number of electrons transferred in the overall reaction is n; $\vec{\gamma}$ electrons are transferred in the steps preceding the rds; one electron is transferred in the rds. Thus, $\overleftarrow{\gamma} \equiv (n - \vec{\gamma} - 1)$ electrons are transferred in the steps after the rds.

1000 CHAPTER 9

Similarly,

$$c_C = K_2 c_B e^{-F\Delta\phi/RT} = K_2 K_1 c_A e^{-2F\Delta\phi/RT}$$

$$c_D = K_3 c_C e^{-F\Delta\phi/RT} = K_3 K_2 K_1 c_A e^{-3F\Delta\phi/RT}$$

and, finally,

$$c_R = \left[\prod_{i=1}^{\vec{\gamma}} K_i\right] c_A e^{-\vec{\gamma} F \Delta\phi/RT} \tag{9.16a}$$

By substituting Eq. (9.16a) in (8.21),

$$\vec{i}_R = F\vec{k}_R \left[\prod_{i=1}^{\vec{\gamma}} K_i\right] c_A e^{-(\vec{\gamma}+\beta)F\Delta\phi/RT} = i'_{0,R} e^{-(\vec{\gamma}+\beta)F\eta/RT} \tag{9.17}$$

where

$$i'_{0,R} = F\vec{k}_R \left[\prod_{i=1}^{\vec{\gamma}} K_i\right] c_A e^{-(\vec{\gamma}+\beta)F\Delta\phi_e/RT} \tag{9.17a}$$

The prime at $i'_{0,R}$ indicates that the rate is now related to the concentration of the initial product A and not R.[†] In complete analogy, the rate of the backward (de-electronation) reaction

$$S \rightarrow R + e^-$$

can be related to the concentration of the final product Z by the equations

$$\overleftarrow{i}_R = F\overleftarrow{k}_R \left[\prod_{i=n-\overleftarrow{\gamma}-1}^{n} K_i\right] c_Z e^{(\overleftarrow{\gamma}+1-\beta)F\Delta\phi/RT}$$

$$\overleftarrow{i}_R = i'_{0,R} e^{(\overleftarrow{\gamma}+1-\beta)F\eta/RT} \tag{9.17b}$$

where

$$i'_{0,R} = F\overleftarrow{k}_R \left[\prod_{i=n-\overleftarrow{\gamma}-1}^{n} K_i\right] c_Z e^{(\overleftarrow{\gamma}+1-\beta)F\Delta\phi_e/RT} \tag{9.17c}$$

Thus, the Butler–Volmer equation for multistep reactions can be written as follows [cf. Eqs. (9.14a), (9.17), and (9.17b)]

$$i = n(\overleftarrow{i}_R - \vec{i}_R) = ni'_{0,R}[e^{(\overleftarrow{\gamma}+1-\beta)F\eta/RT} - e^{-(\vec{\gamma}+\beta)F\eta/RT}]$$
$$= i_0[e^{(\overleftarrow{\gamma}+1-\beta)F\eta/RT} - e^{-(\vec{\gamma}+\beta)F\eta/RT}] = i_0[e^{(n-\vec{\gamma}-\beta)F\eta/RT} - e^{-(\vec{\gamma}+\beta)F\eta/RT}] \tag{9.18}$$

since $\overleftarrow{\gamma} = n - \vec{\gamma} - 1$ and where

$$i_0 = ni'_{0,R} \tag{9.19}$$

and $i'_{0,R}$ is given by Eqs. (9.17a) and (9.17c).

[†] Note that, in Section 9.1.5 and in Eq. (8.43) referred to in this section, $i_{0,j}$ refers to the concentration of the substrate reactant in step j.

In the *high-field* approximation (*cf*. Section 8.2.10), the first exponential term can be neglected for $\eta \ll 0$, i.e., for net electronation, and the second exponential term for $\eta \gg 0$, i.e., for net de-electronation.

In the *low-field* approximation, where both exponential terms in the Butler–Volmer equation can be linearized, Eq. (9.18) becomes

$$i = ni'_{0,R}\left(\frac{nF}{RT}\eta\right) = i_0\left(\frac{nF}{RT}\eta\right) \qquad (9.20)$$

where $n = \overleftarrow{\gamma} + \overrightarrow{\gamma} + 1$.

This treatment remains valid for two other possible reaction sequences; these are sequences in which there are (a) chemical, i.e., non-charge-transfer, steps before and after a charge-transfer rds and (b) charge-transfer steps before and after a chemical rds. In the latter case, where no charge transfer occurs in the rds, the number of electrons transferred after the rds will be $n - \overrightarrow{\gamma}$. There will be no effect of potential on the rate of the rds except from that arising from previous charge-transfer steps; thus, the Butler–Volmer equation for a chemical rds is given as

$$i = i_0[e^{(n-\overrightarrow{\gamma})F\eta/RT} - e^{-\overrightarrow{\gamma}F\eta/RT}] \qquad (9.21)$$

which, when applying the low-field approximation, produces Eq. (9.20).

Equations (9.18) and (9.20) may be written in a general form by including a factor r,[†] e.g.,

$$i = i_0[e^{(n-\overrightarrow{\gamma}-\beta r)F\eta/RT} - e^{-(\overrightarrow{\gamma}+\beta r)F\eta/RT}] \qquad (9.18a)$$

Comparison of Eqs. (9.9a) and (9.20) allows the term σ_r to be identified as

$$\sigma_r = i'_{0,R}\frac{nF}{RT} \qquad (9.22)$$

where $i'_{0,R}$ is given by Eqs. (9.17). Thus, Eq. (9.9a) can be rewritten as

$$i = ni'_{0,R}\frac{nF}{RT}\eta = i_0\frac{nF}{RT}\eta \qquad (9.9b)$$

Note that Eqs. (9.18) and (9.21) pertain to a case where the rds occurs once per one occurrence of the reaction sequence. A more general expression will be given in Section 9.1.7.

[†] When the rds is a charge-transfer step, $r = 1$ and, when the rds is a chemical step, $r = 0$.

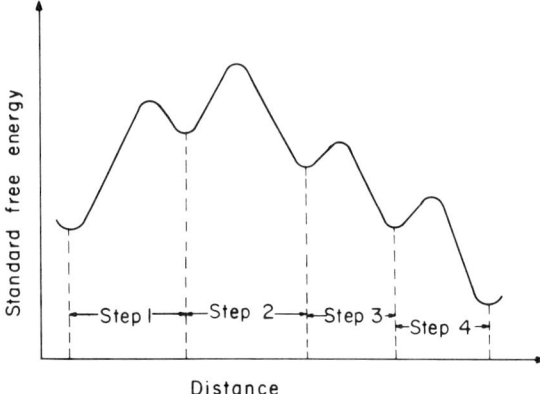

Fig. 9.4. Activation-energy barrier for a multistep reaction.

9.1.6. Rate-Determining Steps and Energy Barriers for Multistep Reactions

Every reaction has an energy barrier associated with it. When, therefore, there are a series of consecutive reactions, one has a series of consecutive barriers (Fig. 9.4). The overall reaction corresponds to the passage in one direction of the point representing the system across all the barriers.

Suppose the standard-free-energy barrier is as shown in Fig. 9.5. It will be noticed that step 1 has a larger standard free energy of activation[†] than step 2, i.e.,

$$\Delta G^{0\ddagger}_{A \to B} > \Delta G^{0\ddagger}_{B \to C}$$

On the other hand, the activated state of step 2, B^*, is higher with respect to the initial state A than step 1's activated state A^*.

The question is: Which step will determine the overall rate of the reaction?

Assume that step 1 determines the overall rate $\vec{v}$. One has (cf. Section 8.2.3)

$$\vec{v}_{\text{step 1}} = \frac{kT}{h} c_A \exp\left(-\frac{\vec{\Delta G}^{0\ddagger}_{A \to B}}{RT}\right) \tag{8.11}$$

[†] The quantity $\Delta G^{0\ddagger}_{A \to B}$, and like quantities, is the standard free energy of activation, which in the theory of absolute reaction rates (Chapter 4, Appendix 1) governs the rate of passage of a representative point in the system from A to B. It is *not* $G_B^0 - G_A^0$, but $G^{0\ddagger}_{A*} - G_A^0$ (see Fig. 9.5).

If, however, step 2 is controlling the overall rate, then

$$\vec{v}_{\text{step 2}} = \frac{kT}{h} c_B \exp\left(-\frac{\vec{\Delta G}^{0\ddagger}_{B \to C}}{RT}\right) \quad (8.11a)$$

One may take into account the fact that steps other than the rds can be considered in virtual equilibrium. Hence, the substance B is in equilibrium with the reactants of A. Therefore, the law of mass action can be used, i.e.,

$$\frac{c_B}{c_A} = \exp\left(-\frac{\Delta G^0_{A \to B}}{RT}\right)$$

where $\Delta G^0_{A \to B}$ is the standard free energy of formation of the substances at B. Hence, by substituting for c_B in Eq. (8.11a), one gets

$$\vec{v}_{\text{step 2}} = \frac{kT}{h} c_A \exp\left(-\frac{\Delta G^0_{A \to B}}{RT}\right) \exp\left(-\frac{\vec{\Delta G}^{0\ddagger}_{B \to C}}{RT}\right)$$

$$= \frac{kT}{h} c_A \exp\left(-\frac{\Delta G^{0\ddagger}_{A \to C}}{RT}\right) \quad (8.11b)$$

On comparing [see (8.11a) and (8.11b)] the expressions for $\vec{v}_{\text{step 1}}$ and $\vec{v}_{\text{step 2}}$, it is clear that, because

$$\Delta G^{0\ddagger}_{A \to C} > \Delta G^{0\ddagger}_{A \to B}$$

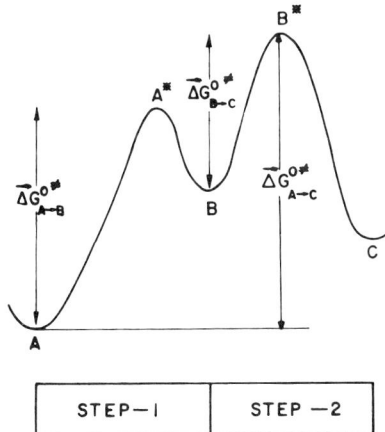

Fig. 9.5. Activation-energy barrier is the total change in energy between the initial state and the activated state with the highest standard free energy.

one has

$$\vec{v}_{\text{step 2}} < \vec{v}_{\text{step 1}}$$

That is, step 2 determines the overall rate $\vec{v}$, or

$$\vec{v} \approx \vec{v}_{\text{step 1}}$$

One concludes that, to qualify as the rds, it is not which step has the highest activation standard free energy with respect to the energy of the previous state that is important, but which step has the highest standard free energy of the activated state compared with that of the initial state.

9.1.7. How Many Times Must the Rate-Determining Step Take Place for the Overall Reaction to Occur Once? The Stoichiometric Number ν

The essential result of these last sections is that one single step, the rds, out of a sequence of steps can determine the rate of the overall reaction.

It has been quickly assumed, however, that, if the rds occurs once, the overall reaction also occurs once. In other words, it has been assumed that there is a one-to-one correspondence between the occurrence of the rds and the overall reaction. Is this always so?

Consider, e.g., that the electronation of hydrated protons, i.e.,

$$M(e) + H_3O^+ \rightarrow MH + H_2O$$

is the rds in the hydrogen-evolution reaction. If this rds occurs only once, then only one adsorbed hydrogen atom has been produced. But it takes two hydrogen atoms to produce a hydrogen molecule

$$2MH \rightarrow 2M + H_2$$

Hence, in this example, two occurrences of the rds are required to produce one occurrence of the overall reaction. One says that the *stoichiometric number ν* of such a reaction scheme is 2. Thus, a stoichiometric number of ν, as introduced by Horiuti in 1939, indicates that there is a ν-to-one correspondence between the occurrences of the rds and the overall reaction.

In situations where the stoichiometric number is ν, what is the theoretical relationship between the current density i and the overpotential η? It has been shown [Eq. (9.9a)] that, when $\nu = 1$, the overall current density is given by

$$i = \frac{nF}{RT} i_0 \eta \qquad (9.9b)$$

under conditions where the use of the linear law is justified. Thus, the conductivity for the overall reaction is determined by $(nF/RT)i_0$.

Is this so when the rds occurs more than once, i.e., v times, for each occurrence of the overall reaction?

Up to now we have dealt with n electrons out of which one was transferred in the once-occurring rds and $n - 1$ were transferred in $n - 1$ other, faster steps. The situation is changed now in that v electrons are now transferred in v times repeated rds and $n - v$ electrons in the remaining, faster steps.

The rate determining step, $R + e^- \rightarrow S$ would have to be repeated v times if (1) more than one R particle is formed by the preceding $\vec{\gamma}$ steps or (2) more than one particle S is required for the following sequence involving now $\overleftarrow{\gamma} = n - \vec{\gamma} - v$ charge-transfer steps (since, not one, but v electrons are now transferred in the v times repeated rds).

Consider now a more general case of a multistep reaction $A + ne^- \rightarrow Z$.

$$\begin{array}{ll} A + e^- \rightleftharpoons B & \text{[Step 1]} \\ B + e^- \rightleftharpoons C & \text{[Step 2]} \\ \vdots & \vdots \\ P + e^- \rightleftharpoons R & \text{[Step } \vec{\gamma}] \\ \underline{v(R + re^- \rightarrow S)} & \text{[rds repeated } v \text{ times]} \\ vS + e^- \rightleftharpoons T & \text{[Step } n - \vec{\gamma} - rv \equiv \overleftarrow{\gamma}] \\ \vdots & \vdots \\ Y + e^- \rightleftharpoons Z & \text{[Step } n] \end{array} \quad (9.15a)$$

Applying the law of mass action for the steps in quasi equilibrium, one has, in analogy to those equations related to (9.16),

$$c_B = K_1 c_A e^{-F\Delta\phi/RT}$$
$$c_C = K_2 c_B e^{-F\Delta\phi/RT} = K_1 K_2 c_A e^{-2F\Delta\phi/RT}$$
$$c_D = K_3 c_C e^{-F\Delta\phi/RT} = K_1 K_2 K_3 c_A e^{-3F\Delta\phi/RT}$$

Taking now the products of all terms to c_R and remembering that this rds occurs v times, one finds that $c_R{}^v$ and c_R are equal to

$$c_R{}^v = \left[\prod_{i=1}^{\vec{\gamma}} K_i\right] c_A^{1/v} e^{-\vec{\gamma}F\Delta\phi/RT} \quad (9.16b)$$

and

$$c_R = \left[\prod_{i=1}^{\vec{\gamma}} (K_i)\right]^{1/v} c_A e^{-(\vec{\gamma}/v)F\Delta\phi/RT} \quad (9.16c)$$

The rate of the rds [cf. Eq. (8.21)] is expressed as

$$\vec{i}_R = F\vec{k}_R c_R e^{-r\beta F \Delta\phi/RT}$$

which, when substituting Eq. (9.16c) for c_R, becomes

$$\vec{i}_R = F\vec{k}_R \prod_{i=1}^{\vec{\gamma}} (K_i c_A)^{1/\nu} e^{-(\vec{\gamma}/\nu)F\Delta\phi/RT} e^{-r\beta F \Delta\phi/RT} \tag{9.23}$$

$$\vec{i}_R = i'_{0,R} e^{-[(\vec{\gamma}/\nu)+r\beta]F\eta/RT}$$

and, hence, using Eq. (9.19), one can obtain for the total forward current $\vec{i}$

$$\vec{i} = i_0 \exp\left[-\left(\frac{\vec{\gamma}}{\nu} + r\beta\right)\frac{F\eta}{RT}\right]$$

since $i'_{0,R} = i_0/n$ and $\vec{i} = n\vec{i}_R$.

The same reasoning can be applied to the backward reaction

$$\nu(S \to R + re^-)$$

the result being

$$c_S = \prod_{n-\vec{\gamma}-r\nu}^{n} (K_i c_z)^{1/\nu} e^{(\vec{\gamma}/\nu)F\Delta\phi/RT}$$

and, further,

$$\overleftarrow{i}_R = i'_{0,R} e^{[(\overleftarrow{\gamma}/\nu)+r-r\beta]F\eta/RT} \tag{9.23a}$$

and, finally,

$$\overleftarrow{i} = i_0 \exp\left[\left(\frac{\overleftarrow{\gamma}}{\nu} + r - r\beta\right)F\eta/RT\right] \tag{9.23b}$$

The total current $i = \vec{i} - \overleftarrow{i}$ is then found to be

$$i = i_0 \left\{\exp\left[\left(\frac{n-\vec{\gamma}}{\nu} - r\beta\right)\frac{F\eta}{RT}\right] - \exp\left[-\left(\frac{\vec{\gamma}}{\nu} + r\beta\right)\frac{F\eta}{RT}\right]\right\} \tag{9.24}$$

or, alternatively,

$$i = i_0 \left\{\exp\left[\left(\frac{\overleftarrow{\gamma}}{\nu} + r - r\beta\right)\frac{F\eta}{RT}\right] - \exp\left[-\left(\frac{\vec{\gamma}}{\nu} + r\beta\right)\frac{F\eta}{RT}\right]\right\}$$

Both these equations are general forms of the Butler–Volmer equation; when $\nu = 1$, these equations reduce to (9.18).

In order to obtain the low-field approximation, both exponential terms in (9.24) are linearized, which yields

$$i = i_0 \left(\frac{nF\eta}{\nu RT} \right) \tag{9.9c}$$

and the general expression for the conductivity of the reaction [*cf.* Eq. (9.22)]

$$\sigma_r = i'_{0,R} \frac{nF}{\nu RT} \tag{9.22a}$$

where

$$i'_{0,R} = \frac{i_0}{n} \tag{9.19}$$

With reference again to Eq. (9.24), the terms $[(n-\vec{\gamma})/\nu] - r\beta$ and $(\vec{\gamma}/\nu) + r\beta$ are called *transfer coefficients* and are denoted by

$$\frac{n - \vec{\gamma}}{\nu} - r\beta = \overleftarrow{\alpha}$$

and

$$\frac{\vec{\gamma}}{\nu} + r\beta = \vec{\alpha} \tag{9.25}$$

It can be easily seen that

$$\vec{\alpha} + \overleftarrow{\alpha} = \frac{n}{\nu} \tag{9.25a}$$

These are the coefficients which determine the slope of the log i versus η curve, i.e., the Tafel slope of a multistep reaction, and are of primary importance in mechanism determinations. This is to be discussed further in Section 9.4.2.

In terms of the transfer coefficients, $\overleftarrow{\alpha}$ and $\vec{\alpha}$, Eq. (9.24) can be written thus:

$$i = i_0 [e^{\vec{\alpha} F\eta/RT} - e^{-\overleftarrow{\alpha} F\eta/RT}] \tag{9.26}$$

Equation (9.26) is the most general form of the Butler–Volmer equation; it is valid for a multistep overall electrodic reaction in which there may be electron transfers in steps other than the rds and in which the rds may have to occur ν times per occurrence of the overall reaction. This generalized equation is seen to be of the same form as the simple Butler–Volmer equation for a one-step, single-electron transfer reaction:

$$i = i_0 [e^{(1-\beta)F\eta/RT} - e^{-\beta F\eta/RT}] \tag{8.31}$$

In comparing the general and the simple equations, it is seen that the transfer coefficients play the same role in a multistep, n-electron-transfer

TABLE 9.1

Tabulation of the Transfer Coefficients $\vec{\alpha}$ and $\overleftarrow{\alpha}$ for Several Mechanisms (assuming $\beta = 0.5$), Where $\vec{\gamma}$, r, ν, $\overleftarrow{\gamma}$, and n Are Known[†]

$\vec{\gamma}$	r	ν	$\overleftarrow{\gamma}$	n	$\vec{\alpha}$	$\overleftarrow{\alpha}$
0	0	1	1	1	0	1.0
1	0	1	2	3	1.0	2.0
1	0	2	1	2	0.5	0.5
2	0	2	2	4	1.0	1.0
0	1	1	1	2	0.5	1.5
1	1	1	2	4	1.5	2.5
1	1	2	1	4	1.0	1.0
2	1	2	2	6	1.5	1.5

[†] For example, $\vec{\gamma} = 0$, $r = 1$, $\nu = 1$, $\overleftarrow{\gamma} = 1$, and $n = 2$ is a two-step, two-electron transfer overall reaction, a reaction involving two electrochemical steps, the first of which is the rds, which occurs once per one act of the overall reaction.

reaction as the symmetry factor does in one-step, one-electron transfer reaction, i.e., the α's determine how the input electrical energy ($F\eta$) affects the reaction rate.

Table 9.1 shows the tabulation of values for $\vec{\gamma}$, r, ν, $\overleftarrow{\gamma}$, and n, from which $\vec{\alpha}$ and $\overleftarrow{\alpha}$ have been evaluated.

9.1.8. The Order of an Electrodic Reaction

In the kinetics of chemical reactions, the *order* of a reaction is a straightforward concept. One simply observes the exponents of the concentration terms in the expression for the reaction rate, e.g.,

$$-\frac{dc_A}{dt} = kc_A{}^a c_B{}^b \cdots c_N{}^n \tag{9.26}$$

Each exponent is termed the *order of reaction* in respect to the species concerned, while the sum of the exponents of the concentration terms defines the *overall order* of a reaction.

Individual reaction orders are often expressed as derivatives of the log of the rate in respect to the log of concentration of the particular species, at constant concentrations of all other species, for it follows from (9.26) that

$$\left(\frac{\partial \log \text{rate}}{\partial \log c_A} \right)_{c_B \cdots c_N} = a$$

or, in a general case,

$$\left(\frac{\partial \log \text{rate}}{\partial \log c_i}\right)_{c_{j \neq i}} = p_i \qquad (9.27)$$

In electrodics, the reaction rate is expressed in terms of current density i (Section 8.6).

Thus, one would expect, by analogy, the electrochemical order of the reaction to be given by an expression similar to (9.27), which should result from the Butler–Volmer expression

$$i = n(\overleftarrow{i_r} - \overrightarrow{i_r}) = nF(k_r c_A^a c_B^b \cdots e^{\bar{x}F\Delta\phi/RT} - k_r c_{A'}^{a'} c_{B'}^{b'} \cdots e^{-\bar{x}F\Delta\phi/RT}) \qquad (9.28)$$

where A^*, B^*, ... are the products of charge-transfer reactions involving A, B, ..., respectively. The exponents a, b, ... and a', b', ... in (9.28), which relates the rate of reaction (current density) to the concentration of various species, are termed *the electrochemical-reaction orders*. It is stressed here that these electrochemical-reaction orders can only be related to equations such as (9.27) when $\Delta\phi$ is constant (see below) and, hence, constant $\Delta\phi$ becomes an essential part of the definition of electrochemical-reaction orders as given above.

It follows from Eq. (9.28) that each reactant A, B, ... has a cathodic- and an anodic-reaction order, e.g., a^*, b^*, ..., a, b, ... At potentials sufficiently anodic to neglect the cathodic reaction, Eq. (9.28) can be expressed in the form of Eq. (9.27), e.g.,

$$\left(\frac{\partial \log i}{\partial \log c_A}\right)_{c_B, c_C, \ldots, \Delta\phi} = a$$

$$\left(\frac{\partial \log i}{\partial \log c_B}\right)_{c_A, c_C, \ldots, \Delta\phi} = b$$

At potentials sufficiently cathodic to neglect the anodic reaction, the electrochemical-reaction orders a' and b' are defined as

$$\left(\frac{\partial \log i}{\partial \log c_{A'}}\right)_{c_{B'}, c_{C'}, \ldots, \Delta\phi} = a'$$

$$\left(\frac{\partial \log i}{\partial \log c_{B'}}\right)_{c_{A'}, c_{C'}, \ldots, \Delta\phi} = b'$$

In a general form,

$$\left(\frac{\partial \log i}{\partial \log c_i}\right)_{c_{j \neq i}, \Delta\phi} = p_{i,\text{an}}.$$

and

$$\left(\frac{\partial \log \vec{i}}{\partial \log c_i}\right)_{c_{i \neq j}, \Delta\phi} = p_{i,\text{cath}} \qquad (9.29)$$

Note that, in all these equations for electrochemical-reaction orders, $\Delta\phi$ has been stipulated as a constant.

A study of the rates of reaction with respect to concentration of the ith species *at constant overpotential* η produces equations which are related to Eqs. (9.29) but are not, obviously, identical. Consider the relation between an anodic current density $\vec{i}$ and p_i, $\Delta\phi_e$ and η (i.e., consider the anodic part of a Butler–Volmer equation in which the reaction order is p_i). This would be Eq. (9.23b). Writing $\vec{\alpha}$ for $(\vec{\gamma}/\nu + r - r\beta)$, and expressing i_0 via (9.17a), one obtains

$$\log \vec{i} = \log nF\vec{k}_r + p_{i,\text{an}} \sum_{i=j}^{i=i} \log c_i + \frac{\vec{\alpha}F\Delta\phi_e}{RT} + \frac{\vec{\alpha}F}{RT}\eta$$

Differentiating now with respect to constant η instead of constant $\Delta\phi$, the result is

$$\left(\frac{\partial \log \vec{i}}{\partial \log c_i}\right)_{c_{j \neq i}, \eta} = p_{i,\text{an.}} + \frac{\vec{\alpha}F}{RT}\left(\frac{\partial \Delta\phi_e}{\partial \log c_i}\right)_{c_{j \neq i}} \qquad (9.29a)$$

This shows that the dependence of the current density on the concentration at constant η (namely, at constant overpotential and not at constant potential) does not produce the electrochemical-reaction order since the expression involves $\Delta\phi_e$, which is itself a function of concentration c_i.

An equation identical to (9.29a) can be obtained from the dependence of the exchange-current density on the concentration c_i; thus,

$$\log i_0 = \log nF\vec{k}_r + p_{i,\text{an.}} \sum_{i=j}^{i=i} \log c_i + \frac{\vec{\alpha}F\Delta\phi_e}{RT}$$

$$\left(\frac{\partial \log i_0}{\partial \log c_i}\right)_{c_{j \neq i}} = p_{i,\text{an.}} + \frac{\vec{\alpha}F}{RT}\left(\frac{\partial \Delta\phi_e}{\partial \log c_i}\right)_{c_{j \neq i}} \qquad (9.30)$$

which is identical with

$$p_{i,\text{an.}} = \left(\frac{\partial \log i_0}{\partial \log c_i}\right)_{c_{j \neq i}} - \frac{\vec{\alpha}F}{RT}\left(\frac{\partial \Delta\phi_e}{\partial \log c_i}\right)_{c_{j \neq i}} \qquad (9.30a)$$

Exactly analogous to this Eq. (9.30a) is the expression for $p_{i,\text{cath}}$,

$$p_{i,\text{cath}} = \left(\frac{\partial \log i_0}{\partial \log c_i}\right)_{c_{j \neq i}} - \frac{\vec{\alpha}F}{RT}\left(\frac{\partial \Delta\phi_e}{\partial \log c_i}\right)_{c_{j \neq i}} \qquad (9.31)$$

It is necessary to examine the terms in these equations more closely, particularly the concentration terms and the condition that $\Delta\phi$ is a constant.

The concentration terms in a rate equation for a multistep reaction usually have the following characteristics:

1. They refer to species other than the reactants in the rds.
2. Reaction orders do not necessarily reflect the molecularity of the rds, since reaction orders may be affected by preceding steps in quasi equilibrium.
3. Reaction orders do not indicate the stoichiometry of the overall reaction.
4. The concentration of the given species may appear with different exponents in the rate equations for the cathodic and anodic reactions.
5. Reaction rates may be influenced by the concentration of a species which does not appear in the overall reaction.

The last point is well illustrated in the case of iron dissolution and deposition (Section 9.4.2) and can be generalized for all cases in which a species is formed in the reaction sequence *before* the rds and consumed *in* or *after* the rds. Very often, such species are OH^- or H^+, which produce a pH dependence of the reaction rate and yet are not involved in the overall reaction.

It has been stressed that the measurement of the reaction order must be made at a constant potential difference across the electrode–solution interface at which the reaction occurs.

It will be recalled that the condition $\Delta\phi =$ constant is tantamount to the condition that the potential of the electrode relative to a standard reference electrode [e.g., SHE (standard hydrogen electrode)] is constant (Section 7.2.5e). Thus, in order to obtain in practice the reaction order of, say, species A, one would measure current densities obtained at a certain potential E referred to a standard electrode potential, in solutions containing various concentrations of A and constant concentrations of all other reactants (*cf.* Fig. 9.6).

The potential E must be chosen sufficiently far from the reversible potential E_e so that the exponential law (Section 8.2.10) applies even at the highest concentrations used of the given species in deelectronation reactions and at the lowest concentrations in electronation reactions.

One additional word of caution has to be added in this regard. In the above derivation, it was tacitly assumed that the change of concentration of the species whose reaction order was determined did not affect the po-

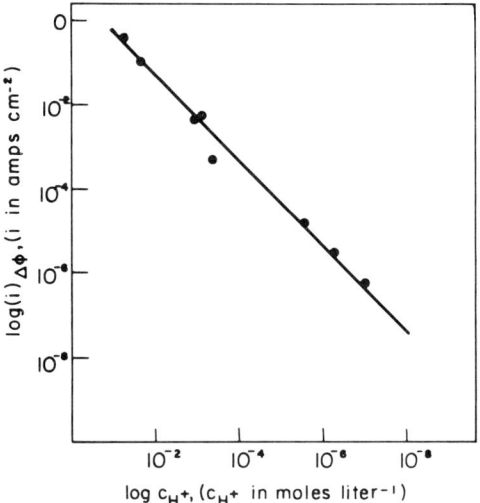

Fig. 9.6. The reaction order of the hydrogen-evolution reaction (on mercury) in respect to hydrogen ion is equal to 1, as seen from the slope of the straight line.

tential distribution in the double layer, i.e., that $\Delta\phi = {}^M\Delta^{\mathrm{OHP}}\phi = $ constant (*cf.* Section 8.3.2). This is true only if the concentration of ions in the double layer remains high and unchanged with varying concentrations of the ionic species investigated. This condition can be closely approximated if the ionic strength of the solution is kept high and constant by the addition of foreign ions, i.e., of a supporting electrolyte which does not participate in the reaction.

A second approach to the determination of reaction orders is by the concentration dependence of the exchange-current density, Eqs. (9.30a) and (9.31). However, herein lies a drawback. In order to determine the concentration dependence of the reversible potential on the concentration of a given species, it is necessary to know the overall reaction. When the overall reaction is unknown, reaction rates can only by determined from the concentration dependence of the current density at constant $\Delta\phi$. This same drawback also lies in the determination of reaction orders by the concentration dependence of the current density at constant overpotential since only by knowledge of the overall reaction can the second term on the right-hand side of Eq. (9.29a) be calculated. This then can be seen as the reason why electrochemical-reaction orders are defined under constant $\Delta\phi$ and not under constant η.

Let us now consider a reaction which will demonstrate these equations; consider the overall reaction

$$I_3^- + 2e^- \rightleftharpoons 3I^- \qquad (9.32)$$

The reaction path, as shown by Vetter, consists of three steps with the third step being rate determining:

1. $\quad I_3^- \rightleftharpoons I_2 + I^-$
2. $\quad I_2 \rightleftharpoons 2I$
3. $\quad 2(I + e^- \to I^-)$

The rate equation is then written as

$$i = 2F[\vec{k}_3 c_{I} e^{(1-\beta)\Delta\phi F/RT} - \overleftarrow{k}_3 c_I e^{-\beta\Delta\phi F/RT}] \qquad (9.33)$$

From the preceding equilibriums in steps 1 and 2,

$$c_I = (K_2 c_{I_2})^{\frac{1}{2}} = \left(K_1 K_2 \frac{c_{I_3^-}}{c_{I^-}}\right)^{\frac{1}{2}} \qquad (9.34)$$

where

$$K_1 = \frac{\vec{k}_1}{\overleftarrow{k}_1} \quad \text{and} \quad K_2 = \frac{\vec{k}_2}{\overleftarrow{k}_2} \quad \text{[equilibrium constants]}$$

Substituting for c_I in Eq. (9.33) produces

$$i = 2F[\overleftarrow{k}_3 c_{I^-} e^{F(1-\beta)\Delta\phi/RT} - \vec{k}_3 (K_1 K_2)^{\frac{1}{2}} c_{I_3^-}^{\frac{1}{2}} c_{I^-}^{-\frac{1}{2}} e^{-\beta\Delta\phi F/RT}] \qquad (9.35)$$

Thus the electrochemical-reaction orders are

$$p_{\text{an.},I^-} = 1$$
$$p_{\text{cath},I^-} = -\tfrac{1}{2}$$
$$p_{\text{cath},I_3^-} = \tfrac{1}{2}$$

This example clearly illustrates points 1 to 4 mentioned earlier concerning the characteristics of the concentration terms in a rate equation. Thus:

1. The species in the cathodic-rate equations are I^- and I_3^-, whereas the reactant in the rds is I.
2. The reaction is unimolecular; the cathodic-rate equation, however, contains concentration terms for two species with reaction orders $-\tfrac{1}{2}$ and $\tfrac{1}{2}$.

3 and 4. The stoichiometry of the overall reaction involves threee I^- and one I_3^- particles, whereas the reaction orders are 1 and $-\frac{1}{2}$ for I^- and 0 and $\frac{1}{2}$ for I_3^- for the anodic and cathodic reactions, respectively.

Experimentally determined reaction orders belong to the most important mechanism-indicating criteria. They may confirm or eliminate an assumed mechanism, depending on whether the rate equation corresponding to the latter postulates the same or different values of p_i as those experimentally established.

9.1.9. Blockage of the Electrode Surface during Charge Transfer: The Surface-Coverage Factor

In writing out the Butler–Volmer equation, it has been assumed that, apart from factors concerning the potential-energy barrier, the current density depends only on the concentrations of reactants on the *solution* side of the interface. The metal surface was always considered empty, i.e., not blocked with any species, intermediate radical, or products.

The electrode, however, is fairly completely covered, at least by water molecules (see Section 7.4.12). Further, there may also be contact-adsorbed ions and organic molecules populating the region in between the metal surface and the OHP.

What effect do these radicals have upon the current density? A large number of scientists in the field have devoted their attention to this problem. In particular, with respect to the role of adsorption on the electrode and its effect on the hydrogen-evolution reaction, very significant contributions have come from Frumkin, Butler, Breiter, Gilman, and Will. There is no simple and general answer to the question: What effect do adsorbed radicals have upon surface-reaction rate? It depends on the radical and the reaction. For instance, the water molecules which cover most of the metal surface do not have direct effects on the reaction

$$H_3O^+ + M(e) \rightarrow MH + H_2O.$$

On the other hand, adsorbed hydrogen atoms do block the surface for the same reaction. If θ is the *fraction* of the surface covered (i.e., the coverage) with adsorbed hydrogen atoms, the reaction can proceed only on the *free* surface, i.e., on $1 - \theta$ of the electrode surface.

So, in writing out the exponential form of the Butler–Volmer equation, one has to introduce a correction. This is done as follows:

In the expression for the exchange-current density, one reckons with the equilibrium coverage θ_e

$$i_0 = \frac{kT}{h} e^{-\vec{\Delta G}^{0\ddagger}/RT} e^{-\beta F \Delta\phi_e/RT} c_{H_3O^+}(1-\theta_e)$$

When the overpotential goes into the exponential i versus η region, the surface coverage may change from θ_e to θ, where θ is the (overpotential-dependent) coverage at η. So one has to write

$$\begin{aligned}
\vec{i} &= \frac{kT}{h} e^{-\vec{\Delta G}^{0\ddagger}/RT} e^{-\beta F \Delta\phi_e/RT} c_{H_3O^+}(1-\theta) e^{-\beta F \eta/RT} \\
&= [kT/h e^{-\vec{\Delta G}^{0\ddagger}/RT} e^{-\beta F \Delta\phi_e/RT} c_{H_3O^+}(1-\theta_e)] \frac{(1-\theta)}{(1-\theta_e)} e^{-\beta F \eta/RT} \\
&= i_0 \frac{(1-\theta)}{(1-\theta_e)} e^{-\beta F \eta/RT}
\end{aligned} \quad (9.36)$$

and, under the special circumstance where $\theta_e \to 0$, Eq. (9.36) reduces to

$$\vec{i} \approx i_0(1-\theta)e^{-\beta F \eta/RT} \quad (9.37)$$

The procedure for correcting for the departure from equilibrium to nonequilibrium surface coverage consists in (1) writing down the actual concentration in the Butler–Volmer equation or its relevant special case and (2) transforming this expression into one involving the equilibrium exchange-current density i_0, which contains the bulk concentration.

In the case of the *de*-electronation of adsorbed hydrogen, it is precisely the fraction of the surface covered which comes into the Butler–Volmer equation. One has

$$\cev{i} = i_0 \theta e^{(1-\beta)F\eta/RT}$$

It may be asked why the water can be neglected as a blocking agent, i.e., no $1-\theta_{\text{water}}$ term was taken into account, whereas one has to take into account a coverage term for the adsorbed hydrogen. The answer is simple if one considers the energies with which water and adsorbed hydrogen atoms are adsorbed on the electrode. Water is bound relatively lightly (10 to 20 kcal mole^{-1}) for most electrode surfaces. In contrast, hydrogen atoms and many other substances adsorb on the electrode surface with much greater binding energies ($\sim$50 kcal mole^{-1}) than that of water. Hence, if there is to be a competition for the surface, as there is when the products of charge transfer have to be adsorbed, water loses out. It gets desorbed during reaction because the charge-transfer products, e.g., adsorbed hydrogen atoms, knock the water off the surface and are able to land even on the sites formerly occupied by water. Water therefore is seldom a blocking agent by itself (*cf.*, however, Section 7.4.29). The strongly bound particles (e.g., H_{ads}), however, cannot be knocked off easily by particles which wish to take their place on the electrode, and, in these cases, one must include a $1 - \theta$ term.

Thus, by thinking about the concentration of the *adsorbed intermediates* during an electrodic reaction—the hydrogen atoms in the above example—one obtains some modified version of the Butler–Volmer equation (which was originally derived in its simplest form). This allowance for θ, the fraction of the surface blocked, can be numerically important because θ can sometimes approach unity. Under such circumstances, $1 - \theta$ is quite small, and its variation with potential has a significant effect on the current density.

The introduction of θ in the equations for current density need by no means refer only to the adsorbed *intermediates* in the electrode reaction. What of other entities which may be adsorbed on the surface? For example, suppose one adds to the solution an organic substance (e.g., aniline) and this becomes adsorbed on the electrode surface. Then, the θ for the adsorbed organic substance must also be allowed for in the electrode kinetic equations. So, in Eq. (9.37), the value of θ would really have to become a $\Sigma \theta$, where the summation is over all the entities which remain upon the surface and block off sites for the discharging entities. Many practical aspects of electrodics arise from this aspect of the Butler–Volmer equation. For example, the action of organic corrosion inhibitors partly arises in this way (adsorption and blocking of the surface of the electrode and hence reduction of the rate of the corrosion reaction per apparent unit area).

In addition, it must not be forgotten that particles which intrude into the crucial region between the metal and the OHP affect the structure of the electrified interface and hence the double-layer potential difference which enters into the Butler–Volmer equation. For example, molecules with dipole moments start contributing dipole-, or $\Delta\chi$-, potential differences (Section 7.3) when they stick on the surface, or their $\Delta\chi$ contributions diminish when they are driven off. These are more subtle points, which are mentioned here only to reveal how important it is to know to what degree (and with what) an electrode surface is blocked during the charge-transfer reaction.[†]

[†] One should note that the surface coverage can affect the standard free energy of intermediate states and thus also the activation-energy barrier and, in this indirect way, the current density of the reaction.

Further Reading

1. A. N. Frumkin and A. I. Shlygin, *Acta Physicochim. URSS*, **3**: 791 (1935).
2. J. D. Pearson and J. A. V. Butler, *Trans. Faraday Soc.*, **34**: 1163 (1938).
3. J. Horiuti and M. Ikusima, *Proc. Imp. Acad. (Tokyo)*, **15**: 39 (1939).
4. K. J. Vetter, *Z. Physik. Chem. (Leipzig)*, **194**: 284 (1950); also, *Z. Elektrochem.*, **55**: 121 (1951).
5. R. Parsons, *Trans. Faraday Soc.*, **147**: 1332 (1951).
6. E. C. Potter, *J. Chem. Phys.*, **20**: 614 (1952).
7. M. Breiter, C. A. Knorr, and W. Völkl, *Z. Elektrochem.*, **59**: 681 (1955).
8. W. Mehl, *Can. J. Chem.*, **37**: 190 (1959).
9. F. G. Will and C. A. Knorr, *Z. Elektrochem.*, **64**: 258 and 270 (1960).
10. A. Despic, *Electrochim. Acta*, **4**: 325 (1961).
11. A. N. Frumkin, "Hydrogen Overvoltage and Adsorption Phenomena," in: P. Delahay and C. W. Tobias, eds., *Advances in Electrochemistry and Electrochemical Engineering*, Vol. III, Chapter 5, Interscience Publishers, Inc., New York, 1963.
12. S. Gilman, *J. Electroanal. Chem.*, **7**: 382 (1964).
13. M. V. Simonova and A. L. Rotinyan, *Russ. Chem. Rev. (English Transl.)*, **34**: 318 (1965).
14. V. G. Levich, *Russ. Chem. Rev. (English Transl.)*, **34**: 792 (1965).
15. A. Damjanovic, *Electrochim. Acta*, **11**: 376 (1966).
16. K. J. Vetter, "Determination of the Electrochemical Reaction-Orders," in: *Electrochemical Kinetics*, Chapter 3c, Academic Press, Inc., New York, 1967.

9.2. THE TRANSIENT BEHAVIOR OF INTERFACES

9.2.1. The Interface under Equilibrium, Transient, and Steady-State Conditions

If one looks back on the treatment of electrodic reactions which has been presented, one cannot fail to notice an important point. In the Butler–Volmer equation which describes the electrodic events at an interface for a single charge-transfer reaction

$$i = i_0[e^{(1-\beta)F\eta/RT} - e^{-\beta F\eta/RT}]$$
$$= F\overleftarrow{k}c_D e^{(1-\beta)F\Delta\phi/RT} - F\overrightarrow{k}c_A e^{-\beta F\Delta\phi/RT}$$

the quantities c_D, c_A, and $\Delta\phi$ have been taken to be independent of time. One has a picture of an interface with a potential difference $\Delta\phi$ and with a structure (i.e., the concentrations of the various species in the interface) which maintain constant values as time passes and a current flows across the interface. How correct is this picture of lack of change and constancy at the interface? How justifiable is it to ignore the time dependence of concentrations and potentials at the interface?

A general answer can easily be put forward. There are two conditions under which one can omit time in an analysis. The first condition is that of *equilibrium*, the condition when nothing net happens. If one writes down the Nernst equation for a reaction $A + e \rightarrow D$[†]

$$\Delta \phi_e = \Delta \phi_e^0 + \frac{RT}{nF} \ln \frac{c_A}{c_D}$$

one has completed the macroscopic description of the condition of equilibrium.

There is, however, another condition under which things seem not to change at an interface. This is the condition of *steady state*. Here, too, the phenomena are time independent, but a net current is flowing and, therefore, something net *is* happening. The concentrations of the intermediates involved in the reaction remain constant with time. The charge flowing into the interface is equal to that flowing out; nothing builds up. Thus, the structure of the interface and the potential difference across it do not change. The description of the steady-state interface, the behavior of which is usually controlled by one step, the rds, of the electrodic reaction, is contained in the Butler–Volmer equation.

But how long does the interface take to change from an equilibrium state to a steady state? Is there an instantaneous transition between the two time-independent states? It is unlikely that the values of c_A, c_D, and $\Delta \phi$ change instantaneously from those characterizing equilibrium to those corresponding to a steady state. For example, if c_A represents the concentration of electron-acceptor ions at the OHP, any change of this concentration requires ionic movements, which are centainly not instantaneous (see Section 4.6.8). The interface, therefore, will take a certain time to move from an equilibrium state to a steady state.

Hence, one has to consider the time variation of the concentrations at the interface. One has to explore the laws of the changing interface, i.e., its *transient* behavior.

[†] The D is the electron donor; A, the electron acceptor.

9.2.2. How an Interface Is Stimulated to Show Time Variations

There is a general procedure for studying the time behavior of any system. One starts off with the system in a state of constancy. Then one perturbs it or applies a stimulus to it and observes the response of the system as a function of time.

What kind of stimuli can be applied to an electrode–electrolyte interface? Examine the Butler–Volmer equation for a single-step reaction

$$i = i_0[e^{(1-\beta)F\eta/RT} - e^{-\beta F\eta/RT}]$$

It can be seen that two quantities which provide convenient levers on the dynamics of the interface are the current density and the potential difference across the interface. Thus, all studies of the changing interface—called the *transient*, *perturbation*, or *relaxation methods*—are based on starting off with the interface in a time-independent state (either equilibrium or steady state) and then applying some planned, or programmed, current or potential stimulus.

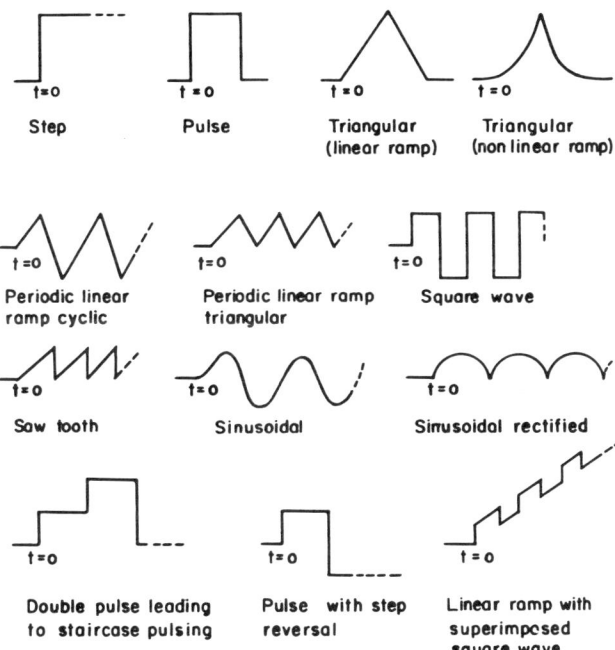

Fig. 9.7. Various types of stimuli that can be applied, as either current or potential, in order to stimulate transient behavior from an electrode–electrolyte interface.

Only imagination and the availability of suitable instrumentation limit the type of stimulus or perturbation chosen for application to the interface. Examples of possible stimuli are shown in Fig. 9.7.

Using the current density as a lever on the electrodic reaction, one can stimulate the interface with a current step, a current pulse, or a periodically varying current (Fig. 9.7) and observe how the potential difference across the interface varies with time. Or one can do the converse. That is, the potential difference across the interface can be changed by the application of a potential step, potential pulse, or a linearly varying or a periodically varying potential difference, and the resultant current density flowing across the interface can be watched (and recorded) as a function of time.

A whole new vocabulary has developed to distinguish the different types of stimuli. However, all one requires (to see what is being done) is two plots, one giving the stimulus as a function of time and the other showing the response of the interface as a function of time (Fig. 9.8).

9.2.3. Some Ideas on the Understanding of Transients

Throwing the interface into a time-dependent behavior is a simple matter. Understanding the laws which govern the transient behavior of it is something else again.

Current electrochemical literature shows that all sorts of stimuli are being used and new ones devised. However, there has been far less corresponding development of interpretative ability. Indeed, the analysis of responses in the transient interface is a frontier of electrodic research. Compared with the understanding of the equilibrium ($i = 0$ and $\Delta\phi = \Delta\phi_e$ = constant) and the steady-state (i and $\eta \neq 0$ but constant) interfaces, the understanding of the time-varying interface involves many new problems. On the other hand, it will be shown that transients provide important insights into electrodic behavior. This is what makes it worthwhile to study the transient interface. The main point is that, if one understands how time responses occur, one can often find out the concentration of some of the radicals blocking the surface. A second point is that sometimes—as, e.g., in studies of metal deposition (where the surface is constantly changing)—it is practically only by means of the use of transients that significant information can be obtained.

To get a feel for some of the problems of the time-varying interface, consider a simple one-step electron-transfer reaction

$$A + e \to D$$

ELECTRODICS: MORE FUNDAMENTALS 1021

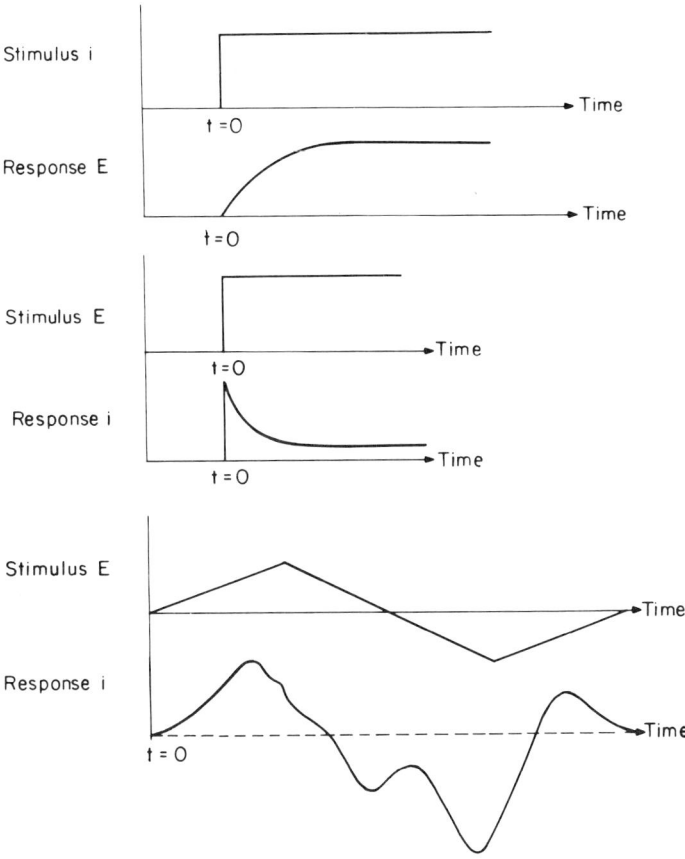

Fig. 9.8. The time dependence of the stimulus and of the response are closely related.

in which an electron acceptor is electronated. Further, consider that one starts off with the interface at equilibrium, i.e., $i = 0$ and $\Delta\phi = \Delta\phi_e$. Now, suppose that, at time $t = 0$, one switches on a constant current density i_g (the subscript g arises from the term *galvanostatic* sometimes used to describe this constant current density technique). The current density *versus* time plot reveals a current step at $t = 0$ (Fig. 9.9). Eventually (e.g., after perhaps a few tens of milliseconds), the interface attains a steady-state condition corresponding to the current density i_g. There will then be a steady-state potential difference across the interface, $\Delta\phi_{t\to\infty} = \Delta\phi$. The question is: What processes control the change of potential from $\Delta\phi_{t=0}$ ($= \Delta\phi_e$) to $\Delta\phi_{t\to\infty}$ ($= \Delta\phi$)?

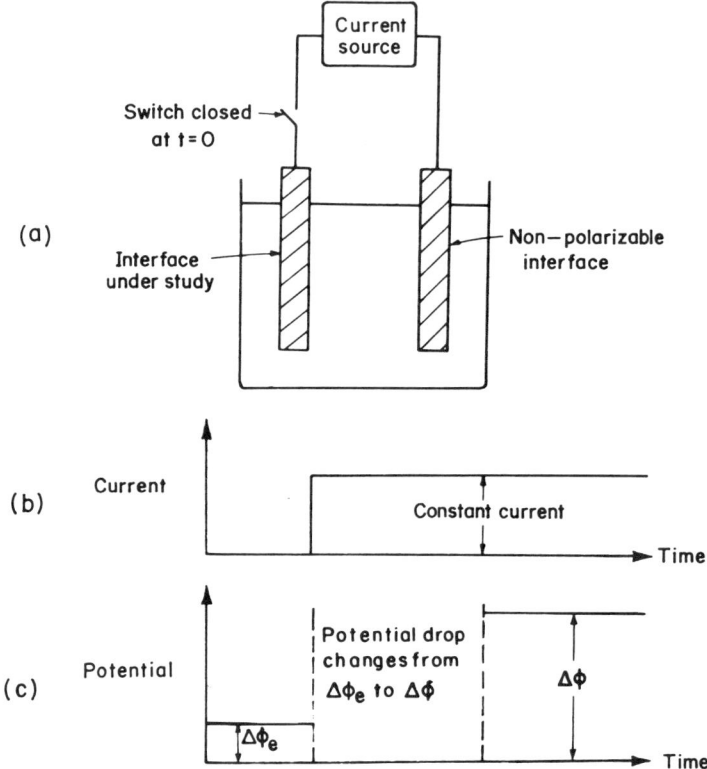

Fig. 9.9. In a galvanostatic arrangement (a), upon closing the switch, a constant current pulse (b) is forced through the cell, which causes the change in potential (c) at the polarizable interface under study.

One can resort to a conceptual artifice at this point. The total time between $t = 0$ and the attainment of steady state can be broken up into smaller and smaller intervals until an infinitesimally small interval dt is considered. Within this time interval dt, one can assume that the interface does not change significantly and is in a steady state (Fig. 9.10). What effects result from the current density i_g that flows during the time interval dt? Or what happens to the charge $q = i_g \, dt$ that flows?

The first effect to be reckoned with is that concerning the double-layer aspect of the interface. Under steady-state conditions, a given potential difference at an electrified interface corresponds to a given structure of the interface (see Section 7.4). This means that the charge density q_M on the metal and the (equal and opposite) density q_S in the solution have to change

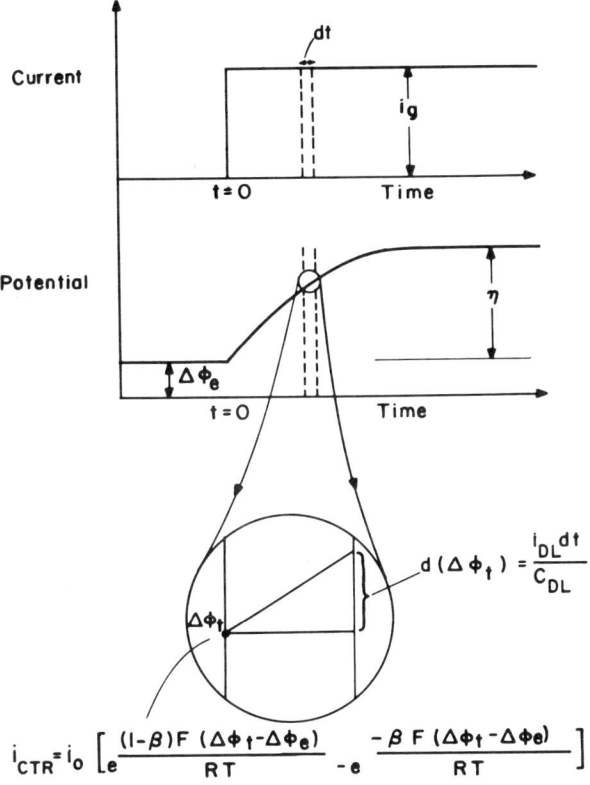

Fig. 9.10. At an infinitesimal time interval, the equation for the steady-state current density of the electrode process, i_{CTR}, can be used, the relevant potential being $\Delta\phi_t$. Also, the change in potential $d\Delta\phi_t$ will involve charging up the double-layer capacity C_{DL} with another part of the current i_{DL}.

from values characterizing the equilibrium potential $\Delta\phi_{t=0}$ to those characterizing the steady-state potential $\Delta\phi_{t\to\infty}$. But, if one has to change the charge density stored on the two sides of the interface, one has to flow in or flow out charge; in other words, part of the current density will be used to charge up the double layer. This is a consequence of the electrified interface's ability to store charge, i.e., to behave as a capacitor.

One can state this argument quantitatively. If C_{DL} is the differential capacity of the interface (remember that, in general, C varies with the potential), then

$$C_{\mathrm{DL}} = \frac{dq}{d\Delta\phi} = i_{\mathrm{DL}} \frac{dt}{d\Delta\phi}$$

or

$$i_{\mathrm{DL}} = C_{\mathrm{DL}} \frac{d\Delta\phi}{dt} \qquad (9.38)$$

The second effect of the current density is to promote the electronation, or charge-transfer reaction. Since, within the time interval dt, the interface can be considered in a steady state, one can use the Butler–Volmer equation, taking care to write down the value of the potential or the excess potential (i.e., the overpotential) obtaining in the time interval dt, i.e., $\Delta\phi_t$ or $\eta_t = \Delta\phi_t - \Delta\phi_e$. Hence, the current density i_{CTR} going toward the charge-transfer reaction is

$$i_{\mathrm{CTR}} = i_0 [e^{(1-\beta)F\eta_t/RT} - e^{-\beta F\eta_t/RT}] \qquad (9.39)$$

Summing up these two contributing equations [Eqs. (9.38) and (9.39)] to the total current density i_g, one has

$$i_g = i_{\mathrm{DL}} + i_{\mathrm{CTR}}$$

or, since the change of potential is the same as the change of overpotential

$$i_g = C_{\mathrm{DL}} \frac{d(\eta_t)}{dt} + i_0 [e^{(1-\beta)F\eta_t/RT} - e^{-\beta F\eta_t/RT}] \qquad (9.40)$$

The total current density i_g is being kept constant; it does not change after $t > 0$. But what about its components i_{DL} and i_{CTR}? When η_t is small compared with the steady-state value $\eta_{t\to\infty}$, the magnitude of i_{CTR} (which depends on η_t) must be small compared with the particular value of the charge-transfer current density which flows in the steady state. This means that, in the early stages of the transient, current density i_g is mainly utilized for double-layer charging, i.e., when $t \approx 0$, $\eta_t \approx 0$, $i_{\mathrm{CTR}} \approx 0$, and, therefore, $i_g \approx i_{\mathrm{DL}}$ [cf. Eq. (9.40)]. (A quantitative treatment of how η changes as a function of time is given in Section 10.2.7.) But, with every bit of electricity used to charge up the electrified interface. $\Delta\phi_t$ increases. This means, however, that η_t increases, and this leads to an increase of i_{CTR} [from (9.39)]. Hence, with the passage of time, less and less of the total *constant* current density i_g is used for double-layer charging; and more and more, for the charge-transfer reaction (Fig. 9.11).

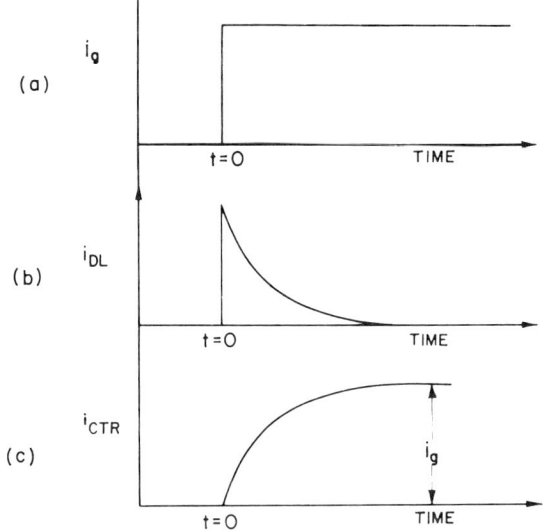

Fig. 9.11. The total current density (a) is used at the beginning mostly as the double-layer charging current (b), while, at longer times, it tends to be used entirely for the electrode reaction (c).

Finally, the steady state $\Delta\phi_t = \Delta\phi_\infty$ is reached; no further changes in the double-layer charge or structure are required. Hence, $d(\eta)/dt = 0$, and therefore $i_{DL} = 0$, from (9.38). That is, all the constant current density i_g goes toward the charge-transfer reaction, i.e., $i_{CTR}[t \to \infty] = i_g$. The current density i_g then becomes related to the familiar Butler–Volmer equation for the steady state

$$i_g = i[e^{(1-\beta)F\eta/RT} - e^{-\beta F\eta/RT}] \tag{9.41}$$

where η is written for $\eta_{t\to\infty}$. Hence, a simple picture emerges for the transition of an interface from its equilibrium state to the steady state when this transition is stimulated by a constant current density. The input current is used partly to change the interface structure by increasing the excess charge on the electrode surface—this is double layer charging—and partly to serve the charge-transfer reaction. In the initial stages of the transient, double-layer charging i_{DL} uses most of the input-current density i_g, but, as the structure of the interface tends more and more to the steady-state structure, i_{DL} becomes smaller and smaller. Finally, i_{DL} becomes zero in the steady state, and all the constant input-current density goes toward charge transfer.

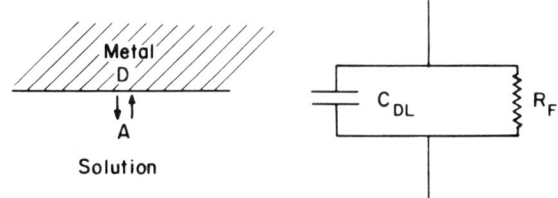

Fig. 9.12. The electrode in its transient behavior can be represented by an equivalent electrical circuit consisting of a resistor and a capacitor connected in parallel.

A certain type of transient behavior of the interface has just been described; it is analogous to the behavior of an electrical circuit consisting of a resistor and a capacitor in parallel (Fig. 9.12). Thus, a metal–solution interface at which there occurs a simple electrodic reaction may be usefully represented by an equivalent circuit in so far as its transient electrical behavior is concerned. Thereby, calculations of some of the properties of the interface (e.g., separation of the contributions of charge transfer and capacitance to the impedance) are facilitated.

9.2.4. Intermediates in Electrodic Reactions and Their Effects on Potential–Time Transient

Now consider a slightly more complicated reaction in which the electrodic reaction proceeds by an adsorbed intermediate which is formed by a charge-transfer reaction

$$A^+ + e \rightarrow AI^\dagger$$
$$AI \rightarrow D$$

How will this electrodic reaction respond to the stimulus of a current step, i.e., to a constant current density i_g switched on at $t = 0$?

A good discussion of the problem has been developed by Conway and Gileadi. The argument for the initial stages of the transient is similar to that used in the previous section [9.2.3]. One says that for $t \ll t \rightarrow \infty$, $\eta_t \ll \eta_\infty = \eta$, and hence $i_{\text{CTR}} \ll i_{\text{CTR}} [t \rightarrow \infty] = i_g$. The balance of the current, i.e., $i_g - i_{\text{CTR}}$, goes toward changing the structure of the interface.

But what does the double-layer structure imply in the present case? It implies not only the charge densities q_{DL} on the electrode and the solution

---

† The AI represents adsorbed intermediate.

sides of the interface but also the surface concentration of adsorbed intermediates. Let the surface concentration of AI be expressed in terms of the fraction θ of the surface covered with AI. Then, to change the structure of the double layer, one has to change θ as well as the charge density q_{DL}.

Why should special attention be given to the adsorbed species? When one says that an inflow of charge dq_{DL} results in a change of potential difference across the interface of $d\Delta\phi$ such that

$$dq_{DL} = i_{DL}\, dt = C_{DL}\, d\Delta\phi \qquad (9.42)$$

has the change in the surface concentration of AI been included or not? It has not. This is because Eq. (9.42) describes how double layers are charged *without electron transfer*. That is, it describes the charging of the polarizable (no leakage of charge, see Section 7.2.5d) part of the interface. Since, however, electron transfer is an essential requirement for the formation of adsorbed intermediates in the case considered, the charge utilized for this process has to be given separate treatment.

If a charge of dq_{AI} is required to change the coverage by $d\theta$, then

$$dq_{AI} = i_{AI}\, dt = k\, d\theta \qquad (9.43)$$

where k is the amount of charge to form a monolayer of adsorbed intermediate, i.e., $\theta = 1$. If the adsorption is potential dependent, one can write

$$d\theta = \frac{\partial \theta}{\partial \Delta\phi}\, d\Delta\phi \qquad (9.44)$$

and, substituting (9.44) in (9.43), one has

$$dq_{AI} = i_{AI}\, dt = k\, \frac{\partial \theta}{\partial \Delta\phi}\, d\Delta\phi \qquad (9.45)$$

However,

$$\frac{dq_{AI}}{d\Delta\phi} = \frac{i_{AI}\, dt}{d\Delta\phi} = k\, \frac{\partial \theta}{\partial \Delta\phi} \qquad (9.46)$$

But what is $dq_{AI}/d\Delta\phi$? It satisfies the definition of a differential capacity. Thus, it looks as if the ability of the interface to store adsorbed intermediates *formed by charge transfer* is somewhat[†] equivalent to an electrical

---

[†] The *name* used for C_{AI} is *pseudo capacity*. This is because it acts electrically as though it were a capacity. But, again, it is *not* a capacity, because the latter holds the charge which is added to it (whereupon its electrical potential changes, simply, proportionately to the flow-in of charge) whereas in a pseudo capacity the charge leaks out across the double layer. The name fits well.

capacity. It shall be symbolized C_{AI} to indicate that it is a capacity[†] arising from the formation of adsorbed intermediates and distinct from the polarizable aspects of the interface. It is given by

$$C_{AI} = k \frac{\partial \theta}{\partial \Delta \phi} \quad (9.47)$$

which permits one to express Eq. (9.45) in the form

$$dq_{AI} = i_{AI} \, dt = C_{AI} \, d\Delta\phi \quad (9.48)$$

This is interesting. It appears that the current density $i_g - i_{CTR}$ which is utilized for changing the structure of the interface is the sum of two current densities. One of these, i_{DL}, charges the double layer by a *non-electron-transfer* process, and the other i_{AI} is used in an electron-transfer process leading to the storage of adsorbed intermediates in the interface. That is,[‡]

$$\begin{aligned} i_g - i_{CTR} &= i_{DL} + i_{AI} \\ &= C_{DL} \frac{d\Delta\phi}{dt} + C_{AI} \frac{d\Delta\phi}{dt} \\ &= (C_{DL} + C_{AI}) \frac{d\Delta\phi}{dt} \end{aligned} \quad (9.49)$$

or

$$i_g = (C_{DL} + C_{AI}) \frac{d\Delta\phi}{dt} + i_{CTR} \quad (9.50)$$

As in the case of a simple, direct electrodic reaction without intermediate formation (Section 9.2.3), the total constant current density i_g has to be divided between the current density $(C_{DL} + C_{AI}) \, d\Delta\phi/dt$ going toward changing the interfacial structure and the current density i_{CTR} which results in the complete overall electrodic reaction.

The similarity between the present case (with intermediate formation) and the previous case of Section 9.2.3 extends further. In the initial stages of the transient, the main portion of i_g goes toward changing the interfacial structure, but, as η_t climbs toward $\eta_{t \to \infty} = \eta$, the current density i_{CTR}, arising from the charge-transfer reaction, increases more and more. In the steady state, when $\eta_t = \eta_\infty = \eta$, $i_{CTR}[t \to \infty] = i_g$ and $(C_{DL} + C_{AI}) \, d\Delta\phi/dt = 0$.

[†] It is customary to refer to this as an *adsorption pseudo capacity*.
[‡] It will be noted that no attempt has been made to introduce explicit expressions for the charge-transfer current density i_{CTR} because such expressions will depend on the detailed mechanism of the reaction.

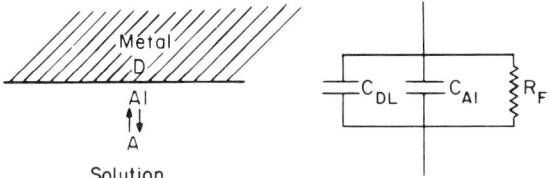

Fig. 9.13. In a multistep electron-exchange reaction, the pseudo capacitance due to the change in the concentration of intermediates acts as a capacitor C_{AI} in parallel with the double-layer capacitor C_{DL} and faradaic resistance R_F.

When there is formation of adsorbed intermediates in an electrodic reaction, the electrical behavior of the interface cannot be represented by the simple equivalent circuit used in a direct electrodic reaction (namely, Fig. 9.12). The relationship between the potential difference $\Delta\phi$ across the interface and the current density $i_g - i_{CTR} = i_{DL} + i_{AI}$ utilized for changing the interfacial structure is determined not only by the double-layer capacitor C_{DL} but also by the pseudo capacitor C_{AI}. Since, however, the relationship (9.49) between $\Delta\phi$ and $i_{DL} + i_{AI}$ is determined by an effective capacity which is equal to the sum $C_{DL} + C_{AI}$ of the double layer and pseudo capacities, this effective capacity may be represented in the equivalent circuit for an electrodic reaction occurring through adsorbed intermediates by a *parallel* combination of the double-layer capacitor and pseudo capacitor (Fig. 9.13).

9.2.5. Experimental Methods for the Determination of Partial Coverage, with Adsorbed Entities, of the Surface of Electrocatalysts

The effect of the presence of adsorbed entities upon transients through the addition of a pseudo capacitance has been outlined in principle above. However, one must discuss this matter a little further because of the great importance of the determination of the coverage of an electrode surface with reactants and also with adsorbed products which may inhibit the electrode process.

Several methods are available for the determination of surface coverage, each method having its own drawbacks for interpretation of the results. These methods can be generally divided into two groups, those which depend upon the measurement of radiations from an adsorbed entity containing a tagged isotopic atom and those methods which depend upon the influence of the adsorbed entity on the electrochemical transient or sweep measurement.

The applicability of any one method depends of course upon the electrode reaction one wishes to study; e.g., clearly, the radiotracer method can be used only when the adsorbed entity can contain a labeled atom, and, for this reason, it could not be applied to O_2 adsorption since O^{18}, the most readily available oxygen isotope, is not radioactive. Correspondingly, certain electrochemical sweep methods will not be useful when, in the potential range under investigation, the substrate undergoes reaction, e.g., dissolution in anodic regions, as would often be the case except for the noble metals. Clearly, then, one must examine each method closely to determine its suitability and then decide which one or, preferably, two methods are the most suitable.

Let us examine the following four methods in more detail.

9.2.5a. Radiotracer Method. The essence of this method is that the amount of radiation emitted is a measure of the *total surface concentration* of tagged atoms. Note that this will not be a measure of the adsorbed species containing tagged atoms when the latter undergoes a surface reaction that produces fragments containing tagged atoms.

It might be possible to get information concerning the radicals upon the surface by examining the adsorbed species containing two labeled atoms, e.g., C^{14} and H^3. If the organic species dehydrogenates upon adsorption, the adsorbed hydrogen will ionize and dissolve off the surface; the number of tritium atoms indicated will now be less than the required number of hydrogen atoms per molecule of organic, i.e., the C^{14} to H^3 ratio for C^{14}- and H^3-labeled benzene will be greater than unity. If the major fraction of the adsorbed species on the electrode is the undissociated molecule, then the C^{14} to H^3 ratio should be as in the molecule itself, i.e., C^{14} to H^3 ratio of unity for benzene. But such possible radiotracer approaches have to be made with caution because of the possibilities of exchange between solution and electrode.

9.2.5b. Galvanostatic Transient Method. In this method, the electrode is maintained at the potential at which knowledge of the coverage with adsorbed species is required, for a time long enough to attain an equilibrium coverage. The electrocatalyst is then subjected to a constant current pulse (*cf.* Fig. 9.9), and the resulting potential–time transient is recorded on an oscilloscope. Figure 9.14 shows a typical transient in which the time $\tau_B - \tau_A$ multiplied by the constant current applied is representative of the charge required to oxidize (de-electronate) the adsorbed species. The

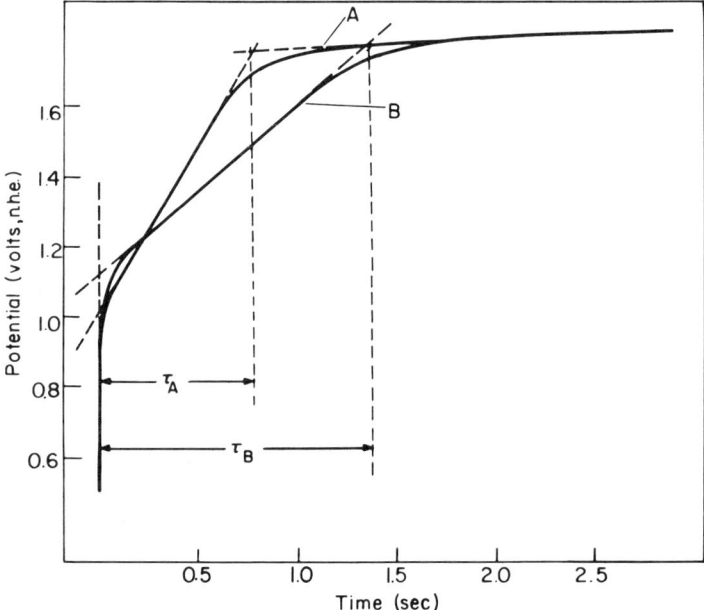

Fig. 9.14. Potential–time relations after a constant current step i is applied to a platinized platinum electrode in N H_2SO_4 in the absence A and in the presence B of benzene. The charge required to oxidize the adsorbed surface species being given as $i(\tau_B - \tau_A)$.

fractional coverage is obtained by dividing this charge by the charge required to completely oxidize a monolayer coverage of the adsorbed species, and, to know this, one requires knowledge of the number of electrons involved in the oxidation of 1 mole of the organic species, i.e., one requires knowledge of what the overall reaction is. Further, one needs confirmation that, under the transient conditions, the reaction *is* going through all the way from its initial state of adsorbed molecule to CO_2. If it were adsorbed as a radical, the number of electrons by which one would have to divide the coulombs per square centimeter would be reduced and uncertain.

Problems that can arise from this method include the inadequate definition of the potential arrest and the time τ, complications due to diffusion of further organic species from the solution to the surface when the rate of oxidation is comparable to the rate of diffusion, and the consumption of a certain number of coulombs in the formation of an oxide film.

An alternative galvanostatic method is an indirect method involving the determination of hydrogen coverage in the presence and in the absence of adsorbed organic species. If the hydrogen coverage in the absence of adsorbed organic species is θ_1 and in its presence, θ_2, then the coverage due to the organic species is sometimes taken simply as $\theta_1 - \theta_2$.

This indirect method, however, is open to a number of questionable assumptions: (1) The coverage of adsorbed species, $\theta_1 - \theta_2$, may not be the *fractional* surface coverage; it will be so only when $\theta_1 = 1.0$, i.e., for monolayer hydrogen adsorption. (2) It is assumed that the organic species is still adsorbed unchanged during the transient, even at the high negative potentials required for hydrogen adsorption. Since organic adsorption is potential dependent (*cf.* Fig. 7.101 and Section 7.5.5), some desorption would be expected even during the short lifetime of a transient. (3) It is assumed that the occupation of hydrogen atoms per adsorption site is equal in both the presence and absence of organic adsorbed species; there is evidence from the influence of organic adsorption on hydrogen-evolution kinetics to suggest that this is not a valid assumption. (4) It is assumed that the current during the transient is used only for the deposition of hydrogen upon the available surface and not for the reduction of the adsorbed organic species.

Finally, there may be, under certain circumstances of potential and time, diffusion of hydrogen into the metal substrate which is still counted in the transient but does not contribute to θ.

Thus, the suitability of the galvanostatic transient method depends upon the validity of these assumptions to the system under investigation.

9.2.5c. Potentiostatic Transients.

As in the galvanostatic transient method, the electrode is first held at a known potential to allow adsorption equilibrium. The potential is then extremely rapidly changed to a new potential, upon which organic oxidation occurs, and the resulting current response is recorded, with respect to time, on an oscilloscope. The charge required to oxidize the adsorbed species, assuming that organic oxidation is the only reaction, is then simply the area under the current–time curve. One further assumption is necessary to relate this charge to fractional surface coverage: it must be assumed that there is no readsorption of the organic species during the lifetime of the transient.

In the presence of a competitive reaction, e.g., oxide formation of the substrate metal, it can no longer be assumed that organic oxidation is the only reaction. Thus, the practice is to determine two transients, one in the presence and one in the absence of the organic species. Subtraction of the area under the current–time curve for the latter from the area (i.e., the charge) of the former produces the charge required to oxidize the adsorbed organic species. Figure 9.15 illustrates this procedure, the shaded

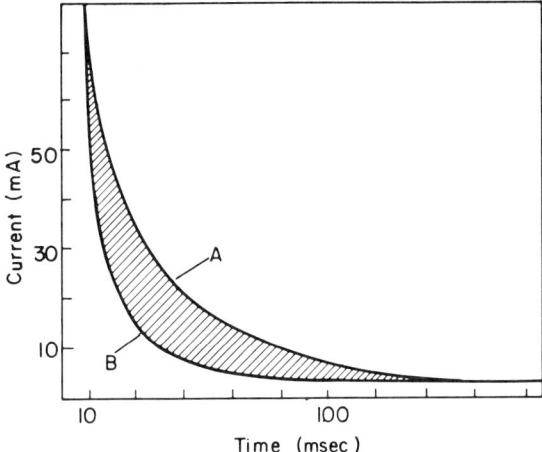

Fig. 9.15. Current–time relations after a constant potential is applied to a platinized platinum electrode in N H_2SO_4 in the presence A and in the absence B of benzene. The charge required for the oxidation of the adsorbed species is represented by the shaded area between the two transients.

area in the figure being the charge representative of the organic surface coverage.

Generally, this method is more useful than galvanostatic transients since fewer assumptions are necessary; also, since potential is the controllable function (as against current in the galvanostatic case), the problems of competitive reactions of the metal substrate may sometimes be overcome. The instrumentation, i.e., to alter the potential within a very short time (10 μsec), is more complex than that needed for the galvanostatic method.

9.2.5d. The Potential-Sweep, or Potentiodynamic, Method. This method is distinct from the previous methods in that the potential is varied linearly with time at some known rate and the current is observed with respect to time.

As for the potentiodynamic method in the presence of competitive substrate reactions, the i versus t curves for a known potential-sweep rate are obtained in the presence and in the absence of adsorbable organic. That the difference in the under-the-curve area, i.e., the charge, is representative of the surface coverage of adsorbed organic species is subject, however, to a number of assumptions, the applicability of which defines the experimental validity of this method. Typical curves for the adsorption of benzene on platinum from acid solution is shown in Fig. 9.16, the shaded area being representative of the charge required to oxidize the adsorbed organic species.

The assumptions involved in this method are: (1) Readsorption of the organic species from solution is negligible, and the oxidation which gives the coulombs counted is that due to complete reaction of the *molecular species* to CO_2 and not the oxidation of *radicals* (see Section 9.5.4). The validity of the first assumption can be checked by the charge determination of several sweep rates. (2) The organic species is removed from the surface by oxidation only, i.e., no desorption by electrostatic effects (Section 7.5). (3) The parallel reaction of substrate oxide formation is the same in the presence as in the absence of the organic species.

The three electrochemical methods mentioned here have several common features. They all determine the charge required to oxidize (de-electronate) whatever is the predominant species on the surface (it has to be *assumed* what this species is, as in the radiotracer method), and they all are limited in their usefulness by the applicability of a number of specific but often implicit assumptions. It may be that the potentiostatic method gives results that are less subject to ambiguity than those of the other two

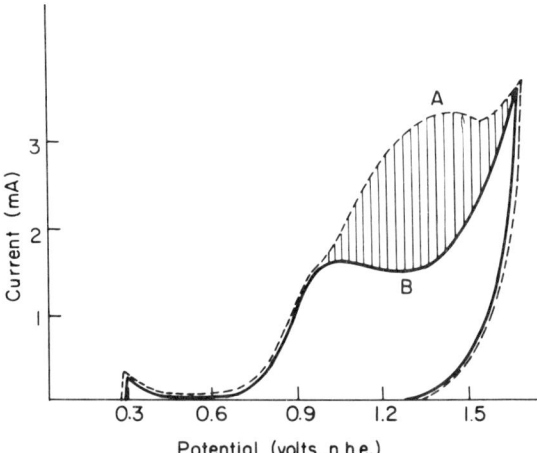

Fig. 9.16. The potential–current response of a platinized platinum electrode in N H_2SO_4 in the presence A and in the absence B of benzene when a constant potential-sweep rate of 0.5 V sec^{-1} is applied. The shaded area is representative of the charge required to oxidize the adsorbed species from the surface.

methods. Thus, the constancy of potential during the transient means that confusion of the coulombs used to oxidize an entity on the surface with some other process (e.g., substrate dissolution or oxide formation) is reduced.

Until recently, it was thought that, when disagreement occurred between the radiotracer method and the electrochemical methods, one method must be incorrect. It is now realized that each gives different information (the labeling method gives the number of labeled atoms, and the electrochemical methods give the quantities of electricity required to carry to completion oxidation reactions removing what is on the surface). The way in which these two approaches can be complementary to each other in the determination of adsorbed surface radicals, by which some knowledge as to the radicals present on the surface can be obtained, is briefly explained in Section 9.5.4).

In conclusion, it is seen that each method has its own drawbacks and areas of applicability, and, therefore, when one determines the surface coverage of a particular organic species, the more complementary methods one uses, the better.

Further Reading

1. F. P. Bowden and E. K. Rideal, *Proc. Roy. Soc. (London)*, **A120**: 59 and 80 (1928).
2. J. A. V. Butler and G. Armstrong, *Proc. Roy. Soc. (London)*, **A137**: 604 (1932); also, *Trans. Faraday Soc.*, **29**: 1261 (1933); and *J. Chem. Soc.*, 743 (1934).
3. A. N. Frumkin and A. I. Shlygin, *Acta Physicochim. U.R.S.S.*, **3**: 791 (1935); and **4**: 991 (1936).
4. H. Gerischer and W. Mehl, *Z. Elektrochem.*, **59**: 1049 (1955).
5. F. Will and C. Knorr, *Z. Elektrochem.*, **63**: 1008 (1959).
6. M. A. V. Devanathan and W. Mehl, *J. Electroanal. Chem.*, **1**: 143 (1959–1960).
7. M. A. V. Devanathan and K. Selvaratnam, *Trans. Faraday Soc.*, **56**: 1820 (1960).
8. M. Breiter, *Electrochim. Acta*, **8**: 447 and 925 (1963).
9. S. Gilman, *J. Electroanal. Chem.*, **7**: 382 (1964); and *Electrochim. Acta*, **9**: 1025 (1964).
10. E. Gileadi and B. E. Conway, "The Behavior of Intermediates in Electrochemical Catalysis," in: J. O'M. Bockris and B. E. Conway, eds., *Modern Aspects of Electrochemistry*, Vol. III, Chapter 5, Butterworths Publications, Ltd., London, 1964.
11. S. Schuldiner, "Electrode Processes," in: J. F. Danielli, K. G. A. Pankhurst, and A. C. Riddiford, eds., *Recent Progress in Surface Science*, p. 160, Academic Press, New York, 1964.
12. F. Will, *J. Electrochem. Soc.*, **112**: 451 (1965).

13. B. E. Conway, "Adsorbed Intermediates," in: *Theory and Principles of Electrode Processes*, Chapter 7, The Ronald Press, New York, 1965.
14. S. Schuldiner and T. B. Warner, *J. Electrochem. Soc.*, **112**: 212 (1965).
15. S. S. Beskorovainaya, Yu. B. Vasil'ev, and V. S. Bagotsky, *Soviet Electrochemistry*, **1**: 916 (1965).
16. M. Breiter, *Electrochim. Acta*, **11**: 905 (1966).
17. N. Pangarov, I. Christova, M. Atanasov, and V. Kertov, *Electrochim. Acta*, **12**: 717 (1967).
18. M. Ya. Popereka and V. N. Lebedeva, *Soviet Electrochemistry*, **3**: 381 (1967).
19. D. I. Leikis and D. P. Aleksandrova, *Soviet Electrochemistry*, **3**: 763 (1967).

9.3. TRANSPORT IN THE ELECTROLYTE EFFECTS CHARGE TRANSFER AT THE INTERFACE

9.3.1. Ionics Looks after the Material Needs of the Interface

The electrodic events at an interface have been the center of attention so far. Quite justifiably, the view has been interface-centered (electrodics) for charge transfer is the essence of electrochemistry. It should not be forgotten, however, that, for an electrodic reaction to keep going, the electron acceptors and electron donors have to move to the interface. It is through this transport of matter from the electrolyte bulk to the interface that ionics (i.e., the behavior of ions in solution) figures in the scheme of things. Thus, ionics looks after the logistics of charge transfer and links up with electrodics to make electrochemical systems run (Fig. 9.17).

The preoccupation with the interface which has characterized the discussion so far is based on an important assumption: the transport aspects of ionics are playing their supply role so well that one has not been aware of the logistic problems of charge transfer. Except for some preliminary indications (*cf.* Section 8.3.3), the interface has been assumed never to fall short of its needs (of electron acceptors and donors). But there are situations where the charge-transfer reaction is inadequately supplied with its material requirements (of electron acceptors, e.g.). Here, a supply problem arises. The *transport* of electron acceptors and donors in the solution becomes the important event. Ionic transport begins to control

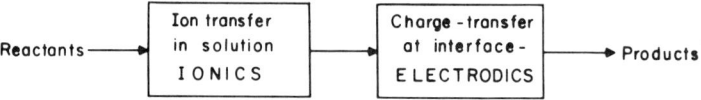

Fig. 9.17. The behavior of electrochemical systems is dependent on both the ionics and the electrodics.

the rate of charge transfer across the interface; then the viewpoint has to become electrolyte-centered.

There is another way of looking at these situations where a lack of a sufficient supply of the needs of the interface compels one to look at the transport processes in solution. The broader view demands that one see the charge-transfer reaction as preceded by a drift of the reaction participants to the interface. In fact, one must think of transport in solution and charge transfer as consecutive steps of an overall reaction.

For example, consider an electron-transfer reaction

$$A_{x=0} \xrightarrow{+e} D$$

where the subscript $x = 0$ emphasizes that, in the initial state of the reaction, the electron acceptors are lined up at the OHP $x = 0$ before being serviced by the electrons in a single or multistep reaction. The step preceding the charge transfer is the transport of the electron acceptors from the electrolyte bulk $x \to \infty$ to the interface $x = 0$, i.e.,

$$A_{x \to \infty} \to A_{x=0}$$

Hence, the overall process is

$$A_{x \to \infty} \xrightarrow[\text{process}]{\text{transport}} A_{x=0} \xrightarrow[\text{transfer}]{\text{charge}} D$$

$$\uparrow \qquad\qquad \uparrow$$
$$\text{Ionics} \qquad \text{Electrodics}$$

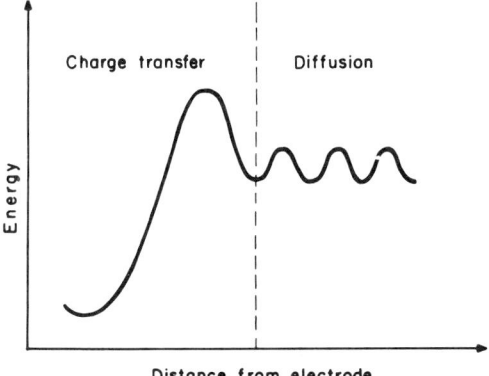

Fig. 9.18. The transport processes are interpretable in terms of the concept of energies of activation (which are, however, relatively very low) and concentration gradients.

1038 CHAPTER 9

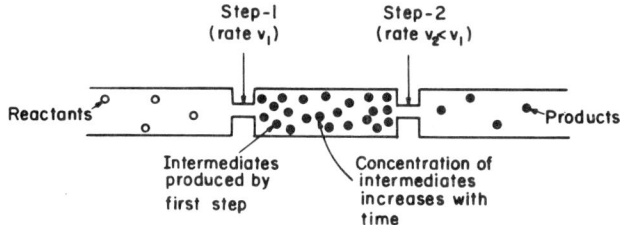

Fig. 9.19. The steady state cannot be established if some intermediates go on accumulating with time.

Now, as in any consecutive reaction, one can pose the question: Which step controls the overall rate of the reaction? Or, which is the rds? Does the transport process or the charge-transfer reaction determine the overall rate? It has been argued (*cf.* Section 4.2.18) that the transport processes of diffusion and migration are rate processes. Hence, they are also connected with activation-energy barriers. These, however, are usually lower than the energy barriers of the electrodic reactions. If diffusion is to be rate controlling, does that mean that its activation-energy barriers have to be higher than those for charge transfer? It does not (Fig. 9.18). One should remember that rates of processes are determined also by concentration. Hence, if the concentration becomes small, the diffusion process *can* become the rds even though its energy barriers remain low.

9.3.2. How the Transport Flux Is Linked to the Charge-Transfer Flux: The Flux-Equality Condition

A simple idea is used to relate the current density across an interface to the rate at which electron acceptors (or electron donors) arrive at (or move away from) the interface.

One starts off with the definition of steady-state (see Section 9.2.1), according to which the concentration of all the intermediates in the reaction must be constant with time. This condition can be achieved if the products of one step are used up in the succeeding step as fast as they are produced. If the first of two consecutive steps proceeded at a faster rate than the second, then the products of the first step would start accumulating (Fig. 9.19) and this would contradict the definition of steady state.

Hence, *in the steady state*, all steps in a consecutive reaction are proceeding at the same rate.[†] If the steps involve charge transfer or charge

---

[†] This idea has been used previously (*cf.* Section 9.1.5) to illustrate the concept of the rds.

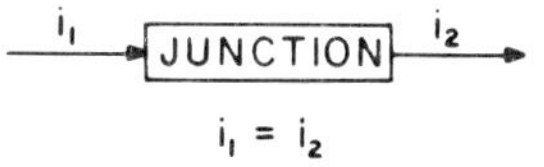

Fig. 9.20. Kirchhoff's law says that the sum of the inflowing currents at a junction must be equal to the sum of outflowing currents.

transport, one says that the net[†] current densities corresponding to all the steps are equal. This is an example of Kirchhoff's first law for electrical currents (Fig. 9.20), which says that the sum of the steady inflowing currents at a junction must be equal to the sum of outflowing currents. Thus, under steady-state conditions, the charge-transfer current density i_{CTR} must be set equal to the current density due to transport,

$$i = i_{\text{CTR}} = i_{\text{transport}}, \quad (9.51)$$

(The junction at which this current equality holds is the $x = 0$, or OHP.) But the transport-current density is the charge transported per mole of ions (i.e., nF) times the transport flux $J_{\text{transport}}$ (i.e., the number of moles of ions transported across 1 cm² of a transit plane per second). If the electron acceptors are not charged species (ions), one can still state the equality of currents in terms of fluxes by expressing (9.51) in terms of moles per unit area and time arriving at, and being reacted in, the interphase

$$\frac{i}{nF} = \frac{i_{\text{CTR}}}{nF} = J_{\text{transport}} \quad (9.52)$$

If diffusion is the transport mechanism, the flux equality condition becomes

$$\frac{i}{nF} = \frac{i_{\text{CTR}}}{nF} = J_D \quad (9.53)$$

where J_D is the diffusion flux.

The flux continuity or flux equality condition can be applied even when the current density i and the transport flux $J_{\text{transport}}$ are changing with time. One simply structures into small time intervals dt and says that the condition is valid within this infinitesimal time.

Once this flux equality condition (9.52) is formulated, one simply

---

[†] The word *net* is used to emphasize that one is not talking of the component forward and backward current densities $\vec{i}$ and $\overleftarrow{i}$ but of the resultant current density $i = \overleftarrow{i} - \vec{i}$.

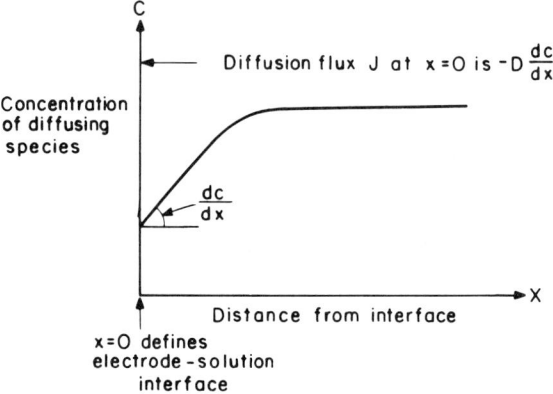

Fig. 9.21. The diffusion flux at $x = 0$ depends on the concentration gradient at this point.

works out $J_{\text{transport}}$ as a pure transport problem and equates it to $1/nF$ times the current density across the interface since n Faradays per mole are required for the transported material to be electronated. If the transport process consists of pure diffusion, i.e., there is no contribution from either migration or hydrodynamic flow, then the flux is given by Fick's first law (*cf.* Section 4.2.2), i.e.,

$$J_D = - D \frac{dc}{dx} \qquad (9.54)$$

so one has to know the concentration gradient in the region where the flux is being considered, i.e., in the region about $x = 0$ (Fig. 9.21). This requires a knowledge of the variation of concentration as a function of distance from the interface, and, if the diffusion is in response to either a constant stimulus switched on at $t = 0$ or a time-varying stimulus (*cf.* Section 9.2.2), then one must also know the *time* variation of the concentration at a particular distance from the interface. To obtain the space and time variation of the concentration, it is necessary to solve Fick's second law under the appropriate initial and boundary conditions, as has been discussed in Section 4.2.9.

9.3.3. Appropriations from the Theory of Heat Transfer

The Laplace transformation method of solving non-steady-state diffusion problems has been briefly treated in Chapter 4. Thus, one can study all sorts of problems by using various types of current or potential stimuli (as in researches using transients, *cf.* Section 9.2.2) and analyzing how trans-

TABLE 9.2
Comparison of the Analogous Parameters Involved in the Analogous Systems, Heat Transfer and Mass Transfer

Mass-transfer-controlled electrodic reaction	Heat-transfer systems, i.e., heat transfer involved in cooling molten metals
Current	Heat flow
Diffusion coefficient	Thermal diffusivity
Concentration gradient	Temperature gradient
Diffusion-layer thickness	Thermal-boundary layer

port in solution influences the response of the system. For example, a sinusoidally varying current density can be used with

$$i = I \sin \omega t \qquad (9.55)$$

(I and ω are the amplitude and angular frequency of the current wave), and the corresponding variation of the potential difference across the interface with time can be measured. It will be recalled, however (cf. Section 4.2.10), that, once one has obtained the solution of the problem involving a constant stimulus (e.g., a current) switched on at $t = 0$, then one can get the solutions for other types of stimuli by using the simple property of Laplace transforms treated in Section 4.2.8.

Another approach can also be rewarding. A common practice in finding solutions to problems of the diffusion of ions in solution is, in fact, to look for analogous heat-flow problems. Since the basic laws of diffusion and heat flow are mathematically similar, a solution of a given heat-flow problem can be used as a solution for the analogous diffusion problem—of course, after changing temperature to concentration and thermal diffusivity to diffusion coefficient (Table 9.2). Solutions to a large number of heat-flow problems have been given in the classic treatise on *Heat Conduction in Solids* by Carslaw and Jaeger, a book which has become a bible for electrochemists studying diffusion-controlled reaction rates.

9.3.4. A Qualitative Study of How Diffusion Affects the Response of an Interface to a Constant Current

The best way to acquire a feel for what happens when the transport of ions determines the overall rate of a reaction at an interface is to consider a specific problem in detail.

However, before tackling such a problem, it is essential to point out that transport processes in electrochemical systems have been analyzed with clarity and adequate detail in many excellent treatises.[†] The present treatment, therefore, is elementary in approach and restricted in content. All that is intended is to sketch in a connected way some of the main concepts relevant to transport-controlled electrodics. Caution must be exercised before extending the ideas to more complex situations.

For example, *the treatment of diffusion that is to follow is solely restricted to semi-infinite linear diffusion*, i.e., diffusion which occurs in the region between $x = 0$ and $x \to +\infty$, *to a plane of infinite area*. Thus, diffusion to a point sink—called *spherical diffusion*—is not treated, though it has been shown to be relevant to the particular problem of the electrolytic growth of dendritic crystals from ionic melts. Further, it is only toward the end (*cf.* Section 9.3.12) of this presentation that deelectronation reactions are considered. Initially, therefore, attention is confined, in this discussion of the effects of diffusion, to electronation reactions.

Consider the electronation of the species M^{n+}

$$M^{n+} + ne \to M$$

where M is deposited on the metal electrode. The electrolyte is assumed to be one with a small concentration of the electron acceptor M^{n+} compared with a large concentration of another positive ion N^+, which is "indifferent" to the charge-transfer reaction, i.e., it does not accept electrons at the given electrode at the potential concerned. Because of the large excess of the indifferent ions N^+, these assume the major part of the burden of carrying the conduction current. It will be recalled (*cf.* Section 4.5.2) that the particular fraction of the conduction current carried by an ionic species depends on its transport number, which in turn depends upon its concentration. Hence, as the electron acceptor M^{n+} is present at relatively low concentrations, it has to reach the interface predominantly by diffusion[‡] rather than by electrical migration. Suppose that, at a time $t = 0$, a con-

---

[†] It is of interest to note that transport to and from electrodes was understood at a sophisticated level many decades before charge-transfer theory began to enter the literature. Some of the frustrations of electrochemists (outside those in the USSR) before the 1950's stemmed, indeed, from attempting to deal with the theory of interfacial charge-transfer reactions with an overstress on diffusion and transport and a neglect of considerations of the energy barrier and charge-transfer theory.

[‡] At this stage, the complicating factor of mass transport by a hydrodynamic flow of the electrolyte is not considered (see, however, Section 9.3.7).

stant current density i_g is switched on, i.e., the stimulus is similar to that considered in the analysis of transients (Section 9.2.3). What will happen?

Initially, the current will largely go toward changing the structure of the interface; this is the double-layer charging discussed earlier. The basic principle in all situations where an external power supply forces a constant current through the interface is that the inflowing charge is consumed in the fastest available process. At the beginning, after switching on a current, this charging of an interface transfer without electron is the easiest process; all that has to be done is to change the charge density on both sides of the interface. At the metal surface, this means moving in excess electrons, and, on the solution side, it means making the excess charge density, due to an unequal distribution of ions, more positive by moving some positive ions toward the interface or negative ions away from it.

After the current has passed for a time, the reaction soon gets going at a considerable rate. In the present section, for simplicity of treatment, one is considering transport-controlled reactions with the charge transfer in virtual equilibrium, i.e., the charge-transfer reaction is assumed to have a high exchange-current density. Hence, one can legitimately use the equilibrium case of the Butler–Volmer relation (Section 8.2.13), i.e., the Nernst equation

$$\Delta\phi = \Delta\phi^0 + \frac{RT}{nF} \ln c_{x=0} \qquad (9.56)$$

where $c_{x=0}$ is the concentration of M^{n+} *at* the interface.

It is in relating $c_{x=0}$ to the c in the bulk that one applies the theory of diffusion. As the electron acceptor M^{n+} gets consumed by the charge-transfer reaction, its concentration at the interface $x = 0$ departs from its initial value, which is the bulk concentration c^0. This is a question of logistics. The diffusion process does not replenish all the electron acceptors which are used up. Thus, the concentration M^{n+} at $x = 0$, namely, $c_{x=0}$, will start diminishing. But this means that the potential difference across the interface will start falling [cf. Eq. (9.56)]. Finally, after a certain time called the *transition time*, the concentration of M^{n+} is almost zero, which implies [Eq. (9.56)] that $\ln c_{x=0}$ and therefore the potential difference $\Delta\phi$ should sink toward $-\infty$.

Thus, when a constant current is imposed on an interface at which the rate of the electronation process is determined by diffusion (i.e., the other reaction steps, particularly electronation, in the overall reaction sequence are relatively fast), it is principally the interfacial concentration of M^{n+} which determines how the potential difference across the interface varies with time. The variation of $c_{M^{n+}}$ with time must therefore be analyzed.

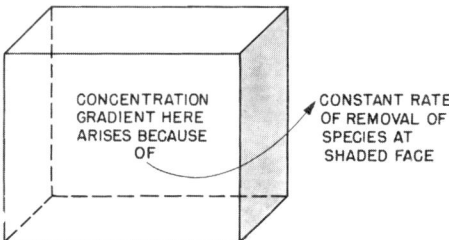

Fig. 9.22. A constant flux at one face of a volume element makes the concentration of the diffusing species vary with distance and time.

9.3.5. A Quantitative Treatment of How Diffusion to an Electrode Affects the Response with Time of an Interface to a Constant Current

In considering the time dependence of a concentration at a given point when this is determined by diffusion (Section 4.2.7), it has been shown that the variation of concentration with time and distance in a rectangular volume with one side parallel to the electrode depends on how the semi-infinite linear diffusion process is stimulated. Diffusion occurs only when a concentration gradient is set up. If the gradient arises from a constant rate of removal of the species across one face of the volume, then this constant flux J_D makes the concentration of the diffusing species vary with distance and time (Fig. 9.22) according to an expression derived in Section 4.2.12 by solving the partial differential equation for diffusion (Fick's second law) under the appropriate initial and boundary conditions. Concentration gradients also result when the interface, instead of removing a species, produces one (Fig. 9.23). From Eq. (4.75) and considering the case where $c < c^0$

$$c_{x,t} = c^0 - \frac{2J_D t^{\frac{1}{2}}}{D^{\frac{1}{2}}\pi^{\frac{1}{2}}} e^{-x^2/4Dt} + \frac{J_D x}{D} \operatorname{erfc} \frac{x}{\sqrt{4Dt}} \qquad (9.57)$$

In this equation, $c_{x,t}$ is the concentration of M^{n+} at a distance x from the $x = 0$ plane at a time t after the switching on of the constant electronation current density, c^0 is the bulk concentration of M^{n+}, and the other terms have their usual significance.

To link the constant-flux problem (of Section 4.2.12) to the constant-current problem discussed here, one can assume that the constant flux arises only from the imposed constant current i_g. Thus, one considers that the boundary of the diffusion problem is the electrified interface $x = 0$ at

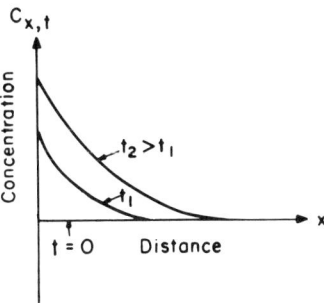

Fig. 9.23. As the substance is being generated, its concentration at the surface increases with time and a positive concentration gradient is established.

which there is equality of the charge transfer (from electrode to ion) and diffusion fluxes (from solution to electrode), i.e.,

$$\frac{i_g}{nF} = J_D \tag{9.53}$$

This expression of J_D must be inserted back into the expression for the concentration $c_{x=0}$ at the interface, i.e., into the expression obtained by setting $x = 0$ in Eq. (9.57), whereupon one obtains

$$c_{x=0} = c^0 - \frac{2i_g}{nF}\sqrt{\frac{t}{\pi D}} \tag{9.58}$$

By writing

$$P = +\frac{2i_g}{nF}\sqrt{\frac{1}{\pi D}} \tag{9.59}$$

Eq. (9.58) becomes

$$c_{x=0} = c^0 - Pt^{\frac{1}{2}} \tag{9.60}$$

This expression, known as *Sand's equation*, gives the variation of the interfacial concentration of M^{n+} with time, after application of a constant current density. But one seeks also to know the time variation of the potential difference across the interface at which the electronation reaction $M^{n+} + ne \rightarrow M$ is occurring. To obtain this information, one recalls that the charge-transfer reaction across the interface is assumed in the present treatment to be virtually in equilibrium and, therefore, the Nernst equation

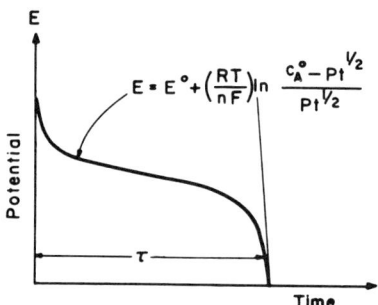

Fig. 9.24. The potential–time transient in a redox system with no reduced species at the beginning is an S-shaped curve.

(9.56) can be used to relate the potential difference to the concentration at the interface. That is, by substituting (9.60) in (9.56),

$$\Delta\phi = \Delta\phi^0 + \frac{RT}{nF} \ln (c^0 - Pt^{\frac{1}{2}}) \tag{9.61}$$

The shape of such potential–time transients as the one represented by (9.61) is interesting (Fig. 9.24). It appears that, after the lapse of a *certain time*, the potential starts falling very rapidly. One must understand what is happening here. Before moving to examine this certain time more closely, consider, instead of a metal-deposition reaction $M^{n+} + ne \to M$, an electronation reaction in which both the electron acceptor A and electron donor D are in solution ($A + ne \to D$), i.e., a *redox* reaction (an example of such a reaction is the electronation of ferric ions to form ferrous ions, $Fe^{+++} + e \to Fe^{++}$), then one would also have had to consider the diffusion of the electron donor *away from* the electrode and the variation of its interfacial concentration with time. Since the electron donor D is being continuously generated, by charge transfer, from the electron acceptor A, the interfacial concentration of D will increase (Fig. 9.23) with time (not decrease, as in the case of the A which is being depleted). If 1 mole of D is *formed* from 1 mole of A and their diffusion coefficients are the same, one will obtain, not the minus sign, but

$$c_{D,x=0} = c_D{}^0 + Pt^{\frac{1}{2}} \tag{9.62}$$

Further, the Nernst equation will take the form [see Eq. (8.61)]

$$\Delta\phi = \Delta\phi^0 + \frac{RT}{nF} \ln \frac{c_{A\,x=0}}{c_{D\,x=0}} \tag{9.63}$$

i.e.,

$$\Delta\phi = \Delta\phi^0 + \frac{RT}{nF} \ln \frac{c_A{}^0 - Pt^{\frac{1}{2}}}{c_D{}^0 + Pt^{\frac{1}{2}}} \tag{9.64}$$

Now, suppose that one starts off the constant current with a zero or negligibly small concentration of electron donor D in the electrolyte. Then $c_D{}^0 \simeq 0$, and Eq. (9.64) reduces to

$$\Delta\phi = \Delta\phi^0 + \frac{RT}{nF} \ln \frac{c_A{}^0 - Pt^{\frac{1}{2}}}{Pt^{\frac{1}{2}}} \tag{9.65}$$

By combining the interface under study with a nonpolarizable interface, i.e., a reference electrode, the potential difference across the system, or cell, will change with time [cf. Fig. 9.24] according to the expression

$$E = E^0 + \frac{RT}{nF} \ln \frac{c_A{}^0 - Pt^{\frac{1}{2}}}{Pt^{\frac{1}{2}}} \tag{9.66}$$

9.3.6. The Concept of Transition Time

Consider the expression (9.65) for the time variation of the potential difference across an interface at which a diffusion-controlled electronation reaction is stimulated by a constant current switched on at $t = 0$

$$\Delta\phi = \Delta\phi^0 + \frac{RT}{nF} \ln \frac{c_A{}^0 - Pt^{\frac{1}{2}}}{Pt^{\frac{1}{2}}} \tag{9.65}$$

The product $Pt^{\frac{1}{2}}$ is zero at $t = 0$. Hence, $c_A{}^0/Pt^{\frac{1}{2}}$ tends to infinity and so does the log term in (9.65), i.e., the potential is supposed to start from plus infinity. Now, $Pt^{\frac{1}{2}}$ grows with the passage of time, and, at some value of time—let it be represented by τ (Fig. 9.24)—, it must become equal to $c_A{}^0$. As this time τ is approached, $c_A{}^0 - Pt^{\frac{1}{2}}$ tends to zero, $\ln(c_A{}^0 - Pt^{\frac{1}{2}})$ tends to minus infinity, and the potential changes very much. In fact, it sinks till it has become sufficiently negative so that some other charge-transfer reaction, e.g., the electronation of the indifferent ions which were inert to electron acceptance at less negative potentials, can utilize the current.

What is the physical meaning that $c_A{}^0 - Pt^{\frac{1}{2}}$ is equal to zero? This is easy to see from

$$c_{A,x=0} = c_A{}^0 - Pt^{\frac{1}{2}} \tag{9.60}$$

At $t = \tau$, $c_A = P\tau^{1/2}$, and hence the interfacial concentration of the electron acceptors has fallen to zero (Fig. 9.25)

$$t = \tau \quad \text{when} \quad c_{A,x=0} = 0 \tag{9.67}$$

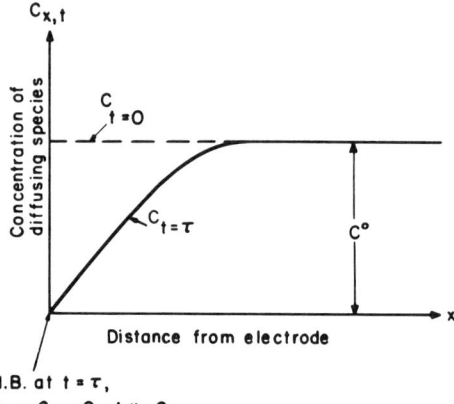

N.B. at $t = \tau$,
$C_{x,t} = C_{0,\tau} = 0$ at $x = 0$

Fig. 9.25. As the transition time is reached, the concentration of the diffusing species at the OHP becomes virtually zero.

This time τ at which the interfacial concentration attains a value of zero is known as the *transition time*.

The transition time has been operationally defined in terms of the rapid variation of potential with time, i.e., it is the time corresponding to the potential jump shown in Fig. 9.24. But one can easily get an explicit expression for it. By making use of the fact that, at τ, $c_{A,x=0} = 0$, one has, using Eqs. (9.59) and (9.60),

$$c_{A,x=0} = 0 = c_A{}^0 - P\tau^{\frac{1}{2}} = c_A{}^0 - \frac{2i_g}{nF}\sqrt{\frac{\tau}{\pi D}} \tag{9.68}$$

or

$$\tau^{\frac{1}{2}} = \frac{nF}{2i_g}c_A{}^0\sqrt{\pi D} = \frac{c_A{}^0}{P} \tag{9.69}$$

Incidentally, with this expression for $\tau^{\frac{1}{2}}$, one has

$$c_A{}^0 = P\tau^{\frac{1}{2}} \tag{9.70}$$

which, in combination with (9.66), gives

$$E = E^0 + \frac{RT}{nF} \ln \frac{\tau^{\frac{1}{2}} - t^{\frac{1}{2}}}{t^{\frac{1}{2}}} \tag{9.71}$$

The equation for $\tau^{\frac{1}{2}}$ [i.e., (9.69)] indicates two main conclusions. Firstly, at a particular concentration $c_A{}^0$, the larger the constant current used, the shorter is the transition time (Fig. 9.26 and Table 9.3). Secondly,

TABLE 9.3
Transition Time for Fe^{2+} Discharge in $0.062M$ $FeSO_4$ + $0.1M$ Na_2SO_4 Solution at pH = 3.4

Current density, amp cm^{-2}	Experimental $\tau_{Fe^{2+}}$, sec × 10^{-3}	Calculated $\tau_{Fe^{2+}}$, sec × 10^{-3}
0.35	9	6
0.32	10	8
0.30	10	9
0.26	13	12
0.24	14	14
0.20	24	20
0.15	40	38
0.08	140	125
0.06	290	222

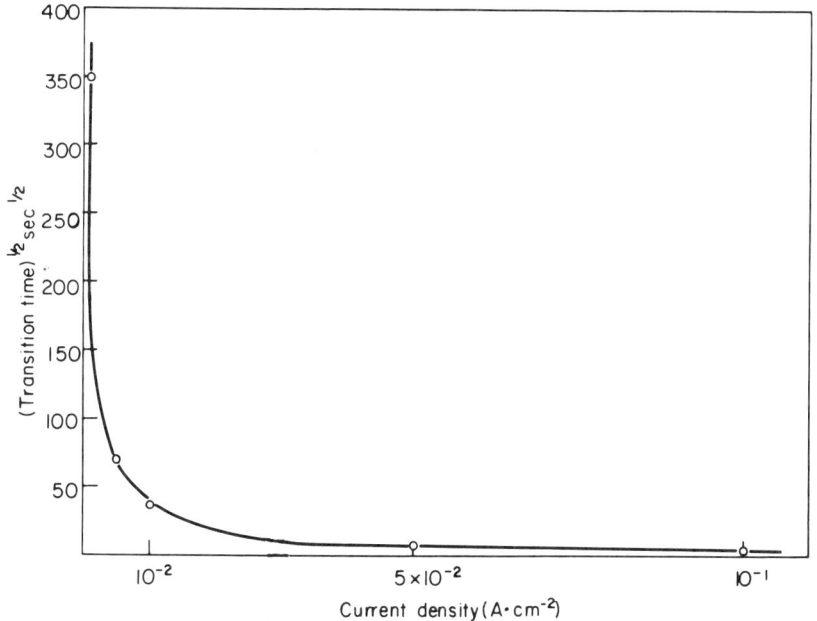

Fig. 9.26. Illustration of the hyperbolic relationship which exists between the square root of the transition time and the current density.

for a fixed constant current i_g, the square root of the transition time is proportional to the bulk concentration; the higher the bulk concentration is, the longer is the transition time.

9.3.7. Convection Can Maintain Steady Interfacial Concentrations

It follows from Eq. (9.69) that, for relatively small concentrations of the reacting species and for not too small currents, transition times are relatively small, of the order of seconds or less (*cf.* Table 9.1). With both increasing concentration and decreasing current density, the calculated transition times increase, reaching, e.g., for a solution with $1M$ concentration of the reactant and for a current density of 10^{-6} amp cm^{-2}, a value of the order of 10 hr. Are there any practical limitations in observing the long transition times which Eq. (9.70) predicts?

A problem does indeed arise when the calculated transition time becomes more than a few seconds. The constant depletion of the electron acceptor makes the electrolyte density near the interface different from that in the bulk. These density differences in different regions of the electrolyte upset the initial condition of hydrostatic equilibrium, and the electrolyte begins to flow to compensate the density changes. This type of hydrodynamic flow is known as *natural convection* (Fig. 9.27).

Though there is fluid flow in the bulk of the electrolyte, it is found that there is a layer adjacent to the electrode in which the electrolyte is stationary, or stagnant. Thus, the electron acceptors travel by convection from the bulk up to the stagnant layer and then cross the layer by diffusion.

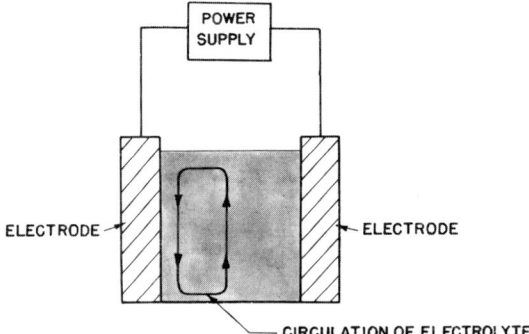

Fig. 9.27. The change in electrolyte concentration produces a change in the density of the electrolyte, a flow of electrolyte results from natural convection, and the hydrostatic equilibrium is thus upset.

This transport by a convection-with-diffusion mechanism has not been taken into account so far. The equations for the time and space variation of concentration [i.e., Eq. (9.57)], for the transition time [Eq. (9.69)], and for the time variation of potential [Eq. (9.71)] have been derived for convection-free conditions, and they break down when convection becomes significant. The first approximation theory given above, therfore, deviates from experiment if the constant current is applied sufficiently long for convection to be important.

The most significant effect of a convective–diffusive transport mechanism is to counteract the tendency of the electronation-current density to reduce the interfacial concentration of electron acceptors to zero. Further, since the interfacial concentration of electron acceptors then remains at a value above that given by the diffusion-based equations, a transition time, indicated by a rapid potential variation, need not be attained.

Thus, the phenomenon of convection (which sets in significantly soon —seconds—after switching on a current) radically alters the picture of the potential–time transient resulting from the switching on of a constant current. When a constant current density is switched on to provoke semi-infinite linear diffusion, a rapid variation of potential is observed, provided the transition time is attained before the onset of natural convection. If, however, the current density is too low and the electron-acceptor concentration is too high, the initial transport process of pure diffusion is soon replaced by a process of convection with diffusion. This convective-diffusion process prevents the electron-acceptor concentration from sinking to zero, and a transition time, marked by a rapid fall of potential, is not attained. Instead, the potential difference across the interface remains at a steady value for an indefinite time even though a constant current density is flowing.

The matter under discussion is of great practical importance. The potential jump associated with the transition time is both the basis of an electroanalytical technique for measuring concentration and also a cause for a lowering of efficiency in substance-producing or energy-converting electrochemical devices (*cf.* Section 11.3.11). Thus, the direct proportionality [Eq. (9.69)] between the square root of the transition time and the bulk concentration of electron acceptors suggests the use of transition-time measurements for analytical purposes. The technique is called *chronopotentiometry*.[†] Workers using this technique must ensure that they do measure-

---

[†] Chronopotentiometry: Chrono- is derived from the Greek word chronos meaning time, so chronopotentiometry is the study of the changes in the potential difference across an interface with respect to time. These potential variations (Fig. 9.24) can be

ments under conditions of pure diffusion, i.e., that the transition time is unaffected by convection. They must choose the current density i_g so that the transition time in the concentration range concerned is reached before the onset of natural convection. The question of how long it takes for natural convection to commence can be answered only by a detailed hydrodynamic analysis of the system, e.g., the concentration of reacting species, current density, viscosity, diffusion coefficient, electrode reaction, and geometry and orientation of the reaction interface. The transition time must be less than a certain time (seconds) to avoid convection effects.

There are, however, many situations in which the potential variations (Fig. 9.24) associated with the approach to a transition time must be scrupulously avoided. Reference is made here to energy-conversion and substance-producing devices in which a departure of the potential from the equilibrium value represents a wastage of electrical energy (Section 11.3.11). Thus, workers interested in such devices (in contrast to those interested in electroanalysis) try to avoid conditions which lead to a transition time. They use appropriately high values of concentration of the ion reacting at the interface to ensure that migration augments mass transport and, above all, they utilize natural convection aided by forced convection (stirring by mechanical means, bubbling of gases, or rotation of the electrode). In this way, the exhaustion of ions at the electrode interface and the rise in overpotential (and corresponding loss in energy) associated with it are avoided.

But what is the quantitative relationship between the steady-state, convection-with-diffusion current density and the potential difference across the interface? How is the steady-state potential difference at a steady current density related to the zero-current, or equilibrium, potential difference? These questions are the relevant ones for steady passage of current in convection-aided situations.

9.3.8. The Origin of Concentration Overpotential

What will be discussed, again for simplicity, is a situation where the charge-transfer reaction is in virtual equilibrium but the interfacial concentration $c_{x=0}$ of the electron acceptor M^{n+} is not the bulk value c^0 but less than that, i.e., $c_{x=0} < c^0$. If a current is passing through the interfaces, the question is: What is the value of the potential difference across the interface?

---

made by either a constant current or a current which varies in some defined manner, e.g., a sinusoidal wave form.

Experiment shows that when the transport of reactants cannot keep pace with the charge-transfer reaction, the potential $\Delta\phi$ observed at the current density i is not equal to the zero-current, or equilibrium potential difference $\Delta\phi_{i=0} = \Delta\phi_e$. If an electronation reaction is considered,

$$M^{n+} + ne \to M$$

the potential sinks to more negative values than that corresponding to equilibrium, although the exchange-current density i_0 has been assumed to be very high (negligible departure from $\Delta\phi_e$ caused by electron transfer). A simple explanation for this phenomenon can be given.

Since the charge transfer is assumed to be virtually at equilibrium, one can again use the Nernst equation to express the potential difference across the interface. Thus, when the current is zero,

$$\Delta\phi_e = \Delta\phi_{i=0} = \Delta\phi^0 + \frac{RT}{nF} \ln c^0 \qquad (9.72)$$

But, what concentration to use in the Nernst equation for the potential difference corresponding to a current density of i? This cannot be the bulk concentration c^0 because it is known that, owing to diffusional hold-up, the interfacial concentration is less than the bulk value. One has to write

$$\Delta\phi = \Delta\phi^0 + \frac{RT}{nF} \ln c_{x=0} \qquad (9.73)$$

This means that the passage of the current has made the potential depart from the zero current value $\Delta\phi_e$. The $\Delta\phi$ has directly resulted from the departure of the interfacial concentration of electron acceptors from the initial bulk value c^0 to a new value $c_{x=0} \neq c^0$. Thus, $\Delta\phi - \Delta\phi_e$ is a potential difference produced by a concentration change at the interface. This concentration-produced[†] potential difference is often known as a *concentration overpotential* η_c to distinguish it from the usual overpotential[‡] η_a, which

---

[†] The change of concentration is the *immediate* cause of the change $\Delta\phi - \Delta\phi_e$ of potential difference across the interface, but the concentration change itself is the result of a current. So, when one refers to concentration-produced overpotential η_c, as opposed to the current-produced overpotential of Section 8.2.8, one is stressing immediate causes.

[‡] To keep the distinction clear, a suffix "a" is hereafter attached to the overpotential arising from the fact that in the charge-transfer reaction there is an activation process necessary. Hence, η_a is often known as the *activation overpotential*.

results from the charge-transfer reaction and was treated at length in Chapter 8. Hence, one writes

$$\eta_c = \Delta\phi - \Delta\phi_e = \frac{RT}{nF} \ln \frac{c_{x=0}}{c^0} \tag{9.74}$$

Since $c_{x=0}$ in cathodic reactions is always smaller than c^0, the concentration polarization has a negative sign, which adds to the activation overpotential in causing the electrode to depart from the equilibrium potential in the negative direction for an electronation reaction.

At this stage, it is worthwhile pointing out a feature of the simplified treatment adopted here. During the discussion of the Butler–Volmer equation (8.31) and the current-produced or activation overpotential η_a, it was assumed that there were no transport limitations on the charge-transfer reaction ($c_{x=0} = c^0$, bulk concentration). Correspondingly, in the present very simple version of transport-controlled electrodics, it has been assumed that charge-transfer limitations are completely absent, i.e., the charge-transfer reaction has such a high exchange-current density that the activation overpotential η_a tends to zero for the current density used [as it would do with a sufficiently high i_0, cf. Eq. (8.31)].

It is now necessary to take a more unified view by considering situations in which the rate of the electrodic process at the interface is subject both to activation and to transport limitations. One refers to a *combined activation-transport control* of the electrodic reaction. Under such conditions, there will be, in addition to the overpotential η_c produced by the concentration change (from c^0 to $c_{x=0}$) at the interface, an activation overpotential η_a because the charge-transfer reaction is not at equilibrium. The total overpotential η is the difference between the interfacial-potential difference $\Delta\phi$, corresponding to a current density i, and the equilibrium-potential difference $\Delta\phi_e$

$$\eta = \Delta\phi - \Delta\phi_e \tag{9.75}$$

but now it is possible to resolve (Fig. 9.28) this total overpotential into two portions: (1) a portion η_a arising from the fact that the charge-transfer reaction must be electrically driven or activated to make it go at a particular rate; and (2) another portion, η_c, arising from the shift in equilibrium potential produced by the transport-induced fall in interfacial concentration

$$\eta = \Delta\phi - \Delta\phi_e = \eta_a + \eta_c \tag{9.76}$$

It must be mentioned here that the activation overpotential η_a, as given by the Butler–Volmer equation (8.31), contains implicit concentration terms

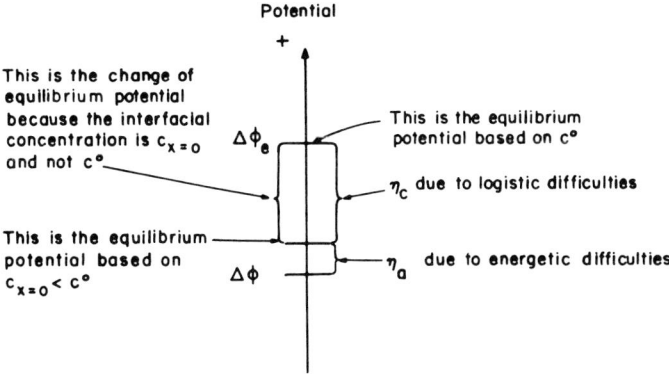

Fig. 9.28. The change in potential during the passage of a current in a cathodic process is negative and due to logistic and energetic difficulties.

hidden in i_0; these concentration terms refer to the concentrations at the OHP and not to bulk values. Only in certain circumstances can the concentration at the OHP be placed equal to the bulk concentration, e.g., when $^{\text{OHP}}\Delta^s\phi = 0$ and $\eta_c = 0$.

9.3.9. The Diffusion Layer

When transport is not able to do its job adequately and there is a change in the interfacial concentrations of electron acceptors and donors from the bulk values, there is a variation of concentration with distance from the interface toward the bulk of the solution. What matters, however, as far as the charge-transfer reaction is concerned, is the gradient of concentration *at the interface* because it is this gradient which drives the diffusion flux J_D. Even when there is convection with a laminar flow[†] of electrolyte, the transport in the (assumed) stagnant layer adjacent to

---

[†] *Laminar* is the name given to the flow of a liquid when it is honey-like. Laminar flow occurs when it is *slow* enough. As the rate of flow increases, a condition arises in which *vortices* begin to form behind a cylinder placed in the path of the flow, and if a further rate increase occurs, these vortices pass out into the rest of the liquid, after it has passed the cylinder, and the surface roughens. This post-honey-like, post-laminar, flow is called *turbulent*. There is an equation for the turnover point from honey-like to rough, laminar to turbulent, namely,

$$(\varrho/\eta)vd > 1$$

where ϱ is the density, η the viscosity, v the velocity, and d the diameter. If this value is *greater* than 1, turbulence begins.

the electrode is by diffusion and the flux J is governed by the concentration gradient in the layer. Thus, using Fick's law of diffusion (*cf.* Section 4.2.2), one has

$$\frac{i}{nF} = J_D = - D \left(\frac{dc}{dx} \right)_{x=0} \qquad (9.77)$$

Now, in general, this concentration profile is such that there is a linear variation of concentration over small distances from the interface and then the concentration asymptotically approaches the bulk value. In this context, Nernst put forward a simplifying suggestion. One might extrapolate (Fig. 9.29) the linear part of the concentration *versus* distance curve until it intersects the bulk value of the concentration at some distance δ from the interface. Then, the gradient of the concentration at $x = 0$, i.e., $(dc/dx)_{x=0}$, can be replaced by $(c^0 - c_{x=0})/\delta$ to give [from (9.77)]

$$\frac{i}{nF} = J_D = - D \frac{c^0 - c_{x=0}}{\delta} \qquad (9.78)$$

In this approximation, therefore, one can consider that the diffusion occurs across a region parallel to the interface, i.e., across a *Nernst diffusion layer* of effective thickness δ.

The diffusion-layer concept is an artifice for handling the flux arising from what would be, if treated in a proper hydrodynamic way, a complicated space variation of concentration at the interface. There is always some gradient of concentration at the interface; there is an initial region in which the concentration changes linearly with distance, but there

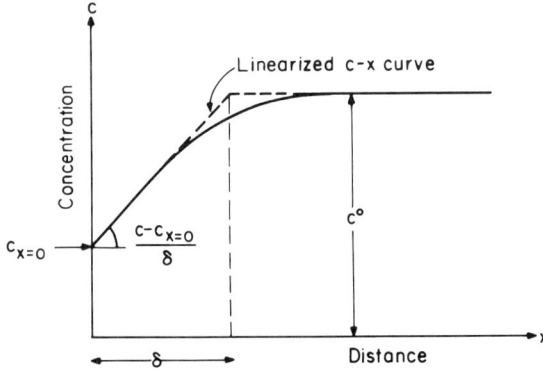

Fig. 9.29. The Nernst diffusion-layer thickness is obtained by extrapolating the linear portion of the concentration change to the bulk concentration value.

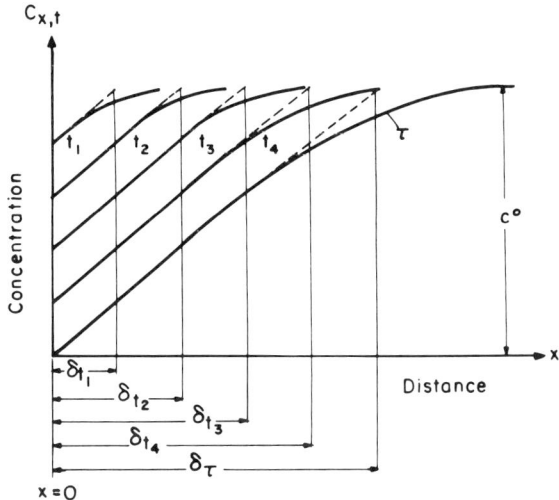

Fig. 9.30. When diffusion is not disturbed by convection, it can be shown that the diffusion-layer thickness δ varies with the square root of time.

is, in the real case, no sharply defined layer of definite thickness, even when convection (natural or forced) produces a steady-state concentration profile. If the concentration profile is not stabilized by forced convection, the effective diffusion-layer thickness varies with time in the case of semi-infinite linear diffusion [*cf.* Eq. (9.80)] until natural convection sets in and "pegs" the concentration profile. In the case of forced convection, e.g., stirring, where the convective transport of species to and from the electrode is much faster than for natural convection, the concentration gradients extend over a much shorter distance δ. The precise value of the Nernst diffusion-layer thickness δ depends largely on the effectiveness of forced convection, being smaller, the greater the effectiveness is. Generally, the convection is laminar where the value of δ and hence of the concentration gradient is governed by the electrode geometry, the kinematic viscosity,[†] the diffusion coefficient, and velocity of the liquid caused by stirring; turbulent flow is also usually involved when the electrolyte is stirred. Both types of convection have been described mathematically by Levich.

Since δ represents only an approximate and simplifying property, it is difficult to evaluate it numerically from fundamental theory. However, it

---

[†] The kinematic viscosity is the ratio of the electrolyte coefficient of viscosity and its density.

TABLE 9.4

The Influence of a Rotating-Disc-Electrode Condition, i.e., Whether Stationary or Rotating, on the Diffusion-Layer Thickness and the Limiting Current Density for the Reaction $I_3^- + 2e \to 3I^-$, assuming that the Concentration of KI_3 is $6.6 \times 10^{-4} M$ and the Diffusion Coefficient of I_3^- is 1.14×10^{-5} cm^2 sec^{-1}

Electrode condition	Limiting current density i_L, μamp cm^{-2}	Diffusion-layer thickness δ, cm
Stationary	28.9	0.05
Rotated at 50 rpm.	134.1	0.011
Rotated at 240 rpm.	292.1	0.005

turns out that, in a very rough and order-of-magnitude sense, a numerical value for δ of about 0.05 cm in solutions in which no forced convection has been artificially introduced is useful for calculations pertaining to transport-controlled electrodics.[†] As artificial convection (stirring, electrode rotation) is introduced, the value of δ which must be introduced into (9.82) depends entirely on the degree of stirring. The value can be reduced to some 10^{-3} cm and even smaller values with sufficiently high stirring. A hydrodynamic theory of δ is available if the hydrodynamic conditions are sufficiently well defined, as, e.g., they are for the rotating-disc electrode (Section 9.3.10a). Table 9.4 indicates the value of δ under three conditions, calculated by assuming a rotating disc of a geometry shown in Fig. 9.33 and a value of δ of 0.05 cm in stationary solutions.

The time variation of δ *before* the onset of natural convection depends on how the diffusion process is provoked. If a constant current density is switched on at $t = 0$, then the time variation of the effective diffusion-layer thickness (Fig. 9.30) can be obtained from Eqs. (9.78) and (9.58)

$$c^0 - c_{x=0} = -\frac{i_g}{nF}\frac{\delta}{D} = -\frac{2i_g}{nF}\sqrt{\frac{t}{\pi D}} \qquad (9.58)$$

or

$$\delta = \frac{2}{\sqrt{\pi}}\sqrt{Dt} \qquad (9.79)$$

[†] For example, $\delta \approx 0.05$ cm yields the correct order of magnitude for the maximum current—the so-called limiting current, i_L [cf. Eq. (9.82)]—which a particular charge-transfer reaction can support. This maximum is determined by the maximum transport flux of reactants.

showing that the effective diffusion-layer thickness increases as the square root of time. It can be shown equally well that, if one considers a diffusion-controlled electronation process occurring under a constant potential switched on at $t = 0$, the diffusion-layer thickness is given as

$$\delta = \sqrt{\pi D t} \qquad (9.80)$$

which is also a square-root dependence of δ on time.

Very much more is known about the theory of concentration gradients at electrodes than has been mentioned in this brief account. Experimental methods for observing them have also been devised, based on the dependence of refractive index on concentration (the Schlieren method). Nevertheless, the basic concept of an effective diffusion-layer thickness, treated here as varying in thickness with $t^{\frac{1}{2}}$ until the onset of natural convection and as constant with time after convection sets in (though decreasing in value with the degree of disturbance, Table 9.4), is a useful aid to the simple and approximate analysis of many transport-controlled electrodic situations. A few of the uses of the concept of δ will now be outlined.

9.3.10. The Limiting Current Density and Its Practical Importance

When an electronation reaction is occurring at an interface, the equality of the charge-transfer flux and the transport flux requires that

$$\frac{i}{nF} = -D\left(\frac{dc}{dx}\right)_{x=0} \qquad (9.77)$$

In terms of the diffusion-layer concept (Section 9.3.9), this condition becomes

$$\frac{i}{nF} = -D\left(\frac{dc}{dx}\right)_{x=0} = -D\frac{c^0 - c_{x=0}}{\delta} \qquad (9.78)$$

It is obvious that the concentration gradient has a maximum value for $c_{x=0} = 0$. Placing this limit upon Eq. (9.78) produces

$$\lim_{c_{x=0} \to 0}\left(\frac{dc}{dx}\right)_{x=0} = \lim_{c_{x=0} \to 0} \frac{c^0 - c_{x=0}}{\delta} = \frac{c^0}{\delta} \qquad (9.81)$$

This maximum-concentration gradient corresponds to a maximum or *limiting current density* denoted by i_L and given from (9.78) and (9.81) by

$$i_L = -\frac{DnFc^0}{\delta} \qquad (9.82)$$

TABLE 9.5

Typical Experimental Limiting Current Density (per Geometric External Square Centimeter of a Porous Electrode) for Four Energy Producers

Cell	Limiting current density, amp cm^{-2}
Propane–oxygen	0.6
Hydrogen–oxygen	0.8
Hydrogen–air	0.13
Hydrazine–oxygen	0.64

For a given electronation reaction, this is the maximum attainable current density. The reaction cannot go faster than i_L because the transport process in the electrolyte bulk is incapable of supplying the electron acceptor to the interface at a faster rate.

The concept of a limiting current density is of great practical importance. Table 9.5 shows some typical experimental limiting current densities for four energy producers employing oxygen (where the calculation of limiting current density is more complicated than that of (9.82); see Chapter 11).

9.3.10a. Polarography: The Dropping-Mercury Electrode. The proportionality between i_L and c^0 [*cf.* Eq. (9.82)] constitutes the basis of *polarography*, a powerful electroanalytical technique introduced by Heyrovsky.[†] The essential part of the polarographic setup (Fig. 9.31) is a dropping-mercury electrode consisting of a glass capillary tube out of which mercury converges at the rate of a drop every few seconds. To the solution is added a large excess (~1 M) of a substance which is termed an *indifferent electrolyte* because its ions do not participate in, or are indifferent to, the charge-transfer reaction at the drop–solution interface. In contrast, the species A under polarographic analysis exists in the solution at a very small concentration (~10^{-3} M). This wide disparity in the concentrations of A and the indifferent electrolyte ensures that the ions of the indifferent electrolyte carry the migration current and that any transport of the species A occurs by a process of pure diffusion.

In a polarographic experiment, a potential difference E is applied across the cell consisting of the dropping-mercury electrode and a nonpolarizable interface (e.g., a calomel electrode). In response to this potential difference,

---

[†] The Czechoslovak electrochemist Heyrovsky was awarded the Nobel prize for this contribution to electrochemical methods in 1959.

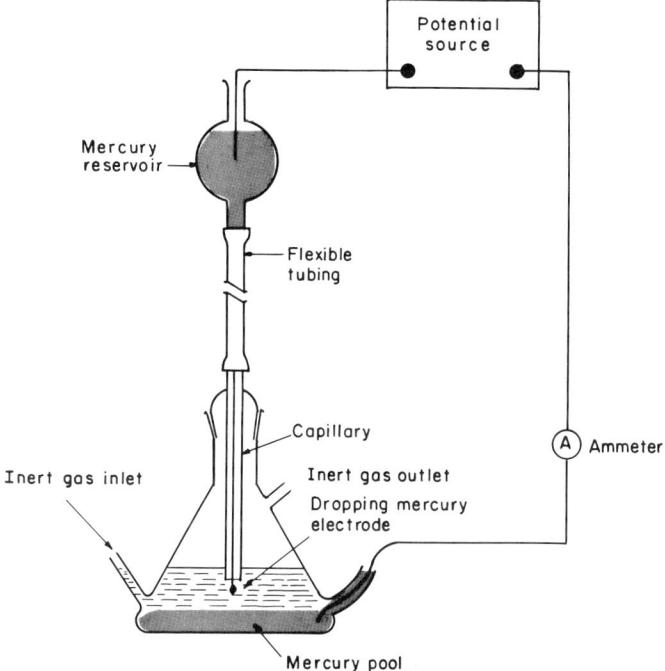

Fig. 9.31. The dropping-mercury electrode arrangement.

a current density i flows across the drop–solution interface. As each drop grows and falls, however, the surface area of the drop also grows, and then becomes effectively zero when the drop falls. Thus, the instantaneous current (current density times surface area) shows fluctuations, but the mean current is a unique function of the potential difference across the drop–solution interface, and therefore of that across the cell.

The relationship between the mean current and the potential will now be derived.

Suppose that, at the drop–solution interface, an electronation reaction: $A + ne \rightarrow D$ is driven by the imposition of a constant potential, E. The reaction results in the depletion of A in the interfacial region, and, therefore, in the diffusion of A toward the drop–solution interface. Let it be assumed that the species D produced by the electrode reaction is soluble either in the electrolyte or in mercury (i.e., D is an amalgam-forming metal). Then, since there is generation of D in the interfacial region, there will be a diffusion of D away from the drop–solution interface either toward the electrolyte or into the mercury.

1062 CHAPTER 9

It is clear that, since the mercury drop approximates a sphere, the theory of *spherical*, and *not* linear, diffusion might have to be used. However, detailed considerations accessible in monographs show that if the electrodic reaction is driven for a sufficiently short time ($t <$ a few seconds) and if the mercury-drop radius is not too small ($r > 0.05$ cm), then the equations of linear diffusion can be used with validity. Thus, the partial differential equations for the diffusion of A and D are[†]—see Fick's second law [*cf.* Eq. (4.32)]—

$$\frac{\partial c_A(x, t)}{\partial t} = D_A \frac{\partial^2 c_A(x, t)}{\partial x^2} \quad \text{and} \quad \frac{\partial c_D(x, t)}{\partial t} = D_D \frac{\partial^2 c_D(x, t)}{\partial x^2} \quad (9.83a)$$

The initial conditions are

$$c_A(x, 0) = c^0 \quad \text{and} \quad c_D(x, 0) = 0 \quad (9.83b)$$

i.e., before the imposition of the potential difference E, the concentrations of A and D for any x are c^0 and zero, respectively.

The first boundary condition to be satisfied by solutions of the differential equations (9.83a) arises from the fact that the only source for the material D is the electrodic transformation of A by the reaction $A + ne \rightarrow D$. For each mole of A reacting at the interface (i.e., at $x = 0$), one mole of D is produced; hence, the sum of the fluxes of A and D at $x = 0$ must be equal to zero:

$$D_A \left(\frac{\partial c_A(x, t)}{\partial x} \right)_{x=0} + D_D \left(\frac{\partial c_D(x, t)}{\partial x} \right)_{x=0} = 0 \quad (9.83c)$$

The second boundary condition involves a relation between the concentrations of A and D and the potential E. The simplest relation is obtained by (initially) assuming that there is *charge transfer equilibrium at the interface* ($x = 0$), in which case the Nernst equation (8.61) can be applied:

$$E = E^0 + \frac{RT}{nF} \ln \frac{a_A(0, t)}{a_D(0, t)} = E^0 + \frac{RT}{nF} \ln \frac{c_A(0, t) f_A}{c_D(0, t) f_D} \quad (9.83d)$$

where a_A, f_A and a_D, f_D are the activities and activity coefficients of the species A and D, respectively.

Finally, the situation far away from the drop can be assumed to be unperturbed by the reaction $A + ne \rightarrow D$. Thus, one has

$$c_A(\infty, t) = c^0 \quad \text{and} \quad c_D(\infty, t) = 0 \quad (9.83e)$$

[†] The notation $c_A(x, t)$ is used to represent the concentration of A at a time t in the life of a drop and at a distance x from the drop–solution interface.

Before attempting to solve the differential equations (9.83a) in the context of the initial and boundary conditions (9.83b)–(9.83e), two variables, θ and $c_1(x, t)$, will be defined thus:

$$\theta = \frac{c_A(0, t)}{c_D(0, t)} = \frac{f_D}{f_A} \exp\left[\frac{nF}{RT}(E - E^0)\right] \tag{9.84}$$

and

$$c_1(x, t) = c^0 - c_A(x, t) \tag{9.85}$$

In terms of these two variables, the differential equations (9.83a) and the initial and boundary conditions (9.83b)–(9.83e) become

$$\frac{\partial c_1(x, t)}{\partial t} = D_A \frac{\partial^2 c_1(x, t)}{\partial x^2} \quad \text{and} \quad \frac{\partial c_D(x, t)}{\partial t} = \frac{\partial^2 c_D(x, t)}{\partial x^2} \tag{9.86a}$$

$$c_1(x, 0) = 0 \quad \text{and} \quad c_D(x, 0) = 0 \tag{9.86b}$$

$$-D_A\left(\frac{\partial c_1(x, t)}{\partial x}\right)_{x=0} + D_D\left(\frac{\partial c_D(x, t)}{\partial x}\right)_{x=0} = 0 \tag{9.86c}$$

$$\frac{c^0 - c_1(0, t)}{c_D(0, t)} = \theta \tag{9.86d}$$

$$c_1(\infty, t) = 0 \quad \text{and} \quad c_D(\infty, t) = 0 \tag{9.86e}$$

The corresponding Laplace transforms of equation (9.86a) are

$$p\bar{c}_1(x, p) - c_1(x, 0) = D_A \frac{d^2\bar{c}_1(x, p)}{dx^2}$$

and

$$p\bar{c}_D(x, p) - c_D(x, 0) = D_D \frac{d^2\bar{c}_D(x, p)}{dx^2}$$

These total differential equations can be combined with the initial condition (9.86b) and solved. The result is

$$\bar{c}_1(x, p) = A_1 \exp[-x\sqrt{p/D_A}] + A_2 \exp[+x\sqrt{p/D_A}] \tag{9.87a}$$

$$\bar{c}_D(x, p) = B_1 \exp[-x\sqrt{p/D_D}] + B_2 \exp[+x\sqrt{p/D_D}] \tag{9.87b}$$

The integration constants A_2 and B_2 must be zero in order to satisfy the boundary conditions (9.86e), in which case

$$\bar{c}_1(x, p) = A_1 \exp[-x\sqrt{p/D_A}] \tag{9.88a}$$

$$\bar{c}_D(x, p) = B_1 \exp[-x\sqrt{p/D_D}] \tag{9.88b}$$

CHAPTER 9

To evaluate A_1 and B_1, it is necessary to use boundary conditions (9.86c) and (9.86d), which after Laplace transformation are

$$\bar{c}_1(0, p) + \theta \bar{c}_D(0, p) = c^0/p \qquad (9.89a)$$

$$D_A\left(\frac{d\bar{c}_1(x, p)}{dx}\right)_{x=0} - D_D\left(\frac{d\bar{c}_D(x, p)}{dx}\right)_{x=0} = 0 \qquad (9.89b)$$

If equations (9.88a) and (9.88b) are used in (9.89a) and (9.89b), the integration constants A_1 and B_1 turn out to be

$$A_1 = \frac{c^0}{1 + \theta m} \frac{1}{p} \qquad (9.90a)$$

$$B_1 = mA_1 \qquad (9.90b)$$

where

$$m = \sqrt{D_A/D_D} \qquad (9.90c)$$

Thus, the solutions of the total differential equations are

$$\bar{c}_1(x, p) = \frac{c^0}{1 + \theta m} \frac{\exp[-(x/\sqrt{D_A})\sqrt{p}\,]}{p} \qquad (9.91a)$$

and

$$\bar{c}_D(x, p) = \frac{c^0 m}{1 + \theta m} \frac{\exp[-(x/\sqrt{D_D})\sqrt{p}\,]}{p} \qquad (9.91b)$$

It is known, however, that the Laplace transform of $\mathrm{erfc}(k/2\sqrt{t})$ is $e^{-k\sqrt{p}}/p$, i.e.,

$$\mathscr{L}[\mathrm{erfc}(k/2\sqrt{t}\,)] = \int_0^\infty e^{-pt}\,\mathrm{erfc}(k/2\sqrt{t}\,)\,dt = e^{-k\sqrt{p}}/p$$

hence

$$c_1(x, t) = \frac{c^0}{1 + \theta m}\,\mathrm{erfc}\left(\frac{x}{2\sqrt{D_A t}}\right) \qquad (9.92)$$

or

$$c_A(x, t) = c^0\left[\frac{\theta m + \mathrm{erf}(x/2\sqrt{D_A t}\,)}{1 + \theta m}\right] \qquad (9.92a)$$

and

$$c_D(x, t) = \frac{c^0 m\,\mathrm{erfc}(x/2\sqrt{D_D t}\,)}{1 + \theta m} \qquad (9.93)$$

TABLE 9.6
Error Functions and Error Function Complements[†]

y	erf (y)	erfc (y)
0	0	1
0.05	0.05637	0.94363
0.1	0.11246	0.88754
0.2	0.22270	0.77730
0.4	0.42839	0.57161
0.8	0.74210	0.25790
1.0	0.84270	0.15730
2.0	0.99532	0.00468
3.0	0.99998	0.00002

[†] Equation (9.92) includes the term erfc $[x/2(D_A t)^{\frac{1}{2}}]$, which can be rewritten as $1 - \text{erf}\,[x/2(D_A t)^{\frac{1}{2}}]$. The error function can be written generally as erf $(y) = (2/\sqrt{\pi})\int_0^y e^{-u^2}\,du$, where u is simply a "dummy" variable; u does not appear in the final answer since integrating between 0 and y produces a final result dependent on y. Let $y = x/2(D_A t)^{\frac{1}{2}}$, then erf $[x/2(D_A t)^{\frac{1}{2}}] = (2/\sqrt{\pi}) \int_0^{x/[2(D_A t)^{\frac{1}{2}}]} e^{-u^2}\,du$. Tabulations of the erf (y) as a function of the upper limit of integration y are available, examples of which are given.

At the time t the instantaneous current i_t is given by Fick's first law:

$$i_t = nFA_t D_A \left(\frac{\partial c_A(x,\,t)}{\partial x}\right)_{x=0} \tag{9.94}$$

where A_t is the surface area of the drop at the time t. It is necessary, therefore, to get an explicit expression for the concentration gradient $(\partial c_A(x, t)/\partial x)_{x=0}$ at the interface. This expression is obtained by differentiating Eq. (9.92a) with respect to x and then setting $x = 0$ in the result. Thus, with

$$\frac{d\,\text{erf}[\lambda(x)]}{dx} = \frac{2}{\sqrt{\pi}} e^{-[\lambda(x)]^2} \frac{d[\lambda(x)]}{dx}$$

the result of the differentiation is

$$\left(\frac{dc_A(x,\,t)}{dx}\right)_{x=0} = \frac{c^0}{\sqrt{\pi D_A t}} \frac{1}{1 + \theta m} \tag{9.95}$$

Combining Eqs. (9.94) and (9.95), one obtains

$$i_t = \frac{nFA_t \sqrt{D_A}\, c^0}{\sqrt{\pi t}} \frac{1}{1+\theta m} \quad (9.96a)$$

$$= i_L \frac{1}{1+\theta m} \quad (9.96b)$$

since the current, when $\theta = c_A(0,t)/c_D(0,t) = 0$, i.e., when $c_A(0,t) = 0$, is equal to the limiting current, i_L,

$$i_L = \frac{nFA_t \sqrt{D_A}\, c^0}{\sqrt{\pi t}} \quad (9.96c)$$

By eliminating θ between Eqs. (9.96b) and (9.84) one obtains the required current–potential relation:

$$E = E^0 - \left[\frac{RT}{nF} \ln \left(\frac{f_D}{f_A}\right)\left(\frac{D_A}{D_D}\right)^{\frac{1}{2}}\right] + \frac{RT}{nF} \ln \frac{i_L - i_t}{i_t} \quad (9.97a)$$

$$= E_{\frac{1}{2}} + \frac{RT}{nF} \ln \frac{i_L - i_t}{i_t} \quad (9.97b)$$

where $E_{\frac{1}{2}}$ is termed the *half-wave potential* because the i_t versus E curve—the so-called *polarographic wave*—is of the form shown in Fig. 9.32 and $E = E_{\frac{1}{2}}$

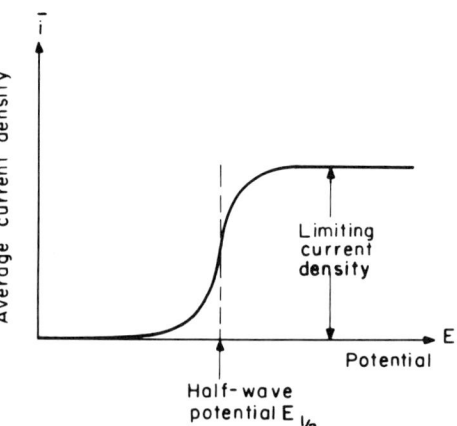

Fig. 9.32. The polarographic wave is the current–potential dependence in which, at some sufficiently high potential values, the current becomes entirely diffusion controlled.

when $i_t = \frac{1}{2}i_L$, i.e., when i_t has attained half the wave height i_L. The potential $E_{\frac{1}{2}}$ also corresponds to an inflection point in the i_t versus E curve, for it may be shown that the second derivative of the i_t versus E curve is equal to zero—the criterion for an inflection point—when $i_t = \frac{1}{2}i_L$.

The half-wave potential $E_{\frac{1}{2}}$ (Table 9.7) might be called a pseudo-fundamental quantity in polarographic analysis since, in the absence of disturbing factors, particularly the heat of amalgam formation, it should be equal to the standard electrode potential E_0 [Eq. (9.97)]. It was once thought that, by determination of the half-wave potential and the assumption that this potential corresponded to the standard thermodynamic potential of the reaction at the mercury–solution interface, a qualitative identification of the species in solution could be made. However, the approximations involved in Eq. (9.97) reveal this as untrue; thus, the identification of the half-wave potential as the standard thermodynamic electrode potential depends not only upon the neglect of amalgam formation between the product D and

TABLE 9.7

Polarographic Half-Wave Potentials $E_{\frac{1}{2}}$, in Volts (versus SCE), for Certain Metal Cations in the Presence of Either Ammonia–Ammonium Chloride Mixture or Tetraethylammonium Hydroxide as the Indifferent Electrolyte

Indifferent electrolyte present in excess	Metal cation	$E_{\frac{1}{2}}$, V, SCE
Ammonia–ammonium chloride mixture	Mn^{2+}	-1.66
	Ga^{3+}	-1.60
	Fe^{2+}	-1.49
	Zn^{2+}	-1.35
	Co^{2+}	-1.29
	Ni^{2+}	-1.10
	Cd^{2+}	-0.81
	Cu^{2+}	-0.51
	Tl^+	-0.48
Tetraethylammonium hydroxide	Li^+	-2.31
	Mg^{2+}	-2.30
	Ca^{2+}	-2.22
	Sr^{2+}	-2.11
	K^+	-2.10
	Na^+	-2.07
	Ba^{2+}	-1.92
	Cr^{2+}	-1.58

the mercury but also upon the assumption made in Eq. (9.97) that the activation overpotential is negligible. It may well be true that, for a number of reactions, η_A is negligible at the current densities obtained in polarographic analysis. However, it is too gross an assumption to allow polarography anything but a historical claim to applications in qualitative analysis.

Quantitative analysis is, however, possible since, in Eq. (9.97), the existence of a limiting current density and its relationship to the concentration of species undergoing electronation are unaffected by the approximations involved. However, as Eq. (9.82a) shows, i_L is a function of time and, as t rises, the diffusion-controlled current decreases toward zero. This is one disadvantage of working at a stationary electrode. Another disadvantage is the effect of impurities which accumulate at the surface. In order to avoid these difficulties, a dropping-mercury electrode is used where each drop extends only over a short time which is determined by the flow rate v. The same law of nonstationary linear diffusion can be applied; however, a correction must be made for the variation of the surface area with time. The surface area A_t can be calculated, on assuming perfectly spherical drops, as follows. The weight of the drop at time t is given by

$$vt = \frac{4\pi r^3}{3} d$$

where r and d are the radius and density of the mercury drop, respectively.
Then,

$$A_t = 4\pi r^2 = 4\pi \left(\frac{3}{4\pi d} vt\right)^{\frac{2}{3}}$$

and, at 25°C,

$$A_t = 0.8515(vt)^{\frac{2}{3}} \tag{9.98}$$

The thickness of the diffusion layer, as shown first by Ilkovic, will be changed, in that, instead of Eq. (9.80), one has now for an expanding spherical drop

$$\delta = (\tfrac{3}{7}\pi\, Dt)^{\frac{1}{2}} \tag{9.99}$$

From Eqs. (9.82), (9.98), and (9.99), the limiting current at time t is given by

$$i_L = i_{\lim} = (7.082 \times 10^4) n v^{\frac{2}{3}} t^{\frac{1}{6}} D_A^{\frac{1}{2}} c^0 \tag{9.100}$$

The above equation—in which i is in amperes; v, in grams per second; t, in seconds; D_A, in square centimeters per second; and c^0, in moles per cubic centimeter—is known as the *Ilkovic equation*.

The mean current $\bar{i}_L$ during the lifetime of the drop, τ, is given by

$$\bar{i}_L = \frac{1}{\tau}\int_0^\tau i_L\,dt = (6.07 \times 10^4)nv^{\frac{2}{3}}\tau^{\frac{1}{6}}D_A^{\frac{1}{2}}c^0 \tag{9.101}$$

The mean current is thus independent of time at the given flow rate v and equal to six-sevenths of the limiting current i_L.

Equations (9.96) and (9.97) can be rewritten as

$$\bar{i} = \frac{\bar{i}_{\lim}}{1 + (c_A/c_D)_{(x=0,t)}(D_A/D_D)^{\frac{1}{2}}} \tag{9.96a}$$

and

$$E = E_0 - \frac{RT}{F}\ln\left(\frac{f_D}{f_A}\right)\left(\frac{D_A}{D_D}\right)^{\frac{1}{2}} + \frac{RT}{F}\ln\frac{\bar{i}_L - \bar{i}}{\bar{i}} \tag{9.97a}$$

In practice, instead of measuring the absolute values of $\bar{i}_L$ and E to attain the concentration of the species under electronation, one simply compares the polarographic wave (Fig. 9.32) with those obtained under identical conditions where the concentration of the species was accurately known, i.e., compares the unknown concentration polarogram with calibrated standard polarograms. With knowledge of $\bar{i}_L$, the concentration of the species undergoing electronation, c^0, can be obtained from Eq. (9.101) if one knows the value of n—the number of electrons exchanged with mercury—and the diffusion coefficient; the values of v and τ are experimental variables that are known and constant for any one determination.

It is essential that a polarographer arrange his system so that the reactant in the solution, the concentration of which he is interested in determining, reaches the interface only by diffusion. This means that he must try to reduce to negligible the amount of the substance which reaches the interface by electrical migration and by convection as well. He ensures that the electrical migration (as distinct from the diffusion) is negligible by adding a large excess of an *indifferent electrolyte*, namely, one which does not take part in the interfacial charge transfer and yet takes care of the passage of current in the solution. For example, in the measurement of, say, Cu^{++} and Ni^{++} concentrations in the same solution, one must have present a large excess of an electrolyte which does not exchange electrons with mercury in the usual potential range covered in polarography; a large excess means concentrations much above the concentration of either Cu^{++} or Ni^{++} in the solution. Such an indifferent electrolyte would be KNO_3 or NH_4NO_3. Under such conditions, the large majority of the electric current is carried by K^+ and NO_3^-, and, hence, diffusion is the only mode of transport to the mercury–solution interface for the cupric and nickel ions.

Thus, the assumption made in Eqs. (9.97) and (9.101) is realized to a good approximation in practice.

The short lifetime of the mercury drop, a few seconds, avoids conditions which lead to convection.

9.3.10b. The Rotating-Disc Electrode. Another method of obtaining well-defined diffusion conditions was developed by Levich, who first introduced and described the rotating-disc electrode.

The rotating-disc electrode, Fig. 9.33, is a device which permits the use of a solid electrode for measurements analogous to the dropping-mercury electrode. It consists of a disc of metal which has one face exposed to the solution and is rotated about its center. Owing to the rotation, the solution is set into motion and flows past the disc surface.

The mathematics of convective diffusion may be solved analytically, and one obtains an expression for the limiting current which is

$$i_L = 0.62nFAD^{\frac{2}{3}}v^{-\frac{1}{6}}\omega^{\frac{1}{2}}c \tag{9.102}$$

where A is the disc area; D, the diffusion coefficient of the reactant; v, the kinematic viscosity of the solution; ω, the angular velocity of the disc; and c, the concentration.

The advantages which the rotating-disc electrode offers are that the limiting current is time independent and very stable and it may be easily

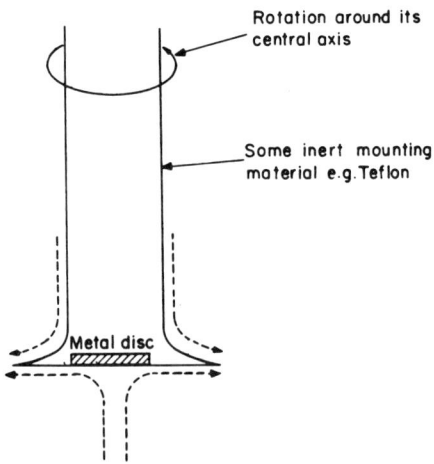

Fig. 9.33. Schematic representation of a rotating electrode. The broken arrows indicate the electrolyte-flow lines produced by rotation.

calculated. Also, it can be made quite large compared with the dropping-mercury electrode, owing to rapid rotation. Another advantage is that both anodic and cathodic reactions may be studied by using an appropriate metal.

A further development by Frumkin and Levich at the Electrochemistry Institute of the Academy of Sciences of the USSR is the rotating disc with ring assembly, Fig. 9.34.

On the disc, a rotating metal plate, a given reaction occurs, e.g., the reduction of O_2 to water. If, in the course of this reaction or any other reaction that is studied, intermediate species are formed that are relatively weakly adsorbed on the disc, the flow lines shown in Fig. 9.33 indicate that at least part of these intermediates will be swept toward the ring. This ring is maintained at various "radical-catching" potentials until a current in the ring circuit is registered. This current is due to the side-swept intermediates of the disc-electrode reaction, undergoing an electrode reaction at the ring.

The importance of this method is that it allows one to get some idea of the radicals present on the electrode surface, at least when they are partially desorbed during their lifetime (this requires weak adsorption). The importance of identifying intermediate radicals at an electrode surface is fairly obvious if one wishes to identify the steps—particularly that which is rate determining—in an electrode reaction.

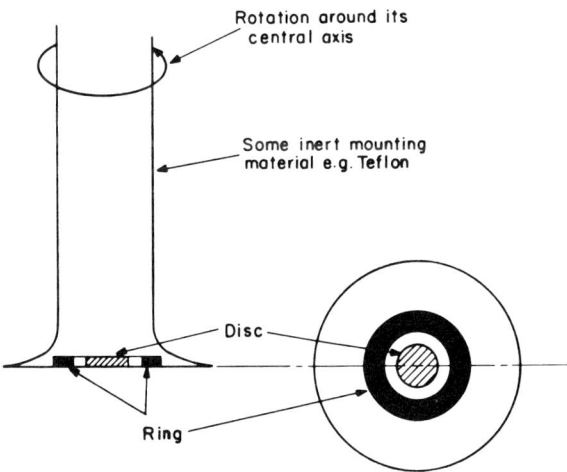

Fig. 9.34. Schematic representation of the rotating disc with ring assembly. The electrolyte-flow lines will be those shown in Fig. 9.33.

9.3.11. The Steady-State Current–Potential Relation under Conditions of Transport Control

The concept of limiting current density permits a simple derivation of a relation between the steady-state concentration overpotential η_c and the current density i if the reaction is such that other forms of overpotential are negligible. One starts from the expression for the concentration overpotential η_c [cf. Eq. (9.74)]

$$\eta_c = \frac{RT}{nF} \ln \frac{c_{x=0}}{c^0} \tag{9.74}$$

the electronation-current density is given by

$$i = -DnF \frac{c^0 - c_{x=0}}{\delta} \tag{9.78}$$

and one can write

$$c^0 - c_{x=0} = \frac{-\delta}{DnF} i$$

or

$$\frac{c_{x=0}}{c^0} = 1 + \frac{\delta}{DnFc^0} i \tag{9.78}$$

But, from (9.82),

$$\frac{\delta}{DnFc^0} = -\frac{1}{i_L} \tag{9.82}$$

hence,

$$\frac{c_{x=0}}{c^0} = 1 - \frac{i}{i_L} \tag{9.103}$$

Thus, (9.103) and (9.74) produce

$$\eta_c = \frac{RT}{nF} \ln \left(1 - \frac{i}{i_L}\right) \tag{9.104}$$

which can also be written in the form

$$i = i_L (1 - e^{nF\eta_c/RT}) \tag{9.105}$$

To understand the significance of (9.105), it is important to bring out the background assumptions. If one is considering semi-infinite linear diffusion to a planar cathode, the switching on of a constant current density leads to a value of the concentration which depends on the distance, increasing from the electrode surface out toward the bulk of the solution.

The effective diffusion-layer thickness δ, the useful simplification of the real situation (Fig. 9.29), varies with time according to (9.79) or (9.80) until natural convection sets in. The effect of convection is to stabilize the concentration profile and permit the use of a steady diffusion-layer thickness in the development of the concept of limiting current density (Section 9.3.9). It may be mentioned here that, under conditions of *spherical* diffusion, a stable diffusion-layer thickness and therefore a constant limiting current density are inherent in the solution of diffusion equations and observed even in the absence of convection.

In terms of the upper limit i_L for the magnitude of the current density possible with a given electrode reaction, the relationship between steady-state current density and concentration overpotential in the absence of significant η_a is given by Eqs. (9.103) and (9.105) (*cf.* also Fig. 9.35). These equations indicate several interesting points. Firstly, it is clear from Eq. (9.104) that $1 - (i/i_L)$ is a fraction and, therefore, $\eta_c = \Delta\phi - \Delta\phi_e$ is a negative quantity. Hence, the $\Delta\phi$ corresponding to an electronation-current density i is always more cathodic than the equilibrium potential $\Delta\phi_e$.

Secondly, if the experimental setup is designed to vary the current density, then, the larger the value of i is, the more $1 - (i/i_L)$ tends to zero and, therefore, the nearer η_c approaches $-\infty$. If, on the other hand, the potential difference at the electrode is being controlled by an external electronic circuit, then, in order to produce a net electronation-current density,

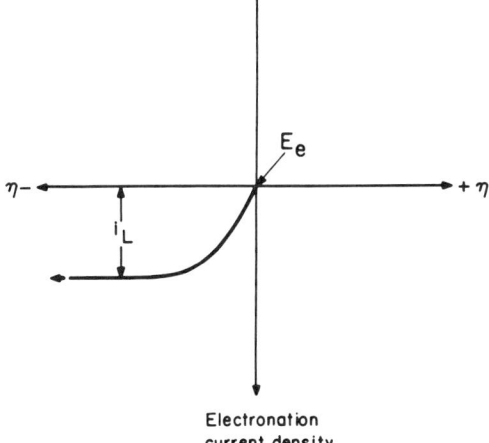

Fig. 9.35. The cathodic limiting current appears as a horizontal straight line limiting the current that can be achieved at any large negative value of overpotential.

it is necessary to make $\Delta\phi < \Delta\phi_e$, i.e., η_c is negative. The more negative η_c is, the smaller is the value of $e^{(nF/RT)\eta_c}$ and the nearer is $1 - e^{(nF/RT)\eta_c} \to 1$; therefore, $i \to i_L$ [Eq. (9.105)]. In the limit, the limiting value of i is reached (Fig. 9.35).

A comparison between the expressions relating current density to η_c [Eq. (9.104)] and to η_a [by assuming that concentration overpotential is negligible, this is given by the Butler–Volmer equation (8.31)] shows an important point. In the cases of both activation and concentration overpotentials (the latter for electronation purposes), the current density is an exponential function of the overpotential, but in the activation case the slope of i versus η_a increases with η_a, while in the transport-limitation case the slope of i versus η_c decreases with increasing η_c and becomes effectively zero for sufficiently negative values of overpotential (see Figs. 8.32 and 9.35).

9.3.12. Transport-Controlled De-electronation Reactions

The above treatment has been solely concerned with electronation reactions, but it is equally valid for de-electronation reactions in which the electron donor is in solution and has to be transported to the interface. It is necessary only to introduce some changes in signs in the equations.

However, for de-electronation reactions in which the electron donor is the solid electrode itself, (as in the dissolution of a metal), some changes must be made in the analysis. The main point is that the electron donor (the metal) is in inexhaustible supply and can keep on building up (Fig. 9.36); the concentration gradient therefore keeps on increasing. Thus, the

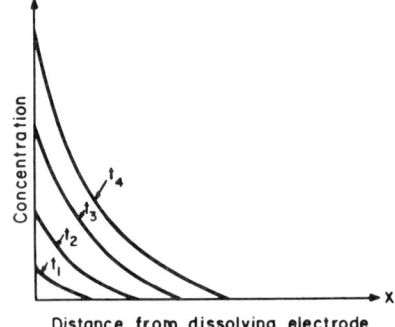

Fig. 9.36. In an anodic dissolution process, the concentration of the metal ions at the electrode surface increases with time from t_1 to t_4.

diffusion away gets faster and faster. Strictly speaking, therefore, there should be no limiting current for such a process. In practice, however, the product of the dissolution, i.e., the metal ion, often starts to precipitate out at some limiting value of the anodic current density because the solubility product of one of the salts is exceeded. For example, the dissolution of nickel metal may lead to the precipitation of nickel hydroxide.

9.3.13. What Is the Effect of Electrical Migration on the Limiting Diffusion-Current Density

Unless there is a large excess of indifferent ions which assume the burden of carrying the current (as indeed was assumed above), the electron acceptors and donors do not move only by diffusion or convection; they also move under the influence of the electric field. In fact, this is generally the case unless one has diminished the fraction of the current in the solution which reactants need for carrying, by adding an excess of ions of another kind which do not undergo electrodic reaction, e.g., the indifferent electrolyte. How must the current–potential equations be modified?

In the steady state, the situation is simple. The basic condition is that, at the $x = 0$ plane, the number of moles per square centimeter per second being transformed by charge transfer is equal to the number of moles arriving per square centimeter per second by electrical migration and by diffusion. The material coming by migration is given by $t_A(i/nF)$, where t_A is the transport number or fraction of the current density carried by the electron-acceptor species A undergoing charge transfer. The diffusion flux is of course given by Fick's law (Section 4.2). Hence,

$$\frac{i}{nF} = \frac{t_A i}{nF} - D\frac{c^0 - c_{x=0}}{\delta}$$

or

$$i = -\frac{DnF}{1-t_A}\frac{c^0 - c_{x=0}}{\delta} \qquad (9.106)$$

It is seen from Eq. (9.106) that the current density i is always greater than in the case of pure diffusion [Eq. (9.78)], in which case $t_A = 0$ and Eq. (9.106) reduces to (9.78). Similarly, the limiting current density must be greater for migration plus diffusion than for pure diffusion [Eq. (9.82)] and is given by

$$i_L = -\frac{DnF}{1-t_A}\frac{c^0}{\delta} \qquad (9.107)$$

This is good; when one wants to make substance-producing or energy-producing devices, one wants to maximize charge transfer and one does not like to be limited by transport in the bulk of the electrolyte. Hence, in such systems, the electron-acceptor species must carry as much of the current as possible; the larger its transport number t_A, the greater is its limiting current density. As i_L increases, η_c decreases [*cf.* Eq. (9.104)] and there is a minimum wastage of electrical energy utilized to combat transport limitations. This is of course only the case when migration helps the transport *toward* the electrode.

When migration enters the transport picture, it is important to realize that the direction of electrical migration of a charged species depends on the sign of the charge. Negatively charged species (anions) migrate to the positive, or electron-sink, electrode (the anode), and positively charged species (cations) migrate to the electron-source electrode (the cathode). It is known, however, that ions which accept electrons do not always have to be positively charged, as might at first be thought, e.g., the negatively charged $[Ag(CN)_2]^-$ is electronated at a negative electron-source electrode. The question is: If such ions migrate under the applied field gradient away from the electron-source electrode, how are they transported to it? This question has been analyzed in Section 4.4.14. It will be referred to here by pointing out that it is the diffusion flux, i.e., the term $DnF(c^0 - c_{x=0})/\delta$ of Eq. (9.78), which has to sustain now not only the charge-transfer reaction but also the loss of ions from the reacting layer due to electrical migration away. Negatively charged particles can be electronated at the negative electrode so long as the diffusion flux *toward* the electrode is greater than the electrical migration *away* from the electrode.

9.3.14. Some Summarizing Remarks on the Transport Aspects of Electrodics

Earlier in this book (Chapters 4 and 6) a fairly detailed treatment of the movements and transport of ions was presented; qualitative pictures and quantitative accounts were given of the diffusion and electrical migration of ions in the bulk of the electrolyte. No mention was made at first, in the treatment of electrodic processes, of a connection between the transport in solution and processes at electrodes. It was then realized that this neglect of ion transport in solution (ionics) was tantamount to assuming that at no stage in the course of a charge-transfer reaction did the interfacial concentrations of electron acceptors and donors depart from their bulk values.

This section commenced with the realization that the supply of the material requirements of the interface may not be sufficient to meet the demands of charge transfer and therefore, one has to be able to analyze such supply problems. The transport of particles through the solution is one of the essential steps that join with the step (or steps) of the charge-transfer reaction to constitute the overall reaction. Hence, the rate of the transport may in some systems determine the overall rate. Thus, one began to think of current densities which may be transport controlled in general. It turned out that diffusion control, in particular, one type of transport process, is easy to describe in a very simple physical way.

The treatment of non-steady-state diffusion is a question of solving Fick's second law of diffusion. In many cases, however, solutions can be taken from the treatments of the analogous problems in heat flow in solids; the point is: Heat flow and diffusion are described by mathematically similar problems.

The analysis of one problem, namely, how semi-infinite linear diffusion affects the response of an interface to the switching on of a constant current showed that, after a certain time known as the *transition time*, the potential difference at the interface undergoes a rapid change. This rapid variation is due to the fact that the interfacial concentrations of the particles diffusing to the interface tend to zero. Application of the fact to a simple equation relating the potential difference at the interface to concentration shows that, when $c_{x=0}$ tends to zero, the corresponding potential difference tends to highly negative values.

However, the outset of natural convection or a deliberate resort to stirring the solution may hold the interfacial concentrations at steady values. Under these steady-state conditions, one may use Fick's first law, and the equality of the flows of particles up to and across the interface for an electronation reaction can be written as

$$\frac{i}{nF} = -D\left(\frac{dc}{dx}\right)_{x=0} \tag{9.77}$$

In reality, the concentration gradient is constant for only a short distance from the interface and then becomes asymptotic[†] to zero in the bulk. But one can resort to a linearization of the concentration profile, and then one can use the artifice of an imagined, i.e., simplified, diffusion layer in

---

[†] In time-dependent semi-infinite linear diffusion, the approach to zero is due to the error function complement in Eq. (9.57); in the steady-state of plane electrodes, convection ensures the zeroing of the concentration gradient at sufficient distance from the electrode, at times greater than that at which diffusion only determines the transport.

which the concentration changes in a linear fashion from the interfacial value $c_{x=0}$ to the bulk value c_0. The effective thickness δ of the diffusion layer, which can be taken as a constant independent of time only under steady-state conditions in which natural convection occurs, proved a useful quantity. With its aid, one can write out the flux-equality condition in the form

$$\frac{i}{nF} = \frac{D}{\delta}(c^0 - c_{x=0}) \quad (9.78)$$

The change of the interfacial concentrations from the bulk values at zero current to different values at finite currents produces an extra potential difference $\eta_c = \Delta\phi_i - \Delta\phi_{i=0}$. This concentration overpotential can be obtained by inserting into the Nernst equation the interfacial concentrations c^0 at $i=0$ and $c_{x=0}$ at i

$$\eta_c = \frac{RT}{nF} \ln \frac{c_{x=0}}{c^0} \quad (9.74)$$

The link between the current density and the concentration overpotential under steady-state conditions for systems in which the exchange-current density is relatively large compared with the limiting current density (hence, the activation overpotential is negligible) was established through the concept of a limiting current i_L arising from the fact that there is a maximum rate at which electron acceptors can move to an interface. In terms of the limiting current density i_L, an exponential current–potential law was obtained for diffusion-controlled current densities involving electronation reactions:

$$i = i_L(1 - e^{nF\eta_c/RT}) \quad (9.86)$$

Electric migration of electron acceptors to the interface aids the transport process. The electron acceptors are driven by concentration gradients but also by the electrical field in the bulk of the electrolyte between the electrodes. Here, their transport number, the fraction of the concentration current carried by the electron acceptors compared with that carried by the other ions, plays a role.

These are only a few elementary ideas concerning the influence of transport processes in solution on the overall rate of charge transfer across an interface. This aspect of electrochemistry has received considerable attention; it is part of classical electrochemistry and dates from about 1900. Detailed knowledge of it is important in the application of electrochemistry to the analysis of systems where diffusion control is deliberately encouraged by the use of low concentrations of electron acceptors and donors and large excesses of species which shun charge transfer, so that complicating electrical migration of the species which react at the interface is avoided.

When it comes, however, to designing devices which produce substances or yield energy, or to understanding what makes a charge-transfer reaction function, the influence of diffusion control is a nuisance. There, one must minimize transport limitations and turn away from consideration of the drift of ions in solution to consideration of charge transfer and its mechanisms, with the central objective of reducing activation overpotential.

Further Reading

1. H. J. S. Sand, *Phil. Mag.*, **1**: 45 (1900).
2. D. Ilkovic, *J. Chim. Phys.*, **35**: 129 (1938).
3. V. G. Levich, *Acta Physicochim. URSS*, **17**: 257 (1942); also, *Zh. Fiz. Khim.*, **18**: 335 (1944).
4. I. M. Kolthoff and J. J. Lingane, *Polarography*, Vol. I, Interscience Publishers Inc., New York, 1952.
5. P. Delahay, *New Instrumental Methods in Electrochemistry*, Interscience Publishers, Inc., New York, 1954.
6. P. Delahay, "Polarography and Voltammetry," in: *Instrumental Analysis*, Chapter 4, The MacMillan Company, New York, 1956.
7. A. N. Frumkin and L. N. Nekrasov, *Dokl. Akad. Nauk SSSR*, **126**: 115 (1959).
8. Yu. B. Ivanov and V. G. Levich, *Dokl. Akad. Nauk SSSR*, **126**: 1029 (1959).
9. V. G. Levich, "Passage of Current Through Electrolytic Solutions," in: *Physicochemical Hydrodynamics*, Chapter 6, Prentice Hall, Inc., Englewood Cliffs, N. J., 1962.
10. A. C. Riddiford, "Rotating Disc System," in: P. Delahay and C. W. Tobias, eds., *Advances in Electrochemistry and Electrochemical Engineering*, Vol. IV, Chapter 2, Interscience Publishers, Inc., New York, 1966.
11. M. Paunovic, *J. Electroanal. Chem.*, **14**: 447 (1967).
12. A. Damjanovic and M. A. Genshaw, *J. Electrochem. Soc.*, **114**: 466 (1967); also, *J. Electroanal. Chem.*, **15**: 173 (1967).
13. V. I. Tikhomirova, V. I. Luk'yanycheva, and V. S. Bagotsky, *Soviet Electrochemistry*, **3**: 673 (1967).
14. A. J. Arvia, S. L. Marchiano, and J. J. Podesta, *Electrochim. Acta*, **12**: 259 (1967).
15. M. C. Giordano, J. C. Bazan, and A. J. Arvia, *Electrochim. Acta*, **12**: 723 (1967).
16. V. V. Losev, A. I. Molodov, and V. V. Gorodetzki, *Electrochim. Acta*, **12**: 475 (1967).

9.4. DETERMINING THE STEPWISE MECHANISM OF AN ELECTRODIC REACTION

9.4.1. How One Tries to Determine the Reaction Mechanism

Though one can present mechanism determination in general and therefore abstract terms, it is preferable, at the outset, to consider a specific electrodic reaction. The reaction which will be considered is the deposition and dissolution of iron in acid solutions.

One must be clear as to what one is seeking in a mechanism determination. What knowledge must be gained before it can be said that the mechanism of an electrodic reaction is understood? Firstly, one must know the *overall reaction*: What are the reactants and what are the products and how many electrons are transferred in one act of this overall reaction? Secondly, one must try to know by what *path* or sequence of steps the reactants are transformed into products. Finally, one must know which particular step determines the overall rate of the reaction, i.e., one must know which is the rds within the reaction sequence.

The simplest part of mechanism determination is to find out what is the overall reaction. The reactants and products can be identified by standard methods of analysis. A whole arsenal of such techniques is available—all the classical methods of inorganic and organic analysis, in particular, the various types of spectroscopy, chromatography, and, of course, electroanalytical methods.

For the iron reaction, however, it is found that the solution accumulates ferrous ions, i.e., $Fe \to Fe^{2+}$. The dissolution must therefore involve two electrons, $Fe \to Fe^{2+} + 2e$. One can check this easily by observing that two faradays of electricity pass in the external circuit per mole of iron dissolved. Thus, the overall reaction is

$$Fe \to Fe^{2+} + 2e \qquad (9.108)$$

for the dissolution (or de-electronation) for iron. The reaction for the deposition is

$$Fe^{2+} + 2e \to Fe \qquad (9.109)$$

although in this case there are complications because in acid solutions, there may be a simultaneous deposition of protons to form adsorbed hydrogen and, eventually, hydrogen gas.

To determine the individual consecutive steps in a reaction is much more tricky. There are several diagnostic criteria which help in this task;

among the first of these are the determination of the transfer coefficients $\vec{\alpha}$ and $\overleftarrow{\alpha}$ and the electrochemical reaction orders. However, one cannot just make measurements, consult tables, and then state the reaction path. Mechanism determination is no dull routine; it is an exciting search often involving inferences and even intuitions rather than firm evidence and data. The experimental evidence is rarely in a form that points directly to the path and the rds. Rather, one must be sensitive to the hints of the experimental trends; one imagines several—perhaps dozens—of conceivable reaction paths and rds's, analyzes what would be the behavior of each hypothetical reaction model, compares the predicted with the observed behavior, and thus singles out a model which yields the closest match with experiment.

This procedure is similar to that taken in other fields. For example, in the determination of the structure of complex crystals by X-ray methods, one has to posit hypothetical structures and compare the X-ray patterns predicted on the basis of them with the experimental observations.

To come back to the iron reactions, the observed dependence of current on potential yields, for the anodic Tafel slope,

$$\frac{dE}{d \log i} = \frac{2.3RT}{\overleftarrow{\alpha} F} = 0.04 \text{ V} \tag{9.110}$$

from which $\overleftarrow{\alpha} = \frac{3}{2}$ is obtained.

The cathodic Tafel slope cannot be obtained directly from the same type of measurement since negative polarization of the iron electrode leads not only to the deposition of iron but also to hydrogen evolution. However, the cathodic Tafel slope for iron deposition was deduced from such measurements as

$$\frac{dE}{d \log i} = \frac{2.3RT}{\vec{\alpha} F} = 0.12 \text{ V} \tag{9.111}$$

from which $\vec{\alpha} = \frac{1}{2}$ is obtained.

Since

$$\nu = \frac{n}{\vec{\alpha} + \overleftarrow{\alpha}} = \frac{2}{\frac{3}{2} + \frac{1}{2}} = 1 \tag{9.25a}$$

the rds occurs once per one occurrence of the overall reaction.

The electrochemical-reaction order for the concentration of ferrous-iron ions was obtained as

$$p_{(\text{Fe}^{2+})} = \left(\frac{d \log i}{d \log c_{\text{Fe}^{2+}}} \right)_{\Delta\phi, \text{pH}} = 1$$

It was also found by Despic and Drazic, greatly to their surprise, that the reaction rate was dependent not only on the Fe^{2+} concentration but also on the pH of the solution. The reaction order for the OH^- ion concentration was obtained as

$$p_{(OH^-)} = \left(\frac{d \log i}{d \log c_{OH^-}}\right)_{\Delta\phi, c_{Fe^{2+}}} = 1$$

Thus, the reaction mechanism postulated for the iron reaction must be consistent with the following

$$\overleftarrow{\alpha} = \tfrac{3}{2}, \quad \overrightarrow{\alpha} = \tfrac{1}{2}, \quad r = 1, \quad p_{Fe^{2+}} = 1, \quad p_{OH^-} = 1$$

One's first thoughts in postulating the mechanism would be that the path consists of a simple one-step two-electron transfer reaction (9.108). In fact, it at first seems likely that the overall reaction is also the path and the rds. This is the simplest guess. In electrochemical—as well as in chemical and biochemical—reactions, what looks like a simple path on paper may not be the most advantageous energetically. What is being demanded in Eq. (9.108) is that two electrons be transferred in one shot. The probability of two electrons tunneling *simultaneously* between an ion and the metal is so low that it has become well accepted in electrodics to *exclude paths which would involve multiple electron transfers in one reaction step*. In fact, if one meets a statement in the literature that an *n*-electron transfer occurs in a reaction, it is likely that what is under discussion is—or is part of—the *overall* reaction and *not* the rate-determining charge-transfer step in a number of consecutive reactions constituting the overall reaction.

If a two-electron transfer step is barred, what about two successive one-electron transfer reactions? Thus,

$$Fe^{2+} + e \rightarrow Fe^+$$
$$Fe^+ + e \rightarrow Fe$$

This sequence, however, would result in $p_{(OH^-)} = 0$ and not the observed value of 1. The latter value is an important clue indicating that either H^+ or OH^- must be involved somewhere in the sequence of the reaction, a conclusion by no means obvious from the overall reaction.

Perhaps a one-electron transfer to an intermediate species formed by the interaction between ferrous ions and either H^+ or OH^- occurs? Then, the concentration of the intermediate species would be pH dependent, and the rate of the electrodic reaction would also become pH dependent. What can $Fe^{2+} + H_2O$ form? One can postulate the species $FeOH^+$ since its existence in solution is known. Thus one thinks of the formation of the intermediate $FeOH^+$ by the following step

$$Fe^{2+} + H_2O \rightarrow FeOH^+ + H^+ \qquad (9.112)$$

and then the electron-transfer reaction

$$FeOH^+ + e \rightarrow FeOH \qquad (9.113)$$

At this stage, the FeOH is on the iron surface, and it must further react in a second one-electron transfer reaction to form iron. What this reaction must be can easily be ascertained. One simply substracts the sum of the two reactions (9.112) and (9.113), namely, $Fe^{2+} + H_2O + e \rightarrow FeOH + H^+$ from the overall electronation reaction (9.109) $Fe^{2+} + 2e \rightarrow Fe$. One gets

$$FeOH + H^+ + e \rightarrow Fe + H_2O \qquad (9.114)$$

In this way, one has devised a possible three-step reaction scheme for the deposition of iron, and one knows that it contains within it the possibility of predicting the unexpected pH variation. Thus,

1. $\qquad Fe^{2+} + H_2O \rightarrow FeOH^+ + H^+ \qquad (9.112)$

2. $\qquad FeOH^+ + e \rightarrow FeOH \qquad (9.113)$

3. $\qquad FeOH + H^+ + e \rightarrow Fe + H_2O \qquad (9.114)$

9.4.2. Which Is the Rate-Determining Step in the Iron Deposition and Dissolution Reaction?

The overall reaction for iron deposition and dissolution has been identified, and a model conceived for the multistep reaction pathway. Now comes the question of the rds. The procedure is to work out the theoretical predictions of several, hypothetically possible, assumed rds's and compare them with experiment.

Let the first hypothesis be that step 3 [Eq. (9.114)] is the rds. The assumed reaction scheme becomes

1. $\qquad Fe^{2+} + H_2O \underset{v_{-1}}{\overset{v_1}{\rightleftharpoons}} FeOH^+ + H^+ \qquad (9.112a)$

2. $\qquad FeOH^+ + e \underset{v_{-2}}{\overset{v_2}{\rightleftharpoons}} FeOH \qquad (9.113a)$

3. $\qquad FeOH + H^+ + e \rightarrow Fe + H_2O \qquad (9.114a)$

1084 CHAPTER 9

The current-density–potential relationship for this will now be worked out.

Instead of using the final Butler–Volmer equation (9.24) developed in Section 9.1.7 for multistep reactions, the procedure leading to Eq. (9.24) will be repeated here on the practical example since it offers certain features not discussed in detail previously. One proceeds as follows:

Firstly, the i versus $\Delta\phi$ relation is written out *for the particular rds*, then the unknown concentration terms in this relation are evaluated by making use of the law of mass action for the other steps assumed to be in virtual equilibrium. Since one has to work with the concentration terms, it is convenient to write the Butler–Volmer equation for the rds in terms of rate constants k's rather than in terms of exchange-current densities i_0's.

On this basis, one has for the assumed rds, i.e., step 3 [reaction (9.114a)].[†]

$$i = 2F(v_{-3} - v_3)$$
$$= 2F[k_{-3}(1 - \theta_{\text{FeOH}})e^{(1-\beta)F\Delta\phi/RT} - k_3\theta_{\text{FeOH}}c_{\text{H}^+}e^{-\beta F\Delta\phi/RT}] \quad (9.115)$$

where v_3 is the reaction rate in the electronation (cathodic) direction and v_{-3} the reaction rate in the de-electronation (anodic) direction and where θ_{FeOH} is the fraction of the total surface covered with adsorbed FeOH. One must now get θ_{FeOH} from the assumed equilibria of the other steps 1 and 2. Consider step 2, i.e.,

$$\text{FeOH}^+ + e \underset{v_{-2}}{\overset{v_2}{\rightleftharpoons}} \text{FeOH} \quad (9.113a)$$

Since it is considered to be in virtual equilibrium, the forward and backward rates must be almost equal, i.e., approximately

$$v_{-2} = v_2 \quad (9.116)$$

or

$$k_{-2}\theta_{\text{FeOH}}e^{(1-\beta)F\Delta\phi/RT} = k_2 c_{\text{FeOH}^+}(1 - \theta_{\text{FeOH}})e^{-\beta F\Delta\phi/RT}$$

or

$$\frac{\theta_{FeOH}}{1 - \theta_{FeOH}} = K_2 c_{FeOH^+}\, e^{-F\Delta\phi/RT} \quad (9.117)$$

[†] Note that the total current density must be double that of the rds since $n = 2$; in other words, each act of the rds must be preceded by one act of reaction (9.113a) for the overall reaction to occur once.

But from the law of mass action for step 1, i.e.,

$$\text{Fe}^{2+} + \text{H}_2\text{O} \underset{v_{-1}}{\overset{v_1}{\rightleftharpoons}} \text{FeOH}^+ + \text{H}^+ \tag{9.112a}$$

one has

$$k_1 c_{\text{Fe}^{2+}} = k_{-1} c_{\text{FeOH}^+} c_{\text{H}^+}$$

or

$$c_{\text{FeOH}^+} = K_1 \frac{c_{\text{Fe}^{2+}}}{c_{\text{H}^+}} \tag{9.118}$$

From Eqs. (9.118) and (9.117), one obtains

$$\frac{\theta_{\text{FeOH}}}{1 - \theta_{\text{FeOH}}} = K_1 K_2 c_{\text{Fe}^{2+}} (c_{\text{H}^+})^{-1} e^{-F \Delta \phi / RT}$$

$$= k' c_{\text{Fe}^{2+}} (c_{\text{H}^+})^{-1} e^{-F \Delta \phi / RT} \tag{9.119}$$

At this stage comes an important assumption in the equational development of $i - \Delta\phi$ equations corresponding to a particular model. One must know θ_{FeOH}, the surface concentration of an adsorbed intermediate. If θ_{FeOH} is assumed to be small in comparison with unity, i.e., $\theta_{\text{FeOH}} \ll 1$, then $(1 - \theta_{\text{FeOH}}) \approx 1$ and hence Eq. (9.119) reduces to

$$\theta_{\text{FeOH}} \approx k' c_{\text{Fe}^{2+}} (c_{\text{H}^+})^{-1} e^{-F \Delta \phi / RT} \tag{9.120}$$

Introducing this expression for θ_{FeOH} into the expression (9.115) for the current density, one gets

$$i = 2F[k_{-3} e^{(1-\beta)F \Delta\phi/RT} - k_3 k' c_{\text{Fe}^{2+}} e^{-F \Delta\phi/RT} e^{-\beta F \Delta\phi/RT}] \tag{9.121}$$

This is the steady-state current-density–potential relation which is predicted on the basis of step 3 as the rds [in the sequence (9.112) to (9.114)]. The experimental results, however, are generally plotted as Tafel plots, i.e., $\Delta\phi$ versus $\log i$ plots. Hence, it is desirable to write Eq. (9.121) in a form suitable for comparison with this kind of experimental plot.

Suppose $\Delta\phi$ is sufficiently *positive* with respect to the equilibrium potential, i.e., $\Delta\phi > \Delta\phi_e$, or $\eta = \Delta\phi - \Delta\phi_e > 0$. Then one can effectively neglect the rate of the electronation reaction. This is tantamount to dropping off the second term in Eq. (9.121). One gets (with $\overleftarrow{k}$ written instead of k_{-3}) the de-electronation-current density

$$\overleftarrow{i} = 2F\overleftarrow{k} e^{(1-\beta)F\phi/RT} \tag{9.122}$$

or

$$\Delta\phi = \text{Constant} + \frac{2.3RT}{(1-\beta)F} \log \overleftarrow{i} \tag{9.123}$$

1086 CHAPTER 9

If β is taken to be equal to one-half (see Section 8.3.4), then the model predicts that the $\Delta\phi$ versus log i plot for the de-electronation should have a slope of $+2(2.3RT/F)$ or $+0.12$ V at 25°C. Now the experimental slope turns out to be $+0.04$ V [Fig. 9.37 and Eq. (9.110)]. The discrepancy between this value and that predicted if step 3 were rate-determining is marked and unambiguous. Consequently, one has to try again with another possible reaction scheme.

Let it now be assumed that step 2 is the rds, with the preceding and succeeding steps virtually in equilibrium. Hence, the assumed electrodic-reaction mechanism can be written thus

1. $$Fe^{2+} + H_2O \underset{v_{-1}}{\overset{v_1}{\rightleftharpoons}} FeOH^+ + H^+ \tag{9.112a}$$

2. $$FeOH^+ + e \overset{v_2}{\rightarrow} FeOH \tag{9.124}$$

3. $$FeOH + H^+ + e \underset{v_{-3}}{\overset{v_3}{\rightleftharpoons}} Fe + H_2O \tag{9.125}$$

The Butler–Volmer equation for this rds is

$$i = 2F(v_{-2} - v_2)$$
$$= 2F[k_{-2}\theta_{FeOH}e^{(1-\beta)F\Delta\phi/RT} - k_2 c_{FeOH^+}(1 - \theta_{FeOH})e^{-\beta F\Delta\phi/RT}] \tag{9.126}$$

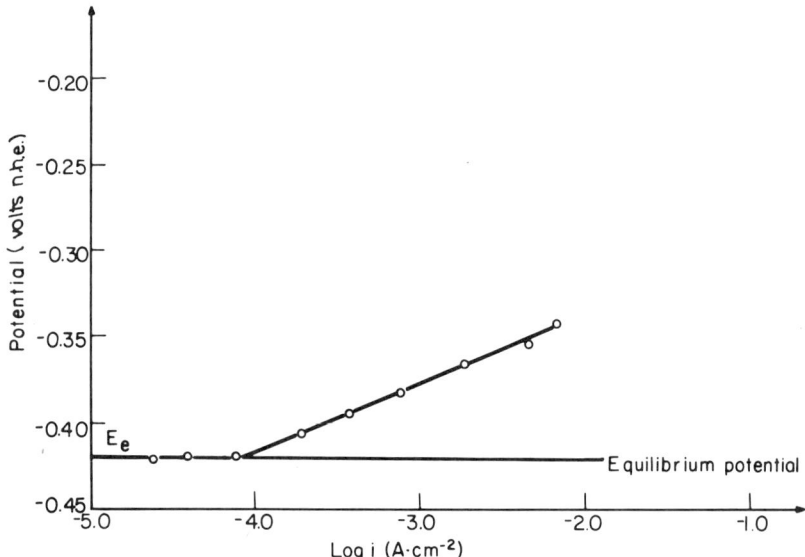

Fig. 9.37. The Tafel slope for iron de-electronation reaction is found to be 0.04 V decade^{-1} of current density, i.e., $\bar{\alpha} = \tfrac{3}{2}$.

The quantity θ_{FeOH} is then evaluated from the equilibrium law for step 3, i.e.,

$$v_{-3} = v_3$$

or

$$k_{-3}(1 - \theta_{\text{FeOH}})e^{(1-\beta)F\Delta\phi/RT} = k_3 \theta_{\text{FeOH}} c_{\text{H}^+} e^{-\beta F\Delta\phi/RT} \quad (9.127)$$

One obtains

$$\frac{\theta_{\text{FeOH}}}{1 - \theta_{\text{FeOH}}} = \frac{1}{K_3}(c_{\text{H}^+})^{-1} e^{F\Delta\phi/RT} \quad (9.128)$$

and, with the same assumption used previously, namely, $1 - \theta_{\text{FeOH}} \approx 1$, one has

$$\theta_{\text{FeOH}} = \frac{1}{K_3}(c_{\text{H}^+})^{-1} e^{F\Delta\phi/RT} \quad (9.129)$$

Introducing the equilibrium constant K_w for the formation of water from its constituents, i.e.,

$$c_{\text{H}^+} c_{\text{OH}^-} = K_w \quad (9.130)$$

one has

$$\theta_{\text{FeOH}} = \frac{1}{K_3 K_w} c_{\text{OH}^-} e^{F\Delta\phi/RT} \quad (9.131)$$

There is another concentration term in Eq. (9.126) which needs substitution, i.e., c_{FeOH^+}. From the equilibrium expression for step 1 [cf. Eq. (9.118)], one has

$$c_{\text{FeOH}^+} = \frac{K_1}{K_w} c_{\text{Fe}^{2+}} c_{\text{OH}^-} \quad (9.132)$$

Expressions (9.131) and (9.132) for θ_{FeOH} and c_{FeOH^+} can be inserted into (9.126) to give a predicted i versus $\Delta\phi$ law of the form

$$i = 2F\left[k_{-2} \frac{1}{K_3 K_W} c_{\text{OH}^-} e^{F\Delta\phi/RT} e^{(1-\beta)F\Delta\phi/RT} - k_2 \frac{K_1}{K_W} c_{\text{Fe}^{2+}} c_{\text{OH}^-} e^{-\beta F\Delta\phi/RT} \right] \quad (9.133)$$

Placing $\overleftarrow{k} = k_{-2}/K_3 K_W$ and $\overrightarrow{k} = (k_2 K_1/K_W)$

$$i = 2F[\overleftarrow{k} c_{\text{OH}^-} e^{(2-\beta)F\Delta\phi/RT} - \overrightarrow{k} c_{\text{OH}^-} c_{\text{Fe}^{2+}} e^{-\beta F\Delta\phi/RT}] \quad (9.134)$$

This second model for the reaction mechanism predicts an exponential i versus η law of the form

$$\overleftarrow{i} = 2F\overleftarrow{k} c_{\text{OH}^-} e^{(2-\beta)F\Delta\phi/RT} \quad (9.135)$$

at potentials sufficiently positive (about 50 mV positive) with respect to the equilibrium potential so that the rate of the electronation reaction becomes negligible compared with that of the de-electronation reaction. At negative overpotentials, where the rate of the de-electronation reaction is negligible, Eq. (9.134) can be written as

$$\vec{i} = 2F\vec{k}c_{OH^-}c_{Fe^{2+}}e^{-\beta F \Delta\phi/RT} \tag{9.136}$$

The predicted $\Delta\phi$ versus $\log i$ slopes are (assuming $\beta = \frac{1}{2}$, see Section 8.3.4)

$$\frac{d\Delta\phi}{d\log i} = \frac{2.3RT}{F}\frac{1}{2-\beta} = \frac{2}{3}\frac{2.3RT}{F} \tag{9.137}$$

$$= 0.04 \text{ V}$$

at 25°C for the de-electronation reaction and

$$\frac{d\Delta\phi}{d\log i} = -\frac{2.3RT}{F}\frac{1}{\beta} = -2\frac{2.3RT}{F} \tag{9.138}$$

$$= 0.12 \text{ V}$$

at 25°C for the electronation reaction, both of which are in agreement with the experimental evidence presented earlier. From the assumed mechanism, one can also predict reaction orders. By differentiating Eq. (9.135) with respect to OH^- ions, maintaining $\Delta\phi$ as a constant, one has

$$p_{OH^-} = \left(\frac{d\log i}{d\log c_{OH^-}}\right)_{\Delta\phi} = 1$$

These three examples of consistency between the predictions of a mechanism and the experimental findings, are examples of evidence on the way to a mechanism determination. They are supplemented by the fact that iron is a well-known subject of numerous studies in *corrosion*, which is in fact a form of *electrochemical dissolution* (*cf*. Section 11.2). The theoretical expressions for the dependence of the corrosion properties, e.g., the rate of corrosion, and the dependence of the potential of the corroding metal upon the pH, all involve the mechanism of the dissolution (and deposition) reaction. The hypothesis [path (9.112) through (9.114), with rds (9.113)] made above receives, then, further support by being able to predict the two aspects of the corrosion behavior to which reference has been made (Table 9.9, p.1092).

9.4.3. The Transfer Coefficient α and Reaction Mechanisms

The transfer coefficients $\overleftarrow{\alpha}$ and $\overrightarrow{\alpha}$ for the postulated rds in the iron deposition–dissolution reaction may be obtained directly by writing the Butler–Volmer equation (9.24) developed in a general form for a multi-step reaction. For this purpose, since (*cf.* Section 9.1.7)

$$\overleftarrow{\alpha} = \left(\frac{\overleftarrow{\gamma}}{\nu} + r - r\beta\right)$$

and (9.25)

$$\overrightarrow{\alpha} = \left(\frac{\overrightarrow{\gamma}}{\nu} + r\beta\right)$$

one must only identify the values of $\overleftarrow{\gamma}$, $\overrightarrow{\gamma}$, r, and ν. They are, assuming $\beta = \frac{1}{2}$,

for step 3 (9.114) as the rds,

$$\overleftarrow{\gamma} = 0, \quad \overrightarrow{\gamma} = 1, \quad \nu = 1, \quad r = 1, \quad \overleftarrow{\alpha}_3 = \tfrac{1}{2}, \quad \overrightarrow{\alpha}_3 = \tfrac{3}{2}$$

for step 2 (9.113) as the rds,

$$\overleftarrow{\gamma} = 1, \quad \overrightarrow{\gamma} = 0, \quad \nu = 1, \quad r = 1, \quad \overleftarrow{\alpha}_2 = \tfrac{3}{2}, \quad \overrightarrow{\alpha}_2 = \tfrac{1}{2}$$

for step 1 (9.112) as the rds,

$$\overleftarrow{\gamma} = 2, \quad \overrightarrow{\gamma} = 0, \quad \nu = 1, \quad r = 0, \quad \overleftarrow{\alpha}_1 = 2, \quad \overrightarrow{\alpha}_1 = 0$$

Thus, the Butler–Volmer equation (9.24) provides a shortcut to the deduction of mechanisms since it leads immediately to the values of α for the postulated mechanism. The values of α thus obtained are then inserted into the expression

$$\frac{d\Delta\phi}{d\log i} = \frac{2.3RT}{\alpha F}$$

In the above case, this procedure yields the following Tafel slopes

$$0.12 \text{ V}, \quad 0.04 \text{ V}, \quad \text{and} \quad 0.03 \text{ V}$$

for anodic rds 3, 2, and 1, respectively, and

$$0.04 \text{ V}, \quad 0.12 \text{ V}, \quad \text{and} \quad \infty$$

(i.e., i independent of $\Delta\phi$) for cathodic rds 3, 2, and 1, respectively.

These results are obviously the same as those obtained previously (*cf.*

Section 9.4.2) since it was the same procedure which led to the development of the generalized Butler–Volmer equation (9.24).

There is no shortcut procedure, however, which can supply the reaction order of species *not appearing in the overall reaction* (*cf.* Section 9.1.8), in this case H^+ (or OH^-). For this purpose, the postulated reaction path has to be considered in detail, as was done for the example of the iron deposition–dissolution reaction. The latter reaction is a good example of the importance of reaction order as a diagnostic criterion in mechanistic determinations. Often, the transfer coefficients for various postulated mechanisms are such that either they are the same, or the difference between their values falls within the experimental error in the determination of transfer coefficients. If, in such cases, the electrochemical reaction orders differ considerably, then a discrimination between the various mechanisms can be made.

The Butler–Volmer equation for the multistep reaction (9.24) was developed without taking into account significant coverages with intermediates; in other words, it was implicitly assumed that θ_i was very small, as was done for the iron deposition dissolution reaction (Section 9.4.2). Thus, application of Eq. (9.24) is limited to cases when $\theta \ll 1$, and modifications must be introduced if coverage with intermediates cannot be neglected.

The consistencies of Tafel slopes and reaction order with respect to OH^- ions encourage one to adopt the model involving step 2 as rds [*cf.* Eqs. (9.112) to (9.114)]. They make the model seem likely. But they do not have much strength until other models have been tested in a similar way. Table 9.8 shows five possible mechanisms for the iron deposition and dissolution reaction; mechanism *E* is the mechanism that was discussed earlier in Section 9.4.2.

One must obviously guard against the possibility that there may be a number of mechanistic possibilities which give consistency equal to that of the reaction sequence proposed here. Such a comparison of theory and experiment is shown in Table 9.9. In fact, the other mechanistic possibilities shown do not give such a degree of consistency as the one treated between (9.126) and (9.138). In the absence of some as yet unsuggested mechanism which does give an equal consistency with that of the present hypothesis, the latter is the one acceptable as the "probable mechanism" (of course, with the obvious reservations of limitation to pH, temperature, and anions with which the experiments were made).

9.4.4. Summarizing Remarks Concerning Mechanistic Studies

Here, then, is a very simple example of how one finds out, for a particular electrodic reaction, which mechanism hypothesis is most consistent

TABLE 9.8

Five Possible Mechanisms for the Deposition and Dissolution of Iron, from Which the Diagnostic Criteria Shown in Table 9.9 Can Be Calculated and Compared with the Experimental Values

Mechanism A	Mechanism B
$Fe + OH^- + FeOH \rightleftharpoons (FeOH)_2 + e$	$Fe + H_2O \rightleftharpoons FeOH + H^+ + e$
$(FeOH)_2 \xrightarrow{rds} 2FeOH$	$FeOH \rightleftharpoons FeOH^+ + e$
$FeOH \rightleftharpoons FeOH^+ + e$	$FeOH^+ + Fe \xrightarrow{rds} Fe_2OH^+$
$FeOH^+ \rightleftharpoons Fe^{2+} + OH^-$	$Fe_2(OH)^+ \rightleftharpoons Fe^{2+} + FeOH + e$
	$FeOH + H^+ \rightleftharpoons Fe^{2+} + H_2O + e$

Mechanism C	Mechanism D
$Fe + OH^- \xrightarrow{rds} Fe(OH)^+ + 2e$	$Fe + OH^- \rightleftharpoons FeOH + e$
$Fe(OH)^+ \rightleftharpoons Fe^{2+} + OH^-$	$FeOH + OH^- \xrightarrow{rds} FeO + H_2O + e$
	$FeO + OH^- \rightleftharpoons HFeO_2^-$
	$HFeO_2^- + H_2O \rightleftharpoons Fe(OH)_2 + OH^-$
	$Fe(OH)_2 \rightleftharpoons Fe^{2+} + 2OH^-$

Mechanism E [Eqs. (9.112a), (9.124), and (9.125)]

$Fe + H_2O \rightleftharpoons FeOH + H^+ + e$
$FeOH \xrightarrow{rds} FeOH^+ + e$
$FeOH^+ + H^+ \rightleftharpoons Fe^{2+} + H_2O$

with the facts about the kinetics of the reaction concerned. Such determinations of mechanism are no routine affairs; speculative invention of possible reaction pathways with an experienced eye on the reasonableness (or estimated standard free energy) of the radicals assumed, working out of detailed predictions, careful collection of the right kind of experimental quantities, i.e., those permitting the testing of reaction models, and then the comparison between the numerical predictions of various models and the experimentally obtained parameters—all these play a role in unraveling the overall reaction, the path, and particularly the rds of an electrodic reaction. The main criterion of how probable a mechanism is relates to the number of experimental parameters it is able to rationalize and to relatively how much poorer is the predictive ability of the other models. It is the pattern of theoretical predictions which must match the pattern of experimental behavior.

TABLE 9.9
Comparison of Experimental Behavior for Iron Deposition and Dissolution with Prediction from Various Mechanisms, $\beta = 0.5$

Quantity	Mechanism					Experimental results, coefficients $\times$ 2.303
	A	B	C	D	E	
$\dfrac{\partial V_{Fe}}{\partial \ln i_a}$	$\dfrac{RT}{2F}$	$\dfrac{RT}{2F}$	$\dfrac{RT}{F}$	$\dfrac{2}{3}\dfrac{RT}{F}$	$\dfrac{2}{3}\dfrac{RT}{F}$	0.042 ± 0.008
$\dfrac{\partial V_{Fe}}{\partial \ln i_c}$	$-\dfrac{RT}{2F}$	$-\dfrac{RT}{2F}$	$-\dfrac{RT}{F}$	$-\dfrac{2RT}{F}$	$-\dfrac{2RT}{F}$	-0.116 ± 0.006
$\left(\dfrac{\partial \ln i_c}{\partial \ln a_{Fe^{2+}}}\right)_{a_{OH^-}}$	2	2	1	1	1	0.8
$\dfrac{\partial \ln i_0}{\partial \ln a_{OH^-}}\bigg|_{a_{Fe^{2+}}}$	2	1	1	2	1	0.9 ± 0.05
$\dfrac{\partial V_{Fe}}{\partial \ln a_{OH^-}}\bigg|_{a_{Fe^{2+}},i}$	$\dfrac{RT}{F}$	$-\dfrac{RT}{2F}$	$\dfrac{RT}{F}$	$-\dfrac{4}{3}\dfrac{RT}{F}$	$-\dfrac{2}{3}\dfrac{RT}{F}$	
$\dfrac{\partial \ln i_0}{\partial \ln a_{Fe^{2+}}}\bigg|_{a_{OH^-}}$	1	1	$\dfrac{1}{2}$	$\dfrac{3}{4}$	$\dfrac{3}{4}$	0.8 ± 0.1
$\dfrac{\partial V_{corr}}{\partial \ln a_{OH^-}}\bigg|_{a_{Fe^{2+}}}$	$-\dfrac{6}{5}\dfrac{RT}{F}$	$-\dfrac{4}{5}\dfrac{RT}{F}$	$-\dfrac{4}{3}\dfrac{RT}{F}$	$-\dfrac{3}{2}\dfrac{RT}{F}$	$-\dfrac{RT}{F}$	-0.060 ± 0.003
$\left(\dfrac{\partial \ln i_{corr}}{\partial \ln a_{OH^-}}\right)_{a_{Fe^{2+}}}$	$-\dfrac{2}{5}$	$-\dfrac{3}{5}$	$-\dfrac{1}{3}$	$-\dfrac{1}{4}$	$-\dfrac{1}{2}$	-0.5 ± 0.01

Further Reading

1. J. Horiuti and M. Ikusima, *Proc. Imperial Acad. Tokyo*, **15**: 39 (1939).
2. R. Parsons, *Trans. Faraday Soc.*, **47**: 1332 (1951).
3. E. C. Potter, *J. Electrochem. Soc.*, **99**: 169 (1952).
4. H. Mauser, *Z. Elektrochem.*, **62**: 419 (1958).
5. A. R. Despic, *Electrochim. Acta*, **4**: 325 (1961).
6. D. M. Drazic, *Electrochim. Acta*, **7**: 293 (1962).
7. P. Delahay, "Kinetics of Electrode Processes Involving more than One Step," in: *Double Layer and Electrode Kinetics*, Chapter 8, Interscience Publishers, Inc., New York, 1965.
8. B. E. Conway, "Applications to Selected Problems," in: *Electrode Processes*, Chapter 8, The Ronald Press Company, New York, 1965.
9. J. J. Podesta and A. J. Arvia, *Electrochim. Acta*, **10**: 159 and 171 (1965).
10. K. J. Vetter, "Experimental Results of Electrochemical Kinetics," in: *Electrochemical Kinetics*, Chapter 4, Academic Press, Inc., New York, 1967.
11. J. O'M. Bockris and S. Srinivasan, "Thermodynamic Aspects of Electrochemical Energy Conversion," in: *Fuel Cells, Their Electrochemistry*, Chapter 3, McGraw Hill Book Company, New York, 1969.

9.5. MORE ON MECHANISM DETERMINATION

9.5.1. Why Review Mechanism Determination?

The business of determining the mechanism of electrodic reactions has been treated above with a light touch in order to provide a gentle introduction to the concepts. It is desirable, however, to have a second look at this important aspect of electrochemistry. Why is mechanism determination important? The reason arises from the concept of an rds in a consecutive reaction. If the rds in a reaction is known, then one's ability to think about the reaction—to consider what factors will change its velocity in a particular direction—is much improved. One knows what one has to try to tamper with in order to make the reaction go faster or slower. For example, it may be that the substrate can be changed to speed up the reaction, and what characteristic of the substrate (e.g., MO bond strength) has to be changed, and in which way (e.g., an increase in MO bond strength), can be known only if the rds is known.

There is another reason why, even in the elementary discussions of this book, attention must be focused on the matter of electrodic mechanisms and how to determine them. The literature in the field shows that there is not yet an entirely valid consensus on the question of what mechanism determination constitutes—that, then, is what must be discussed at this stage.

9.5.2. What Is Mechanism Determination?

There are three sorts of answers implicitly given at the present time to the question: What is the mechanism of an electrodic reaction?

The first type of answer consists of simply referring to the *overall reaction* as the mechanism. For example, in the electrodic de-electronation (or oxidation) of ethylene, the overall reaction would be

$$C_2H_4 + 4H_2O \rightarrow 12H^+ + 2CO_2 + 12e$$

If one terms the overall reaction as the mechanism, one is content with knowledge which is inadequate for the control of the rate of the reaction. The knowledge of the overall reaction is a matter of thermodynamics, which deals with the initial and final state of the reaction but not with the intermediate steps. It is these latter steps which must be known if one is to attain a rational control on the velocity of the electrode reaction, e.g., by change of substrate. It is this possibility of rationally altering the rate that underlies much of the interest in the mechanism of an electrodic reaction.

Hence, in mechanism determination, one must look beyond the mere determination of the overall reaction, knowledge of which must, nevertheless, always be attained as a first step.

The second type of possible approach is to equate mechanism determination with the determination of the *consecutive steps* which make up an overall reaction. One seeks the steps of the reaction pathway along which reactants travel and end up transformed into products. If each one of these steps is known, one knows much more than one knew from the thermodynamic information contained in the overall reaction, but one is still ignorant about why the reaction has a particular rate for a given driving force (i.e., a given overpotential η).

In the steady state, all the steps are going at the same *net* rate (the net velocity manifested to the external observer), but all the individual steps are not responsible to the same extent for the overall rate. The equality of the net rates of all the consecutive steps conceals the fact that, usually, one single step is the dominating cause, under given conditions, of the electron queue and can be held responsible for the overall rate. This is the rds (rate-determining step, *cf.* Section 9.1.5). Only when one knows the rds does one understand what controls the overall rate.

The most important aspect of an electrodic reaction, therefore, is the rds which controls the rate at which charge is pumped across an interface. The most essential stage in the mechanism determination is the stage of singling out the rds from a series of consecutive steps which constitute a multistep overall reaction. If the rds is known, one can say that an essential aspect of the mechanism of the reaction is known. If one were restricted either to knowing the sequence of steps making up a reaction path or to knowing the crucial step which determines the overall rate, it would be more useful to acquire knowledge of the rds. One has then a grasp on the handle which controls the overall reaction rate.

9.5.3. Stages in the Elucidation of a Reaction Mechanism

A number of scientists have contributed to this important matter of mechanism determination, with special reference to the elucidation of the rds. In particular, mention must be made of the contributions of Frumkin, Parsons, Vetter, Gerischer, Conway, Bagotsky, and Wroblowa. These workers have developed clear diagnostic criteria, and a knowledge of these criteria taken together contributes much to the elucidation of the mechanism of charge-transfer reactions at interfaces.

The above discussion suggests a stage-by-stage procedure in the determination of the mechanism of an electrodic reaction. Some of these stages are described below.

1. The elucidation of the *overall reaction* is the first objective. This is usually easy. It comprises a coulombic analysis of the reaction, i.e., finding out the number of coulombs necessary to accomplish the reaction of one formula weight of the substance concerned as well as a chemical (or other) analysis of products.

2. The second stage is the identification of the *entities in solution*. This information is needed for the following reason: The reaction path generally consists of an initial transport process in which a number of particles from the bulk of the solution diffuse to the electrode, undergo charge transfer there in a single- or multistep reaction, and then produce particles on the electrode or move back into solution. One cannot begin to think about the mechanism of the reaction unless one knows what particles are there to start with. The point is that the ion may be complexed. Thus, if aqueous solutions of silver salts containing cyanide ions are considered, the particles which take part in the electronation reaction leading to the deposition of silver are not Ag^+. The particles which participate in the deposition reaction are $Ag(CN)_2^-$, i.e., the silver ions are complexed. One must know to what extent and how the particles are complexed.

3. The third stage is the determination of the extent to which the electrode surface is covered with adsorbed species which may influence the reaction rate. This is not as simple and straightforward as determining the

overall reaction involving the particles in the solution. In fact, until a few years ago, it was difficult to obtain data on this problem, except for data obtained by methods connected with the measurement of surface tension on mercury (see Section 7.3.8). However, there are several ways in which the surface coverage θ can be obtained at the present time (e.g., by the study of transients; see Section 9.2). One seeks to know how the coverage of the various entities changes with potential and the concentration of some stable entity in solution from which they are derived.

4. Having confirmed that there is a given overall reaction, known initial species in the solution, and an electrode surface which is covered with adsorbed species to an extent which varies with potential and concentration in a known way, one embarks upon the task of *mechanism determination* with the prospect of some degree of success. The objective, as stated earlier, is the elucidation of the rds and, as far as it is feasible, the other steps which make up the sequence in the reaction.

9.5.4. The Elucidation of the Overall Reaction, the Entities in Solution, and the Surface Coverage

Often, the first stage, namely, the determination of the overall reaction, need not be carried out at all. The overall reaction may be obvious. This is so when all the current goes toward one electrode reaction and there is an easily determined product. Thus, by knowing how many moles of the product are formed per Faraday of electricity (*cf.* Faraday's laws, Section 4.3.5), one can easily determine how many electrons are transferred in the overall reaction and what the overall reaction is. A good example here would be the deelectronation, or oxidation, of water to form oxygen. The amount of oxygen formed can easily be analyzed by standard methods, and the overall reaction is deduced to be

$$2H_2O \rightarrow O_2 + 4H^+ + 4e$$

Similarly, by following the CO_2 formed at a platinum electrode, it can be shown that the deelectronation of ethylene is given by the overall reaction

$$C_2H_4 + 4H_2O \rightarrow 2CO_2 + 12H^+ + 12e$$

Sometimes, however, two electrodic reactions occur at the same electrode. They share the total current and form different products in different overall reactions. Thus, in the oxidation of ethylene on gold, two products (acetaldehyde in addition to CO_2) are produced at the same time, and these products have to be followed and analyzed in the solution as a function of potential, concentration, etc. A great number of methods can be used to do this. For example, chromatographic analysis would be used to distinguish between acetaldehyde and CO_2, which are products of the de-electronation of ethylene on gold. One might also pass the gases evolved into a mass spectrograph to determine the various species which are still alive at a few centimeters away from the electrode.

The identification of the species in the solution sometimes gives positive surprises. Many solutes remain as they are when they are put into the solvent. For example, except in very concentrated solutions, ethylene remains as ethylene when dissolved in acids. This need not always be the case. The entities existing in solution may be quite different from what would be expected on the basis of what one has added to the solvent. Thus, in molten salts, solutes often form complex ions with the solvent. For example, $CdCl_3^-$ is formed in a solution of cadmium chloride and potassium chloride (see Section 6.6.4); this complex formation shows up in the Raman spectrum of the resulting system. Spectroscopic methods are the most important ones for determining what complex ionic species are present in the solution. This part of the analysis of the electrode reaction will become more important in the future because the electrochemistry of nonaqueous solutions will probably grow in importance in the next decade, and, in these solutions particularly, identification of the species present will require spectroscopy.

The determination of the surface coverage by adsorbed species can be carried out by several methods. One of the less ambiguous ones involves the use of radiotracers (see Sections 7.2.6e and 9.2.5). The electrode material (in the form of a very thin foil) is placed over the end of a gas proportional counter, and this assembly is then immersed in a solution containing the substances concerned. These must, of course, contain a tagged or radioactive species. It is then possible by the reading on the counter tube to obtain the amount which has adsorbed on the electrode.

There are other methods of determining coverages on electrodes (Section 9.2.5). Each has its own difficulty. For example, the radioactive method has the difficulty that it measures only the concentration of tagged atoms, regardless of whether the tagged atom is residing in an adsorbed substance or residing in some species resulting from dissociation of this adsorbed substance. Hence, at least two different types of measurements should be used when adsorption is being determined, each type of determination having different assumptions and approximations.

Possible alternative methods that can be used include various types of transient techniques (*cf.* Section 9.2.5), which permit one to calculate the charge required for the oxidation of an adsorbed species, e.g., the potentiostatic transient method. Calculation of the surface concentration requires the knowledge of the number of electrons required to oxidize the gram-molecular weight of the material predominantly adsorbed to CO_2; e.g., in a determination of the coverage with benzene, one would use in the calculations 30 electrons, according to the overall reaction

$$C_6H_6 + 12H_2O \rightarrow 6CO_2 + 30H + 30e$$

However, under certain conditions, i.e., adsorption at more anodic potentials, benzene may not be the main entity on the surface; if benzene undergoes some fast reactions to some intermediate radical whose surface concentration is high compared with that of benzene and the oxidation of which requires less than 30 electrons per molecular entity to reach CO_2, then the surface concentration given as

$$\Gamma = \frac{Q}{30F} \text{ [moles cm}^{-2}] \tag{9.139}$$

would be lower than that really existing (Q is the number of coulombs per square centimeter required to oxidize the adsorbed species). The value of Q can be obtained by calculating the shaded area bounded by the two potential step transients shown in Fig. 9.15, this being the charge required to oxidize the adsorbed substance.

Comparison of results obtained under similar conditions by radiotracer and electrochemical-transient methods may thus give information which neither technique can furnish separately. Agreement between the results of both (in cases similar to the benzene adsorption) would indicate that the adsorbate is predominantly the undissociated original species (in the above case, benzene). If, however, various electrochemical methods consistently yield results lower than radiotracer techniques, it is possible to attribute the difference to the fact that the species in question is predominantly adsorbed in the form of radicals. Assuming that one type of radical predominates, one may attempt a more quantitative description, i.e., to calculate its molecular formula. Great caution is indicated here, since other assumptions have also to be made. Taking benzene again as an example, one would have to assume that no breakup of C—C bonds, but only dehydrogenation, occurred to yield C_6H_x radicals. Then the radiotracer method yields the amount of moles, Γ, of C_6H_x per square centimeter and the voltage sweep, the amount of charge Q required for their complete oxidation.

Then the amount of electrons used in oxidation of one radical is

$$n = \frac{Q}{F\Gamma} \tag{9.139}$$

and, obviously, $(6 - x) = 30 - n$.

Once the coverage of material which is reacting on the electrode, the substances in the solution which are giving rise to the reaction, and the overall reaction are known, one can proceed to the various techniques used to obtain indication of an rds and to obtain an indication of path of the reaction.

9.5.5. Some Techniques for Mechanism Determination

There are three main methods for the study of the mechanism of electrodic reactions.

9.5.5a. The Determination of Reaction Order. This determination is nearly always carried out in a mechanism study. It consists (*cf.* Section 9.1.8) in maintaining the potential at the electrode constant and the concentration of all the reactants and products except one (say, A_i) constant. Then, according to Eq. (9.29), the variation, $(\partial \log i/\partial \log c_{A_i})_{A_{j \neq i}, \Delta\phi}$, of the current density with respect to the concentration of the component A_i gives the reaction order of the reaction with respect to that component (Fig. 9.6). This determination can be a quite lengthy one in practice, particularly for reactions in which there are several reactants and products. One must try to determine the reaction order for all the known reactants and products, i.e., a series of determinations must be made by varying only one concentration at a time. To do this, the electrodic reaction must be run, if possible,[†] in both the de-electronation (anodic) and electronation (cathodic) directions. One ends up with a series of reaction orders for each component of the overall reaction.

When these reaction orders have been determined, one looks at the overall reaction and writes down various seemingly possible consecutive

---

[†] One of the difficulties in mechanism determination is that one sometimes cannot in practice reverse a given reaction. Thus,

$$C_2H_4 + 4H_2O \rightarrow CO_2 + 12H^+ + 12e^-$$

However, CO_2, upon reduction, does not yield hydrocarbons but formaldehyde, or, if one tries to reduce it on electrocatalysts other than the noble metals, e.g., on mercury or cadmium, one may obtain some methanol.

reaction sequences in which the overall reaction might be brought about. But, what do these guesses say on the question of the reaction orders? Usually, they make different predictions, and the ones which do not check with the experimentally determined reaction orders can be rejected. In this fashion, the possibilities available for the sequence and rds are narrowed down.

9.5.5b. The Determination of the Transfer Coefficients. The experimental transfer coefficients $\vec{\alpha}$ and $\overleftarrow{\alpha}$ (Section 9.1.7) can often be determined simply. If the electrodic reaction is relatively slow, i.e., has a low exchange-current density, then, for a given applied current density, a steady-state potential can be obtained without diffusion (*cf.* Section 9.3.8) setting in and generating a concentration overpotential. Under these conditions, passage of a steady current density produces only an activation overpotential, which can be easily measured. A series of applied current densities are used, and a $\Delta\phi$ versus log i or η versus log i plot is drawn. The slope of such a plot (Fig. 9.38) at a constant concentration of reactants gives the

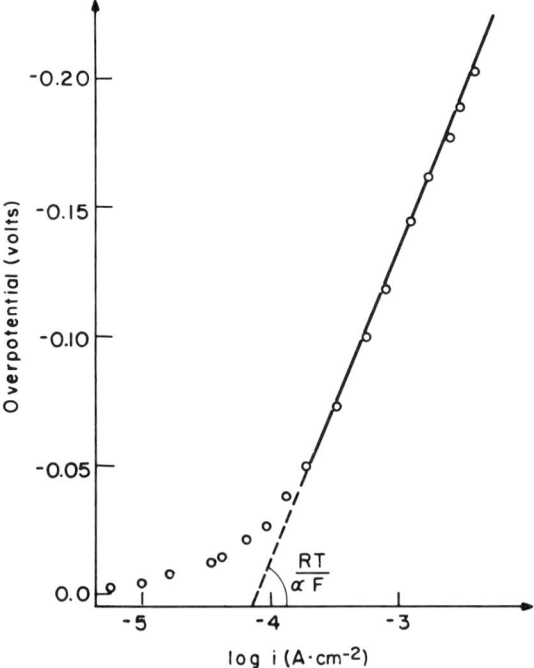

Fig. 9.38. The plot of η versus log i is usually a straight line, which can render transfer coefficients from the slope.

transfer coefficients through equations such as (9.110). These coefficients are invariably simple numbers, e.g., 2, $\frac{3}{2}$, or $\frac{1}{2}$. They are usually simply related to $\frac{1}{2}$.

For simple reactions, the transfer coefficient alone may occasionally give some information concerning the mechanism of the reaction.

Thus, as an example, consider the hydrogen-evolution reaction, the overall reaction of which is

$$2H^+ + 2e \rightarrow H_2$$

One model for the reaction mechanism is that the rds is the electronation, or discharge, of protons to form adsorbed hydrogen atoms and the following reaction (in which the adsorbed hydrogen atoms combine and desorb as hydrogen molecules) is in equilibrium. One can then write the reaction mechanism in the form

$$2(H^+ + e \rightarrow H_{ads})$$
$$2H_{ads} \rightleftharpoons H_2$$

The same two steps (discharge and combination) can be used in an alternative model. Here, the discharge of the protons is taken as at equilibrium; and the combination of the hydrogen atoms, as rate determining.

The predicted cathodic transfer coefficients for these two mechanisms (on assuming $\theta_H \rightarrow 0$) can be shown to be $\frac{1}{2}$ and 2, respectively,[†] on the assumption that $\beta = 0.5$ (see Table 9.10), and constitute, therefore, a particularly simple criterion for deciding between these two possible rds's.

9.5.5c. The Determination of the Stoichiometric Number.

It will be noticed that the two assumed mechanisms of the preceding section corresponded to different stoichiometric numbers 2 and 1, respectively (*cf.* Section 9.1.7). Thus the stoichiometric number ν may sometimes be an aid to the determination of the reaction mechanism. This quantity can be determined in several ways. One way, e.g., is to note that, by taking the sum of the transfer coefficients for the deelectronation, $\overleftarrow{\alpha}$ and electronation, $\overrightarrow{\alpha}$, reactions [*cf.* Eq. (9.25)], one gets (*cf.* Fig. 9.39).

$$\nu = \frac{n}{\overleftarrow{\alpha} + \overrightarrow{\alpha}} \tag{9.25a}$$

[†] For rate-determining proton discharge $\overrightarrow{\alpha} = (\overrightarrow{\gamma}/\nu) + r\beta = (0/2) + \frac{1}{2} = \frac{1}{2}$ and $\overrightarrow{\alpha} = (2/1) + 0 = 2$ for rate determining hydrogen-atom recombination [*cf.* Eq. (9.25), Section 9.1.7].

TABLE 9.10
Tabulation of the Cathodic-Transfer Coefficients for Four Possible Mechanisms for the Evolution of Hydrogen (cf. Section 10.3)

Path	Mechanism Steps	rds	$\vec{\nu}$	$\overleftarrow{\nu}$	r	ν	Cathodic-transfer coefficient $\overleftarrow{\alpha}$
1. Chemical recombination	1. $H^+ + M + e \rightleftarrows MH$	1	0	0	1	2	β
	2. $2MH \rightleftarrows H_2 + 2M$	2	2	0	0	1	2
2. Electrodic desorption	1. $H^+ + M + e \rightleftarrows MH$	1	0	1	1	1	β
	2. $H^+ + MH + e \rightleftarrows M + H_2$	2	1	0	1	1	$1 + \beta$

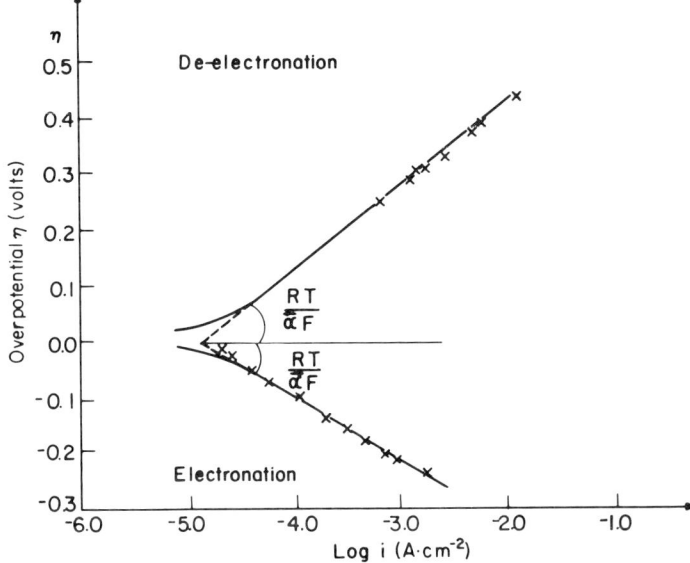

Fig. 9.39. Two transfer coefficients can be obtained from Tafel plots for de-electronation and electronation reactions.

In the case of the oxygen evolution reaction in acid solutions on platinum electrodes, the experimental values of $\overleftarrow{\alpha}$ and $\overrightarrow{\alpha}$ are $\tfrac{1}{2}$ and $\tfrac{1}{2}$, respectively, and, since $n = 4$, it follows from Eq. (9.25a) that the experimental stoichiometric number is 4. Consider one possible reaction path, called the *oxide path*, for the oxygen reaction. It runs thus

1. $\quad\quad\quad 4Pt + 4H_2O \rightarrow 4PtOH + 4H^+ + 4e \quad\quad\quad (9.140)$

2. $\quad\quad\quad 4PtOH \rightarrow 2PtO + 2PtH_2O \quad\quad\quad (9.141)$

3. $\quad\quad\quad 2PtO \rightarrow O_2 + 2Pt \quad\quad\quad (9.142)$

Depending upon whether step 1, 2, or 3 is the rds, the rds has to go four times, twice, or once, respectively, per act of the overall reaction. That is, the predicted stoichiometric numbers are 4, 2, and 1, and, hence, if the correct path is the oxide path, then only a rate-controlling step of water deelectronation, step 1, is consistent with the observed stoichiometric number.

Thus, determinations of the reaction orders of the transfer coefficients and of the stoichiometric number are some of the more direct ways in which one can try to approach the elucidation of the rds in an electrodic reaction.

In addition, if one knows the overall reaction, the coverage, and the nature of the particles in the initial state, it is often possible to build up a case in support of a particular electrodic reaction mechanism, particularly if its ability to predict the reaction orders, transfer coefficients, and stoichiometric numbers has been studied comparatively with those of alternative possibilities. Above all, one has to take several criteria together and let each contribute something to narrowing down the possibilities of path and rds toward one. It may be perhaps less difficult to do this for electrodic reactions than for heterogeneous chemical reactions. In the latter case, the surface is more difficult to control. One cannot obtain the degree of surface coverage easily; radicals on the surface are virtually impossible to identify; lastly (but most importantly) one does not have the extra criterion ($\vec{\alpha}$ and $\overleftarrow{\alpha}$) given by the current–potential reactions. There is no easy road to mechanism determination but even if the electrochemical road is torturous and has confusing sidetracks, there *is* such a road—often more than can be said for the same reaction carried out in a gas-phase situation.

9.5.5d. Auxiliary Methods. The above methods are by no means the only ones which can be used in the determination of the reaction mechanisms at an electrode. There are several more, the use of which is dependent on the circumstances. Some examples of auxiliary methods will be quoted.

i. *Steady-state mixed potentials as a function of certain concentration variables*. It must not be assumed that when the net current density across an interface is zero, the potential difference observed is *ipso facto* a thermodynamic equilibrium potential [*cf*. Eq. (8.61)]. The latter (Section 8.2.13) arises when the electronation and de-electronation directions of the single given electrode reaction occur at the same rate (i.e., net $i = 0$). But there is another situation at which net $i = 0$. It may be that, at a certain potential, the de-electronation current of the given reaction may be equaled by the electronation current of *another* reaction, the mechanism of which is known (Fig. 9.40). Under these conditions, what is observed is not an equilibrium potential, but a steady-state mixed potential set up by two reactions.

An example of a mixed potential has already been mentioned in the study of the mechanism of the dissolution of iron. In corrosion processes (Section 11.2.8), one reaction is that of anodic metal dissolution; the other, that of hydrogen deposition, or oxygen reduction. Study of, say, the pH dependence of the corrosion potential, $\Delta\phi_{\text{corr}}$ (the potential at which iron dissolution and hydrogen evolution currents are equal) gives equations which depend on the mechanism assumed for iron dissolution. Therefore, the abilities of the various hypotheses to yield the experimental value of,

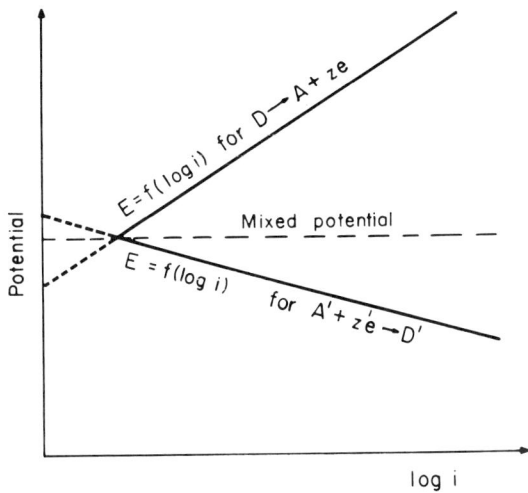

Fig. 9.40. The intersecting potential *versus* log i relations for two different electrode reactions define a steady-state potential—"mixed potential"—for that electrode.

say, $\partial \Delta \phi_{\text{corr}}/\partial \text{pH}$ contribute to one's estimate of the probability of a mechanism.

ii. *The nature of radicals on the surface.* Organic-reaction mechanisms involve certain radicals on the electrode surface. If the nature of the radical can be determined, a contribution to the determination of the mechanism has been made. Indeed some electrochemists (Breiter, Brummer) have tried to develop this approach as a kind of royal road to the elucidation of the consecutive steps (with a relative neglect of determination of the rate-determining step).

Such determinations are not easy. But suppose one subjects the surface undergoing this organic reaction to a cathodic pulse, introducing onto it a high concentration of adsorbed hydrogen. One observes (say, by the use of a mass spectrograph) that a stable hydrocarbon, perhaps H_3C-CH_3, is evolved. The inference is then that entities containing C—C are on the surface. This kind of knowledge, coupled with knowledge of the overall reaction and the type of data discussed earlier in this section, may help distinguish among alternate possibilities (Grubb).

iii. *The dependence of the reaction rate on substitution of isotopic atoms.* When atoms of different masses and the same atomic number are used in studies of reaction mechanisms, it transpires that the different isotopes undergo reaction at different rates. Apart from this difference of rates, the reactions are the same, having the same path and the same rds. Thus, hydrogen evolves at a higher rate than deuterium and tritium (Fig. 9.41). In fact, this is how deuterium (heavy hydrogen) was first obtained in appreciable quantity; a deuterium-enriched electrolyte was left behind after the gases were evolved at the electron-source electrode. Now, the point is that the difference of rate between the various isotopes can be calculated quite accurately, and the result of the calculation depends on what mechanism is assumed. Hence, if one makes up various hypotheses concerning the mechanism, one can calculate various ratios of the rates at which hydrogen and its isotope are evolved, each ratio corresponding to one of the assumed reaction mechanisms. In this way, one can distinguish between alternate mechanisms, both of which fit other experimental facts equally well. This method is particularly useful in determining the mechanism of hydrogen evolution (*cf.* Section 10.3.4g) because of the significant difference in zero-point energies between the isotopes, which may be shown to lead to a two- to tenfold difference in the evolution rate of hydrogen and tritium, according to alternative mechanism assumptions. This method has been used in the investigation of the hydrogen-electrode kinetics, particularly by Horiuti, by Conway, and by Srinivasan.

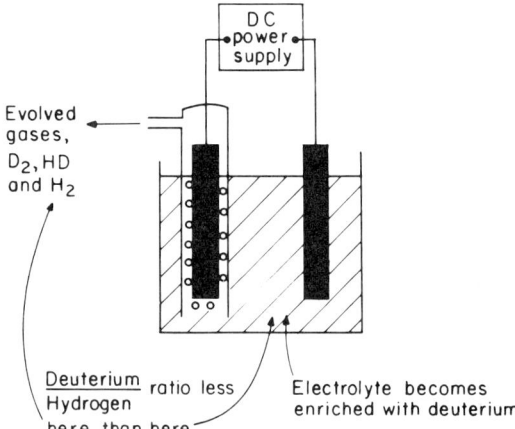

Fig. 9.41. When deuterated water is electrolyzed, hydrogen evolves at a higher rate than deuterium relative to their concentrations, and this results in an increased deuterium to hydrogen ratio in the remaining water.

The isotopic method does not apply only to the mechanism of hydrogen evolution. It can be used, e.g., in the study of hydrocarbon oxidation, where comparison of rate between reactions that are carried out in, e.g., H_2O and D_2O contributes evidence for distinguishing one mechanism from another, as does also a study of the effect that the position of the isotopic atoms in the radicals and reactants has upon the products yielded.

9.5.6. Mechanism Determination for Saturated Hydrocarbons

The saturated hydrocarbons form the most important class of possible fuels for use in electrochemical energy converters for the simple reason that they have the most electricity in them per unit price and unit weight of any substance known except hydrogen. In fact, the higher hydrocarbons are little more than hydrogen carriers, and it would be advantageous if one were able to burn methane, propane, etc., in an electrochemical energy converter, the other electrode in which would be an air electrode (Chapter 11).

Conversely, the saturated hydrocarbons are by no means among the most reactive substances, and many electrochemists at first thought that their electrochemical burning to give electricity directly would not be achievable at sufficiently low overpotentials to make possible practical fuel cells.

The principal need, of course, is that of electrocatalysis, and the principal point of interest is the catalyst's activity per unit of monetary value. This active area of electrochemical research depends to a large extent upon some knowledge of the sequence of reactions which occur during the oxidation of the hydrocarbons, and, in particular, the rds. Knowledge of this would be a kind of contour map acting as a basis for the search for good catalysts.

It is therefore particularly important to evaluate the rds in the electrochemical oxidation of hydrocarbons, and a brief summary will be given here of the present stage to which this investigation has come for hydrocarbons of low molecular weight.

The oxidation of propane, the example taken, takes place with an overall reaction which needs 20 electrons. The reaction is

$$C_3H_8 + 6H_2O \rightarrow 3CO_2 + 20H^+ + 20e$$

There are three diagnostic coefficients which are known for this reaction. The first is the transfer coefficient for the anodic reaction in the range of about 300 to 500 mV on the reversible hydrogen scale, that part of the relation which is linear and simple.

The Tafel relation gives

$$\left(\frac{d\eta}{d\ln i}\right)_{H_2O, C_3H_8} = \frac{RT}{F}$$

Thus, $\alpha_a = 1$.

The desirable and complementary cathodic transfer coefficient cannot be determined because the reversal of the above reaction on the same electrode is essentially impossible. It is possible to reduce carbon dioxide to formaldehyde or even to methanol, but not to propane. This is a trouble. Correspondingly, it is found experimentally that the propane-oxidation reaction cannot be run in the vicinity of the reversible potential because, when one tries to do this, the other reactions (possibly the reduction of carbon dioxide to formaldehyde) take over and spoil the measurement of reversibility.

The second diagnostic criterion that is known is the reaction order with water. It is given by

$$\left(\frac{d\log i}{d\log c_{H_2O}}\right)_{\Delta\phi} = 0$$

and the third criterion is the reaction order with propane

$$\left(\frac{d\log i}{d\log c_{C_3H_8}}\right)_{\Delta\phi} = 1$$

These three pieces of information are by no means a very full list of diagnostic criteria. But there are some other points which help one to get to a mechanism. Thus, when we place the propane and the electrode firstly in solution, it is observed that a great deal of current is evolved and then the current settles down to a low value. It is difficult to avoid the interpretation that, when the propane first strikes the electrode, it does not adsorb as a molecule but splits up into something which gives rise to charge transfer. One of the possibilities here would be

$$C_3H_8 \rightarrow C_3H_{7\,ads} + H^+ + e$$

and, when one compares this possibility with some other charge-transfer modes of adsorption, it turns out that the one given above is the most likely; so this dissociation mode of adsorption is assumed for propane on platinized platinum. There is a way of treating the transient current that one gets upon contacting the solution with the electrode, which gives rise to a rough measurement of the concentration of the radicals upon the surface, and it may be concluded that it is less than about 25%. The coverage, two reaction orders, and the transfer coefficient are therefore the data which one has to use to try to deduce an rds for the propane oxidation.

Now let the various possibilities be considered. In Table 9.11 are nine selected mechanisms, all of which have some closeness to the criteria concerned, from a much larger list which may be devised for this reaction if the rudimentary criteria (e.g., Tafel slope) are neglected. Of those reactions chosen in Table 9.11, as many as three satisfy all the criteria evolved; these are reactions 3, 4, and 5.

How may we decide what is the most probable of these three fitting mechanisms? Strike out firstly mechanism 5. It has been argued elsewhere

TABLE 9.11

Nine Possible Rate-Determining Steps (with Preceding Quasi-Equilibrium Steps Where Present) for the Oxidation of Propane on Platinum in Phosphoric Acid, 80 to 150°C and 300 to 500 mV (Reversible Hydrogen Scale)

1. $C_3H_{8_{soln}} \rightarrow C_3H_{7_{ads}} + H^+ + e$

2. $C_3H_8 \rightleftharpoons C_3H_{7_{ads}} + H^+ + e$
   $C_3H_{7_{ads}} \rightarrow C_3H_{6_{ads}} + H^+ + e$

3. $C_3H_8 \rightleftharpoons C_3H_{7_{ads}} + H^+ + e$
   $C_3H_{7_{ads}} \rightarrow C_3H_{6_{ads}} + H$

4. $C_3H_8 \rightleftharpoons C_3H_{7_{ads}} + H^+ + e$
   $C_3H_{7_{ads}} \rightarrow C_2H_{4_{ads}} + CH_{3_{ads}}$

5. $C_3H_8 \rightarrow C_3H_{6_{ads}} + 2H^+ + 2e$

6. $C_3H_8 \rightleftharpoons C_3H_{8_{ads}}$
   $C_3H_{8_{ads}} \rightarrow C_3H_{7_{ads}} + H^+ + e$

7. $C_3H_8 \rightleftharpoons C_3H_{7_{ads}} + H^+ + e$
   $H_2O \rightleftharpoons OH_{ads} + H^+ + e$
   $C_3H_{7_{ads}} + OH_{ads} \rightarrow C_3H_7OH_{ads}$

8. $C_3H_8 \rightleftharpoons C_3H_7 + H^+ + e$
   $C_3H_7 + H_2O_{soln} \rightarrow C_2H_4OH_{ads} + CH_{3_{ads}} + H^+ + e$

9. $C_3H_8 \rightleftharpoons C_3H_{7_{ads}} + H^+ + e$
   $C_3H_7 \rightleftharpoons C_3H_{6_{ads}} + H^+ + e$
   $C_3H_{6_{ads}} + H_2O_{soln} \rightarrow C_3H_7OH_{ads}$

(Section 9.4.1) that the probability of two electrons tunneling simultaneously is so small that the presence of multiple-electron transfers in single steps at electrodes can be neglected. This means that mechanism 5 can be rejected.

Can we distinguish between mechanisms 3 and 4? Close inspection reveals that mechanism 3 involves the breaking of a C—H bond. Mechanism 4 involves the breaking of a C—C bond. Which of these mechanisms, one asks, is the more energetically favorable? It has been argued that the C—C bond breaking with subsequent bond formation with the substrate produces a transition state whose standard free energy with respect to the initial system is lower than that produced by the breaking of a C—H bond. Therefore, upon these energetic grounds, the last selection is made, i.e., mechanism 4 is preferred over mechanism 3.

Thus, the conclusion is that the rds in the oxidation of propane is a carbon–carbon bond breaking, and the first two steps in the mechanism must run like this

$$C_3H_8 \rightleftharpoons C_3H_{7_{ads}} + H^+ + e$$
$$C_3H_{7_{ads}} \rightarrow C_2H_{4_{ads}} + CH_{3_{ads}}$$

Thus propane is oxidized to carbon dioxide on platinized platinum in phosphoric acid over the temperature range of 80 to 150°C (and for the potential range of 300 to 500 mV) by a charge-transfer step in quasi equilibrium, followed by a chemical rds. The rds probably involves a carbon–carbon bond rupture of an adsorbed C_3H_7 radical. The reactions following the rds would be relatively rapid in the sense used in reaction kinetics, and there is at present little knowledge concerning them (however, as far as catalyst thinking is concerned, this lack of knowledge is not of major concern so long as the rds is securely known).

Further Reading

1. A. N. Frumkin, *Discussions Faraday Soc.*, **1**: 57 (1947).
2. K. J. Vetter, *Z. Physik. Chem. (Leipzig)*, **194**: 284 (1950).
3. R. Parsons, *Trans. Faraday Soc.*, **47**: 1332 (1951).
4. H. Gerischer, *Z. Elektrochem.*, **62**: 256 (1953).
5. M. A. V. Devanathan and W. Mehl, *J. Electroanal. Chem.*, **1**: 143 (1959–1960).
6. B. E. Conway and P. L. Bourgault, *Can. J. Chem.*, **40**: 1690 (1962).
7. A. J. Krazilshchikov, *Russ. J. Phys. Chem. (English transl.)*, **37**: 273 (1963).
8. H. Wroblowa and B. Piersma, *J. Electroanal. Chem.*, **6**: 401 (1963).
9. S. Srinivasan, *Electrochim. Acta*, **9**: 31 (1964).

10. J. W. Johnson and H. Wroblowa, *Electrochim. Acta*, **9**: 639 (1964).
11. J. W. Johnson and H. Wroblowa, *J. Electrochem. Soc.*, **111**: 863 (1964).
12. H. Wroblowa and A. T. Kühn, *Trans. Faraday Soc.*, **61**: 2531 (1965).
13. V. S. Bagotsky, L. N. Nekrasov, and N. A. Shumilova, *Russ. Chem. Rev.* (*English Transatl.*), **34**: 717 (1965).
14. A. Damjanovic, *Electrochim. Acta*, **11**: 376 (1966); also, (with A. Dey) *J. Electrochem. Soc.*, **113**: 739 (1966); and (with M. A. Genshaw) *J. Chem. Phys.*, **45**: 4057 (1966).
15. V. S. Bagotsky and Yu. B. Vassiliev, *Electrochim. Acta*, **11**: 1439 (1966).
16. A. Damjanovic and M. A. Genshaw, *J. Electrochem. Soc.*, **114**: 1107 (1967).
17. E. Yeager and R. W. Zurilla, *The 153rd National Meeting*, Am. Chem. Soc. (*Div. of Fuel Cell Chemistry*), **11**(1): 66 (Apr., 1967).
18. E. Gileadi and L. Duic, *Electrochim. Acta* **13**: 1915, (1968).

9.6. CURRENT–POTENTIAL LAWS FOR ELECTROCHEMICAL SYSTEMS

9.6.1. The Potential Difference across an Electrochemical System

In the introduction to the fundamentals of electrodics (Chapter 8), it was pointed out that the only electrochemical systems of practical interest are those that consist of at least a *pair* of electrode–electrolyte interfaces. In such electrochemical systems, or cells, one electrode can function as an electron sink (or anode) and the other as an electron source (or cathode).

But, before trying to understand the behavior of electrochemical systems, or cells, it was considered tactical to disassemble, or *analyze*, them conceptually into two isolated electrode–electrolyte interfaces and then to study *single* interfaces. This has been done. The whole treatment so far has concerned itself with a single electrode–solution interface[†] and with the current–potential laws that govern its behavior. The Butler–Volmer equation is the key equation for a single interface. The behavior of an *electrochemical system, or cell*, must be conceptually synthesized from the behavior of the individual interfaces that combine to form a cell.

Consider two electrode–electrolyte interfaces (Fig. 9.42), M/S and M'/S', which are assembled to form an electrochemical system or cell. Recalling that potential differences are always *measured* between two metals of the same composition (*cf.* Section 7.2.5a), a metal M'' which is identical in composition to M is attached to M'. Under these circumstances, the

---

[†] Or, if the treatment considered a cell, the second interface was considered nonpolarizable, i.e., its potential difference was taken to be a constant.

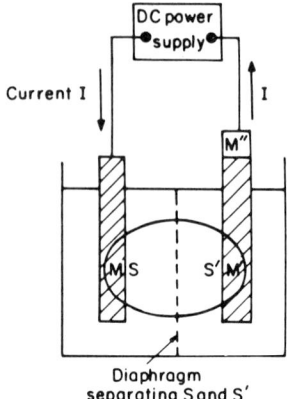

Fig. 9.42. An electrochemical system consists of a series of potential drops that constitute the voltage measured between the two wires of equal composition M and M''.

potential difference V across the whole system, or cell, has been shown [Eq. (7.22)] to be given by the inner potential of the electrode on the right minus the inner potential of a wire of the same composition connected to the electrode on the left

$$V = {}^M\!\Delta^{M''}\!\phi$$
$$= (\phi_M - \phi_S) + (\phi_S - \phi_{S'}) + (\phi_{S'} - \phi_{M'}) + (\phi_{M'} - \phi_{M''})$$
$$= {}^M\!\Delta^S\!\phi + {}^S\!\Delta^{S'}\!\phi + {}^{S'}\!\Delta^{M'}\!\phi + (\phi_{M'} - \phi_{M''}) \tag{9.143}$$

where ${}^S\!\Delta^{S'}\!\phi$ is the potential difference between the bulk of the solution S near the M/S interface and the bulk of the solution S' near the M'/S' interface.

The ${}^S\!\Delta^{S'}\!\phi$ will include (Fig. 9.43) the liquid-junction potential that always arises whenever two solutions of different composition are in contact. But it will be assumed throughout this treatment that the liquid-junction potential has been minimized by the use of salt bridges so that it can be taken as zero. In addition, ${}^S\!\Delta^{S'}\!\phi$ will also include the potential drop in *solution*. This, in the absence of liquid-junction potential, is given by

$$ {}^S\!\Delta^{S'}\!\phi = IR \tag{9.144}$$

where I is the current through the solution, and R is its resistance. If the resistance is negligible, then the quantity ${}^S\!\Delta^{S'}\!\phi$ tends to zero.

Initially, it will be assumed that this special case is applicable. On this basis, Eq. (9.143) reduces to

$$V = {}^{M}\!\Delta^{S}\phi + {}^{S'}\!\Delta^{M'}\phi + \phi_{M'} - \phi_{M''} \tag{9.145}$$

Since, however, the convention is to express the potential difference across an interface as the inner potential of the *electrode* minus the inner potential of the *electrolyte*, one can use the relation that

$$ {}^{S}\!\Delta^{M'}\phi = - {}^{M'}\!\Delta^{S}\phi \tag{9.146}$$

to write

$$V = {}^{M}\!\Delta^{S}\phi - {}^{M'}\!\Delta^{S'}\phi + (\phi_{M'} - \phi_{M''}) \tag{9.147}$$

This, then, is the potential difference across the whole cell, or electrochemical system, on assuming that there is zero potential difference in the solution, i.e., ${}^{S}\!\Delta^{S'}\phi = 0$. The potential difference V is not independent of the current passing through the system because the potential differences across the two individual interfaces M/S and M'/S' are functions of the current densities; in fact, that is what the Butler–Volmer equation is all about. (Note that it has been assumed that ${}^{S}\!\Delta^{S'}\phi \approx 0$. Otherwise, as the current I changes, ${}^{S}\!\Delta^{S'}\phi = IR$ will also change—another contribution to the variation of V with I.) Hence, the cell potential V depends, in general, on the current flowing through the electrochemical system. The problem therefore is to understand the laws relating the potential difference across an electrochemical cell to the current I flowing through it.

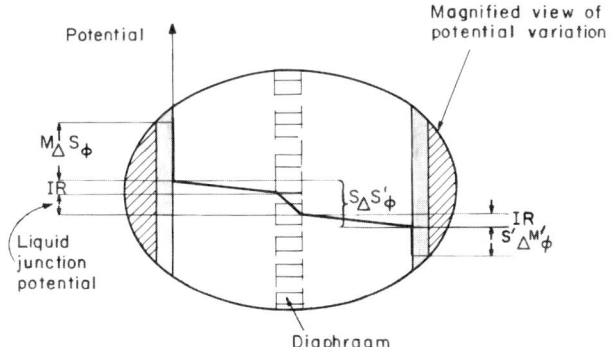

Fig. 9.43. A magnified view of the situation in Fig. 9.42 shows that, besides the two potential changes at the two metal–solution interfaces, there is also a liquid-junction potential as well as an ohmic drop in both solutions.

9.6.2. The Equilibrium Potential Difference across an Electrochemical Cell

Before treating cells with currents flowing across them, an expression will be developed for the zero current or equilibrium potential difference across a cell.[†] Since there is zero cell current, the cell is not connected to either an external current source or an external current sink (or load); one says the cell is on *open circuit*. It is neither a driven cell nor a self-driving system. Each interface therefore must be at equilibrium because the net current is zero across both interfaces.

Hence, both $^M\!\varDelta^S\phi$ and $^{M'}\!\varDelta^{S'}\phi$ in Eq. (9.147) are equilibrium potential differences and may be represented as $^M\!\varDelta^S\phi_e$ and $^{M'}\!\varDelta^{S'}\phi_e$, respectively, in which case one has for the equilibrium potential difference V_e across the cell

$$V_e = {}^M\!\varDelta^S\phi_e - {}^{M'}\!\varDelta^{S'}\phi_e + (\phi_{M'} - \phi_{M''}) \tag{9.148}$$

Even though one is considering the simplified situation of equilibrium, one is still faced with the fact that the absolute potential differences across single interfaces are experimentally inaccessible (Section 7.2.5b). However, one can always resort to the convention (see Sections 7.2.5e and 8.2.14) of referring all potentials to a SHE (standard hydrogen electrode), which would consist, e.g., of a platinized platinum electrode in contact with hydrogen gas at 1-atm pressure and a solution of hydrogen ions at unit activity.

So one can adopt the following procedure[‡]: The quantities $\varDelta\phi_{\text{SHE}}$ and ϕ_{Pt} are added to and subtracted from the right-hand side of the expression (9.148) for the equilibrium cell potential V_e. Thus,

$$V_e = [{}^M\!\varDelta^S\phi_e - \varDelta\phi^0_{\text{SHE}} + (\phi_{\text{Pt}} - \phi_{M''})] - [{}^{M'}\!\varDelta^{S'}\phi_e - \varDelta\phi^0_{\text{SHE}} + (\phi_{\text{Pt}} - \phi_{M'})] \tag{9.149}$$

Examine the terms in the square brackets. What do they represent? They are the potential differences across cells consisting of an SHE coupled with the M/S and M'/S' interfaces. In other words, they are the potential dif-

---

[†] Corrosion cells (*cf.* Sections 11.2.3 and 11.2.8) are excluded from this analysis because even when a corroding metal is not connected into a circuit, i.e., even when it is on open circuit, the metal is not at equilibrium.

[‡] Compare Eq. (9.149) with (7.11). The potentials which are being added to and subtracted from each side here are the potential which was described as constant and arising from the arbitrary designation as zero of the potential of the reference electrode (Section 7.2.5e).

ferences of the M/S and M'/S' interfaces relative to the SHE. It is customary to call them *relative potentials* or, simply, *potentials* of the electrodes M and M'. Thus, Eq. (9.149) becomes

$$V_e = E_{e,M/S} - E_{e,M'/S'} \qquad (9.150)$$

Since, however, it has been stipulated that the equilibrium situation is being considered, one can use the Nernst expression (8.61) for the equilibrium potentials and write[†]

$$V_e = \left(E^0_{M/S} + \frac{RT}{nF} \ln \frac{a_A}{a_D}\right) - \left(E^0_{M'/S'} + \frac{RT}{nF} \ln \frac{a_{A'}}{a_{D'}}\right) \qquad (9.151)$$

where A and D are the electron acceptors and donors involved in the electron-transfer reaction at the M/S interface; and A' and D', the corresponding quantities at the M'/S' interface.

Thus, the equilibrium-potential differences across cells can be predicted for known a_A/a_D and $a_{A'}/a_{D'}$ ratios by making use of tabulated values of the standard electrode potentials $E^0_{M/S}$ and $E^0_{M'/S'}$.

9.6.3. The Problem with Tables of Standard Electrode Potentials

In consulting tables of standard electrode potentials (see Table 9.12), it is necessary to be aware of an (unfortunate) *difference in conventions* for the *sign* of the E^0 values.

Consider, e.g., the E^0 values for zinc electrodes dipped in Zn^{++} solutions of unit activity and for copper electrodes dipped in Cu^{++} solutions of unit activity. Tables which follow the *zinc-minus* and *copper-plus* convention report (Table 9.12) that the E^0 values for the Zn/Zn^{++} and Cu/Cu^{++} electrodes are -0.76 and $+0.34$ V, respectively. Other tables, particularly those in American textbooks on physical chemistry, follow the *zinc-plus* and *copper-minus* convention and state that the E^0 values of Zn/Zn^{++} and Cu/Cu^{++} are $+0.76$ V and -0.34 V respectively. The situation at first appears unlikely because one would think that a standard electrode potential must be an objective fact, not a matter of convention. The situation therefore needs some analysis.

Consider that a cell is set up consisting of an SHE and another electrode, whose standard potential is to be measured. To measure the open-circuit potential of the cell, one can use the Pogendorff compensation

---

[†] To avoid cumbersome notation, it is taken for granted that the standard electrode potential is an *equilibrium* potential and that, therefore, one can write E^0 instead of E_e^0.

TABLE 9.12

Standard Potentials of Certain Electrode Reactions at 25°C†

Electrode reaction	$E°$, V	Electrode reaction	$E°$, V
$Li^+ + e \rightleftharpoons Li$	-3.01	$Cd^{2+} + 2e \rightleftharpoons Cd$	-0.40
$Rb^+ + e \rightleftharpoons Rb$	-2.98	$In^{3+} + 3e \rightleftharpoons In$	-0.34
$Cs^+ + e \rightleftharpoons Cs$	-2.92	$Tl^+ + e \rightleftharpoons Tl$	-0.34
$K^+ + e \rightleftharpoons K$	-2.92	$Co^{2+} + 2e \rightleftharpoons Co$	-0.27
$Ba^{2+} + 2e \rightleftharpoons Ba$	-2.92	$Ni^{2+} + 2e \rightleftharpoons Ni$	-0.23
$Sr^{2+} + 2e \rightleftharpoons Sr$	-2.89	$Sn^{2+} + 2e \rightleftharpoons Sn$	-0.14
$Ca^{2+} + 2e \rightleftharpoons Ca$	-2.84	$Pb^{2+} + 2e \rightleftharpoons Pb$	-0.13
$Na^+ + e \rightleftharpoons Na$	-2.71	$D^+ + e \rightleftharpoons \frac{1}{2}D_2$	-0.003
$Mg^{2+} + 2e \rightleftharpoons Mg$	-2.38	$H^+ + e \rightleftharpoons \frac{1}{2}H_2$	0.000
$Ti^{2+} + 2e \rightleftharpoons Ti$	-1.75	$Cu^{2+} + 2e \rightleftharpoons Cu$	0.34
$Be^{2+} + 2e \rightleftharpoons Be$	-1.70	$\frac{1}{2}O_2 + H_2O + 2e \rightleftharpoons 2OH^-$	0.40
$Al^{3+} + 3e \rightleftharpoons Al$	-1.66	$Cu^+ + e \rightleftharpoons Cu$	0.52
$V^{2+} + 2e \rightleftharpoons V$	-1.5	$Hg^{2+} + 2e \rightleftharpoons 2Hg$	0.80
$Mn^{2+} + 2e \rightleftharpoons Mn$	-1.05	$Ag^+ + e \rightleftharpoons Ag$	0.80
$Zn^{2+} + 2e \rightleftharpoons Zn$	-0.76	$Pd^{2+} + 2e \rightleftharpoons Pd$	0.83
$Ga^{3+} + 3e \rightleftharpoons Ga$	-0.52	$Ir^{3+} + 3e \rightleftharpoons Ir$	1.00
$Fe^{2+} + 2e \rightleftharpoons Fe$	-0.44	$O_2 + 4H^+ + 4e^- \rightleftharpoons 2H_2O$	1.23

† Although reactions are written as electronation (as recommended by the IUPAC), the sign of the standard potential is the same whether electronation or de-electronation is considered (sign invariant).

method, which consists in applying a potential difference exactly equal and opposed in sign to that produced by the cell itself (cf. Fig. 9.44). This is effected by adjusting the potentiometer till the galvanometer G shows zero current. The potentiometer reading in such a balanced cell shows then the magnitude of the potential difference across the cell as well as the sign of the charge on the electrode.

These items of information can also be obtained by using a high-input-impedance voltmeter, which, when connected, e.g., across the cell,

$$Pt/H_2 \text{ [1 atm]}, H^+[a_{H^+} = 1]//Zn^{++}[a_{Zn^{++}} = 1]/Zn$$

shows (1) that the *magnitude* of the potential difference across the cell, i.e., the magnitude of $E^0_{Zn/Zn^{++}}$, is 0.76 V and (2) that the zinc electrode is

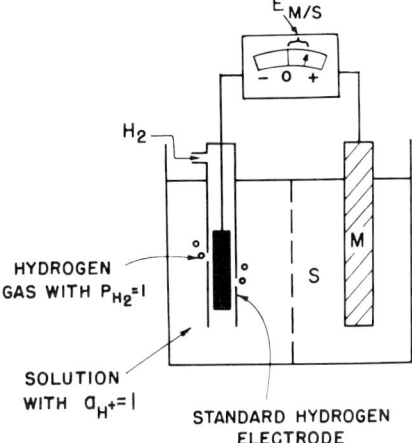

Fig. 9.44. The IUPAC convention ascribes to the standard electrode potential the same sign as that experimentally observed when the electrode in question is connected to a cell with the SHE.

negative.[†] A similar measurement across the following cell

$$\text{Pt/H}_2 \text{ [1 atm]}, \text{H}^+ [a_{\text{H}^+} = 1]//\text{Cu}^{++} [a_{\text{Cu}^{++}} = 1]/\text{Cu}$$

indicates that the magnitude of E^0 is 0.34 V and that the copper electrode is *positive*.

Now suppose that one decides to affix to the measured magnitude of the $E^0_{\text{Zn/Zn}^{++}}$ of a Zn/Zn^{++} electrode the same sign as the observed polarity of the zinc electrode. Then $E^0_{\text{Zn/Zn}^{++}} = -0.76$ V; and, similarly, $E^0_{\text{Cu/Cu}^{++}} = +0.34$ V. It is by this approach that one gets the zinc-minus and copper-plus table. Thus, the rationale behind the zinc-minus–copper-plus convention is based on *observed* polarities and follows consequently the firmly established electrical convention of assigning the minus sign $(-)$ to the electron charge and, thus, of showing the flow of electrons in an electronic conductor from minus $(-)$ to plus $(+)$.

Note also, that, if infinitesimally small currents are allowed to flow (*cf.* Fig. 9.44) through the galvanometer G in one or another direction,

---

[†] A voltmeter indicates, as negative, the electrode *from* which electrons flow into the external circuit toward the voltmeter and, as positive, the electrode *into* which electrons flow from the external circuit.

i.e., when the electrode reactions occurring at both electrodes are reversed from their spontaneous direction, the polarity of the electrodes remains unchanged. Thus, the sign of the electrode potential remains in this convention *invariant*, irrespective of whether the electrode processes proceed in the spontaneous or reverse direction and thus are written as

$$\tfrac{1}{2}H_2 \rightarrow H^+ + e \quad \text{and} \quad M^+ + e \rightarrow M$$

or

$$H^+ + e \rightarrow \tfrac{1}{2}H_2 \quad \text{and} \quad M \rightarrow M^+ + e.$$

The other convention, called the American convention, is based upon the following argument: If a measuring instrument shows the zinc electrode to be negative, it means that the zinc is an electron sink and therefore the site of a spontaneous reaction in which the zinc atoms of the metal transfer electrons to the electrode and become converted into zinc ions. That is, the fact that the zinc electrode is negatively charged implies that the deelectronation reaction $Zn \rightarrow Zn^{++} + 2e$ proceeds spontaneously. But, if a reaction takes place spontaneously, the corresponding free-energy change ΔG must be negative. It is known, however, that $\Delta G = -nFE$ or $\Delta G^0 = -nFE^0$ under standard conditions; hence, $-nFE^0$ must be negative, and E^0, positive. Thus, the observed negative polarity of the zinc electrodes suggests that the standard potential for the deelectronation reaction $Zn \rightarrow Zn^{++} + 2e$ must be positive, i.e., $E^0_{Zn/Zn^{++}} = +0.76$ V. The corresponding argument for the Cu/Cu^{++} electrode suggests that the standard potential for the $Cu \rightarrow Cu^{++} + 2e$ reaction must be negative, i.e., $E^0_{Cu/Cu^{++}} = -0.34$ V. These E^0 values corresponding to de-electronations (oxidations), i.e., standard oxidation potentials, are in accordance with the zinc-plus–copper-minus convention, which now appears reasonable from the free-energy point of view.

Notice, however, that, if the reaction at the Zn/Zn^{++} interface is reversed and written as an electronation (reduction) rather than a de-electronation, then this electronation does not proceed spontaneously and its free-energy change is positive. This positive value of $\Delta G^0 = -nFE^0$ implies that E^0 must be negative. The standard reduction potentials for the zinc and copper systems are, therefore, -0.76 and $+0.34$ V, in contrast to the standard oxidation potentials which are $+0.76$ and -0.34 V respectively.

Thus, an unsatisfactory feature of the zinc-plus–copper minus convention has emerged; the sign of the potential varies depending upon whether the electrode reaction is written as an electronation or a de-electronation. *It is a sign-bivariant convention* with respect to the standard potential.

In contrast, experiment indicates a unique polarity for an electrode, irrespective of the direction in which the infinitesimally small current is flowing and of the way the interfacial charge-transfer reaction is *written*. *The zinc-minus–copper-plus convention is sign invariant* with respect to the potential; this is its distinct advantage.

An important point will now be stressed. If charge-transfer reactions are written as electronations (reductions), e.g., $Zn^{++} + 2e \rightarrow Zn$, the sign of the electrode potential as derived from the free-energy change comes out in agreement with that indicated by the observed polarity of the electrode. This agreement is what prompted the International Union of Pure and Applied Chemistry (IUPAC) to take the following decisions (*cf.* Fig. 9.44):

1. The cell implicit in the measurement of a standard electrode potential should be arranged so that the standard hydrogen electrode is on the *left*

$$Pt/H_2 \text{ [1 atm]} \quad H^+ [a_{H^+} = 1]//M^{z+}/M$$

2. The measured potential difference across such a cell furnishes the *magnitude* of the standard electrode potential.
3. The polarity of the electrode on the *right*, i.e., the sign of the charge on the M electrode, serves to define the *sign* that is affixed to the E^0 value.
4. The charge-transfer reaction implicit in the statement of a standard potential of an M/M^{z+} electrode is an electronation reaction $M^{z+} + ze \rightarrow M$.

Thus, the IUPAC decision supports the zinc-minus–copper-plus table of standard electrode potentials. The first thing to do, therefore, when consulting a table of standard electrode potentials is to examine the E^0 values of the zinc and copper electrodes. If the values are -0.76 and $+0.34$V, respectively, the table can be used forthwith. If, however, the values are $+0.76$ and -0.34V, the convention contravenes the IUPAC decision. To use such a table, one can retain all the magnitudes of the E^0 values but simply change all the signs of the E^0 values; the table will then be in accord with the international convention (Table 9.12).

With this background, consider the calculation of the equilibrium-potential difference V_e across the cell

$$Zn/Zn^{++}//Cu^{++}/Cu$$

According to the international convention, the standard potential of the electrode on the left is always *subtracted* from that of the right-hand electrode. From Eq. (9.151), one has

$$V_e = \left(E^0_{\text{Cu/Cu}^{++}} + \frac{RT}{2F} \ln a_{\text{Cu}^{++}}\right) - \left(E^0_{\text{Zn/Zn}^{++}} + \frac{RT}{2F} \ln a_{\text{Zn}^{++}}\right) \quad (9.152)$$

Suppose that a particular table of standard potentials shows that $E^0_{\text{Zn/Zn}^{++}} = +0.76$ V and $E^0_{\text{Cu/Cu}^{++}} = -0.34$ V. These values are not consistent with the internationally accepted zinc-minus–copper-plus convention, and, therefore, the signs are changed to read $E^0_{\text{Zn/Zn}^{++}} = -0.76$ and $E^0_{\text{Cu/Cu}^{++}} = +0.34$ V. With these values, the equation becomes

$$V_e = 1.10 + \frac{RT}{2F} \ln \frac{a_{\text{Cu}^{++}}}{a_{\text{Zn}^{++}}} \quad (9.153)$$

Now, assume that $a_{\text{Cu}^{++}}/a_{\text{Zn}^{++}} = 1$. Then, $V_e^0 = 1.10$ V, which means that the copper electrode is positive with respect to the zinc electrode; the sign of the potential difference across a cell corresponds to the polarity of the electrode on the right. For ratios of $a_{\text{Zn}^{++}}/a_{\text{Cu}^{++}}$ other than unity, one can calculate V_e from Eq. (9.153).

A simple way of visualizing the procedure is to represent all equilibrium potentials on a single vertical axis (Fig. 9.45). Corresponding to any activity ratio a_A/a_D, there is an equilibrium electrode potential for the interface relative to the SHE. The same is true for the other activity ratio a_A'/a_D'. Thus, the separation between any two points yields the potential difference across a cell with the activity ratios corresponding to the points.

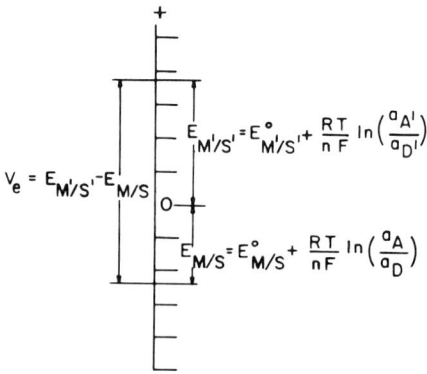

Fig. 9.45. Since each electrode potential is referred to the hydrogen scale, the cell voltage, which is equal to the difference in the two electrode potentials, is given by the separation V_e.

9.6.4. The pH–Potential Diagrams: A General Representation of Equilibrium Potential Differences across Cells

In predicting the equilibrium potential differences across cells, one has so far considered that, at both electrode–electrolyte interfaces, the reactions are purely electron-transfer reactions

$$A_1 + ne \to D$$

and

$$A_2 + ne \to D$$

In such a case, one could represent the electrode potentials on a *single vertical axis*.

Suppose, however, that the electrodic reaction involves not only acceptor and donor but also water and its constituents H^+ and OH^-. That is, suppose one has a reaction of the type

$$xA + mH^+ + ne = yD + zH_2O \tag{9.154}$$

Then the Nernst equation (in terms of relative electrode potentials) assumes the form

$$E_e = E^0 + \frac{RT}{nF} \ln \frac{a_A{}^x a_{H^+}^m}{a_D{}^y a_{H_2O}^z} \tag{9.155}$$

If the activity of water in Eq. (9.155) can be taken as equal to unity, then $a_{H_2O}^z$ can be omitted from Eq. (9.155); thus,

$$E_e = E^0 + \frac{2.303 RT}{nF} \log \frac{a_A{}^x}{a_D{}^y} + \frac{2.303 RT}{nF} \log a_{H^+}^m \tag{9.156}$$

But there is a convenient scale for expressing the activity of hydrogen ions in a solution, the pH scale (*cf.* Section 5.4.6)

$$\text{pH} = -\log a_{H^+} \tag{9.157}$$

Hence, Eq. (9.156) becomes, at 25°C,

$$E_e = E^0 + \frac{2.303 RT}{nF} \log \frac{a_A{}^x}{a_D{}^y} - \frac{2.303 RT}{F} \frac{m}{n} \text{pH} \tag{9.158}$$

Thus, the equilibrium electrode potential in a cell having a reaction involving H^+ or OH^- ions depends not only on the activity ratio of the electron acceptor and donor but also on the pH. The equilibrium electrode potential for such a reaction cannot be plotted on a single vertical axis. It needs a two-dimensional diagram. Thus, assuming a particular concentration ratio $a_A{}^x/a_D{}^y$, one has a plot of the equilibrium potential E against the pH (Fig. 9.46). It will be a straight line with a slope of $(-2.303RT/F)(m/n)$.

Consider, e.g., the reaction

$$O_2 + 4H^+ + 4e = 2H_2O \qquad (9.159)$$

The relative electrode potential for this reaction is given by Eq. (9.158) after inserting $x = 1$, $A = O_2$, $m = 4$, $n = 4$, and $y = 0$. Thus,

$$E_e = E^0 + \frac{2.303RT}{4F} \log P_{O_2} - \frac{2.303RT}{F} \text{pH} \qquad (9.160)$$

where P_{O_2} is the partial pressure of O_2 gas. If this is taken as 1 atm and the numerical value of RT/F at 25°C is inserted, then

$$E_e = E^0 - 0.059 \text{ pH} \qquad (9.161)$$

From Table 9.12, $E^0 = 1.23$ and, hence,

$$E_e = 1.23 - 0.059 \text{ pH} \qquad (9.162)$$

Similar consideration shows that the equilibrium potential of the hydro-

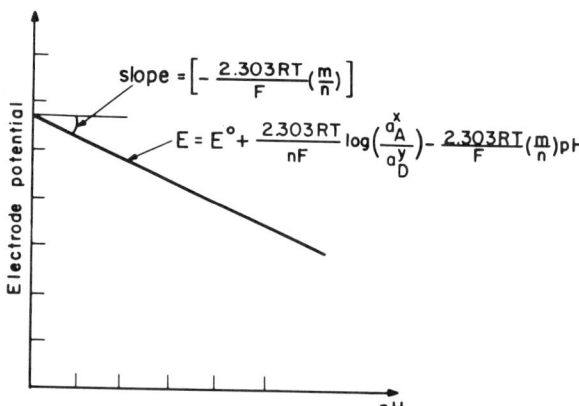

Fig. 9.46. The equilibrium potential for a reaction involving H⁺ or OH⁻ ions for any concentration ratio of acceptors and donors depends linearly on pH.

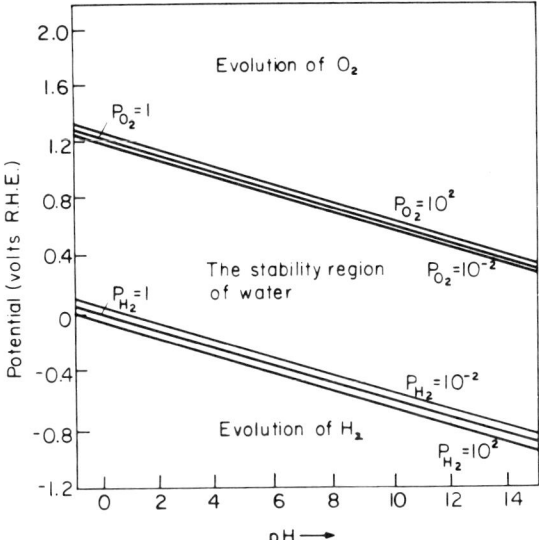

Fig. 9.47. The equilibrium potential difference of an oxygen–hydrogen cell does not depend on the pH and can be obtained at any pH as a difference between the two lines in the figure.

gen electrode is pH dependent since the Nernst equation for the reaction

$$2H^+ + 2e \rightleftharpoons H_2$$

is

$$E_e = E^0 + \frac{2.303RT}{2F} \log P_{H_2} - \frac{2.303RT}{F} \text{pH} \tag{9.163}$$

and, for $P_{H_2} = 1$ at 25°C,

$$E_e = -0.059 \text{ pH} \tag{9.164}$$

A simple pictorial pH *versus* potential presentation can be used for the equilibrium situation in an oxygen–hydrogen cell (Fig. 9.47). Although the potentials of both electrodes are pH dependent, the cell potential (represented by the distance of the points of intersection of a perpendicular erected at any desired pH) remains constant since both electrode potentials change in the same way with varying pH.

Extensive contributions made by Pourbaix to the use of such representations lead to the commonly accepted name of *Pourbaix diagrams*. Figure 9.47 is a very simple example of pH–potential diagrams, which are widely used to determine not only the equilibrium cell potentials but, primarily, the pH regions of stability, corrosion, passivity, etc. (*cf.* Section 11.2.7).

9.6.5. Are Equilibrium Cell Potential Differences Useful?

There is an important piece of information that emerges from the calculation of the equilibrium cell potential. For example, if unit activites of Zn^{++} and Cu^{++} are taken, it has been found (Section 9.6.3) that the potential difference across the Daniell cell is

$$V_e^0 = E^0_{Cu/Cu^{++}} - E^0_{Zn/Zn^{++}} = 1.1 \text{ V}$$

i.e., the right-hand electrode, copper, is positive with respect to the zinc electrode (Fig. 9.48).

What is the implication of the zinc electrode's being negative and the copper electrode's being positive? It means that, relative to the copper electrode, the zinc electrode is negatively charged, or bursting with excess electrons, and, relative to the zinc electrode, the copper electrode is posi-

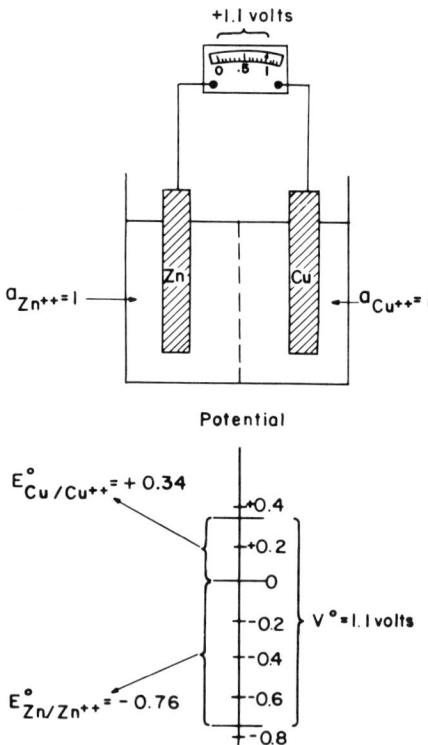

Fig. 9.48. In the Daniel cell, the zinc electrode is 1.1 V negative with respect to the copper electrode.

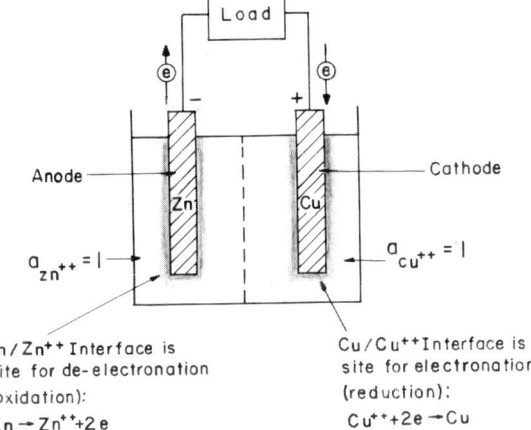

Fig. 9.49. The sign of the voltage on the Daniel cell indicates that, upon placing a load on the cell, a spontaneous de-electronation will occur on the zinc electrode; and electronation, on the copper electrode.

tively charged, or starved of electrons. Hence, when an external electron path (or circuit) is provided, electrons tend to flow out from the zinc electrode, through the external circuit, and into the copper (Fig. 9.49). Thus there tends to be a net de-electronation current $Zn \rightarrow Zn^{++} + 2e$ at the Zn/Zn^{++} interface and a net electronation current $Cu^{++} + 2e \rightarrow Cu$ at the Cu/Cu^{++} interface. Hence, the zinc electrode tends to function spontaneously as an electron sink; and the copper, as an electron source electrode.

An interesting result has emerged. When a Zn/Zn^{++} interface and a Cu/Cu^{++} interface are built into an electrochemical cell, or system, one can proceed from the equilibrium electrode potentials and the zero-current cell potential to predict at which interface there will be a tendency for de-electronation (oxidation) and at which, a tendency for electronation (reduction), i.e., which electrode will function as the electron source and which as the electron sink.

The result can in fact be generalized. Suppose any two electron-transfer reactions taking place at separated interfaces in a cell are considered (Fig. 9.50)

1. $$x_1 A_1 + m_1 H + n_1 e = y_1 D_1 + z_1 H_2 O \qquad (9.165)$$

and

2. $$x_2 A_2 + m_2 H + n_2 e = y_2 D_2 + z_2 H_2 O \qquad (9.166)$$

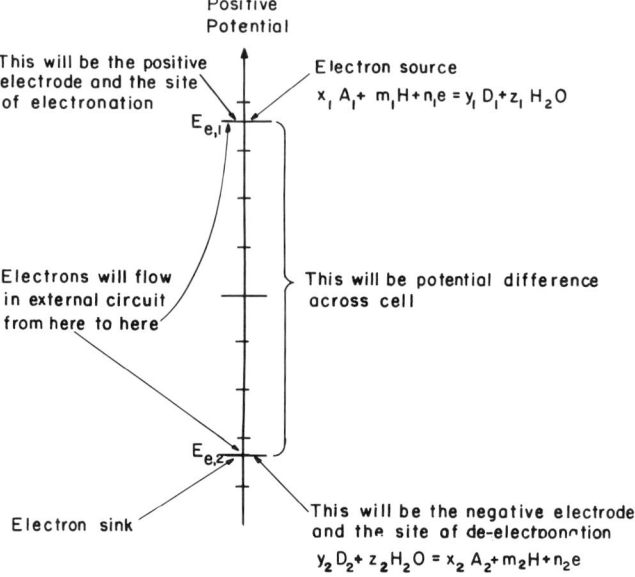

Fig. 9.50. A general situation of the potentials of two electrodes connected to a cell and the sites for spontaneous reactions.

Further, let the equilibrium electrode potentials $E_{e,1}$ and $E_{e,2}$ for the two reactions be such that $E_{e,1}$ is positive with respect to $E_{e,2}$. Then there will be a tendency for de-electronation at the electrode which is the site of reaction 2, i.e., electrode 2 will tend to be the electron sink for D_2 and H_2O. Correspondingly, there will be a tendency for electronation at the positive electrode 1, which will tend to be the electron source for A_1 particles.

Thus, the tables of standard electrode potentials predict those processes which tend to occur spontaneously if any pair of listed interfacial systems are built into an electrochemical cell; that with the lower (algebraically, i.e., more negative) standard potential will spontaneously undergo deelectronation (oxidation), while that with the higher potential (i.e., more positive) will spontaneously undergo electronation (reduction).

In this book, the electrode from which electron acceptors in the solution accept electrons has been termed the *electron-source electrode*, and the electrode which receives electrons from electron donors has been termed the *electron-sink electrode*. The conventional terms, introduced by Faraday upon a suggestion by the Reverend Whewell, for an electron-source electrode and an electron-sink electrode are *cathode* and *anode*, respectively.[†]

[†] Originally Faraday wanted to call them "east-ode" and "west-ode," reasoning thus:

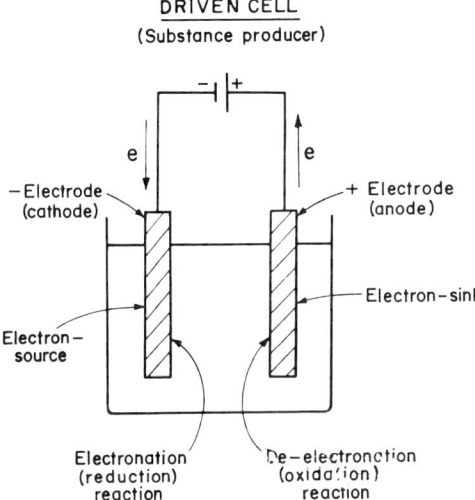

Fig. 9.51. In a substance producer, the cathode must be connected to the negative terminal and the anode to the positive terminal of a power supply.

Consider a driven cell, or substance producer (Fig. 9.51). To make an electronation reaction proceed at a particular electrode, it must function as an electron source for electron acceptors in solution, and must therefore receive an electron flow through the conductor from the power supply. But the terminal of the power supply which pushes out an electron stream is the negative terminal. Thus, to ensure that an electrode functions as an electron source, or cathode, in an electronation reaction, the electrode must be connected to the negative terminal of the power supply. Similarly, to make an electrode function as an electron sink, or anode, in a deelectronation reaction, the electrode must be connected to the positive terminal of the power supply.

It is important to remember that these terms are connected with the

---

"If we admit the magnetism of the Globe as due to Electric currents running...from East to West, and if a portion of water under decomposition by an electric current be placed so that the current through it shall be parallel to that considered as circulating round the earth, then the oxygen will be rendered towards the east...and hydrogen towards the west.... [These, however] are names which a scholar could not suffer," and thus Reverend Whewell, a well-known (among other things) philologist, suggested "anode" and "cathode" as words which also might signify an eastern and western "way."

direction in which the electrode reaction proceeds and *not* with the electrode interface. Thus, e.g., the Zn/Zn^{++} interface in a *self-driven electrochemical cell* is an electron sink (anode) since the reaction that is proceeding there is deelectronation $Zn \rightarrow Zn^{++} + 2e$. By forcing the reaction to proceed in the reverse direction, i.e., $Zn^{++} + 2e^- \rightarrow Zn$, one would make it an electron source (cathode). This can be done by introducing a power supply in the external circuit and building thus a *driven cell*, or substance producer (Fig. 9.51).

On this basis, the terms anode and cathode have often been taken to signify the positive and negative electrodes, respectively. This conclusion is erroneous since it is only in a driven cell that the cathode is the negative electrode and the anode the positive electrode, as may be shown by considering a self-driving cell or energy producer (Fig. 9.52). In such a cell, the negative electrode is that electrode which serves as an electron sink (anode) with respect to electron donors in the solution; and the positive electrode is that electrode which works as an electron source (cathode) for an electronation reaction. Thus, in a self-driven cell, the anode is the negative terminal of the cell and the cathode the positive terminal, a situation which is precisely the opposite of that which obtains in an externally driven cell.

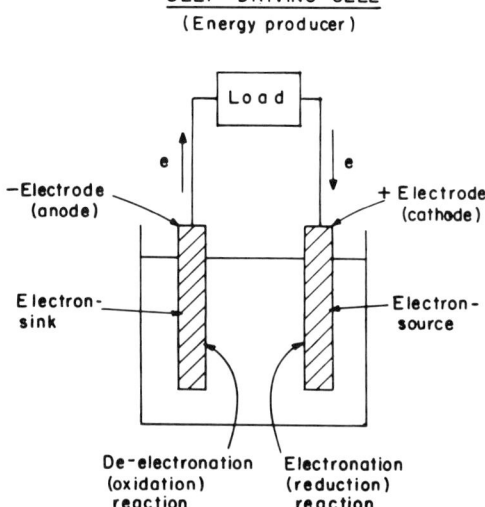

Fig. 9.52. In a self-driving cell, the negative electrode is the anode, or an electron sink for de-electronation, and the positive electrode is the cathode, or an electron source for the electronation reaction.

It is important, therefore, to remember that the terms *anode* and *cathode* are connected with the *nature of the reaction* (de-electronation or electronation) at the electrode, and not with its polarity.

9.6.6. Electrochemical Cells: A Qualitative Discussion of the Variation of Cell Potential with Current

As long as one is making predictions on the basis of *equilibrium* cell potentials, one can talk only of *tendencies* for deelectronation and electronation at the interfaces. Once the system spontaneously drives a current through the external load, the current density at each interface will set up a current-produced potential (i.e., an overpotential) η. What effect will the overpotentials at the two electrodes have on the cell potential V, an increase or decrease in it? To answer that question, one has to abandon the Nernst framework of equilibrium potentials and think in the Butler–Volmer framework of electrodics.

It is easy to see in a qualitative fashion what happens when an energy-producing, spontaneously acting cell drives a current through an external load.

Consider the electron-sink electrode or anode. At its interface with the electrolyte, a net de-electronation-current density will flow. The Butler–Volmer equation for the *overall* reaction at the electron-sink electrode is[†] [*cf.* Sections 9.1 and 9.4 and Eq. (9.24)]

$$i_{\text{si}} = \overleftarrow{i}_{\text{si}} - \overrightarrow{i}_{\text{si}} \tag{9.167}$$

$$i_{\text{si}} = i_{0,\text{si}}(e^{\overleftarrow{\alpha}_{\text{si}} F \eta_{\text{si}}/T} - e^{-\overrightarrow{\alpha}_{\text{si}} F \eta_{\text{si}}/RT}) \tag{9.168}$$

where the subscript si indicates that the electron-sink electrode is being considered. For there to be a *net* de-electronation-current density, the first term $\overleftarrow{i}_{\text{si}}$, arising from the de-electronation-current density (not the *net* current density) at the sink electrode has to be larger than the second term $\overrightarrow{i}_{\text{si}}$ due to the electronation-current density at the same electrode. This requires that η must be positive. But

$$\eta_{\text{si}} = \Delta\phi_{\text{si}} - \Delta\phi_{e,\text{si}} = E_{\text{si}} - E_{e,\text{si}} \tag{9.169}$$

[†] When one is considering a single interface, there is no ambiguity regarding the interface at which there is a net current density $i = \overleftarrow{i} - \overrightarrow{i}$. Here, cells consisting of two interfaces are being discussed. Hence, there are net current densities at two electrodes which must therefore be distinguished by subscripts to indicate which is the electron sink and which is the source.

where $\Delta\phi_{si}$ and $\Delta\phi_{e,si}$ are the absolute Galvani potential differences across the interface when a current density i is passing and when there is equilibrium, and E_{si} and $E_{e,si}$ are the corresponding relative electrode potentials. Hence,

$$E_{si} > E_{e,si} \quad \text{for} \quad i_{si} > i_{si} \tag{9.170}$$

This means that the flow of a current through an external load makes the potential of the electron-sink electrode climb in the positive direction (Fig. 9.53).

Similarly, at the electron-source electrode or cathode, the condition for net electronation is

$$\overleftarrow{i}_{so} < \overrightarrow{i}_{so} \tag{9.171}$$

i.e.,

$$\eta_{so} = -ve \tag{9.172}$$

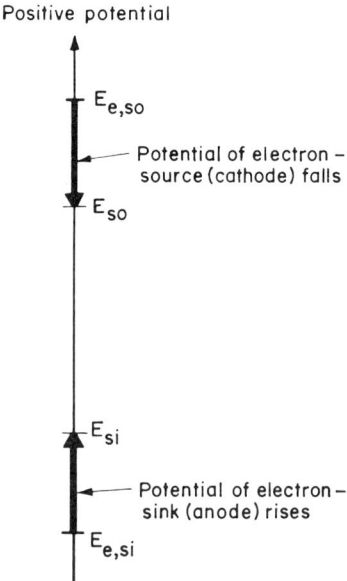

Fig. 9.53. When a self-driving cell delivers current to an external load, the potential of the electron sink shifts in the positive direction, while that of the electron source shifts negative. The net result is a decrease in the cell potential as compared with that at an open circuit.

or
$$E_{so} < E_{e,so} \tag{9.173}$$

Thus, to drive a current through the external circuit, the potential of the electron sink has to become more positive and that of the electron source more negative (Fig. 9.53). But under zero-current, or equilibrium, conditions, the electrode which tends to be a sink is negative with respect to the electrode which tends to be a source. This means that, in the course of driving a current, the potentials of the two electrodes climb toward each other; *the cell potential decreases with cell current in a self-driving cell.*

If it is recalled that, on a purely thermodynamic basis, any cell which has a tendency to drive a current through an external load seems capable of being harnessed as an *energy-producing device*, then the conclusion just reached is serious for it bears the following implication: In the development of an energy-producing device, the variation of cell potential with cell current is as important as, if not more important than, its open-circuit or equilibrium potential. What is the use of a device, the equilibrium potential of which offers big hope but which decays drastically the moment one tries to draw some current from it?[†] *The crucial problem, therefore, is the quantitative relation between cell potential V and the actual cell current passed through a load I*. If it turns out that the fall in potential with current is small, then the actual potential of the cell *in action* at a significant current density is not much different from the theoretical open-circuit potential and the system offers hope of being a good energy producer.

A qualitative understanding of the change of cell potential with current in the case of *driven* electrochemical systems (substance producers) can be developed on similar lines. Here, an external current source has to *oppose* the spontaneous current flow from the cell. That means that it has to promote a net electronation reaction at the electrode which would tend to run spontaneously as a sink and a net de-electronation at the electrode which would tend to be a source.

The thinking takes off from the fact that the equilibrium condition ($i = 0 = \overleftarrow{i} - \overrightarrow{i}$ or $\overleftarrow{i} = \overrightarrow{i}$ when $\eta = 0 = E - E_e$ or $E = E_e$) demarcates the

---

[†] In the pre-electrodic days, essentially before 1950, the attitude of most workers toward electrochemical cells was such that mainly the thermodynamical and diffusional aspects were important. When the cell potentials decreased as the power drawn from them increased, the causes were sought in special phenomena such as *gas layers* on the electrode. The general character of such a decrease, above all, its relation to bonding between substrate and reactant and to electrocatalysis (Section 10.1.4), was not realized.

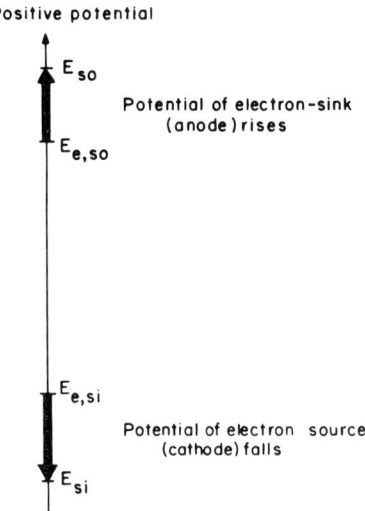

Fig. 9.54. When a substance producer is driven at an appreciable rate, the electron-sink potential becomes more positive and the electron-source potential becomes more negative. The net result is an increase in the cell potential compared with that at an open circuit.

regions of the i versus η curve where net deelectronation occurs ($i > 0$ or $\overleftarrow{i} > \overrightarrow{i}$ when $\eta > 0$ or $E > E_e$) from the region where net electronation occurs ($i < 0$ or $\overleftarrow{i} < \overrightarrow{i}$ when $\eta < 0$ or $E < E_e$)—cf. Fig. 8.32. This means (Fig. 9.54) that the electrode which will function as a source must be driven to more negative potentials ($E < E_e$), and the other electrode which, at equilibrium, sits at more positive potentials must be driven more positive ($E > E_e$). The result is that *the cell potential increases with current in a driven cell,* i.e., it opposes the external cell increasingly as the cell current increases (Fig. 9.55). But how much? One cannot answer this question until one has worked out the quantitative relation between cell potential and current.

9.6.7. Electrochemical Cells in Action: Some Quantitative Relations between Cell Current and Cell Potential

At the outset, consider a self-driving, or energy-producing, cell with two interfaces 1 and 2, and let the equilibrium electrode potentials on the

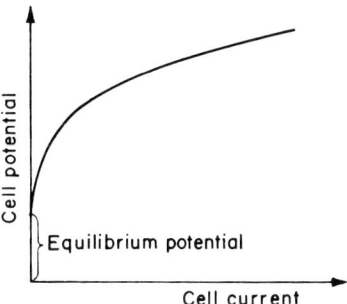

Fig. 9.55. The higher the current driven through a substance producer is, the larger will be the cell potential opposing the driving cell potential.

hydrogen scale be $E_{e,1}$ and $E_{e,2}$. Suppose $E_{e,1}$ is more positive than $E_{e,2}$. Then, if an external load is provided, electrode 2 will tend to be an electron sink for a net deelectronation reaction and electrode 1 will tend to be an electron source for a net electronation reaction.

If interface 2 is far from equilibrium, one can use the simple one-term exponential form of the current-density–potential law [*cf.* Eq. (8.47)]

$$i_2 = i_{0,2} e^{\overleftarrow{\alpha}_2 F \eta_2 / RT} \tag{9.174}$$

or

$$i_2 = i_{0,2} e^{\eta_2 / \overleftarrow{\lambda}_2} \tag{9.175}$$

where $\overleftarrow{\lambda}_2$ is the slope of the η_2 versus $\ln i$ curve and is given by

$$\overleftarrow{\lambda}_2 = \frac{RT}{\overleftarrow{\alpha}_2 F} \tag{9.176}$$

By taking logarithms in Eq. (9.175), the result is

$$\eta_2 = \ln (i_2)^{\overleftarrow{\lambda}_2} - \ln (i_{0,2})^{\overleftarrow{\lambda}_2} \tag{9.177}$$

Similarly, if one assumes that the electronation reaction at interface 1 is far from equilibrium, then, since, by convention, a net electronation current is negative, one has[†]

$$-i_1 = -i_{0,1} e^{-\overrightarrow{\alpha}_1 F \eta_1 / RT} \tag{9.178}$$

[†] Note that, in the electronation reaction, η is a negative quantity.

or
$$i_1 = i_{0,1} e^{-\eta_1/\vec{\lambda}_1} \qquad (9.179)$$

where
$$\vec{\lambda}_1 = \frac{RT}{\vec{\alpha}_1 F} \qquad (9.180)$$

Taking logarithms in Eq. (9.179),

$$\eta_1 = -\ln (i_1)^{\vec{\lambda}_1} + \ln (i_{0,1})^{\vec{\lambda}_1} \qquad (9.181)$$

From these relations (9.181) and (9.177) between the current densities i_1 and i_2 and the overpotentials η_1 and η_2, at the two electrodes, one has to develop a relation between the current I flowing through the cell and the potential difference V across it. This is done in the following way: Since

$$\eta_1 = E_1 - E_{e,1} \qquad (9.182)$$

and
$$\eta_2 = E_2 - E_{e,2} \qquad (9.183)$$

it is clear that
$$\eta_1 - \eta_2 = (E_1 - E_2) - (E_{e,1} - E_{e,2}) \qquad (9.184)$$

But
$$E_1 - E_2 = V \qquad (9.185)$$

and
$$E_{e,1} - E_{e,2} = V_e \qquad (9.186)$$

Hence,
$$\eta_1 - \eta_2 = V - V_e \qquad (9.187)$$

Or
$$V = V_e + \eta_1 - \eta_2 \qquad (9.188)$$

Now one must use Eqs. (9.181) and (9.177) to substitute for η_1 and η_2. The result is

$$\begin{aligned} V &= V_e - \ln (i_1)^{\vec{\lambda}_1} + \ln (i_{0,1})^{\vec{\lambda}_1} - \ln (i_2)^{\vec{\lambda}_2} + \ln (i_{0,2})^{\vec{\lambda}_2} \\ &= V_e - \ln [(i_1)^{\vec{\lambda}_1}(i_2)^{\vec{\lambda}_2}] + \ln [(i_{0,1})^{\vec{\lambda}_1}(i_{0,2})^{\vec{\lambda}_2}] \end{aligned} \qquad (9.189)$$

One still has to transform the current densities to the currents. This is done by recalling that the total *current* I_1 (not current density!) flowing through electrode 1 is equal to the current I_2 through electrode 2, i.e.,

$$I = I_1 = I_2 \qquad (9.190)$$

Further, the current density at each electrode is obtained by dividing the total current by the area A of the interface, i.e.,

$$i_1 = \frac{I}{A_1} \quad \text{and} \quad i_2 = \frac{I}{A_2} \tag{9.191}$$

Combining Eqs. (9.191) and (9.189), one obtains

$$V = V_e - \ln\left[\frac{I^{(\vec{\lambda}_1 + \overleftarrow{\lambda}_2)}}{A_1^{\vec{\lambda}_1} A_2^{\overleftarrow{\lambda}_2}}\right] + \ln\left[(i_{0,1})^{\vec{\lambda}_1}(i_{0,2})^{\overleftarrow{\lambda}_2}\right] \tag{9.192}$$

This is the general I versus V relation for a cell when there is activation overpotential at the two interfaces, and the i versus η relation is taken to be exponential in form. A special case of (9.192) results from assuming electrodes of unit area, i.e.,

$$A_1 = A_2 = 1 \tag{9.193}$$

Then, if one uses the following notation,

$$\vec{\lambda}_1 + \overleftarrow{\lambda}_2 = q \tag{9.194}$$

and

$$i_{0,\text{cell}}^q = (i_{0,1})^{\vec{\lambda}_1}(i_{0,2})^{\overleftarrow{\lambda}_2} \tag{9.195}$$

Eq. (9.192) reduces to

$$V = V_e - \ln I^q + \ln (i_{0,\text{cell}})^q \tag{9.196}$$

or

$$I = i_{0,\text{cell}} e^{-(V - V_e)/q} \tag{9.197}$$

This is an interesting result. Firstly, when the i versus η relations at the two electrodes are of the exponential form, then the I versus V relation for the whole cell is also of the exponential form. Secondly, under these conditions, a plot of what might be called the cell overpotential $V - V_e$ versus log I should be a straight line (Fig. 9.56). That is, when both the individual interfaces show Tafel behavior, the whole cell shows Tafel behavior. Finally, there is, in the expression for the current–potential relation (9.197) for the whole cell, a quantity $i_{0,\text{cell}}$ analogous to the equilibrium exchange-current density i_0 for a single interface. The quantity $i_{0,\text{cell}}$ is obtained by extrapolating the $V - V_e$ versus log I curve back to the equilibrium cell potential V_e, i.e., $V - V_e = 0$.

The form of the I versus V relation (9.197) for a cell, which has just been derived, depends upon the assumption that the activation overpotentials η_1 and η_2 at the two interfaces have pushed the i versus η curves into the exponential region. If, instead, the two interfaces are showing

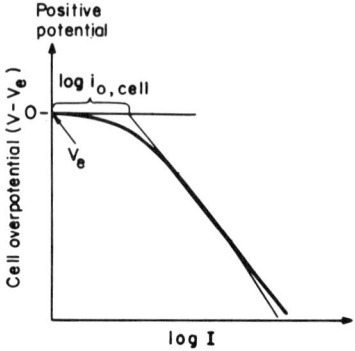

Fig. 9.56. In a self-driving cell, the plot of cell overpotential *versus* log cell current density should be a straight line if the charge transfer at both electrodes are both rate controlling and valid under the high-field approximation. An apparent i_0 for the cell as a whole can be deduced.

ohmic behavior, then one has from Eq. (9.188) and linear i versus η relations (8.43)

$$V = V_e + \eta_1 - \eta_2$$
$$= V_e - \frac{RT}{\vec{\alpha}_1 F}\frac{i_1}{i_{0,1}} - \frac{RT}{\overleftarrow{\alpha}_2 F}\frac{i_2}{i_{0,2}}$$
$$= V_e - I\left(\frac{\vec{\lambda}_1}{A_1 i_{0,1}} + \frac{\overleftarrow{\lambda}_2}{A_2 i_{0,2}}\right) \tag{9.198}$$

Thus, when the i versus η relations at the two interfaces are linear, the cell potential V is a linear function of the cell current I.

There are many ways in which one can go on to make more realistic the above I versus V laws for cells. At the outset, it is necessary to free the treatment from the assumption that there is no potential drop in the solution due to the passage of current. The potential difference $^{S_1}\Delta^{S_2}\phi$ in the solution is introduced in a straightforward way

$$V = V_e + \eta_1 - \eta_2 - {^{S_1}\Delta^{S_2}\phi} \tag{9.199}$$

and, since

$$^{S_1}\Delta^{S_2}\phi = IR \tag{9.200}$$

where R is the resistance of the electrolyte,

$$V = V_e + \eta_1 - \eta_2 - IR \tag{9.201}$$

Now the overpotentials η_1 and η_2 are the total overpotentials at the two interfaces. They must include the activation overpotentials η_a and concentration overpotentials η_c. Thus, Eq. (9.201) can be written

$$V = V_e + \eta_{a,1} + \eta_{c,1} - \eta_{a,2} - \eta_{c,2} - IR \qquad (9.202)$$

So, if there are mass-transport limitations on the concentrations of the species involved in the reactions at the two electrodes, expressions for $\eta_{c,1}$ and $\eta_{c,2}$ must be introduced. Thus, one can write

$$V = V_e + \eta_{a,1} - \frac{RT}{zF} \ln\left[1 - \frac{i_1}{i_{L,1}}\right] - \eta_{a,2} - \frac{RT}{zF} \ln\left[1 - \frac{i_2}{i_{L,2}}\right] - IR \qquad (9.203)$$

These are matters of detail which will also be discussed more thoroughly in the sections on energy conversion (Section 11.3).

The fundamental point is that in a self-driving cell (Fig. 9.56)—the case treated above—all the terms on the right-hand side of Eq. (9.203) make the cell potential V at a current I less than the equilibrium potential V_e. In a driven cell with (Fig. 9.55)

$$V = V_e + \eta_{a,1} + \eta_{c,1} - \eta_{a,2} - \eta_{c,2} + IR \qquad (9.204)$$

all the terms on the right-hand side make V greater than V_e. Thus, in a self-driving cell,[†] or energy producer, the equilibrium potential is the *maximum* cell potential, and, in a driven cell, or substance producer, the equilibrium potential is the *minimum* cell potential—compare Fig. 9.55 (a driven cell) with Fig. 9.56 (a self-driving cell). These basic relations will be discussed in a more detailed and practical way in Section 11.3.11.

Further Reading

1. W. M. Latimer, *Oxidation States of the Elements and Their Potentials in Aqueous Solutions*, Prentice-Hall, Inc., Englewood Cliffs, N. J., 1964.
2. J. M. West, "Surface Films," in: *Electrodeposition and Corrosion Processes*, Chapter 4, D. Van Nostrand Co., Inc., London, 1965.
3. M. Pourbaix, *Atlas of Electrochemical Equilibria in Aqueous Solutions*, Pergamon Press, New York, 1966.

[†] The existence of self-driving electrochemical mechanisms (i.e., chemical systems which spontaneously produce electric power) is a concept which has been hitherto completely neglected in chemistry. It may find significant application in biochemistry.

9.7. THE GRAND DIVIDE

The introduction to electrodics (Chapter 8) commenced with a question: What are the consequences of charge transfer at electrified interfaces? The consequences were seen to be twofold: Electrical currents flow across the interface, and chemical transformations of substances occur. These consequences were held to be of tremendous significance. They opened up the new vistas of electrodics, electrical power generation directly from chemicals and the electrical production of substances. To make these devices, however, it was seen to be necessary to use electrochemical systems as cells with *pairs* of interfaces. Before analyzing pairs, however, the single interface was taken up for study. How does it "tick"?

The central question which clamored for an answer was: What is the law relating the current density flowing across the interface (i.e., the rate of reaction occurring there) and the potential difference across it? Very elementary arguments regarding the passage of particles across potential-energy barriers at an interface led to the development of the Butler–Volmer equation

$$i = i_0[e^{(1-\beta)F\eta/RT} - e^{-\beta F\eta/RT}]$$

The Butler–Volmer equation was heralded as the central equation of electrodics. The thought was that, if one understands the basis of the Butler–Volmer equation and its exponential, linear, and equilibrium forms, one can think about many electrodic situations.

There were, however, a number of quantities in the Butler–Volmer equation which required elucidation. The potential had to be understood in terms of the actual potential difference between the metal surface and the plane from which the particles start their climb of the potential-energy hill. The concentrations were the interfacial concentrations of the electron acceptors and donors.

In analyzing the symmetry factor β, one had to go into how electrons leak out from the metal when the electron acceptors are in their proper welcoming state. The symmetry factor β was seen to be a peculiarly electrodic quantity; it governs the flow of chemical into electrical energy at the interface and *vice versa*.

Once single-step reactions were understood, one began to realize that many of the common electrodic reactions are multistep in character.

How are multistep reactions to be handled? It turned out that there was a simple rule. Though all the consecutive steps which link together to make up the overall reaction proceed in the steady state with the same net velocity, *one* of these reaction steps, the *rate-determining step*, tends to control the overall rate much more than the other partial reactions. This rds was pictured as tending to be the step with the highest standard free energy of activation with respect to the standard free energy of the particles before the reaction begins.

The concept of an rds proved of great utility; one simply[†] wrote out the Butler–Volmer equation for the rds, and that was the expression for the reaction of the overall reaction. Of course, the concentration terms figuring in the expressions for the rds may not be those that correspond to the initial reactants or products. It is easy, however, to express the concentrations of hypothetical intermediate radicals which occur in the supposed rds, in terms of the experimentally known concentrations of reactants.

The last step consisted in using the understanding of the laws of single interfaces to synthesize the laws of electrochemical systems, or cells. It is eventually the control of the functioning of these systems which constitutes the principal challenge, which is the goal of getting at the fundamentals.

At this stage, therefore, one can say that a scaffolding for electrodics has been erected. The "continental divide" of electrodics has been reached. Now one tries to look at the fundamentals of electrodic situations which occur outside the laboratory.

There are many interesting questions on which to exercise one's basic electrodic knowledge. Do electrodes differ from catalysts? What is the mechanism of the hydrogen evolution and oxygen dissolution reactions, which are so important to the conversion of chemical energy into electrical power in the hydrogen–oxygen fuel cell? What are the basic mechanisms of the growth of electrodeposits? How do materials spontaneously destroy themselves? Can the products of electrodic reactions (e.g., hydrogen) seep into the material on which they are produced (perhaps by means of "local currents") and, if so, with what effect? How do films build up on surfaces and affect their stability? For example, what stabilizes the surface of the aluminum panels of a modern building or of stainless steel kitchen pans, or of the mirror-bright chrome on the fender of a car?

It is to some of the electrodic mechanisms, which lie behind such questions, that attention is now turned.

[†] Simply, i.e., if one is dealing with a reaction which has *one* path and contains *one* rds. This is the assumption usually made. The question here is: Are multistep reactions examples of reactions having more than one path at the same potential? The answer is yes, but, in most reactions (less than a dozen at the present time) to which detailed mechanism attention has been given, it seems that *one* reaction *path* is much faster than all the others over most potential ranges.

CHAPTER 10
ELECTRODIC REACTIONS OF SPECIAL INTEREST

10.1. ELECTROCATALYSIS

10.1.1. A Chemical Catalyst and an Electrocatalyst

A catalyst is a substance which alters the rate of a chemical reaction without itself being either consumed or generated in the process. Some examples of heterogeneous catalysis are the hydrogenation of unsaturated organic compounds on nickel catalysts or the combination of nitrogen and hydrogen on iron substrates to form ammonia. In all such examples, the solid catalyst acts as a meeting place for the reactants and promotes their union.

Now, an electrode, too, acts as the site or substrate for the *electrodic* reaction. Further, it survives unchanged during the process of the reaction.[†] Hence, an electrode is a catalyst for charge-transfer reactions; it is a *charge-transfer catalyst*, or an *electrocatalyst*.

The catalytic aspect of an electrode is not immediately obvious in the Butler–Volmer equation for a multistep reaction. Thus, one recalls in this equation the presence of the transfer coefficient α, which, as Eq. (9.25) shows, involves the symmetry factor β, the stoichiometric number ν, and

---

[†] Except, of course, when it grows or dissolves (electrodeposition or dissolution) or when it "feeds on itself" (corrosion).

the number of electrons transferred in steps other than the rds. From this general version of the Butler–Volmer equation (9.24), one writes the current-density–overpotential relationship in the nonlinear region for an electron-ation reaction

$$i = i_0 e^{-\vec{\alpha} F \eta / RT} \tag{10.1}$$

The potential-dependent aspect of the reaction rate appears to be the most important part. One can however reveal the catalytic role of the electrode material by writing the Butler–Volmer equation in a different way, as follows:

$$i = nFc\frac{kT}{h} e^{-\vec{\Delta}G^{0\ddagger}/RT} e^{-\vec{\alpha} F \Delta\phi/RT}$$

$$= nFc\vec{k} e^{-\vec{\alpha} F \Delta\phi/RT} \tag{10.2}$$

Any value of the inner potential $\Delta\phi$ can be *conceptually* split into a charge-dependent $\Delta\psi$ portion and a dipole-dependent $\Delta\chi$ portion (*cf.* Section 7.2.5m); thus,

$$\Delta\phi = \Delta\psi + \Delta\chi \tag{10.3}$$

$$= \frac{4\pi q_M}{\varepsilon} + \frac{4\pi N\mu}{\varepsilon} \tag{10.4}$$

At the pzc (potential of zero charge) (*cf.* Section 7.3.7), $q_M = 0$ and the $\Delta\psi$ contribution vanishes; thus,

$$\Delta\phi_{\text{pzc}} = \Delta\chi = \frac{4\pi N\mu}{\varepsilon} \tag{10.5}$$

Hence, in terms of the $\Delta\phi_{\text{pzc}}$, one can rewrite the Butler–Volmer equation for the overall reaction thus

$$i = nFc\vec{k} e^{-\vec{\alpha} F \Delta\phi_{\text{pzc}}/RT} e^{-\vec{\alpha} F (\Delta\phi - \Delta\phi_{\text{pzc}})/RT} \tag{10.6}$$

This equation makes clear the catalytic role of the electrode. The rate of the electrodic reaction depends, firstly, on the *potential difference* across the electrified interface and, secondly, on a *catalyst*, the $\vec{k}$ term. Thus, electrocatalysis consists of the two aspects, an *electrical* one (the term depending on the potential difference) and a *chemical* one (*cf.* the $\vec{k}$ term).

A one-line distinction between chemical catalysis and electrocatalysis can now be suggested. The electrocatalytic reaction rate is potential dependent; the chemical catalytic rate is not. The rate expressions show this distinction clearly. The catalytic rate of a heterogeneous chemical re-

TABLE 10.1
Comparison of the Rate Equation and the Characteristics of the Two Types of Catalysis, Chemical (or Thermal) Catalysis and Electrocatalysis

	Chemical catalysis	Electrocatalysis
Rate depends on	$e^{-\Delta G^{0\pm}/RT}$	$e^{-\Delta G^{0\pm}/RT} e^{-\bar{\alpha} F \Delta\phi/RT}$
Potential dependent	No	Yes
Temperature dependent	Yes	Yes
Operating temperature range	Usually above 150°C	Usually below 150°C
Activation energy, kcal mole^{-1}	10–100	5–35

action is given by the *Arrhenius equation*

$$v = c \frac{kT}{h} e^{-\Delta G^{0\pm}/RT} \tag{10.7}$$

whereas the electrocatalytic rate is given by

$$v = \frac{i}{nF} = \left(c \frac{kT}{h} e^{-\Delta G^{0\pm}/RT} \right) e^{-\bar{\alpha} F \Delta\phi/RT} \tag{10.8}$$

The potential dependence permits one to use enormous *leverage* on the reaction rate. By changing the potential difference across the interface, one can change the reaction rate on a given catalyst in some cases by more than 10 orders of magnitude. In the case of a chemical reaction, no such great variation in its rate can be achieved by altering the temperature in the rate expression within the range usually available for such changes (<500°C).

Another important factor distinguishing electrocatalysis from heterogeneous catalysis is the presence of nonreacting species (ions and solvent molecules) at the interface, which often affect the reaction rate and cannot be neglected since their effect, even in the same solution, may differ on various substrates. Table 10.1 shows the characteristics of the two types of catalysis.

10.1.2. At What Potential Should Electrocatalysts Be Compared?

The potential dependence of the electrocatalytic rate introduces some problems when it comes to a comparison of the relative efficacy of a number of electrocatalysts. This point is illustrated by the experimental current-

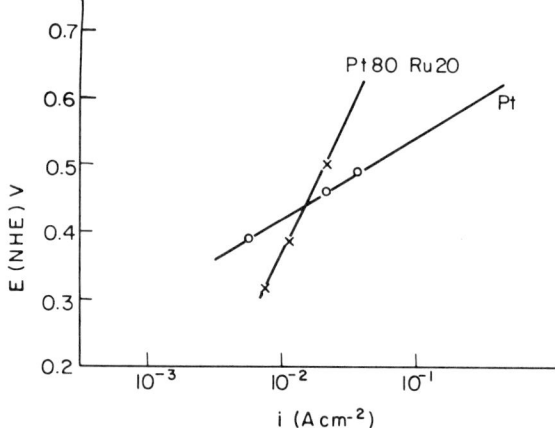

Fig. 10.1. Experimental current-density–potential relationship for the oxidation of ethylene on platinum and 80% Pt–20% Ru alloy.

density–potential plot (Fig. 10.1) for ethylene oxidation on platinum and a platinum–ruthenium alloy (80 to 20). Notice how the alloy is a better catalyst at potentials below about 0.45 V, whereas, at higher potentials, pure platinum is better.

The facts just noted about the relative electrocatalytic abilities of platinum and platinum–ruthenium generate the following question: How are these electrocatalysts to be compared? In other words, at what value of the electrode potential shall the comparison of electrocatalytic rates be made?

The basis for a comparison is straightforward. At a given potential with respect to the reference electrode, the best electrocatalyst for the given reaction is one on which the current density is largest. If i_1 and i_2 are the current densities on electrocatalysts 1 and 2, respectively, then, because $\eta = \Delta\phi - \Delta\phi_e$,

$$i_1 = (i_0)_1 e^{-\vec{\alpha}_1 F(\Delta\phi - \Delta\phi_e)/RT} \tag{10.9}$$

and

$$i_2 = (i_0)_2 e^{-\vec{\alpha}_2 F(\Delta\phi - \Delta\phi_e)/RT} \tag{10.10}$$

If the electrodic reaction has the same mechanism (i.e., path and rds) on both electrocatalysts so that one can reasonably assume that $\vec{\alpha}_1 = \vec{\alpha}_2$, then the exponential terms in the two expressions are equal. Thus, one obtains

$$\frac{i_1}{i_2} \approx \frac{(i_0)_1}{(i_0)_2} \tag{10.11}$$

Hence, when one compares the current densities on two electrocatalysts at a given relative electrode potential (e.g., at the same overpotential), one is really comparing the exchange-current densities on the two substrates.

To try to see what atomistic factors make electrode 1 a better catalyst than electrode 2, one must analyze the situation more closely.

Since an electrocatalyst acquires its separate status from a heterogeneous chemical catalyst by the reaction-accelerating field at the interface, let the electrocatalyst be stripped of its special feature. To what extent would it be possible (*cf.* Section 8.2.2) to make the potential difference across the interface zero? If it were possible to make an experimental arrangement by which the potential differences across various catalyst–electrolyte interfaces are zero, the current densities at that potential would give a clear comparison of the *intrinsic* catalytic powers of the various substrates. One would have succeeded in isolating the chemical catalytic power of the electrocatalysts.

Can such experiments be devised? One knows the pzc for various metal–solution interfaces (Table 8.1), and, when $q_M = 0$, it has been shown [Eq. (10.5)] that $\Delta\phi_{\text{pzc}} = \Delta\chi$.

Now $\Delta\chi$, it will be recalled (*cf.* Section 7.2.5*l*), is the potential difference at an interface due to dipole layers. There are two contributions to the total $\Delta\chi$: When dipolar molecules from the electrolyte adsorb and orient at an interface, there is a $\Delta\chi_{\text{dipole}}$ (see Section 7.2.5*k*); in addition, due to the probability of electrons from the metal's being found outside the confines of the metal surface, there is an electronic contribution $\Delta\chi_e$. That is,

$$\Delta\phi_{\text{pzc}} = \Delta\chi_{\text{dipole}} + \Delta\chi_e \qquad (10.12)$$

Let it be assumed for the moment that $\Delta\chi_e$ is negligible and can be ignored. Even then, the use of the pzc as a comparison potential for electrocatalysts implies either that $\Delta\chi_{\text{dipole}}$ is sufficiently small to make an insignificant effect on the variation of the reaction rate from one substrate to another or that $\Delta\chi_{\text{dipole}}$ is the same on all metals. The experimental evidence is slender at present. What there is of it suggests that variations in $\Delta\chi_{\text{dipole}}$ from metal to metal have a relatively small enough effect. Thus, the pzc would seem to be a fairly satisfactory comparison point for electrocatalysts, at least when the reaction velocities differ by several orders of magnitude over the available catalysts.

However, a much more important theoretical problem is the effect of $\Delta\chi_e$. The understanding of $\Delta\chi_e$ is, at present, very scant indeed (*cf.* Section 7.2.5*c*). What little knowledge exists is indirect, but it suggests that $\Delta\chi_e$ is an uncontrolled factor in electrocatalysis, i.e., there may be widely differing $\Delta\chi_e$ values from metal to metal at the pzc. This uncertain factor could affect electrocatalytic rates by several orders of magnitude. Thus, there is a very serious objection to using the pzc as the potential for comparison of electrocatalysts. A further objection to this choice is that the experimental values of E_{pzc} are far from well established for the majority of metals. The pzc of only a few metals are reasonably well established—see Table 8.1.

A third uncertainty that can arise in the choice of the pzc as the comparison potential is that, for some processes, the rate of reaction at E_{pzc} is accessible only by a lengthy extrapolation. This is so, for example where $E_{pzc} < E_e$ for anodic or $E_{pzc} > E_e$ for cathodic processes.

Another potential at which the rates of a given reaction on various electrocatalysts can be compared is the equilibrium potential E_e. The relation of the current densities observed for a certain reaction on various substrates at the same electrode potential, on a relative scale, can now be discussed in sufficient detail to make the principles understandable. The discussion, however, is restricted to a very simple case.

10.1.3. Electrocatalysis in Simple Redox Reactions

The catalytic effect of an electrode substrate must be examined separately for (1) weak interactions between the reactants and the electrode surface, i.e., when no adsorption occurs at the metal surface; and (2) strong interactions, i.e., where one or more reactants adsorb at the metal surface. The second class is obviously a more complicated one since the reaction rate will depend additionally on the heats of adsorption, on the strength of bonds forming or breaking, and on the properties of the adsorption isotherm.

10.1.3a. How Does the Electrocatalytic Rate Depend upon the Substrate Work Function at the Reversible Potential? Let is be assumed for simplicity that $\beta = \frac{1}{2}$ for the reaction

$$A^{z+} + e \rightleftharpoons A^{(z-1)+} \tag{10.13}$$

and that one is concerned with the simplest type of redox reaction, i.e., no bonding between reactants and electrode surface.

Then, from Eq. (8.26),

$$i_0 = \vec{k} c_{A^{z+}} e^{-\Delta\phi_e F/2RT} \tag{10.14}$$

where $\Delta\phi_e$ is the absolute Galvani potential difference across the interface under conditions of equilibrium.

The term $\vec{k}$ contains the catalyst-dependent factors, those associated with the electronic—and perhaps crystallographic—properties of the substrate. If one takes into account the *lack* of any bonding to the substrate, it is easily appreciated that the principal electronic factor which affects the rate of exit of electrons from the metal is the electronic work function Φ_M. By assuming that the rate of exit attributable to all other factors (e.g., the potential difference at the interface) is kept constant, then the rate constant $\vec{k}$ decreases exponentially as a function of Φ_M. One could write, in analogy to Eq. (A4.3.8),

$$\vec{k} = \vec{k}_0 e^{-\Phi_M/RT} \tag{10.15}$$

It has been stressed before that individual Galvani potential differences are not subject to an experimental determination (Section 7.2.5*b*). One has to work with *cells*, not individual electrodes (or catalysts). Thus, when one makes a measurement of a current (i.e., a rate) on one or several electrode substrates, the actual potential which one refers to as *the potential* is a *cell*-potential difference. It includes the vital $\Delta\phi_e$ of Eq. (10.14), but it also includes a difference in metal–metal junction potential difference. The latter is related to the difference of the work functions of the working electrode, i.e., the catalyst electrode, which varies from experiment to experiment, and the cell's *reference* electrode, which is kept constant. Hence, when one realizes that a reaction rate, measured in terms of a current density at an electrode, is dependent upon a *cell*-potential difference (because it depends on $\Delta\phi_e$ and *this* is obtained by introducing a certain *cell* potential, cf. Section 7.2.5*e*), then one begins to see (Adam, Frumkin) that the work function may enter twice into the relationship between the rate and the effectively measured potential (which is a *cell* potential, the *relative* electrode potential). The first entry originates in a conceptual way, from (10.15). Thus, the electron (in an imagined cathodic reaction) exits with more difficulty at larger Φ [cf. Eq. (10.14)]. The second entry arises because of the nature of electrochemical rate measurements. They are taken at a certain potential *referred to another potential*. In *this* way, the extra Φ gets in.

Thus (*cf.* Sections 7.2.5*c* and 9.6),

$$\Delta\phi_e = V + \Delta\phi_{\text{ref}} - \Phi_M + \Phi_{\text{ref}} \tag{10.16}$$

where V is the cell potential difference, i.e., that potential difference across a cell consisting of a catalyst test electrode and a nonpolarizable reference electrode (together, of course, with a counterelectrode through which the current is made to flow to and from the catalyst test electrode, see Fig. 8.33. The $\Delta\phi_{\text{ref}}$ is the Galvani potential difference at the metal–solution interface formed by the reference electrode; Φ_{ref} and Φ_{M} are the work functions of the reference electrode and the test catalyst electrode M, respectively.

Substituting Eqs. (10.16) and (10.15) in (10.14), one obtains

$$i_0 = \vec{k}_0 c_{A^{z+}} e^{-\Phi_{\text{M}}/RT} e^{-F(V + \Delta\phi_{\text{ref}} - \Phi_{\text{M}} + \Phi_{\text{ref}})/2RT} \qquad (10.17)$$

In Eq. (10.17), therefore, the term Φ_{M} cancels out, and so the rate of an electrocatalytic reaction at the reversible potential becomes independent of the work function of the catalyst. This is true only as long as $c_{A^{z+}}$ is independent of the electrode material, a point about which discussion will follow.

One should stop to ponder on this result, for it is by no means an expected one. Many workers not concerned intimately with the field of electrocatalysis would have predicted an increasing difficulty, i.e., a lessening in rate, of the simplest kind of electronation reaction (the redox type of reaction because there is no bonding to the substrate) as the thermionic work function increased. This thought arises because, the larger the value of Φ, the more deeply are electrons embedded in the potential energy well inside the metal and, hence, the more difficult—it might be thought—to extract the electrons from the metal and transfer them to acceptors in solution. But it is one of the difficulties of understanding electrodics that one is always concerned in actuality with *cell* measurements. It is the terminology which is usually misleading. One refers to the "electrode potential," meaning potential difference across a cell, and much of the time, in considerations other than those connected with the effect of electronic properties upon electrocatalysis, one forgets about this fact and refers to the electrode potential (*cf.* Section 9.6.2) as though it were a metal–solution potential difference. The reason why it is possible to consider many areas in electrodics without bothering to remind oneself that a *cell* is always under measurement is that the metal–metal junction potential difference is constant with change of current density—and, of course, with changes in the solution and electrode surface. Further, the $\Delta\phi_{\text{ref}}$, the Galvani potential difference at the nonpolarizable reference electrode, is also constant, and, therefore, the cell potential varies mostly (*cf.* the discussion of when one can measure changes of Galvani potential difference, Sections 7.2.5c and 7.2.5f) with the variation of the test electrode $\Delta\phi$.

In electrocatalysis, all one is considering are changes of the substrate, say, from M to M_1 or M_2, etc.; consequently, the potential difference

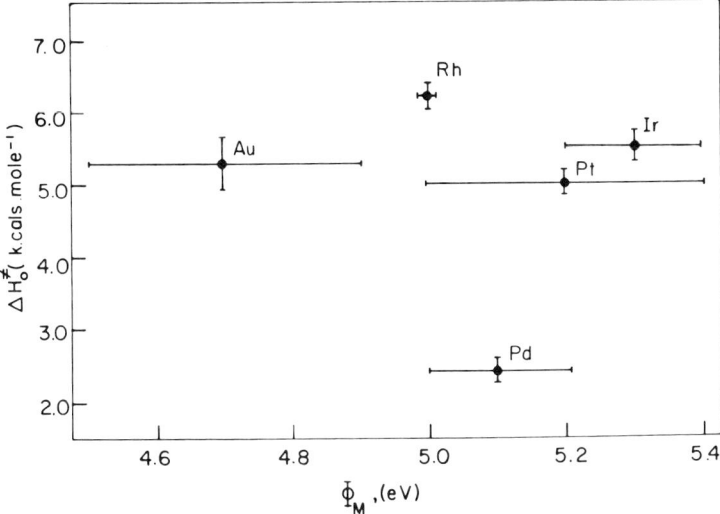

Fig. 10.2. The heat of activation $\Delta H_0^\ddagger$ as a function of the substrate work function Φ_M for the ferric–ferrous redox system on the noble metals (Damjanovic and Mannan).

across the junction between the catalyst electrode and the reference electrode varies from catalyst to catalyst and thereby compensates that part of the Galvani potential-difference variation at the catalyst–electrolyte interface which depends upon Φ_M. Thus, the simple dependence expected of the electron emission rate upon Φ_M does not take place.

These considerations were speculative until the late 1960's when the systematic measurements of the dependence of the heat of activation upon work function for the $Fe^{3+} + e \rightarrow Fe^{2+}$ reaction at various electrodes became available. The results, shown in Fig. 10.2, show that the heat of reaction $\Delta H_0^\ddagger$ is essentially[†] independent of Φ_M.

10.1.3b. Can the Exchange-Current Density Depend upon the Work Function? It was argued in the last section that, when one strips down a consideration of electrode reaction rates to those which are electrocatalytically the simplest, i.e., nonbonding redox reactions, the heat of activation would be independent of the work function and hence (taking into account

---

[†] The result cannot be classed better since one point of the five is sharply deviant from the indications of the other four. It may be that such deviation arises from an experimental error concerning competing H_2 evolution and dissolution on palladium, the deviant metal.

the nonbonding character of the reaction concerned—no partaking of d levels in the substrate, e.g.) would not depend upon the electronic properties of the substrate at all. However, caution has been used. The *heat of activation* has been referred to in the confirming Fig. 10.2, not the exchange-current density, although, in Eqs. (10.14) and (10.17), the argument was begun in terms of the latter.

There is a very good reason for the caution and the limitations in these statements. It has been shown that the heat of activation should be independent of the substrate electronic properties for nonbonding reactions. However, contemplation of Eq. (10.14) will indicate that this means the heat terms implicitly contained in $\vec{k}$. It does not imply the constancy of $c_{A^{z+}}$ with change of electrode potential. Thus, it is in principle conceivable that the value of $c_{A^{z+}}$ could change at various electrode substrates, while the heat of activation for the transfer of electrons from various substrates (*cf.* Fig. 10.2) remains the same. Then [*cf.* Eq. (10.14)], the exchange current density could vary with the substrate. At first, it may seem surprising to suggest that the concentration of ions reacting depends upon the electrode substrate chosen. Such a concentration might be thought to be an independent variable, i.e., essentially the concentration in the bulk of the solution.

However, it must now be admitted that simplification in Eq. (10.14) concerning the identification of $c_{A^{z+}}$ is the source of such a concept. The $c_{A^{z+}}$ term should have been stated clearly as the concentration of A^{z+} species *in the double layer*; this concentration is related to the bulk concentration, and, under some conditions, it is nearly equal to it. Such relations between these two concentration terms and their influence on the double-layer structure in electrode kinetics has been touched upon earlier in Section 8.3 2.

The relationship between the concentration of A^{z+} in the double layer to that in the bulk of solution depends upon the potential $^{\mathrm{OHP}}\varDelta^{S}\phi$, i.e., that potential which exists between the layer of solvated ions nearest the electrode and those in the bulk of the solution; the relevant equation is (8.74). However, there is one point not yet brought out: How does $^{\mathrm{OHP}}\varDelta^{S}\phi$ depend upon the electrode material for a given bulk-solution concentration? Further examination[†] of this question shows that $^{\mathrm{OHP}}\varDelta^{S}\phi$ at constant bulk concentration is a function of how far from the pzc is the $\varDelta\phi_e$ value for the substrate concerned. It is known that the potential of zero charge E_{pzc} varies linearly, with a slope near unity, with the work function of the substrate concerned, see Fig. 10.3. Thus, as the substrate work function changes,

---

[†] The equation which is the basis of this examination is given in Section 8.3.2.

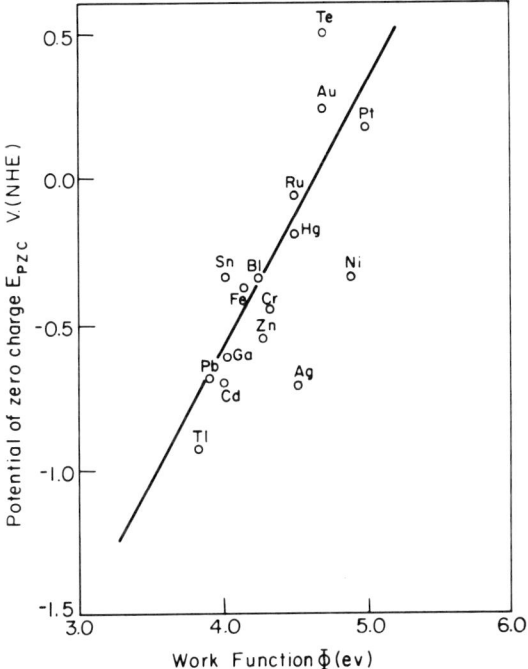

Fig. 10.3. The potential of zero charge E_{pzc} plotted against the work function Φ, showing a linear relationship with a positive slope.

so does E_{pzc}; as E_{pzc} changes, the value of $^{OHP}\Delta^S\phi$ changes; and, therefore, changes in the concentration of $c_{A^{z+}}$ in the double layer occur from one substrate to another. *The double layer has secondary effects*; as its structure changes, so does the concentration of $c_{A^{z+}}$ for a given bulk concentration. The conclusion is that, although the *primary* electrocatalytically determining factors (those which affect the rate constant exponentially) are independent of the work function and hence the electronic properties of the substrate, the latter do come back again into consideration owing to a change in the reactant concentration in the double layer. The double-layer concentration depends upon the substrate work function and upon the value of $\Delta\phi_e$ at the equilibrium potential for the electrode reaction concerned.

Hence, the exchange-current density *does* depend on the work function of the substrate upon which the reaction is carried out, even in the absence of bonding effects, i.e., reactions with no metal-reactant bond formation—see Fig. 10.4.

Two questions must be answered before one can leave these very basic

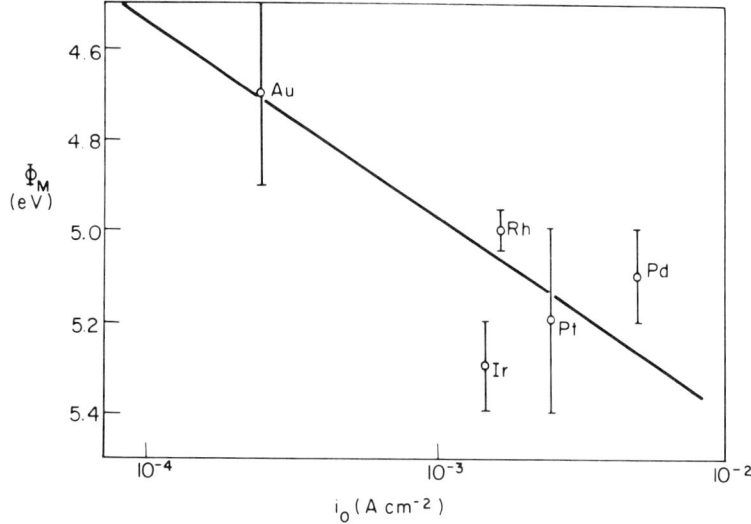

Fig. 10.4. The exchange-current density i_0 for the Fe^{3+}/Fe^{2+} redox couple (0.01 M concentration) plotted against the work function Φ_M of the noble-metal substrate.

considerations of electrocatalysis and approach more practically interesting ones. Firstly, one has to ask why one calls the dependence of i_0 upon Φ_M *secondary*, and, secondly, one has to ask if the *primary* independence, i.e., the independence of i_0 through $\Delta H_0^{\ddagger}$ to Φ_M, is general, or restricted to redox (nonbonding) reactions.

The answers are as follows: The dependence of i_0 on Φ is termed *secondary* because it has a small effect relative to the dependencies of reaction rates upon electronic properties which arise in other reactions where bonding does occur, e.g., in hydrogen evolution; under such conditions, there is a vast variation in the rate ($\sim 10^{10}$ times) at the same overpotential on different substrates. The range of double-layer changes are in the range 10 to 100 and are therefore seen as secondary compared with the effects produced when bonding begins to affect the rates. The second question can be answered by implication; the exchange-current density (and the heat of activation) can be very much dependent on electronic properties, essentially the d-level energy when there is bonding. The background of this will be discussed later in Section 10.1.4.

The basic aspect of electrocatalysis, a negative and surprising one, has been established in Section 10.1.3*a* in that the heat of activation for an electrode reaction is independent of the electronic properties of the surface at which it occurs if it involves only electron transfer from substrate to ions in solution or *vice versa*. This is the ideal case on which the rest of the theory of electrocatalysis may in the next few decades be built up.

10.1.4. Electrocatalysis in Reactions Involving Adsorbed Species

The effect of the substrate on the rates of reactions in which one or more reactants are adsorbed on the electrode is much more pronounced than in the simple case discussed in Section 10.1.3. Here, the substrate affects the heat of activation and, because the exchange-current density is related *exponentially* to the heat of activation, the latter can be changed by many orders of magnitude by a change in the heat of adsorption of one of the reactants or products.

The strength of adsorptive bonds will affect both the concentration terms and the free energy of activation in the rate equation. The extent of the effect depends of course on the reaction mechanism and has to be considered individually in each case. Evaluation of this effect requires knowledge of a number of facts, (1) the reaction path, or at least that part of it which precedes the rds; (2) the rds itself; (3) the adsorption characteristics of the reactants at the electrode surface (i.e., the dependence of coverage on concentration and potential); and (4) the heat of adsorption. These demands are rather high; *prima facie* data are often not yet readily available. Two examples will be discussed below, hydrogen evolution and dissolution and hydrocarbon deelectronation.

10.1.4a. Electrocatalysis in the Hydrogen-Evolution and -Dissolution Reaction. This reaction is chosen for discussion on two grounds. Firstly, there has been much more thinking and experimentation on this reaction than on any other electrodic reaction. Secondly, it is of fundamental importance in hydrogen–oxygen energy producers (called the hydrogen–oxygen *fuel cells*), in corrosion and in electrodeposition.

The hydrogen-evolution reaction shall be considered to consist of two steps, a hydrogen *adsorption* step onto the metal, namely, $M(e) + H_3O^+ \rightarrow MH + H_2O$, followed by an electrochemical *desorption* step, e.g., $MH + H_3O^+ + e \rightarrow M + H_2 + H_2O$.

Consider the first step

$$M(e) + H_3O^+ \rightarrow MH + H_2O$$

A complete understanding of this reaction involves a calculation of the potential energy of the system as a function of the M—H and H^+—H_2O

distances. Such calculations would lead (*cf.* Section 8.3.4*b*) to a potential-energy surface—the three-dimensional and complete presentation of the relationship between the potential energy of the system and the interatomic distances between the particles reacting.

In an electrodic charge-transfer reaction, e.g., $M(e) + H_3O^+ \to MH + H_2O$, the energy changes which occur during the reaction arise principally from movements or vibrations of the transferring particle (e.g., H^+ or H) in a direction normal to the metal surface. Under these conditions, a potential-energy curve, the two-dimensional analog of a three-dimensional potential-energy surface,[†] is a helpful conceptual device. Hence, the influence of an electrocatalyst on the catalytic rate will be discussed in terms of the potential-energy curves for the $M(e) + H_3O^+ \to MH + H_2O$ electronation reaction which were used in Chapter 8.

The reaction rate—and the effect of the substrate on it—is intimately connected with the mechanism of this electronation reaction. The essential act of the reaction which leads to the adsorption of hydrogen on the electrocatalyst has been described (Section 8.5) to be the tunneling of an electron from the electrode to the proton. There is, however, a necessary precondition to this electron tunneling; the H^+—H_2O bond has to be stretched until the tunneling condition is satisfied (Fig. 8.89). For zero field at the interface, this condition for tunneling reads

$$R + A \leq I + L - \Phi \qquad (8.148)$$

To save looking back, these quantities are briefly defined again. The R is the H—H_2O repulsion energy, A is the M—H bond energy, I is the ionization potential of the hydrogen atom, L is the proton-hydration energy, and Φ is the work function of the electrode.

It requires little thought to decide that the quantities R, I, and L do not depend on the nature of the electrode material. For example, L, the hydration energy of the proton, is not affected by a change of the electrode from, say, tin to tantalum. But the M—H interaction energy A and the work function Φ do depend on the nature of the electrocatalyst. Hence, one can rewrite the electron tunneling condition in the form

$$A \leq -\Phi + \text{Constant}$$

or

$$A + \Phi \leq \text{Constant} \qquad (10.18)$$

[†] In fact, a *cut* through the appropriate three-dimensional surface vertically out from the electrode surface.

where the constant is characteristic of the particles involved in the reaction, H^+, H, and H_2O. Since the constant is independent of the substrate, it could be concluded that the tunneling condition[†] and hence the reaction rate vary from one electrode to the other essentially because of the differences in work function Φ and the M—H interaction energy A. However, the apparent dependence of the exchange-current density for an electrode reaction on the electronic work function Φ exists because *the present discussion has been restricted to that of a single interface and not to a cell.* If one measures the rate of an electrocatalytic reaction at *a given overpotential*, i.e., at a known cell potential, the contact-potential difference which exists between the metal M under examination and the metal of the reference electrode contains the work function of the metal M, as shown for the redox case in Section 10.1.3a. This work function, phenomenologically introduced, one might say, cancels with the work function contained in the tunneling condition. Hence, there is no net, direct effect of the work function upon bonding reactions.

Suppose that the rate of the hydrogen-adsorption reaction [i.e., $M(e) + H_3O^+ \rightarrow MH + H_2O$] determines the rate of the overall reaction $2H_3O^+ + 2e \rightarrow H_2 + 2H_2O$. Thus, the activation energy for the adsorption step is greater than that for the succeeding electrochemical-desorption step $MH + H_3O^+ + e \rightarrow M + H_2 + H_2O$. But the activation energy of the adsorption reaction leading to the formation of adsorbed hydrogen atoms is determined by the ease with which electron tunneling occurs according to the condition

$$\Phi + A \leq \text{Constant} \qquad (10.18)$$

Stated in these terms, the larger the M—H adsorption energy is, i.e., the stronger the M—H bond, the more easily is the tunneling condition satisfied and the lower is the activation energy for the adsorption reaction.[‡]

One can easily **visualize** the effects of the M—H adsorption energy in terms of the **analogs** of the energy barrier. Increase in the magnitude of A moves the R versus A curve down (Fig. 10.5) and thus decreases the

---

[†] Here, as in the discussion in Chapter 8, it is *electron* tunneling which is intended. The implicit assumption is that tunneling of protons does not occur, i.e., the protons have to go *over* the energy barrier and not *through* it. In fact, however, there is some evidence to show that a certain fraction of the protons do tunnel to a significant degree on certain metals, particularly on metals for which the overpotential is high. The quantum-mechanical characteristics of the proton have been stressed in equations developed first by Christov and then by Bockris and Matthews.

[‡] Although Φ is still present in the tunneling condition (10.18), its effect on the measured heat of activation cancels out in a cell, as explained in Section 10.1.3a.

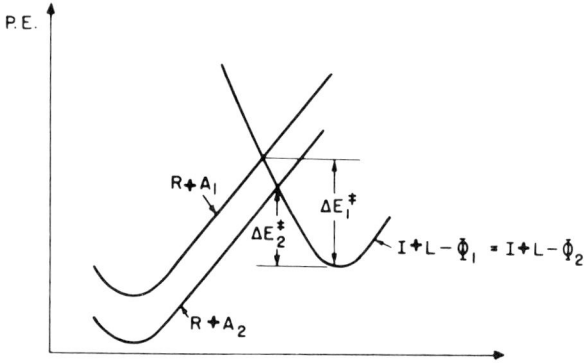

Fig. 10.5. A potential-energy–distance diagram illustrating the effect of an increased M—H bond energy A upon the activation energy ΔE.

activation energy for the achievement of the tunneling condition for the electronation reaction, i.e., the reaction rate increases. Thus, for a series of metals in which the rds for the hydrogen-evolution reaction is the fulfilment of the electron-tunneling condition by deformation of H^+—OH_2, i.e., proton discharge with resulting hydrogen-adsorption reaction, *decrease* of the M—H adsorption energy along the series of metals produces a *decrease* in the reaction rate.

When the mechanism of hydrogen evolution involves rate-determining desorption, $MH + H_3O^+ + e \rightarrow M + H_2 + H_2O$, the M—H adsorption energy A works in the opposite direction, i.e., the more strongly the metal holds onto the hydrogen atom, the more difficult will be the rate-determining desorption to form molecular hydrogen and the slower will the reaction proceed.

Figure 10.6 illustrates clearly that, as the M—H adsorption energy increases for the series of metals cadmium, thallium, mercury, and lead, the rate of reaction, as reflected by an *increase* in i_0, also increases, i.e., for this series of metals, proton discharge to *form* M—H is rate determining. For the series of metals aluminum to platinum (at a low current density), a decrease in i_0 as the M—H adsorption energy increases is observed which, as outlined above, is indicative of rate-determining desorption. Thus, the prediction of the theory and the experimental data are in reasonable accord in this simple case.

10.1.4b. The Electrocatalytic De-electronation of Hydrocarbons. One would perhaps not think that hydrocarbons, particularly unsaturated ones, undergo de-electronation easily. In fact, this is what most electrochemists

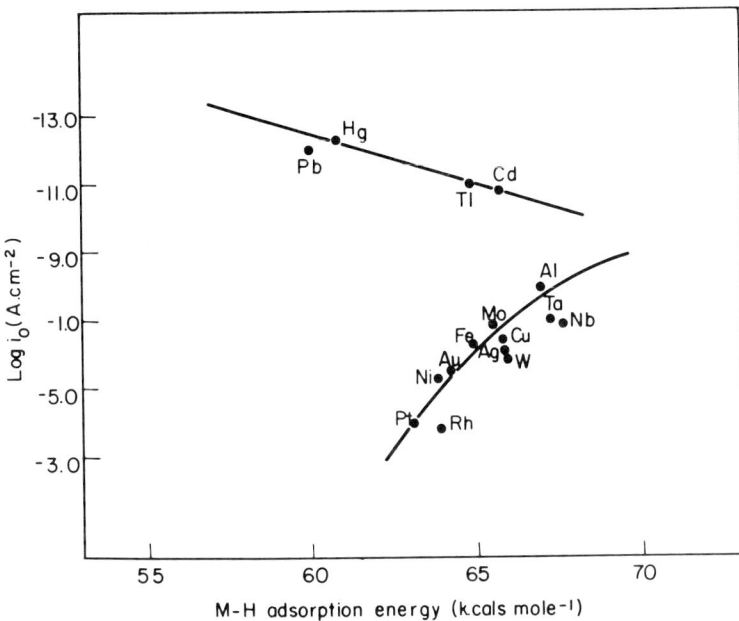

Fig. 10.6. The exchange-current density for the hydrogen-evolution reaction in acid solution as a function of the M—H adsorption energy.

thought until some experiments in 1960 showed that the de-electronation of hydrocarbons at electrodes was possible. More significantly, the experiments on ethane and subsequently on many other hydrocarbons, saturated and unsaturated, aliphatic and aromatic, showed that the electrocatalytic rate on some metals, notably the noble metals, was high enough to make it feasible to incorporate the hydrocarbon-de-electronation reaction into an energy-producing device with, of course, porous electrodes to increase the limiting current density (*cf.* Section 11.3.17). The outstanding importance of the discovery of the electrochemical oxidizability of hydrocarbons will only be touched upon here since a fuller discussion of energy producers is to be given in Chapter 11. The main point is that the hydrocarbon-de-electronation reaction, as an electron-sink electrode, could be combined with the electronation of air, i.e., of atmospheric oxygen, to develop an energy-producing device. Such a device for driving an electric motor would constitute an electrochemical engine. The technological development of electrochemical engines would introduce a revolution in transportation because electric vehicles would be noise-free and vibration-free, cost less to run than gasoline-fueled cars, and, in addition, would not pollute the atmosphere with toxic fumes.

What holds back such a breakthrough? It is partly a matter of the electrocatalysis of hydrocarbons.[†] The not yet sufficient rate of electrocatalyzed oxidation per unit area of electrode necessitates the use of cells which are too heavy for the power that they produce. A more serious problem is that the only satisfactory substrates known at present for the hydrocarbon-de-electronation reaction are the noble metals, in particular, platinum; and the cost of such an electrochemical engine using platinum is prohibitive for all but very specialized uses where the cost of the whole device (e.g., a spaceship) per unit weight is so large that a costly energy conversion device is cheap if it saves weight.

Furthermore, were all cars to be run on fuel cells using platinum catalysts, world resources of this metal would be rapidly exhausted. However, some relief from this gloomy outlook may be obtained by using platinum only at those regions in a porous structure where diffusion and ohmic drops allow activity to be exerted; consequently, the power output per unit weight of platinum would be increased (see Section 11.3.17).

Since the need is for a satisfactory electrocatalyst for hydrocarbon de-electronation, it is necessary to understand the mechanism of the oxidation reaction with the objective of finding the rds. If this understanding is gained and it becomes clear why platinum catalyzes hydrocarbon de-electronation faster than any other catalyst at present known, it may become possible to design a faster and cheaper electrocatalyst.

By no means enough research has been done to permit a firm and general statement on the rds in the de-electronation of saturated hydrocarbons on a representative number of substrates. In fact, this is a research area in which there is intense activity which is likely to generate significant knowledge in the years to come. The de-electronation of ethylene has received detailed attention, so that it will be discussed as an introduction to the mechanism of saturated hydrocarbons.[‡]

[†] One alternative is to arrange the thermal cracking of hydrocarbons and use the resulting hydrogen gas as the electron donor in a de-electronation reaction. However, the weight and volume of such a cracking device is disadvantageous. For intraurban transportation, where range necessary without recharge is small, electrochemical energy *storage* (rather than conversion) may be the best solution (*cf.* Section 11.3).

[‡] Many of the published researches on hydrocarbon oxidation have been carried out by using porous electrodes, the reason for this being that porous electrodes give much greater currents per external, geometric area (*cf.* Section 11.3.17). A maximum current per external area is of course what is required for a practical device; however, the equations which relate current to potential in porous electrodes and other kinetic expressions are much more complex than those for planar electrodes. Thus, little mechanism-indicative data can be obtained with porous electrodes. (Footnote continued on p. 1159.)

After diffusion from the electrolyte bulk to the interface, the ethylene adsorbs on the electrode. At temperatures below 80°C, the ethylene molecules remain undissociated

$$M + C_2H_4 \rightarrow M - C_2H_{4_{ads}}$$

The overall reaction, shown by Faraday's law analysis on the amount of CO_2 produced for a given number of coulombs (*cf.* Section 9.4.2), is

$$C_2H_4 + 4H_2O \rightarrow 2CO_2 + 12H^+ + 12e.$$

Hence, apart from ethylene adsorption, there must be a reaction which introduces oxygen in some form to the electrode surface, and, at least in acid solutions (low concentration of OH^-), this reaction is almost certain to be the de-electronation of water

$$M + H_2O \rightarrow M - OH_{ads} + H^+ + e$$

What next? There must be a reaction between the OH radicals and the molecules of adsorbed ethylene. The product of such a reaction must then enter into a series of consecutive reactions involving OH radicals and the various radicals representing further and further stages in the de-electronation of ethylene. Many full reaction sequences may be formulated, but they would be largely speculative. The lacuna is the absence of direct methods of identifying organic reaction intermediates in the angstrom region from the electrode surface. The information that present-day techniques reveal concerns the rds, i.e., the step in the sequence of reactions which controls the overall de-electronation rate. Since this is the crucial item of information required for considering how knowledge of the mechanism of a reaction may contribute to the direction in which to seek appropriate catalysts, it is not so urgent to know the *complete* sequence of steps leading from ethylene to carbon dioxide; this refers particularly to those steps which *follow* the rds in the *anodic* direction.

The available information suggests that the rds in the de-electronation of ethylene on platinum black is the electron transfer from water to free

---

Ethylene has a somewhat higher exchange-current density at temperatures below 100°C than do saturated hydrocarbons; thus, the current per real unit area on a planar electrode will be greater than that for saturated hydrocarbons. Therefore, mechanism-significant measurements are carried out more easily with ethylene than with, say, propane, where one would have to go, e.g., to 125°C and hence a high-boiling-point electrolyte to get comparable reaction velocities.

sites on the catalyst surface. If water de-electronation is the rds, then the current density should be given by a relation of the form

$$i = \overline{k}(1 - \theta_{C_2H_4}) f(\text{pH}, \Delta\phi) \tag{10.19}$$

where the $1 - \theta_{C_2H_4}$ factor has been introduced to recognize that the electrode has a high coverage ($\theta_{C_2H_4} \to 1$) with ethylene.

The above expression for the current density demands that, since the C_2H_4 in solution is in adsorption equilibrium with $\theta_{C_2H_4}$ on the surface, an increase of the partial pressure of C_2H_4 in solution should increase $\theta_{C_2H_4}$ and thus lead to a decrease in the reaction rate, this is the so-called "negative-pressure" effect measured quantitatively in terms of $(d \log i)/(d \log p_{C_2H_4})$. By radiotracer measurements of marked carbon on the electrode surface, one can determine the adsorption isotherm, i.e., the variation of ethylene surface coverage $\theta_{C_2H_4}$ with the partial pressure $P_{C_2H_4}$ of C_2H_4 in solution. The parameters of the adsorption isotherm can be correlated with the observed negative-pressure effect, a point in favor of the rate-determining water-de-electronation mechanism because the two sets of data, $(d \log i)/(d \log P_{C_2H_4})$, and the change in $\theta_{C_2H_4}$ with ethylene pressure, are obtained in entirely different ways.

There is supporting evidence for the view that water de-electronation is rate determining for ethylene oxidation in that the value of the experimental transfer coefficient is $\frac{1}{2}$; this is in accord with the simple rate-determining

$$M + H_2O \to M\text{---}OH_{ads} + H^+ + e$$

The observed effect on the reaction rate of replacing hydrogen in the electrolyte by deuterium (*cf.* Section 9.5.5*d*) agrees well for platinized platinum with that predicted theoretically for the above rds and is at sharp variance with the theoretical predictions of several other envisaged rds's, e.g., the combination of adsorbed ethylene with adsorbed OH radicals. Furthermore, if one examines the relative rates of the de-electronation of a series of ethylenic hydrocarbons (Table 10.2) and of benzene, it is found that the rates do not differ by more than a *few* times and that the rate decreases as the adsorbability of the hydrocarbon increases, in accordance with Eq. (10.19).

These facts on the relative electrocatalytic rates of various hydrocarbons provide additional qualitative support for the water de-electronation model.

In conclusion, therefore, it seems that the rds in the de-electronation of ethylenic hydrocarbons on platinized platinum at temperatures below 80°C is indeed the water-discharge reaction, as given above.

TABLE 10.2

The Catalytic Activity of Group I and Group VIII Substrates (Metals and Alloys) for the Oxidation of Ethylene

Metal or alloy	$-\log i$ [amp cm^{-2}] at a polarization of +0.60 V (NHE)
Platinum	5.3
Iridium	6.5
Palladium	6.7
Gold	6.7
Rhodium	7.0
Ruthenium	7.0
Silver	7.3
Osmium	>8.0
Cu–Rh	5.4
Rh–Pd	5.8–7.5
Pt–Rh	6.1–6.3
Pd–Au	6.2–6.5
Cu–Au	7.0

The mechanism described is by no means general for the oxidation of unsaturated hydrocarbons on other metals, it pertains to the oxidation of ethylenic hydrocarbons and benzene *on platinized platinum*. Change of the metal can change the rds or even the whole reaction path.

What happens on other metals? A similar diagnostic study (Table 10.3) of the mechanism of ethylene de-electronation on a series of noble metals other than platinum leads to the conclusion that the oxidation of ethylene often proceeds, at least at first, along the same *path* as that on platinum, i.e.,

$$C_2H_4 \rightleftharpoons C_2H_{4_{ads}} \qquad (10.20)$$

$$H_2O \rightleftharpoons OH_{ads} + H^+ + e \qquad (10.21)$$

$$C_2H_{4_{ads}} + OH_{ads} \rightarrow \text{Organic radicals} \qquad (10.22)$$

but with a different rds, (10.22) instead of (10.21).

10.1.4c. *The Dependence of the Rate upon Substrate for the Oxidation of Ethylene.* It is obvious that the rate of the above reaction with (10.22) rate determining will increase with increasing coverage of the electrode by adsorbed ethylenic and hydroxyl radicals and with decreasing strength of M—C and M—O bonds which are broken in step (10.22).

TABLE 10.3
Diagnostic Criteria for the De-electronation of Ethylene on Platinum and Four Other Metals, in Acid Solution at 80°C

Metal	Reaction products	Tafel slope $\left(\dfrac{d\eta}{d\log i}\right)_{pH}$, V	$\left(\dfrac{d\log i}{d\text{pH}}\right)_v$	$\left(\dfrac{dV}{d\text{pH}}\right)_i$, V	$\dfrac{d\log i}{d\log P_{C_2H_4}}$	$\left(\dfrac{i_{H_2O}}{i_{D_2O}}\right)_v$	i (600 mV), amp cm^{-2}
Platinum	CO_2	0.14	0.45	0.07	−0.2	1.5–2.5	5×10^{-6}
Palladium	50% CO_2 + aldehydes	0.19	0.5	0	+0.5		2×10^{-7}
Rhodium	CO_2	0.16	0.5	0.07	+0.5	4–5	10^{-7}
Iridium	CO_2	0.16	0.5	0.075	+0.5		3×10^{-7}
Gold	CO_2 + aldehydes	0.20	0.5	0	+0.5	4–6	2×10^{-7}

The coverage with OH radicals should increase with the amount of unpaired d electrons in the metal, in analogy with the dependence found for oxygen coverage (Table 10.4). By comparison with evidence from the gas-phase studies, the coverage with ethylene might be expected to be low for Group I metals (Silver, gold, and copper) and higher, but of the same order of magnitude, for Group VIII metals (platinum, iridium, osmium, palladium, rhodium, and ruthenium). Radiotracer determination of ethylene coverage from solution has shown, however, that the ratio of coverages $\theta_{Au}^{eth}:\theta_{Pt}^{eth}:\theta_{Rh}^{eth}$ is about 0.2:1:0.02. This result is not entirely unexpected since ethylene adsorbs from solution in competition with hydroxyl radicals, and the more strongly adsorbed species, i.e., the one the heat of adsorption of which is highest, pushes the more weakly adsorbed species off the surface. For example, the heats of adsorption from the gas phase of oxygen and ethylene on rhodium are about 120 and about 60 kcal mole^{-1} for the single Rh—O and Rh—C bonds, respectively; thus, the coverage of rhodium with oxygen or OH radicals would be expected to be much higher than that of ethylene on rhodium.

It may thus be anticipated that the coverage with oxygen or OH radicals will increase with increasing d-band vacancies and hardly be disturbed by ethylene competition, whereas ethylene coverage, low for Group IB metals, increases to a maximum on certain of the Group VIII metals. Thus, $\theta_{Au}^{eth} < \theta_{Pt}^{eth}$. In Group VIII, the metal with the largest heat of adsorption of OH radicals, i.e., the greatest number of d-band vacancies, will tend to reject ethylene, i.e., have the lowest coverage of ethylene; thus

TABLE 10.4

Data for the Oxygen Adsorption on Several Noble Metals and Its Relation to the Degree of d-Band Character of the Metal

Metal	Obs oxygen coverage, $\mu C\ cm^{-2}$	Calc oxygen coverage required for a monolayer, $\mu C\ cm^{-2}$	Fraction of surface covered by oxygen	Number of unpaired d electrons per metal atom
Palladium	110	510	0.22	0.55
Platinum	135	500	0.27	0.6
Rhodium	480	530	0.90	1.7
Iridium	440	525	0.84	1.7
Ruthenium	500	530	0.95	2.2
Gold	<15	–	<0.03	0

$\theta_{Rh}^{eth} < \theta_{Pt}^{eth}$. The Group VIII metal which has the lowest heat of adsorption of OH radicals and the smallest number of unpaired d electrons is platinum, thus, the coverage of ethylene on platinum is the highest among those metals so far investigated.

In terms of the rate of reaction, the ethylene oxidation rate will proceed the faster as the number of d-band vacancies increases from gold to platinum (θ^{OH} and θ^{eth} increasing); the reaction rate among the noble metals will depend on whether the ethylene coverage decreases faster or slower than the coverage with OH radicals.

Another factor which must be considered in the elucidation of the ethylene-oxidation mechanism is the influence of the bond strengths M—O and M—C upon the rate at constant coverage. These, according to equations of the type suggested by Pauling, can be expressed as[†]

$$E_{M-A} = \tfrac{1}{2}(E_{M-M} + E_{A-A}) + 23.06(\chi_M - \chi_A)^2 \qquad (10.23)$$

where χ is the electronegativity. It can be seen that, as E_{M-M} bond strength increases, the E_{M-A} bond strength also increases, which suggests that the reaction rate should decrease with increasing M—M bond strength if the rds is reaction (10.22), i.e., the combination reaction between adsorbed OH and an adsorbed ethylene radical.

For the metals considered here, one may write for the M—M bond strength

$$E_{M-M} = \tfrac{1}{6} L_s \qquad (10.24)$$

where L_s is the latent heat of sublimation of the metal M.

Thus, examination of the rate equation (10.22) would indicate that the rates of ethylene oxidation must depend on such substrate properties as the amount of unpaired d electrons and the latent heat of sublimation. For the metals considered, the order in which both these properties increase is almost the same. This would indicate that, by plotting either of these two properties against the reaction rate (expressed as a current density at a fixed overpotential) should produce plots of a similar nature; this has been shown to be correct by Kuhn *et al.* Figure 10.7 shows that the plot of the reaction rate against L_s possesses a "volcano" shape and appears to be valid for alloys as well as for pure metals. This type of plot is often encountered in heterogeneous catalysis when two principal factors (coverage and bond strength) affect the reaction rate in opposing directions.

[†] In this reaction A represents the atom to which the metal is bound. This will often be C or O for organic reactions.

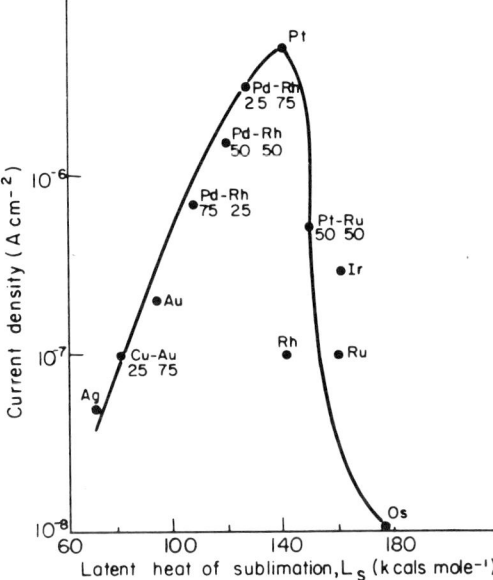

Fig. 10.7. The volcano relationship for the oxidation of ethylene upon various substrates, relating the reaction rate in amperers per square centimeters to the substrate heat of sublimation.

How can one interpret this volcano plot? In view of the previous considerations, it seems reasonable to consider the ascending part of the curve as one in which the dominating factor is the increase of coverage of the reactants. The relatively low bonding energy of radicals to the surface and hence a faster combination between them and a faster overall reaction are less important than the fact that the coverage with ethylene and OH is low. Consequently, as the coverage increases with increasing heat of sublimation, the reaction rate increases. On the descending side of the volcano, further increases in radical coverage are no longer dominant because the coverage is already relatively high and further increases will be small; thus, the increase in the radical–substrate bond energy will dominate, and the reaction rate will decrease.

This interpretation can perhaps also be stretched to rationalize the fact that, on platinum, the rds is the water-discarge step (10.21), while, for substrates of both higher and lower heats of sublimation, the rds is the chemical-recombination step (10.22). Thus, *per site*, the specific rate constant for the discharge of water molecules increases from left to right across the volcano plot shown in Fig. 10.7. However, to the left, e.g., on gold, there is a very low coverage of OH and ethylene, whereas, on platinum, these coverages, particularly that of ethylene, are relatively higher. Although the specific rate constant per site for water discharge would tend to be greater on platinum than on gold, the rate per unit area may well be less because of the greater reduction of free surface area per unit of total electrode area. The result of this is that the water-discharge reaction proceeds slower on platinum than on gold, thus water discharge tends to become the rate-controlling step.

As the heat of sublimation increases to the right of platinum (Fig. 10.7), the effect of increasing M—O bond strength causes an increase in the specific rate per site for water discharge, which causes this process (10.21) to be fast enough for combination (the specific rate of which is undergoing reduction as the strength of the bonding of reactants to the surface increases) to take over the rate control. Correspondingly, θ^{eth} is smaller for rhodium than platinum, and this also tends to increase the water-discharge rate per unit area.

10.1.4d. The Special Position of Platinum as an Electrocatalyst. These interpretations may also be used to discuss why it is that the statement is often made that platinum is the best catalyst in electrochemical-oxidation reactions. Firstly, it must be stated that the statement is probably not exactly true since there is increasing evidence that, at least for methanol oxidation, there are some noble-metal alloys which do give reaction rates at some overpotential, which are much faster than those on platinum (see Table 10.5). Further, the statement has less wide significance than it appears to have because the number of pure-metal substrates with which it can be compared is more or less limited to noble metals and noble-metal-containing alloys, other metals undergoing anodic dissolution in the potential regions in which organic oxidation occurs.

Nevertheless, the statement that platinum is the best electrocatalyst has sufficient basis (indeed, it shows up well in the volcano diagram given here, Fig. 10.7) and therefore requires interpretation. Thus, in a hypothetical situation of a catalyst with limitingly low bond strength to the reactant, the coverage, θ, of the catalyst with this will be negligible and the reaction rate correspondingly tend to zero. As the catalyst is modified and changed from one metal to another, so that the bond strength to the reactant and its radicals is increased, they will increase in their degree of coverage of the electrode; the reaction will consequently have a chance to proceed on the surface and the rate will increase. Such an increase may be expected to

TABLE 10.5

Catalytic Activity of Platinum Black and the Noble-Metal Alloys for the Electrochemical Oxidation of Methanol in 2N H_2SO_4 at 100°C with a Current Density of 20 mamp cm^{-2}

Metal or alloy	Polarization, η
Platinum black	0.44
Binary Platinum alloys	
Pt–Ru	0.24–0.32
Pt–Os	0.30–0.35
Pt–Rh	0.44
Pt–25%Ir	0.33
Binary Platinum-free alloys	
Ru–Rh	0.29–0.38
Ru–Pd, Ir, Rh	0.38–0.46
Ternary alloys	
Pt–Ru–W	0.25–0.37
Pt–Ru–Ta	0.24–0.27
Pt–Ru–Zr	0.30
70% Pt–15% Ru–15% Mo	0.23

continue with increase of coverage. But there are reasons why this increase of θ must come to a halt with increase of bond strength; one of these, of course, is that θ will eventually approach "full coverage." What will happen, then, to the rate of the reaction if there is a change in catalyst which corresponds to a still further increase in bonding to the surface concerned? The answer depends on the nature of the rate-determining step in the reaction. In many electrochemical reactions (e.g., the oxidation of hydrocarbons on substrates apart from platinized platinum), it appears that a *chemical* surface reaction, such as (10.22), governs the rate. But, the rate *constant* of such a reaction clearly *decreases* as the bond by which the reactants are adsorbed to the surface strengthens (the particles have to stretch up off the substrate to mix orbitals, and that gets more difficult the tighter they are bound to their bed). Thus, when increase of bonding no longer increase θ, the decreasing rate constant rules the reaction rate, and (for θ is now nearly constant with increase of bond strength) the reaction *rate* decreases as the bond strength of substrate to reactant increases. The reaction rate, as a function of this varying bond strength to reactant, thus passes through a maximum.

Such ideas—which are sketchy and undetailed—serve to rationalize relations such as that of Fig. 10.7. They suggest that platinum's preeminence as an electrocatalyst arises because its tendency to bond to O and C (and other) atoms is about half-way between that of gold (heat of sublimation, L_{sub} = 92 kcal mole^{-1}) and that of osmium (L_{sub} = 180 kcal mole^{-1}). That of platinum is 125 kcal mole^{-1}. Equation (10.23) shows that if the electronegativity influence does not reverse the linearity between L_{sub} and E_{M-A}, the bond strength of Pt to O and C will be an intermediate one among the noble metals. Thus, platinum's surface optimizes the opposing demands of coverage and rate constant (or free energy of activation). Its surface makes the coverage "quite high," and the free energy of activation "not too high," so that the value of the product of θ and k passes through a maximum when portrayed against some variable such the heats of sublimation of the catalyst materials (on the assumption [*cf.* Eq. (10.23)] that L_{sub} is proportional to the strength of the bond each metal makes with O or C).

10.1.5. Special Features of Electrocatalysis

10.1.5a. The Effect of the Electric Field. The main factor distinguishing electrocatalysis from heterogeneous chemical catalysis is of course the effect of the electric field across the interface on the reaction rate. The rate may change by several orders of magnitude with increasing overpotential. It may achieve values which reduce the heat of activation toward zero, as can be seen from the equation[†] relating the heat of activation at overpotential η, $\Delta H_\eta^{\ddagger}$, with the heat of activation at the equilibrium potential, $\Delta H_0^{\ddagger}$,

$$\Delta H_\eta^{\ddagger} = \Delta H_0^{\ddagger} - \alpha \eta F \qquad (10.25)$$

where α is the transfer coefficient. Thus, at an overpotential given by

$$\eta = \frac{\Delta H_0^{\ddagger}}{\alpha F} \qquad (10.26)$$

[†] In previous deductions, the term $\alpha \Delta \phi F = \alpha \Delta \phi_e F + \alpha \eta F$ has been shown to be the potential-dependent part of the free energy of activation. Now $\Delta G^{0\ddagger} = \Delta H^{0\ddagger} - T \Delta S^{0\ddagger}$. It is generally assumed that $\Delta S^{0\ddagger}$ is independent of potential. If this is so, $\Delta H_\eta^{\ddagger} = \Delta H_0^{\ddagger} - \alpha \eta F$, as in (10.25). The assumption of the potential dependence of ΔS seems reasonable. Only if the activated state were affected by potential in its order differently from the initial state would $\Delta S^{0\ddagger}$ be a function of potential.

the heat of activation becomes zero. No analogous situation exists in heterogeneous catalysis. By changing the overpotential one may effectively adjust the heat of activation of the reaction.

Additional advantages in mechanistic determinations arise from this dependence of rate on the electric field because the relation between the rate and the field strength supplies an extra diagnostic criterion for reaction mechanism compared with those of reaction kinetics, a knowledge of which is a prerequisite in any attempts to introduce theoretical considerations into the development of catalysts.

Investigations of electrocatalysts are now being carried on extensively in many countries. Attention has been focused largely upon the noble metals and noble-metal alloys (Tables 10.2 and 10.5); their present high cost may probably be overcome by the development of porous-electrode systems in which the catalyst is only placed where it is of use (Section 10.1.5e). Possible new materials, useful as electrocatalysts, that are to be investigated include organic semiconductors and metallically conducting oxides, e.g., the possible application of the class of compounds called the *bronzes*.†

10.1.5b. *Reactivity at Low Temperatures*. In heterogeneous catalysis, working temperatures are often high, say, a few hundred degrees Centigrade. With the advent of interest in electrochemical energy conversion, oxidations of many organic compounds have been demonstrated at much lower temperatures electrochemically than those used in the gas phase. For example, the saturated hydrocarbons can be oxidized to carbon dioxide and water at temperatures of less than 100°C. Much of the hydrocarbon fuel-cell research (*cf.* Section 11.3) is however conducted at temperatures nearer 150°C to attain lower overpotentials and hence high energy-conversion efficiencies; this temperature is still well below that used in gas-phase oxidation studies.

Even compounds with a complex structure, such as cellulose, can be

---

† The bronzes are a class of nonstoichiometric compounds of the general formula $M'_x M''_y O_z$, where M'' is a transition metal present in the form of its highest binary oxide $M''_y O_z$, and M' is some other metal. Many transition metals have the ability to form bronzes, e.g., vanadium, niobium, and tungsten (see Dickens and Wittingham in *Further Reading*).

These bronzes, owing to their high electrical conductivity and their chemical inertness, could perhaps have some application as electrode materials for many reactions. For example, sodium tungsten bronze $Na_x WO_3$, in which x lies between 0.2 and 0.8 and which contains <100 ppm of platinum or nickel, has been shown to possess a catalytic activity comparable to or slightly better than that of pure platinum at the reversible potential for the oxygen-reduction reaction (Damjanovic).

oxidized quantitatively to CO_2 by using platinized-platinum electrocatalysts; this is of interest because of the difficulty of a chemical or biochemical oxidation of cellulose. Practical implications of this fact include a cellulose–air fuel cell which would yield about 1 mW cm^{-2} of power and the possibility of the electrochemically accelerated disposal of body wastes (which contain large quantities of cellulose). A feature of this low-temperature reactivity of electrically accelerated reactions is that they tend to give rise to one path and product for given ambient conditions and overpotential, whereas the corresponding gas-phase reactions often go along several paths simultaneously with comparable rates.

10.1.5c. *The Activation of an Electrocatalyst.* Certain electrode materials can be activated either before or during an electrode reaction so as to produce an increased rate of reaction, i.e., a higher current density at some constant value of overpotential. This increased reaction rate can produce gains in power output of 100 to 200% when such electrodes are used in an electrochemical reactor.

Activation can be achieved in several ways, among them (1) by the application of alternate anodic and cathodic pulses by which fresh, clean (oxide-free) electrode surfaces are produced, and (2) by the application of ultrasonic radiation (Fig. 10.8). The mechanism of this effect may be that activation arises out of the removal of an adsorbed, inhibiting layer from the metal surface, introducing new metal atoms into contact with reactants. For example, the α emission from polonium causes the acceleration of the oxygen-dissolution reaction $O_2 + 2H_2O + 4e \rightarrow 4OH^-$ on silver electro-catalysts. The detailed mechanism of this effect, i.e., the effects on the catalyst structure and the secondary effect on ions in solution, is as yet unknown.

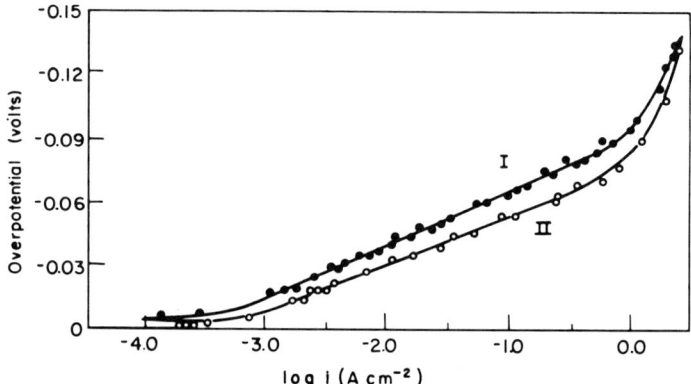

Fig. 10.8. The effect of ultrasonic irradiation (II) for the hydrogen-evolution reaction (I) on a platinum electrode in N H_2SO_4 at 25°C.

10.1.5d. Increasing the Power Output by Changing the Reaction Path. It is sometimes possible to stimulate conditions to promote an electrode process which, although having the same overall reaction, possess a different reaction path to that normally followed. If the reaction path has an rds whose activation energy is lower than that of the normal rds this will be reflected in the increased power output of the energy converter or a reduced power need of an electrochemical reaction. For example, using oxygen as a reactant, one might introduce a suitable redox system, e.g., the nitric acid–nitric oxide couple. The primary process in this example is not the reduction of oxygen but the reduction of nitric acid. The oxygen then oxidizes the formed nitric oxide back to nitric acid while it is itself being reduced. Since the electrochemical reduction of nitric acid and the chemical oxidation of nitric acid are considerably faster than the electroreduction of oxygen, the power output of oxygen-using systems could thus be increased.

10.1.5e. The Use of Porous Electrodes. The maximum reaction rate of a process in which diffusion to a planar electrode surface is rate controlling (cf. Section 9.3) can be calculated from the expression for the limiting current density

$$i_L = \frac{DnF}{\delta} c_{\text{reactant}} \tag{9.82}$$

Suppose (optimistically) that the maximum concentration of reactant that one is likely to have is 1 mole liter^{-1}, and that the value of δ is taken as 0.05 cm (a roughly correct value for unstirred solutions, Section 9.3.9), then, for an order-of-magnitude significant result, one can take $D = 10^{-5}$ cm^2 sec^{-1} and, e.g., $n = 2$; one obtains

$$i_L = 4 \times 10^{-2} \text{ amp cm}^{-2}$$

i.e., a maximum current density of 40 mamp cm^{-2}.

Technological electrode processes are valued eventually in the cost of the products which they produce, not only per unit weight of product but also per unit area taken up by the reactor. A higher current density means a lower specific area and hence a more economical process; however, compensating factors may arise owing to higher current densities, e.g., an overpotential increase with increasing current density. How may higher currents per geometric square centimeter of an electrode be obtained? One way would be to agitate the solution, and, indeed, this is often done. The effect is to reduce δ below 0.05 cm; reductions may be at least one order of magnitude, so that the limiting current density would approach 0.5 amp cm^{-2}.

However, the work done in producing agitation and solution flow may be significant and costly, and the solubility of reactants is sometimes several orders of magnitude below the optimistically chosen value for the solubility assumed in the initial calculation. Agitation would then give relatively poor maximum current densities, and, for this reason, *porous* electrodes are used.

At first, indeed, up till the mid-1960's, the empirical fact that much greater current densities could be obtained per external geometric square centimeter on porous electrodes (e.g., porous carbon with perhaps some particles of catalyst distributed inside the electrode) was attributed simply to the greater available internal surface area per external geometric area. However, in this area, the fundamental equations have tended to take the lead over the inspired gropings of technologists. It has been shown (Cahan) that, at least for systems in which there is wetting, little current is effectively produced inside most of the porous electrode. What is important and makes it work (particularly, when the reactant is a gas) is that it gives a great number of menisci at which there are sections where the layer of solution is very thin, i.e., δ has been made very small ($\sim 10^{-5}$ cm) by physical means and not by stirring. It is in these areas that very high limiting currents per meniscus are found; it turns out that, when they are added together to give the total current per external geometric area, the apparent amperes per square centimeter are also very much, i.e., several orders of magnitude, higher, than for the corresponding situation at a planar electrode. The situation is discussed in detail in Chapter 11.

A difficulty with some electrochemical reactions is that the catalyst (all too often an expensive noble metal), which has been distributed in fine particle form in the pores, has been pretty *uniformly* distributed throughout all of the electrode, and, because of the very local character of the production of high currents— at those areas where the meniscus is very thin— there is a waste of much of the catalyst. A bright picture exists in the future for those who can find devices in which the position of the catalysts coincides with the position of minimum δ; very considerable reduction of the catalyst metal used per external square centimeter of electrode could then ensue while the power produced remained the same.

Further Reading

1. F. P. Bowden and E. K. Rideal, *Proc. Roy. Soc. (London)*, **120A**: 80 (1928).
2. B. E. Conway and J. O'M. Bockris, *J. Chem. Phys.*, **26**: 532 (1957).

3. A. T. Grubb, *Nature*, **198**: 883 (1963).
4. H. Wroblowa and B. Piersma, *J. Electroanal. Chem.*, **6**: 401 (1963); also, **7**: 428 (1964).
5. V. S. Bagotsky and Y. B. Vasilyev, *Electrochim. Acta*, **9**: 869 (1964).
6. R. Parsons, *Surface Science*, **2**: 418 (1964).
7. A. N. Frumkin, *J. Electroanal. Chem.*, **9**: 173 (1965).
8. A. T. Kuhn and H. Wroblowa, *Trans. Faraday Soc.*, **63**: 1458 (1967).
9. A. Damjanovic, *Electrochim. Acta*, **12**: 746 (1967).
10. S. Srinivasan and H. Wroblowa, "Electrocatalysis," *Advan. Catalysis*, **17**: 351 (1967).
11. P. G. Dickens and M. S. Whittingham, "The Tungsten Bronzes and Related Compounds," *Quart. Rev. (London)*, **22**: 30 (1968).
12. A. Damjanovic and R. J. Mannan, *J. Chem. Phys.*, **48**: 1898 (1968).
13. B. D. Cahan and S. Srinivasan, *J. Advan. Energy Conversion*, **6**: 183 (1968).

10.2. THE ELECTROGROWTH OF METALS ON ELECTRODES

10.2.1. The Two Aspects of Electrogrowth

Electrocatalysis has just been described. One important feature of an electrocatalyst is that it goes through the electrodic reaction unchanged. Its sole function is to act as an electron source or sink and as a surface for the adsorption of any intermediates involved in the reaction. Or, if one prefers to think in terms of the crystalline lattice which constitutes the solid electrocatalyst, it is clear that the lattice neither disintegrates by its constituent particles' walking off into solution nor grows by particles from the solution's adding onto the lattice permanently. The surface of the electrocatalyst is a stable frontier; it neither advances nor recedes (Fig. 10.9).

But think what happens when a piece of copper is immersed in a silver nitrate solution (Fig. 10.10) and then made an electron-source electrode. The electronation of Ag^+ ions to silver metal takes place on the copper, and the reddish copper surface becomes coated with a silvery color. A cross section of the electrode shows that the electrode surface has advanced toward the solution (Fig. 10.11). Silver has electrocrystallized on the copper. Thus, the copper electrode has not behaved as an electrocatalyst; it has been altered by electrocrystallization. It is not simply an electron source.

What happens in the electrocrystallization process? How do metals "electrogrow" on other metals? There are, strictly speaking, two aspects to this question. The first (Fig. 10.12) involves the process of *deposition*, i.e., the path taken by an ion in solution to move up to and be incorporated in the lattices of the crystals which make up the electrode. The second aspect

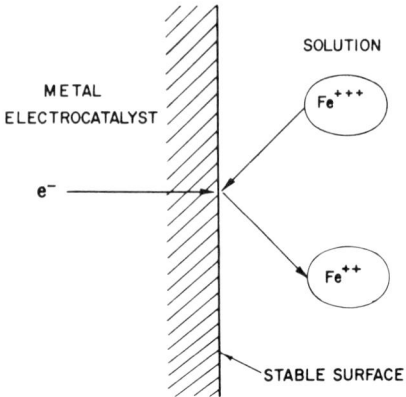

Fig. 10.9. Representation of a stable surface that acts only as an electron source or sink for a chemical reaction.

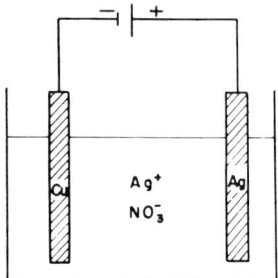

Fig. 10.10. An electrochemical cell involving the electronation of Ag^+ at an electron source (copper cathode).

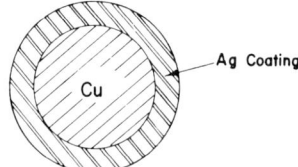

Fig. 10.11. An illustration of how the copper cathode in Fig. 10.10 has changed during the electronation of Ag^+. Unlike the situation in Fig. 10.10, the copper surface has not been maintained in its original form.

ELECTRODIC REACTIONS OF SPECIAL INTEREST 1175

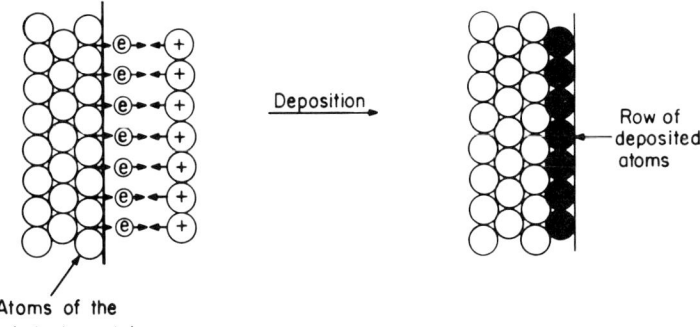

Atoms of the substrate metal

Fig. 10.12. The path taken by ions undergoing deposition and lattice incorporation to form a new row of atoms.

(Fig. 10.13) concerns the process of *crystallization*, or crystal growth, the name given to the cooperative process by which the individual acts of ionic deposition link up to build up old crystals or grow new ones.

10.2.2. The Reaction Pathway for Electrodeposition

The first step in the deposition process is that in which an ion crosses the electrified interface, i.e., the charge-transfer reaction.

Picture the situation (Fig. 10.14). A hydrated ion, e.g., a silver ion, is waiting at the OHP. In the direction of the silver metal electrode, there is the three-dimensional network, or lattice, consisting of silver ions cemented

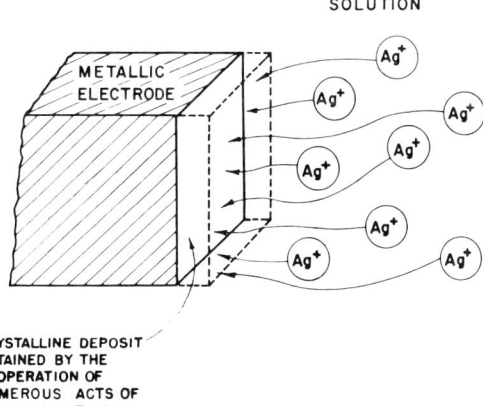

CRYSTALLINE DEPOSIT OBTAINED BY THE COOPERATION OF NUMEROUS ACTS OF IONIC DEPOSITION

Fig. 10.13. The formation of a crystalline deposit involving the deposition of several rows of atoms in a manner identical to that shown in Fig. 10.12.

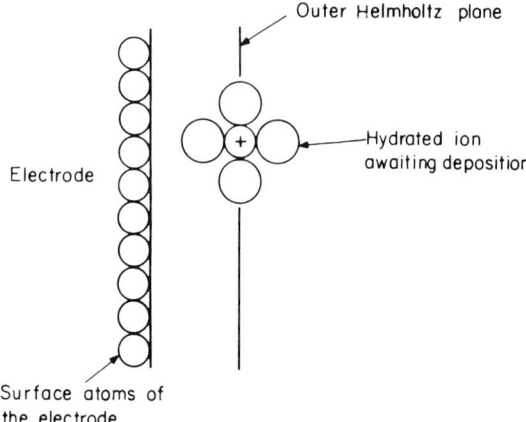

Fig. 10.14. Representation of a hydrated ion at the OHP awaiting deposition and lattice incorporation.

together by an electron gas. The silver ions in the lattice each lay claim to an electron of the electron gas; in this sense, they can be said to be neutral and referred to as metal atoms which are, of course, unhydrated. On the other hand, the silver ions in solution are not only charged but undeniably hydrated (*cf.* Section 2.4).

A simple conclusion follows. Before a silver ion from solution becomes part of the metallic lattice (Fig. 10.15), it has to receive an electron and divest itself of its sheath of hydration water. In short, the deposition of an ion consists of *electronation* and *dehydration*. How the ion goes through electronation and dehydration is an interesting story, only the broad outlines of which will be sketched here.

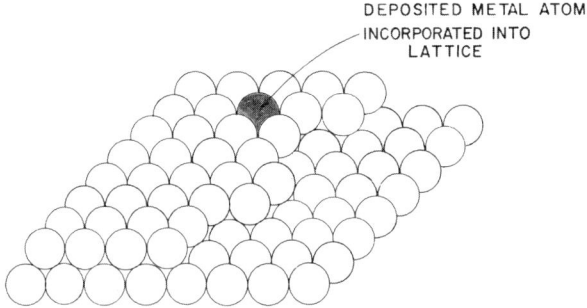

Fig. 10.15. The final stage in a metal deposition process, in which the deposited atom becomes incorporated into the lattice.

The electronation of the metal ion is basically the same as the electronation of the proton. There, the proton was considered closely associated with one water molecule (*cf.* Section 5.2). Here, the metal ion is surrounded by several water molecules. Electron tunneling could only occur when the H_3O^+ ion was deformed sufficiently (Section 8.5); here, too, the electron in the metal, intent on tunneling to the ion, has to wait until, during the constantly changing deformations of the ion–water complex, the water sheath is pushed aside sufficiently and the metal ion comes near enough the electrode to satisfy the tunneling condition of Section 8.5.5. Thus, the distortion of the hydration shell is a precondition for electron tunneling from the electrode to the metal ion in solution awaiting deposition. One can, of course, write down the tunneling conditions in terms of potential-energy surfaces or profile diagrams and become quantitative. The aim here, however, is to present the basic picture.

What happens to the charge on the ion after the electron has tunneled to it from the metal? This charge depends on what the ion does with the electron which it receives into an acceptor energy level. If, after the momentarily neutralized ion (atom) lands on the metal surface, it tosses back the electron as its contribution to the free-electron gas; then the ion has acquired full status as a member of the metallic lattice. It has become a neutral metal atom in the sense that it can lay claim to an electron of the electron gas, i.e., it has an electron associated with it. But this means that the electronated ion has to get rid of all its water molecules at one shot.

Some alternatives arise here. When electrons tunnel to hydrated ions, are the particles which are formed on the electrode surface metal atoms of zero charge? Or are they *adions*, i.e., partially charged adsorbed ions? The evidence on this point is necessarily indirect. It arises largely from comparative calculations of the heat of activation for the electronation of some metal ions on assuming that they become, on the one hand, charge-free surface atoms and, on the other hand, partially charged surface adions. Calculations which assume the direct formation of charge-free surface atoms yield values of the heat of activation that are so much higher than those of experiments that their formation must be regarded as improbable.

However, what are the consequences of the residual charge which calculations indicate must be possessed by surface adions?

10.2.3. Stepwise Dehydration of an Ion; the Surface Diffusion of Adions

To answer this question one had to take a more detailed look at the electrode surface. If one does not do this, one might assume that all sites

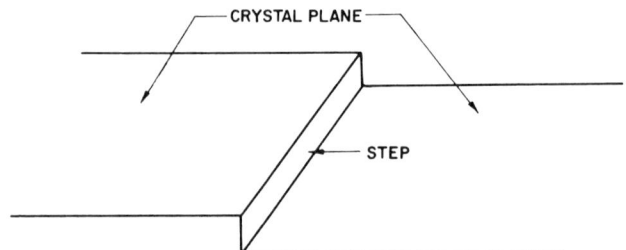

Fig. 10.16. Step site upon a crystal-lattice plane.

on the electrode look alike. The fundamental point is that an electrode presents a richly differentiated array of sites to an ion crossing the interface.

At first, one can consider that the electrode is a *single* crystal. Thus, instead of consisting of small crystals separated by grain boundaries, there is one crystal with an uninterrupted network of atoms extending right through its bulk. The surface of such an ideal crystal is not necessarily a perfect plane. The planes on its surface exhibit steps (Fig. 10.16), kinks (Fig. 10.17), edge vacancies (Fig. 10.18), and holes (Fig. 10.19).

One can do a thought experiment at this stage. Place an ion at each one of these different kinds of sites. It will be partly in contact with metal ions of the substrate, but the remaining space around it can accommodate water molecules. But how many water molecules can be associated with the metal ion on the surface? *That depends on the site.* The maximum number of hydration water molecules with which an adion can associate is available when the adion is sited *on* the plane, and the number progressively decreases as one considers the ion at a step, kink, edge vacancy, and hole (Fig. 10.20). So, if an ion moves from plane to step to kink and then is enveloped in a hole, its surroundings change; it tends to lose a water molecule and acquire a metal atom neighbor with each move.

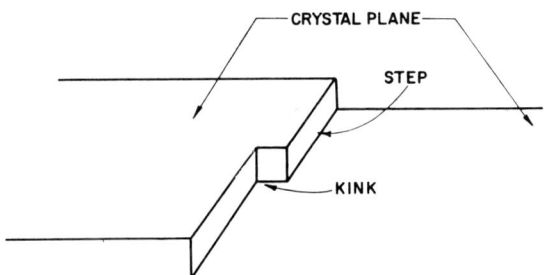

Fig. 10.17. Kink site upon a crystal-lattice plane.

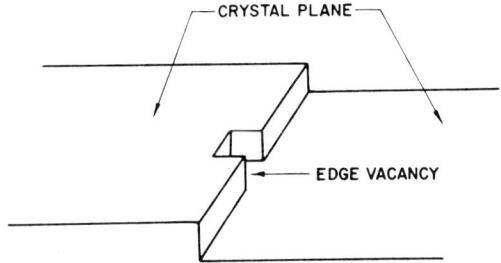

Fig. 10.18. Edge vacancy upon a crystal-lattice plane.

Now, this description of the stepwise replacement of water molecules by metal ions as nearest neighbors to an ion can be linked up with the charge-transfer reaction. To what site is the ion likely to cross the interface?

One possibility is for the ion to wander about on the solution side of the interface, say, in the OHP, till it comes face to face with a hole site. Then, in one shot, the ion could get electronated, divest itself of its solvent sheath, and dive into the lattice. This would be a direct one-step deposition reaction (Fig. 10.21).

Alternatively, the ion could jump onto a plane (Fig. 10.22). Two factors favor such a jump to a plane site compared with a direct jump to a hole. Firstly, sites on a smooth plane are far more numerous than the other sites such as holes, kinks, and steps. Secondly, crossing over to a plane site requires the minimum amount of distortion in the ion–water complex, and hence the minimum energy change from this source. This is an important point because it is connected with basic charge-transfer theory. Before electron tunneling to the ion can occur, the ion–water-sheath bonds have to present themselves in activated form, and it is the stretching and distorting associated with this activation which determines the amount of energy needed for activation (see Section 8.5.5). Since ion transfer to a planar site involves the minimum changes in hydration, such transfers have the least activation energy and go fastest compared with the rates of electron transfer at other sites (Table 10.6), for example, those to kink sites (Fig. 10.20), where the distortion is relatively greater.

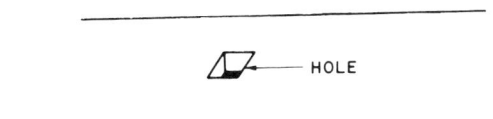

Fig. 10.19. Hole vacancy upon a crystal-lattice plane.

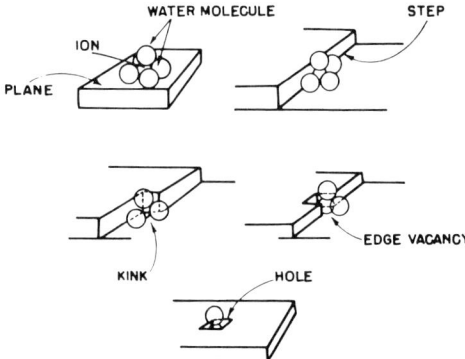

Fig. 10.20. Representation of the progressive dehydration of a hydrated ion as its site changes from surface to step to kink to edge vacancy to hole.

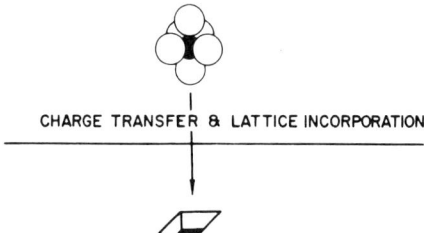

Fig. 10.21. Representation of the direct transfer of a hydrated ion to a hole site upon a crystal-lattice plane.

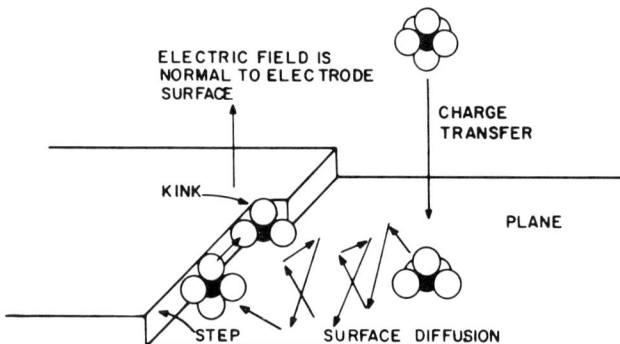

Fig. 10.22. Consecutive stages involved in the incorporation of an ion at a kink site.

TABLE 10.6

Calculated Heats of Activation at the Potential of Zero Charge for Ag⁺ and Cu²⁺ Transfer Directly from the Outer Helmholtz Plane to Various Surface Sites (in kcal mole⁻¹)

Solvated ion	To planar site	To step site	To kink site	To edge-vacancy site	To hole site
Ag^+	10	21	35	>35	$\gg 35$
Cu^{2+}	130	180	>180	>180	$\gg 180$

After having landed on a plane, the ion has become a surface adion which still has some charge and therefore some molecules of hydration water associated with it. It has less than the full ionic charge ze_0 of an ion in solution; hence, it must have less than the number of water molecules which hydrate an ion in solution. But the adion has quite a few things to accomplish before it can get incorporated into the metal lattice. It must move on the surface to a step, where it loses one more water molecule, and then move along the step to a kink site (another water molecule lost, Fig. 10.22). A similar process continues. Further, hydration water molecules are replaced by coordinating metal atoms until, finally, the series of actions ends when this ion, now with "zero" charge, gets embedded in the lattice.

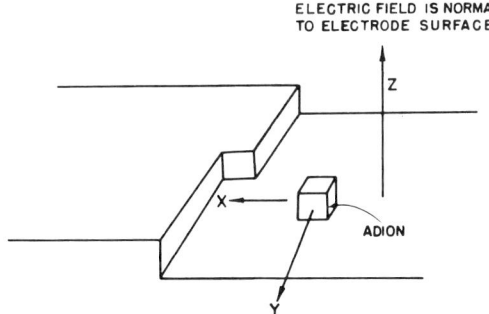

Fig. 10.23. Since the surface adion is unaffected by the electric field normal to the electrode surface, in order to reach the step site, the adion diffuses in a random-walk manner.

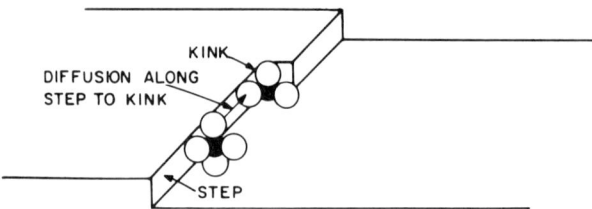

Fig. 10.24. Following surface diffusion to a step site, the adion diffuses along the step to a kink site and to final lattice incorporation.

The ion is now like any other metal "atom" in the lattice. The *deposition*[†] process has ended.

How does the ion move on the surface? It cannot drift under an electric field because the field at an interface is normal to the electrode surface (Fig. 10.23) and what is under discussion here is motion parallel to the surface plane. The movements are by a random-walk diffusion process in two dimensions, surface diffusion.

So, what one is talking about here is the goings-on which take place in getting a hydrated ion out of the solution and into the metal lattice. Initially, charge transfer occurs and an adion is formed; then, several phenomena follow—the zig-zag walk across a planar surface (surface diffusion), the "collision" with a step, a gradual surrendering by the ion of its remaining water molecules as it surrounds itself with other metal atoms. In short, electrodeposition is a multistep reaction of charge transfer followed by surface diffusion to steps (Fig. 10.22), transfer from plane sites to step sites, then diffusion along the step to a kink site (Fig. 10.24), and finally lattice incorporation. Figure 10.25 shows the arrival of several ions at surface sites and the consecutive steps involved in the advance of a step by lattice building.

10.2.4. Mechanism Determination on Surfaces Which Change with Time

The moment one talks of consecutive steps in an overall reaction, one wonders: What step determines the rate of the reaction? In the context of deposition, the question becomes: Of the various steps (Fig. 10.25), namely, ion transfer across the electrified interface, surface diffusion of adions to

---

[†] As distinct from what happens after that, *viz.*, the buildup of the new material formed into certain types of crystal forms.

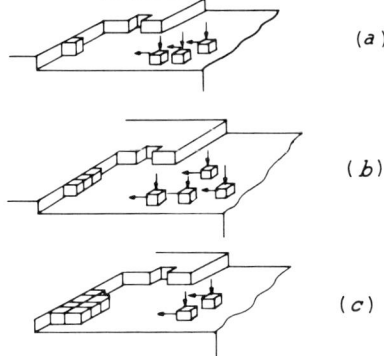

Fig. 10.25. Consecutive steps (charge transfer, surface diffusion, and lattice incorporation) involving several ions and showing the advancement of a step by lattice building.

steps, diffusion along steps to kinks, etc., which one controls the rate of electrodeposition?

So one considers the methods of mechanism determination described earlier in order to pick out suitable methods. But what is particular about electrogrowth? What of the use of steady-state methods, for instance? One could pass a steady current across the interface, measure the steady potential across it, and then determine reaction orders, transfer coefficients, stoichiometric number, etc. (*cf.* Section 9.5.5)? A steady-state procedure may be possible when the electrode is an electrocatalyst so that it can be assumed to be unchanging with time. Then the occurrence of the reaction does not change the substrate.

In the case of electrogrowth, however, there is a special problem. The substrate is growing; its surface is advancing and changing with time. Actually, as will be explained later, even its topographical features change with time. Hence, the measurement of current density and potential at any instant is made on an electrode surface which is different from the surface for a measurement made at an earlier instant. One's experimental data become suspect if the electrode surface is not constant during measurements.

How does one circumvent the difficulty of an electrode surface which is fickle? The key to success is the use of transients (*cf.* Section 9.2). The deposition reaction must be "caught on the wing." The measurement must be done so quickly that the surface, for all intents and purposes, has not changed during the measurement.

The simplest transient technique involves the use of a constant current (a current step) switched on suddenly at a time arbitrarily taken as $t = 0$. The electrode is built into an electrochemical system or cell, and the whole system is connected in series with a high resistance to an external current source (Fig. 10.26). The high resistance is to the circuit what the rds is to an electrodic reaction; it determines the current flowing through the circuit and hence through the electrified interface under study. Because the resistance of the interface, given by $d\eta/di$ (cf. Section 8.2.12), is small, i.e., a few ohms, in comparison with the large series resistance, i.e., a few megohms, the current through the interface is virtually unaffected by changes in the resistance of the interface with time as the overpotential builds up, i.e., current is constant with time. Hence, one can make measurements of the time variation of the potential and arrange things so that the total number of ions depositing onto the electrode during the measurement is so small (e.g., 10% of a monolayer) that the electrode surface is negligibly changed. The use of transients thus permits one to clutch a little bit of permanence out of a rapidly changing situation on the electrode surface.

How can such transients throw light on the rds? A quantitative treatment is given in Section 10.2.5, so a qualitative answer will be indicated here (see also Section 9.2). The point is that the way in which the potential difference across the interface builds up as a function of time depends upon the rds.

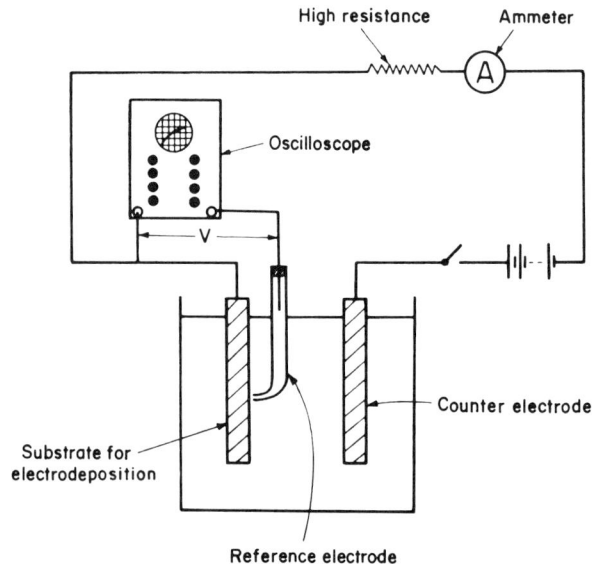

Fig. 10.26. An electrochemical cell and its ancillary equipment used for the observation of transient phenomena.

Suppose the charge transfer of ions is the rds, the succeeding steps of surface diffusion of adions, etc., being in equilibrium. When the external current is switched on at $t = 0$, electrons flow through the metallic conductor to the electrode surface. But the electronation current depends on the overpotential. Hence, for small values of overpotential, only a small fraction of the electrons flowing in from the external circuit get across to the ions, and, therefore, most of them simply mount up on the electrode surface without going across to the ions. The double layer is being charged up, and the potential rises rapidly.

The rate of buildup of potential can be calculated on the basis of a model. Thus, one can assume, for instance, that charge transfer is the process that determines the time for the buildup of potential, the *rise time*.[†] The rise time can also be measured and compared with the calculated value. If the two do not agree, one must conclude that an incorrect model has been assumed and the rise time is governed by a rate-determining process which occurs after charge transfer. The most obvious second model to try is that in which the surface diffusion of adions to steps is rate determining. Of course, if one still does not get agreement between theoretical and measured rise times, one can try further models, e.g., the rate-determining formation of a crystal nucleus.

The interpretation of constant-current transients (Mehl) provided the first indication that, under certain conditions, metal deposition may involve surface diffusion of adions as an rds (Lorenz, Gerischer).

10.2.5. The Time Variation of the Average Adion Concentration in Response to the Switching on of a Constant Current

The problem, therefore, is to develop a quantitative treatment for the rise time of potential in a constant-current transient. To simplify the treatment, it will be assumed that the deposition reaction consists of two steps, the transfer of an ion across the electrified interface to form an adion on a plane—the charge-transfer step—followed by the surface diffusion of adions to steps. The subsequent transfer of adions from plane sites to step sites, the diffusion along these edges to kinks, etc., will not be considered in the quantitative treatment because calculations have suggested that these subsequent processes are energetically easier than the processes of charge transfer and surface diffusion.

Thus, the problem involves the derivation of the expression for the

---

[†] A more quantitative definition of the rise time will be given later (*cf.* Section 10.2.7).

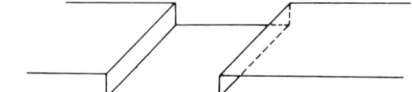

Fig. 10.27. A surface plane bounded on two sides by steps.

time variation of the average adion concentration and thence the time variation of the potential, on the assumption of a two-step reaction—a charge transfer followed by surface diffusion, with double-layer charging being ignored.

Consider an area of the electrode surface in which the surface plane is bounded on two sides by two parallel steps (Fig. 10.27). Let everything be initially at equilibrium—no net deposition and no net dissolution. One-half of the picture is that of ions from the solution crossing the interface and landing on the plane as adions, which then random-walk to the steps (Fig. 10.28). The subsequent movement to kinks, etc., will not be considered. The other half of the picture consists of adions quitting the steps to random-walk on the plane and, from there, leaping off across the electrified interface into solution as hydrated ions (Fig. 10.29). The two halves of the equilibrium picture when juxtaposed yield a view of the interface with charges flitting to and fro across it and of the electrode surface with adions zig-zagging about on it.

What of the concentration of adions on the plane surface of the electrode? At equilibrium, the adion concentration c_0 must be the same everywhere on the surface planes. If this concentration were not the same throughout the surface, a concentration gradient would set up diffusion which would even out the concentrations to the c_0 value.

Consider, now, that a net deposition-current density is switched on at a time $t = 0$. The ions crossing the double layer land principally on the

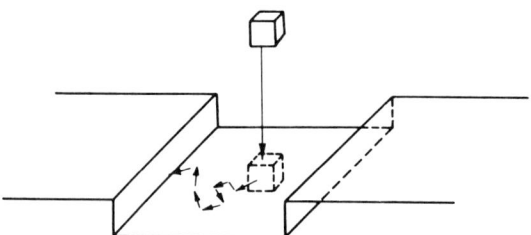

Fig. 10.28. The charge transfer of an ion to the surface followed by a random-walk process to the step: Deposition.

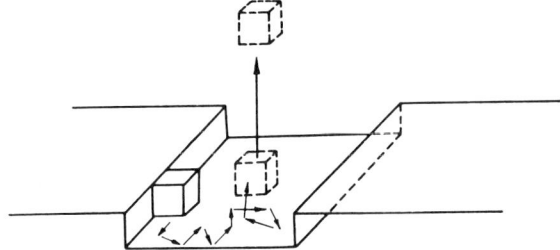

Fig. 10.29. An adion leaving the step site, random-walking out onto the surface, and then undergoing charge transfer into the solution as a hydrated ion: Dissolution.

planes between the steps and, in doing so, increase the adion concentration at any point on the plane to values greater than the equilibrium adion concentration. At the steps, however, where there is an easy exchange of adions between planar sites and step sites, the equilibrium adion concentration is maintained. Hence, a concentration gradient is set up, which forces adions to surface-diffuse to the steps (Fig. 10.30).

The equilibrium situation has been disturbed. A clear view of this perturbation emerges (Fig. 10.31) by analyzing what happens to the adions in a small area on the plane. Despic, by considering only surface diffusion along the x axis normal to the growth steps, has described four processes which affect the equilibrium adion concentration c_0: (1) Adions diffuse into the area, (2) adions diffuse out, (3) adions are formed by the electronation reaction, and (4) adions leap into solution in the de-electronation reaction.

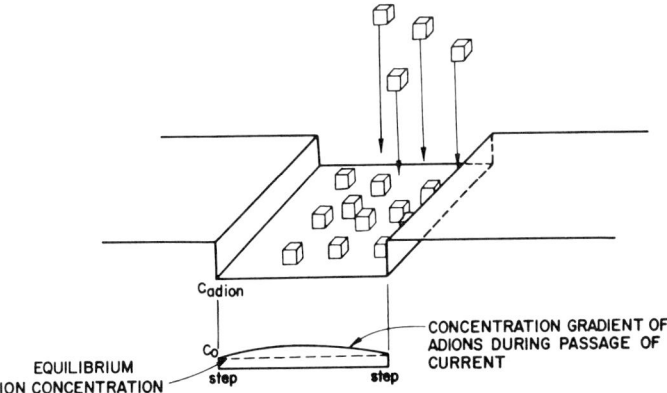

Fig. 10.30. Deposition of surface adions under nonequilibrium conditions, resulting in a concentration gradient from a point between the steps to the step.

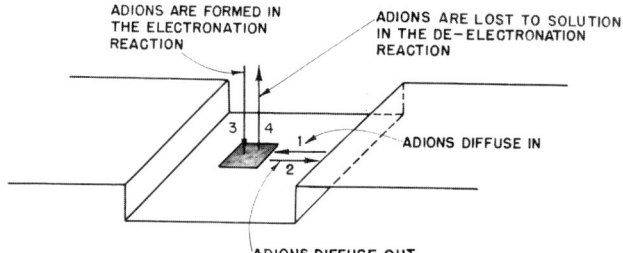

Fig. 10.31. Representation of the four processes which influence the adion equilibrium concentration c_0.

The change of adion concentration with time, dc/dt, depends on the rates of all these four processes.

Instead of handling all these complications, a simple treatment will be adopted. A "blind eye" will be turned to all local variations of the individual electronation and deelectronation currents on the plane and also on variations of adion concentration with distance from the growth steps. It will be assumed that, during the buildup of potential, there is a average current density i on the plane equal to the constant current density i which has been switched on and that the average adion concentration at the time t is $\bar{c}_t$. In an electronation (cathodic) transient, ions are being delivered *to* the surface, and, hence, the value of $\bar{c}_t$ will be greater than the equilibrium value c_0. Thus, the perturbation in the adion concentration due to the current is $\bar{c}_t - c_0$. It is this perturbation which makes the ions move, surface-diffuse, from landing points on the plane to acceptor sites at the steps on the single-crystal surface.

One can assume that, for small perturbations, the surface-diffusion flux J_{SD} (the number of moles crossing 1 cm of growth step per second) is proportional to the perturbation,[†] i.e., equal to $k(\bar{c}_t - c_0)$, where k is the proportionality constant. Now, depending on whether a deposition (electronation) or dissolution (deelectronation) current is applied, adions are formed on or removed from the surface. It will be recalled, however, that the convention is to designate electronation currents as negative and deelectronation currents as positive. If, therefore, the average rate at which adions are formed on the plane is represented by $-i/F$, then positive currents (dissolution) make $-i/F$ negative, i.e., lead to adion removal, and negative currents (deposition) make $-i/F$ positive, i.e., lead to adion formation.

[†] It will be shown later that, if one assumes a linear variation in adion concentration with distance from the growth steps, this proportionality follows automatically.

What happens to the difference between $-i/F$, the average rate at which adions are supplied to the plane, and $k(\bar{c}_t - c_0)$, the rate at which adions are removed from the plane to the steps by surface diffusion? The difference goes to increase the adion concentration with time, i.e.,

$$\frac{d\bar{c}_t}{dt} = -\frac{i}{F} - J_{SD}$$

$$= -\frac{i}{F} - k(\bar{c}_t - c_0) \qquad (10.27)$$

This differential equation can easily be integrated after the following rearrangement

$$\frac{d\bar{c}_t}{dt} = -\left(\frac{i}{F} - kc_0\right) - k\bar{c}_t$$

$$= -(k\bar{c}_t + m) \qquad (10.28)$$

where

$$m = \frac{i}{F} - kc_0 \qquad (10.29)$$

Hence,

$$\frac{d(k\bar{c}_t + m)}{(k\bar{c}_t + m)} = -k\, dt \qquad (10.30)$$

which, upon integration, gives

$$k\bar{c}_t + m = Ae^{-kt} \qquad (10.31)$$

where A is an integration constant. To evaluate A, one recalls that, at time $t = 0$, the adion concentration is c_0. Hence, at $t = 0$, $e^{-kt} = 1$ and

$$A = kc_0 + m \qquad (10.32)$$

whereupon

$$k\bar{c}_t + m = (kc_0 + m)e^{-kt} \qquad (10.33)$$

Substituting for m from Eq. (10.29), one has

$$k\bar{c}_t + \frac{i}{F} - kc_0 = \frac{i}{F}e^{-kt}$$

i.e.,

$$\bar{c}_t - c_0 = -\frac{i}{kF}(1 - e^{-kt}) \qquad (10.34)$$

Here, then, is an expression for the average adion concentration $\bar{c}_t$

at any time t. For a deposition process, $i < 0$, i.e., the current is negative, and, therefore, $\bar{c}_t > c_0$, which shows that the surface concentration of adions builds up with time to a steady-state value given by setting $t \to \infty$ in Eq. (10.34). The rate of buildup is seen to depend on the magnitude of the constant current density used. The higher this current density is, the quicker is the buildup.

10.2.6. The Contributions of Double-Layer Charging and Faradaic Reaction to the Total Deposition-Current Density

An expression (10.34) has been derived for the average surface concentration of adions as a function of time. It will be recalled, however (cf. Section 9.2.3), that, in the study of the transient response of an interface to a constant current, the experimental technique consists in observing the time variation of the potential (or overpotential). Thus, one has to proceed from the expression for $\bar{c}_t$, the average surface concentration of adions at a time t, to derive an equation for η_t, the overpotential at the time t.

The starting point is the argument that the total constant current density can be resolved into a double-layer charging-current density $i_{\rm DL}$ and the faradaic current density, which arises from a charge-transfer reaction $i_{\rm CTR}$ (cf. Section 9.2.3),

$$i_g = i_{\rm DL} + i_{\rm CTR} \tag{10.35}$$

The current density for charging the double layer is determined by the capacity of the interface (cf. Section 9.2.3)

$$i_{\rm DL} = C_{\rm DL} \frac{d\eta_t}{dt} \tag{10.36}$$

The faradaic current density is given by the de-electronation (dissolution) current density minus the electronation (deposition) density (cf. Section 8.2.7)

$$i_{\rm CTR} = \overleftarrow{i} - \overrightarrow{i} \tag{10.37}$$

Now explicit expressions for the two component current densities $\overleftarrow{i}$ and $\overrightarrow{i}$ must be worked out in terms of the model of adion formation.

Consider the dissolution-current density. It depends on the average adion concentration and the overpotential at the time t, in the following way,

$$\begin{aligned}\overleftarrow{i} &= F\overleftarrow{k}\bar{c}_t \exp\left\{\frac{(1-\beta)F\Delta\phi_e}{RT}\right\}\exp\left\{\frac{(1-\beta)F\eta_t}{RT}\right\}\\ &= \left[F\overleftarrow{k}c_0\exp\left\{\frac{(1-\beta)F\Delta\phi_e}{RT}\right\}\right]\frac{\bar{c}_t}{c_0}\exp\left\{\frac{(1-\beta)F\eta_t}{RT}\right\}\end{aligned} \tag{10.38}$$

The term in square brackets is the equilibrium exchange-current density i_0; recall (cf. Section 8.2.6) that it is always defined with respect to the equilibrium concentration of reactants. Hence,

$$\overleftarrow{i} = i_0 \frac{\bar{c}_t}{c_0} e^{(1-\beta)F\eta_t/RT} \tag{10.39}$$

In the charge-transfer step of the deposition (electronation) reaction, ions from the OHP cannot land on top of the adions on the electrode surface; they can only land on areas which are not blocked up. Hence,

$$\overrightarrow{i} = F\overrightarrow{k}(1 - \theta_t)c_{M^{z+}}e^{-\beta F \Delta\phi_e/RT}e^{-\beta F\eta_t/RT} \tag{10.40}$$

where θ_t is the fraction of the surface covered with adions at the time t; and $c_{M^{z+}}$, the bulk concentration of metal ions. By rearrangement,

$$\overrightarrow{i} = F\overrightarrow{k}(1 - \theta_e)c_{M^{z+}}e^{-\beta F \Delta\phi_e/RT} \frac{1-\theta_t}{1-\theta_e} e^{-\beta F\eta_t/RT}$$

$$= i_0 \frac{1-\theta_t}{1-\theta_e} e^{-\beta F\eta_t/RT} \tag{10.41}$$

However, if θ_t and θ_e are both much less than unity, then one can ignore the term $(1 - \theta_t)/(1 - \theta_e)$. This is what will be assumed in the present treatment. Thus,

$$\overrightarrow{i} = i_0 e^{-\beta F\eta_t/RT} \tag{10.42}$$

Substituting Eq. (10.39) for $\overleftarrow{i}$ and Eq. (10.42) for $\overrightarrow{i}$ in Eq. (10.37), one has, for $i_{CTR} = \overleftarrow{i} - \overrightarrow{i}$,

$$i_{CTR} = i_0 \left[\frac{\bar{c}_t}{c_0} e^{(1-\beta)F\eta_t/RT} - e^{-\beta F\eta_t/RT} \right] \tag{10.43}$$

Considering small departures from equilibrium, i.e., the low-field approximation, the exponentials can be linearized thus

$$i_{CTR} = i_0 \left\{ \frac{\bar{c}_t}{c_0} \left[1 + \frac{(1-\beta)F\eta_t}{RT} \right] - 1 + \frac{\beta F\eta_t}{RT} \right\}$$

$$= i_0 \left\{ \frac{\bar{c}_t - c_0}{c_0} + \frac{F\eta_t}{RT} \left[\frac{\bar{c}_t}{c_0}(1-\beta) + \beta \right] \right\} \tag{10.44}$$

But, for small departures from equilibrium, $\bar{c}_t/c_0 \approx 1$, in which case,

$$\frac{\bar{c}_t}{c_0}(1-\beta) + \beta \approx 1 \tag{10.45}$$

and, therefore,

$$i_{\mathrm{CTR}} = i_0 \left(\frac{\bar{c}_t - c_0}{c_0} + \frac{F\eta_t}{RT} \right) \quad (10.46)$$

Introducing the expressions (10.36) and (10.46) for i_{DL} and i_{CTR}, respectively, in Eq. (10.35), it is clear that the constant current density switched on at $t = 0$ is given by

$$i = C_{\mathrm{DL}} \frac{d\eta_t}{dt} + i_0 \left(\frac{\bar{c}_t - c_0}{c_0} + \frac{F\eta_t}{RT} \right) \quad (10.47)$$

10.2.7. The Time Variation of the Overpotential and the Rate-Determining Step in Electrodeposition

The time variation of the overpotential expected on the basis of two assumed rds's will now be worked out.

In the first instance, assume that the charge-transfer step is rate determining. This implies that the surface diffusion of adions from planes to steps is in quasi equilibrium and does not, therefore, control the overall deposition rate. In other words, the average adion concentration $\bar{c}_t$ at the time t is assumed to be virtually equal to the equilibrium concentration, i.e.,

$$\bar{c}_t \approx c_0 \quad \text{and} \quad \frac{\bar{c}_t - c_0}{c_0} \approx 0 \quad (10.48)$$

Under these conditions of rate-determining charge transfer, the $(\bar{c}_t - c_0)/c_0$ term in Eq. (10.47) can be set equal to zero, in which case one has

$$i = C_{\mathrm{DL}} \frac{d\eta_t}{dt} + i_0 \frac{F\eta_t}{RT} \quad (10.49)$$

If one recalls that near-equilibrium conditions are being considered, it is clear that, when steady state is reached at $t \to \infty$, the steady-state overpotential η_∞ for rate-determining charge transfer is given by the low-field law (cf. Section 8.2.11)

$$\eta_\infty = \frac{RT}{F} \frac{i}{i_0} \quad (10.50)$$

On combining this relation with Eq. (10.49), the result is

$$\frac{d\eta_t}{dt} = \frac{Fi_0}{RTC_{\mathrm{DL}}} (\eta_\infty - \eta_t) \quad (10.51)$$

By using the notation

$$\tau_{CT} = \frac{RTC_{DL}}{Fi_0} \qquad (10.52)$$

Eq. (10.51) becomes

$$\frac{d(\eta_\infty - \eta_t)}{\eta_\infty - \eta_t} = -\frac{1}{\tau_{CT}} dt \qquad (10.53)$$

By integration,

$$\eta_\infty - \eta_t = A e^{-t/\tau_{CT}} \qquad (10.54)$$

and the integration constant A is obtained by noting that, at $t = 0$, $e^{-t/\tau_{CT}} = 1$ and $\eta_t = 0$ (i.e., the system is at equilibrium—zero overpotential—before switching on the constant current), which means that

$$A = \eta_\infty \qquad (10.55)$$

Thus, the variation of the overpotential with time is given by

$$\eta_t = \eta_\infty (1 - e^{-t/\tau_{CT}}) \qquad (10.56)$$

This equation is of the same form as the equation describing the time variation of potential difference V_t across a parallel capacitor–resistor network when it is charged with a constant current

$$V_t = V_\infty (1 - e^{-t/\tau}) \qquad (10.57)$$

From the mathematical point of view, Eqs. (10.56) and (10.57) indicate that the final steady-state values η_∞ and V_∞, respectively, will only be reached after an infinite time $t \to \infty$. In practice, therefore, the progress of the process of potential variation from the initial equilibrium value to the final steady-state value is judged on an arbitrary basis. In the case of condenser charging, the *rise time* is defined as the time at which the magnitude of the exponent in the exponential term is equal to unity. Setting $t/\tau = 1$ in Eq. (10.57) implies that

$$V_t = V_\infty (1 - e^{-1}) \qquad (10.58)$$

or

$$\frac{V_t}{V_\infty} = 63\% \qquad (10.59)$$

By adopting a similar criterion in the case of the buildup of the overpotential in the metal-deposition process, it is clear from Eq. (10.56) that

$$\frac{\eta_t}{\eta_\infty} = 63\% \qquad (10.60)$$

when

$$t = \tau_{\text{CT}} = \frac{RTC_{\text{DL}}}{Fi_0} \qquad (10.61)$$

Alternatively, one can consider in what time the overpotential η_t is virtually equal to the steady-state overpotential η_∞, i.e., say, 98.2% of η_∞. The precise figure 98.2 is chosen because

$$\frac{\eta_t}{\eta_\infty} = 98.2\% \qquad (10.62)$$

when

$$\eta_t = \eta_\infty(1 - e^{-4}) \qquad (10.63)$$

or

$$t = 4\tau_{\text{CT}} = \frac{4RT}{Fi_0} C_{\text{DL}} \qquad (10.64)$$

With the use of Eqs. (10.64) and (10.61), it is possible to predict in what time interval the overpotential builds up to either 63 or 98.2% of the final value. In other words, one can calculate either the rise time of the deposition process or the time to attain a value which is virtually equal to the steady-state value.

When, however, these predictions are compared with the experimental variation of potential, it turns out that the observed buildup of overpotential is about ten times *slower* that that expected on the basis of a rate-determining charge-transfer step in the electrodeposition process (see Table 10.7 later in this section). This puzzled the researchers who carried out the basic experiments on the mechanism of electrodeposition. In fact, it was this discrepancy between the predicted and observed rise times that led them to think that electrodeposition is not a one-step reaction and there may be other steps which may be rate determining. It was in this context that the concepts of adions and their surface diffusion were developed (although the concept of adions was born in calculations of the likely path which an ion would take during deposition, Section 10.2.3).

To return to the analysis, now assume that surface diffusion is rate determining (Fig. 10.32), i.e.,

$$\bar{c}_t \not\approx c_0 \quad \text{and} \quad \frac{\bar{c}_t - c_0}{c_0} \not\approx 0 \qquad (10.65)$$

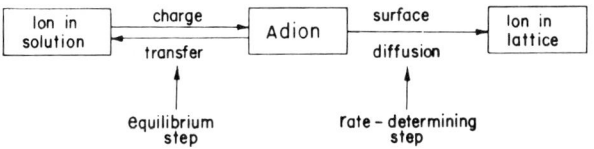

Fig. 10.32. Schematic representation of the processes involved in the incorporation of an ion into a lattice position if surface diffusion is assumed to be the rds.

Thus, the total current density is given by Eq. (10.47), i.e.,

$$i = C_{\rm DL}\frac{d\eta_t}{dt} + i_0\left(\frac{\bar{c}_t - c_0}{c_0} + \frac{F\eta_t}{RT}\right) \quad (10.47)$$

Since $i_{\rm DL} = C_{\rm DL}(d\eta_t/dt)$ is a maximum immediately after switching on the constant current subsequent to which it decays asymptotically to zero, it is only in the initial stages of the potential-time transient that the double-layer charging current must be considered. If, therefore, one considers times longer than the time for charging the double layer, the double-layer current can be considered virtually zero, i.e.,

$$C_{\rm DL}\frac{d\eta_t}{dt} \approx 0 \quad (10.66)$$

Under these conditions, from Eq. (10.47),

$$i = i_0\left(\frac{\bar{c}_t - c_0}{c_0} + \frac{F\eta_t}{RT}\right) \quad (10.67)$$

and by substituting for $\bar{c}_t - c_0$ from Eq. (10.34), the result is

$$\frac{i}{i_0} = -\frac{i}{Fkc_0}(1 - e^{-kt}) + \frac{F}{RT}\eta_t \quad (10.68)$$

or

$$\eta_t = \frac{RT}{F}\frac{i}{i_0} + \frac{RT}{F^2}\frac{i}{kc_0}(1 - e^{-kt}) \quad (10.69)$$

As in the case of the analysis of rate-determining charge transfer, the steady-state overpotential η_∞, when surface diffusion controls the deposition rate, is obtained by setting $t \to \infty$ in Eq. (10.69), i.e., the result is

$$\eta_\infty = \frac{RT}{F}\frac{i}{i_0} + \frac{RT}{F^2}\frac{i}{kc_0} \quad (10.70)$$

Combining this relation with Eq. (10.69), one finds that

$$\eta_t = \eta_\infty - \frac{RT}{F^2} \frac{i}{kc_0} e^{-kt} \tag{10.71}$$

or

$$\ln [\eta_t - \eta_\infty] = -\ln \left(\frac{RT}{F^2} \frac{i}{kc_0}\right) - kt \tag{10.72}$$

This is the basic equation for the time variation of the overpotential, assuming that deposition consists of a two-step charge-transfer plus surface-diffusion reaction and the rate-determining surface diffusion is under near-equilibrium conditions. From this equation, the rise time τ_{SD} is easily found by setting the magnitude of the exponent in Eq. (10.71) equal to unity. Thus,

$$\tau_{SD} = \frac{1}{k} \tag{10.73}$$

To compare the rise time τ_{SD} calculated on the basis of rate-determining surface diffusion with the observed rise time of potential, it is necessary to evaluate k. This quantity, it will be recalled [cf. Eq. (10.27)], is the proportionality constant relating the surface-diffusion flux of adions to the difference between the average surface concentration $\bar{c}_t$ of adions at the time t and the equilibrium adion concentration c_0.

There is however a simpler test of the theory. Equation (10.72), which is based on the rate-determining surface diffusion of adions, predicts that, when $\log (\eta_t - \eta_\infty)$ is plotted against the time t, a straight-line plot should be obtained. Some experimental results in conformity with this prediction are shown in Fig. 10.33. This agreement indicates the basic validity of the

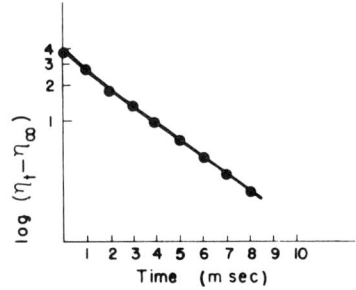

Fig. 10.33. Log $(\eta_t - \eta_\infty)$ against time t for a two-step reaction mechanism involving surface diffusion as the rds.

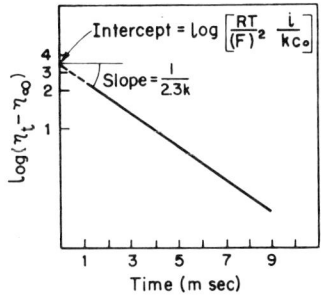

Fig. 10.34. Log ($\eta_t - \eta_\infty$) against time t from which values of rise time τ_{SD} and the equilibrium adion concentration c_0 can be evaluated.

model as a two-step reaction consisting of a charge-transfer step followed by a rate-determining surface-diffusion step.

The rise time $\tau_{SD} = 1/k$ can also be obtained from the log ($\eta_t - \eta_\infty$) versus t plot; it is the reciprocal of the slope (Fig. 10.34). Rise times obtained in this way are much longer than the rise times τ_{CT} calculated on the basis of rate-determining charge transfer (see Table 10.8).

Another check on the theory consists of an extrapolation of the log ($\eta_t - \eta_\infty$) versus t plot to zero time. Since the intercept thus obtained (Fig. 10.34) is equal to ln $[(RT/F^2)(i/kc_0)]$—cf. Eq. (10.72)—and the value of $[(RT/F^2)(i/k)]$ is easily calculated by using the k given by the reciprocal slope of log ($\eta_t - \eta_\infty$) versus t, one can determine the equilibrium adion concentration c_0.[†] Table 10.7 shows the value of c_0 as a function of current density i, and Table 10.8 shows c_0 as a function of the type of electrode surface.

---

† The value of c_0 can be obtained in a more convenient way by the use of an approximation. Equation (10.69) can be written as

$$\eta_t = \frac{RT}{F} \frac{i}{i_0} + \frac{RT}{F^2} \frac{i}{kc_0} (1 - 1 + kt)$$

when the exponential e^{-kt} can be linearized (cf. Section 8.2.11); this is valid only when t is very much less than the rise time τ_{SD}. Differentiating the above equation with respect to time produces

$$\frac{d\eta_t}{dt} = \frac{RT}{F^2} \frac{i}{c_0}$$

Thus, the value of equilibrium adion concentration c_0 can be evaluated.

TABLE 10.7

Equilibrium Adion Concentration c_0 in the Electrodeposition of Silver at Near-Equilibrium Conditions

Current density i, mamp cm^{-2}	Equilibrium adion concentration c_0, moles cm^{-2}	Rise time τ_{SD}, μsec
4.5	3.3×10^{-11}	342
6.3	2.1×10^{-11}	187
8.9	2.6×10^{-11}	156
15.0	2.3×10^{-11}	96
22.6	3.2×10^{-11}	75
44.7	5.8×10^{-11}	53
62.6	1.1×10^{-11}	22

Introducing the value of c_0 into the expression (10.70) for the steady-state overpotential, one can calculate the exchange-current density. The values of i_0 derived in this manner are in agreement with those obtained by methods which do not involve the theory of surface diffusion, e.g., extrapolation to $\eta = 0$ from conditions of much higher overpotentials, where (as can be shown) surface diffusion is no longer rate controlling, i.e., rate control is electron charge transfer.

TABLE 10.8

Equilibrium Concentration c_0 as a Function of the Initial Electrode-Surface Preparation for Silver

Type of electrode	Equilibrium adion concentration c_0, moles cm^{-2}
Electrode quenched in H_2 atmosphere	90×10^{-11}
Electrode quenched in a helium atmosphere	3×10^{-11}
Electrode anodically pulsed	160×10^{-11}
Electrode prepared from an electrodeposition in AgCl–KCl–LiCl melt	90×10^{-11}
Electrode whose surface is abraded in solution	15×10^{-11}

10.2.8. The Contribution of Charge Transfer and Surfaces Diffusion to the Total Overpotential for Electrodeposition at Steady State

The steady-state overpotential η_∞ expected for a two-step charge-transfer plus surface-diffusion reaction has been shown to be given (for low overpotentials) by [*cf.* Eq. (10.70)]

$$\eta_\infty = \frac{RT}{F}\frac{i}{i_0} + \frac{RT}{F^2}\frac{i}{kc_0} \qquad (10.70)$$

Consider the factor kc_0. It is the product of the proportionality constant determining the surface-diffusion flux and the equilibrium adion concentration. If this product is large, the second term in (10.70) becomes negligible compared with the first. Under these conditions,

$$\eta_\infty = \frac{RT}{F}\frac{i}{i_0} \qquad [\text{for } k \to \infty] \qquad (10.74)$$

Since this is the familiar low-field approximation for the overpotential in a single-step charge-transfer reaction, it means that, when k is large, the total overpotential for the electrodeposition process is determined by the charge-transfer step. Hence, $(RT/F)(i/i_0)$ is the charge-transfer contribution η_{CT} to the total overpotential, i.e.,

$$\eta_{CT,\infty} = \frac{RT}{F}\frac{i}{i_0} \qquad (10.75)$$

The second term in Eq. (10.70) contains the surface-diffusion parameter k and must therefore be the surface-diffusion contribution η_{SD} to the total overpotential. That is,

$$\eta_{SD,\infty} = \frac{RT}{F^2}\frac{i}{kc_0} \qquad (10.76)$$

and

$$\eta_\infty = \eta_{CT,\infty} + \eta_{SD,\infty} \qquad (10.77)$$

It is seen, therefore, that the total steady-state overpotential η_∞ consists of a charge-transfer contribution and a surface-diffusion contribution and their relative magnitudes depend upon the surface-diffusion parameter kc_0. When low total overpotentials are used for electrodeposition, it has been established that, for certain systems, surface diffusion is rate determining. Hence, at low overpotentials, the parameter kc_0 must be small enough

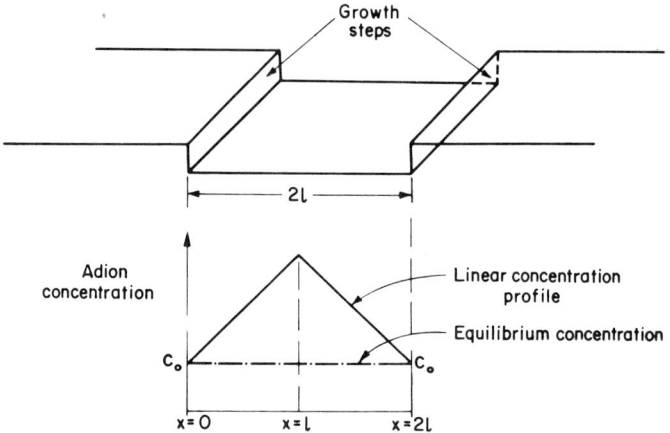

Fig. 10.35. Representation of a linearized adion-concentration profile under nonequilibrium conditions between a pair of growth steps at a distance $2l$ apart.

[cf. Eq. (10.76)] to make it impermissible to neglect $\eta_{\text{SD},\infty}$ in comparison with $\eta_{\text{CT},\infty}$.

What about at high overpotentials? Is the situation unchanged? The quantity c_0 must be independent of overpotential because it is an equilibrium quantity. In contrast, the diffusion parameter k may become dependent on overpotential. The form of this dependence may be intuited by the following simple picture.

Assume a linear variation of adion concentration with distance from the growth steps (Fig. 10.35). From Fick's first law, the surface-diffusion flux is given by

$$J_{\text{SD}} = -\frac{D}{l}(c_l - c_0) \tag{10.78}$$

where c_l is the adion concentration at the midpoint l between the growth steps, c_0 is the equilibrium adion concentration *at* the growth step.

Equation (10.78) can be rewritten in the form

$$J_{\text{SD}} = -\frac{D}{l/2}\left(\frac{c_l + c_0}{2} - c_0\right) \tag{10.79}$$

In the case of a linear variation of adion concentration with distance from the growth steps,

$$\frac{c_l + c_0}{2} = \bar{c}_t \tag{10.80}$$

where $\bar{c}_t$ has been defined as the average adion concentration. Hence,

$$J_{SD} = -\frac{D}{l/2}(\bar{c}_t - c_0) \tag{10.81}$$

from which it follows by writing

$$k = -\frac{D}{l/2} \tag{10.82}$$

that

$$J_{SD} = k(\bar{c}_t - c_0) \tag{10.83}$$

which is the expression used earlier in Eq. (10.27).

Now suppose that the assumption of a linear concentration profile is relaxed (Fig. 10.36). Then it means that $l/2$ is not a quarter of the distance between steps but an effective distance δ (cf. the diffusion layer δ, Section 9.3.9) over which the concentration varies linearly

$$k = -\frac{D}{\delta} \tag{10.84}$$

So the question is: What happens to the effective thickness δ of the surface-diffusion layer with an increase in overpotential? An exact analysis of the variation in adion concentration shows that δ decreases with increase in overpotential, and, consequently, the diffusion parameter k increases. Hence, there must come a stage when k is so large that the second term in Eq. (10.70) becomes negligible compared with the first, i.e., the rate of the overall electrodeposition reaction is determined by the charge-transfer step. Table 10.9 shows how the values of k and i_0 determine the rate-controlling

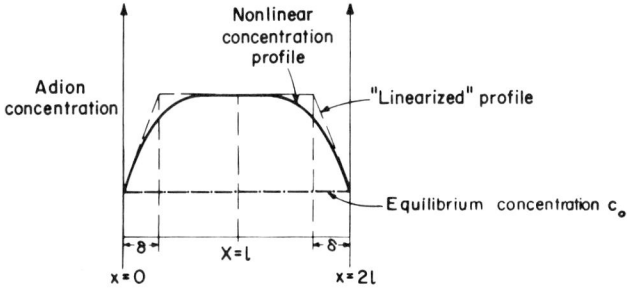

Fig. 10.36. An adion-concentration profile showing a linearized profile only over the distance δ (the diffusion-layer thickness).

TABLE 10.9
Exchange-Current Density, Surface-Diffusion Parameter k, and the Rate-Controlling Step, as a Function of the Electrode Surface for the Cu–Cu^{2+} System

Type of electrode surface	Exchange-current density i_0	Low current density		High current density	
		Surface-diffusion parameter k	Rate-control step	Surface-diffusion parameter k	Rate-control step
Electrodeposited	Low	High	CT+SD	Very high	CT
Helium quenched	High	Low	SD	Very high	CT
H$_2$ quenched	High	Low	SD	Very high	CT
Oxide film	Low	High	CT+SD	Very high	CT

step at high and low current densities for the Cu–Cu^{2+} system for different preparations of the surface.

The deposition of metals such as silver or copper may therefore be summarized as follows: The process involves a multistep reaction consisting of a charge-transfer step to form adions on the planes, surface-diffusion of adions to steps, transfer of adions from plane sites to step sites, diffusion along steps to kinks, followed by lattice incorporation. Table 10.9 shows that, at low overpotentials, surface diffusion (SD) is rate determining for some substrates (of copper) but, at high overpotentials, the charge-transfer (CT) step tends to become rate controlling.

10.2.9. From Deposition to Crystallization

A very elementary treatment has been presented of the deposition aspect of the process of electrogrowth. Many approximations have been made. For example, it has been assumed that the surface adion concentration is small not only at equilibrium but also during the passage of current. Consequently, in considering the transfer of ions from the OHP to the surface, a $1 - \theta$ factor was not included to ensure that ions land on free sites. Another major approximation is that of near equilibrium. Thus, the treatment would need modification for metal-deposition processes with low equilibrium exchange-current densities because such processes are relatively polarizable and therefore show large overpotentials, i.e., departures from the equilibrium potential even at very low current densities.

Ideas about the mechanism of deposition are vital to the understanding of the process of electrogrowth. But the sequence of steps (consisting of charge transfer to form adions, surface diffusion of adions to steps, transfer to step sites, diffusion along steps to kinks, and then lattice incorporation) constitute *one* individual act of deposition, and it takes many deposition acts to result in the observed electrocrystallization of metals. One can conclude, therefore, that, though the electrochemical growth of crystals is founded on acts of ionic deposition, understanding of the *electrogrowth* of metals does not end with the picture of deposition that has been sketched. Indeed, that part of the matter is only the beginning of the overall process of electrocrystallization, although perhaps it is the part which, at the present time, is known best. One now has to treat the crystallization aspect of the process of the electrogrowth of metal.

The remainder of this section will therefore be devoted to a sketch of some of the key features of the theory of crystal growth at electrodes in contact with an electrolyte. Since some of the basic aspects of this theory derive from considerations of crystal growth from the vapor phase, one may question the inclusion of this topic in a book on electrochemistry. But the point is that there are specifically electrochemical features to crystal growth at electrodes. One may mention, e.g., the effect on ionic transport in solution, the role of the electroadsorption of constituents of the electrolyte, and the dependence of the surface free energy upon potential.

10.2.10. Some Devices for Building Lattices from Adions: Screw Dislocations and Spiral Growths

The steps by which ions from solution are incorporated into the lattice have been pictured. The next question is obvious. What happens when many ions travel the deposition path, i.e., the path of charge transfer to a plane, surface diffusion to steps, then movement to kinks, and finally lattice incorporation?

As more and more adions join a step, it advances. Electrogrowth is occurring. The more adions incorporated into a step, the farther it advances, but also the closer it comes to the edge of the electrode and, eventually, a stage must come when there is no step on the surface. The step has disappeared (Fig. 10.37). But, according to the deposition mechanism sketched in Section 10.2.3, the existence of steps is a necessary condition. Without steps, the adion concentration on the plane builds up, θ increases and

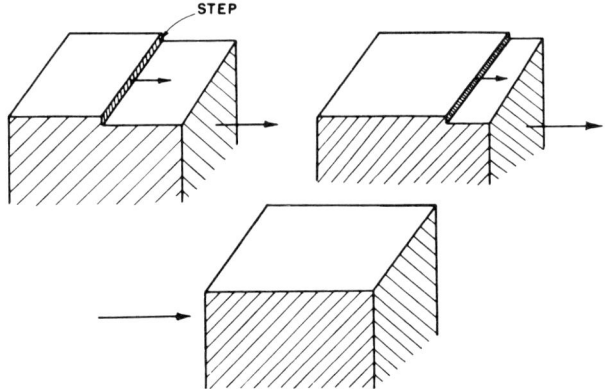

Fig. 10.37. Growth-step advancement by lattice building until the electrode edge is reached, where the step disappears.

makes it more and more difficult for charge transfer to occur (remember that electronation requires a bare surface), and, thus, further deposition should stop.[†]

In practice, however, the deposition current just keeps flowing on and the electrogrowth does not cease. Nature seems to have some trick up her sleeve by which the surface of a crystal is perpetually provided with steps. What is this device?

It turns out to be simple. Nature rarely works with crystals as ideal as the ones considered in Section 10.2.3. Crystals are grown, e.g., from a melt, and, in the general rush of crystallization, the majority of crystals grow with built-in defects and imperfections in the way in which their atoms are arranged. It is one type of these imperfections which contains the secret of nonvanishing, self-perpetuating steps. The mechanism by which this type of defect arises has yet to be understood in its details and complications; so what will be done here is to describe a model of a crystal with a defect which is both nonvanishing and self-perpetuating.

Imagine that a perfect crystal is cut, not right through, but only up to a point, and then the part of the crystal on one side of the cut is pushed up or down through one interatomic distance relative to the part of the

---

[†] Actually, there is an out to this situation. When the overpotential is fairly high, two-dimensional *nucleation* can occur (as suggested by Stranski, and proved by Kaischew and Budewski), i.e., adions condense together without random walking separately to steps. But this process requires current densities—rates of deposition—much higher than those normally used. So (under normal circumstances) one has to think about the buildup of lattices without dragging in the nucleation mechanism.

crystal on the other side of the cut. What is meant is shown in Fig. 10.38. A *dislocation* has been produced in the once-perfect crystal.

The mismatch of atomic layers (arising from the process which formed the single crystal) has made a ledge emerge on the surface. The defect has been advertised on the surface. But think of the atomic layers beneath the surface. The mismatch penetrates (Fig. 10.39) right through the crystal; remember that one whole side of the crystal has been pushed up relative to the other side. A view from above the surface looking down the axis, shown in Fig. 10.39, resembles what one would see if one looked down upon a spiral staircase from above, but the staircase is spiraling all the way down.

The most important point about the type of defect described above is that it gives rise to a step on the crystal surface. Now consider what happens if adions keep adding on to the step. In the first instance, think about the addition of a whole row of adions starting from the point X where the step originates at the surface and ending at the edge M (Fig. 10.40). What is the result of the addition of this row? Has the step disappeared? No. The point X is still anchored to the same axis normal to the surface, but the step, XM', is at an angle to its former position XM (Fig. 10.40).

Further, however many uniform rows of adions are added to the surface, the step still remains on the crystal surface; all that happens is that its orientation to the surface changes. When the orientation of the

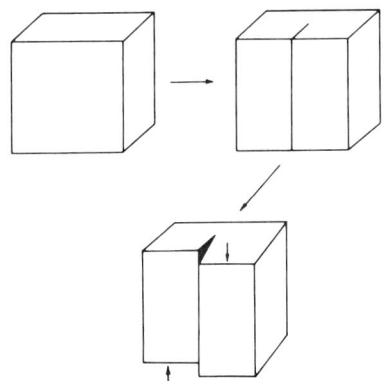

Fig. 10.38. Representation of a crystal-lattice dislocation. The left half of the crystal has been pushed up, while the right half has been pushed down.

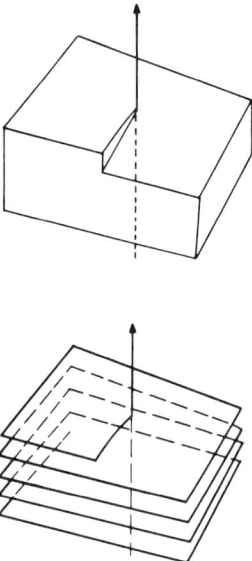

Fig. 10.39. A block and atomic-layer representation of a lattice dislocation illustrating the spiral of atomic layers.

step changes by one complete revolution (i.e., an angle of 2π radians), the crystal has added on a new layer of atoms in its growth upward. Thus, as the crystal grows, the step rotates about the axis at X going through the crystal; it winds like a screw (Fig. 10.41), which is why the type of defect has been described by Frank as a *screw dislocation*.

What happens, however, if adions do not add on in complete rows

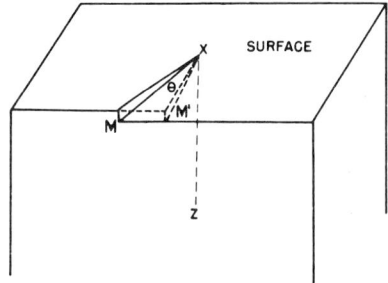

Fig. 10.40. The initial stage in the formation of a screw dislocation; the addition of adions to edge XM moves the step to position XM'.

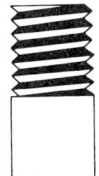

Fig. 10.41. Screw-thread spiral analogy to a screw dislocation.

all along the step from the screw dislocation axis to the edge of the crystal? Suppose that they add on only up to a fraction XY of the step length XM (Fig. 10.42). Then, if this happens several times, another small step PQ has formed on the surface. This new step, too, can be the recipient of adions, and it can advance. If this process goes on, one obtains an interesting sort of growth. In plan, the growth looks like a spiral; and, in elevation, like a mountain which has been terraced (Fig. 10.43). This is known as a *microspiral* growth.

10.2.11. Microsteps and Macrosteps

The steps that have been described so far are microsteps. They are one atomic layer in height and therefore too small to be seen in an optical microscope. But sometimes steps *are* clearly visible in an ordinary optical microscope (Fig. 10.44). Such steps must therefore have a height of the

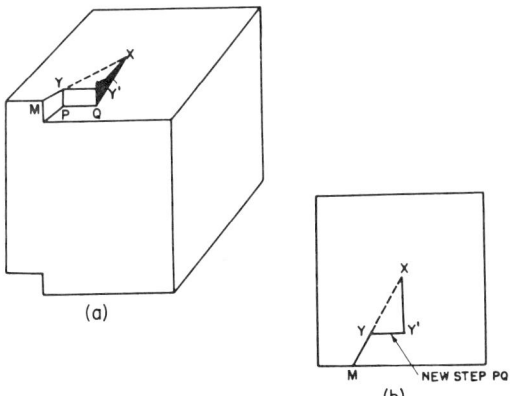

Fig. 10.42. An incomplete step XM movement; lattice building occurring only along XY, the first stage in the formation of a terraced microspiral growth.

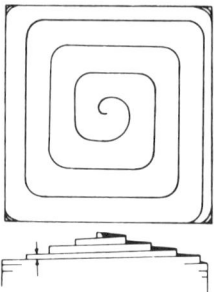

Fig. 10.43. Plan and elevation of a microspiral growth. The arrows indicate the thickness of one atomic layer.

order of a wavelength of light, several thousand angstroms: These are known as *macrosteps*. How do they form?

Numerous reasons for the formation of macrosteps have been suggested, including those by Frank and by Cabrera and Vermilyea. It was considered that the velocity of a microstep was dependent upon two factors, the proximity of the microsteps to each other—the closer the steps, the more slowly they move since the flux corresponding to a given current density is distributed among *all* steps—and the presence of adsorbed impurities. *Ideally*, one can consider what will happen when one microstep stops advancing, al-

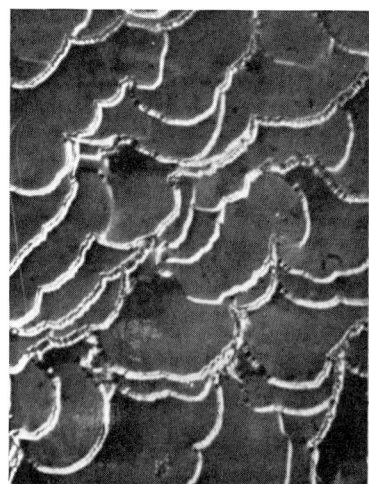

Fig. 10.44. Layer type of growth for copper deposition showing macrosteps with irregular edges.

though the result of this, i.e., microstep bunching, does not depend upon the microstep's coming to a complete halt but only on the fact that its rate of advance has become slightly less than the steps following it.

Let it be supposed then, that an advancing microstep suddenly stops advancing. The movement may cease, e.g., owing to the adsorption of impurities from solution at the step. On a solid surface with its hierarchy of sites, there will be a hierarchy of free energies of adsorption (see Section 7.6), and it may happen that impurities seek adsorption at steps in preference to adsorption on flat planes.

So think of a microstep which, for a reason such as that given above, has stopped advancing somewhere within the boundary of the crystal (Fig. 10.45). Now imagine that a layer B of atoms is growing on top of the layer A. The step B will keep advancing until it comes to the point where the advance of the step A was blocked. The layer B will then act as though it has reached the edge of the crystal. If the same process is repeated with another layer C on top of B, and then another layer on C, and so on, then there is a pile-up of layer upon layer. Microsteps bunch into macrosteps. Sometimes the pile-up reaches such proportions that it can be seen in a microscope as a macrostep.

The development of the macrostep through the bunching of microsteps has been described. Now suppose that this bunching mechanism occurs at all the steps of a microspiral. Then, instead of the difference in

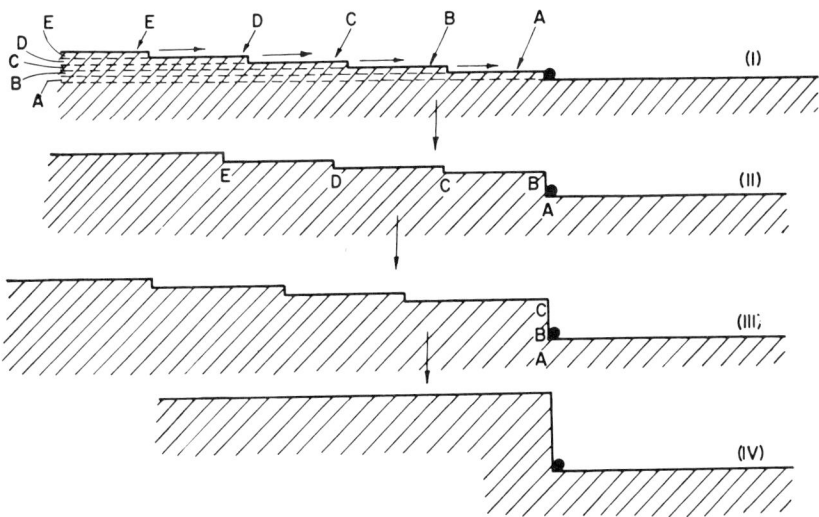

Fig. 10.45. Representation of the four successive stages in the formation of a macrostep by microstep bunching.

height between the steps of a microspiral's being one atomic layer (*cf.* Fig. 10.43), it will be of the same order as in a macrostep, i.e., several thousand angstroms. In short, the result is a macrospiral growth which is clearly visible in a microscope. The observation of macrospiral growths is a clear verification of the role of screw dislocations in sustaining crystal growth.

10.2.12. How Steps from a Pair of Screw Dislocations Interact

It must not be imagined that a single-crystal surface is allotted only one isolated screw dislocation and, therefore, there will be only one growth spiral. Except in the case of special kinds of very thin rodlike crystals called *whiskers*, even so-called "perfect" single crystals are richly endowed with screw dislocations. Now each screw dislocation generates a rotating step which can in principle span the whole surface. If, therefore, there is more than one screw dislocation, there will be an interaction of the steps generated from each dislocation. In other words, the step rotating from one screw dislocation can collide with the step rotating from another screw dislocation. What happens when these steps collide?

This problem will be side-stepped for a moment, and a simpler one tackled. Consider two steps which are parallel to each other, the type of steps considered in the analysis of the constant-current transient (*cf.* Sections 9.2.3 and 10.2.5). As ions transfer across the electrified interface and the adions thus formed surface-diffuse and become incorporated in the steps, there is an advance of the steps toward each other (Fig. 10.46). Eventually the two steps approach each other, some closely, so that all one is left with is a one-atom-wide and one-atom-deep chasm. The moment this is filled in, the two steps disappear. The collision of the two steps moving toward each other has resulted in their mutual annihilation.

Now back to the original problem. How can the steps emanating from two screw dislocations move toward each other? One way is for the steps to rotate in opposite directions. This will happen if one screw disloca-

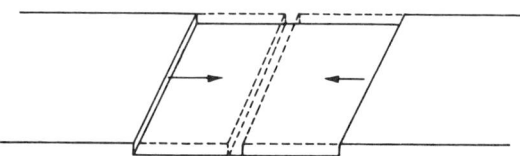

Fig. 10.46. The movement of two parallel growth steps toward each other, resulting in their mutual annihilation.

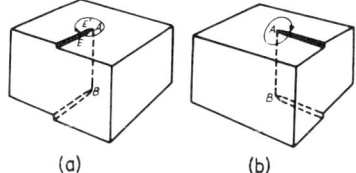

Fig. 10.47. (a) Right- and (b) left-handed screw dislocations.

tion is right-handed and the other is left-handed (Fig. 10.47); they bear the same relation to each other as a right screw thread to a left screw thread.

Consider, therefore, two screw dislocations emerging on the surface (Fig. 10.48) and forming two steps. Suppose that, in an ideal case, adions are adding on to the steps uniformly. There comes a stage when the two steps collide.

This collision will generally occur at a crystal edge because steps originate at a dislocation axis and extend to the edge of the crystal. A V-shaped step is formed (Fig. 10.49), the region inside the V being one atom layer lower than the outside. Now adions will be incorporated to the inside of the V, but this means that the angle inside the V keeps increasing and eventually it becomes a straight line joining the axes of the two screw dislocations. The step now runs from one axis to another, and further growth must be based on such a step. One-half of this step continues to spiral left and the other half spirals right, the final result being the formation of spiral growths with closed loops (Fig. 10.50).

This is only a highly simplified version of the interaction of screw dislocations. The situation is more complicated where there is nonuniform

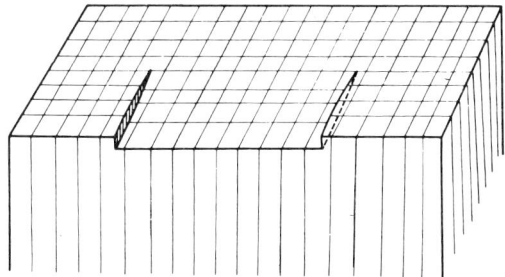

Fig. 10.48. Parallel growth steps each at a screw dislocation, one left-handed spiral and the other right-handed spiral.

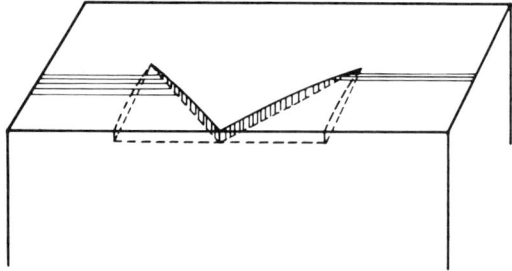

Fig. 10.49. The resultant merging of the two growth steps shown in Fig. 10.48, assuming uniform growth on both steps.

growth along the steps of the two screw dislocations, where both steps are rotating in the same direction, etc.

10.2.13. Crystal Facets Form

One has, therefore, a picture of ions from solution's being transferred onto the electrode surface as adions, of adions' joining steps, kinks, etc., of steps advancing on the surface, of screw dislocations' yielding growth spirals, of the surface advancing and occupying more of the solution. But, apart from the macrospiral growths which require special blockage and bunching mechanisms, the other types of growths (step advance, microspirals, etc.) lead to surface irregularities which are of *atomic* dimensions. The description is too *flat*. In practice, electrodeposits do not consist of smooth-faced, single crystals decorated with an occasional macrospiral. They

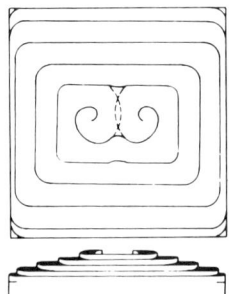

Fig. 10.50. Plan and elevation of the final result of two merging screw dislocations—growth spirals similar to those in Fig. 10.43 but with closed loops.

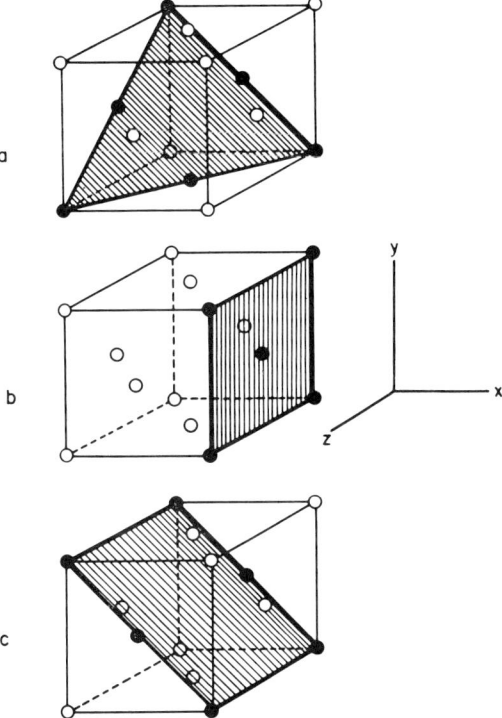

Fig. 10.51. Three faces of a cubic lattice showing how the atomic arrangement varies with the direction of the plane chosen.

display the forms and shapes characteristic of crystals; they exhibit facets and also nonuniformities of various kinds, e.g., the formation of complicated and beautiful growths such as dendrites. How do these forms and shapes develop? One has to see things on a grosser scale than steps, kinks, etc., otherwise, one will miss the forest for the trees.

Consider the three-dimensional arrangement of ions in a metal crystal. The ions are close packed. If one imagines a plane cutting the assembly, then, depending on the direction of cut, characteristic arrangements of ions are exposed at the surface (Fig. 10.51). Each arrangement is generated by the repetition of a unit pattern. In silver, e.g., the unit patterns might consist (Fig. 10.52) of silver ions in the center of other silver ions arranged in hexagons or squares. The silver ions could also be arranged in rectangles, the long side being $\sqrt{2}$ times the short side. The regular internal arrangement of ions in a silver crystal is advertising itself at the surface

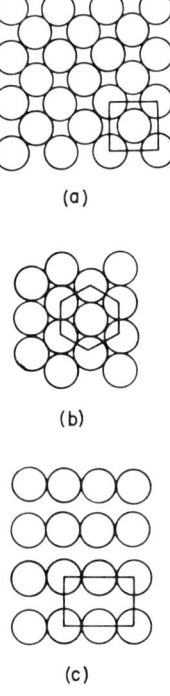

Fig. 10.52. Three crystallograhic arrangements as seen for the close-packed cubic lattice of silver.

through the characteristic unit pattern which represents different faces of the crystal. This is why they are known as crystal faces.[†] (A small, e.g., 1000 Å square, face is known as a *facet*).

[†] Every face in a crystalline lattice can be identified precisely in terms of *Miller indexes*. For cubic lattices, e.g., silver, to describe exactly each face, plane, or direction in the lattice, three space vectors x, y, and z are required, one vector y in the vertical plane intersecting at right angles two vectors x and z in the horizontal plane. Figure 10.51 shows three crystal faces of the cubic lattice, (a) (111), (b) (100), and (c) (110), crystal planes. For the hexagonal lattice, e.g., zinc, to describe lattice faces and planes exactly, one requires four space vectors because this lattice structure is now six sided compared with the four-sided cubic lattice, one vector y in the vertical plane intersecting at right angles three vectors u, x, and z, each describing a 60° angle to the other, in the horizontal plane. The Miller index is then the reciprocal of the length each plane describes along each vector and expressing the ratio of each vector length in integral numbers. For example, Fig. 10.51(b) describes unit-cell length along the x vector and infinity

TABLE 10.10

Exchange-Current Densities and the Total Deposition Overpotentials When $i = 10^{-2}$ amp cm^{-2} for Copper on Copper Single Crystals

Crystal face	i_0, amp cm^{-2}	Deposition overpotential [mV] when $i = 10^{-2}$ amp cm^{-2}
(110)	2×10^{-3}	-85
(100)	10^{-3}	-125
(111)	4×10^{-4}	-185

The relevance of crystal faces to the subject of electrocrystallization comes up as follows: Each of the crystal faces just described contains all the microfeatures which have been described in previous sections, i.e., steps, kinks, etc. Further, the same phenomena of deposition—the ions crossing the electrified interface to form adions, the surface diffusion, lattice incorporation of adions, screw dislocation, growth spirals, etc.—occur on *all* the facets.

What, then, is the difference between electrogrowth on one face compared with that on another? The *rates* of electrogrowth are (Razumney) different on different faces. Table 10.10 shows that, for the electrodeposition of copper on copper single crystals, the exchange-current densities increase from the (111) face through the (100) face to the (110) face. The rates of deposition, at constant overpotential, on these single crystals increases in the order of $(110) > (100) > (111)$.

This phenomenon should not be surprising for it is generally found that the surface properties of crystals depend on the atomic arrangements which are exposed at the surface.

The explanation of the difference in the rates of electrogrowth on different crystal faces is quite complicated. The differing energetics of two-dimensional nucleation upon different crystal faces has been suggested to account for the different growth rates (Pangarov), but, as mentioned earlier in Section 10.2.10, since nucleation is not likely to be involved in crystal growth at low current densities and yet preferential growth is experienced as such conditions, this proposal is not often applicable. The explanation probably lies more in connection with the energies with which adions bond

---

along the y and z vectors; taking the reciprocal of these lengths and expressing it as a ratio, one to another, produces for xyz the (100) plane.

onto the various types of crystal planes. The number of underlying metal atoms which are in contact with an adion depends on the pattern of atomic arrangement on the surface, i.e., on the particular crystal plane. An adion sitting on the (111), (100), and (110) planes will have 3, 4, and 5 close lattice-atom neighbors, respectively, to which it will be bonded. A situation analogous to this exists when a surface adion diffuses from one lattice site to another and so gains additional lattice-atom neighbors at each step, which finally culminates in lattice incorporation (*cf.* Section 10.2.3). Figure 10.20 illustrates that an adion has 3, 4, and 5 close atom lattice neighbors when it resides at a kink, edge vacancy, and hole, respectively, and Table 10.6 shows that the strength of bonding to the adion increases as its number of close lattice-atom neighbors increases.

Thus, the larger the number is of atoms of any crystal plane that are contiguous with the adion, the stronger is the bonding and, therefore, the faster is the charge-transfer step (*cf.* Fig. 10.5). Another possible factor affecting the electrogrowth rate on a particular face is the work function, which is known to be different on different crystal planes. Since the work function helps to determine the ease with which an electron tunnels to the depositing ion, it has an influence on the rate of the charge-transfer reaction (and, if this is a local effect, it may not completely cancel out at the metal–metal junction in a cell, as it would otherwise; *cf.* Section 10.1.2).

This differential deposition rate onto different faces has an important consequence; fast-growing faces tend to grow out of existence and disappear, and slow-growing faces tend to survive. This, perhaps at first, seemingly contradictory assertion follows from simple geometric arguments best grasped from a diagram (Fig. 10.53). The function of crystal faces or diminutive

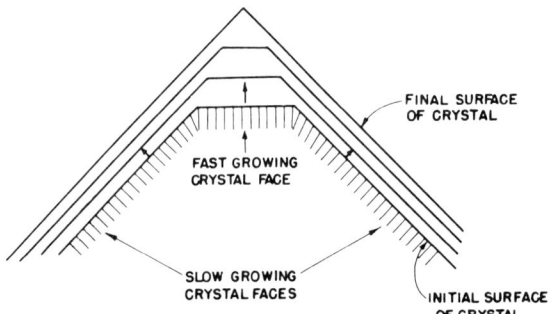

Fig. 10.53. The growth of a crystal, illustrating how a fast-growing face grows out of existence, while the slow-growing crystal faces remain.

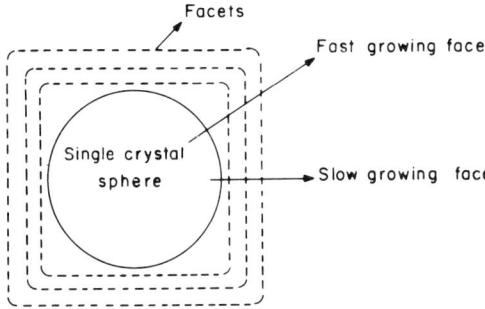

Fig. 10.54. Fast- and slow-growing faces as applied to a single crystal sphere. The facets which develop and are observed are those crystal faces with the slowest growth rates.

faces, or facets, arises, then, as a result of the different rates of deposition on different crystal faces of the substrate. This means that, even if one carries out deposition onto a single crystal sphere, it soon breaks out (Fig. 10.54) into a rash of those facets that grow slowest.

At this point, an objection may be raised. The above argument about faceting may be valid on a sphere where all possible faces are present (the tangent to the surface at any point may be considered a hypothetical face), but, if deposition is carried out on a flat single-crystal face, where are the different faces to grow at different velocities? There is only one face, it may be said.

The single flat face may be valid as a starting condition, but, with continued deposition, all sorts of things happen. The microsteps bunch into macrosteps, and nonuniformities appear on the surface. At the projections, the electric field becomes concentrated and causes faster growth (Fig. 10.55). In this way, the surface of the electrode starts showing sufficient nonflatness for the operation of the law of differential growth velocities of different crystal faces. Then facets start developing on the electrodeposit. This is the crystallographic stage of growth.

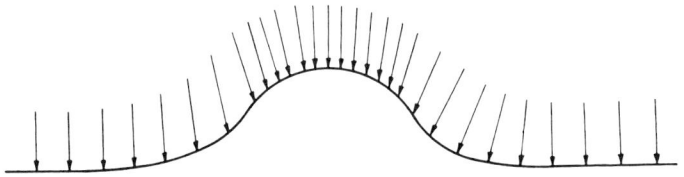

Fig. 10.55. A representation of the concentrated electric field, at a projection, producing a faster growth than at the flat surface.

10.2.14. Deposition on Single-Crystal and Polycrystal Substrates

In the elementary treatment of the phenomena of electrogrowth being presented here, it is inevitable that several simplifications and idealizations are adopted. Thus, it has been assumed so far that the electrodes used were single crystals. Such electrodes, however, have to be specially prepared by techniques such as the Bridgman technique.[†] The usual piece of metal encountered in everyday life is not one single crystal; X-ray analysis would show that it is a polycrystal, i.e., an agglomerate resulting from many single crystals (sometimes called *grains*) meeting at grain boundaries.

So the question arises: How valid is the picture of deposition, developed above, when the electrodes are polycrystalline metals? The answer is simple. One can consider the surface exposed to the solution by each grain as a single-crystal *micro*substrate and describe the deposition on this microsubstrate in the same terms as those used for single-crystal macrosubstrates. That is, one would have charge transfer followed by surface diffusion, transfer to steps, then to kinks, etc., and one would also have rotating steps resulting from screw dislocations, growth spirals, faceting, etc. In addition, however, at the grain boundaries where the single-crystal microsubstrates meet and the periodic atomic arrangement of each grain is interrupted, the deposition and growth processes will be abnormal. But the actual area of an electrode surface occupied by the grain boundaries is so negligible that the abnormal processes occurring there can be largely ignored.

In conclusion, therefore, the basic picture of deposition and growth developed for single crystals is valid as a basis for understanding the electrogrowth of polycrystals.

10.2.15. How the Diffusion of Ions in Solution May Affect Electrogrowth

What next? The situation is replete with possibilities. If the growing electrodeposit is inadequately supplied with metal ions, the nature of the

---

[†] The Bridgman technique is essentially the controlled solidification of a molten metal in a slowly moving temperature gradient. The apparatus used consists of a sealed glass tube drawn into a constriction at one end; the metal, when molten, is allowed to flow through this constriction, and, when slow cooling is begun, crystallization commences at this point and then proceeds at a controlled rate from there throughout the molten metal. Other methods used in single-crystal preparation are vapor deposition onto some suitable substrate, electrodeposition on a suitable substrate, and crystal-pulling techniques from molten metals.

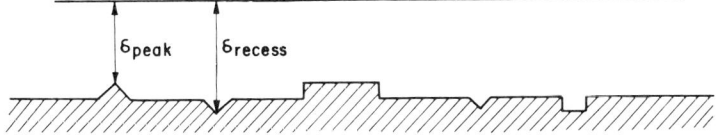

Fig. 10.56. A representation of a microrough surface where δ_{peak} is less than δ_{recess} and, hence, the growth rate when diffusion-controlled will be greater at the peak than in the recess.

further growth depends on how easily different parts of the electrode secure the supply of ions used to build up the crystal surface. One is talking of the logistical differences between different parts of the advancing crystal front.

One case is where the ions are traveling to the electrode by a process of diffusion. Then the steady-state diffusion problem can be looked at from the diffusion-layer point of view (Section 9.3). The variation of concentration with distance can be approximated to a linear variation, and the linear concentration gradient can be considered to occur over an effective distance of δ, the diffusion-layer thickness. Then the diffusion current is given by (cf. Section 9.3.9)

$$i = -DnF\frac{c^0 - c_{x=0}}{\delta} \qquad (9.78)$$

If the heights between peaks and recesses on the electrode are small compared with the diffusion-layer thickness δ (Fig. 10.56), then δ_{peak} will be less than δ_{recess} and, therefore, i_{peak} will be greater than i_{recess}. Hence, there will be greater amounts of deposition on the parts of the substrate which stick out. The rich become richer, and the poor become poorer. The nonuniformity increases, and the formation of "macrorough" deposits can be understood.

There are other ways in which ion transport leads to nonuniform growths. Consider, e.g., the following elementary theory of dendrite formation. Suppose that a *macro*spiral growth develops on a flat substrate surface. The tip of the spiral has quite a small radius of curvature ($r \sim 10^{-6}$ cm) and should not therefore be considered a plane sink that stimulates linear diffusion. It is virtually a *point* sink with the radius of curvature's being much less than the diffusion-layer thickness ($r \ll \delta$). Under these conditions, there is spherical diffusion[†] to the point sink (Fig. 10.57), and the limiting

---

[†] *Spherical diffusion* unlike linear diffusion where the reaction coordinate along which diffusion occurs is considered to reach vertically out from the electrode surface, has a reaction coordinate which is hemispherical. This hemispherical diffusion layer is de-

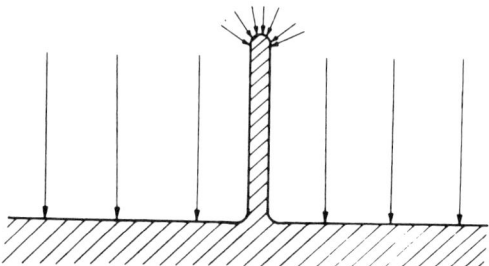

Fig. 10.57. The elementary dendrite theory in which the tip of a macrospiral has a radius of curvature $r \gg \delta$ and hence the growth rate is much greater at the tip than at a flat surface.

current density (*cf.* Section 9.3.10) is given by $DnFc^0/r$ and not $DnFc^0/\delta$ as is the case for linear (or planar) diffusion. Since $r \ll \delta$, it is obvious that the limiting current density to the spiral tip is much higher than to a projection with a radius of curvature of the order of the diffusion-layer thickness.

Another feature of the spiral tip is that it has an abnormally high step and kink density and perhaps the tip has a higher exchange-current density for deposition than the corresponding planar surface. If this were so, the activation overpotential would be much less at the tip of the spiral than around its base.

Arguments have been presented by Barton (and also by Hamilton) for both the concentration and activation overpotentials being much less at the tips of macrospirals than on the planar surface. It follows, therefore, that electro-growth tends to become concentrated at the spiral tip. The tip tends to grow faster than the rest of the substrate. This is part of the basis of the theory of the growth of long, thin, fast-growing faceted rods which some-

---

scribed, from geometrical considerations, as having a radius equal to the radius of curvature r of the surface from which it extends. Therefore, in the limiting current density and other diffusion equations, δ, which is a distance along the vertical coordinate, is replaced by r, the distance along a hemispherical coordinate.

Since, in stationary solutions, a diffusion-layer thickness of roughly 0.05 cm is present (*cf.* Section 9.3.9), to reduce the influence of diffusion in these stationary solutions, a spherical electrode of radius less than 0.05 cm could be used. Spherical diffusion will not become operative if the spherical electrode is larger than 0.05 cm in radius when used in stationary solutions. In highly stirred solutions where δ decreases toward approximately 10^{-3} cm, spherical diffusion will not be operative till the electrode has a radius of curvature of less than 10^{-3} cm.

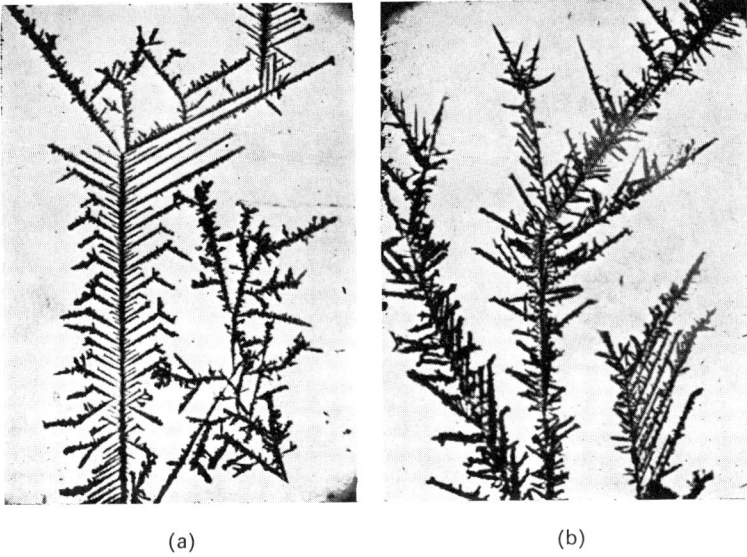

Fig. 10.58. (a) Two-dimensional (left) and three-dimensional (right) silver dendrites (after Wranglen); (b) three-dimensional cadmium dendrites (after Wranglen).

times shoot out from the electrode surface. These *dendrites* usually grow side arms, ending up like mini-Christmas trees (Fig. 10.58).

It must be emphasized here that the study of dendrite formation has important practical implications. In energy-storage devices (batteries), dendrites often rupture the membranous *separators* and go over to touch the other, electron-sink electrode (anode), which leads to a diastrous short circuit of the cell. In substance producers designed for the preparation of metal powders by electrodeposition, dendrite formation is to be avoided because it does not give the desired type of deposit.

10.2.16. Organic Additives and Electrodeposits

The effect of organic substances in solution on the nature of electrocrystallization is an area in which there has been a vast number of facts with little theory explaining them. Here one cannot do more than hint at some of the factors.

In the first place, the adsorption of organic substances is generally dependent on the charge of the electrode. It will be recalled that the relation of the coverage to the electric charge is bell shaped (see Section 7.5.3, Fig. 7.101). A model in which the organic competes for the electrode against water permits a simple view of the process. As a very crude first approximation (see, however, Section 7.5.5 and Table 7.13), the potential at which there is a maximum adsorption of uncharged organics may be taken as the pzc.

All this only helps one to understand whether the organic adsorbs or not at the potential difference prevailing at the interface during electrodeposition. But where does it adsorb? Uniformly all over the electrode? This would only lead to an all-round slowing down of growth. Preferentially on some crystal faces but not so much on others? In this case, the ratio of growth velocities of the different faces will be altered from that obtained in the absence of the organic additive, and perhaps the growth rates will get evened out, which will result in smooth, even deposits. Selectively, on the planes? This will cause a stopping of step movement, perhaps the nucleation of new crystals (fine-grained deposits). Preferentially on micropeaks owing to the greater diffusion of organic to these peaks (Fig. 10.56)? This would lead to smoothing and brightening because the deposit would tend to grow where there is *no* adsorbed, blocking organic additive, i.e., to fill in the valley (Kardos). All these effects are in fact observed by the

TABLE 10.11

Crystal Growth Morphology for the Electrodeposition of Copper in the Presence and Absence of 10^{-8} mole liter^{-1} *n*-Decylamine

Deposition-current density, mamp cm^{-2}	Crystal-growth morphology	
	In the presence of *n*-decylamine	In the absence of *n*-decylamine
5	Layers	Layers but with larger distances between steps
10	Layers + truncated pyramids	Layers + pyramids, again with larger distances between steps
15	Layers + truncated pyramids + blocks	Layers + pyramids (little tendency for truncated pyramids to form)
20	Polycrystalline	Layers + truncated pyramids + blocks

addition of organic solutes to solutions from which electrocrystallization is to occur. But the detailed understanding of how they are related to the structure of the additive is as yet at a very low level.

The effects made upon crystal-growth forms by very small concentrations of additives can be seen in Table 10.11, where the additive is *n*-decylamine.

10.2.17. The Simultaneous Deposition of More Than One Metal: Alloy Deposition

It has been assumed thus far that the electrode–electrolyte interface is the seat of one single charge-transfer process. For example, in the discussion of electrocrystallization, it has been considered that the only electrodic reaction is the deposition of metal ions of the species M_i^{z+}. The theory may be applied, e.g., to the electrodeposition of copper or of zinc. It is known, however, that it is possible to carry out the electrodeposition of brass (an alloy of zinc and copper). This implies that two electrodic reactions are occurring at the electrode, i.e., there is a simultaneous deposition of zinc and of copper resulting in the electrogrowth of a zinc–copper alloy on the electron-source electrode. It is necessary, therefore, to learn to handle situations in which the interface is the scene of more than one electrodic reaction.

The treatment of simultaneous reactions is based on an important fact: though the current densities $i_1, i_2, \ldots, i_n$ for n simultaneous reactions may be different from each other, they are all producing (or being produced by) the same absolute Galvani potential difference $\Delta\phi$ at the interface. Thus, in writing out expressions for the current densities $i_1, i_2, \ldots, i_n$, one must use the same value $\Delta\phi$ for the potential difference across the interface irrespective of what particular reaction is under consideration. If all the reactions are proceeding under conditions of *activation* overpotential and the high-field approximation of the Butler–Volmer equation is valid, one can write[†]

[†] Equations containing $\Delta\phi_e$ involve the activity of the alloy component in the surface, from the Nernst relation,

$$\Delta\phi_e = \Delta\phi_e^\circ + \frac{RT}{F} \ln \frac{a_{M^{z+}}}{a_A}$$

where a_A is the activity of component A in the alloy. It should be an equilibrium quantity, but sometimes one component may build up on the surface, at least temporarily before bulk diffusion has smoothed out the concentration. However, there is some evidence that diffusion from the bulk of the metal to the surface in the dissolution of

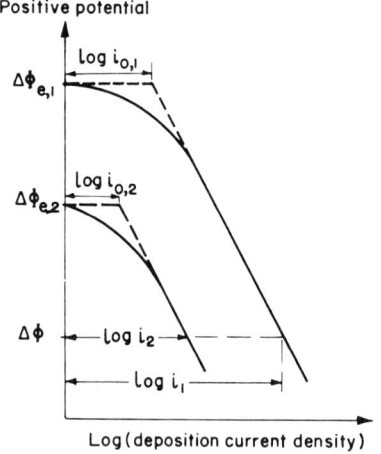

Fig. 10.59. The alloy deposition of metals M_1 and M_2 in which (1) $\Delta\phi_{e,1}$ is more positive than $\Delta\phi_{e,2}$ and (2) $i_{0,1} > i_{0,2}$ but $\vec{\alpha}_1 = \vec{\alpha}_2$.

$$i_1 = i_{0,1} e^{-\vec{\alpha}_1 F \eta_1 / RT} = i_{0,1} e^{-(\vec{\alpha}_1 F / RT)(\Delta\phi - \Delta\phi_{e,1})} \tag{10.85}$$

$$i_2 = i_{0,2} e^{-\vec{\alpha}_2 F \eta_2 / RT} = i_{0,2} e^{-(\vec{\alpha}_2 F / RT)(\Delta\phi - \Delta\phi_{e,2})} \tag{10.86}$$

$$i_n = i_{0,n} e^{-\vec{\alpha}_n F \eta_n / RT} = i_{0,n} e^{-(\vec{\alpha}_n F / RT)(\Delta\phi - \Delta\phi_{e,n})} \tag{10.87}$$

For simplicity, suppose that there are only two reactions 1 and 2 and, further, $\vec{\alpha}_1 = \vec{\alpha}_2$, then

$$\frac{i_1}{i_2} = \frac{i_{0,1}}{i_{0,2}} e^{(\vec{\alpha} F / RT)(\Delta\phi_{e,1} - \Delta\phi_{e,2})} \tag{10.88}$$

Thus, if one is considering the simultaneous deposition of two metals M_1 and M_2, the ratio of the current densities with which the two metals are deposited depends on their exchange-current densities and the difference of their equilibrium potentials. The larger the value of $i_{0,1}$ is compared with $i_{0,2}$ and the more positive the equilibrium potential $\Delta\phi_{e,1}$ is compared with $\Delta\phi_{e,2}$, the greater will be the amount deposited of metal M_1 compared with the amount deposited of M_2 (Fig. 10.59).

---

alloys is relatively fast, so that equilibrium between bulk and surface is usually maintained. The high diffusion rate, unusual in the solid state, arises because of the large number of vacancies introduced into the structure from the surface. Some of these vacancies are paired, and the heat of activation for the migration of these is particularly low.

Of course, the above approach is a highly simplified one with many approximations which can be removed at the expense of the simplicity of the algebra. For instance, generally the transfer coefficients and hence their Tafel slopes for the two reactions are unequal. In such a case, the current density ratio becomes

$$\frac{i_1}{i_2} = \left[\frac{i_{0,1}}{i_{0,2}} e^{(F/RT)(\vec{\alpha}_1 \Delta \phi_{e,1} - \vec{\alpha}_2 \Delta \phi_{e,2})} \right] e^{(F/RT)(\vec{\alpha}_2 - \vec{\alpha}_1) \Delta \phi} \quad (10.89)$$

Thus, the value of the current-density ratio depends not only upon the exchange-current densities, the transfer coefficients, and the equilibrium potentials but also, in a exponential way, upon the potential difference across the interface (Fig. 10.60). Thus, even if the deposition of metal 2 is handicapped with a lower exchange-current density (i.e., $i_{0,2} < i_{0,1}$) and a more negative equilibrium potential (i.e., $\Delta \phi_{e,2} < \Delta \phi_{e,1}$), it can have a higher deposition-current density i_2 than that of another metal i_1 (i.e., $i_2 > i_1$) if its transfer coefficient $\vec{\alpha}_2$ is greater than that of the other (i.e., $\vec{\alpha}_2 > \vec{\alpha}_1$).

In many cases of alloy deposition, the predominance of the deposition of one metal constituent over another is strongly potential dependent owing to converging or intersecting Tafel lines. Figure 10.61 illustrates that, at potentials more positive than $\Delta \phi$, metal 1 is predominantly deposited, whereas, at potentials more negative than $\Delta \phi$, metal 2 is predominantly deposited.

Another way in which a metal M_2 with a more negative equilibrium potential (i.e., $\Delta \phi_{e,2} < \Delta \phi_{e,1}$) can be deposited to a greater extent is if the

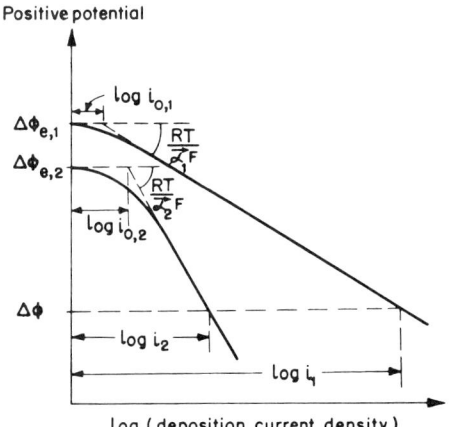

Fig. 10.60. Alloy deposition of metals M_1 and M_2 in which $\vec{\alpha}_2 > \vec{\alpha}_1$—diverging Tafel lines.

deposition of metal M_1 is limited by transport. Under such conditions,

$$i_1 = i_L = -\frac{DnFc_1^0}{\delta} \qquad (9.82)$$

whereas the current density for the reaction which is not limited by transport continues to be given by

$$i_2 = i_{0,2} e^{-(\vec{\alpha}_2 F/RT)(\Delta\phi - \Delta\phi_{e,2})} \qquad (10.86)$$

Thus,

$$\frac{i_1}{i_2} = \frac{i_L}{i_{0,2}} e^{(\vec{\alpha}_2 F/RT)(\Delta\phi - \Delta\phi_{e,2})} \qquad (10.90)$$

$$= \frac{DnFc_1^0}{\delta i_{0,2}} e^{(\vec{\alpha}_2 F/RT)(\Delta\phi - \Delta\phi_{e,2})} \qquad (10.91)$$

which shows that metal M_2 is deposited to a greater extent than M_1, i.e., $i_2 > i_1$, if i_2 exceeds the limiting current density for the deposition of M_1 (Fig. 10.62). Further, the more negative the potential $\Delta\phi$ is with respect to the equilibrium potential $\Delta\phi_{e,2}$, the smaller becomes the value of i_1/i_2, i.e., the greater is the fraction $i_2/(i_1 + i_2)$ of the total current density going toward the deposition of M_2.

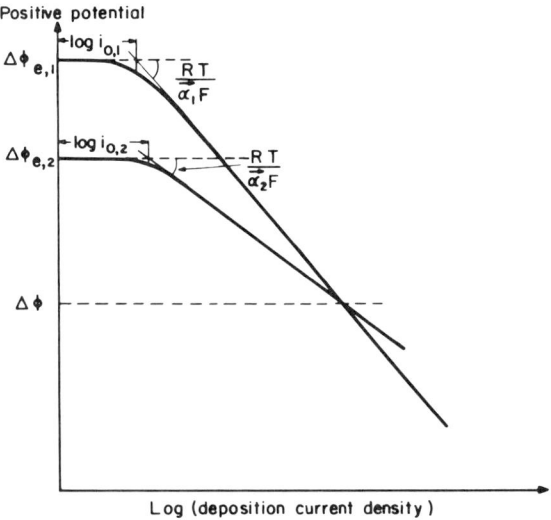

Fig. 10.61. Alloy deposition of metals M_1 and M_2 in which $\vec{\alpha}_2 < \vec{\alpha}_1$—converging Tafel lines.

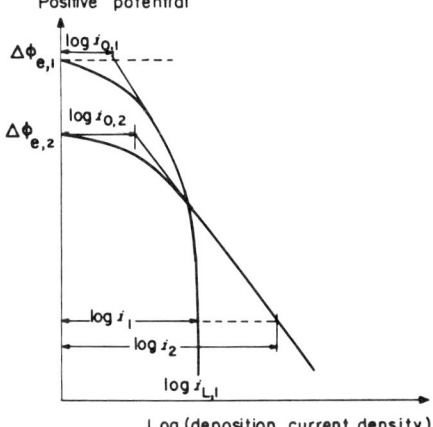

Fig. 10.62. Alloy deposition of metals M_1 and M_2 in which the deposition of metal M_1 is limited by transport, i.e., diffusion control.

10.2.18. The Sometimes Unavoibable Complication: Hydrogen Codeposition

The above elementary discussion of simultaneous reactions was aimed at giving the first essentials of how alloys are deposited, i.e., the simultaneous deposition of two or more metals; but it is equally valid for a situation in which metal deposition occurs along with any other electronation (cathodic) reaction.

The need for this generalization arises from the fact that one sometimes finds that there is less metal deposited than there should be on the basis of Faraday's laws, which require the deposition of one gram equivalent of metal per faraday of electricity passed through the system if the depositing ion is univalent. In fact, in some cases, there is almost no deposit at all. What has happened to the missing coulombs? In such cases, hydrogen evolution is occurring. From a technological point of view, however, one seeks to deposit a metal without a wastage of electrical energy on hydrogen evolution.

It is important, therefore, to understand the current-density ratio i_H/i_M or the percentage of current efficiency $[i_M/(i_M + i_H)] \times 100$, and to try to use all the current only for metal deposition, i.e., to make $i_H/i_M \to 0$ or $i_M/(i_M + i_H) \to 100\%$.

The basic factors determining current ratio or current efficiency are easily revealed by considering that reaction 1 of Eq. (10.85) is hydrogen evolution and reaction 2 of Eq. (10.86) is metal deposition. Thus, assuming equality of the transfer coefficients for the hydrogen-evolution and metal-deposition reactions, i.e., $\vec{\alpha}_H = \vec{\alpha}_M = \alpha$, one has

$$\frac{i_H}{i_M} = \left[\frac{i_{0,H}}{i_{0,M}} e^{(\vec{\alpha}F/RT)(\Delta\phi_{e,H} - \Delta\phi_{e,M})}\right] e^{(F/RT)(\vec{\alpha}_M - \vec{\alpha}_H)\Delta\phi} \quad (10.92)$$

which indicates that the fraction of the total current going toward hydrogen evolution (Fig. 10.63) depends upon the ratio of the exchange-current densities, the equilibrium potentials, and the transfer coefficients of hydrogen-evolution and metal-deposition reactions. In general, the deposition of metal is favored by a more positive equilibrium potential with respect to that of hydrogen, a higher exchange-current density, and a lower value of the transfer coefficient.

Even if the factors of equilibrium potential, exchange-current density, and transfer coefficient are unfavorable for the deposition of metal, i.e., $i_H > i_M$ when $i_{0,M} < i_{0,H}$, $\Delta\phi_{e,M} < \Delta\phi_{e,H}$, and $\vec{\alpha}_M > \vec{\alpha}_H$, it may still be possible to get high current efficiencies for metal deposition if it is possible to ensure that the current density for hydrogen evolution is limited by transport. In the latter case, the current ratio is given by [cf. Eq. (10.91)]

$$\frac{i_H}{i_M} = \frac{DnFc^0_{H^+}}{\delta i_{0,M}} e^{(\vec{\alpha}_M F/RT)(\Delta\phi - \Delta\phi_e)} \quad (10.93)$$

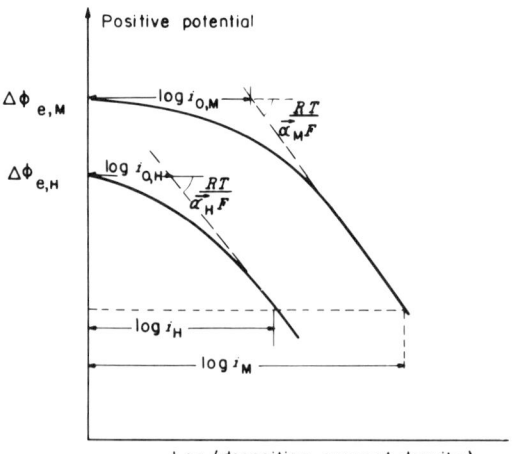

Fig. 10.63. The deposition of metal M with concurrent hydrogen evolution, on assuming $\vec{\alpha}_M = \vec{\alpha}_H$.

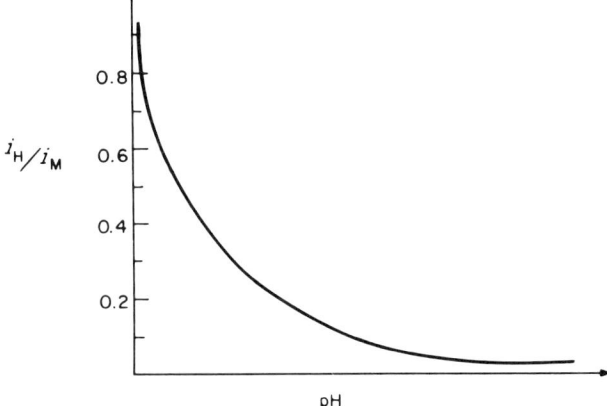

Fig. 10.64. The current efficiency (expressed as i_H/i_M) for the deposition of metal M, in the presence of concurrent hydrogen evolution, as a function of pH.

which shows that the current ratio i_H/i_M decreases as $c^0_{H^+}$ decreases, i.e., as the pH increases (Fig. 10.64).

Conversely, if the limiting current density for the deposition of the metal is exceeded, hydrogen evolution will replace it. In aqueous solutions, hydrogen will always be evolved if the total deposition current is made sufficiently high; the hydrogen may come not only from H_3O^+ (of limited concentration) but also from the water itself.

The effects of hydrogen codeposition on the crystallography of metal deposition are widespread and rather complicated; an attempt to relate them would take one into specialized reaches. Several starting paths to new effects and mechanisms may be noted, however. Hydrogen may adsorb more freely on certain crystal planes and block them so that the metal grows preferentially on others. The development of a preferred orientation of the deposit crystals can sometimes be ascribed to the preferential adsorption of hydrogen on certain crystal facets (Reddy). Hydrogen may also permeate into the metal and *change its mechanical properties.* Finally, removal of H_3O^+ ions from the diffusion layer near the electrode changes its properties, too; it makes the solution alkaline. If the solution produced is sufficiently alkaline, it may cause the solubility product of a hydroxide of the metal ion present to be exceeded, which would cause precipitation and therefore a relatively thick (10^2 to 10^3 Å) film on parts or on all of the electrode. Thus, hydrogen codeposition produces many things and they are usually evil.

These are only some elementary ideas on the current efficiency of metal deposition, but they lead one on from the story of electrodeposition to that of hydrogen evolution.

Further Reading

1. I. N. Stranski, *Z. Physik. Chem. (Leipzig)*, **136**: 259 (1928).
2. F. C. Frank, *Discussions Faraday Soc.*, **5**: 48 (1947).
3. G. I. Finch, *Discussions Faraday Soc.*, **1**: 144 (1947).
4. W. K. Burton, N. Cabrera, and F. C. Frank, *Proc. Roy. Soc. (London)*, **A243**: 299 (1951).
5. W. Lorenz, *Z. Naturforsch.*, **9a**: 716 (1954).
6. W. Mehl, *J. Chem. Phys.*, **27**: 818 (1957); also *Can. J. Chem.*, **37**: 190 (1959).
7. H. Gerischer and R. P. Tischer, *Z. Elektrochem.*, **61**: 1159 (1957).
8. N. Cabrera and D. A. Vermilyea, "The Growth of Crystals from Solution," in: R. H. Doremus, R. W. Roberts, and D. Turnbull, eds., *Growth and Perfection of Crystals*, John Wiley and Sons, Inc., New York, 1958, p. 393.
9. F. C. Frank, "On the Kinematic Theory of Crystal Growth and Dissolution Processes," in R. H. Doremus, R. W. Roberts, and D. Turnbull, eds., *Growth and Perfection of Crystals*, John Wiley and Sons, Inc., New York, 1958, p. 411.
10. G. Wranglen, *Electrochem. Acta*, **2**: 130 (1960).
11. A. Despic, *J. Chem. Phys.*, **32**: 389 (1960).
12. J. L. Barton, *Proc. Roy. Soc. (London)*, **A268**: 485 (1962).
13. O. Kardos and D. G. Foulke, "Applications of Mass Transfer Theory: Electrodeposition on Small Scale Profiles," in: P. Delahay and C. Tobias, eds., *Advances in Electrochemistry and Electrochemical Engineering*, Vol. 2, 1962, p. 145.
14. N. A. Pangarov, *Electrochim. Acta*, **7**: 139 (1962).
15. A. K. N. Reddy and S. R. Rajagopalan, *J. Electroanal. Chem.*, **6**: 141 (1963).
16. A. Brenner, *Electrodeposition of Alloys*, Vols. I and II, Academic Press, Inc., New York, 1963.
17. A. K. N. Reddy, *Electrochim. Acta*, **8**: 831 (1963).
18. M. Fleishmann and H. R. Thirsk, "Metal Deposition and Electrocrystallization," in: P. Delahay and C. Tobias, eds., *Advances in Electrochemistry and Electrochemical Engineering*, Vol. 3, 1963, p. 123.
19. A. Damjanovic, "The Mechanism of the Electrodeposition of Metals," in: J. O'M. Bockris and B. E. Conway, eds., *Modern Aspects of Electrochemistry*, Vol. III, p. 224, Butterworths Publications, Ltd., London, 1964.
20. H. R. Thirsk, *J. Electrochem Soc.*, **113**: 1120 (1966).
21. K. M. Gorbunova and Yu. M. Polukarov, "Electrodeposition of Alloys," in: P. Delahay and C. Tobias, eds., *Advances in Electrochemistry and Electrochemical Engineering*, Vol. 5, 1966, p. 249.
22. A. Damjanovic, *J. Electrochem. Soc.*, **113**: 429 (1966).

23. E. Budewski, R. Kaischew, *et al.*, *Electrochim. Acta*, **11**: 1697 (1966).
24. N. N. Balashova and L. I. Klimenkova, *Soviet Electrochemistry*, **3**: 38 (1967).
25. J. O'M. Bockris and G. Razumney, *Fundamental Aspects of Electrocrystallization*, Plenum Press, New York, 1967.

10.3. THE HYDROGEN-EVOLUTION REACTION

10.3.1. A Reaction with a Special History

The reaction in which hydrogen is evolved from aqueous solution is a *very* special one in the history of electrodics.

There are many reasons for this. First of all, it is a famous reaction. Every chemist knows something about the evolution of hydrogen, and most of them have heard about *hydrogen overvoltage*.[†] In acid solutions, the overall reaction is

$$2H_3O^+ + 2e \rightarrow H_2 + 2H_2O \qquad (10.94)$$

and, in alkaline solutions, it is

$$2H_2O + 2e \rightarrow H_2 + 2OH^- \qquad (10.95)$$

A second reason for the special status of this reaction is its technological importance. Not only is the electrodic evolution of hydrogen a standard method of producing the gas, but it is of vital relevance to many other processes. Consider, e.g., the industrial production of heavy water, D_2O. The principle of the method is that hydrogen H_2 gas evolves faster at the electron-source electrode than deuterium D_2 gas, and this leaves behind a D_2O-enriched solution. Obviously, the efficiency of the operation depends on the ratio of the electronation currents i_{H_2}/i_{D_2}, which in turn is a function

---

[†] In the days of classical electrochemistry, i.e., until the middle of the twentieth century, it was thought that overpotential—or "overvoltage" as it was then termed—was a peculiarity of a few electrode reactions, in particular, of hydrogen evolution. Special causes of overvoltage were sought. It was generally thought that an electrode–solution system which exhibited overvoltage was sick in some way, e.g., had a gas film which impeded the current. The picture today is clear; overpotential is a necessary concomitant of the passage of a net current density across any interface. Reactions differ only in the magnitude of overpotential associated with a particular current density. To have a zero activational overpotential, the exchange-current density would have to be infinity; to avoid concentration overpotential, the limiting current density and therefore the transport flux must be infinity; both these conditions are realizable only in thought experiments.

of the ratio of the exchange-current densities $i_{0,H_2}/i_{0,D_2}$. The higher this ratio is, the faster is the separation of isotopes. Or consider the electrodeposition of metals such as nickel or chromium. The deposition is accompanied by simultaneous hydrogen evolution. The efficiency of deposition—called the *current efficiency*—depends upon the ratio i_M/i_{H_2}, and the lower the ratio, the lower the efficiency is. One sets out to deposit the metal only, and one ends up also evolving hydrogen gas (*cf.* Section 10.2.18). In all cases, a selective control over the rate of the hydrogen-evolution reaction is vital; hence the need to understand the reaction.

Thirdly, the gas evolves on a great number of metals. This fact permits the rate of hydrogen evolution to the studied upon a much larger number of electrocatalysts than any other reaction. For example, when one tries to evolve oxygen on certain metals, there is also the possibility of a competitive dissolution of the metal or a formation of an oxide film. It is difficult to learn about oxygen reduction or evolution in such systems. In the case of hydrogen evolution, however, results are available on more than 20 metals and can therefore reveal trends on the reaction rate as a function of the substrate properties.

Fourthly, as will be seen later (Section 11.2.12), the hydrogen-evolution reaction plays a dominant role in the theory of the corrosion of metals in acidic mediums. The rate of hydrogen evolution often determines the extent of the corrosion of a metal and therefore the stability of the surface of the metal. But events at the metal surface, particularly the amount of hydrogen permeating into it, have powerful repercussions on the inner strength of the metal; hence, it would not be an exaggeration to say that the stability of technologically structures in moist air may depend upon the kinetics of the hydrogen-evolution reaction at the metal–moist-film-layer interface (*cf.* Section 11.2.2).

Finally, one should mention a rather different kind of reason why the hydrogen-evolution reaction is so special to the subject of electrodics. It is the prototype of an electrodic reaction. Name almost any feature of a charge-transfer reaction at an electrified interface, and the hydrogen-evolution reaction has it, generally in its quintessential form. In fact, this is why the reaction has already been used so much in the exposition of the fundamental principles of electrodics. Thus, the hydrogen-evolution reaction involves charge transfer, consecutive steps, the formation of an adsorbed intermediate, an ionic reactant in acid solution, a neutral reactant in alkaline solutions, and a gaseous product. In addition, it may involve rate control by the diffusion of hydrogen ions in solution or a chemical reaction succeeding charge transfer. On some electrodes, e.g., mercury, it is a very slow reaction, and, on other electrodes, e.g., platinum, it is a very fast one. With this one reaction as an example, one could, if pressed, develop a great deal of the whole theory of electrodics. It is to electrodics what germanium and silicon are to semiconductor physics or what water is to radiation chemistry.

10.3.2. What Are the Possible Paths for the Hydrogen-Evolution Reaction?

There is a long history to the study of the hydrogen-evolution reaction, dating from the beginning of the century, and, in the course of this history, all sorts of reaction steps have been proposed as part of the overall process. But today, owing to the work of many electrochemists and, in particular, to Frumkin, Conway, and Parsons, only two reaction paths are regarded as likely.

Common to both these reaction paths is the description of how hydrogen gets onto an electrode surface. Hydrated protons are electronated to form neutral hydrogen atoms upon those areas of the surface which are unoccupied; one says that protons (or hydrogen ions) are discharged onto free sites on the electrode to form adsorbed hydrogen atoms

$$M(e) + H_3O^+ \rightarrow MH + H_2O \text{ [acid solutions]} \qquad (10.96)$$

A step of this type, which was developed by Volmer (after a suggestion by Smits) is, in fact, an old friend; it was used as the archetype of a charge-transfer reaction in discussing the fundamental theory of charge transfer (*cf.* Section 8.5).

It must not be concluded that the discharge reaction will not proceed unless hydrogen ions are available. Water is an invariably present source of protons. So, in alkaline solutions (with negligible concentrations of hydrogen ions), there is an electronation of water molecules. The discharge of water molecules also yields adsorbed hydrogen atoms

$$M(e) + H_2O \rightarrow MH + OH^- \text{ [alkaline solutions]} \qquad (10.97)$$

Now one must think how the adsorbed hydrogen gets off the surface. If the first step in the hydrogen evolution consists of *adsorption*, the second step must be a *desorption* step.

Tafel (the discoverer, at an experimental level, of exponential i versus η relations) suggested a desorption step which does not involve charge transfer

$$MH + MH \rightarrow 2M + H_2 \qquad (10.98)$$

This is a *chemical-desorption step* and is also known under various aliases, *catalytic-recombination reaction, Tafel recombination, atom–atom step*, etc. In this step, the adsorbed hydrogen atoms formed from the discharge step diffuse about on the surface, either among the adsorbed water molecules (*cf.* Section 9.1.9) or by pushing them out of the way, until two adsorbed hydrogen atoms collide with each other, whereupon they combine to form hydrogen molecules.

Kobosew and Nekrassow suggested that the adsorbed hydrogen atoms escape from the surface in another way, by means of an *electrodic-desorption* step:

$$\text{MH} + \text{H}_3\text{O}^+ + \text{M}(e) \rightarrow \text{H}_2 + \text{H}_2\text{O} + 2\text{M} \qquad (10.99)$$

(This step is also known as *electrochemical desorption* or *ion–atom recombination*). In electrodic desorption, one starts off with adsorbed hydrogen atoms, and then protons discharge on top of the adsorbed hydrogen atoms with the simultaneous formation of hydrogen molecules.

The two basic reaction paths are therefore (a) discharge D followed by chemical desorption CD

$$\text{M}(e) + \text{H}_3\text{O}^+ \rightarrow \text{MH} + \text{H}_2\text{O} \qquad (10.96)$$

$$2\text{MH} \rightarrow 2\text{M} + \text{H}_2 \qquad (10.98)$$

and (b) discharge followed by electrodic desorption ED

$$\text{M}(e) + \text{H}_3\text{O}^+ \rightarrow \text{MH} + \text{H}_2\text{O} \qquad (10.96)$$

$$\text{MH} + \text{H}_3\text{O}^+ + \text{M}(e) \rightarrow 2\text{M} + \text{H}_2\text{O} + \text{H}_2 \qquad (10.99)$$

These two paths differ in an important way which has not come out in the above description. One can see that, in the electrodic-desorption path, the surface concentration of adsorbed hydrogen atoms must tend to be high; otherwise, there will be little chance of the protons discharging onto the hydrogen atoms on the surface. This point can be seen from the equation for the rate of the electrodic-desorption step. Since this step depends on collisions between hydrogen ions and adsorbed hydrogen atoms, the concentrations of both these species must appear in the rate expression. Hence,

$$v_{\text{ED}} = k_{\text{ED}} \theta c_{\text{H}^+} e^{-\beta F \Delta\phi/RT} \qquad (10.100)$$

where θ is the fraction of the surface covered with adsorbed hydrogen. It is obvious that $v_{\text{ED}} \rightarrow 0$ as θ becomes small. This condition, however, does not prevent the chemical-desorption path which can proceed at low coverage.

10.3.3. What Mechanisms Are Possible in Hydrogen Evolution?

In the treatment of multistep consecutive reactions (*cf.* Section 9.1.5), the view has been taken that one step, namely, the rds, controls the overall reaction rate. The condition for the existence of a single rds is that the activated state corresponding to the rds is much higher with respect to the initial state for the overall reaction than the activated state corresponding to any other step. Suppose, however, that the activated states of more than one step are within 1 to 2 kcal mole^{-1} of the same energy with respect to the initial state. Then there will be a multiple control of the rate of the overall reaction. For two steps which have activated states at almost the same level of energy with respect to the initial state, the two steps will exercise dual control over the rate of the overall reaction. In such a case, the overall reaction is said to have a *dual mechanism*. For example, if the activated states of the discharge and chemical-desorption steps have almost equal free energies of activation (with respect to the initial state, *cf.* Section 9.2), the two steps will effect a dual control over the rate of hydrogen evolution. However, since the potential only affects the discharge step, it is only within certain regions of potential that a dual mechanism can operate.

There is another point on which the earlier treatment of consecutive reactions (*cf.* Sections 9.1.5 and 9.1.6) can be made more realistic. Steps other than the rds were considered earlier always to be in quasi equilibrium, meaning that their forward and backward velocities were taken to be virtually equal. Suppose, however, that the reverse velocities of all steps are negligible compared with the forward velocities. Then one cannot use the approach of treating the other steps as in virtual equilibrium. In such a situation there must be still an equality of the net velocities of each consecutive step. But, because of the negligible velocity of the reverse reaction, equal velocity of the consecutive reactions means equal velocities of their *forward* partial reactions. Increase the forward velocity of the first step, and the forward velocity of the second step has to increase to satisfy the steady-state hypothesis (concentration of intermediate—adsorbed H—invariant with time). Under these conditions (namely, negligible backward velocities), the forward velocities of the two steps are coupled to each other (respond to the same extent to variables), and the overall reaction is said to be a *coupled reaction*.

Now, two reaction paths have been suggested for hydrogen evolution (Section 2.3.2) and each of these is a two-step reaction. Within each path, either of the consecutive steps can be rate determining.

Consider, e.g., the chemical-desorption path; either the discharge step

D or the chemical-desorption step CD can determine the rate of the overall reaction. Steps other than the rds can be assumed to be in equilibrium (*cf.* Section 9.1.5) if there is no coupled mechanism present, and one follows the usual convention of representing a mechanism by indicating which step is in equilibrium ($\rightleftharpoons$) and which step is rate determining ($\rightarrow$). Thus, if the discharge step is rate determining, one can represent the mechanism thus

$$H_3O^+ \xrightarrow{D} MH \underset{}{\overset{CD}{\rightleftharpoons}} H \qquad (10.101)$$

and, if the chemical-desorption step is rate controlling, the mechanism is

$$H_3O^+ \underset{}{\overset{D}{\rightleftharpoons}} MH \xrightarrow{CD} H_2 \qquad (10.102)$$

Frequently, the rds is referred to as the *slow* step, and the other steps are called *fast* steps. For example, the mechanism $H_3O^+ \xrightarrow{D} MH \rightleftharpoons H_2$ is known as a *slow-discharge–fast-chemical-desorption mechanism*. The terms slow and fast must be understood in a technical way. They do not refer to the net rates of the two consecutive steps because, *in a consecutive reaction, all steps must proceed at the same net velocity in the steady state* (otherwise, the intermediate concentration would change with time in contradiction to the definition of steady state, *cf.* Section 9.2.1).

What, then, is the meaning of the term "*slow* step"? One must imagine what happens when a constant current is switched on at $t = 0$. The question is: Which step is responsible for the piling up of charge on either side of the interface, and, therefore, which step causes the buildup of overpotential across the interface? It will be the step which runs slow.

Suppose that proton discharge is the bottleneck, i.e., the rds, then the formation of adsorbed hydrogen atoms occurs with difficulty and those hydrogen atoms which do arrive on the surface are whipped away by a rapid chemical-desorption step. (For this reason, the surface concentration of hydrogen atoms will tend to be very low, less than, say, 1% of the surface.) So, the cause of the buildup of overpotential is the slow discharge step. Of course, once a steady state is attained, both the slow discharge step and the following "fast" step run at the same net rate which, however, is determined by the rate of the discharge step.

On the other hand, suppose that, during the transient, the chemical-desorption step is slow. What will happen? When the current is turned on, the discharge step starts taking place with ease, but the hydrogen atoms will begin to accumulate on the surface because the rate of hydrogen combination is too slow to whip them away. It needs a large fraction of surface coverage θ before it can encourage the desorption-step to take place at a rate equal to the rate at which the protons are being delivered by the relatively fast proton-discharge step. According to this mechanism, the surface coverage of adsorbed hydrogen atoms will mount up to a relatively high value before it tends to reach steady state.

Similarly, in the case of the electrodic-desorption path, either the discharge step D or the electrodic-desorption step ED can be rate determining. These two possible mechanisms can be represented thus

$$H_3O^+ \xrightarrow{D} MH \underset{}{\overset{ED}{\rightleftharpoons}} H_2 \tag{10.103}$$

and

$$H_3O^+ \underset{}{\overset{D}{\rightleftharpoons}} MH \xrightarrow{ED} H_2 \tag{10.104}$$

The above mechanisms may also run as coupled mechanisms, i.e., mechanisms in which the backward reaction of the first step is negligible in rate. Thus, in the chemical-desorption path, there can be a coupled-discharge–chemical-desorption reaction

$$M(e) + H_3O^+ \xrightarrow[\overleftarrow{v}_1 \to 0]{v = \vec{v}_1} MH + H_2O$$

$$2MH \xrightarrow[\overleftarrow{v}_2 \to 0]{v = \vec{v}_2} 2M + H_2 \tag{10.105}$$

where v is the rate of the overall reaction, $\vec{v}_1$ and $\overleftarrow{v}_1$ are the forward and backward velocities of the discharge step, and $\vec{v}_2$ and $\overleftarrow{v}_2$ are the corresponding quantities for the chemical-desorption step. For the electrodic-desorption path also, there can be a coupled-discharge–electrodic-desorption mechanism

$$H_3O^+ + M(e) \xrightarrow[\overleftarrow{v}_1 \to 0]{v = \vec{v}_1} MH + H_2O$$

$$MH + H_3O^+ + M(e) \xrightarrow[\overleftarrow{v}_3 \to 0]{v = \vec{v}_3} 2M + H_2O + H_2 \tag{10.106}$$

where v, $\vec{v}_1$, and $\overleftarrow{v}_1$ have the same significance as in (10.105), and $\vec{v}_3$ and $\overleftarrow{v}_3$ are the forward and backward velocities of the electrodic-desorption step.

10.3.4. How One Determines the Path and Rate-Determining Step of the Hydrogen-Evolution Reaction

In general, the determination of a reaction mechanism has been shown (*cf.* Sections 9.4 and 9.5) to require a number of observations. Which are the most worthwhile ones for diagnosing the mechanism depends upon the reaction and the substrate. Some of the techniques useful for the study of the hydrogen-evolution reaction will now be looked at and a few remarks made upon what help they give in determining the reaction mechanism.

TABLE 10.12
The Exchange-Current Density i_0 for the Hydrogen-Evolution Reaction

Metal	$-\log i_0$ [amp cm^{-2}] in $\sim 1M$ H$_2$SO$_4$
Palladium	3.0
Platinum	3.1
Rhodium	3.6
Iridium	3.7
Nickel	5.2
Gold	5.4
Tungsten	5.9
Niobium	6.8
Titanium	8.2
Cadmium	10.8
Manganese	10.9
Thallium	11.0
Lead	12.0
Mercury	12.3

10.3.4a. The Determination of the Exchange-Current Density. The measurement of the exchange-current density gives a general idea of the reaction rate in the standard state of equilibrium and permits the classification of the substrate upon which the reaction is occurring as a good or bad electrocatalyst for the hydrogen-evolution reaction. For example, mercury, lead, and thallium are poor catalysts for hydrogen evolution. From Table 10.12, it can be seen that the exchange-current density is very small upon these three metals, of the order of 10^{-10} amp cm^{-2} or less; this would correspond to the renewal of about one monolayer of adsorbed atomic hydrogen per second. The noble metals, however, have values of exchange-current densities in the milliampere range: about ten million monolayers of adsorbed hydrogen are renewed per second per square cm when the electrode is at the equilibrium potential.

10.3.4b. The Determination of the Transfer Coefficient. This can sometimes be useful as an indication of mechanism. Consider, e.g., the chemical desorption path.

In the first instance, let the discharge step be assumed to be rate determining. What is the transfer coefficient appearing in the current-density–

potential relation? This can easily be calculated for the assumed mechanism, namely,

$$H_3O^+ + M(e) \xrightarrow{v_1} MH + H_2O \qquad (10.107)$$

$$2MH \underset{v_2}{\overset{v_2}{\rightleftharpoons}} 2M + H_2O \qquad (10.108)$$

The current density for the discharge step (10.107) far from equilibrium is given by the Butler–Volmer equation

$$\vec{i} = 2Fk_1 c_{H_3O^+}(1-\theta)e^{-\beta F \Delta\phi/RT} \qquad (10.109)$$

Assuming low coverage with adsorbed hydrogen, i.e., $\theta \to 0$ or $1 - \theta \approx 1$, one has

$$\vec{i} = 2Fk_1 c_{H_3O^+} e^{-\beta F \Delta\phi/RT} \qquad (10.110)$$

In this case, therefore, the transfer coefficient $\vec{\alpha}$ is simply equal to the symmetry factor, i.e., $\vec{\alpha} = \beta$.

Now consider the discharge step to be in equilibrium followed by a rate-determining chemical-desorption step

$$H_3O^+ + M(e) \underset{v_{-1}}{\overset{v_1}{\rightleftharpoons}} MH + H_2O \qquad (10.111)$$

$$2MH \xrightarrow{v_2} 2M + H_2 \qquad (10.112)$$

The electronation-current density is given by

$$\vec{i} = 2F\vec{v} = 2F\vec{v}_2 \qquad (10.113)$$

But the rate of the *chemical* desorption is

$$\vec{v}_2 = k_2(k'\theta)^2 \qquad (10.114)$$

where $(k'\theta)$ is the concentration of adsorbed hydrogen atoms at a coverage of θ, i.e., k' is the concentration when $\theta = 1$.

Combining Eqs. (10.113) and (10.114),

$$i = 2Fk_2(k')^2\theta^2 \qquad (10.115)$$

Now, the quasi equilibrium assumption can be used for the discharge step

$$k_1 c_{H_3O^+}(1-\theta)e^{-\beta F \Delta\phi/RT} \approx k_{-1}k'\theta e^{(1-\beta)F\Delta\phi/RT} \qquad (10.116)$$

or

$$\frac{\theta}{1-\theta} = \frac{k_1}{k_{-1}k'} c_{H_3O^+} e^{-F\Delta\phi/RT} \qquad (10.117)$$

1240 CHAPTER 10

i.e.,

$$\theta = \frac{k_1 c_{H_3O^+} e^{-F\Delta\phi/RT}}{k_1 c_{H_3O^+} e^{-F\Delta\phi/RT} + k_{-1} k'} \tag{10.118}$$

If $k_1 c_{H_3O^+} e^{-F\Delta\phi/RT} \ll k_{-1} k'$ (a situation that corresponds to low overpotential), it follows from Eq. (10.118) that

$$\theta \approx \frac{k_1 c_{H_3O^+} e^{-F\Delta\phi/RT}}{k_{-1} k'} \tag{10.119}$$

Substituting for θ in Eq. (10.115), one has

$$\vec{i} = 2F k_2 k'^2 \frac{k_1^2 c_{H_3O^+} e^{-2F\Delta\phi/RT}}{k_{-1} k'^2}$$

$$= 2F k_2 \left(\frac{k_1}{k_{-1}}\right)^2 c_{H_3O^+} e^{-2F\Delta\phi/RT} \tag{10.120}$$

Thus, under conditions of low overpotential, the transfer coefficient for rate-determining chemical-desorption preceded by a discharge step in quasi equilibrium turns out to be $\vec{\alpha} = 2$.

It is seen, therefore, that a mechanism involving rate-determining discharge has an $\vec{\alpha} = \frac{1}{2}$ (assuming $\beta = \frac{1}{2}$), whereas rate-determining chemical-desorption has an $\vec{\alpha} = 2$. Thus, in this instance, a determination of $\vec{\alpha}$ will distinguish between the two possibilities given. However, one must not pretend that a mere determination of $\vec{\alpha}$ is particularly diagnostic by itself alone. There are other mechanisms for which $\vec{\alpha} = \frac{1}{2}$ as well as reactions (10.107) and (10.108).

Consider a discharge–electrodic-desorption mechanism. This is probably a mechanism which is often applicable at high overpotentials when the reverse currents of the two steps are negligible. The electronation-current density for the discharge step is

$$i = 2F k_1 c_{H_3O^+} (1 - \theta) e^{-\beta F \Delta\phi/RT} \tag{10.109}$$

For the electrodic-desorption step, it is

$$i = 2F k_1 c_{H_3O^+} k' \theta e^{-\beta F \Delta\phi/RT} \tag{10.121}$$

According to the steady-state hypothesis, the expressions (10.109) and (10.121) can be set equal to one another, and, thus,

$$\frac{\theta}{1-\theta} = \frac{k_1}{k_3 k'} \tag{10.122}$$

or

$$\theta = \frac{k_1}{k_1 + k_3 k'} \tag{10.123}$$

Now suppose that $k_1 \ll k_3 k'$. It follows from (10.123) that

$$\theta \approx \frac{k_1}{k_3 k'} \tag{10.124}$$

or

$$\theta \ll 1 \tag{10.125}$$

If, however, for $\theta \ll 1$, Eq. (10.109) becomes

$$\vec{i} = 2Fk_1 c_{H_3O^+} e^{-\beta F \Delta \phi / RT} \tag{10.126}$$

$$= i_0 e^{-\beta F \eta / RT} \tag{10.127}$$

where

$$i_0 = 2Fk_1 c_{H_3O^+} e^{-F\beta \Delta \phi_e / RT} \tag{10.128}$$

Thus, again, it has turned out that $\vec{\alpha} = \beta = \frac{1}{2}$.

Even for a rate-determining discharge followed by an electrodic-desorption step in quasi equilibrium,

$$M(e) + H_3O^+ \rightarrow MH + H_2O \tag{10.129}$$

$$M(e) + H_3O^+ + MH \rightleftharpoons 2M + H_2O + H_2 \tag{10.130}$$

one can show that $\vec{\alpha} = \beta = \frac{1}{2}$, e.g., by using the general formula (for $\theta \rightarrow 0$) given as Eq. (9.25) but rewritten here as

$$\vec{\alpha} = \frac{\beta(n - \vec{\gamma} - \overleftarrow{\gamma}) + \vec{\gamma}}{\nu}$$

with $n = 2$, $\vec{\gamma} = 0$, $\overleftarrow{\gamma} = 1$, and $\nu = 1$.

Thus, irrespective of whether the mechanism is rate-determining discharge followed by a quasi-equilibrium step of chemical or electrodic desorption or whether it is a discharge step coupled with a chemical- or electrodic-desorption step, the transfer coefficient is $\beta = \frac{1}{2}$. In such a situation ($\vec{\alpha} = \frac{1}{2}$), the value of $\vec{\alpha}$ is an inadequate criterion for diagnosing the reaction mechanism.

On the other hand, the rate-determining chemical desorption is almost the only mechanism which gives a transfer coefficient of 2. There is only one other mechanism which gives the same transfer coefficient; this is a rate-determining diffusion of the evolved hydrogen molecules away from the electrode. Here one considers that the observed overpotential is a concentration overpotential, in which case

$$i = i_L(1 - e^{-2F\eta/RT}) \qquad (10.131)$$

and once again the η versus ln i curve has a slope of $RT/2F$, which is tantamount to a transfer coefficient of 2. However, one can always tell when this diffusion control is present and distinguish it from the case where the chemical desorption is rate determining. One simply increases mass transfer, e.g., by using a rotating electrode. If diffusion is rate determining, the rate of rotation makes a large difference in the rate of the reaction, but it would not have any effect if the chemical-desorption step is rate determining. Thus, when the experimental value of the transfer coefficient does turn out to be 2 and there is no effect of stirring, it is likely that the chemical-desorption step is rate controlling.

One definitely should not stress the value of the transfer coefficient, especially in the study of hydrogen evolution, because, as has been pointed out, four different mechanisms give rise to the same value for the transfer coefficient (see Table 10.13). On the other hand, one always records the experimental value of α from the η versus log i (Tafel) slope and considers its diagnostic implications because it permits one to eliminate some mechanistic possibilities and thus narrows down the number of mechanisms which have to be considered.

TABLE 10.13

A Summary of the Transfer Coefficients Associated with Some Hydrogen-Evolution Reaction Mechanisms for the Condition $\theta \to 0$

Mechanism	Transfer coefficient α, assuming $\beta = 0.5$
Discharge-rate determining followed by chemical desorption	0.5
Discharge followed by rate-determining chemical desorption	2.0
Coupled discharge–chemical desorption	0.5
Discharge-rate determining followed by electrodic desorption	0.5
Discharge followed by rate-determining electrodic desorption	1.5
Coupled discharge–electrodic desorption	0.5

10.3.4c. The Determination of Reaction Order with Respect to Hydrogen Ions in Solution. The order of reaction is often a good criterion for determining the reaction mechanism, but when one calculates the reaction orders for hydrogen ions which one would expect on the basis of the various reaction paths and rate-determining steps, one finds that many of them indicate the same dependence of current density upon the hydrogen ion, $(\partial \ln i / \partial \ln c_{H_3O^+})_{\Delta\phi}$. Hence, reaction orders are generally not very helpful in determining the mechanism of a hydrogen-evolution reaction.

From one point of view, however, reaction orders are helpful. In strong acid solutions, it is to be expected that hydrated hydrogen ions H_3O^+ will be the source of the protons for the discharge step. In alkaline solutions, where there are hardly any hydrated hydrogen ions in the solution, one would expect water molecules to be the proton source for the discharge reaction.

To illustrate principles and not details, let a relatively simple assumption be made. Strong, neutral salt solutions will be assumed, and hence $^{OHP}\Delta^S\phi$ will tend to zero (*cf.* Section 8.3.2). Then, if the source of the protons is H_3O^+ and the coverage θ_H is taken as being low,

$$i = k_1 c_{H_3O^+} e^{-\beta \Delta\phi F/RT} \tag{10.132}$$

or

$$i = k_1 c_{H_3O^+} e^{-\beta \Delta\phi_e F/RT} e^{-\beta F \eta/RT} \tag{10.133}$$

From the cathodic form of Eq. (9.29a), one may write

$$\left(\frac{\partial \ln i}{\partial \ln c_{H_3O^+}}\right)_\eta = 1 - \frac{\beta F}{RT}\left(\frac{\partial \Delta\phi_e}{\partial \ln c_{H_3O^+}}\right)_\eta \tag{10.134}$$

The term $(\partial \Delta\phi_e / \partial \ln c_{H_3O^+})_\eta$ can be obtained from the Nernst equation (*cf.* Section 8.2.14) as

$$\left(\frac{\partial \Delta\phi_e}{\partial \ln c_{H_3O^+}}\right)_\eta = \frac{2RT}{2F} = \frac{RT}{F} \tag{10.135}$$

Substituting Eq. (10.135) in (10.134) produces the result

$$\left(\frac{\partial \ln i}{\partial \ln c_{H_3O^+}}\right)_\eta = 1 - \beta \simeq +\tfrac{1}{2} \tag{10.136}$$

However, if, with the same assumptions concerning mechanism, salt concentration, and coverage, the proton source is now water,

$$i = k_2 c_{H_2O} e^{-\beta \Delta\phi F/RT} \tag{10.137}$$

or

$$i = k_2 c_{H_2O} e^{-\beta\eta F/RT} e^{-\beta \Delta\phi_e F/RT} \tag{10.138}$$

Following an identical procedure as before, the dependence of current density on the concentration of H_3O^+ ions is

$$\left(\frac{\partial \ln i}{\partial \ln c_{H_3O^+}}\right)_\eta = -\beta \simeq -\tfrac{1}{2} \tag{10.139}$$

Corresponding calculations can be made for other reaction mechanisms and also for less-simple assumptions about the value of β. They tell one [by comparison of the experiment with the predictions of Eqs. (10.136) and (10.139)] which particle delivers the proton from the solution to an adsorbed site on the surface.

10.3.4d. The Stoichiometric Number ν. The stoichiometric number ν (*cf.* Section 9.1.7) is the number of times the rds takes place per act of the overall reaction. Experimental determinations of the stoichiometric number are quite helpful in thinking about the reaction mechanism so long as there is (as there usually seems to be) a clearly marked rds.

Consider that the discharge step is determining the overall rate of the hydrogen-evolution reaction. Then, if the reaction is following the chemical-desorption path,

$$M(e) + H_3O^+ \rightarrow MH + H_2O$$
$$2MH \rightleftharpoons 2M + H_2 \tag{10.140}$$

it is obvious that the discharge step has to proceed twice for the overall reaction, i.e., $\nu = 2$. If, instead, the reaction is following the electrodic-desorption path,

$$M(e) + H_3O^+ \rightarrow MH + H_2O$$
$$M(e) + MH + H_3O^+ \rightleftharpoons 2M + H_2O + H_2 \tag{10.141}$$

then only one occurrence of the rds is required per act of the overall reaction, i.e., $\nu = 1$.

Thus, on the basis of stoichiometric-number measurements, a distinction can be made between the two desorption paths. (Such a distinction, it will be recalled, could not be made on the basis of transfer coefficients.) Further, it can be shown that each of the possible mechanisms for the hydrogen-evolution reaction has a characteristic value of stoichiometric number. Hence, a knowledge of the stoichiometric number does provide information

TABLE 10.14

Stoichiometric Number for Some Hydrogen-Evolution Mechanisms

Mechanism	Stoichiometric number ν
Discharge-rate determining followed by chemical desorption	2
Discharge followed by rate-determining chemical desorption	1
Discharge-rate determining followed by electrodic desorption	1
Discharge followed by rate-determining electrodic desorption	1

on the path and rds. Table 10.14 shows that the stoichiometric number resembles the transfer coefficient in that it does not of itself always uniquely characterize the reaction path and rds. For example, if one considers the $H_3O^+ \underset{}{\overset{D}{\rightleftharpoons}} MH \overset{CD}{\rightarrow} H_2$ mechanism and the $H_3O^+ \overset{D}{\rightarrow} MH \underset{}{\overset{ED}{\rightleftharpoons}} H_2$ mechanism, the stoichiometric number is unity in both cases. Conversely, if $\nu = 2$, there is strong indication of a rate-determining proton discharge followed by a fast chemical-desorption step.

10.3.4e. *The Determination of Hydrogen Coverage.* As has been shown by Devanathan, the criterion of hydrogen coverage is helpful in the elucidation of the mechanism of the hydrogen-evolution reaction. Thus, suppose that, in the steady state, the coverage of the surface with adsorbed hydrogen is high (i.e., virtually a monolayer). Then it is improbable that the rds in the proton–discharge reaction is the chemical-desorption path because it has been seen that this is associated with a low coverage. Thus, if stoichiometric-number data were not available and one could not distinguish between the two mechanisms $H_3O^+ \overset{D}{\rightarrow} MH \underset{}{\overset{CD}{\rightleftharpoons}} H_2$ and $H_3O^+ \underset{}{\overset{D}{\rightleftharpoons}} MH \overset{ED}{\rightarrow} H_2$ determination of the hydrogen coverage would permit a distinction because small coverages are incompatible with the latter mechanism.

But, again, one must not be too sanguine about this criterion because the coverage is often difficult to determine. The total amount of coverage per square centimeter on a surface is only of the order of magnitude of 10^{-10} moles cm^{-2}, which is a very small quantity indeed as far as chemical analysis is concerned.

The determinations of coverage can be done by transient, or sweep, techniques. For example, consider an electrode on which hydrogen evolution is occurring at a steady rate. Suppose that the potential is switched rapidly ($\approx 10^{-5}$ sec) to a value such that the adsorbed hydrogen dissolves and the current is measured as a function of time. The adsorbed hydrogen is rapidly dissolved off by the deelectronation current and the course of the current–time line curve, which corresponds to the dissolution process—and not, e.g., to the charging of the double layer—can be picked out from the screen of the cathode-ray oscilloscope. An integral of the current-time curve obviously gives the total number of coulombs used to ionize the hydrogen dissolved from the surface at the moment at which the potential was switched from that at which steady cathodic evolution occurs to that at which anodic dissolution occurs (a monolayer of adsorbed hydrogen is equivalent to about 100 microcoulombs). Thus, the charge can be used as an index of the coverage at the given conditions of current density and potential for the hydrogen-evolution reaction.

Sometimes the dependence of this hydrogen coverage on the potential is a useful criterion in identifying the reaction mechanism. Suppose, e.g., that a rate-determining discharge–chemical-desorption mechanism is operating, then the rate at which protons leave the solution phase, undergo deelectronation, and become adsorbed on the metal is [*cf.* Eqs. (10.109) and (10.115)]

$$\vec{i} = 2Fk_1 c_{H_3O^+}(1 - \theta_H)e^{-\beta F \Delta\phi/RT} \qquad (10.142)$$

The rate at which such adsorbed hydrogen atoms are being removed by chemical desorption is

$$\vec{i} = 2Fk_2 k'^2 \theta_H^2 \qquad (10.143)$$

Equating these two equations and evaluating θ_H, with the simplification that θ_H is assumed to be negligible compared with unity,

$$\theta_H^2 = \left(\frac{k_1}{k_2} \frac{c_{H_3O^+}}{k'^2} e^{-\beta F \Delta\phi_e/RT}\right) e^{-\beta F \eta/RT} \qquad (10.144)$$

Taking $\beta = \tfrac{1}{2}$ produces

$$\left(\frac{\partial \ln \theta_H}{\partial \eta}\right) = -\frac{F}{4RT} \qquad (10.145)$$

which is an expression for the rate of increase of θ_H with increasingly negative values of overpotential. Such a value is a diagnostic characteristic of the mechanism and, if θ_H as a function of η can be experimentally obtained, serves therefore as an extra mechanism-indicating criterion.

TABLE 10.15

Dependence of the Hydrogen Coverage θ_H upon the Overpotential η for Four Hydrogen-Evolution Mechanisms, Assuming $\theta_H \ll 1$

Mechanism	$\dfrac{\partial \ln \theta}{\partial \eta}$
Discharge-rate determining followed by chemical desorption	$\dfrac{-F}{4RT}$
Discharge followed by rate-determining chemical desorption	$\dfrac{-F}{RT}$
Discharge-rate determining followed by electrodic desorption	0
Discharge followed by rate-determining electrodic desorption	$\dfrac{-F}{RT}$

In the case of hydrogen evolution on silver in alkaline solutions, the experimental slope of the η versus $\log \theta_H$ curve is -0.31 V. Hence, in alkaline solutions and on silver electrodes, the hydrogen evolution reaction seems to follow a rate-determining discharge followed by a chemical-desorption mechanism. From Table 10.15, which tabulates the values of $(\partial \ln \theta_H / \partial \eta)$ for four possible reaction mechanisms for the case where $\theta_H \ll 1$, it is clear that the value of -0.31 for $(\partial \eta / \partial \ln \theta_H)$ appears consistent only with the mechanism stated above.

10.3.4f. The Heat of Adsorption of Atomic Hydrogen on the Electrode. The change in heat content associated with the gas-phase adsorption of atomic hydrogen upon a metal is available from experiment. This heat of adsorption on most metals is approximately equal to the strength of the bond between the metal and hydrogen, which in turn is related to the *d*-band character of the metal. Thus, as the *d*-band character of the metal increases, the adsorption energy may be expected to decrease and the coverage with hydrogen atoms to increase.

It is, of course, unreasonable to assume that the gas-phase heat of adsorption is applicable to a situation in which the metal is made an electrode in an electrolytic solution. One must not ignore the fact that atomic hydrogen and also water molecules adsorb on the electrode surface. Nevertheless, if a series of metals are arranged in the order of their gas-phase heats of adsorption, it seems likely that the *same order* (in the sense of a succession of values) is valid when the metals are used as electrodes immersed in an electrolyte.

The trend of the heat of adsorption of atomic hydrogen or of the metal–hydrogen bond strength can be used in the following way to give a clue to the mechanism of hydrogen evolution. Using various substrates, the current density for hydrogen evolution is noted at a certain overpotential. Then this current density is plotted against the heat of adsorption of hydrogen or against the M—H bond strength. It was found (Fig. 10.6) by Conway that the metals fall into two groups. In one group, which contains mercury and thallium, the current density increases with increasing bond strength; in the other group, which contains the transition metals, the current density decreases as the M—H bond strength increases. It can readily be shown (*cf.* Section 10.1.4a) that the mercury group behaves as though proton transfer were rate determining, and, in the transition-metal group, the electrodic desorption of hydrogen appears to be rate determining.

10.3.4g. Isotopic Separation Factors. A standard method used in chemical kinetics for the elucidation of reaction mechanisms is based on the study of isotope effects on reaction rates. What is done is to examine the change in reaction rate resulting from a change in the isotope of one of the atoms taking part in the reaction. For example, consider a reaction involving hydrogen. Suppose that, instead of hydrogen, one uses deuterium or tritium. Since hydrogen evolves faster than deuterium, which in turn evolves faster than tritium, one can consider the isotope effect in terms of what is called the *separation factor*, a quantity that is proportional to the relative rates of evolution of H_2 and HT, e.g. The experimental isotopic effect on the rate of the particular reaction can then be compared with the isotopic effect calculated from the theory of rate processes.

The *important part* is that the calculation involves an assumption concerning the mechanism and the rds of the reaction. By varying the assumptions that are made, one can obtain a series of predicted values, and these predicted numerical values differ rather greatly (up to several hundred percent) with the assumptions. Since the experimental determination of the separation factor can be made to within a few percent, comparison of theory and experiment can indicate the mechanism. The theoretical prediction of the isotopic effect for each assumed mechanism is based on a consideration of the potential-energy surface for a reaction, of the change in the activation-energy surface for a reaction, and of the change in the activation energy which would result from a change in the isotope of one of the reactants.

Since the isotopic effect is based on the mass differences of the isotopes, it is at its best when the mass differences are large. Thus, the largest isotope effects are obtained with hydrogen, deuterium, and tritium when the masses are in the ratio 1:2:3. It follows that the hydrogen-evolution reaction is particularly suited for the use of the isotopic effect method.

The application of this isotopic separation factor to the hydrogen-evolution reaction was attempted by Horiuti and later by Srinivasan with considerable success. Assuming the various mechanisms for the hydrogen-evolution reaction, one then calculates the separation factor for each mechanism by adopting a model of the activated complex and working out the relative activation energies for the different isotopes. Since one is only aiming at the *ratio* of the rates of hydrogen and tritium evolution, many of the uncertainties in the calculation disappear. It turns out that the separation factors for hydrogen and tritium are typically 6 for one mechanism and about 20 for another. Table 10.16 illustrates the value of the hydrogen–tritium separation factor in the elucidation of the hydrogen-evolution reaction mechanism. Hence, it looks as if the mechanism can easily be determined. Separation-factor measurements indicate that there are two groups

TABLE 10.16

Correlation of Calculated and Experimental Separation Factors for Hydrogen Evolution from Acid Solution

Metal	Separation factor		Probable mechanism
	Calc value	Exper value	
Platinum	9.4	9.6 ± 0.4	Fast discharge, slow chemical desorption
Rhodium	9.4	10.7 ± 0.4	
Tungsten	6.2	6.0 ± 0.2	Coupled discharge–electrodic-desorption reaction, which, at high hydrogen-atom coverage, becomes fast discharge, slow electrodic desorption
Nickel	19.8	18.0 ± 0.9	Fast discharge, slow electrodic desorption
Mercury	6.2	5.8 ± 0.3	Slow discharge, fast electrodic desorption
Cadmium	6.2	9.2 ± 0.5	
Lead	6.2	6.7 ± 0.7	

TABLE 10.17
Mechanism for the Hydrogen-Evolution Reaction on Various Metals in Acid Solution

Metal	Mechanism
Mercury, lead, and cadmium	Proton-discharge-rate determining followed by electrodic desorption
Nickel, tungsten, and gold	Proton discharge followed by rate-determining electrodic desorption
Platinum and rhodium	Proton discharge followed by rate-determining chemical desorption; on particularly activated metals, rate-determining diffusion of molecular hydrogen away from the electrode surface

of metals, the "soft" metals (mercury, thallium, etc.), for which the rate of hydrogen evolution is determined by the rate of proton discharge, and the transition metals, for which electrodic desorption is rate determining. An analogous conclusion resulted from consideration of the dependence of the reaction rate upon the bond strength of hydrogen on the substrate (*cf.* Section 10.1.4*a*).

10.3.4h. What Are the Probable Mechanisms for Hydrogen Evolution? One can state the mechanism of hydrogen evolution on several substrates with a high degree of probability, and, although some workers in this field still argue about the degree of certainty of the mechanism on some metals, the reaction is certainly one of the better ones in the matter of degree of confidence in the mechanism determination. So long as one sticks to acid solutions (curiously, not much mechanistic work has been done in alkaline solutions), the mechanisms are now reasonably well accepted. Examples of some of the more certain ones are given in Table 10.17.

Further Reading

1. J. Tafel, *Z. Physik. Chem.* **50A**: 641 (1905).
2. N. I. Kobosew and N. Nekrassow, *Z. Elektrochem.* **36**: 529 (1930).
3. J. Horiuti and G. Okamoto, *Sci. Papers Inst. Phys. Chem. Res. (Tokyo),* **28**: 231 (1936).
4. A. N. Frumkin, *Acta Physicochim. U.R.S.S.*, **7**: 474 (1937).
5. A. N. Frumkin, *Acta Physicochim. U.R.S.S.*, **18**: 23 (1943).
6. A. N. Frumkin, *Discussions Faraday Soc.*, **1**: 57 (1947).
7. E. C. Potter, *J. Chem. Phys.*, **20**: 164 (1952).
8. B. E. Conway, *J. Chem. Phys.*, **26**: 532 (1957).

9. A. N. Frumkin, *Acta Physicochim. U.R.S.S.*, **207**: 321 (1957).
10. M. A. V. Devanathan and W. Mehl, *J. Electroanal. Chem.*, **1**: 143 (1959–1960).
11. A. N. Frumkin, "Hydrogen Overvoltage and Adsorption Phenomena, Part I," in: P. Delahey and C. Tobias, eds., *Advances in Electrochemistry and Electrochemical Engineering*, Vol. 1, 1961, p. 65.
12. A. N. Frumkin, "Hydrogen Overvoltage and Adsorption Phenomena, Part II," in: P. Delahey and C. Tobias, eds., *Advances in Electrochemistry and Electrochemical Engineering*, Vol. 3, 1964, p. 287.
13. S. Srinivasan, *J. Electrochem. Soc.*, **111**: 844, 853, and 858 (1964); also, *Electrochim. Acta*, **9**: 31 (1964).
14. B. E. Conway, *Theory and Principles of Electrode Processes*, The Ronald Press Company, New York, 1965.
15. D. B. Matthews, *J. Chem. Phys.*, **44**: 298 (1966); also, *Discussions Faraday Soc.*, **39**: 239 (1965).
16. K. J. Vetter, *Electrochemical Kinetics*, Academic Press, Inc., New York, 1967.
17. J. O'M. Bockris and S. Srinivasan, *Full Cells: Their Electrochemistry*, Chap. 2, McGraw–Hill Book Company, New York, 1968.

10.4. THE ELECTRONATION OF OXYGEN

10.4.1. The Importance of the Oxygen-Electronation Reaction

One cannot overemphasize the importance of the reaction involving the electronation or reduction of oxygen molecules to give water

$$O_2 + 4H^+ + 4e \rightarrow 2H_2O \tag{10.146}$$

To justify this statement, one has only to list some of the principal areas where the reaction plays a role.

First, the reaction is the basis of *electrochemical combustion*, which bears an interesting relation to chemical combustion. When a substance reacts chemically with oxygen, it is said to burn or undergo chemical combustion. Thus, hydrogen burns with the explosive release of thermal energy to form water

$$2H_2 + O_2 \rightarrow 2H_2O \tag{10.147}$$

There is, however, an electrochemical method of producing water from hydrogen and oxygen. In the same electrochemical system, hydrogen is deelectronated at the electron-sink electrode, and oxygen is electronated at the electron-source electrode. The two charge-transfer reactions are

$$2H_2 \rightarrow 4H^+ + 4e \tag{10.148}$$

$$O_2 + 4H^+ + 4e \rightarrow 2H_2O \tag{10.149}$$

Four hydrogen ions and four electrons are produced in the hydrogen deelectronation reaction, but the same numbers are used up for oxygen reduction. So the overall reaction in the electrochemical system (not the reaction at each interface) is the reaction

$$2H_2 + O_2 \rightarrow 2H_2O \qquad (10.150)$$

i.e., hydrogen is burned electrochemically. Hydrogen undergoes *cold combustion*.

If, therefore, one wants to burn a substance electrochemically, it must be deelectronated at one electrode and oxygen must be electronated at the other electrode; then the overall cell reaction will represent the electrochemical combustion of the substance. Whereas chemical burning is done by direct interaction, electrochemical burning is done in two separate charge-transfer reactions at two separate electrode–electrolyte interfaces.

It will be seen later on (Chapter 11) that most electrochemical energy converters are air breathing; some fuels, e.g., decane, the principal constituent of diesel oil, gives up electrons to one electrode, and an equivalent number of electrons are used up at the other electrode for reducing oxygen. Thus, the oxygen-electronation reaction is of crucial importance to the production of electrical energy by electrochemical means, the most promising technique of direct energy conversion.

The reaction also plays an important role in the instability of metal surfaces (as will be shown in Section 11.1.5). Consider, e.g., a steel bridge in a saline atmosphere. It continuously corrodes, or consumes itself, because oxygen[†] from the air diffuses to the metal, where it is electronated by electrons arising from the dissolution of iron atoms in the steel. To protect the bridge from catastrophic failure, one must try to reduce the rate of oxygen electronation, and hence, as a result, metal dissolution.

The third area in which the oxygen-electronation reaction is of tremendous significance is less often emphasized. Life itself depends on the electronation of oxygen. In air-breathing organisms, the vital energy is derived by the oxidation, or deelectronation, of foods (carbohydrates, e.g.), and, to keep the reaction going, the electrons so obtained must be consumed by an electron acceptor. Oxygen is the electron acceptor and, hence, plays the essential role of the oxygen-reduction reaction in life producing electrochemical reactions.

[†] The other common electronation reaction involved in corrosion is hydrogen evolution. However, for a metal in contact with moist air, oxygen reduction (i.e., electronation) is a more likely reaction.

These thoughts seem strange because one wonders where the electrocatalysts are in living organisms. They are there under the name of enzymes. Of course, enzymes are not strips of metal, like iron, but consist of metal ions attached to giant protein molecules. In cytochromes (iron-containing enzymes), it is likely that the metal ions are the electronation and de-electronation sites analogous to the metallic electrodes of an electrodic system like an energy producer.

It is an interesting thought to realize that the oxygen electronation (or reduction) reaction is giving one the energy and the curiosity to wonder how the reaction works.

10.4.2. The Evaluation of One of the Mechanisms of Oxygen Electronation

Despite the crucial role of the oxygen reduction reaction, it has hardly been studied compared with the much-examined hydrogen-evolution reaction. This delay has probably occurred because it is only in the last few years that the importance of the reaction has been realized. Hence, generalizations concerning which criteria are best for mechanism determination cannot be discussed as soundly as in hydrogen evolution. Further, far fewer substrates can be used for the study of oxygen electronation than for hydrogen evolution. This is because oxygen begins to be electronated (reduced) at pH = 0 at potentials about 1 V more positive than the hydrogen equilibrium potential, i.e., at about 1.23 V on the standard hydrogen scale, and, at these potentials, many of the possible substrates for oxygen electronation would already, themselves, anodically dissolve. Thus, the potential of the electrode would be determined by two reactions, oxygen electronation and metal de-electronation; the electrode is, in fact, a corroding system, and the extraction of the parameters of the oxygen reaction for such a complex situation is none too easy. One of the major aims of electrochemical technological research is to produce an electrode in which the substrate is less expensive than platinum and the other noble metals, possesses good chemical resistance to the reaction environment, and is metallically conducting. The tungsten and possibly other bronzes appear to be promising in this respect (*cf.* Section 10.1.5*a*) when prepared under conditions where less than 100 ppm of platinum or nickel is incorporated.

What will be done, therefore, is to present only *one example* of the determination; this example involves the use of iridium[†] as substrate.

[†] Oxygen electronation is relatively rapid on this substrate.

First, consider the experimental facts. In both acidic and alkaline solutions, the slope of the η versus log i curve is about 0.11 V for the oxygen-electronation reaction $O_2 + 4H^+ + 4e \to 2H_2O$ and about 0.04 V ($\eta < \sim 0.25$ V) for the oxygen-evolution reaction $2H_2O \to O_2 + 4H^+ + 4e$. In other words, the transfer coefficients are $\overleftarrow{\alpha} \approx \frac{3}{2}$ for the oxygen-evolution reaction and $\vec{\alpha} \approx \frac{1}{2}$ for the oxygen-electronation reaction. The stoichiometric number, as given by (*cf.* Section 9.1.7)

$$\nu = \frac{n}{\overleftarrow{\alpha} + \vec{\alpha}} \tag{9.25}$$

turns out thus to be 2.

From the electronation transfer coefficient $\vec{\alpha} = \frac{1}{2}$, the stoichiometric number $\nu = 2$ and n, the number of electrons transferred in the overall oxygen reduction $n = 4$, one has, assuming $\beta = \frac{1}{2}$,

$$\vec{\alpha} = \frac{\frac{1}{2}(4 - \overleftarrow{\gamma} - \vec{\gamma}) + \vec{\gamma}}{2} = \frac{1}{2} \tag{10.151}$$

or

$$\frac{(\overleftarrow{\gamma} + \vec{\gamma})}{2} - \gamma = 1 \tag{10.152}$$

Now $(\overleftarrow{\gamma} + \vec{\gamma})$, which is the total number of electrons transferred in all steps except the rds, must have an integral value. Similarly, $\vec{\gamma}$ also must be integral. Hence, the only way to satisfy Eq. (10.152) is to have $(\overleftarrow{\gamma} + \vec{\gamma}) = 2$ and $\vec{\gamma} = 0$ or to have $(\overleftarrow{\gamma} + \vec{\gamma}) = 4$ and $\vec{\gamma} = 1$.

By a similar argument, from $\beta = \frac{1}{2}$, $n = 4$, $\overleftarrow{\alpha} = \frac{3}{2}$, and $\nu = 2$, one has for the oxygen-evolution reaction (*cf.* Section 9.1.7)

$$\overleftarrow{\alpha} = \frac{2 - (\overleftarrow{\gamma} + \vec{\gamma}/2) + \overleftarrow{\gamma}}{2} = \frac{3}{2} \tag{10.153}$$

or

$$\overleftarrow{\gamma} - \frac{\overleftarrow{\gamma} + \vec{\gamma}}{2} = 1 \tag{10.154}$$

To satisfy this condition, $\overleftarrow{\gamma} + \vec{\gamma} = 2$ and $\overleftarrow{\gamma} = 2$ or $\overleftarrow{\gamma} + \vec{\gamma} = 4$ and $\overleftarrow{\gamma} = 3$.

Hence, there are two possibilities, (1) $\vec{\gamma} = 0$ and $\overleftarrow{\gamma} = 2$ or (2) $\vec{\gamma} = 1$ and $\overleftarrow{\gamma} = 3$. If the former possibility is correct, then the rds is a charge-transfer step; if, however, the latter possibility is true, then the rds must be a chemical step. Several paths have been suggested for the oxygen-reduction reaction, but it has not yet proved possible to formulate a path which has a rate-controlling *chemical* step and yet has $\overleftarrow{\gamma} = 3$ and $\vec{\gamma} = 1$.

Three possible paths have been suggested by Damjanovic for the oxygen-reduction reaction in which a charge-transfer step is rate controlling and the conditions $v = 2$, $\vec{\gamma} = 0$, $\overleftarrow{\gamma} = 2$, $\overleftarrow{\alpha} = \frac{3}{2}$, and $\vec{\alpha} = \frac{1}{2}$ are satisfied. These are shown in Table 10.18. The mechanisms involving rds's $A2$, $B3$, and $C2$ in the table are all in accord with the experimental facts.

Hence, to discriminate between these three paths would require further criteria. One such criterion may be the pH dependence of the overpotential. It may be shown that $\partial \eta / \partial$ pH is $+RT/3F$ for path A and $-RT/3F$ for the other two paths in Table 10.18. Further, in the case of path A, the pH dependence of η changes sign in going from acid to alkaline solutions; this sign reversal is not expected for the other paths. The experimental behavior appears to favor path A with the step $A2$ as rate-determining.

Several more complex reaction mechanisms have been suggested for the oxygen reduction on other metals; e.g., the reaction at gold electrodes is considered to involve hydrogen peroxide either as an intermediate or as a side product, depending on the overpotential and the electrolyte in which the reaction takes place.

TABLE 10.18

Three Possible Reaction Paths for the Oxygen-Electronation Reaction Which Satisfy the Known Diagnostic Criteria on Iridium Substrates

Reaction path	Mechanism
A	1. $O_2 + 2M \rightleftharpoons 2MO$
	2. $MO + H^+ + e \rightarrow MOH$
	3. $MOH + H^+ + e \rightleftharpoons M + H_2O$
B	1. $O_2 + 2M \rightleftharpoons 2MO$
	2. $MO + H_2O \rightleftharpoons MO - H - OH$
	3. $MO - H - OH + e \rightarrow MO - H - OH^-$
	4. $MO - H - OH^- + H^+ \rightleftharpoons MOH + H_2O$
	5. $MOH + H^+ + e \rightleftharpoons M + H_2O$
C	1. $O_2 + 2M \rightleftharpoons 2MO$
	2. $MO + e \rightarrow MO^-$
	3. $MO^- + H^+ \rightleftharpoons MOH$
	4. $MOH + H^+ + e \rightleftharpoons M + H_2O$

10.4.3. Catalysis and the Oxygen Reaction

Although the number of metals useful as substrates for the oxygen reduction is much smaller than for the hydrogen-evolution reaction, it is possible to study the electrocatalytic aspects of the oxygen reaction. For example, it is possible to change the surface of the noble metals from those in which there is an oxide on the surface to those in which the free metal is bare. Then, at the same potentials, the rate of the oxygen-reduction reaction can be compared on the two surfaces. Thus, one can clearly pick out what happens to the reaction rate with a known change of substrate.

First, it is necessary to find out how much oxygen is upon the surface and then to find out something about the properties of oxygen upon the surface of metals. This is done by carrying out electronation transients in which the current is kept constant and the change of potential noted (one could also keep the potential constant and note the change in the current). A typical constant-current transient is shown in Fig. 10.65 for oxygen evolution upon platinum, with an indication of the region in which oxygen is reduced. In this way, it is possible to relate the amount of oxygen on an electrode to the concentration of oxygen molecules in the solution (the variation of oxygen pressure being by the use of gaseous mixtures including oxygen and nitrogen). Thus, one obtains relations between the fraction of the surface covered with oxygen and the oxygen pressure; such relations are known as *isotherms*. By doing this on various metals and various metal

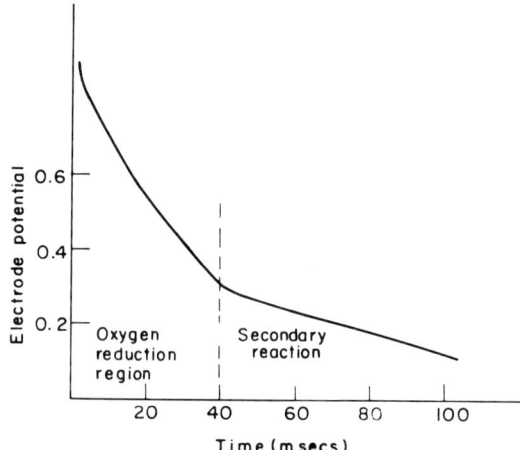

Fig. 10.65. A typical constant-current transient for oxygen reduction on platinum, the oxygen reduction occurring prior to the sharp change in slope at 0.3 V.

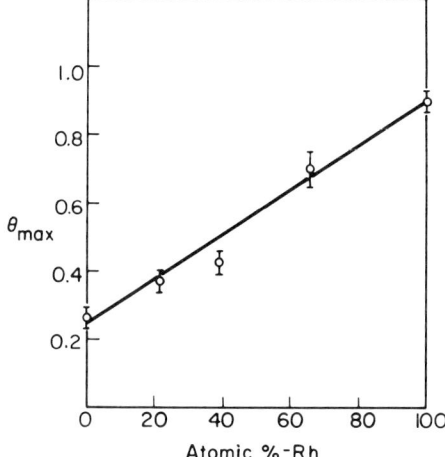

Fig. 10.66. The linear variation of maximum oxygen coverage θ_{max} with Pt–Rh alloy composition (the number of unpaired d electrons increases from pure platinum to pure rhodium).

alloys, it is possible to determine the saturation oxygen coverage (where the isotherm reaches its limiting value as pressure increases). One can obtain (Bhaskara Rao) a relation between this constant amount of adsorbed oxygen and some characteristic of the surface, e.g., the number of unpaired electrons in the d band of the metal (Table 10.4 and Fig. 10.66).

It is seen that there is a relation between the number of unpaired d electrons in the metal and the amount of oxygen coverage; this is information on the electrocatalytic abilities of different substrates in the oxygen-evolution reaction.

Just as the hydrogen-electronation reaction mechanisms can be related to the d-band character of the substrate (*cf.* Section 10.3.4*f*), so can the oxygen electronation. For example, gold, which has no unpaired d electrons (Table 10.4, Section 10.1.4*c*), has a very low oxygen coverage and the Tafel slope under such a condition has been given as $-2RT/F$; palladium, however, shows a Tafel slope of $-RT/F$ owing to an intermediate oxygen coverage and a different rds step. Alloys of gold and palladium show Tafel slopes of either $-RT/F$ or $-2RT/F$ depending upon the alloy composition. The change from $-2RT/F$ to $-RT/F$ as palladium is added to gold has been related, by Damjanovic, to the occurrence of sufficient palladium in the alloy to create d orbital vacancies and consequently a change from low to intermediate oxygen coverage.

A clear case of catalysis is revealed when one changes from the bare metal substrate to one on which there is an oxide film. For many years electrochemists had difficulty in distinguishing between *adsorbed oxygen* on a surface and *oxide film*, but this difficulty has been removed by the use of modern optical methods, which reveal the characteristics of surface films through a study of polarized light reflected from the substrate surface. It is now possible to say at what potential there is a definite beginning of the formation of an oxide film. An oxide film may be conceptually distinguished from a film of adsorbed oxygen in the following way: In the latter, the oxygen and metal atoms are in separate planes, unlike those in an oxide lattice, where the oxygens and metal atoms are interspersed. The potential dependence of the rate of the oxygen-reduction reaction is different on a bare platinum substrate compared with that on a substrate on which there is an oxide; Fig. 10.67, illustrating this point, shows that the bare platinum is a better catalyst than the oxide-covered surface by about one-hundred times, at least at the potential of 0.85 V. It is premature to attempt an explanation of this change in rate with the formation of an oxide film. These relations have been given here as an example of clear and quantitative evidence of changes of oxygen-reduction rates with substrate.

Another example is that of the change which occurs when one has an

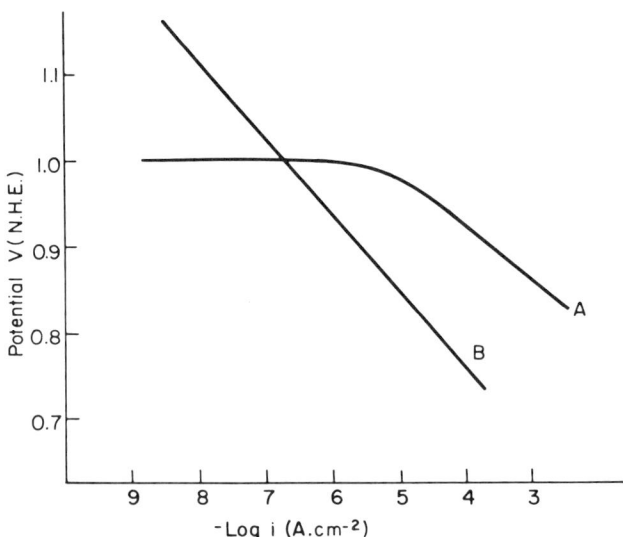

Fig. 10.67. Comparison of the catalytic activity of oxide-free platinum A and oxide-covered platinum B for the reduction of oxygen.

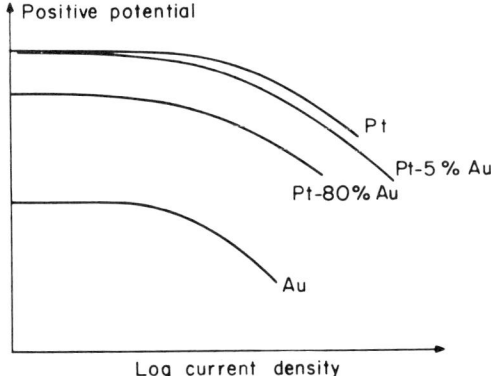

Fig. 10.68. Current–potential relationship, as a function of alloy composition for the system Au–Pt, for the oxygen-reduction reaction.

alloy substrate. How does the rate now depend upon the constitution of the surface? Here a rather peculiar thing takes place. One can see the essential result in Fig. 10.68 for the Pt–Au system and in Table 10.19 for the Pt–Rh system. It is seen that the substrate does not behave as though it were following the properties of the alloy but as though the reaction takes place always on that metal which, before alloying, gave the higher catalytic rate.

10.4.4. Some Special Difficulties with Electrodic Reactions Having Small Exchange-Current Densities

One of the problems in electrode-kinetic studies is that some of the mechanistically useful quantities, e.g., the stoichiometric number, can best be determined under near-equilibrium conditions. It will be recalled that,

TABLE 10.19

Exchange-Current Density for Several Noble Metals and a Platinum-Rhodium Alloy in the Reduction of Oxygen from Perchloric Acid Solution

Metal or alloy	Exchange-current density i_0, amp cm^{-2}
Platinum	10^{-9}
Platinum and 40 atomic % rhodium	10^{-9}
Rhodium	6×10^{-9}
Iridium	10^{-11}

under these conditions, there is a linear relation between the current density and the overpotential

$$\eta = \Delta\phi - \Delta\phi_e = \frac{RT}{\eta F}\frac{i}{i_0} \tag{8.53}$$

Thus, for a given value of η, the smaller the value of i_0 is, the smaller is the value of i corresponding to η. In oxygen reduction, the exchange-current density is about 10^{-10} amp cm^{-2}; hence, under near-equilibrium conditions, i.e., $\eta \leq 10^{-2}$ V, the oxygen-reduction current density would be about 10^{-10} amp cm^{-2}.

It has been seen, however, from the theory of simultaneous reactions at an interface (cf. Section 10.2.17) that there is no guarantee that the potential $\Delta\phi$ promotes only the reaction under study. Any electron acceptor present will be electronated provided that the equilibrium potential for this reaction is more positive than the potential difference $\Delta\phi$ across the interface. Similarly, any electron donor present in the solution or the electrode itself will be deelectronated provided that the equilibrium potential for this reaction is more negative than $\Delta\phi$.

If the reaction under study is oxygen electronation and the electrode consists of a metal that does not dissolve at the potential $\Delta\phi$, one has to bother with electron acceptors and electron donors in solution which are simultaneously electronated or deelectronated along with the oxygen molecules. Now, if the objective is to study oxygen reduction, the only electron acceptors and electron donors in solution that need be considered are those which have crept into the solution as unintended impurities. The impurity content of an *ordinarily purified* solution may be in the region of about 10^{-6} mole liter^{-1}. At such low concentrations, the current arising from the electronation or deelectronation of the impurities is likely to be controlled by their transport to the interface. Thus, at the potential $\Delta\phi$, the current density i_{imp} utilized by the impurities will be equal to the limiting current density i_L for their transport, i.e.,

$$i_{\text{imp}} = i_L = \frac{DnFc}{\delta} \tag{10.155}$$

and, by setting $c = 10^{-6}$ mole liter^{-1} = 10^{-9} mole cm^{-3}, $D = 10^{-5}$ cm^2 sec^{-1}, $F \approx 10^5$, $\delta = 10^{-2}$ cm, and $n = 4$,

$$i_{\text{imp}} \approx 4 \times 10^{-7} \text{ amp cm}^{-2}$$

Since this impurity current is 10^2 to 10^3 times the oxygen-reduction current density (cf. Table 10.19) under near-equilibrium conditions, the bulk of the current density at the interface will be used for an electrodic reaction involving impurities and not for oxygen electronation. Clearly, this situation must be avoided; to study oxygen reduction, the major part of the current density corresponding to a potential $\Delta\phi$ must be used for the electronation of oxygen molecules, i.e., the impurity current i_{imp} must be a negligible fraction of the total current density i. The obvious way of doing this is to reduce the impurity content of the solution until the limiting current density for impurity transport becomes negligible compared with i. To keep the impurity-current density less than the oxygen-reduction current density of 10^{-10} amp cm^{-2} under near-equilibrium conditions, the impurity concentration must be less than 10^{-10} mole liter^{-1}. This extremely low impurity becomes essential only because of the low exchange-current density for oxygen electronation and because of the desire to work under near-equilibrium conditions which require $i \approx i_0$.

One must conclude, therefore, that, in order to work near the equilibrium potential of a reaction, the lower its exchange-current density is, the higher will be the purification of the solution necessary to avoid interference from impurities in the solution. It is clear that only under conditions of extremely high purity will the oxygen equilibrium potential of 1.23 V be observed. The first experimental verification of this value was reported by Hoar (1.20 $\pm$ 0.03 V), obtained by the intersection of extrapolated cathodic and anodic Tafel lines (see also Section 10.4.6), and by Huq by direct determination in highly purified systems.

10.4.5. An Electrodic Method of Purifying Solutions

It has been argued that the impurity content of a solution must be reduced by about four powers of ten from 10^{-6} mole liter^{-1} (the impurity level of an ordinarily purified solution) to less than 10^{-10} mole liter^{-1} before one can secure credible data on oxygen evolution and reduction in the vicinity of the equilibrium potential. To obtain such ultrapure solutions, conventional purification procedures are inadequate. Special methods have to be used.

One of the best methods (called *pre-electrolysis*) is to deposit the impurities on a scavenger electrode and, after deposition has occurred for a sufficiently long time, to remove the scavenger electrode and introduce the electrode that is to be used for the reaction under study. All this shows that many fine details must be observed to get good results at very low current densities in electrodic systems where competition with impurities becomes important.

It is easy to calculate for what period of time one must electronate or de-electronate with a scavenger electrode of area A cm^2 before one can be confident that impurities have been successfully eliminated. Suppose that the solution initially contains an impurity at a concentration of c moles cm^{-3}, and let the total volume of solution be V cm^3. Since the decrease in the number of moles of impurity in the solution must be equal to the number of moles deposited on the scavenger electrode,

$$\frac{dVc}{dt} = -\frac{i_{\text{imp}}A}{nF} \qquad (10.156)$$

or

$$\frac{dc}{dt} = -\frac{i_{\text{imp}}A}{nFV} \qquad (10.157)$$

since V is a constant.

If, however, the impurity is being supplied to the electrode at the maximum diffusion rate (i.e., at the limiting current density), then

$$i_{\text{imp}} = i_L = \frac{DnFc}{\delta} \qquad (10.155)$$

Hence,

$$\frac{dc}{c} = -\frac{DA}{\delta V}dt \qquad (10.158)$$

which can be integrated to give

$$c_t = c_0 e^{-(DA/\delta V)t} \qquad (10.159)$$

where c_t and c_0 are the impurity concentrations at a time t and at the start of the scavenging process.

Electrodic methods for the purification of solutions are in extensive use today; among the first to use these methods were Lewina, in the purification of alcohols, and Jaffe, in the cleansing of water and other solutions utilizing very high potentials (2000 V). Their rational introduction into studies of electrode kinetics at solid electrodes was first carried out by Conway, who showed that hydrogen evolution on nickel electrodes was sensitive to solution concentrations of 10^{-10} mole liter^{-1} of As_2O_3 and other "poisons."[†]

[†] It is a most intriguing point that the classical Agatha Christie poisons are just those substances which seem most able to slow down interfacial electron-transfer reactions. This point must be placed alongside the observations of Srinivasan *et al.* that the energy-conversion efficiency situation in the body is such that *only* electrochemical mechanisms of the conversion of food to power are consistent with all the facts. Are there electrodic mechanisms to many life processes? What makes arsenic, cyánide, etc.,

10.4.6. Observing Very Slow Reactions near Equilibrium

On the basis of what has just been said, subtle and careful techniques are required to observe slow or low exchange-current-density reactions which are in the neighborhood of the equilibrium potential. One has to have very clean solutions and make a great deal of fuss concerning the purification. This means, in effect, that, to observe the equilibrium or Nernst potential, the electronation- and de-electronation-current densities of the reaction intended must predominate over the electrode surface. If other reactions involving impurities enter the picture, one is, in fact, observing some mixed potential (see Section 9.5.5d) which does not refer to any thermodynamic condition. In past years, this caused a great deal of trouble for classical electrochemists, who often did not realize what was happening—a stimulus to the breaking away from the Nernstian hiatus of half a century of electrochemistry.

The oxygen-reduction reaction is a case in point here. Its equilibrium potential was easily upset by impurities, and its low exchange-current density made any attempt to determine it very difficult. When, as Huq first showed, great care was taken to purify the solution, both electronation and de-electronation being carried out on scavenger electrodes, it became possible to make an experimental determination of the equilibrium potential. Also, the pressure relation, which proved the reaction to be oxygen reduction and evolution at the equilibrium potential, was obtained.

Why is it important to be able to attain the equilibrium potential? An operational answer is that, in an energy producer, the maximum possible potential difference between the electrodes is the difference of the equilibrium potentials. If oxygen reduction is to be the charge-transfer reaction at one of the electrodes, any fall of the potential from the equilibrium value represent a reduction in the electrical energy available from the converter at low rates of working.

Further Reading

1. G. Jaffé, *Ann. Physik*, **28**: 326 (1909).
2. T. P. Hoar, *Proc. Roy. Soc. (London)*, **A142**: 628 (1933).
3. S. Lewina and M. Zilberfarb, *Acta Physicochim. U.R.S.S.*, **4**: 275 (1936).

---

so effective as poisons in such small concentrations? Is it entirely coincidental that such poisons in the electrodic situation often accelerate reactions in extremely small concentrations and reduce their rates at small concentrations, while some well-known biological poisons—strychnine and arsenic—are also substances which act as tonics if taken in sufficiently small quantities?

4. A. K. M. S. Huq, *Proc. Roy. Soc.* (*London*), **A237**: 1733 (1956).
5. A. Damjanovic and M. L. Bhaskara Rao, *J. Phys. Chem.*, **67**: 2508 (1963).
6. A. I. Krasilshchikov, *Zh. Fiz. Khim.*, **37**: 531 (1963).
7. A. Damjanovic, *J. Electrochem. Soc.*, **113**: 739 (1966); also, *Electrochim. Acta*, **11**: 791 (1966).
8. A. Damjanovic and V. Brusic, *Electrochim Acta*, **12**: 615 (1967); also, *Electroanal. Chem.*, **15**: 29 (1967).
9. J. P. Hoare, "The Oxygen Electrode on Noble Metals," *Advan. Electrochem. Eng.*, **6**: 201 (1967).
10. J. P. Hoare, *The Electrochemistry of Oxygen*, Interscience, New York, 1968.
11. A. K. N. Reddy and M. A. Genshaw, *J. Chem. Phys.*, **48**: 671 (1968).
12. A. Damjanovic, "The Kinetics of the Mechanism of the Oxygen Reduction," in: J. O'M. Bockris and B. E. Conway, eds., *Modern Aspects of Electrochemistry*, No. 5, Plenum Press, New York. 1969.
13. J. O'M. Bockris and S. Srinivasan, *Fuel Cells: Their Electrochemistry*, Chap. 8, McGraw–Hill Book Company, New York, 1969.

CHAPTER 11
SOME ELECTROCHEMICAL SYSTEMS OF TECHNOLOGICAL INTEREST

11.1. TECHNOLOGICAL ASPECTS OF ELECTROCHEMISTRY

At the outset of this book (*cf.* Sections 1.3, 1.4, and 1.9) the role of electrochemistry in the development of science and technology and thus in the material standard of civilization was mentioned. To recall two milestones, the whole science of electricity with all its technological consequences was associated with the discovery and development of the first electrochemical energy source by Alessandro Volta (March 20, 1800).[†] The voltaic pile was the first source of large quantities of stable electric current and it enabled studies of the elementary properties of the new form of energy to be made. Secondly, aluminum metal is obtained by electrodeposition from a melt containing a product from the chemical processing of a naturally occurring

---

[†] Very remarkable discoveries have been made more recently concerning the apparent existence of a type of crude electrochemical storage device at least 1400 years before the time of Volta (1800). The first discovery was made during excavations in Khujut Rabuah, near Bagdad, by König in 1936. The apparatus (estimated to have originated between 300 B.C. and 300 A.D.) was essentially an iron–copper element. There is evidence that gold plating was carried out in 2500 B.C. (*cf.* Winkler, *Elektrie*, Heft 2, p. 71, 1960). Thus, it appears that flowing electricity was only *re*discovered in the 18th century. It is an interesting speculation that this may apply to a number of other parts of scientific knowledge, a thesis developed in detail by the French physicist Louis Pauwels.

ore. The consumption of that metal in this present civilization has taken proportions such that it may be considered along with steel in degree of importance in practical metallurgy. Together with plastics, it has extended the path of engineering beyond the iron era.

Presently, one can enumerate six basic fields of technology and everyday life in which electrochemistry plays a major part. These are:

Chemical industry. The production of chlorine and caustics by electrolysis of salt brine is of such a volume as to put these products close to the top of the list of the most important industrial products. The electrosynthesis of organic compounds is increasing; for example, it is used in a stage of the synthesis of nylon.

Metallurgy. Besides aluminum, electrolysis of melts or solutions is the basic or the only way of winning or refining a large number of other metals such as alkali metals, alkaline earths, zinc, copper, uranium, and others. Increasingly, controlled electrodissolution is used to machine metals at high speed.

Surface decoration and protection. In the case of metals, electroplating and electroless plating (nevertheless, electrochemical in nature) are basic ways of obtaining decorative and corrosion- and wear-resistant metallic finishes. Electropolishing is extensively used to obtain smooth surfaces.

Corrosion. Electrochemical reactions are the main cause of processes of degradation and deterioration of metallic constructions.

Energy conversion and storage. Electricity is stored in batteries. The development of fuel cells as primary sources of electricity and high-power-density batteries as storers has remarkable potentialities not only in the field of electricity production, by converting directly to electric power the considerable energy in the natural gas and coal reserves, but particularly in vehicle propulsion as the vitally needed replacement for the pollution-causing internal-combustion motor. Indeed, fuel cells and the new batteries can be the solution to the pollution problems of the cities.

Biological mechanisms. A number of these, particularly that of the transfer of currents through nerves, are certainly electrochemical. It is possible that biological cells consume the breakdown products of food and oxygen in an electrodic way (Del Ducca).

The fundamental bases of all of the enumerated applications are contained in the preceding chapters. It is felt, however, that more must be said before the basic aspects of corrosion are understood equally well and there are many fundamental aspects of electrochemical energy conversion and storage which are still to be explained. It is to these objectives that this chapter is devoted.

11.2. CORROSION AND THE STABILITY OF METALS

11.2.1. Civilization and Surfaces

In the early stages of human civilization, man preserved himself in a hostile environment by functioning as a bioelectrochemical machine, converting the solar energy stored in food via electrochemical reactions into muscle power. But recently man has become increasingly a *cyborg*;[†] he has linked himself more and more with machines that harness nuclear energy and the solar energy stored in coal and oil and has thereby satisfied his needs with increasing efficiency. Thus, the progress of civilization has been marked by an increasing use of machine power and a decreasing use of muscle power. What of the future? The trend is clearly that man makes minimal use of his own biochemical energy converters and turns to insentient machines to effect the conversion of energy into convenient forms. Man is bound to lean increasingly on computerized mechanisms programmed to make energy derived from atomic fission and fusion reactions in the form of electricity do the work which he wants.

To make this vision a reality, the machines that do man's bidding and thus become the basis of the material aspects of civilization must be able to function without decay over years in the terrestrial atmosphere. Materials, mainly metals, used in fabrication must be stable. If the metals become unstable, then the machines fabricated partly from these metals undergo an undesired obsolescence. An industrial civilization depends in a crucial way upon the stability of metals in its moist (and often impurity-containing) atmospheres.

It is an interesting fact that a piece of metal remains stable for an almost indefinite period of time provided that it is stored in vacuum. It appears that metals acquire stability when their surfaces are isolated from the normal terrestrial environment. If this isolation is not achieved, metals become unstable in various ways. They develop cracks and break upon strain with catastrophic suddenness. They suffer fatigue, i.e., loss of strength, when subjected to periodic stress. They undergo a process of embrittlement. Their surfaces are transformed into oxides which peel off, or they just dissolve away. With the exception of the (hence) expensive noble metals, all metals are unstable to varying degrees in a terrestrial atmosphere.

[†] A cyborg, or cybernetic organism, is a human being functioning in association with a machine. A man writing with a pen is a trivial example of a cyborg, but the visionary purpose of the concept is best brought out by thinking of a hypothetical writing machine connected to and operated directly by electrical impulses from the brain.

The most widely used metals, namely, iron, aluminum, copper, nickel, and alloys of these metals, all decay and lose good mechanical properties in unprotected contact with air.

One conclusion is obvious. The stability of metals is determined by the events at the interface between these metals and their environment. The internal strength of a metal (particularly a metal under stress) is influenced in the long run by happenings at its surface. If the surface of a metal is stable, its interior tends to remain so. The detrimental transformation of the bulk properties of a metal begins at its surface.

This, then, is a link between civilization, surfaces, and, as will be seen, electrochemical reactions.

11.2.2. Charge-Transfer Reactions Are the Origin of the Instability of a Surface

An important feature of the terrestrial environment must now be noted. The atmosphere is essentially moist air containing dissolved carbon dioxide. (Marine atmospheres consist of moist air often containing in suspension sodium chloride.) Moisture in contact with the terrestrial atmosphere becomes an ionically conducting medium, an electrolyte.

Since metals become unstable (undergo the happenings named above) when allowed to come into contact with the moist atmosphere, it is reasonable to conclude that this instability of metals results from charge-transfer reactions at their interfaces. This is why the rate of corrosive destruction of a metal's surface is greatly reduced by removal of moisture from the atmosphere. Keeping a metal in vacuum is equivalent to removal of the electrolyte in contact with the metal and therefore to the prevention of charge-transfer reactions. Thus, the spontaneous instability (or corrosion) of metals results from the charge-transfer reactions at the electrified interface between the metal and the moist, CO_2- or NaCl-containing, air (Wollaston, de la Rive).

Such a view has been confirmed by many detailed experiments in the first half of this century. These experiments included direct studies of the rate and products of the corrosion of a metal as a function of the electrolytic conduction of the moisture film. They also involved an imaginative extrapolation from experiments on energy-producing electrochemical cells, in which, e.g., separate pieces of zinc and copper were immersed in an electrolytic solution, to a situation in which an actual piece of impure (copper-containing) zinc decayed when brought into contact with a film of moisture that contained dissolved electrolyte.

11.2.3. A Corroding Metal Is Analogous to a Short-Circuited Energy-Producing Cell

Charge-transfer reactions are the basis of electrochemical substance-producing cells driven by an external current source and of electrochemical energy-producing cells driving an external load (*cf.* Sections 8.1.5 and 8.1.6). Metallic corrosion, too, it has been stressed, arises from the electrodic charge-transfer reactions at the interface between a metal and its electrolytic environment. But, in the case of a corroding metal, where is the external source driving the charge-transfer reactions, or where is the external load consuming the current produced by the charge-transfer reactions? The conceptual relationship between electrochemical cells and corroding metals must be developed (*cf.* Section 8.1.7).

Suppose that a piece of zinc and a piece of copper are immersed in an electrolyte containing Zn^{++} and Cu^{++} ions (Fig. 11.1). It has been argued that, because the equilibrium potential of the $Zn^{++} + 2e = Zn$ reaction is

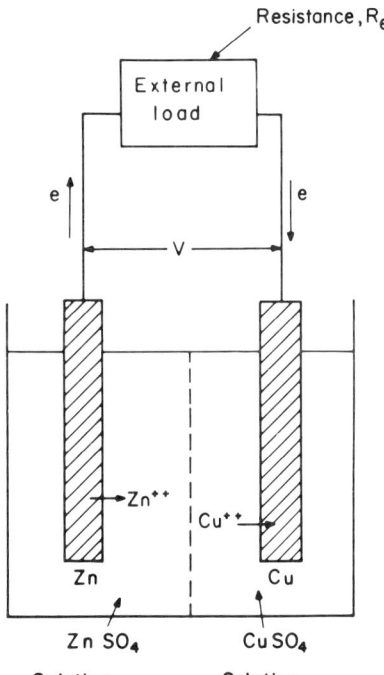

Fig. 11.1. In a Daniel cell, the current is kept going by the dissolving zinc and depositing copper ions.

negative with respect to that for the $Cu^{++} + 2e = Cu$ reaction, the zinc electrode is negatively charged with respect to the copper electrode. When an external electron path is provided by connecting the zinc and copper electrodes through an *external* load of resistance R_e, electrons flow through this external circuit from the zinc to the copper. To keep this current going, the zinc electrode dissolves to form Zn^{++} ions, and the copper deposits on the copper electrode. Hence, a Zn–Cu *electrode couple* acts as an energy producer. The potential difference across such cells has been analyzed (*cf.* Section 9.6.1), and it has been shown that the potential difference decreases with the cell current (Fig. 11.2). But this cell current is decided by the external load; make its resistance R_e lower, and the cell current increases.

What happens when the external resistance is made zero, i.e., when the copper and zinc electrodes are brought into electrical contact, or short-circuited (Fig. 11.3)? Of course, the copper continues to deposit and the zinc continues to dissolve at a certain current, but the potential difference across the cell will become zero. This thought experiment is equivalent to what happens when a bar of copper and a bar of zinc are welded together and put into an electrolyte containing cupric ions (Fig. 11.4). The zinc dissolves as the copper deposits.

Similarly, if, e.g., iron is welded together with some other metal and placed in an electrolytic solution, whether it dissolves or not will depend on whether its equilibrium potential is more negative or more positive than that of the other metal.

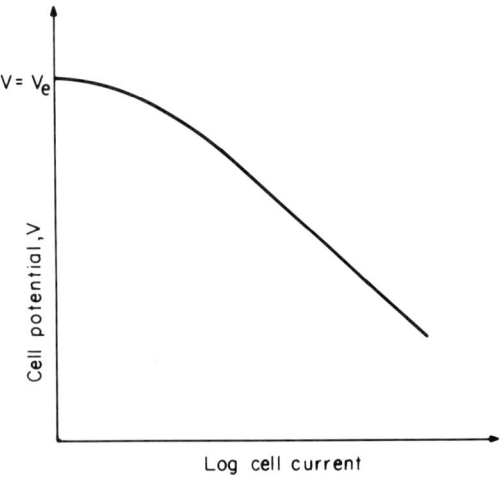

Fig. 11.2. The cell potential decreases as increasing current passes from the cell through the load.

SOME ELECTROCHEMICAL SYSTEMS OF TECHNOLOGICAL INTEREST 1271

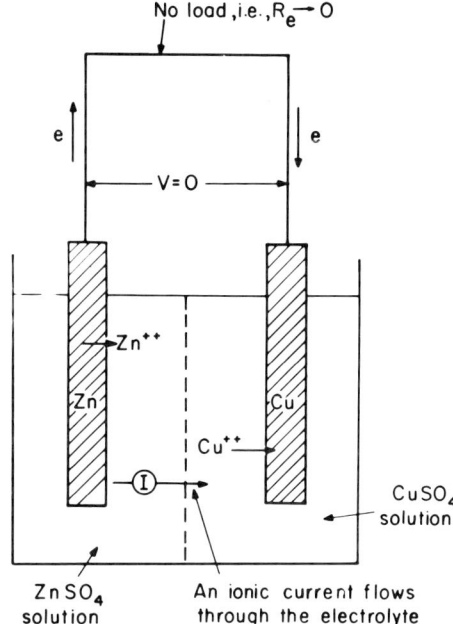

Fig. 11.3. The maximum current given by the cell is attained when the load is removed and the two electrodes are short-circuited.

The next step in the thought experiment consists in taking a large number of strips of copper and zinc and joining them so that there are alternate strips of the two metals, a multiband arrangement. If this assembly is immersed in solution containing cupric ions, the copper strips will be the sites for copper deposition and the zinc strips will be the sites for zinc de-electronation. Once again, the net result is copper deposition and zinc dissolution.

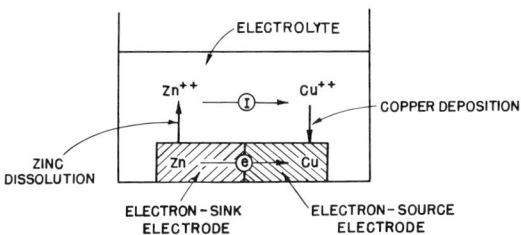

Fig. 11.4. The case of short-circuited electrodes is obtained also when two welded bars, one of zinc and one of copper, are immersed in an electrolyte containing cupric ions.

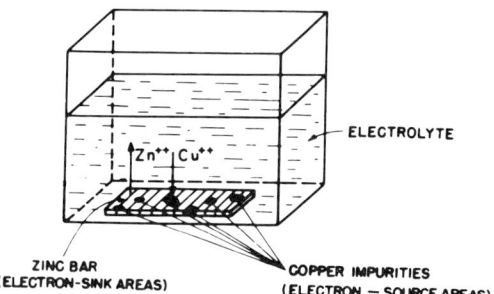

Fig. 11.5. Even without copper strips, copper will deposit on a zinc bar from a cupric-ion-containing electrolyte, preferably at some microscopic inclusions of copper impurities.

Finally, think of a bar of zinc with microscopic inclusions of copper, i.e., with copper impurities (Fig. 11.5). If this zinc bar is immersed in a solution containing Zn^{++} and Cu^{++} ions, the result is that the zinc will dissolve out and copper will deposit preferentially on the already existing copper areas or even form such areas by crystallization.

But notice that, in all these thought experiments to keep the zinc dissolving, it is not essential that the deposition of copper should be the electronation reaction. Even if the aqueous electrolyte has no Cu^{++} ions but consists of an ionically conducting moisture film, other electronation reactions are possible, e.g., hydrogen evolution $2H_3O^+ + 2e = H_2 + 2H_2O$ or oxygen reduction $O_2 + 4H^+ + 4e = 2H_2O$, and, as long as these electronation reactions take place, zinc dissolution will continue (Fig. 11.6). The bar of zinc is undergoing corrosion; it becomes unstable and eventually

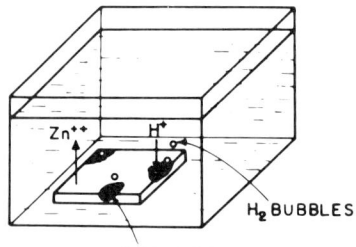

Fig. 11.6. The hydrogen-evolution reaction can be the electronation reaction necessary for the short-circuited cell reaction to occur and for zinc to corrode (i.e., anodically to dissolve).

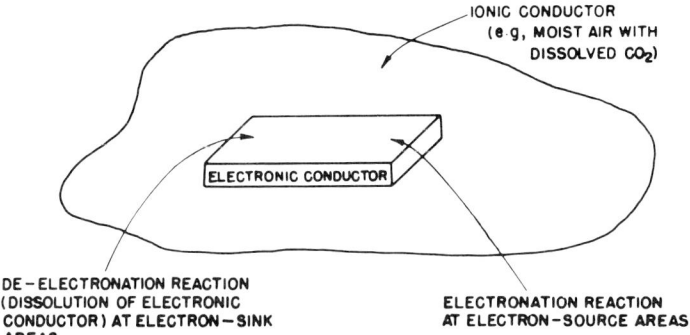

Fig. 11.7. According to the local-cell theory, the corrosion can take place if the piece of metal is in a moist atmosphere. There are separate sites for the de-electronation and electronation reactions.

destroys itself as a consequence of the electrochemical reactions occurring at the interface of the metal and ionically conducting moisture films or actual solutions.

Starting from the familiar Zn–Cu cell, the above discussion has shown that, by short-circuiting the cell and altering the spatial location of the electron source and sink, one is able to understand the corrosion of a piece of zinc. A corroding metal consists of an electron-sink area at which a de-electronation reaction (i.e., metal dissolution) occurs, an electronic conductor to carry the electrons to the electron-source area where an electronation reaction occurs, and an ionic conductor to keep the ion current flowing and to function as a medium for the electrodic reaction (Fig. 11.7). This model of corrosion is often termed the *local-cell theory of corrosion*.

11.2.4. The Mechanism of the Corrosion of Ultrapure Metals

On the basis of the local-cell theory, an ultrapure metal without impurity inclusions would be expected to be incorrodible. In general, the purer a metal, the more stable it is in an aqueous environment. But even an ultrapure metal does corrode. Why?

The basic mechanism for the instability of ultrapure metals was suggested by Wagner and Traud in a classic paper of 1938.[†] The essence of their view is that, for corrosion to occur, there need not exist *spatially separated* electron-sink and -source areas on the corroding metal. Hence, impurities or other heterogeneities on the surface are not essential for the

---

[†] The decomposition of amalgams was first discussed by Schultin in a sense to which Wagner and Traud's suggestion bore a resemblance.

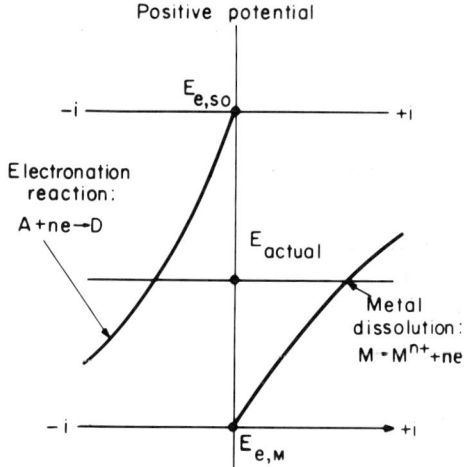

Fig. 11.8. Corrosion can occur always when the reversible potential of the metal dissolution is more negative than the actual potential of the metal and the potential of the electronation reactions is more positive.

occurrence of corrosion. The necessary and sufficient condition for corrosion is that the metal dissolution reaction and some electronation reaction proceed simultaneously[†] at the metal–environment interface. For these two processes to take place simultaneously, it is necessary and sufficient that the potential difference across the interface be more positive than the equilibrium potential of the $M^{n+} + ne = M$ reaction and more negative than the equilibrium potential of the electronation reaction $A + ne \to D$ involving electron acceptors contained in the electrolyte (Fig. 11.8).

Hence, the present view is a unified one. When the electronsink and -source areas are distinct in space and stable in time, one has the local-cell, or heterogeneous, theory of corrosion (Fig. 11.9a). On the other hand, when the metal-dissolution and electronation reactions occur randomly over the surface with regard to both space and time, one has the Wagner–Traud, homogeneous theory of corrosion (Fig. 11.9b). The Wagner–Traud mechanism with its random and dynamic de-electronation and electronation

---

[†] This possibility of simultaneous reactions occurring on one and the same electrode surface has already been considered in the discussion of how the study of the oxygen-reduction reaction is affected by the presence of impurities. It was pointed out that the electronation or deelectronation of impurities can occur simultaneously with the reduction of oxygen (*cf.* Section 10.4.4).

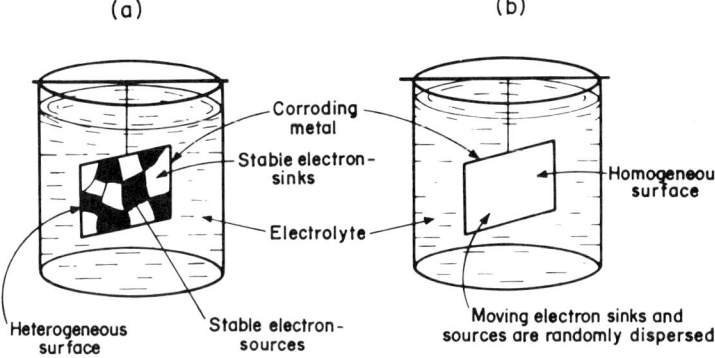

Fig. 11.9. A metal can corrode (a) by a heterogeneous mechanism if it contains areas of different electrodic properties or (b) by a homogeneous one if the surface is uniform.

sites requires a homogeneous metal surface. This is because heterogeneities tend to fix the de-electronation and electronation reactions to stable sink and source areas.

In some practical situations, however, there are heterogeneities of one type of another. Impurities are the most obvious type of heterogeneities, but there are other types, e.g., different phases of an alloy, or a metal with a nonuniform stress distribution or with a nonuniform access to electron acceptors. Thus, the local-cell, or heterogeneous, theory of corrosion has a wide scope of applicability. The homogeneous theory of corrosion emphasizes that, irrespective of the presence or absence of impurities, *metals become unstable because of different electrodic charge-transfer reactions occurring simultaneously and in opposite directions at the surface.*

11.2.5. What Is the Electronation Reaction in Corrosion?

A very important aspect of the corrosion of metal has been only touched upon so far. This aspect concerns the electronation reaction required to complete the corrosion circuit by consuming the electrons transferred to the metal from the metal-dissolution reaction. The question is: What is the electronation (cathodic) reaction?

Theoretically, it can be any reaction with an equilibrium potential which is more positive than the equilibrium potential of the metal-dissolution reaction. In practice, it is a reaction of the type of $A + ne = D$, where A is an electron-acceptor species present in the electrolyte which is in contact with the corroding metal. In aqueous electrolytes, the electron acceptors invariably present are H_3O^+ ions and dissolved oxygen, the corresponding electronation reactions being

$$2H_3O^+ + 2e \rightarrow 2H_2O + H_2 \quad \text{[acid solutions]} \quad (11.1)$$

and

$$O_2 + 4H^+ + 4e \rightarrow 2H_2O \quad \text{[acid solutions]} \quad (11.2)$$

or

$$O_2 + 2H_2O + 4e \rightarrow 4OH^- \quad \text{[alkaline solutions]} \quad (11.3)$$

The electrolyte may also contain species such as Fe^{3+} ions or nitric acid, in which case there can be additional electronation reactions of the type of

$$Fe^{3+} + e \rightarrow Fe^{2+} \quad (11.4)$$

or

$$3H^+ + NO_3^- + 2e \rightarrow HNO_2 + H_2O \quad (11.5)$$

If several electronation reactions are possible, i.e., their equilibrium potentials are positive with respect to the metal-dissolution equilibrium potential, then the one which yields the highest corrosion current is preferentially adopted. There is, of course, no new principle here; when parallel reactions can occur, the current is controlled by that reaction which yields the largest current corresponding to the given potential. This point is brought out clearly by the increase in the corrosion rate of iron in oxygenated solution compared with that in a deoxygenated one; the rate in the latter is decided by the $2H_3O^+ + 2e = 2H_2O + H_2$ electronation reaction, whereas, in the former, it is determined by oxygen reduction. The higher the pressure is of oxygen in the gas phase, the higher is the corrosion rate since the solubility of oxygen in the electrolyte is proportional to

TABLE 11.1

Effect of Oxygen Pressure on Corrosion Rate of Iron in $3\frac{1}{2}$% Sodium Chloride Solution[†]

Oxygen pressure, atm	Corrosion rate, mm yr^{-1}
0.2	2.2
1	9.3
10	86.4
61	300

[†] F. L. La Que and H. R. Copson, eds., *Corrosion Resistance of Metals and Alloys*, Reinhold Publishing Corp., New York, 1963.

pressure (Table 11.1). Upon addition of dilute nitric acid, the corrosion rate is increased even more because of the occurrence of reaction (11.5) involving nitrate ions.

11.2.6. Thermodynamics and the Stability of Metals

Suppose that one were faced with the task of deciding whether a particular metal would be suitable as a material of construction or fabrication in a given environment. The problem, e.g., may consist of approving or rejecting the use of a mild-steel reaction vessel in a technologically important process involving an aqueous medium. The real criterion for making the decision on the stability of the iron vessel is the magnitude of the rate of its dissolution; if it has a negligible rate of corrosion and sufficient strength, it is suitable for the purpose.

But suppose that, even before one calculates or measures the corrosion rate, one requires a yes or no answer regarding the stability of the steel vessel. The question is: Will the $Fe \rightarrow Fe^{++} + 2e$ de-electronation reaction and the electronation reaction, which together constitute the corrosion process, proceed spontaneously or not? Such questions concerning the spontaneous occurrence of reactions fall within the scope of equilibrium thermodynamics.

Now, there are several ways in which thermodynamics can be used to answer the question at hand. For instance, one can make use of the relation between free-energy change and equilibrium potential (*cf.* Section 8.2.13) to obtain the free-energy changes for the de-electronation and electronation reactions. The sum of the two free-energy changes yields the total free-energy change for the corrosion process

$$\Delta G = -nFV \qquad (11.6)$$

If this total free-energy change is negative, then the corrosion of the metal will proceed spontaneously. Table 11.2 shows the calculated free-energy change for the corrosion reactions of different metals with hydrogen evolution and oxygen reduction as the electronation reactions.

There is, however, a shortcut approach based on the potential *versus* pH representation of equilibrium potentials (*cf.* Section 9.6.4). The approach is as follows: Suppose the $M^{n+} + ne = M$ reaction does not involve proton transfer. Its equilibrium potential is then independent of pH and can therefore be represented on the potential–pH diagram as a straight line parallel to the pH axis (Fig. 11.10). Next, one considers the electron acceptor A present in the solution which is in contact with the metal M and calculates

TABLE 11.2

Products and Overall Energetics of Spontaneous Corrosion Reactions[†]

Metal	Solid product	Hydrogen type, $p_{H_2} = 1.0$ atm	Oxygen type, $p_{O_2} = 0.21$ atm
		Free-energy change at 25°C, cal g-mole^{-1} metal	
Silver	Ag$_2$O	+ 27,000	− 1,080
Copper	CuO	+ 24,800	− 37,450
	Cu(OH)$_2$	+ 27,800	− 28,300
	Cu$_2$O	+ 9,500	− 18,600
Lead	PbO (red)	+ 11,500	− 44,600
Nickel	Ni(OH)$_2$	+ 7,800	− 48,500
Cadmium	Cd(OH)$_2$	+ 600	− 55,600
Iron	Fe(OH)$_3$	+ 4,700	− 80,000
	Fe(OH)$_2$	− 2,260	− 58,500
	Fe$_3$O$_4$	− 5,020	− 80,000
Zinc	Zn(OH)$_2$(?)	− 19,240	− 75,200
Chromium	Cr(OH)$_3$	− 32,500	−117,000
Aluminum	Al(OH)$_3$(?)	−102,570	−180,700
Magnesium	Mg(OH)$_2$	− 84,000	−140,000

[†] F. L. La Que and H. R. Copson, eds., *Corrosion Resistance of Metals and Alloys*, Reinhold Publishing Corp., New York, 1963.

the equilibrium potential for its reactions. Suppose it does involve a proton transfer as well, i.e., $xA + mH^+ + ne = yD + zH_2O$. Since this reaction involves both electron and proton transfer, its equilibrium potential will vary with pH and can be represented as a straight line sloping downward in the potential–pH diagram.

Once one has a pH–potential diagram with lines drawn for the $M^{n+} + ne = M$ reaction and for the $xA + mH^+ + ne = yD + zH_2$ reaction, all one has to do is to draw a line perpendicular to the pH axis at the particular value of pH corresponding to that of the solution (Fig. 11.10). If that line intersects the $M^{n+} + ne = M$ line at a more negative value of potential than the $xA + mH^+ + ne = yD + zH_2O$ line, then a simple conclusion follows. The $M^{n+} + ne = M$ reaction will tend to run spontaneously in the de-electronation direction and produce M^{n+} from M, i.e., dissolution, and the other reaction will tend to proceed spontaneously as an electronation reaction (and thus absorb electrons supplied during

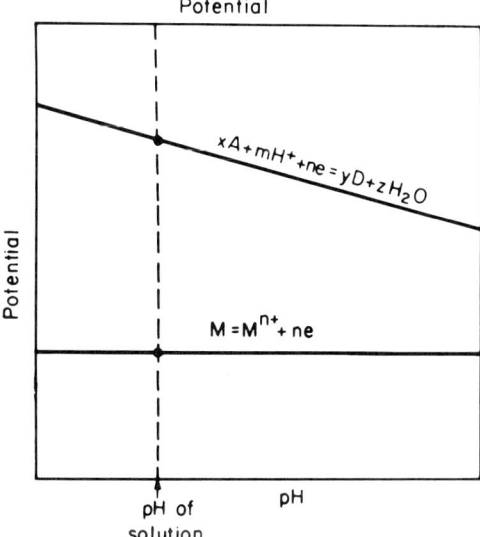

Fig. 11.10. The potential–pH diagram for the given system of the metal-dissolution and electronation reaction reveals the tendency of the metal to corrode.

the de-electronation of the metal) if a path is provided for the electron flow from the sink for the de-electronation reaction to the source for the electronation reaction. The metal M will be said to tend to corrode spontaneously.

On this basis, it is clear from Fig. 11.11 that, if the solution in the mild-steel reaction vessel contains Fe^{++} ions at a concentration of unit activity and if the pH $= 2$, the material of the vessel must tend to dissolve.[†] It must, therefore, be rejected as an unsuitable material for holding a pH $= 2$ solution.

Suppose, however, that the solution in the reaction vessel does not contain any ferrous ions. Then what is the concentration or activity that must be inserted in the Nernst expression for the equilibrium potential of the $Fe^{++} + 2e = Fe$ reaction? What value should be used for $c_{Fe^{++}}$ in

$$E_{e,\text{Fe}^{++}/\text{Fe}} = E^0_{\text{Fe}^{++}/\text{Fe}} + \frac{RT}{2F} \ln c_{\text{Fe}^{++}} \tag{11.7}$$

[†] Note that, at higher pH values, the potential of iron becomes pH dependent. This is because, with increasing OH^- ion concentration, the $Fe(OH)_2$ species is formed and controls the potential. Its concentration is of course pH dependent, and the excess ferrous ions originally present are precipitated.

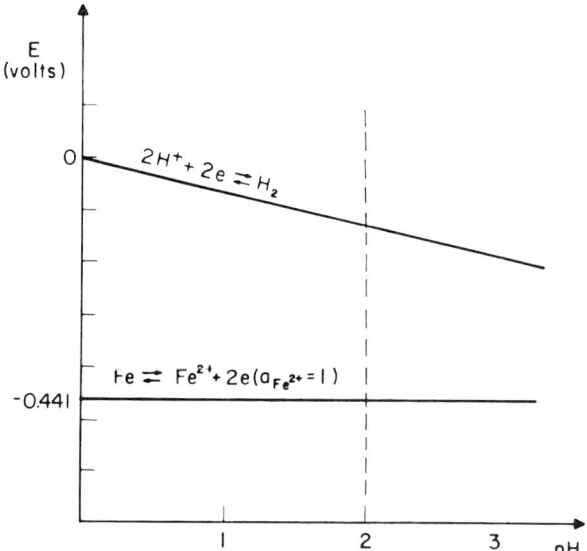

Fig. 11.11. The potential–pH diagram for iron immersed in a ferrous-ion solution of unit activity.

The obvious value to insert for $c_{Fe^{2+}}$ is zero, in which case Nernst's equation for the electrode potential of the $Fe^{++} + 2e = Fe$ equilibrium indicates a value which would be highly negative. Since this potential is negative with respect to the equilibrium potential of any possible electronation reaction, the iron will start dissolving and building Fe^{++} concentration in the solution layer which is in contact with the electron-sink area of the iron. If this layer is stagnant and the Fe^{++} ions are not removed by a chemical reaction, e.g., precipitation, the Fe^{++} concentration will climb up from zero. Clearly, the Fe^{++} ion concentration adjacent to the metal is determined by the amount of metal which has dissolved, and how fast this diffuses away.

How much iron must dissolve to attain a given Fe^{++} concentration? A ferrous ion concentration of 10^{-6} mole liter^{-1} corresponds to the dissolution of about 0.06 mg of iron per liter of solution. Hence, a concentration of less than 10^{-6} mole liter^{-1}, e.g., 10^{-8} mole liter^{-1}, corresponds to the dissolution of around a microgram of iron per liter of solution in contact with the metal. On the other hand, a ferrous ion concentration of more than 10^{-6} mole liter^{-1}, e.g., 10^{-4} mole liter^{-1}, requires the dissolution of a few milligrams of iron per liter of solution in contact with it, i.e., the dissolution of a significant quantity of iron.

In view of these considerations, one can adopt a practical and reasonable, though arbitrary, criterion: A ferrous-ion concentration of 10^{-6} mole liter^{-1} and higher implies the occurrence of "considerable" dissolution, i.e., of corrosion. With these considerations as background, it is conventional, in using a potential–pH diagram for deciding whether a metal can possibly corrode or not, to calculate the equilibrium potential for the $M^{n+} + ne = M$ reaction for a metal-ion concentration of 10^{-6} mole liter^{-1}.

Coming back to the question of whether the mild-steel vessel will corrode in a solution not initially containing Fe^{++} ions, the answer can now be found by examining the position of the $Fe^{++} + 2e = Fe$ line drawn for a concentration of 10^{-6} mole liter^{-1} of Fe^{++}. This line will be $(RT/2F) \times \ln 10^{-6}$ volts below that shown in Fig. 11.11. When this line then gets into a region more negative than the hydrogen line, iron will corrode.

The corrosion of an iron vessel has been treated here only to make the discussion less abstract. A similar approach can be used to inspect the potential–pH diagrams of other metals and decide whether they tend to corrode spontaneously or not in solutions of a given pH.

11.2.7. Potential–pH (or Pourbaix) Diagrams: Uses and Abuses

It must not be imagined that the *ultimate* product of the metal-dissolution reaction is always an ionic species, e.g., $M \rightarrow M^{n+} + ne$. Often, it is a solid oxide or hydroxide.

From the free-energy considerations, one can calculate the reversible potentials for a metal that is in equilibrium with its simple hydrated ions or with its soluble product of hydrolysis or with its insoluble oxide. Such a calculation provides the set of data shown in Table 11.3. Under the given conditions, the preferred state is that which gives the most negative values of potential, and any other existing state would spontaneously turn into that preferred one.

Potential–pH diagrams are useful in this respect, too. They indicate the potential and pH conditions under which a solid product is thermodynamically stable (Fig. 11.12). The regions of the potential–pH diagram in which oxide or hydroxide formation receives thermodynamic approval arise as follows:

Consider the case of iron, and assume, for the sake of argument, that the *immediate* product of iron dissolution is ferrous ions. Now, the solution in contact with iron can dissolve ferrous ions only up to the limit which is given by applying the law of mass action to the reaction

$$Fe(OH)_2 + 2H^+ = Fe^{++} + 2H_2O \tag{11.8}$$

According to the law of mass action, for a constant concentration of

TABLE 11.3

Potentials of Some Metals in Equilibrium with Unimolar Solutions of Their Aquo- and Hydrolyzed Ions Compared with the Potentials of the Formation of Corresponding Oxides, in Volts, N-Hydrogen Scale[†]

Aquo-ions		Soluble hydrolyzed ions			Oxides–hydroxides		
			pH 0	pH 14		pH 0	pH 14
Al^{3+}	-1.33	$Al(OH)_6^{3-}$	-1.20	-2.84	$Al(OH)_3$	-1.49	-2.31
		$Al(OH)_4^-$	-1.25	-2.34	Al_2O_3	-1.69	-2.51
Cr^{3+}	-0.51	$Cr_2O_7^{2-}$	$+0.65$	-1.27	Cr_2O_3	$+0.27$	-0.55
Cr^{2+}	-0.56	CrO_4^{2-}	$+0.40$	-0.69	$CrO(OH)$	?	?
					$Cr(OH)_3$	-0.47	-1.29
		$Cr(OH)_4^-$	-0.14	-1.23			
		$Cr(OH)_2^+$	-0.30	-0.84			
Cu^+	$+0.54$	$Cu(OH)_4^{2-}$	$+1.53$	-0.11	$Cu(OH)_2$	$+0.61$	-0.21
Cu^{2+}	$+0.34$	$Cu(OH)_3^-$	$+1.13$	-0.10	CuO	$+0.57$	-0.25
					Cu_2O	$+0.47$	-0.35
Fe^{3+}	-0.04	FeO_4^{2-}	$+0.63$	-0.46	Fe_2O_3	-0.05	-0.87
Fe^{2+}	-0.44	$Fe(OH)_3^-$	$+0.49$	-0.74	$Fe(OH)_2$		
		$Fe(OH)^{2+}$	$+0.01$	-1.22	$Fe(OH)_3$	-0.06	-0.88
					Fe_3O_4	-0.08	-0.90
Ni^{2+}	-0.25	$Ni(OH)_3^-$	$+0.65$	-0.58	NiO_2	$+0.67$	-0.15
					Ni_2O_3	$+0.42$	-0.40
					Ni_3O_4	$+0.31$	-0.51
					NiO	$+0.12$	-0.70
					$Ni(OH)_2$	$+0.11$	-0.71
Zn^{2+}	-0.76	$Zn(OH)_4^{2-}$	$+0.44$	-1.20	$Zn(OH)_2$	?	?
					ZnO	-0.65	-1.47

[†] J. M. West, *Electrodeposition and Corrosion Processes*, D. Van Nostrand Co., Inc., New York, 1965.

$Fe(OH)_2$ in equilibrium with a solid phase,

$$\frac{c_{Fe^{++}}}{c^2_{H^+}} = K = 10^{+13.29} \tag{11.9}$$

or

$$\log c_{Fe^{++}} = 2 \log c_{H^+} + 13.29$$
$$= 13.29 - 2\,\text{pH} \tag{11.10}$$

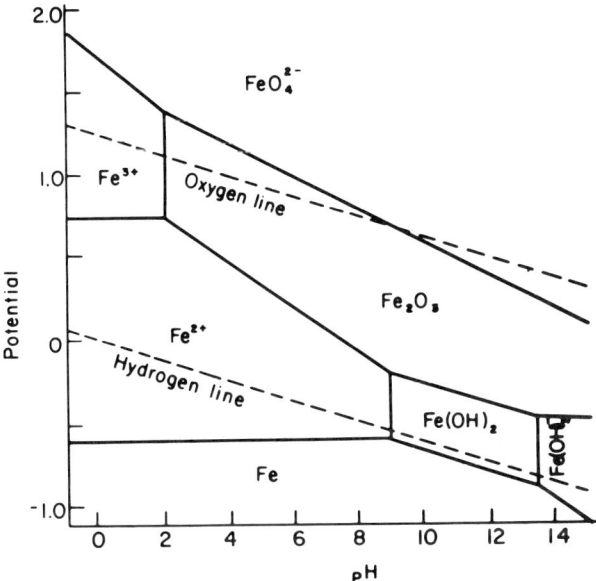

Fig. 11.12. Example of potential–pH diagram for a system with a solid phase as a dissolution product. Iron has a tendency to corrode at all pH's, but at pH $\gtrsim 9$ it forms $Fe(OH)_2$.

If, as before, the minimum concentration of Fe^{++} ions corresponding to a piece of corroding iron is arbitrarily taken as 10^{-6} mole liter^{-1}, then

$$\text{pH} = \frac{13.29 + 6}{2} = 9.6$$

which means that, above a pH of 9.6, $Fe(OH)_2$ is stable. Since the $Fe(OH)_2/Fe^{++}$ equilibrium depends only on pH and not on potential, i.e., it is a pure proton-transfer reaction and does *not* involve electron-transfer, the $Fe(OH)_2/Fe^{++}$ equilibrium is shown on the potential–pH diagram as a vertical line parallel to the potential axis (see Fig. 11.12).

Above the indicated pH value, the concentration of Fe^{2+} ions is governed by Eq. (11.10), and, hence, the potential changes with increasing pH with a slope of RT/F.

The question of how the corrosion of a metal is affected by the formation of a solid product of the dissolution reaction is a rather complex matter, which will be considered in due course. It is important, however, to stress one important point which is of relevance not only to oxide formation. It often turns out that, whereas the potential–pH diagram indicates that a particular hydroxide, e.g., can be formed only above a certain pH value, it is experimentally observed that the hydroxide is formed when the electrode is immersed in a solution with much lower pH value. This apparent contradiction arises (1) because *the values of pH in a potential–pH diagram always refer to the solution in the immediate vicinity of the electrode* and (2) because the local pH near an electrode can increase well above the bulk pH of the solution if the electronation reaction taking place at the electrode consumes hydrogen ions (e.g., $2H^+ + 2e \rightarrow H_2$) or generates hydroxyl ions (e.g., $O_2 + 2H_2O + 4e \rightarrow 4OH^-$).

In conclusion, therefore, potential–pH diagrams can be used to yield yes or no answers on whether a particular corrosion process is thermodynamically possible or not. The diagrams provide a compact pictorial summary of the electron-transfer, proton-transfer, and electron-and-proton-transfer reactions which are favored on thermodynamic grounds when a metal is immersed in a particular solution. Yet, they should be used with caution. On the one hand, when a potential–pH diagram indicates that a particular metal is immune to corrosion, it is immune *provided* the pH in the close vicinity of the surface is what it is assumed to be. On the other hand, when the diagram indicates that a particular corrosion process can occur spontaneously, it does not mean that significant corrosion must in practice be observed. For this to be so, the *rate* of corrosion must be appreciable, and one must at this stage refrain from making any predictions

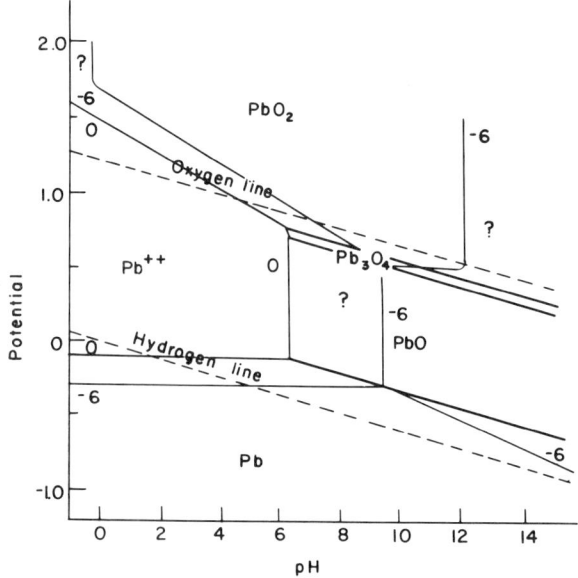

Fig. 11.13. The potential–pH diagram for lead.

regarding the rate of corrosion since this cannot be done from a knowledge of the thermodynamics of the system alone. If one does not observe this caution, one can be led into serious errors. A classical example of the point under discussion is the case of lead in contact with aerated water. The potential–pH diagram (Fig. 11.13) shows the equilibrium potential for the $Pb^{++} + 2e = Pb$ reaction for a Pb^{++} concentration of $\sim 10^{-6}$ g liter^{-1} to be negative with respect to the equilibrium potential for the hydrogen-reduction reaction at a pH less than about 5. This implies a tendency on the part of lead to corrode in an aerated aqueous environment. In actual fact, however, the rate of corrosion is so negligible that lead is often used in pipes for carrying water.

Thermodynamics, therefore, defines a necessary, vital precondition for corrosion; it determines the direction in which an overall corrosion reaction will *tend*. But the determination of the rate and control of a corroding system can emerge only by a study of the *electrodics of corrosion*.

11.2.8. The Corrosion Current and the Corrosion Potential

Consider a system consisting of a metal corroding in an electrolyte. The corrosion process involves a metal-dissolution de-electronation reaction at electron-sink areas on the metal and an electronation reaction at electron-source areas. (This picture is applicable to a metal's corroding by a Wagner–Traud mechanism provided one imagines the sink and source areas shrunk to atomic-sized dimensions and considers the situation at one instant of time.)

The corroding metal, it has been pointed out, is equivalent to a short-circuited energy-producing cell with the following specifications: The electron-sink (anodic) and electron-source (cathodic) areas of the equivalent energy-producing cell are chosen equal to the corresponding areas on the corroding metal. Thus, the total metal-dissolution *current* I_M and electronation *current* I_{so} (not current densities) on the corroding metal are equal in magnitude but opposite in sign, just as they are in an energy-producing cell,

$$I_M = -I_{so} \tag{11.11}$$

The rate of corrosion of the metal is obviously given directly by the rate of metal dissolution; hence, the corrosion current I_{corr} is equal to the metal-dissolution current I_M

$$I_{corr} = I_M = -I_{so} \tag{11.12}$$

Apart from this important feature of a corroding system, there is another characteristic which arises from the *short-circuit* condition of the corrosion cell and of the equivalent cell. (It will be recalled that the electron sources and sinks in the corroding metal are *internally* short-circuited; the two electrodes in the equivalent cell are *externally* short-circuited.) The total potential difference V across the equivalent cell is zero. But this cell potential is composed of[†] the absolute potential differences across the interfaces at the two electrodes and the potential drop IR in the electrolyte

$$V = 0 = \Delta\phi_{so} - \Delta\phi_M + IR \qquad (11.13)$$

where $\Delta\phi_{so}$ is the metal–solution potential difference at the electron-source electrode (cathode); and $\Delta\phi_M$, the corresponding quantity at the electron-sink electrode (anode).

Now assume that $IR \approx 0$. The assumption requires that the interelectrode distance be negligibly small, that the electrolyte be sufficiently conducting, and that there be no high-resistance oxide films on the electrodes. Under these circumstances,

$$\Delta\phi_{so} - \Delta\phi_M \approx 0 \quad \text{or} \quad \Delta\phi_{so} \approx \Delta\phi_M \qquad (11.14)$$

Thus, when $IR \approx 0$, the potential difference across the metal–electrolyte interface at the electron-source electrode of the short-circuited equivalent cell is virtually equal to that at the electron-sink electrode.

What is the validity of the assumption $IR \approx 0$ in the case of a corroding metal?

If the metal is homogeneous and is corroding by a Wagner–Traud mechanism, the sink and source areas are separated at any one instant by a distance of the order of a few angstroms. Further, the sink and source areas are shifting around with time and therefore smearing out the negligible potential differences in the solution adjacent to these areas. Thus, $IR = 0$ is almost exactly true.

If the metal has heterogeneities and is corroding by local-cell action, the validity of $IR \approx 0$ depends upon the separation of the sink and source areas and upon the conductivity of the electrolyte. Now, there are special circumstances in which the distance apart of the sink and source areas is considerable (of the order of centimeters). For these situations, $IR \not\approx 0$ and $\Delta\phi_{so} \not\approx \Delta\phi_M$, which implies a difference in the metal–electrolyte potential difference at electron-source and -sink areas. In general, however, the sink-to-source distance is of the order of microns or less, in which case

---

[†] Since both electrodes consist of the same metal, no potential difference arises owing to metal–metal contact.

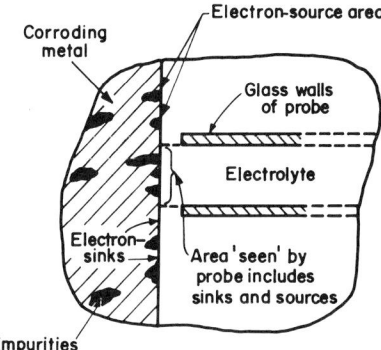

Fig. 11.14. In heterogeneous systems, the distances between source (e.g., metallic inclusions) and sink points are so small that a probe cannot usually detect any potential difference in the electrolyte between them.

the conducting path in the solution and therefore IR becomes negligible. Thus, the $\Delta\phi_{so}$ is virtually equal to $\Delta\phi_M$, and any negligible difference that exists occurs over distances which are too small to be resolved by a probe used to measure the potential difference between the metal and the solution (Fig. 11.14).

This uniform potential difference across the interface between a corroding metal and its electrolytic environment may be termed the *corrosion potential* $\Delta\phi_{corr}$; it is considered to be given by

$$\Delta\phi_{corr} = \Delta\phi_{so} = \Delta\phi_M \qquad (11.15)$$

It follows that the corrosion potential on a heterogeneous metal corroding by local-cell action is *virtually equal* to the mixed potential at an electrode on which electronation and de-electronation reactions are occurring on spatially separated sinks and sources and is *identical* to a mixed potential when the metal is corroding homogeneously by a Wagner–Traud mechanism.

The concept of the corrosion current I_{corr} and the corrosion potential $\Delta\phi_{corr}$ will now be treated quantitatively.

11.2.9. The Basic Electrodics of Corrosion in the Absence of Oxide Films

Two fundamental ideas have been developed. Firstly, the rate of corrosion, a quantity of great practical significance, is given by the corrosion current I_{corr}, which is equal to the metal-dissolution, de-electronation current

I_M and to the negative of the electronation (cathodic) current I_{so} at the electron-source areas, i.e.,

$$I_{corr} = I_M = -I_{so} \tag{11.12}$$

Since the metal-dissolution current is equal to the product of the corresponding current density i_M times the sink area A_M, one can write

$$I_{corr} = I_M = A_M i_M \tag{11.16}$$

and, similarly,

$$I_{corr} = -I_{so} = -A_{so} i_{so} \tag{11.17}$$

Secondly, there is a uniform potential difference, namely, the corrosion potential $\Delta\phi_{corr}$, all over the surface of the corroding metal. It is this corrosion potential which is associated with both the metal-dissolution and electronation currents, i.e.,

$$\Delta\phi_{corr} = \Delta\phi_M = \Delta\phi_{so} \tag{11.15}$$

To obtain quantitative expressions for the corrosion current and the corrosion potential, one has to substitute the proper expression for the metal-dissolution- and electronation-current densities. If no oxide films form on the surface of the corroding metal and neither of the current densities is controlled by mass transport, i.e., there is no concentration overpotential, one can insert the Butler–Volmer expression for the de-electronation- and electronation-current densities. Thus,

$$I_{corr} = I_M = A_M i_M \tag{11.16}$$

$$= A_M i_{0,M} \left[\exp\left(\frac{\vec{\alpha}_M F}{RT} \eta_M\right) - \exp\left(-\frac{\overleftarrow{\alpha}_M F}{RT} \eta_M\right) \right] \tag{11.18}$$

Now, the overpotential η_M is equal to the potential difference at the electron-sink areas, i.e., the corrosion potential $\Delta\phi_{corr}$ minus the equilibrium potential $\Delta\phi_{e,M}$ for the metal-dissolution reaction $M^{n+} + ne = M$, i.e.,

$$\eta_M = \Delta\phi_{corr} - \Delta\phi_{e,M} \tag{11.19}$$

Further, since $i_{0,M}$ is the exchange-current density for the $M^{n+} + ne = M$ reaction and A_M is the area over which this reaction occurs, the product of $i_{0,M}$ and A_M must be the exchange *current*, i.e.,

$$I_{0,M} = A_M i_{0,M} \tag{11.20}$$

Finally, as was done in the treatment of electrochemical cells (*cf.* Section 9.6.7), one can use the following notation

$$\overleftarrow{\lambda}_{M} = \frac{RT}{\overleftarrow{\alpha}_{M}F} \quad \text{and} \quad \overrightarrow{\lambda}_{M} = \frac{RT}{\overrightarrow{\alpha}_{M}F} \quad (11.21)$$

where $\overleftarrow{\lambda}_{M}$ and $\overrightarrow{\lambda}_{M}$ are the η_{M} − log i_{M} Tafel slopes for the deelectronation and electronation directions of the $M^{n+} + ne = M$ reaction.

In view of Eqs. (11.19) to (11.21), the expression (11.18) for the corrosion current becomes

$$I_{\text{corr}} = I_{0,M}\left[\exp\frac{\Delta\phi_{\text{corr}} - \Delta\phi_{e,M}}{\overleftarrow{\lambda}_{M}} - \exp\left(-\frac{\Delta\phi_{\text{corr}} - \Delta\phi_{e,M}}{\overrightarrow{\lambda}_{M}}\right)\right] \quad (11.22)$$

Similarly, one can write for the relation between the corrosion current and the electronation current at the electron-source area

$$I_{\text{corr}} = -I_{\text{so}}$$
$$= -A_{\text{so}}i_{0,\text{so}}\left[\exp\frac{\overleftarrow{\alpha}_{\text{so}}F\eta_{\text{so}}}{RT} - \exp\left(-\frac{\overrightarrow{\alpha}_{\text{so}}F\eta_{\text{so}}}{RT}\right)\right]$$
$$= I_{0,\text{so}}\left[\exp\left(-\frac{\Delta\phi_{\text{corr}} - \Delta\phi_{e,\text{so}}}{\overrightarrow{\lambda}_{\text{so}}}\right) - \exp\frac{\Delta\phi_{\text{corr}} - \Delta\phi_{e,\text{so}}}{\overleftarrow{\lambda}_{\text{so}}}\right] \quad (11.23)$$

From Eqs. (11.22) and (11.23), it is clear that the corrosion current depends upon the exchange currents (i.e., available areas and exchange-current densities), Tafel slopes, and equilibrium potentials for both the metal-dissolution and electronation reactions. To obtain an explicit expression for the corrosion current [*cf.* Eq. (11.22)], one has first to solve Eqs. (11.22) and (11.23) for $\Delta\phi_{\text{corr}}$. If, however, simplifying assumptions are not made, the algebra becomes unwieldy and leads to highly cumbersome equations.

One such simplifying assumption is

$$\overleftarrow{\alpha}_{M} = \overrightarrow{\alpha}_{M} = \overleftarrow{\alpha}_{\text{so}} = \overrightarrow{\alpha}_{\text{so}} = \tfrac{1}{2} \quad (11.24)$$

i.e.,

$$\overleftarrow{\lambda}_{M} = \overrightarrow{\lambda}_{M} = \overleftarrow{\lambda}_{\text{so}} = \overrightarrow{\lambda}_{\text{so}} = \frac{2RT}{F} \quad (11.25)$$

If, making use of this assumption, one divides the two expressions (11.22) and (11.23) for I_{corr} by $\exp(-F\Delta\phi_{\text{corr}}/2RT)$ and equates them, the result is

$$I_{0,M}(e^{F\Delta\phi_{\text{corr}}/RT}e^{-F\Delta\phi_{e,M}/2RT} - e^{F\Delta\phi_{e,M}/2RT})$$
$$= I_{0,\text{so}}(e^{F\Delta\phi_{e,\text{so}}/2RT} - e^{F\Delta\phi_{\text{corr}}/RT}e^{-F\Delta\phi_{e,\text{so}}/2RT}) \quad (11.26)$$

An evaluation of $\Delta\phi_{corr}$ from this expression shows that

$$\Delta\phi_{corr} = \frac{RT}{F}\ln\left[\frac{I_{0,so}\exp(F\Delta\phi_{e,so}/2RT) + I_{0,M}\exp(F\Delta\phi_{e,M}/2RT)}{I_{0,so}\exp(-F\Delta\phi_{e,so}/2RT) + I_{0,M}\exp(-F\Delta\phi_{e,M}/2RT)}\right] \quad (11.27)$$

Although obtained under the simplifying assumption of $\vec{\alpha}_M = \vec{\alpha}_M = \vec{\alpha}_{so} = \vec{\alpha}_{so} = \frac{1}{2}$, this equation brings out a simple characteristic of the corrosion potential; it approximates being near the equilibrium potential for the metal-dissolution reaction or near the equilibrium potential of the electronation reaction, depending upon whether the exchange current at the sink areas is much greater than the exchange current at the source areas or *vice versa*. In symbols, if $I_{0,M} \gg I_{0,so}$, then $\Delta\phi_{corr} \approx \Delta\phi_{e,M}$, and, if $I_{0,so} \gg I_{0,M}$, then $\Delta\phi_{corr} \approx \Delta\phi_{e,so}$.

The expression (11.27) for the corrosion potential can be introduced into Eq. (11.22), and, thus, an explicit result for the corrosion current can be obtained. But the resulting equation is quite cumbersome, and, therefore, a simpler equation will be derived by assuming that overpotentials are sufficiently large so that the high-field approximation of the Butler–Volmer equation can be used for the electronation- and de-electronation-current densities. Thus, Eqs. (11.22) and (11.23) become

$$I_{corr} = I_{0,M}\exp\frac{\Delta\phi_{corr} - \Delta\phi_{e,M}}{\vec{\lambda}_M} = I_{0,so}\exp\left(-\frac{\Delta\phi_{corr} - \Delta\phi_{e,so}}{\vec{\lambda}_{so}}\right) \quad (11.28)$$

Hence,

$$\exp\frac{\Delta\phi_{corr}}{\vec{\lambda}_M} = \left(\frac{I_{0,so}}{I_{0,M}}\right)^{\vec{\lambda}_{so}/(\vec{\lambda}_M + \vec{\lambda}_{so})}\exp\left[\frac{\vec{\lambda}_M\Delta\phi_{e,so} + \vec{\lambda}_{so}\Delta\phi_{e,M}}{\vec{\lambda}_M(\vec{\lambda}_M + \vec{\lambda}_{so})}\right] \quad (11.29)$$

and, therefore,

$$I_{corr} = I_{0,M}^{\vec{\lambda}_M/(\vec{\lambda}_M + \vec{\lambda}_{so})} I_{0,so}^{\vec{\lambda}_{so}/(\vec{\lambda}_M + \vec{\lambda}_{so})}\exp\left[\frac{\Delta\phi_{e,so} - \Delta\phi_{e,M}}{\vec{\lambda}_M + \vec{\lambda}_{so}}\right] \quad (11.30)$$

The dependence of the corrosion current (the rate at which a metal destroys itself) on the exchange currents, Tafel slopes, and equilibrium potentials of the metal-dissolution and electronation reactions is clearly brought out in this expression. In general, the more positive the equilibrium potential of the electronation reaction is with respect to the equilibrium potential for the $M^{n+} + ne = M$ reaction and the larger the exchange currents (areas times exchange-current densities) are, the greater is the rate of corrosion. The Tafel slopes also enter the picture; high slopes diminish the enhancing effect which the exponential term has on the rate of the corrosive attack on the metal.

A simpler form of Eq. (11.30) arises by setting

$$\overleftarrow{\lambda}_M = \overleftarrow{\lambda}_{so} = \frac{2RT}{F} \quad (11.25)$$

in which case one gets

$$I_{corr} = (I_{0,M}I_{0,so})^{\frac{1}{2}} \exp \frac{F(\Delta\phi_{e,so} - \Delta\phi_{e,M})}{4RT} \quad (11.31)$$

and, if the potentials are written as relative potentials on the standard hydrogen scale, Eq. (11.31) becomes

$$I_{corr} = (I_{0,M}I_{0,so})^{\frac{1}{2}} \exp \frac{F(E_{e,so} - E_{e,M})}{4RT} \quad (11.32)$$

This approximate and special-case equation brings out the role of the exchange currents and the equilibrium potentials in determining the corrosion rate.

What has been presented above is a very elementary account of corrosion under superideal conditions. In a few cases, it does give a fairly good agreement with the observed rates of corrosion. Yet, in real systems, corrosion is nearly always too complex a phenomenon for the above simple treatment to be directly applicable. The simple version would be valid *if* there were no oxide films, *if* there were a negligible *IR* drop in the solution, *if* the corrosion potential $\Delta\phi_{corr}$ settled down to a value such that the high-field approximations [*cf.* Eq. (11.28)] could be applied, and *if* the transfer coefficients of the metal-dissolution and electronation reactions were $\frac{1}{2}$ [*cf.* Eq. (11.25)]. But, the point of an introductory treatment is not to treat the details, and the complex realities, but to present the idealized essence about an electrochemical mechanism which has substantial effects in the everyday world.

11.2.10. An Understanding of Corrosion in Terms of Evans Diagrams

Most of the factors affecting the rate of corrosion can be understood from a graphical superposition of the current–potential curves for the metal-dissolution and electronation reactions. The principle of the graphical superposition method is straightforward.

Consider the metal-dissolution reaction $M^{n+} + ne = M$. One can construct a curve (Fig. 11.15) for the variation of the potential of an M electrode with the de-electronation current crossing the electrode–electrolyte interface. This curve can be obtained either experimentally or from a knowledge of

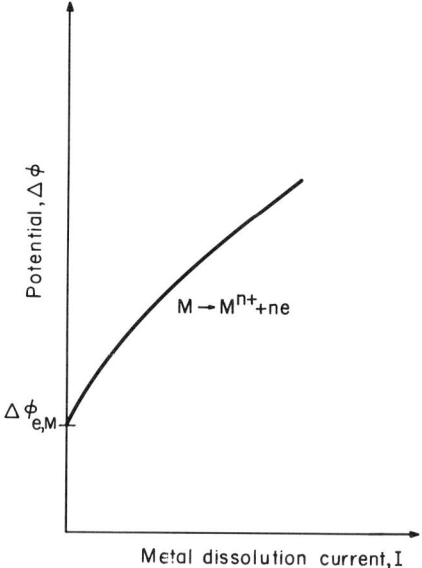

Fig. 11.15. The potential–current relation for the metal-dissolution reaction.

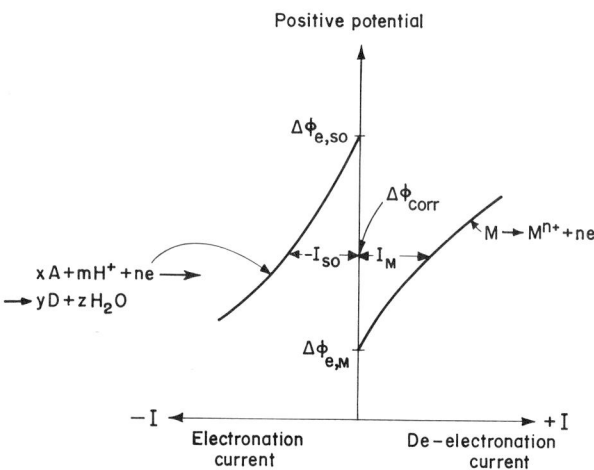

Fig. 11.16. The potential–current relation for the two reactions occurring at a corroding interface. The corrosion current and corrosion potential are defined by the point on the diagram at which the two currents I_M and I_{so} are equal.

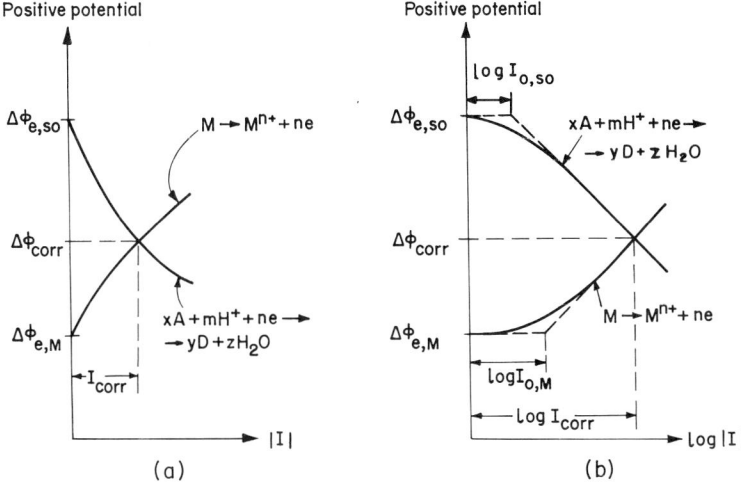

Fig. 11.17. The Evans diagrams are plots of the potentials of the two reactions (a) *versus* the magnitude of the two currents or (b) *versus* their logarithms. The intersections of the curves define the corrosion current and corrosion potential.

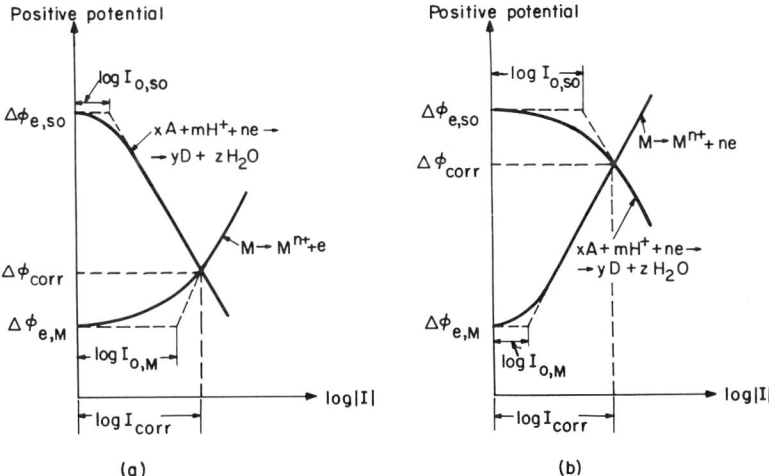

Fig. 11.18. Changing the exchange-current densities produces a shift of the corrosion potential from a medium value [Fig. (11.7)] toward the equilibrium potential of (a) the metal-dissolution or (b) the electronation reaction.

the parameters that determine the overpotential associated with the de-electronation-current density. For concentration overpotential, this parameter is the limiting current density, and, for activation overpotential, the parameters are the exchange-current density and the transfer coefficients. On the same diagram, one can then superpose a curve (Fig. 11.16) for the variation of the potential of the M electrode with the current associated with the electronation of electron acceptors present in the electrolyte. The current at which the metal dissolution and electronation are equal is, in fact, the corrosion current (*cf.* Fig. 11.16). The potential corresponding

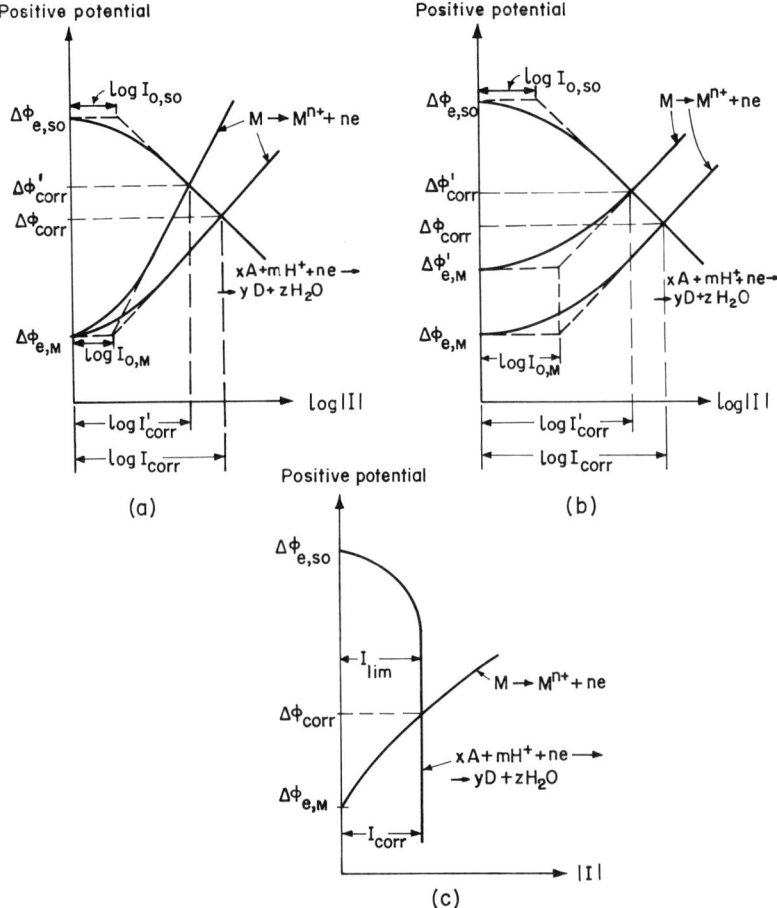

Fig. 11.19. The effect of (a) the Tafel slope, (b) the equilibrium potential of the metal dissolution, or (c) the transport difficulties of the electron acceptor are reflected in the Evans diagram.

to the corrosion current is the corrosion potential. If one uses only the *magnitude* of the de-electronation and electronation currents in the construction of the $\Delta\phi$ versus I curves, one has what is known as an *Evans* type of diagram (Fig. 11.17).

The particular form of the Evans diagram obtained depends upon the current–potential curves for the metal-dissolution and electronation reactions. Some of the common diagrams are shown in Figs. 11.18 to 11.20. They cover situations in which the exchange current for the metal-dissolution reaction is much greater than that for the electronation, i.e., $I_{0,\mathrm{M}} \gg I_{0,\mathrm{so}}$ [Fig. 11.18(a)] or in which $I_{0,\mathrm{so}} \gg I_{0,\mathrm{M}}$ [Fig. 11.18(b)]. Evans diagrams can also be used to bring out the influence of Tafel slopes [Fig. 11.19(a)], the influence of equilibrium potentials [Fig. 11.19(b)], the effect of mass-transport control on the electronation current [Fig. 11.19(c)]. The effect of an *IR* drop in the electrolyte between the electron-sink (anodic) and electron-source (cathodic) areas can also be represented in an Evans diagram (Fig. 11.20), which then shows the inequality of the metal–solution potential difference at the two areas, i.e., the anodic (or de-electronation) and the cathodic (or electronation) areas.

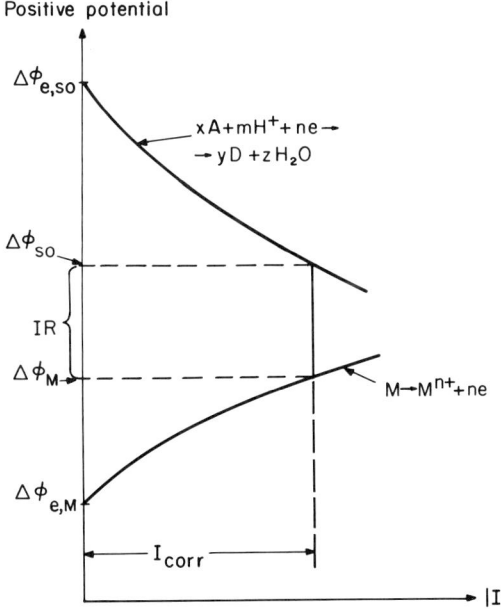

Fig. 11.20. If there is some potential drop in the solution (*IR*), the two potential–current curves do not intersect at the corrosion current.

11.2.11. Which Step in the Corrosion Process Controls the Corrosion Current?

If one considers the Evans diagram [Fig. 11.18(a)] for the situation in which $I_{0,M} \gg I_{0,so}$, it is obvious that the magnitude of the corrosion current I_{corr}, i.e., the position of the intersection of the two curves, depends essentially on the curve for the electronation (cathodic) reaction. That is, the corrosion current is controlled by the electronation (cathodic) reaction. One says that the corrosion process is under *electronation (cathodic) control*. By the same token, it follows that the diagram [Fig. 11.8 (b)] for the $I_{0,so} \gg I_{0,M}$ situation indicates *de-electronation (anodic) control*. It is also possible to have situations in which there is a *mixed control* (Fig. 11.17) over the corroding system (i.e., the corrosion current is controlled by both the de-electronation and electronation reactions) or what is called an *ohmic control* (Fig. 11.20), i.e., the corrosion current is determined by the *IR* drop in the electrolyte.

There is another way of looking at these various types of control over the corroding system. It has been shown in Chapter 9 that the rate of a multistep, consecutive reaction may be controlled by *one* of the steps, the rds (the rate-determining step). Why not therefore look upon the corrosion process as a consecutive reaction consisting of the following steps: (1) de-electronation, (2) electron flow in metal, (3) electronation, and (4) ion flow in the electrolyte. When a consecutive reaction attains a steady state, the net currents associated with the various steps are equal to each other and also equal to the overall current. In the case of the corroding metal, the overall current is the corrosion current, and, therefore,

$$I_{corr} = I_M = I_{so} = I_e = I_i \qquad (11.33)$$

where I_e and I_i are the currents due to electron flow in the metal and ion flow in the electrolyte. Despite the equality of the currents in the various steps, the overall current in the consecutive reaction may be controlled by one of the four steps. Since the electronic conductivity of the metal will be high enough to prevent electron flow in the metal from controlling the corrosion current, one is left with the possibility of three rate-controlling steps; it is in this way that the corroding system becomes subject to de-electronation control or electronation control or control by ion flow in the electrolyte (ohmic control). Further, just as, in the case of hydrogen evolution, the possibility was considered of two steps' exercising dual control over the reaction rate, there are situations in which the de-electronation and electronation steps exercise dual or mixed control over the corrosion current.

The importance of identifying the type of control over the corrosion current arises from the same reason that one should know the rds in a multistep electrodic reaction. One can ignore the non-rds's and concentrate on *tampering with the parameters of the rds to achieve desired results* (in the case of corrosion reactions, the diminution of the rate).

11.2.12. Metals, pH, and Air

The basic (very simplified) quantitative theory for the corrosion of a metal in the absence of oxide films has just been presented. Now a few qualitative inferences can be drawn regarding the effect of various factors in determining corrosion.

First and foremost, corrosion obviously depends a great deal upon the particular metal involved. Sodium, e.g., corrodes in aqueous solutions so vigorously that considerable heat is evolved, sparks fly, and the evolved hydrogen explodes. Gold, on the other hand, does not corrode at all (its stability is the main point of its use as a monetary standard).

The electrochemical origin for this difference in behavior is clear. The standard equilibrium potential for the $Na^+ + e = Na$ reaction is extremely negative (-2.71 V) with respect to the equilibrium potential of either the hydrogen-evolution or oxygen-reduction reaction. Hence, the $E_{e,\text{so}} - E_{e,\text{Na}}$ term in [*cf.* Eq. (11.32)]

$$I_{\text{corr}} = (I_{0,\text{Na}} I_{0,\text{so}})^{\frac{1}{2}} \exp \frac{F(E_{e,\text{so}} - E_{e,\text{Na}})}{4RT}$$

has a large positive value, and, therefore, the corrosion current is very high. For gold, the standard equilibrium potential for the $Au^{+++} + 3e = Au$ reaction is $+1.50$ V, which is more positive than the equilibrium potentials of either the hydrogen-evolution or oxygen-reduction reactions. Hence, if there is unit activity of ions in solution, gold *cannot* corrode.

A table of standard electrode potentials (*cf.* Table 9.12) thus yields what is called the *electrochemical series* when the metals are arranged according to decreasing standard electrode potentials. All metals with E^0 values negative with respect to hydrogen, which has an E^0 of zero, are known as "base metals." These metals can be expected to corrode spontaneously (the corrosion rate is a question of kinetics and may prove to be negligible, *cf.* lead). Metals with positive E^0 values—the "noble" metals—are incorrodible in deoxygenated solutions of pH = 0.

The influence of air and pH upon the rate of corrosion of a metal is linked up with the fact that hydrogen evolution and oxygen reduction are the most common electronation reactions in corrosion. When more than one reaction can be the electronation reaction, the reaction which yields the largest corrosion current becomes the predominant electronation reaction. The question of whether oxygen reduction or hydrogen evolution becomes the predominant electronation reaction depends upon several factors.

1. Since the equilibrium potential for oxygen reduction $O_2 + 2H_2O + 4e = 4OH^-$ is given by

$$E_{e,O_2/OH^-} = 1.23 - 0.059 \text{ pH} \qquad (11.34)$$

whereas that for hydrogen evolution $2H^+ + 2e = H_2$ is given by

$$E_{e,H^+/H_2} = -0.059 \text{ pH} \qquad (11.35)$$

the former is always 1.23 V more positive than the latter (cf. Fig. 9.46) in a solution of the same pH. Since one of the factors determining the corrosion rate is the difference between the equilibrium potentials of the electronation reaction at the electron-source area and of the metal-dissolution reaction, one would expect that the oxygen-reduction reaction would always be preferred. The point is: Thermodynamics is on the side of oxygen reduction.

2. Kinetics, however, generally favors the hydrogen-evolution reaction. The exchange-current densities for hydrogen evolution on the corrodible metals are generally many orders of magnitude higher than the exchange-current densities for oxygen reduction. In the case of iron, e.g., the i_0 for hydrogen evolution is about 10^{-6} amp cm^{-2}, whereas that for oxygen reduction is about 10^{-14} amp cm^{-2}. Thus, a quantitative inspection of Eq. (11.32) (insert value in I_0 for the two electronation reactions) indicates that, in highly acidic solutions, hydrogen evolution is the preferred reaction (Fig. 11.21).

3. But, as the pH increases, the situation may be reversed. The main reason for this reversal is that the exchange-current density for the hydrogen-evolution reaction depends on the hydrogen-ion concentration so long as the solution is sufficiently concentrated (see Sections 8.3.1 and 10.3.4a). As the pH increases, the hydrogen-ion concentration decreases and so does the exchange-current density for hydrogen evolution. For example, if it is assumed that the discharge step is the rds in hydrogen evolution on a particular metal, the exchange-current density on a particular metal is given by (cf. Section 8.2.6)

$$i_{0,H} = 2Fk_1 e^{-\beta F \Delta\phi_e/RT} c_{H_3O^+} \qquad (11.36)$$

Taking into account the Nernst equation, $\Delta\phi_e = (RT/F) \ln c_{H_3O^+}$, one finds

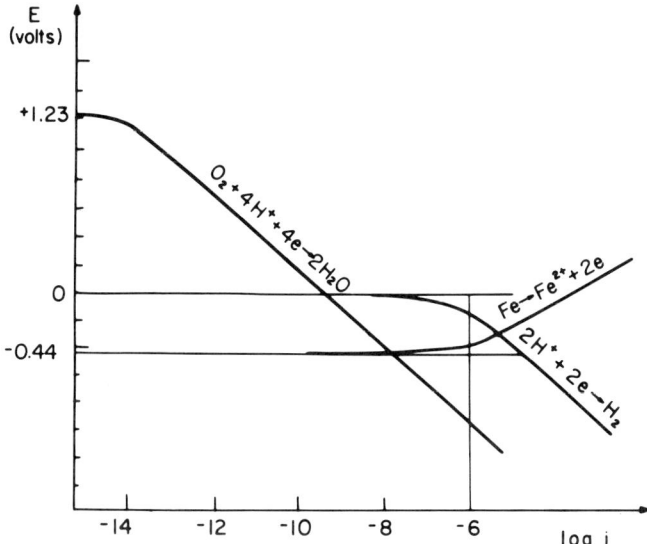

Fig. 11.21. The hydrogen evolution reaction is the preferred electronation reaction even though its equilibrium potential is more negative than the corresponding value for the oxygen reduction reaction, because the exchange current for hydrogen is larger at a pH=0.

that $i_{0,H}$ in this instance is proportional to $c_{H_3O^+}^{1-\beta}$, and, if $\beta = \frac{1}{2}$, increase of pH by 6 (decrease of H^+ by 10^6) decreases $i_{0,H}$ by a factor of 10^3.

Another way in which a decrease of pH decreases the corrosion rate is by the reduction of the difference between the equilibrium potentials for hydrogen evolution and metal dissolution. For example, in the corrosion of iron,

$$E_{e,H} - E_{e,Fe} = -0.059 \text{ pH} - (-0.44 + 0.029 \log c_{Fe^{++}}) \quad (11.37)$$

For $c_{Fe^{++}} = 1$, this reduces to

$$E_{e,H} - E_{e,Fe} = (-0.059 \text{ pH}) - (-0.44) \quad (11.38)$$

Thus, if pH = 0, then $E_{e,H} - E_{e,Fe} = +0.44$ V, and, if pH = 7, then $E_{e,H} - E_{e,Fe} = -0.42 + 0.44 = +0.02$ V. Since it is this difference of equilibrium potentials that provides the thermodynamic driving force for corrosion, it follows that, as the pH increases, the driving force for corrosion with hydrogen evolution decreases (but *cf.* kinetic effects which occur, e.g., the influence of pH changes upon $i_{0,H}$ or $i_{0,Fe}$).

In conclusion, therefore, the predominating electronation reaction depends upon the pH; in highly acid reactions, hydrogen evolution generally has a substantial role, but, in neutral or alkaline solution, oxygen reduction tends to become the predominant reaction.

4. The oxygen molecules required for the oxygen-electronation reaction originate from dissolved air. Now, the solubility of oxygen in water is of the order of 10^{-4} mole liter^{-1}. This is quite a low concentration and implies that the limiting current density for oxygen reduction is of the order of 10^{-5} amp cm^{-2} in unagitated aqueous solutions. Hence, when the electronation of oxygen molecules is the reaction associated with metal dissolution in a corrosion process, the transport of oxygen to the metal causes the maximum corrosion rate to be the value given by the limiting current density for oxygen reduction (Fig. 11.22). Consequently, in this situation of oxygen-reduction control, *agitation of the solution* in which corrosion is occurring leads to an increase of the limiting current for oxygen reduction and therefore to an *increase of the corrosion rate*.

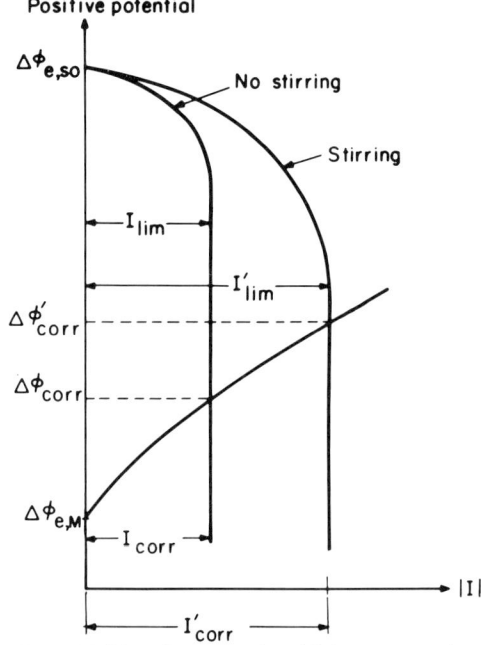

Fig. 11.22. Corrosion in which oxygen reduction is the de-electronation reaction, the low solubility of oxygen makes the oxygen limiting current and the degree of stirring the rate-controlling factors.

TABLE 11.4
Effect of Dissolved Oxygen on the Corrosion of Some Metals by Acids at Atmospheric Temperatures[†]

Material	Acid	Normality, N	Hydrogen-saturated acid (no oxygen)	Oxygen-saturated acid
			Corrosion rate, mm yr^{-1}	
Mild steel	Sulfuric	1.2	31	358
Lead	Hydrochloric	1.1	17	163
Copper	Hydrochloric	1.1	17	1380
Tin	Sulfuric	1.2	9	1100
Nickel	Hydrochloric	1.1	6	440

[†] F. L. La Que and H. R. Copson, eds., *Corrosion Resistance of Metals and Alloys*, Reinhold Publishing Corp., New York, 1963.

Several examples given in Table 11.4 illustrate the extent of the effect that hydrogen- and oxygen-electronation reactions have on the rate of corrosion of different metals in different acid media.

11.2.13. Some Common Examples of Corrosion

It is intended to present in this section the electrodic principles underlying some familiar instances of corrosion.

Automobiles are painted to protect them from corrosion, but often there may be small regions where the steel is exposed to the atmosphere because the paint has been chipped or scratched off. One might at first expect it to be the unprotected metal beneath the broken-paint spot which corrodes. In fact, it turns out that the exposed metal is not the electron-sink area where metal dissolution, or corrosion, occurs. The exposed metal has a better access to oxygen than the metal still covered with paint and is therefore the electron-source area; it is the adjoining metal underneath the paint coating which corrodes. *The situation is much worse than it may appear* (Fig. 11.23). Having a break in paint coatings leads therefore to a spreading of the corroded area rather than to a restriction of corrosion to the exposed spot.

This example of corrosion provides a general reason why paint and coatings of various kinds are only a partial answer to the corrosion-pre-

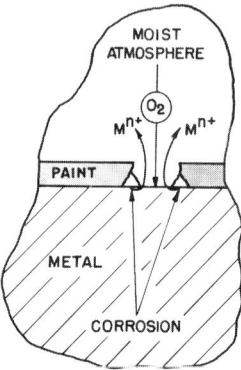

Fig. 11.23. As the paint coating is damaged, the corrosion couple is established with the metal dissolving at the edges *underneath* the coating, while the exposed part is the place for the electronation of oxygen (and thus not that which dissolves).

vention problem. They often tend to develop cracks or pinholes, and these exposed spots provide access to oxygen for the electronation reaction, which result in unseen corrosion of the surrounding areas.

The above example of corrosion brings out an important consequence of oxygen reduction being the electronation reaction. This consequence is known as the *principle of differential aeration*. The principle can be stated as follows: If a part of a metal surface has greater access to oxygen, i.e., is in contact with a higher oxygen concentration, than an oxygen-starved area, then oxygen electronation tends to occur at the oxygen-rich area and metal dissolution tends to occur at the oxygen-poor area (Fig. 11.24). In other words, oxygen-rich areas act as electron-sources (cathodes); and oxygen-starved regions, as electron sinks (anodes). Thus, owing to the differential accessibility of various parts of a metal surface to the diffusion of oxygen, a corrosion cell is produced with spatially separated electron-source and -sink areas. The exclusion of air (oxygen) from any particular part of a metal system leads to a localized attack of the metal precisely in the oxygen-starved regions.

One can provide several practical examples of localized corrosion occurring by differential aeration. Crevice attack is a common phenomenon (Fig. 11.25), or, one may mention the corrosion of partially immersed metals in seawater (Fig. 11.26). The region near the waterline provides easy access to oxygen and thus becomes an electron-source area for the lower

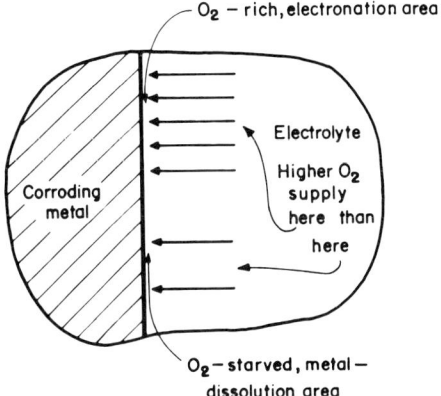

Fig. 11.24. As the availability of oxygen varies on a metal, metal dissolution takes place at oxygen-starved areas.

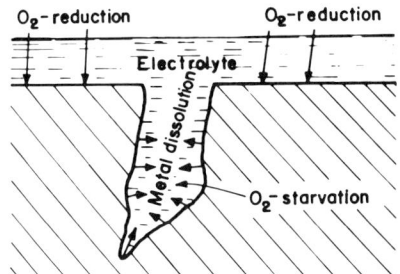

Fig. 11.25. A crevice in the metal surface is a region of limited aeration. Hence, it has a tendency to corrode farther and expand.

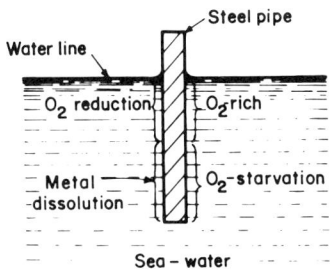

Fig. 11.26. The concentration of oxygen decreases with depth, and this makes the metal dissolution favored in regions farther away from the waterline.

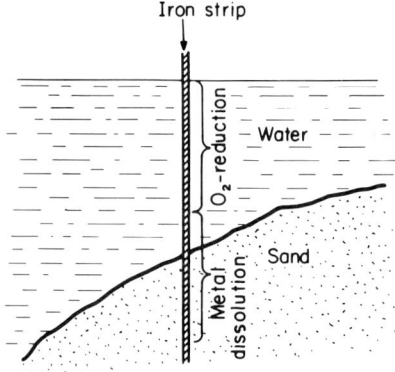

Fig. 11.27. When iron strips are imbedded in sand, it is the imbedded part which undergoes intensive dissolution.

part of the metal, which becomes an electron sink because of its relative oxygen starvation.

A similar situation prevails when a strip of iron is partly embedded in moist sand underneath water (Fig. 11.27). Contrary to more naïve expectations that the sand-covered metal would be protected, it is just this part of the metal strip which dissolves because of its relative oxygen starvation.

The differential-aeration principle can also be exemplified by the underground corrosion of an iron pipe which runs partly through sand with high oxygen permeability and partly through clay soil with low oxygen per-

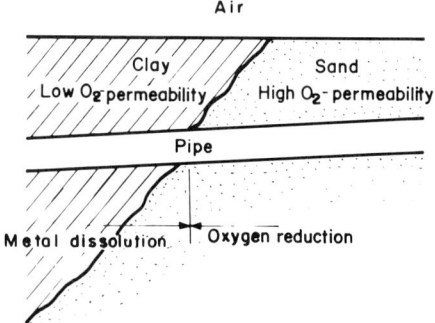

Fig. 11.28. A pipe passing through sand and clay will undergo a more intensive corrosion in clay since this has a lower permeability to oxygen than sand.

SOME ELECTROCHEMICAL SYSTEMS OF TECHNOLOGICAL INTEREST 1305

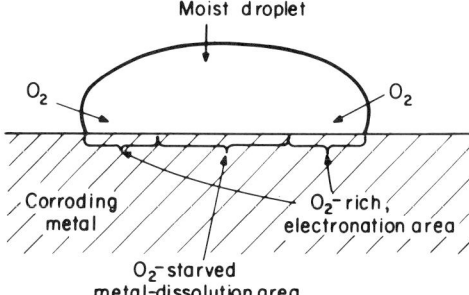

Fig. 11.29. When a droplet of moisture condenses at a metal surface, its center tends to become the point of intensive dissolution activity.

meability (Fig. 11.28); the portion of the pipe in the clay corrodes considerably in comparison with the portion in the sand.

Another illustrative example of a differential-aeration corrosion cell is an iron sheet with a drop of moisture on it (Fig. 11.29). The central region of the drop is oxygen starved compared with the peripheral regions, which therefore become electron-source areas and corrosion is observed at the central electron-sink section.

Apart from corrosion due to differential aeration, corrosion of underground metal structures and pipelines may also arise from stray currents. How this comes about can be seen in the accompanying diagram (Fig. 11.30). The presence of a current-carrying cable in conducting soil results in stray currents passing through the soil. These stray currents may set up

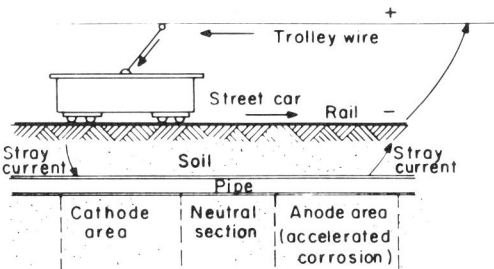

Fig. 11.30. If an underground metal construction comes into a field of stray currents from bare electric lines (e.g., rails), the potential difference established along the construction may set up a corrosion process.

a potential difference between two portions of a pipeline, which, therefore, develops electron-source (cathodic) and -sink (anodic) areas. Thus, pipelines tend to corrode when they pass near electric lines.

11.2.14. Electrodic Approaches to Increasing the Stability of Metals

The control and prevention of corrosion is a subject with tremendous technological significance. It is not a surprise, therefore, that the literature on this subject contains vast amounts of empirical information. No attempt is made here to survey these details. Rather, it is intended to present in a very much simplified way the essence of the electrochemical approaches to corrosion prevention.

A good starting point is the basic picture of corrosion by local-cell action, according to which a corroding metal consists of electron-sink areas at which metal dissolution takes place and electron-source areas at which an electronation reaction occurs. Under conditions where there is an exponential current–potential relationship for both the metal-dissolution and electronation reactions and the transfer coefficients are equal to $\frac{1}{2}$, the corrosion current has been shown to be given by

$$I_{\text{corr}} = (I_{0,\text{M}} I_{0,\text{so}})^{\frac{1}{2}} \exp \frac{F(E_{0,\text{so}} - E_{e,\text{M}})}{4RT} \qquad (11.32)$$

Two fundamental ways in which the magnitude of I_{corr} can be reduced may be seen from (11.32). The first method is based on diminishing the product $I_{0,\text{M}} I_{0,\text{so}}$; this is the method called *corrosion inhibition*. The second method is based on making the relative potential of the corroding metal, E, equal to or less than the equilibrium potential $E_{e,\text{M}}$ for the metal-dissolution reaction; this is the method called *cathodic protection*.

The basic principles of these two methods of corrosion control and prevention will now be presented.

11.2.14a. Corrosion Inhibition by the Addition of Substances to the Electrolytic Environment of a Corroding Metal. Consider the ways in which the term $I_{0,\text{M}} I_{0,\text{so}} = (A_\text{M} i_{0,\text{M}})(A_{\text{so}} i_{0,\text{so}})$ can be reduced.

Firstly, one can try to reduce the exchange-current densities of the metal-dissolution and electronation reactions. For instance, if hydrogen evolution is the electronation reaction, the addition of phosphorus, arsenic, or antimony compounds (e.g., As_2O_3) reduces the exchange-current density for hydrogen evolution (Fig. 11.31). Or, if oxygen electronation is the reaction at the electron-source areas, then, by adding substances which react with dissolved oxygen, the O_2 concentration is reduced and therefore,

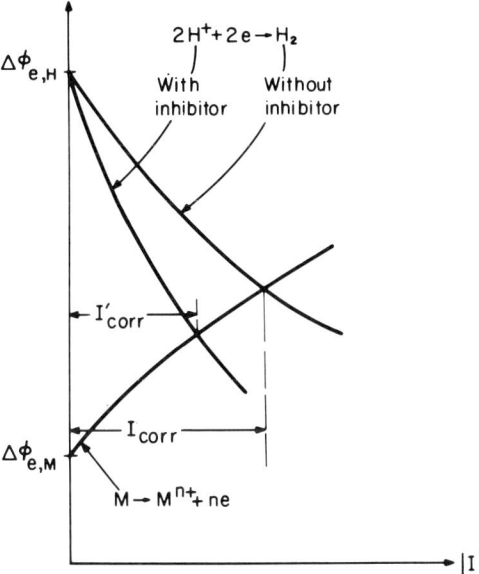

Fig. 11.31. If hydrogen-evolution reaction is inhibited, the potential–current relation changes in such a way that the corrosion current is reduced.

also, the exchange-current density for oxygen reduction. Two substances which act in this way are hydrazine N_2H_4 and sulphite ions SO_3^{2-}

$$2N_2H_4 + 5O_2 \rightarrow 4NO_2^- + 4H^+ + 2H_2O \tag{11.39}$$

and

$$2SO_3^{2-} + O_2 \rightarrow 2SO_4^{2-} \tag{11.40}$$

Alternatively, or in addition, the exchange-current density for the metal-dissolution reaction can be reduced by the addition of compounds which adsorb on the electron-sink areas of the corroding metal and slow down the metal-dissolution reaction (Fig. 11.32). The compounds most often used for this purpose are nitrogen-containing organic compounds (aliphatic and aromatic amines), sulphur-containing compounds (thiourea and its derivatives), and various oxygen-containing compounds (aldehydes).

Now it will be recalled that the adsorption of a particular constituent of the electrolyte depends not only on its chemical nature (i.e., on the chemical part, ΔG_c^0, of its free energy of adsorption) but also upon the electrode charge (remember the parabolic θ_{org} versus q_M curves in Chapter 7).

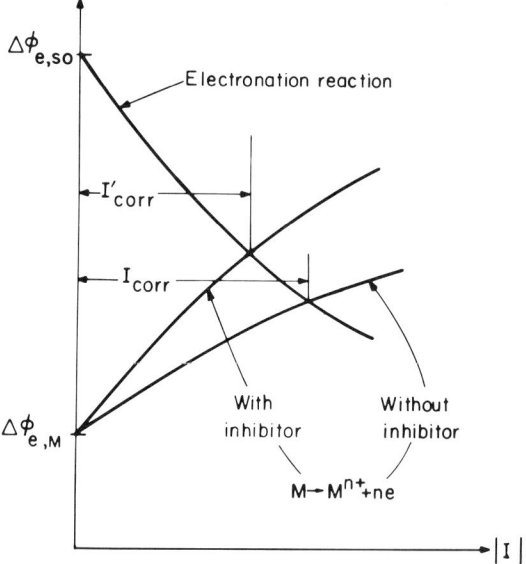

Fig. 11.32. Metal-dissolution reaction can be inhibited by adsorbing organics at electron-sink areas, in which case the corrosion current is reduced.

So the corrosion inhibitor must not only be highly adsorbable in a chemical sense; it must also adsorb in the range of potentials which includes the potential at which the corrosion reactions occur. Correspondingly, differences in the degree of coverage arise when a metal is polarized cathodically or anodically in respect to the corrosion potential, as exemplified in Table 11.5.

A second way of reducing the product $(A_M i_{0,M})(A_{so} i_{0,so})$ is to cut down on the areas of the corroding metal which function as electron sinks or electron sources. The reduction of the electron-sink area is achieved usually by means of the solid products of metal dissolution. This, however, is a process which will be discussed later on (*cf.* Sections 11.2.17 to 11.2.19). Here, reference will be made to methods of producing solid films which reduce the electron-source areas. What is done is to add a film-forming inhibitor which causes the precipitation of a solid film over the electron-source areas. An example of a film-forming inhibitor is the HCO_3^- ion, which interacts with the OH^- ions produced by oxygen reduction to precipitate a carbonate film over the electron-source areas. Another example is the PO_4^{3-} ion, which causes the precipitation of a mixture of ferrous and ferric phosphates over steel.

Film-forming inhibitors which affect the electron-source areas may

TABLE 11.5

Approximate Per Cent of Coverage of Certain Inhibitors (When Anodic and Cathodic) on Mild Steel in 5% Sulfuric Acid at 70°C[†]

Solution concentration of inhibitor $\times 10^3$, moles liter^{-1}	Tolythiourea		β-Naphthoquinoline	
	Anodic	Cathodic	Anodic	Cathodic
0.003	3	3	4	
0.01	22	22	20	
0.03	70	70	40	
0.1	92	90	70	
0.3	99	90	85	<5
1	>99	90	95	20
3	>99	90	98	35
10	–	–	99	60

[†] J. M. West, *Electrodeposition and Corrosion Processes*, D. Van Nostrand Co., Inc., New York, 1965.

be distinguished from those which affect the electron-sink areas by the fact that the former alter the corrosion potential in the negative direction (Fig. 11.33), and the latter, in the positive direction (Fig. 11.34). However, although an inhibitor may start its action on the electron-source areas, it may continue causing the precipitation over the whole surface of the corroding metal and produce a general blockage of the metal surface. With inhibitors which produce a general coverage of the surface, the corrosion potential may move either way, depending on which reaction is affected more.

Table 11.6 exemplifies some commonly used corrosion inhibitors for different situations.

11.2.14b. Corrosion Prevention by Charging the Corroding Metal with Electrons from an External Source. A metal corrodes because the potential difference across the interface at the electron-sink areas is positive with respect to the equilibrium potential for the metal-dissolution reaction. If, by some means, the potential difference could be made negative with respect to the equilibrium potential, metal dissolution would not occur. This depression of the potential difference between a metal and its corrosive

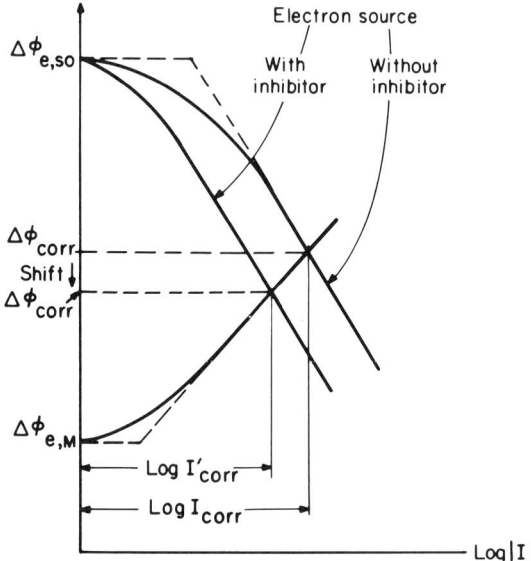

Fig. 11.33. If an inhibitor adsorbs on electron-source areas, the corrosion potential shifts in the negative direction.

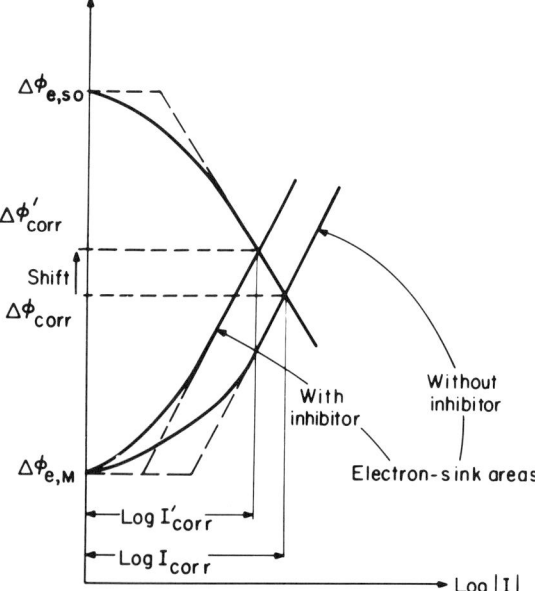

Fig. 11.34. If an inhibitor adsorbs on electron-sink areas, the corrosion potential shifts in the positive direction.

TABLE 11.6
Some Examples of Practical Organic and Inorganic Corrosion Inhibitors Used in Acidic, Neutral, and Alkaline Solutions

| Metal | Medium | Inhibitor | |
		Organic	Inorganic
Iron	Acidic	Aniline, ethyl- and diethylamine, pyridine, quinoline, β-naphthoquinoline, tolylthiourea, thiodiglycol, and formaldehyde	As_2O_3 $NaAsO_2$ $K_2Cr_2O_7$
	Neutral	Sodium benzoate and hydrazine	Na_2SO_3
Steel	Acidic	The same as for iron	
	Neutral	Sodium benzoate	
Copper	Acidic	Thiourea	$K_2Cr_2O_7$ $KMnO_4$
	Neutral	—	$Ca(HCO_3)_2$ $K[Fe(CN)_6]$ Chromates
Aluminum and zinc	Neutral	—	$Ca(HCO_3)_2$ Sodium and calcium Hexametaphosphate
	Alkaline	Glucose	S^{--} and SiO_3^{--}

environment can be achieved by arranging for electrons to be pumped into the corroding metal. These electrons will make the metal more negatively charged and thus lower the potential difference in the negative direction. To prevent dissolution of the metal, it is necessary to pump in an adequate number of electrons.

One method of pumping electrons into the corrodible metal is based on a well-known electrodic fact. When the ions inside a suitably selected metal pass into solution, they leave behind excess electrons, which, if provided with an electronically conducting path, can be made to flow into the corrodible metal. Suppose that an auxiliary metal having an equilibrium potential negative to that of the corrodible metal is immersed in the corrosive environment and connected by a short-circuiting wire to the metal to be

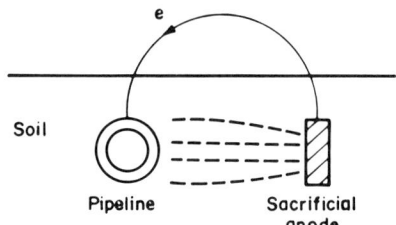

Fig. 11.35. Cathodic protection can be accomplished if an auxiliary metal of a more negative potential is short-circuited with the metal construction to be protected.

protected (Fig. 11.35). Then the auxiliary metal will function as an electron sink (anode) and sacrificially dissolve (hence the term *sacrificial anode*). Further, the corrodible metal will act as the electron-source electrode for the electronation reaction, which would otherwise have produced its corrosion. Hence, what is done is to set up a new corrosion cell in which an auxiliary metal is made to corrode in place of the metal to be protected and in which the entire surface of the latter metal is converted into an electron-source area. For example, if a steel structure has to be protected, one can use zinc or magnesium as a sacrificial electron sink and save the structure from corrosion.

The electrons pumped into the corrodible metal have come, in the above method, from the dissolution of a sacrificial auxiliary metal. Instead, they can come from an external current source (i.e., an electrical power supply). The electrical circuit, however, has to be completed, and, toward this end, an auxiliary *inert* electrode can be immersed in the corrosive electrolyte to provide a return path for the electron current (Fig. 11.36). The external source can then be adjusted so that the potential difference between

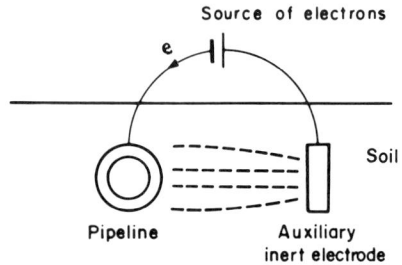

Fig. 11.36. Cathodic protection can be accomplished also by using an inert auxilliary electrode and sending current into the circuit from an external current source.

the corrodible metal and its environment becomes negative with respect to its equilibrium potential. Under these circumstances, the whole of the metal to be protected against corrosion will function as an electron source for the electronation reaction, and the second electrode will serve as an electron sink for some de-electronation reaction (Hoar).

The above two methods of preventing corrosion can be understood easily with an Evans diagram (Fig. 11.37; *cf.* Section 11.2.10). (These diagrams, it will be recalled, result from the superposition of the potential–current curves of the electronation and de-electronation reactions that occur during corrosion.) In spontaneous corrosion reactions, the electronation current is equal to the metal-dissolution current at the corrosion potential. With an external current source for adjusting the potential difference between the corroding metal and its electrolytic environment, it follows that, as the metal–solution potential difference becomes more negative, the metal-dissolution current decreases, whereas the electronation current increases, so that

$$I_{so} > I_M \qquad (11.41)$$

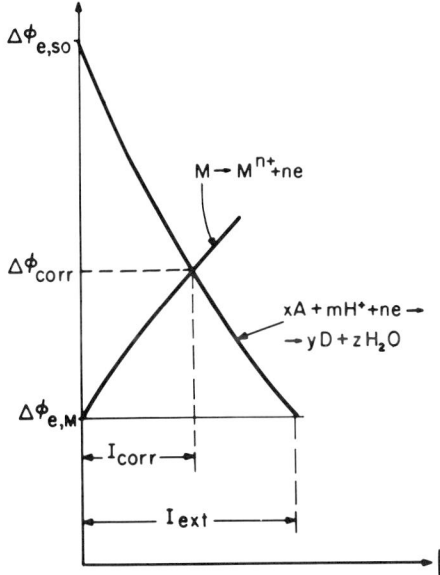

Fig. 11.37. The change in potential from $\Delta\phi_{corr}$ to $\Delta\phi_{e,M}$ *results in a decrease of the metal-dissolution current to zero* but also in an increase in the electronation current, and this increase in current has to be supplied from an external source.

When the potential of the previously corroding metal becomes equal to the equilibrium potential for the dissolution reaction,

$$I_{\mathrm{M}} = 0 \quad \text{and} \quad I_{\mathrm{ext}} = (I_{\mathrm{so}})_{\Delta\phi_{e,\mathrm{M}}} \qquad (11.42)$$

where I_{ext} is the current that has to be supplied by the external source to protect the corroding metal from corrosion.

The stabilization of metal surfaces by the superimposition of an adequately negative potential difference across the interface between the metal and its environment appears to be an ideal method of corrosion prevention. There are, however, some less favorable aspects of the method.

Firstly, when external current sources are used, the power consumption may be impractically large. It all depends on the electrodic parameters of the electronation reaction—the larger its exchange-current density and the lower its Tafel slope, the larger will be the external protection current which must be used to achieve protection.

Secondly, it is important that the potential difference across the *entire* interface between the metal to be protected and its environment is shifted below the equilibrium potential. Suppose that the current through the electrolyte is not distributed uniformly over the corroding metal (e.g., the current lines may pass through greater distances to reach some parts of the metal and hence introduce an *IR* drop near those parts); there may then be localized areas at which the potential difference remains insufficiently cathodic and metal dissolution occurs. Under these circumstances, one may be deceived into thinking that an adequate protection against corrosion has been set up, whereas, in fact, localized corrosion is occurring. Also, it is often better, e.g., in a pipeline, to suffer a very slow, uniform dissolution than localized attack and puncturing.

Conversely, under conditions where the surface of the metal is made *excessively* negative with respect to the hydrogen equilibrium potential, hydrogen evolution may occur and one has to reckon with a hazardous consequence of such cathodic protection. While the metal is successfully protected from dissolving, its surface is being covered with adsorbed hydrogen atoms which are intermediates in the hydrogen-evolution reaction. It will be shown later that some of those adsorbed hydrogen atoms may dissolve into the metal and that this hydrogen which is thus pumped into the interior of the metal can *undermine* the internal strength of the metal by what is called *hydrogen embrittlement* (Section 11.2.24). One may end up, therefore, by losing the strength of the body of the metal.

11.2.15. Passivation: The Transformation from a Corroding and Unstable Surface to a Passive and Stable Surface

One of the ways of protecting a metal from corrosion has been shown to consist in pumping electrons into it and thereby *decreasing* the potential difference across the metal–environment interface. In other words, the double-layer field is altered in such a way as to *diminish* the metal-dissolution reaction that results in corrosion. Suppose, therefore, it is suggested that a corroding metal can be made stable by superimposing a double-layer field which *accelerates* the metal-dissolution reaction rather than hinders it. The suggestion is likely to be considered unreasonable. Yet the fact is that the unstable surface of an actively corroding metal can sometimes[†] be made passive and stable by increasing the potential in the positive direction. This phenomenon of *enforced passivation* (Tomashov) is unexpected, and its study reveals many interesting and novel aspects of the electrochemical interaction of substances with their environments.

Before seeking the mechanisms underlying the enforced passivation of the metal, a fuller description of the phenomenon will be presented.

A piece of corrodible metal (called the *test* electrode) is immersed in an electrolyte along with two other electrodes, one of which is a nonpolarizable reference electrode and the other, an auxiliary electrode. The potential of the test electrode relative to that of the reference electrode can be held at any preselected value by an electronic device known as a *potentiostat*.

This instrument—one of the benefits acquired by electrodics from electronics—consists of a source of potential, an electronic voltmeter and a current source all organized in a particular way (Fig. 11.38). The potentiostat measures the potential V of the test electrode under study and compares this with the preselected value V^* from the potential source (in the potentiostat). If there is a difference $\delta V = V^* - V$ between the measured potential and the chosen potential, the potentiostat tells its current source to send a current I between the auxiliary electrode and the test electrode. The direction and magnitude of this current is electronically chosen to keep the potential of the test electrode at the desired value, i.e., to make $\delta V = V^* - V = 0$. In effect, the potentiostat controls the rate of the charge-transfer reactions at the metal–solution interface and, by increasing the de-electronation or electronation current, pushes the electrode potential in the positive or negative direction to the preselected value.

The study of enforced passivation is facilitated by the use of a potentio-

---

[†] Depending on the metal, solution conditions, etc.

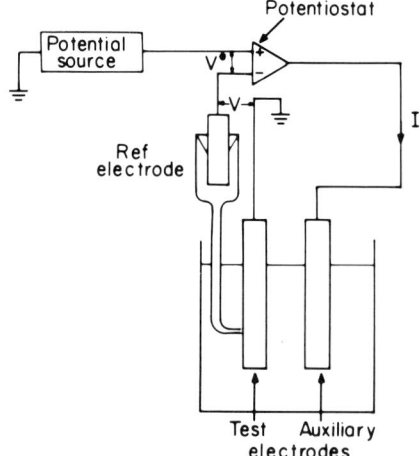

Fig. 11.38. In a potentiostatic circuit, an imposed potential difference V^* drives the current from the auxiliary electrode to the test electrode till the potential difference V becomes equal to V^*, so that the total input into the potentiostat falls to zero.

stat because it enables one to hold the potential of the electrode at chosen values and make measurements, e.g., of the steady-state current density at each value of the potential.

If iron is taken as the corrodible metal, the experimental current–potential plots (Fig. 11.39) show an interesting pattern. As the potential is made more positive, starting from the corrosion potential, the dissolution current increases; this is what one feared. But, at a certain potential, the current–potential curve changes direction and starts decreasing rapidly with potential. This decrease of dissolution with increase of potential is a trend opposite to that expected. Apparently, the dissolution of the metal is undergoing a sudden inhibition at the potential corresponding to the maximum in the current–potential curve, the *passivation potential*. Different metals exhibit different passivation potentials, as shown in Table 11.7.[†]
Once the metal is passivated, the currents which flow are negligible, usually

---

[†] The values given in this table are not precisely those of the passivation potential but *Flade potentials*. These are potentials very near the passivation potential of Fig. 11.39. They are defined in a slightly different way by various authors. For example, "the potential corresponding to the inflection point on the i versus V curve [Fig. 11.39] anodic to the passivation potential."

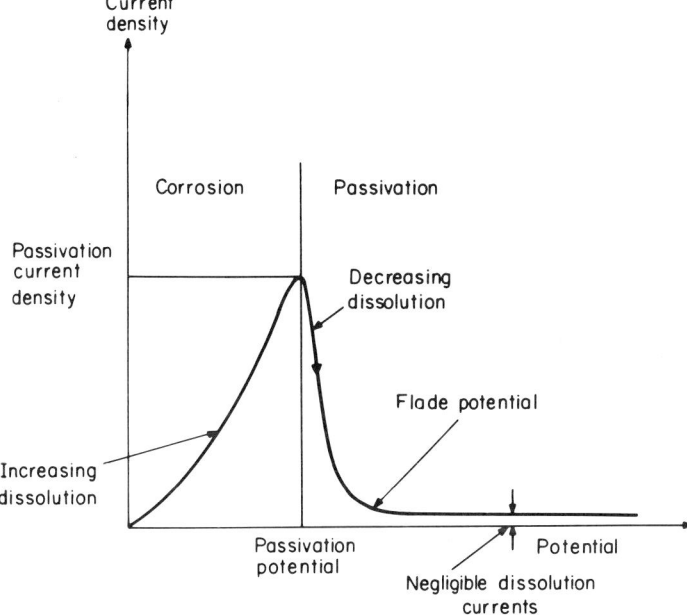

Fig. 11.39. As the potential of steel is increased, the current initially increases, reaches a maximum value, and starts sharply decreasing to negligible values.

TABLE 11.7

Approximate Passivation Potentials of Some Metals at pH = 0 and 25°C

Metal	Passivation potential, V, N-Hydrogen Scale
Gold	+1.36
Platinum	+0.91
Iron	+0.58
Silver	+0.40
Nickel	+0.36
Chromium	−0.22
Titanium	−0.24

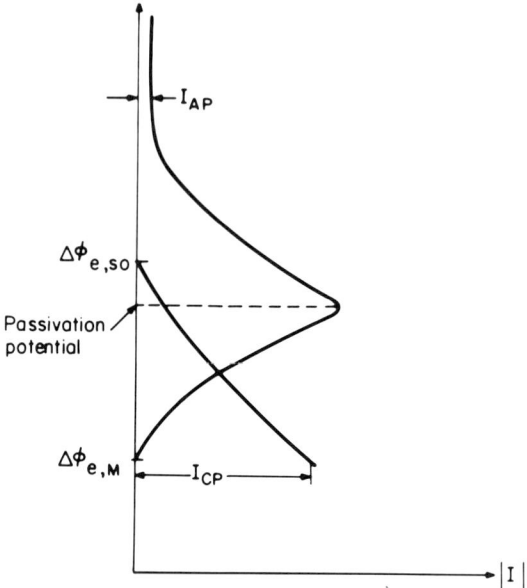

Fig. 11.40. The metal-dissolution current in the passive region, I_{AP}, to be supplied for anodic protection, is usually much smaller than the cathodic protective current. Also, the potential region is such that there can be no hydrogen evolution.

100 to 1000 times less than the dissolution current at the maximum in the current–potential region in which the corrosion current increases with increasing potential. To all intents and purposes, corrosion has ceased, and the metal has been *anodically* protected by pumping electrons away from the metal instead of pumping electrons into the metal as in cathodic protection. The point, however, is that anodic protection can be conferred on a corrodible metal by the passage of a very small total number of coulombs, compared with that needed in cathodic protection, and cannot result in hydrogen entering and damaging the metal, as can cathodic protection (Fig. 11.40).

This method of stabilizing a metal can be used on a large scale, e.g., large tanks used to store oil, which may contain water and be acidic, are sometimes protected from corrosion in this way.

11.2.16. The Mechanism of Passivation

A model for the mechanism of the enforced passivation of nickel has emerged by making a light beam incident on the metal surface and studying the information contained in the reflected light. The point is that the reflection of light is intimately connected with the characteristics of the reflecting surface. Hence, by arranging matters so that a metal acts as an electrode in an electrodic experiment and simultaneously as a mirror in an optical experiment, one can get simultaneous information not only on the current and potential but also on the state of the surface (*cf.* Fig. 11.41).

Consider the enforced passivation of nickel in acid solutions. It turns out that, as the potential of the electrode is increased in the positive direction, there comes a critical potential at which a film of $Ni(OH)_2$ suddenly forms on the metal surface. This film formation, however, cannot explain passivation because it occurs at potentials negative to the passivation potential. The film that is formed is therefore called a *precursor* or *prepassive film*. What extra happenings occur at the passivation potential to transform an unstable corroding surface into a stable surface?

Here, again, the reflected light comes back with a dramatic answer (Fig. 11.42). It indicates that the prepassive film is to all intents and purposes a nonabsorber of light, but, at the passivation potential, the surface film becomes an absorber. What, however, is the connection between the absorption of light and the problem of the electrodic stability of the surface? There can be two explanations of change in absorption properties. On the one hand, this can indicate that some change in the physical properties of the surface film made it just capable of absorbing the light quanta of the

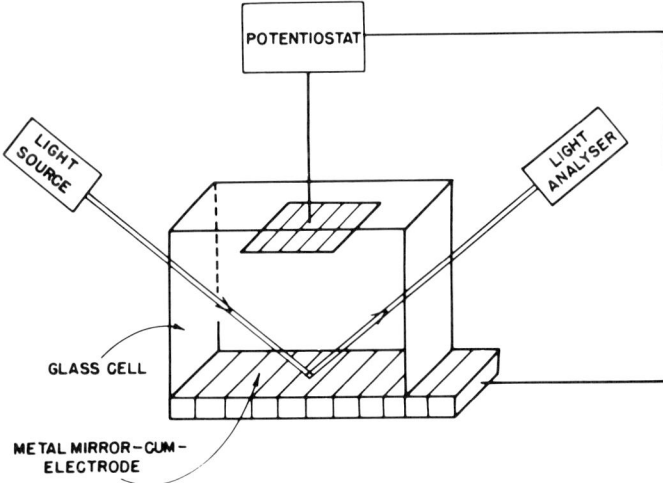

Fig. 11.41. The mechanism of passivation can be studied by using a piece of metal both as an electrode and as a reflector for a light beam.

particular wavelength used. On the other hand, the change of absorption properties of the film may indicate something much more important to the understanding of passivation, *viz.*, that, at this point, the material has all of a sudden acquired free electrons which can then absorb light of any wavelength (Reddy and Bhimasina Rao). But, when it has free electrons, it functions as an *electronic* conductor. On the other hand, the absence of free electrons makes it an electronic insulator. Thus, the change of the prepassive layer from a nonabsorber to an absorber of light may mean that, while the prepassive surface film was an electrical insulator, it was converted at the passivation potential into an electronic conductor. If so, the phenomenon of passivity would find a very straightforward explanation.

Thus, once this electronic conductivity develops in the oxide film (in the prepassive film), and thus its resistance falls, the potential drop across the erstwhile electronically insulating film collapses, and, without a potential gradient to drive the ions, they do not drift through the film from the metal surface to the solution. Hence, dissolution (and, therefore, corrosion) ceases; the metal becomes passive (Fig. 11.42), and the surface is stabilized.

One can conclude, therefore, that the mechanism of enforced passivation of nickel in somewhat acid solutions consists of the formation of a prepassive film at potentials negative to the passivation potential. This film

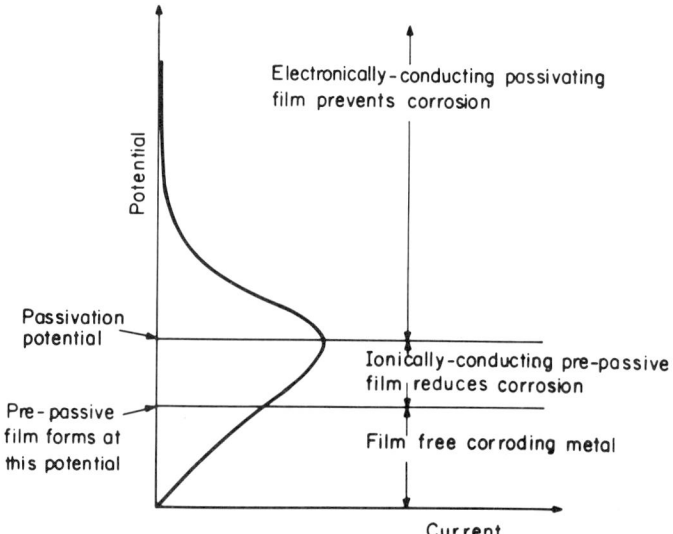

Fig. 11.42. The investigation of the reflected light reveals that the drastic reduction of corrosion current and passivity appear as the film that is formed at the surface becomes electronically conducting.

is a necessary condition of passivation, but not a sufficient condition. The prepassive film has to be, or become, an electronic conductor (Vetter).

11.2.17. The Dissolution-Precipitation Model for Film Formation

A precondition for the enforced passivation of a metal appears to be the formation of a surface film. But how does this precursor film form?

Once again, optical monitoring of the electrode surface is extremely useful. It permits one to know when the film is formed. What can be done, therefore, is to start off with an initial condition in which the metal is kept free of films by potentiostatic control of the potential difference across its interface. Then the potentiostat can be switched off and a constant current switched on. The metal will dissolve, and, as it is dissolving, the light will be watching the surface. As soon as the film forms, a change in the properties of the reflected light signals the event of film formation.

It turns out that there is an induction time τ_i for the film to form, i.e., τ_i is the time interval between the switching-on of the constant current and the appearance of the prepassive film (Fig. 11.43). This induction time increases with a decrease in the current density used to produce the film (Fig. 11.44). Further, stirring the solution during the passage of the constant current increases the induction time. This fact demonstrates that mass transport in the solution is a deciding factor in the formation of the precursor film.

These observations fit into a simple and coherent picture. When the constant current is switched on, the dissolution of the metal tends to make

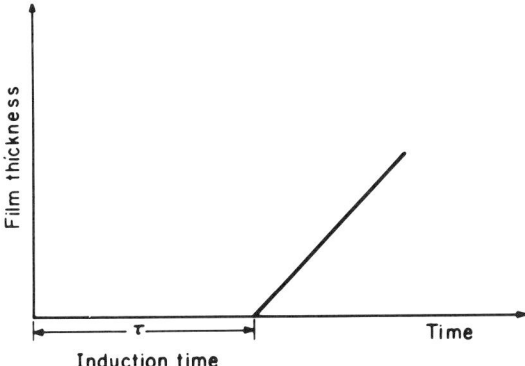

Fig. 11.43. It takes some time after switching on the passivating current before the film formation is observed.

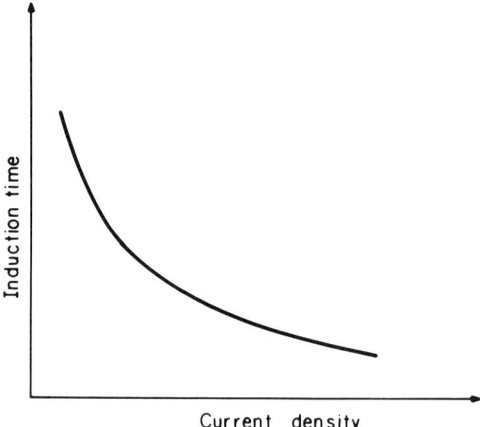

Fig. 11.44. The induction time decreases with increasing current density of passivation.

the metal-ion concentration near the electrode increase. In fact, the variation of concentration with time and distance is given by solving the constant-flux-diffusion problem (Sections 4.2.10 and 9.3.5). In a deposition process under diffusion control, the interfacial concentration of depositing ions is given by

$$c_{x=0} = c^0 - \frac{2i_g t^{\frac{1}{2}}}{nF\pi^{\frac{1}{2}}D^{\frac{1}{2}}} \tag{9.57}$$

i.e., the concentration near the electrode decreases with time and may reach a value of zero at the transition time (in the absence of convection, etc.).

In a dissolution process, however, the interfacial concentration of ions produced is given by

$$c_{x=0} = c^0 + \frac{2i_g t^{\frac{1}{2}}}{nF\pi^{\frac{1}{2}}D^{\frac{1}{2}}} \tag{11.43}$$

which means that the concentration is increasing with time (Fig. 11.45). This interfacial concentration cannot increase indefinitely because there is a limit to the solubility of an ionic species. Hence, a stage is always reached when a precipitate is formed—perhaps $Ni(OH)_2$ in the case of nickel dissolution. Thus, the induction time τ_i is to a metal-dissolution process what the transition time τ is to a deposition process. The former is the time when the interfacial concentration of the metal ions attains a value approximately determined by the solubility product of its insoluble salt; the latter

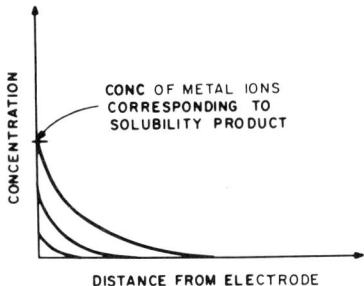

Fig. 11.45. The concentration of ions near the electrode surface increases as the dissolution current is switched on.

corresponds to the time when the interfacial concentration reaches a value of zero. Both the induction time and the transition time depend upon the current density used to produce them.

11.2.18. Spontaneous Passivation: Nature's Method of Stabilizing Surfaces

Passivation has been shown, at least in the example of nickel, to result from superimposition of a double-layer field which enforces conditions for film formation, etc. But does it always have to be a human agency which produces enforced passivation? Cannot nature do the job? In other words, can a metal spontaneously passivate?

Examine the induction-time *versus* current-density curve. It is seen that the induction time increases as the current density decreases. In fact, for all practical purposes, the induction time is infinitely long below a certain critical current density; i.e., there is no film formation, and hence no passivity. What happens is that the rate of dissolution is so slow that diffusion transports away the ions as they are formed and does not allow their concentration to build up sufficiently for precipitation.

If, therefore, a metal is corroding spontaneously at a rate less than the minimum current density (Fig. 11.46), it will corrode—and corrode. But suppose the dissolution-current density during corrosion is greater than the minimum current density; then a film will form by a dissolution–precipitation mechanism, and the metal will passivate spontaneously. Here lies an explanation for the well-known fact that iron corrodes to destruction in *dilute* nitric acid but becomes protected by a passive film in *concentrated* nitric acid. Nature can perform the passivation trick, too.

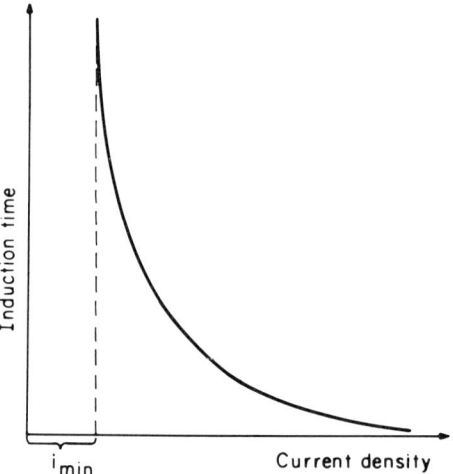

Fig. 11.46. Below a certain minimum current density, the induction time is so long that passivity practically never occurs.

11.2.19 A Competition in Models for Passivation?

The ideas presented here are by no means the only ones which have been suggested for explaining why it is that, on certain metals, as the potential is made more anodic, the current suddenly falls, instead of continuing to rise as it should do according to the Butler–Volmer relation. If one neglects subtleties, the models can be divided into two main groups. In one (Bonhoeffer, Frank, Okamoto, and Reddy), the dissolution of the metal (which does at first occur according to the Butler–Volmer relation) provokes a situation in which the solubility of a salt or hydroxide is exceeded and a film (10–100 Å in thickness) forms. The essential cause of passivation is then some change of property of this film, e.g., its turn-on of electronic conduction, as described above. One might describe such a model for passivation as the solid state mode; it is what happens *in* the film which causes the passivity.

Another, quite different, view (Uhlig, Kolotyrkin, Schwabe) regards the essential cause of passivation as arising from the formation of a monolayer of adsorbed oxygen, or oxygen which is "in" and not "on" the metal surface. In some versions of this view, the sudden fall of current (Fig. 11.42) characteristic of passivation is supposed to occur as a result of the presence on the surface of much less than a monolayer of O^{2-}, OH, or other anion.

The reasons why such small amounts of material, adsorbed on an electrode, should cause such dramatic happenings as shown in Fig. 11.42, are rationalized in various ways. For example, the adsorbed ion may block a kink site in the dissolving metal, and this may lower the free energy of the initial state of the atom in its dissolution reaction, so that it no longer dissolves with the former rate—its i_0 for dissolution has been dropped by several orders of magnitude. This latter model could be called the electrodic one, because there is no need in it to have a solid oxide phase, and the changes are supposed to come about entirely by happenings at the electrode–solution interface.

For a long time, these two types of models existed side by side with very little to distinguish between them. In fact, the various protagonists seemed rather evenly matched. The distinction between the models depended upon the presence or absence of a film, which, according to the electrodic model, was a monolayer thick, and, according to the solid state model, many monolayers thick. But even though the film of the latter model is much thicker than that of the former, it is still very thin, and for a long time was difficult to detect, and its thickness hence uncertain.

From the early 1950's on, however, it gradually became clear that in the region of potential *after* the current has fallen (*cf*. Fig. 11.42), and while it remains so low ("passive region"), the film is indeed (relatively) "thick," i.e., typically some 50 Å in thickness. This became an accepted fact—it is not now a doubted one—and the solid state theories seemed to have gained the day.

However, all was not at rest, because, if the thickness of the oxide in the anodic passive region is measured, it is found to decrease as the potential becomes more negative. The protagonists of the monolayer theory consequently rallied. They advanced the view that, although there was indeed a thick film present in the more anodic region, that was not the *cause* of passivity, but only an incidental after-growth. The *cause*, they said, in their electrodic way, is the effect of a less-than-monolayer situation which occurs from before the peak of Fig. 11.42, the coverage increasing from this peak, thus making the dissolution current fall.

With the comeback of monolayer theories, a diagnostic test became more difficult but more necessary. Now, it was not sufficient to find out the thickness of the film *after* the act of passivation is complete, i.e., after the current *had* fallen in Fig. 11.42. Something more refined was required, namely, to find out how the film thickness varied with potential *around the maximum* of current against potential, and, moreover, to have some measure of the electronic conductance of the film as a function of potential.

Thus, in the electrodic model, at the peak of the current, on the current–potential curve for passivation, there should be very little material present—the adsorbing O^{2-} ions (or whatever it is which in very small quantities might cause the fall of current) have only just begun to adsorb. At the positive end of the hump of Fig. 11.42, something like a monolayer may be expected, and the electronic conductance of the film in such a situation will not change in a profound and concomitant way with respect to the changes of current. But, in the solid state view, a substantial film (about as thick as it is going to get) has already formed *before* the peak. The changes which are to occur later, namely, at increasingly positive potentials, will be the change of electronic conductance; it is the increase of this which will cause cause the fall of the anodic current after the peak of Fig. 11.42.

The distinction criteria are very clear, but what of the methods of measuring, simultaneously as functions of potential, the thickness and conductance of films as little as 1 Å thick? Further, how does one measure the electronic conductance of a film 50 Å or less thick in solution? These are not easy measurements to make. No application of any form of electron diffraction or microscopy can be made, because the electrons are absorbed in the solution and, therefore, do not reflect the image of the surface. Removal of the metal from the solution means that electron diffraction done "in the dry" (and perhaps even after contact with air) may refer to a different entity from that in the solution. This is where the boundary of knowledge lies, except for the case of nickel described above. The future lies certainly in *optical methods* of examination of the electrode surface *in situ*, just through the peak region of the passivation curve. Visible light must be used because it is not (as electrons are) absorbed by the solution. And the method may not be a simple interferometric one, but some more sophisticated form of spectroscopy which measures conductance, refractive index (hence, composition), and, of course, thickness, all at the same condition. After such a method has been applied extensively to metals and alloys of practical importance, a lot more will be known about the reasons for the fortunate stability of the thermodynamically unstable technological world.

11.2.20. The Thermodynamics of Passivation

The discussion of the phenomenon of passivity has led to an important conclusion; the surface of a metal can under the appropriate environmental conditions be covered with oxide films.

Is it possible to tell approximately whether the environmental conditions are ripe for oxide formation on a metal surface? This is a thermodynamic question, and, as indicated previously, answers are obtained by looking up the potential–pH diagrams. Such complete potential–pH diagrams (Fig. 11.47) for metals indicate the conditions for the stability of a metal (immunity from corrosion), of its ions (corrosion), and of its oxides (the possibility of passivation). It can be seen that there are often large ranges of pH and potential in which passive layer formation is the thermodyamically sanctioned process.

There is an important caution that must be borne in mind while using potential–pH diagrams to understand oxide-film formation on electrodes. As stressed earlier, it is only the pH of the solution adjacent to the electrode which is of relevance and not the pH of the bulk of the solution. Thus, e.g., the pH adjacent to an electron-source electrode may be increased to 9 by the removal of hydrogen ions ($2H^+ + 2e \rightarrow H_2O$), whereas the bulk pH remains at 3. Then the potential–pH diagram must be examined at a pH of 9 and not at a pH of 3.

Further, *oxide formation does not necessarily imply passivation*. Passivating oxide films must have the right properties (e.g., electronic conductivity). Then they can serve as excellent inhibitors of corrosion. Of course, the films must satisfy several secondary requirements. The protective oxide films must be mechanically stable; they must not flake off or crack. The films must also be continuous. If they are porous or full of pinholes, corrosion cells can be set up. One reaction (perhaps the cathodic reduction of oxygen) can occur *on* the oxide; and metal dissolution, in the pores.

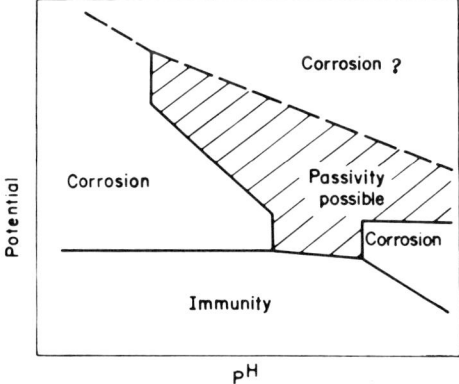

Fig. 11.47. Complete potential–pH diagrams indicate the conditions of oxide formation at the surface which is a precursor to passivity.

11.2.21. Hydrogen Diffusion into a Metal

It has been sufficiently emphasized that the *instability of metal surfaces arises from an electrodic mechanism*; an electronation reaction teams up with the metal-dissolution reaction to keep numerous micro *corrosion cells* running.

What has all this to do with the inside of the metal? One would think that the inside is sufficiently isolated from the surface to remain safe and stable. It will be shown, however, that events at the borders of a metal may have internal repercussions and eventually cause even the inside of the metal to decay, i.e., to lose its mechanical properties. Thus, electrodic charge-transfer reactions at the surface have far-reaching implications for the strength of bulk metals. It is intended here to indicate briefly how the surface instability can be propagated to the inside of the metal.

Consider a corroding metal, and let hydrogen evolution be the electronation reaction. The formation of hydrogen atoms adsorbed on the metal surface is an essential intermediate step in the electrodic evolution of hydrogen. What happens to these adsorbed hydrogen atoms? They can get desorbed in either a chemical or electrodic reaction as hydrogen molecules which diffuse out into the solution or collect in bubbles of hydrogen gas. This is the visible *way out* from the metal surface.

But there is also a *way in* from the surface; the *adsorbed* hydrogen atoms can dissolve into the metal to form *absorbed* hydrogen. Since one has started off with zero concentration of absorbed hydrogen inside the bulk of the metal, a gradient of hydrogen concentration develops between the surface where hydrogen enters and the interior of the metal. This concentration gradient makes the adsorbed hydrogen diffuse into the metal. The extent of this phenomenon is exemplified in Table 11.8 for different types of iron. The diffusion coefficients are seen to be of the same order of magnitude as

TABLE 11.8

Diffusion Coefficients and Critical Concentrations of Hydrogen for Diffusion in Iron at 25°C

Material	$D_{298 \cdot K}$, 10^{-5} cm^2 sec^{-1}	c_k, 10^{-8} mole cm^{-3} Fe
Polycrystalline Armco iron	6.3	14.3
Single-crystal iron	8.3	12.0
Zone-refined iron	6.1	13.9

those for diffusion of ions in aqueous solution. Thus, diffusion of hydrogen in the bulk of the metal can be considered a fairly fast process.

The permeation of hydrogen into the interior of a metal can be shown in a simple way (Fig. 11.48). A thin, metal membrane separates two vessels containing an electrolyte. Hydrogen is electrodically evolved on one side.

The potential difference across the other membrane–electrolyte interface is adjusted for the de-electronation (or ionization) of any absorbed hydrogen, $H_{ads} \to H^+ + e$, coming through the metal from the entry side of the membrane. Thus, the ionization current is the manifestation of the hydrogen permeating through the metal membrane from the surface on which hydrogen is evolved. Quantitative correlations can be made, as shown below, between the permeation current and the diffusion coefficient and the flux of hydrogen through the metal.

The steady-state de-electronation current density i_d is related by the condition of flux equality (cf. Section 9.3.2) to the flux J_p of hydrogen permeating through the metal (Fig. 11.48)

$$\frac{i_d}{F} = J_p \tag{11.44}$$

and, using Fick's law for the steady-state permeation flux, one has

$$\frac{i_d}{F} = J_p = \frac{D(c_{\text{entry}} - c_{\text{exit}})}{l} \tag{11.45}$$

where c_{entry} and c_{exit} are the concentrations of absorbed hydrogen on the

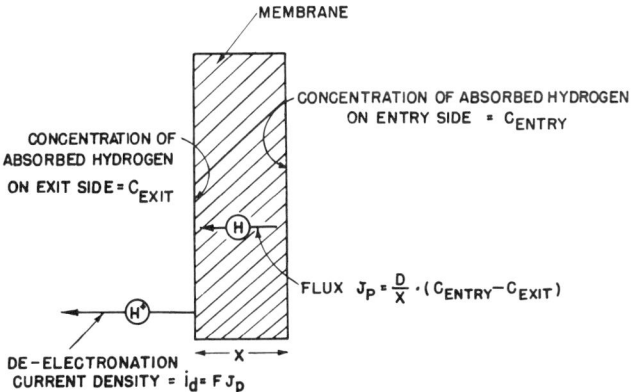

Fig. 11.48. The penetration of hydrogen into the bulk of a metal can be detected if the metal is sliced into a membrane and anodic dissolution of hydrogen is observed on the side opposite to that at which the evolution takes place.

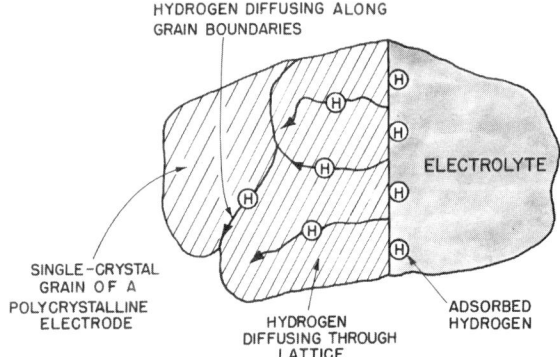

Fig. 11.49. Hydrogen atoms adsorbed at the surface may diffuse into the bulk of the metal by intercrystalline, or interstitial, diffusion (diffusion *through* the crystal lattice).

entry and exit sides[†] of the membrane of thickness l. The potential difference across the membrane–electrolyte on the exit side can be adjusted so that $c_{\text{exit}} = 0$, i.e., all the hydrogen coming through is immediately ionized. Thus,

$$\frac{i_d}{F} = \frac{Dc_{\text{entry}}}{l} \tag{11.46}$$

Hence, by measuring the permeation current, it is possible to study the diffusion coefficient of the hydrogen inside the metal.

11.2.22. The Preferential Diffusion of Absorbed Hydrogen to Regions of Stress in a Metal

In a polycrystalline material, one might have considered that the grain boundaries are *irrigation channels* for the diffusing hydrogen. Yet, it is found that the diffusion coefficient for hydrogen is the same for polycrystalline and single-crystal iron. Thus, the hydrogen must be diffusing through the lattice—interstitial diffusion (Fig. 11.49).

When a hydrogen atom occupies an interstitial site, there is a certain displacement of the atoms around the interstitial atom; it is as though some atoms moved apart a little to accommodate the interstitial hydrogen

---

[†] *Entry side* refers to a plane *just inside* the surface at which the absorbed hydrogen atoms enter the membrane, and *exit side* refers to a plane *just before* the hydrogen leaves the membrane and becomes ionized.

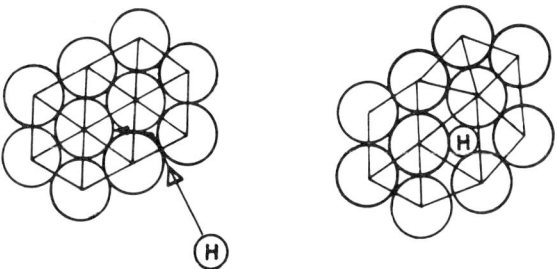

Fig. 11.50. As a hydrogen atom penetrates into the lattice, slight distortion takes place to accommodate it.

atom (Fig. 11.50). Hence, one may consider that there is a certain change in volume due to the entry of a hydrogen atom into the lattice. The net change in volume so resulting, due to one gram atom of hydrogen, is the partial molar volume $\bar{V}_H$, of hydrogen in the metal.

It is possible to obtain the partial molar volume of hydrogen in a metal provided one knows the solubility of hydrogen in it, corresponding to a constant pressure or overpotential, *as a function of applied stress*. For the applied stress to be thermodynamically significant, it should be within the Hooke's-law region for the metal (Fig. 11.51). Proceeding from the

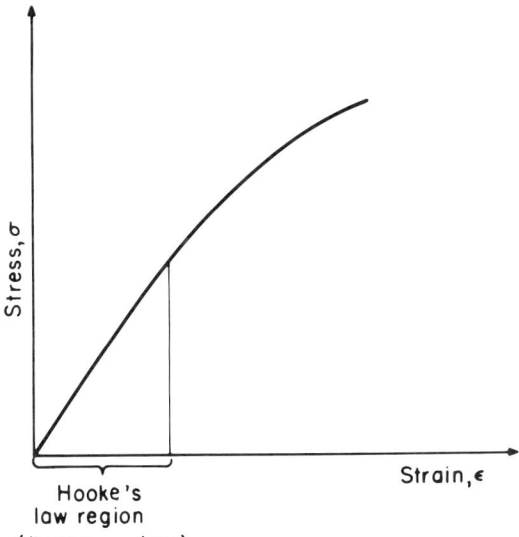

Fig. 11.51. Hooke's law relates the strain produced in a body under stress (the linear stress–strain laws). Past a certain stress, the law is not obeyed any more and a plastic deformation of the metal takes place.

thermodynamic relations $(\partial \mu/\partial P)_T = \bar{V}$ and $\mu = \mu^0 + RT \ln c$ (when c is small), one has:

$$\left[\frac{\partial \ln(c/c_0)}{\partial P}\right]_T = \frac{\bar{V}_H}{RT} \qquad (11.47)$$

Here c is the solubility of hydrogen (g atom H/cc metal) when the applied uniaxial stress (equivalent to pressure, see below) is σ and c_0 the solubility when the applied stress is zero. The general relation between the applied stress and pressure or the equivalent hydrostatic stress σ_h, can be written as

$$P = \tfrac{1}{3}(\sigma_x + \sigma_y + \sigma_z) = \sigma_h \qquad (11.48)$$

where σ_x, σ_y and σ_z are the components of the applied stress. For uniaxial stress condition, $\sigma_x = \sigma_z = 0$ and hence, $P = \tfrac{1}{3}\sigma_y = \tfrac{1}{3}\sigma = \sigma_h$. Substituting for P in terms of σ_h in equation (11.47) and integrating one obtains

$$c = c_0 \, e^{\sigma_h \bar{V}_H/RT} \qquad (11.49)$$

Measurements of c and c_0 (the hydrogen solubility in the presence and absence of stress) can be made by determining the permeation of H, at a series of stresses. Equation 11.49 was used by Beck, Nanis, and McBreen to determine $\bar{V}_H$ for H in pure iron. Their value (numerically corrected for the absence of a $(2.303)^2$ factor) is 2.6 cc/mole.

Thus, from 11.49, and taking σ_h as positive when the stress is tensile, and negative when it is compressive, one seen that hydrogen accumulates in regions of compressive stress. The stress need not be externally applied

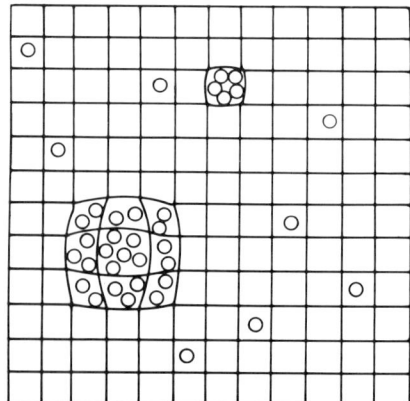

Fig. 11.52. Hydrogen atoms are accumulating at those points in the lattice at which there are larger interstices produced by stretching the specimen.

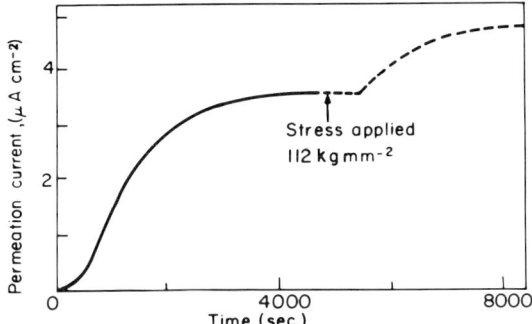

Fig. 11.53. As the metal membrane is strained, the hydrogen-permeation current increases.

stress. The above equation is true also for residual stresses in metal. The latter kind of stress usually will have tensile *and* compressive stress fields associated with it. As far as the solubility of hydrogen is concerned, the effect of the tensile stress field (which *increases* solubility) overwhelms the counter effect due to the compressive stress field (which tends only to decrease the already small solubility). Therefore, the larger the lattice strain or distortion, the larger the concentration of hydrogen (Fig. 11.52). All imperfections in crystals are regions of distortion or strain. Hence, absorbed hydrogen finds its way to, and concentrates at, such imperfections.

There is a simple way of experimentally demonstrating that stretching a metal leads to increased hydrogen absorption (McBreen). The permeation currents, measured by the membrane technique, should show an increase when the membrane is stretched. This is, in fact, what happens (Fig. 11.53).

11.2.23. Interstitial Hydrogen Can Crack Open a Metal Surface

It is known that the imperfections in a metal include voids which are larger than atomic dimensions, say about 100 Å across. On reaching these regions, the *ab*sorbed hydrogen atoms feel they have reached an exposed surface. They become *ad*sorbed hydrogen atoms and combine to form hydrogen molecules; a chemical desorption reaction takes place, i.e., $2H_{ads} \rightarrow H_2$ (Fig. 11.54). Thus, a pressure of hydrogen gas builds up inside the void. Calculations show that the pressures can become enormous,[†] indeed so

---

[†] Pressures of more than 10^{10} atm ought thermodynamically to be obtained inside some electrodes, but the highest pressures reported in experimental hollow cathodes range between 200 and 300 atm. These, however, are not equilibrium values. Indirect measurements based upon the measurement of the expansion of the specimens suggest an effective H_2 pressure of 10^5 to 10^7 atm (Smialowski).

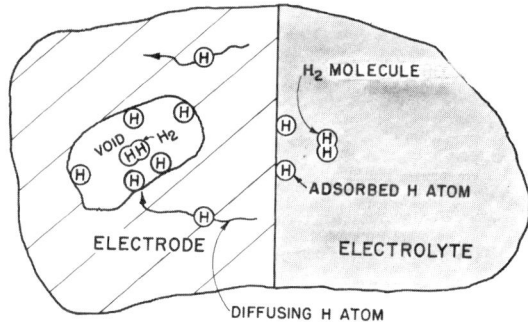

Fig. 11.54. Hydrogen atoms absorbed in the metal recombine when they reach the voids and form molecular hydrogen inside the metal.

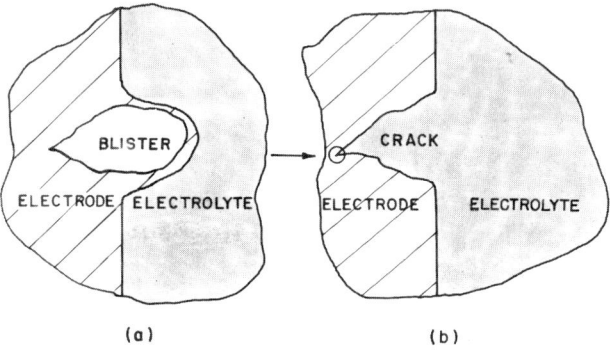

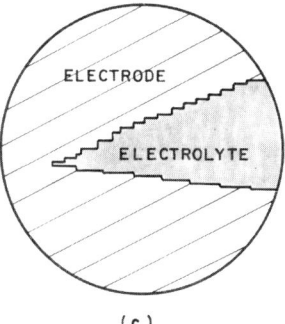

Fig. 11.55. The metal near the surface yields under the pressure of the accumulated molecular hydrogen forming a blister (a), which breaks up and so initiates a crack in the metal (b). At the bottom of the crack (c), the stress is very great and may produce further yielding.

high that the surrounding metal is stressed beyond its elastic limit. The metal yields; the void becomes a cavity (supervoid) as discussed in Section 11.2.25.

If all this happens near the surface of the metal from which hydrogen is entering the metal, a *blister* may be formed near the surface. Eventually, the walls of the blister collapse and this rupture allows the gas to escape. In the process, however, a crack has been initiated at the metal surface (Fig. 11.55). The whole process of crack initiation is of course facilitated if an outside stress is applied to the metal. Then, if the metal structure is such that there are specially high stresses at some points, it is precisely at these points that there is the greatest likelihood of crack initiation because hydrogen permeates preferentially into the stressed region and enters the voids nearest the stressed region. The cracks thus *initiate* there.

11.2.24. Surface Instability and the Internal Decay of Metals: Stress-Corrosion Cracking

So far, corrosion has not come into the picture except insofar as it stimulates the $H^+ + e \to H_{ads}$ electronation reaction and is therefore responsible for the hydrogen accumulation inside the metal. Now consider a metal which is simultaneously being corroded and some parts of which are subjected to a tensile stress. The permeating hydrogen will tend to initiate† a crack in the region where the stress is great by the mechanism described in the previous section (see also Section 11.2.20), and the electrolytic solution (the corrosive environment) comes into contact with the inside of the crack (Fig. 11.55).

Once the crack is initiated, the metal surface inside the crack may be quite different from the normal surface of a metal. For example, the outside surface may be covered by a passive film, while the inside is a fresh and oxide-free metal surface. Or, in the course of plastic deformation, the metal could have developed slip steps [*cf.* Fig. 11.55(c)] which contain crystallographic planes of high Miller index at which the specific dissolution rate (or exchange-current density) may be larger than that at the normal metal surface. Anodic current densities of some 10^4 times those at a passive surface have been shown to appear at a metal surface yielding under stress (Despic and Raicheff).

---

† Hydrogen initiation of stress-corrosion cracking is indeed the probable mechanism. But what has been given here is rather overgeneral. For example, the stress corrosion of alloys shows specificities which hint at unexplained factors. Perhaps passive films form at the bottom of pits and it is the breaking of these upon stress that is sometimes determinative of crack spreading.

One conclusion is clear. The instability of a metal with surface cracks will tend to be greater than that on a surface without such cracks. The metal-dissolution and hydrogen-evolution reactions tend to occur indiscriminately on the normal surface of a homogeneous *single* crystal. When, however, there is a crack, the metal dissolution will occur preferentially inside the crack; and the hydrogen evolution, on the surface outside the crack (Fig. 11.56). But this implies that the electron-source area A_H is very large compared with the area A_M inside the crack, i.e., compared with the area over which there is metal dissolution. It is essential, however, that the corrosion current (not the current *density*) be equal to the electronation current

$$I_{\text{corr}} = I_M = |I_H| \quad (11.50)$$
$$= A_M i_M = A_H |i_H| \quad (11.51)$$

Hence,

$$i_M = \frac{A_H}{A_M} |i_H| \quad (11.52)$$

and, since $A_H \gg A_M$, it follows that

$$i_M \gg |i_H| \quad (11.53)$$

i.e., even though the hydrogen-evolution current density is small (say, 10^{-4} amp cm^{-2}), the metal-dissolution current density may be greater by a factor of A_H/A_M, which can be of the order of 10^3. Such high current densities can be sustained by the metal-dissolution reaction inside the crack because of the abnormally high exchange-current densities possible there.

A model has been given for high dissolution rates inside cracks in terms of the oxide-free and highly kinked state of the surface there. What

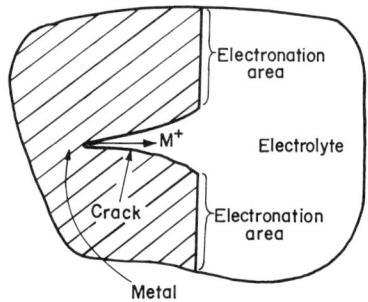

Fig. 11.56. When a stress crack appears at the surface, this becomes a *locus* of an increased dissolution activity, the electrons being drawn away to the rest of the surface as the electronation area.

happens when the metastable kinks are dissolved off or when the passive film is formed to cover the walls of the crack? The surface inside the crack becomes more normal, and so does the current density in that region. Thus, the dissolution rate should become normal inside the crack.

Now consider that the whole corroding metal is being stretched by a tensile stress. The gross average applied stress is not sufficient to make the metal yield; the stress is within the elastic limit. Release it, and the metal will spring back to its original dimensions. This does not mean, however, that the local stresses are equal to the average stress. An abnormally high stress concentration at the apex of the crack need not arise from externally applied stresses; the high stress may be due to residual stresses left behind in the metal at the time of its incorporation into a fabricated structure (e.g., the region around the rivets or welds in a steel boiler).

Irrespective of how the abnormal stress concentration arises, it is possible that the material at the crack apex is locally stressed into the plastic-deformation region of the stress–strain curve. What is the result of the yielding of the metal near the crack apex? The result is that, as anodic dissolution dissolves away the kinky surface, further plastic yielding creates fresh kinky surface inside the crack, and, thus, the yielding helps along the metal dissolution at a rate (e.g., of millimeters per hour) which turns out to be far greater than what would be expected at the overpotential concerned from measurements on the normal surface (Hoar and West).

This is not all. What is happening is that the crack is propagating into the interior of the metal with the advancing edge of the area serving as the electron-sink area for the metal-dissolution reaction. Superficially, everything is normal; if one measures the potential difference between the solution and the apparent surface of the metal, one gets almost the usual corrosion potential. Then microcracks begin to join up with other microcracks, macrocracks are produced, and the piece of metal ceases to be a stable structural material; an axle cracks, or a part of an aircraft disintegrates when only normally stressed.

What has been described is what is called the *stress-corrosion cracking*. Some common examples of systems that tend to undergo this type of corrosion are given in Table 11.9. But perhaps one should call it *yield-assisted corrosion* (an electrochemical-plus-mechanical phenomenon) in contrast to normal *field-assisted dissolution* (an electrochemical phenomenon).

At this point, one may feel a lurking doubt. Maybe the crack-propagation process has nothing to do with electrodic dissolution, and the stress by itself does the damage. This view can be tested simply by superimposing a double-layer field and adjusting the metal–solution potential difference so

TABLE 11.9
Some Common Examples of Stress-Corrosion Systems[†]

Alloy	Medium	Type of cracking
18 Cr–8 Ni	Cl$^-$	Transgranular
Steels	OH$^-$	Transgranular
70 Cu–30 Zn	NH$_4^+$ and some amines	Transgranular at low pH, intergranular in neutral solutions
Al–4% Cu alloys	Cl$^-$	In regions adjacent to grain boundaries
Al–7% Mg alloys	Cl$^-$	Intergranular
Magnesium alloys	Cl$^-$	Transgranular or intergranular
Titanium alloys	Cl$^-$	Transgranular and intergranular
Mild steel	NO$_3^-$ and OH$^-$	Intergranular
Cu–P and Cu–Al	NH$_4^+$ and some amines	Intergranular
β brasses	Cl$^-$	Transgranular
	NH$_4^+$	Intergranular
Cu$_3$Au	FeCl$_3$	Intergranular

[†] J. C. Scully, *The Fundamentals of Corrosion*, Pergamon Press, New York (1966).

that metal dissolution stops. Despite the continued presence of the stress, the crack propagation stops—clear evidence that, for stress-corrosion cracking to occur, both the stress and the abnormally high corrosion rate inside the crack are essential. Without the stress, the dissolution rate inside the crack becomes normal and the crack ceases to advance into the metal; without the electrochemical dissolution at the crack apex, the stress cannot of itself make the crack advance into the body of the metal.[†]

11.2.25. Surface Instability and Internal Decay of Metals: Hydrogen Embrittlement

It has been known for quite some time that some very strong metals may suddenly lose their strength and become brittle even though there is no indication of an applied or initial stress. Thus, a piece of iron sheet can under certain circumstances become so brittle that it can easily be torn in the hands like paper; the metal has become embrittled.

[†] This is not to say, of course, that all cracking of materials has an electrochemical step. It is clear, e.g., that the cracking of glass (a nonconductor) is unconnected with anodic dissolution. Thus, if the energy associated with stressing a material is greater than the appropriate surface free energy, spontaneous cracking will occur.

What is the mechanism of this phenomenon? Very early during investigations of this field, it was realized that metals become embrittled because, at some stage of their career, their surface was the scene of a hydrogen-evolution reaction either because the metal was deliberately used as an electron-source electrode in a substance-producing cell or because parts of the metal became electron-source areas in a corrosion process. In fact, the phenomenon has come to be known as *hydrogen embrittlement*.

An approximate picture of what happens during hydrogen embrittlement can be sketched. The process commences with hydrogen's diffusing into the metal and accumulating in distorted regions of the lattice. Any voids (tiny cavities) in the lattice permit the accumulation of hydrogen gas by the chemical desorption of hydrogen atoms (supplied by diffusion from the surface). If the pressure of the gas becomes sufficiently high, cracks or large cavities are initiated. The result of all these events is that the metal ends up with plenty of cracks inside it (Fig. 11.57). When stretched, it does not yield like a ductile material; it fractures along the cracks. The hydrogen has embrittled the metal.

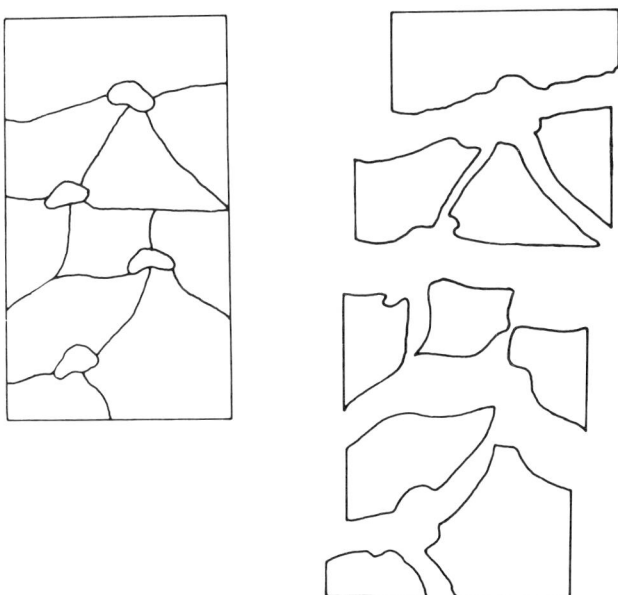

Fig. 11.57. Accumulated hydrogen (originating in electrochemical formation at the surface) produces cracks inside the metal. Hence, when strained, the metal fractures along the cracks rather than yields. It has become brittle.

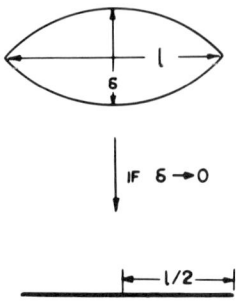

Fig. 11.58. The Griffith crack is a disc-shaped crack.† When in equilibrium, the strain energy expanding it is equal to the surface energy contracting it. For a quantitative argument, it can be represented by a flat disk of small thickness, δ, and relatively considerable radius $l/2$.

It is possible to write down approximate conditions for crack propagation in a very simple way. Consider the *Griffith crack*, as it is called, a disk-shaped crack with a length l (Fig. 11.58).

The strain energy U_{strain} is the work done to increase the strain of the crack from $\varepsilon = 0$ to $\varepsilon = \varepsilon$ under the action of the stress σ that arises from the pressure of the hydrogen gas, i.e.,

$$U_{\text{strain}} = \int_0^\varepsilon \sigma \, d\varepsilon \tag{11.54}$$

But, by Hooke's law,

$$\sigma = Y\varepsilon \tag{11.48}$$

where Y is Young's modulus and ε is the strain. Hence,

$$U_{\text{strain}} = \int_0^\varepsilon Y\varepsilon \, d\varepsilon = \frac{Y\varepsilon^2}{2}$$

$$= \frac{\sigma^2}{2Y} \tag{11.55}$$

---

† The Griffith crack of Fig. 11.58 is a hypothetical entity, necessary in the present model for embrittlement. However, small cavities do actually form inside metals due to a piling up of dislocations in specific regions. Such areas are wedge-shaped rather than disk-shaped. In the very crude analysis given here, the shape of the crack will be taken as somewhere between wedge and sphere, and the lack of definition covered by a numerical constant.

This is the strain energy per unit volume. Hence, for the crack considered, the strain energy is

$$-\frac{\sigma^2}{2Y} V_{crack} \quad (11.55a)$$

If the crack were a sphere (cf. Fig. 11.58), $V_{crack} = (4/3)\pi(l/2)^3$. Let it be taken as $K_1 \pi l^3/8$, where K_1 is a dimensionless constant close to 1.

But, also, as the crack expands under the influence of the H_2 within it, it absorbs more surface energy. The total surface energy of the crack is

$$\gamma A_{crack} \quad (11.56)$$

where γ is the surface tension of the metal–hydrogen interface. If the Griffith crack were spherical, A_{crack} would equal $4\pi(l/2)^2$, and if it were a disk, it would be $2\pi(l/2)^2$. Let it be $K_2 \pi l^2/4$, where $2 < K_2 < 4$. Thus, $\gamma A = K_2 \pi (l^2/4) \gamma$.

Hence,

$$U_{crack} = -\frac{\sigma^2}{2Y} K_1 \pi \frac{l^3}{8} + K_2 \pi \frac{l^2}{4} \gamma \quad (11.57)$$

If the H_2 pressure is large enough, the crack will grow, i.e., l will increase. One can obtain the critical value of σ (the stress in dynes cm^{-2}), and also the H_2 pressure in the crack, by finding what σ value corresponds to

$$\frac{\partial U}{\partial l} = 0 \quad (11.58)$$

Using (11.57) in (11.55), one finds (assuming $K_2/K_1 \simeq 2$)

$$\sigma_{critical} = p_{H_2 \text{ in crack for embrittlement}} > \left[\frac{16}{3} \frac{Y\gamma}{l}\right]^{\frac{1}{2}} \quad (11.59)$$

One may take some likely values of Y, γ, and l to get a feeling of what sort of p_{H_2} values are implied. They are

$$Y = 10^{12} \text{ dynes cm}^{-2}$$
$$\gamma = 10^3 \text{ dynes cm}^{-1}$$
$$l = 10^{-5} \text{ cm}$$

Then,

$$p_{H_2} = \left(\frac{16}{3} \times 10^{20}\right)^{\frac{1}{2}} \text{ dynes cm}^{-2} \quad (11.60)$$

$$\simeq 3 \times 10^4 \text{ atm}$$

Thus, when hydrogen atoms from the metal surface have diffused inside the metal and reached the voids within the metal, they deabsorb from the surface of these as H_2 molecules. If their pressure inside the voids is high enough (3×10^4 atm, as deduced above), the crack will spread: collapse is on the way.

One can now think: What has all this got to do with electrodics, with current density and overpotential? Suppose that H_2 is being evolved on the surface concerned (in aqueous solutions, even with moisture films, metal surfaces are seldom free from adsorbed H), and that the overpotential is η. One might conceive of an equivalent H_2 pressure (*outside* the electrode, now) which would produce *thermodynamically* the same shift in electrode potential as that of the overpotential η. Nernst's equation for a hydrogen electrode in a solution of unit activity of protons but gas pressure P_{H_2} is

$$e_{H_2} = -\frac{RT}{2F} \ln p_{H_2} \qquad (11.61)$$

The H_2 pressure in contact with the outside of the electrode, which would shift its potential as much as the η due to the irreversibility, would be

$$\eta = -\frac{RT}{2F} \ln p_{H_2} \qquad (11.62)$$

or

$$p_{H_2} = e^{-2\eta F/RT} \qquad (11.63)$$

This is the molecular hydrogen pressure outside the electrode, then, to which the electrode's overpotential η is equivalent. It seems, thus, reasonable to say that the η developed on the outside of the electrode could *support* a pressure of p_{H_2} inside a crack, and, if the processes of diffusion of H within the metal and desorption, to form molecular H_2 inside the crack, are in equilibrium,

$$(p_{H_2})_{\text{equivalent}} = e^{-2\eta F/RT} > \left(\frac{16}{3} \frac{\gamma Y}{l}\right)^{\frac{1}{2}} \qquad (11.64)$$

is a first crude condition for embrittlement. Numerical substitution in (11.64) shows that, at η less negative than about 0.1 V at room temperature, no hydrogen embrittlement is possible (of course, assuming the parameters stated above).

It is better to state the result of the above type of theory as a limit of overpotential below which the embrittlement will not occur rather than a statement of the overpotential at which it will commence. This is so because

the argument given is quasi-thermodynamic in nature. The effect of various kinetic hindrances (for example, the diffusion of H atoms through the lattice to the cracks) is neglected. At cathodic overpotentials greater than that of (11.64), embrittlement and destruction of a metal *could* occur. But if the hydrogen solubility or diffusion coefficient is sufficiently small, it will not actually occur in any practical time.

There is clearly a similarity here to the type of information given by Pourbaix diagrams (Section 11.2.7). These devices tell one the conditions under which corrosion is impossible. At what rate it will occur under conditions in which thermodynamics says it is possible has to be investigated kinetically (Section 11.2.9).

How valid is this elementary[†] picture of crack initiation and propagation? Perhaps it can be tested by permeation experiments. One can first make some predictions. The permeation currents are a measure of the concentration of atomic hydrogen in the lattice. When the concentration is below a critical limit, the only sinks for the diffusing hydrogen are interstitial lattice positions; but when the concentration inside the lattice exceeds the critical value, then the propagating cracks will also start consuming the diffusing hydrogen. This means that there will be many hydrogen sinks inside the metal and the switching-on of these sinks should reflects itself in the permeation–time behavior.

Experiment shows that (when the hydrogen concentration is enough to cause embrittlement) the permeation current builds up with time and then, instead of stabilizing to a steady state as it normally does in the absence of a crack-initiation and -propagation process, it *drops down* and only then becomes steady (Fig. 11.59). Thus, the fall in permeation current marks the onset of crack propagation and embrittlement. The permeation–time behavior of an embrittled membrane differs in another fundamental way from membranes which do not suffer cracking. In the case of the latter, the permeation–time transients are reversible; after running one transient, the hydrogen inside the metal can be pumped out by dissolution at the exit side (*cf.* Fig. 11.48), and a second transient can be shown to reproduce the first. In the case of an embrittled membrane, a second transient is entirely different from the first. This is not due to irreproducibility, as was originally thought. The hydrogen-permeation properties of the membrane

---

[†] Apart from the cavalier handling of the shape of the crack, plastic deformation energy and the pV energy of the gas in the crack have been neglected; but both energies are usually much smaller than the energy of the surface and the strain energy which are here taken into account.

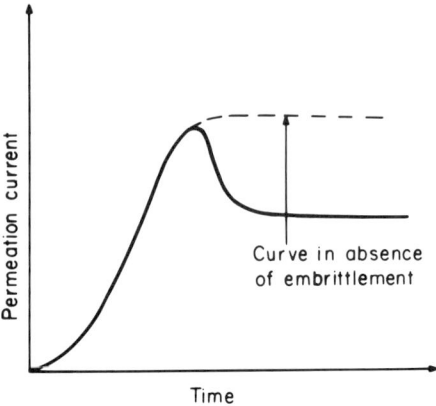

Fig. 11.59. When cracks are initiated and hydrogen starts accumulating inside the cavities, the permeation current falls off.

have been irreversibly changed by embrittlement. More of the hydrogen remains trapped in the metal.

Two modes of the internal disintegration of a metal have been described, namely, stress-corrosion cracking and hydrogen embrittlement. What is the difference between the two mechanisms of decay? Firstly, in stress-corrosion cracking, stress (either applied or residual) is a necessary but not sufficient condition, whereas, in hydrogen embrittlement, an externally applied stress is not a usual part of the phenomenon. Secondly, there is the question of hydrogen. The permeation of hydrogen into the metal is a necessary condition for hydrogen embrittlement. In stress-corrosion cracking, it is the surface crack that is a necessary condition. How this crack is produced is irrelevant. It may be the result of mechanical stresses, or it may arise from the blistering produced by preferential hydrogen entry in regions of stress. Thus, hydrogen may or may not have a role in stress-corrosion cracking. Finally, there is the question of the electrochemical basis of the two modes of the internal decay of metals. The propagation of a stress-corrosion crack in metals is sustained by the electrochemical metal-dissolution reaction, whereas one can conceive of hydrogen embrittlement even though the permeating hydrogen does not arise from an electrochemical hydrogen-evolution reaction at the metal surface. For example, metals embrittle in the presence of hot dry hydrogen. However, under practical conditions, hydrogen is usually introduced into the metal by the electrodic hydrogen-evolution reaction which occurs on the exposed external surface unintentionally as part of a corrosion couple.

11.2.26. Charge Transfer and the Stability of Metals

A bold hypothesis was made earlier to the effect that it is electrochemical phenomena which decide many aspects of the stability of metals. The essential correctness of this view should be clear by now.

The mechanisms of instability are basically simple, and it is appropriate here to develop a perspective. The fundamental cause of instability is that the metal is in contact with an ionically conducting medium. This may be an ionic solution (e.g., saline water) or even condensed moisture from a humid atmosphere.

The juxtaposition of an electronic and an ionic conductor creates all the ingredients of an electrochemical system, or cell—an electron sink for a de-electronation reaction, a path for the flow of electrons to an electron source which supplies the needs of electron acceptors in an electronation reaction, and, finally, an ionic solution for the logistics of charge transfer. Thus, the surface of the metal becomes the scene of two charge-transfer reactions, and its instability originates in the fact that the de-electronation reaction consists of metal dissolution. The electronation reaction, however, depends on the situation, but the common reactions are hydrogen evolution in acid solutions and oxygen reduction in neutral solutions.

The actual rate of corrosion can be derived, for simplified conditions, by a straightforward application of electrodic principles based on two facts: (1) The potential of the metal is uniform over the surface, and (2) the total dissolution current must be equal to the total cathodic electronation current.

In this simplest version of a model for corrosion, the metal dissolves to destruction, layer by layer. In reality, many other effects come in. Even on a single crystal, the different faces dissolve at different rates (etching). In a polycrystalline material, the grain boundaries contain atoms in unorthodox positions, and, hence, they react differently to the corrosive environment. Further, the two oppositely directed electron-transfer reactions may or may not be confined to separate areas of the surface; this only depends on how homogeneous they are. These factors, however, decide how the topography of the surface changes with time, for example, how much it may become rough during corrosion.

Another fundamental possibility is that oxide formation occurs. One mechanism is based on a dissolution of the metal followed by precipitation of an insoluble compound, e.g., a metal hydroxide. If the oxide so formed has the right properties—electronic conductivity, mechanical stability, compactness, and spatial continuity—it may form a protective layer which inhibits further corrosion. This is passivity.

But there are more complex ways in which electrochemical reactions lead to the instability of metals. If hydrogen evolution is the electronation (cathodic) half of corrosion, part of the intermediately produced adsorbed hydrogen dissolves into the metal. Once inside, it prefers to collect in regions of strain where the lattice is distorted. It is preferential distribution of hydrogen in strained regions that causes the internal damage of metals. When the distortion attains void dimensions, the hydrogen atoms combine to produce a gas of hydrogen molecules inside the voids. This pressure buildup acts in different ways depending on the degree of penetration of atomic hydrogen into the metal.

Some metals develop blisters by plastic deformation, and, if these blisters are near the metal surface, they may rupture to leave cracks which thus begin to bring the interior of the metal into contact with the electrolytic environment.† The inside edge of a surface crack is a highly reactive area; metal dissolution can occur there at an abnormally high rate. If the whole metal is under tensile stress, the crack advances by the electrodic dissolution's keeping pace with the continuous creation of the reactive crack edge by the plastic-deformation process (the corresponding electronation process is hydrogen evolution or oxygen reduction on the surface). The metal cracks under conjoint action of the stress and corrosion. This is stress–corrosion cracking.

In other situations, the hydrogen builds up in voids and opens them out into cracks when the hydrogen concentration inside the metal reaches critical values. The metal ends up full of cracks. The hydrogen makes it brittle. The metal fractures easily. This is hydrogen embrittlement.

11.2.27. The Cost of Corrosion

A technological civilization depends upon the stability of machines and structures in diverse industrial, rural, and marine atmospheres. The plain truth, however, is that man has achieved a very limited success in stabilizing metal surfaces largely because the knowledge of the mechanisms concerned is not widespread. One has only to look at the junkyards of industrialized societies to realize the ravages of the electrochemical charge-transfer reactions responsible for corrosion. The loss, in the United States, is estimated at over 5 billion dollars per year, or about 1 cent for every dollar spent. This is a high price to pay for unplanned electron-transfer reactions, especially when they could often be avoided or inhibited.

---

† Moisture films containing ions (e.g., HCO_3^- and H^+ from CO_2 dissolution) may provide this in the absence of bulk solution.

But there is another serious aspect to corrosion. Except for the noble metals, most metals exist in nature in the deelectronated form as ions or ionic compounds. They exist as oxides, sulphides, etc. The reason is simple; thermodynamics recommends the deelectronation (oxidation) of metals, a return to relative disorder. It is only by human enterprise and effort and by the use of energy resources that the pure metals are recovered from their ores. Corrosion and oxide formation are, therefore, expressions of the desire of metals to return to nature. They represent a colossal wastage of resources and human effort.

11.2.28. A Bird's-Eye View of Corrosion

The present civilization depends upon machines and structures which utilize metal parts. But metals in contact with moist oxygen-containing atmospheres are fundamentally unstable. With the exception of the noble metals, all metals display a thermodynamic tendency to revert to the oxides from which most of them came. This reversion would occur by their *chemical* reaction with O_2 or H_2O. But a special electrochemical mechanism obtains to make things worse. The humid atmosphere gives rise to surface moisture films, containing dissolved ions from the CO_2 which air contains, and this makes the environment of such metals conducting, and local electrochemical reactions occur, the essential one being the anodic dissolution of the metal into the moisture film. To complement this anodic dissolution reaction, another reaction (often the reduction of oxygen, sometimes the evolution of hydrogen) occurs on some other sites, which may be quite separated from the sites from which the metal dissolves into the moisture film. When, instead of a moisture film an actual ionic solution, e.g., sea water, is in contact with the metal, the environment may prove even more corrosive. In essence, this is the way corrosion works. The corrosion process represents the electrochemical undevice, the energy waster, and the substance destroyer of Section 8.1.7.

A purely thermodynamic approach reveals much concerning corrosion, but one has to use this approach in a defensive vein. It tells one the situation in which no corrosion can occur. While this is very useful information, thermodynamics is intrinsically incapable of saying anything about the corrosion rate. It is too inadequate; it does not tell one when corrosion could occur in a practical situation at a significant rate. It can mislead. For example, according to thermodynamics, lead should be unstable in oxygen-containing and CO_2-containing water, yet lead is used as piping for holding water.

Corrosion must be interpreted from the standpoint of electrodics, which provides a more informative and practical picture. One can view the corrosion situation by thinking of two currents, both expressed in the Butler–Volmer type of equations, the anodic one literally representing the actual corrosion (the dissolution of the metal) and the cathodic one representing the utilization of the electrons transferred to the metal in the metal dissolution process. Such equations contain, as an unknown, the potential at which the anodic and cathodic currents are equal. This unknown potential, the corrosion potential, sometimes called the mixed or compromise potential, can then be obtained and related to such electrodic parameters as the transfer coefficients and, above all, the exchange-current densities. In fact, with little more algebra, it is possible to get an equation which not only expresses the rate of corrosion of the metal in terms of the exchange-current densities for both the anodic and cathodic reactions but also gives the equilibrium potentials which correspond to the concentrations of ions present in the solution in contact with the corroding interface. The above principles form a basis for handling most of the complications of real corrosion, e.g., the presence of small metallic inclusions in the surface, impurities in the metal, and, above all, the presence of oxide films upon the surface. They permit an electrodic understanding of a large number of everyday situations. For example, they explain why pipe embedded in moist sand will corrode on the part under the sand and not that above it or why, when paint chips off a car, it is the metal under the surrounding paint, *not* that exposed to moist air, which corrodes.

Electrodic principles can also be used to promote an increase of stability. One can try to reduce the exchange-current densities by adsorbing organics upon either cathodic or anodic sites and thus reduce the the exchange-current densities typical of those sites, or one can inject electrons from an outside source to force the mixed potential to more negative potentials where the metal dissolution rate is slowed to zero, i.e., no corrosion occurs.

A most important mechanism for defending the materials of civilization against the ravages of moist air and oxygen-containing solutions is that of passivation. Perhaps, without this mechanism, many uses made of metals before the electrochemistry of corrosion was understood would not have been possible. Passivation is the transformation from an unstable state to a stable one by the presence of a special type of thin oxide film. Special optical measurements show what happens when at least some metals become passive. The metal first dissolves and attains a concentration in the solution near the interface greater than that of the solubility product of the hydroxide or oxide concerned. What then? The latter precipitates and covers the electrodes. This is not enough, however, to cause passivity; some ions can quite easily get through the oxide which is formed if it has ionic conductivity. It is something else which causes the passivation itself; it is the conversion of the oxide film to an electronically conducting film. The potential on the electrode then acts no longer to push *ions* through the film to the solution—corrosion—, but, instead, it pulls *electrons* through the film from some entity in the solution (e.g., from water in oxygen evolution).

The corrosion and its damping down by various artifices or its shutting off by passivation are only part of the battle between man's desire to keep metals stable and the spontaneous tendency of metals to decrepitate, or spoil themselves as useful materials. A still more pervasive danger is that caused by hydrogen. In some metals—and one is iron—it easily diffuses from its atomic form on the metal's surface to the interior. There is a special aspect of that diffusion, one that has only very recently been established. The rate of its permeation depends on the local stress. What does this do? Near points of stress, the hydrogen diffuses particularly well into the metal and reaches small voids which exist anyway inside the metal. Once inside them, it recombines and forms molecular H_2, but now at very great pressure. The cracks spread and meet each other internally, and the metal becomes brittle.

Hydrogen effects are probably as omnipresent as moisture itself. It is even suspected that metal fatigue, a cause of sudden failures under repeatedly applied stress in aircraft may sometimes be hydrogen-caused.

Stress-corrosion cracking has not necessarily to do with hydrogen, although it may be initiated by it. Phenomenologically, it is the spread of cracks in materials which occurs under stress. It need not always be electrochemical in mechanism, but, in metals, it often is. The mechanism is simple. A cathodic current on the surface (oxygen ionization) removes electrons from the metal which originate from deelectronation reactions at the head of the crack, i.e., at the head, metal anodically dissolves and ionizes away. The crack is able to provide fresh atoms when under stress (it actually *slips* under stress and new crystal planes in the interior are exposed), and these new atoms turn out to have a particularly fast dissolution rate. The unseen electrochemical spread of cracks inside metals under stress is one of the more dangerous aspects of the very costly subject of corrosion. Indeed, in this decade, the cost of corrosion is about one unnecessary cent off each dollar.

Further Reading

1. C. Wagner and W. Trand, *Z. Elektrochem.*, **44**: 391 (1938).
2. A. N. Frumkin and A. Aladjova, *Acta Physicochim. U.R.S.S.* **19**: 1 (1944).
3. H. Uhlig, ed., *Corrosion Handbook*, John Wiley and Sons, Inc., New York, 1949.

4. K. J. Vetter, *Z. Elektrochem.*, **62**: 642 (1958).
5. J. M. Kolotyrkin, *Z. Elektrochem.*, **62**: 664 (1958).
6. M. Smialowski, *Hydrogen in Steel*, Pergamon Press, New York, 1962.
7. Ulick R. Evans, *An Introduction to Metallic Corrosion*, Edward Arnold [Publishers], Ltd., London, 1963.
8. H. R. Copson, in: F. L. LaQue and H. R. Copson, eds., *Corrosion Resistance of Metals and Alloys*, Reinhold Publishing Corp., New York, 1963.
9. M. Sato and G. Okamoto, *J. Electrochem. Soc.*, **110**: 605 (1963).
10. J. M. West, *Electrodeposition and Corrosion Processes*, D. Van Nostrand Co., Inc., New York, 1965.
11. G. T. Bakhvalov and A. V. Turkovskaya, *Corrosion and Protection of Metals*, Pergamon Press, New York, 1965.
12. J. C. Scully, *The Fundamentals of Corrosion*, Pergamon Press, New York, 1966.
13. N. D. Tomashov, *Theory of Corrosion and Protection of Metals*, The Macmillan Company, New York, 1966.
14. Hugh L. Logan, *The Stress Corrosion of Metals*, John Wiley and Sons, Inc., New York, 1966.
15. N. D. Tomashov and E. N. Mirolyubov, eds., *Corrosion of Metals and Alloys*, Israeli Program for Scientific Translations, Jerusalem, 1966.
16. A. K. N. Reddy and B. Rao, *J. Electrochem. Soc.*, **113**: 1133 (1966).
17. W. Beck, J. McBreen, and L. Nanis, *Proc. Roy. Soc.*, **290**: 220 (1966).
18. N. D. Tomashov and G. P. Chernova, *Passivity and Protection of Metals Against Corrosion*, Plenum Press, New York, 1967.
19. T. P. Hoar, *Corrosion Science*, **7**: 341 (1967).
20. M. G. Fontana and N. D. Greene, *Corrosion Engineering*, McGraw-Hill Book Company, New York, 1967.
21. A. Despic and R. Raicev, *J. Chem. Phys.*, **49**: 926 (1968).

11.3. ELECTROCHEMICAL ENERGY CONVERSION

11.3.1. The Present Situation in Energy Consumption

About 85% (Fig. 11.60) of the energy consumed today by man comes from the thermal combustion of coal, oil, and natural gas. These hydrocarbon fossil fuels are limited and a once-only gift[†] from natural processes

---

[†]Were man to obtain energy predominantly from other sources, e.g., solar and gravitational energy and fissionable materials, the remaining reserves of oil and natural gas could be used as the raw materials for the manufacture of synthetic food stuffs and textiles. Used in this way and not burned to provide mechanical energy, they would last for centuries even if used extensively to feed the population in those parts of the world in which the application of antiseptics in prenatal care has become out of phase with the acceptance of contraceptives (Fig. 11.61).

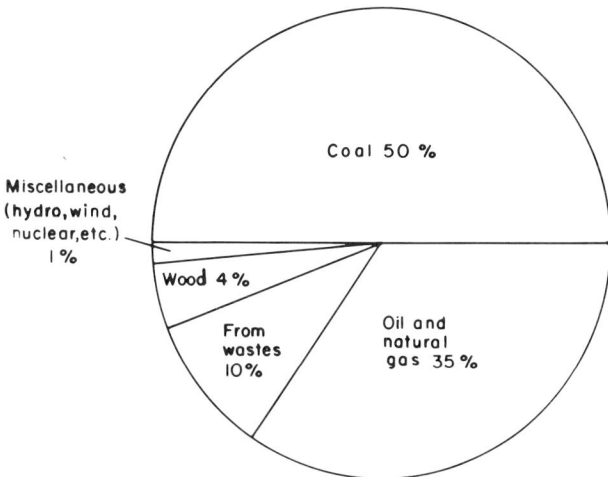

Fig. 11.60. A schematic representation of the total world energy among various sources.

which occurred many millions of years ago. This fact constitutes a starting point for thinking about the world energy situation.

Several conclusions follow clearly. Firstly, what is being consumed is energy *capital*, not energy *income*, and, therefore, the day must come when the reserves of fossil fuels are exhausted, particularly the easily transportable and usable oil and natural gas. Secondly, it is important that, in energy conversion, the efficiency of the process of the thermal combustion of these fossil fuels be critically examined. Is the present limited store being squandered? Or is it being used with the maximum efficiency obtainable for the conversion of chemical energy to mechanical work?

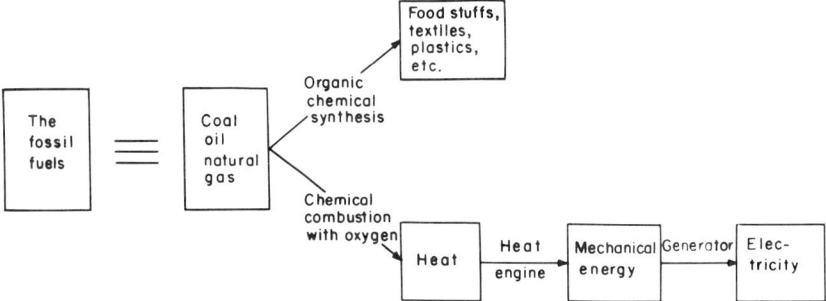

Fig. 11.61. A schematic representation of two possibilities in the utilization of the remaining fossil fuel deposits.

11.3.2. How Are the Hydrocarbon Fuels Used at Present?

The energy produced when fossil fuels undergo chemical reaction with oxygen in burning is at first in the form of heat, which is converted in an engine to mechanical energy. The conversion of mechanical to electrical energy is a process in which the efficiency of conversion is nearly unity. Hence, it is the conversion of the heat energy given out in the chemical reaction of combustion to mechanical energy or work which will be scrutinized.

The essence of a heat engine is a piston or turbine rotor which is forced to move by the expansion of a gas, which occurs as a consequence of a rise of temperature due to heat that is given out in the chemical reaction between the fuel and oxygen. The combustion process can take place either outside the engine or inside the engine. In the former case, i.e., in an external-combustion engine, the fuel is burned outside the engine and the heat generated is transferred to the so-called "working substance," which is then conducted into the cylinder or turbine and allowed to expand, which produces mechanical work. An example of an external-combustion engine is the steam engine, the essential instrument by which man made machines do work for him, beginning in the eighteenth century. In an internal-combustion engine, an appropriate mixture of air and the fuel (e.g., the hydrocarbon mixture gasoline) is either ignited inside the cylinder by an electric spark or compression, or continuously burned inside a gas turbine. The expansion of the gas therein due to the heat given out in the reaction of hydrocarbon and oxygen to carbon dioxide causes the piston or the rotor to move. The internal-combustion engine has been the basis of the transport revolution of the twentieth century.

At present, electricity is generated largely by utilizing mechanical energy from the external and sometimes the internal combustion of fossil fuels[†] to drive generators.

The feature common to both the external- and internal-combustion engines is that the chemical energy contained in the fuel is first converted into *heat* and then this heat energy is used to produce mechanical power through the force associated with the expansion of a heated gas. Heat is essentially an intermediate form of energy in the conversion to mechanical work of the difference between the energy of the reactants and that of products in chemical reactions and thence its conversion by the generator to electricity.

[†] Oil and natural gas are preferred to coal as fossil fuel sources because of the ease of transporting and burning them.

11.3.3. The Pollution of the Atmosphere with Products from Internal-Combustion Reactions and Its Possible Effect on World Temperature and Sea Levels

Apart from questions of rationality raised by a continued burning-up of limited oil and natural gas supplies to give man the energy which runs his machines (instead of using energy from continuous or quasi-inexhaustible sources), some direct negative results arise from the present predominant form of energy conversion.

11.3.3a. Products of Combustion Other than Carbon Dioxide. Gasoline and diesel oil consumed in present internal combustion engines do not undergo 100% conversion to CO_2. The organic compounds remaining incompletely burned are several dozens in number and include particularly unsaturated compounds. Carbon oxides, nitrogen oxides, sulphur oxides, and a lead-containing compound [from the $Pb(C_2H_5)_4$ added to fuels to break up chains in the combustion reactions] are also present in significant amounts.

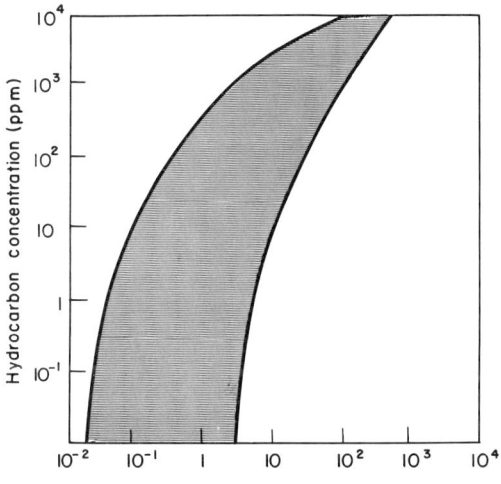

Fig. 11.62. Environment favorable for the production of ozone as a function of the concentrations of hydrocarbons and nitrogen dioxide. The shaded area represents the zone in which ozone is produced. Concentrations of ozone are quite low, 1 ppm, but enough to cause trouble in many ways (cf. Fig. 11.63). The figure shows that attempts to decrease the amounts of hydrocarbons alone may increase the formation of the harmful ozone.

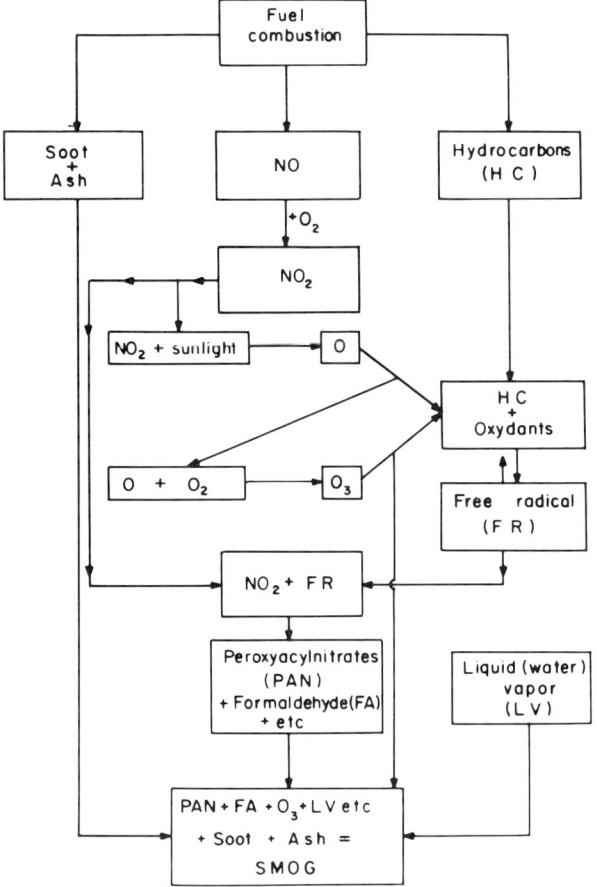

Fig. 11.63. A schematic representation of the formation of smog.

Little is known about the present or the possible future effect of these substances. However, it is known that certain of the organic compounds undergo a photochemical reaction of oxides of nitrogen to produce a complex addition compound which is the origin of smog (Figs. 11.62 and 11.63). There is evidence which suggests that some of the compounds from automobile exhausts are carcinogenic.

Conditions favoring the production of smog would, then, be sunlight and a large concentration of the exhaust gases (nitrogen oxides and the unsaturates) from automobiles together with a geographic situation where there are few winds to sweep the mixture away and hence dilute it. Such circumstances prevail in Los Angeles, California, and it is this city which is well known for the frequent existence there of the smoggy state. A sufficiently high concentration of the necessary mixture of oxides of nitrogen and unsaturates may occasionally occur in any city over which an *inversion layer* occurs, i.e., a layer of air over the city which is warmer than the exhaust gases so that the latter do not tend to rise and dissipate themselves, as they should if they are relatively warm.

The deleterious consequences of the smoggy state are obvious. Other dangers arising from increasing concentrations of products associated with combustion of fuels into the open (e.g., emphysema) are receiving research attention. However, the situation expected in the next two or three decades may be estimated by taking into account the expected increase in the number of cars powered by internal combustion engines. Frequent smog conditions would be expected to threaten many urban areas within the mentioned time period should no effective modification in internal-combustion engines be made which would cause the reaction of combustion to convert more completely to CO_2. Many such modifications have been suggested. In addition to the unknown degree of success which developments may bring, the question arises as to what extra costs or reduced performance may be incurred by modifications which would avoid the drift toward smogginess in all urban areas.

11.3.3*b*. *Carbon Dioxide.* This gas, an inevitable consequence of the method of obtaining mechanical energy by burning coal and oil, absorbs radiant energy in the infrared. The temperature of the earth's atmosphere

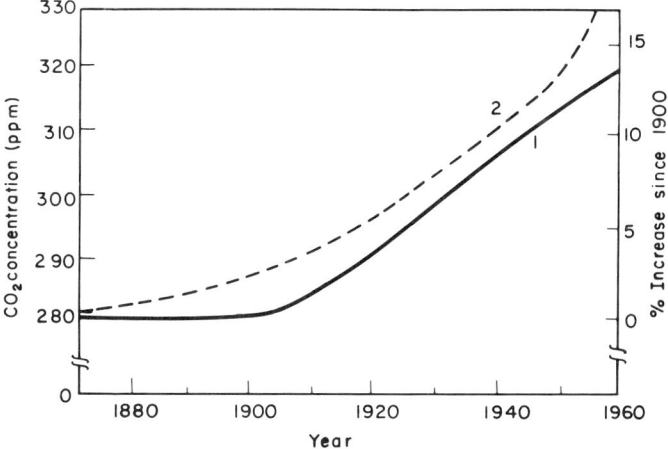

Fig. 11.64. Plots of cumulative amounts of carbon dioxide in the atmosphere *versus* time. Curves 1 and 2 represent, respectively, the observed and calculated amounts added by burning fossil fuel.

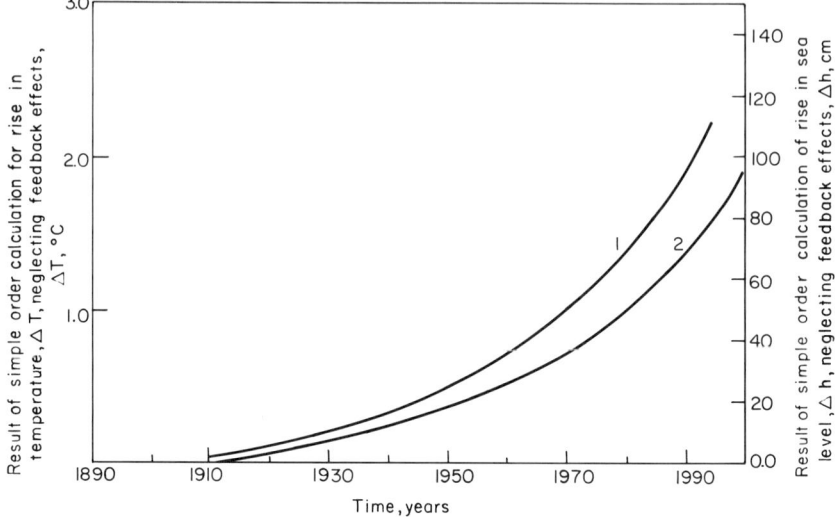

Fig. 11.65. Plots of increase in atmospheric temperature and the consequent rise in sea level *versus* time. Curve 1 represents the rise in atmospheric temperature, and curve 2, the rise in sea level as a result of simple order calculations, feedback effects being neglected.

is the net consequence of a number of influences which include the amount of solar radiation reflected back to space. The greater the concentration of CO_2 in the atmosphere, the less of this reflected energy escapes, i.e., the more is stored in the earth's atmosphere (Fig. 11.64). The energy thus absorbed degrades to heat. All other things being equal, the average temperature of the atmosphere is expected to increase, and there is some tentative evidence which suggests that, so long as one neglects short-time fluctuations, it is doing so. If the present absence of a balance between the production of CO_2 by metabolism and its consumption by vegetation (caused by combustion of fossil fuels) is maintained, the ice at the poles should be increasingly converted to water. As a consequence, world sea levels would be expected to rise[†] (Fig. 11.65).

---

† It is possible to trace through these ideas to an estimation of the rise in sea level which would be expected to result, say, by the year 2000, if the present curve of the concentration of CO_2 in the atmosphere as a function of time is extrapolated to this date. The calculation suggests (Fig. 11.65) that the sea-level rise for continued injection of CO_2 till the time stated would be in the region of 1 m (Haynes). However, such a calculation involves assumptions which, although plausible, tend to give the upper limit of the effect. The calculation neglects complex feedback effects, e.g., the results of increased cloud level, and the fact that, with increasing combustion, an increasing

11.3.3c. Uncertainties in Predicting the Future Pollution of the Atmosphere. An assessment of the problem of the time dependence of the growth of atmospheric pollution is difficult because of a number of gaps existing in the necessary data, e.g., the lifetime of the unsaturated compounds produced in internal-combustion engines and their distribution in the atmosphere. Changes in the design of internal-combustion engines will give a reduction in the amount of unsaturates emitted, but to what extent, at what increase in cost and decrease in engine performance, and how much these factors will offset the counter tendencies (e.g., the expansion of the car-using population and the industrialization of populations at present in a primitive stage of development) is not at present accurately estimable. However, whatever the consequences predicted for the *increase in carbon dioxide* concentration resulting from burning hydrocarbons, they would seem difficult to avoid.†

Although, here, mainly automobile exhaust products are considered (for they are the largest single cause of pollution by unsaturates), analogous statements apply to stationary industrial machines using combustion of fossil fuels to produce energy, in particular, electricity.

Enough has been said to show that increasing difficulties are becoming associated with the method of obtaining energy by the burning-up of the fossil fuel supply and that the time constant for the advent of widespread and serious negative consequences of continued use of this *thermal* approach for energy production seems likely to be some two or three decades.

11.3.4. Thermal-Combustion Engines Waste the Chemical Energy Available from Burning Hydrocarbons in Air

Let it be supposed that combustion engines can be modified in design so that they do not pollute the atmosphere and even the deleterious effects of the absorption of reflected solar energy by CO_2 can somehow be significantly delayed so that internal combustion could continue to be used till

---

concentration of light-obscuring solid particles obtains. Nevertheless, such complexities *can* be taken into account and computer programs devised which will give more significant predictions. Congressional committees have begun to exhibit concern about the matter. In 1966, John Malone ended his testimony before the Doddario Committee thus: "The degree of danger which exists from the warming of the earth is something we must resolve in a matter of decades. The situation could become serious by the end of the century."

† It has been suggested that, if metallic particles appropriately shaped were injected at a high level in the atmosphere, these could reflect solar energy significantly to diminish the "greenhouse effect" caused by continued combustion of the hydrocarbons to obtain mechanical energy and electricity.

the end of the present century. The question remains: With what efficiency is the chemical energy released in the combination of hydrocarbons with oxygen that is being converted to mechanical work in the present indirect process of obtaining mechanical and electrical energy?

This question was answered in principle over a century ago when Carnot showed that all engines which convert heat to mechanical work operate by transferring heat from a source at a temperature T_1 to a sink at a lower temperature T_2 and that the efficiency ε of such an engine is given by (if temperatures are taken in degrees Kelvin)

$$\varepsilon = \frac{T_1 - T_2}{T_1} \qquad (11.65)$$

Since $T_1 - T_2 < T_1$, then $\varepsilon < 1$, i.e., the efficiency is less than 100% (of the change in enthalpy in the reaction).

Hence, for given source and sink temperatures, the maximum possible efficiency is prescribed by what is called the *Carnot limitation* on the conversion of heat into mechanical work. This limitation is *intrinsic*; it cannot be avoided by improvement of the engine design. The physical basis of this Carnot limitation will be discussed in Section 11.3.9. Thus, a steam engine working between 356 and 100°C has a maximum efficiency of 41%.

In a real engine with moving parts, nonideal materials of construction, etc., extrinsic efficiency losses come in and reduce the efficiency to values below the theoretical maximum efficiency for a heat engine as derived by Carnot. Thus, most mobile combustion engines have in practice percentage efficiencies of 10 to 20%.

The upshot of this discussion on the intrinsic limitations of the heat-engine method of obtaining mechanical energy from chemical reactions is that some 60 to 90% of the energy contained in the reaction of the fossil fuels to oxygen is being wasted, i.e., lost in heat and not converted to mechanical work. The present method of providing available mechanical and electrical energy thus not only uses up the limited store of hydrocarbons (a potential source of food and textiles, which would last for centuries) but wastes in doing so some two-thirds of the energy of the thermal-combustion reactions concerned. At the same time, it pollutes the atmosphere and may even within decades cause a significant reduction in land area throughout the world.

11.3.5. Direct Energy Conversion

Apart from the fact that the usual method of producing electricity— the most convenient form in which energy can be made available—is subject

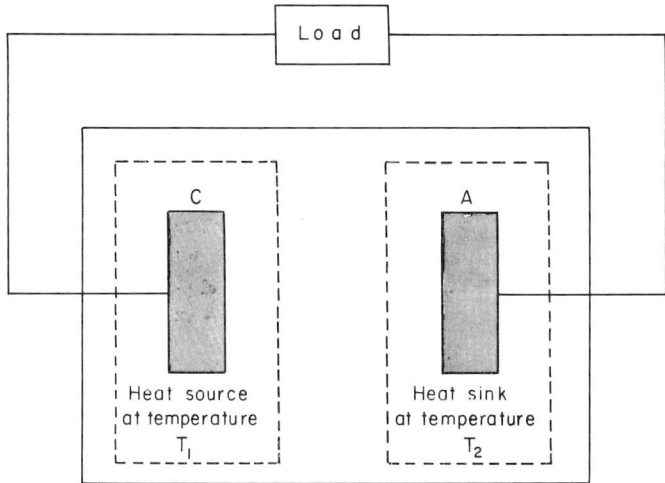

Fig. 11.66. A schematic representation of a thermionic energy converter where C is the emitter cathode and A is the collector anode.

to the intrinsic Carnot limitation and has the other disadvantages described, it is clear that it is an *indirect* method. A chemical reaction produces heat, which causes a gas to expand and thus push forward a part of a machine. This machine then drives another which produces the electricity. It would be preferable to attempt to devise a means of converting directly to electricity the energy released during most chemical reactions without involving a second machine and also preferably without moving parts for these are subject to erosion and mechanical failure.

Several such methods, called *direct energy-conversion methods*, have been known, some since the nineteenth century. In the thermionic converter (Fig. 11.66), an electronic conductor is heated till it emits electrons; these electrons are taken off by a counterelectrode, and a current is made to flow through an external circuit. In thermoelectric devices (Fig. 11.67), two

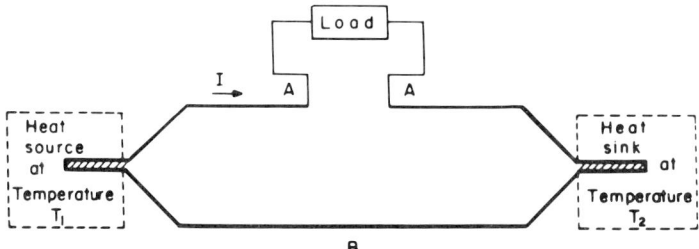

Fig. 11.67. A simple thermoelectric energy converter.

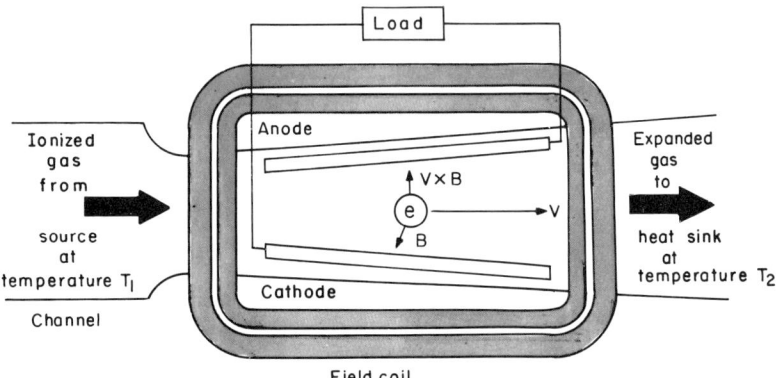

Fig. 11.68. The elements of a magnetohydrodynamic engine. Here, V represents the flow velocity of the ionized gas; B, the strength of the magnetic field (at right angles to the plane of the paper); and $V \times B$, the strength of the induced electric field.

differing materials A and B (metals or semiconductors) are made to form two junctions, each consisting of an A to B contact; one junction is maintained hot and the other cold, and electricity flows between them through a load. In magnetohydrodynamic converters (Fig. 11.68), a hot plasma is circulated past the poles of a magnet. A fuel is used to heat the gas and ionize it. While passing between the magnets, the ions of opposite sign are attracted to the respective magnetic poles and produce a current which can then be led through a load.

In each of these methods, heat is *directly* converted to electricity without an intermediate stage of mechanical work or a machine with moving parts. Compared with the conventional method, the present methods represent an improvement and give rise to much modern interest. However, rather than direct energy-conversion methods, a better term might be *direct heat-to-electricity conversion* for these methods still share one very disadvantageous property with the conventional method; energy in the form of heat is put in at a high temperature and comes out of the converter at a lower temperature. They are therefore still heat engines subject to the Carnot efficiency limitation, i.e., only a fraction of the energy of the chemical reaction can be turned into electricity however well the devices are engineered.

What is needed is a method in which the energy released in chemical reactions is indeed converted *directly* to electricity without moving parts, as with the several methods outlined above, but *also* in a way which *avoids the loss of energy due to the intrinsic Carnot limitation.*

SOME ELECTROCHEMICAL SYSTEMS OF TECHNOLOGICAL INTEREST   1361

11.3.6. Direct Energy Conversion by Electrochemical Means

It will be recalled that, of the three types of electrochemical systems that have been described, the energy producer is a spontaneously working, self-driving electrochemical system (Fig. 11.69). In it, there is the spontaneous occurrence of a de-electronation reaction at the electron-sink electrode and an electronation reaction at the electron-source electrode. If an external load is connected to the two electrodes, a current of electrons flows in the external circuit. Thus, the electrodic reactions at the two electrodes bring about the conversion of the difference in the chemical energy (Gibbs free energy) of the reactants and products directly into a flow of electricity (and thereafter, by a motor, to mechanical energy). There is no intermediate step in which the energy has to bring itself to power by expansion of a gas converting thereby only part of its thermal energy to mechanical work (Section 11.3.9).

11.3.7. The Maximum Intrinsic Efficiency in Electrochemical Conversion of the Energy of a Chemical Reaction to Electric Energy

To obtain this, let an essential thermodynamic equation be recapitulated (*cf.* Section 7.3.4). It is

$$-\Delta G = W_{\text{rev}} - P \Delta V \qquad (11.66)$$

It means the following: The change in free energy in a reaction is equal to the *total reversible work* obtainable from the reaction (this work

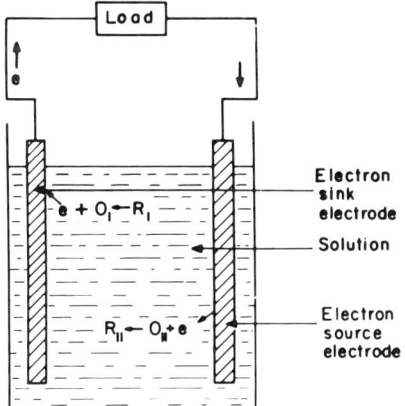

Fig. 11.69. Schematic diagram of an electrochemical energy producer.

to include *all* kinds of work, i.e., gravitational, electrical, surface, etc., and also the work of expansion) diminished by the work of expansion, $P \Delta V$. Hence,

$$-\Delta G = W'_{\text{rev}} \qquad (11.67)$$

where W'_{rev} is all the work obtainable from the reaction exclusive of any work which can be obtained from a possible volume change in the system.

Now, if one carries out a chemical reaction in an electrochemical way, then, in the example of the reaction $2H_2 + O_2 \to 2H_2O$, there will be two partial reactions

$$2H_2 \to 4H^+ + 4e \qquad (11.68)$$

and

$$O_2 + 4H^+ + 4e \to 2H_2O \qquad (11.69)$$

After each has been carried through with the stoichiometric quantities (and simultaneously), one has in fact carried out the overall reaction $2H_2 + O_2 \to 2H_2O$ and, hence, the normal change of free energy associated with this reaction at a given temperature and pressure has occurred.

But something else has occurred, too, namely, the transport of four electrical charges across a total potential difference of V, the cell potential. As thermodynamics calculations are based on the assumption that the chemical change has been carried out near equilibrium, i.e., in a reversible way, the cell potential V to which one here refers is the thermodynamic equilibrium potential $V = V_e$. It is that obtained on an electronic voltmeter with the velocity of the electrode reactions at "infinitely slow" in conformity with the conditions of thermodynamic reversibility. Now, the electrical work of transporting such charges (four electrons per two molecules of water formed, or 4 Faradays, if molar quantities are considered) is the total charge transported multiplied by the potential difference through which it passes, i.e., $4FV_e$. Thus, the general expression for the change of free energy in one act of an electrochemical reaction, in which the number of electrons transported externally for each act of the equivalent electrode reactions is n, is

$$nFV_e \qquad (11.70)$$

Compare this *electrical* work carried out in the reaction, now, with W'_{rev}, the total work obtainable from the reaction excluding volume-change work. What other kind of work than electrical work is obtainable from the reaction $2H_2 + O_2 \to 2H_2O$ *carried out in this electrochemical way?* There is no surface work and no gravitational work. Carried out in this electro-

TABLE 11.10

Theoretical Cell Potentials of Various Oxidation Reactions at 25°C

Fuel	Reaction	ΔG^0, kcal mole^{-1}	V_e^0, V
Hydrogen	$H_2 + \frac{1}{2}O_2 \rightarrow H_2O$	−56.69	1.229
	$H_2 + Cl_2 \rightarrow 2HCl$	−62.70	1.370
Propane	$C_3H_8 + 5O_2 \rightarrow 3CO_2 + 4H_2O$	−503.90	1.093
Methane	$CH_4 + 2O_2 \rightarrow CO_2 + 2H_2O$	−195.50	1.060
Carbon monoxide	$CO + \frac{1}{2}O_2 \rightarrow CO_2$	−61.45	1.333
Ammonia	$NH_3 + \frac{3}{4}O_2 \rightarrow \frac{3}{2}H_2O + \frac{1}{2}N_2$	−80.8	1.170
Methanol	$CH_3OH + \frac{3}{2}O_2 \rightarrow CO_2 + 2H_2O$	−168.95	1.222
Foraldehyde	$CH_2O + O_2 \rightarrow CO_2 + H_2O$	−124.7	1.350
Formic acid	$HCOOH + \frac{1}{2}O_2 \rightarrow CO_2 + H_2O$	−68.2	1.480
Hydrazine	$N_2H_4 + O_2 \rightarrow N_2 + 2H_2O$	−143.9	1.560
Zinc	$Zn + \frac{1}{2}O_2 \rightarrow ZnO$	−76.05	1.650
Sodium	$Na + \frac{1}{2}H_2O + \frac{1}{4}O_2 \rightarrow NaOH$	−71.84	3.120
Carbon	$C + O_2 \rightarrow CO_2$	−94.26	1.020

chemical way and in an ideal manner (i.e., infinitely slowly so that the potential differences in the cell are those characteristic of equilibrium),

$$W'_{\text{rev}} = nFV_e \tag{11.71}$$

Hence, from Eq. (11.67),

$$-\Delta G = nFV_e \tag{11.72}$$

In Table 11.10 are listed the Gibbs free-energy change and the corresponding equilibrium-potential differences for the reactions of the oxidation of some currently used and potential fuels.

It is in this sense that it is said that, in an electrochemical energy converter, the ideal maximum efficiency is 100% for, as in the above idealized situation, if one could carry out reactions in a such way that the electrode potentials were infinitely near the equilibrium values, the electrical energy one could draw[†] from the reaction would be nFV_e and this is all of the free-energy change ΔG, which is the maximum amount of useful work one can obtain from a chemical reaction.

[†] One would draw it out by making it drive a current through an external load, e.g., the armature of an electric motor, which ideally would convert the electrical energy to mechanical energy at 100% efficiency and in fact does carry out such a conversion at more than 95% efficiency.

What has been shown, then, is that the intrinsic maximum efficiency of the electrochemical converter working under ideal conditions is 100% of the ΔG, which is the *useful* or intrinsically *available* work of a chemical reaction. This is an encouraging fundamental result for electrochemical energy conversion. It shows a clear and unique advantage over the efficiency of classical thermal energy converters. It shows also an intrinsic advantage when comparison is made with other, newer direct energy converters because most of these, such as the thermoelectric device in which heat is taken in at one temperature and rejected at a lower temperature are also subject to the debilitating Carnot efficiency limitation. Herein lies, then, the unique and attractive potentiality of the electrochemical method of the conversion of the energy of chemical reactions to energy in the form of electricity.

But, in making a numerical comparison with other types of converters, there is something which has been done here which is not fair. A comparison has been made with the *available* energy in a reaction, ΔG, and it has been shown that the electrochemical method of energy conversion could intrinsically convert to electricity all the energy which is intrinsically available as a result of a chemical reaction's occurring (independently of the method of conversion). Not quite all the energy difference between the reactants and products of an electrochemical reaction can be made available, however, even by the electrochemical method because some of it is wasted in very fundamental processes connected with the ordering and disordering, i.e., the entropy losses and gains, which also occur in chemical reactions. It is the enthalpy change (or change in heat content) ΔH which is equivalent to the total change in energy between the reactants and products of a reaction, *including* the energy lost in entropy increases. It is a more significant standard of comparison, therefore, to base the efficiency of any energy-conversion method on a comparison of how much energy it gives compared with heat-content (enthalpy) change, ΔH, in a reaction because ΔH is the total energy difference between the products and reactants of a reaction. The ΔH is usually larger in magnitude than ΔG, often by 10 to 20% (see Table 11.11). Hence, a second and better expression [see also Eq. (11.72)] for the intrinsic maximum efficiency of an ideal electrochemical converter is[†]

$$\varepsilon_{max} = \frac{\Delta G}{\Delta H} = -\frac{nFV_e}{\Delta H} \qquad (11.73)$$

[†] Of course, the electrical energy nFV_e must be converted from its units of joules to the heat units of calories through the equation joule = 4.18 cal.

TABLE 11.11

Standard Free Energy, Enthalpy, and Maximum (Intrinsic) Efficiency for Some Possible Fuel-Cell Reactions

Reaction	$-\Delta G^0$, kcal	$-\Delta H^0$, kcal	V_e, V	Efficiency, %, $(\varepsilon)_{max} = \dfrac{-nFV_e}{\Delta H} \times 100$
$H_2 + \frac{1}{2}O_2 \to H_2O$	56.69	68.32	1.229	83
$CH_4 + 2O_2 \to CO_2 + 2H_2O$	195.50	212.80	1.060	92
$C_3H_8 + 5O_2 \to 3CO_2 + 4H_2O$	503.90	530.61	1.093	95
$C_{10}H_{22} + \frac{1}{2}O_2 \to 5CO_2 + 11H_2O$	1574.42	1632.33	1.102	97
$CH_3OH + \frac{3}{2}O_2 \to CO_2 + 2H_2O$	168.95	182.61	1.222	93
$C + \frac{1}{2}O_2 \to CO$	32.81	26.42	0.712	124
$CO + \frac{1}{2}O_2 \to CO_2$	61.45	67.63	1.333	91
$H_2 + Br_2 \to 2HBr$	24.57	28.90	1.066	85

Thus, it can be seen from Eq. (11.73) that there is no general single number (e.g., 100%) which one can give for the maximum intrinsic efficiency of an electrochemical energy converter on a heat-content basis. Examples of values for typical overall reactions which are or might be used in fuel cells are given in Table 11.11. The values can depend on whether the cell reaction is carried out in an acid or alkaline electrolyte since, in the latter case, the reaction product is a carbonate with a somewhat different standard free energy than CO_2. Since the activity of the reactants and products depends on the concentration of the electrolyte, so does the potential of the cell reactions and the efficiency of conversion. Still, one may say that the maximum intrinsic efficiency for electrochemical energy conversion on a heat-content comparison basis is in the region of 90%[†] com-

---

[†] Note that, in the hypothetical converter in which one would realize the reaction $C + \frac{1}{2}O_2 \to CO$, the efficiency of conversion on a heat-content basis could be *greater* than 100%. Thus, more electrical energy would be obtained from the system (worked in an ideal reversible way) than is the difference in the heat content of the products and reactants. This is because the entropy change is positive in the reaction quoted, i.e., the disorder of the product, 1 mole of gas, is greater than the disorder of the reactants, ½ mole of gas. The cell would tend to cool upon working, and heat energy would be absorbed from the surroundings and converted to electricity if it were arranged for the converter to continue to work isothermally.

pared with heat engines which, when operating in currently tolerable temperature intervals, were shown to have a maximum intrinsic efficiency of 20 to 40%.

11.3.8. The Actual Efficiency of an Electrochemical Energy Converter

It has been shown (Section 11.3.7), however, that, in an electrochemical energy converter, the maximum cell potential is the value V_e obtainable when the reaction in the cell is electrically balanced out to equilibrium, i.e., when no current is being drawn from the cell. *As soon as the cell drives a current through the external circuit, the cell potential falls from the equilibrium value V_e to V.* The value of the actual potential V at which the cell works when delivering a current i is always *less* than the equilibrium potential V_e (*cf.* Section 9.6.6 and Fig. 9.56). Hence, one has from Eq. (11.73)

$$\varepsilon_0 = -\frac{nFV_e}{\Delta H}\frac{V}{V_e} \quad (11.74)$$

or

$$\varepsilon_0 = \varepsilon_{\max}\varepsilon_p \quad (11.75)$$

where $\varepsilon_{\max}$ is the maximum efficiency given by Eq. (11.73), and ε_p is known as the *voltage efficiency* given by

$$\varepsilon_p = \frac{V}{V_e} \quad (11.76)$$

Of course, this picture is true only if the reactants are completely converted to final reaction products, i.e., if the overall reaction is fully accomplished and none of the electrons take part in some alternative reaction. To allow for the possibility that such a wastage does occur, we must consider the current or faradic efficiency ε_f to take into account the incomplete conversion of reactants into products. The overall efficiency will be

$$\varepsilon_0 = (\varepsilon_{\max}\varepsilon_p)\varepsilon_f \quad (11.77)$$

In many reactions of interest, ε_f is virtually unity.

11.3.9. The Physical Interpretation of the Absence of the Carnot Efficiency Factor in Electrochemical Energy Conversion

At first thought, one might conclude that there is no need for an explanation of why electrochemical energy converters differ both from classical indirect energy converters and also from other direct energy-conversion devices in lacking the intrinsic limitation in the conversion of the energy of a chemical reaction as a consequence of Carnot's theorem. Those with Carnot terms are all *heat engines*. The fuel cells, on the other hand, do not use the heat given out in a reaction but, rather, separate the reactants (which do not collide and react and expand a gas to do $p\,dV$ work) and make them yield equivalent electric charges, which then create magnetic fields which turn the armatures of motors, etc.

However, there is something so important and fundamental in this different way of bringing about the conversion to work of the energy released in a chemical reaction that it is at this moment worthwhile following through a more detailed inquiry, on a molecular level, concerning the difference in intrinsic-conversion-efficiency maxima between thermal and electrochemical reactions.

Thus, in the thermal reaction, the *mechanism* of energy conversion may be visualized on a molecular scale in the way that the change in energy between products and reactants is released in the form of the heat or kinetic energy of the constituent gases. These then impact upon some mechanical and movable object (e.g., the piston in an internal-combustion engine), but, in these collisions, *there is a series of glancing angles of incidence of the molecules on the cylinder and the transfer of momentum to the piston from the* (hot) *molecules is not complete*. The degree of its completeness is measured by the change in the kinetic energy of the particles after they have struck the piston (Fig. 11.70). The temperature of the gases initially—i.e., before striking the piston—is T_1 or the kinetic energy per particle, $\tfrac{3}{2}kT_1$. Afterwards, it is T_2 or the kinetic energy per particle, $\tfrac{3}{2}kT_2$. The point is that only a part of the kinetic energy, $\tfrac{3}{2}kT_1 - \tfrac{3}{2}kT_2$, or heat energy was transferred to the piston; the rest of the energy escaped, i.e., remained as heat in the gas that was rejected to the heat sink (the products of the reaction are emitted from the reaction chamber at a lower temperature than the maximum but still containing much of the reaction's energy, i.e., still *hot*). The efficiency of the conversion process is given by

$$\varepsilon = \frac{\tfrac{3}{2}kT_1 - \tfrac{3}{2}kT_2}{\tfrac{3}{2}kT_1} = \frac{Q_1 - Q_2}{Q_1} = \frac{W}{Q_1} = \frac{T_1 - T_2}{T_1} \qquad (11.78)$$

In the electrochemical converter, there are no collisions between the particles (e.g., H_2 and O_2) which react to form, e.g., H_2O. They simply undergo electron-supplying (for the H_2) and electron-accepting (for the O_2) reactions on the electrodes. These electrons travel round an external circuit and, while doing so, pass through a *load* (e.g., the armature of a

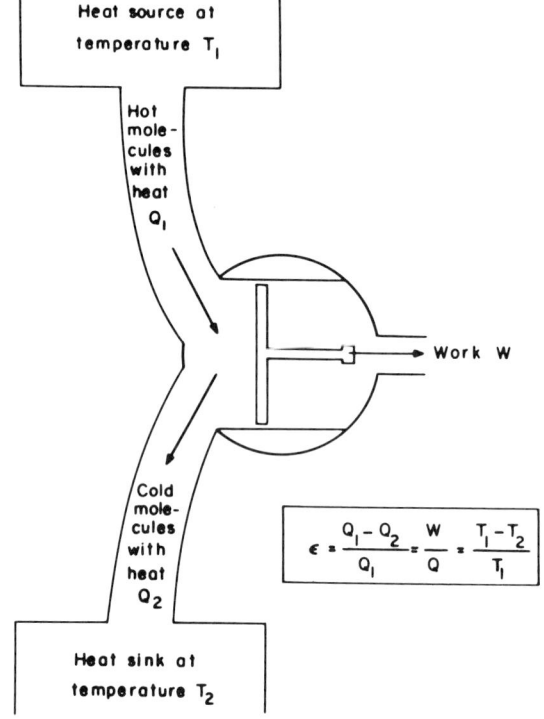

Fig. 11.70. Representation of the efficiency of a heat engine.

motor) and thereby do work. They complete the reaction and allow the product water to be formed. Hence, there is no analog of any process of incomplete transfer of the energy difference between the two sides of a reaction to the outside of the convert, as there is in thermal conversion. The electrons pass through the *entire* potential difference generated by the cell reaction and not through a part of it only.

It may, then, be asked why the electrochemical converter working under ideal conditions does not convert into electric energy *all* the energy released in a chemical reaction, ΔH, but only the free-energy change in the reaction, ΔF. This is of course because the reaction taking place is still the same overall chemical reaction as in the thermal case, i.e., it is still $2H_2 + O_2 \rightarrow 2H_2O$. The *entropy* change will be unaltered. A part of the ΔH is used up in the unavailable energy connected with the differences in order and disorder between the products and the reactants, and this is unavoidably so, independent of any method of energy conversion.

11.3.10. Cold Combustion

These considerations lead to an explanation of why reactions between some substance and oxygen that are carried out in an electrochemical way are called *cold combustion* (Justi and Winsel).

Thus, the *net* cell reaction (the summation of the two electrode reactions $2H_2 \rightarrow 4H^+ + 4e$ and $O_2 + 4H^+ + 4e \rightarrow 2H_2O$ in which the electrons cancel out) is identical to the actual combustion reaction; it gives out heat which may be converted to mechanical work with only an efficiency given by the Carnot expression (11.65). The rest of the heat is evolved—as heat. But, in the electrochemical reaction, the heat ΔH is not given out. What is given out is not hotter molecules (made hotter by transfer of the difference of the potential energy of the reactants and products to kinetic energy) but a stream of electrons, the total energy of which is ΔG (per Avogadro number of act of the overall reaction). The combustion reaction has occurred, but *cold*.

It has been said that the same reaction has occurred as in combustion, and the energy ΔG has been electrically drawn off. The total energy change in the reaction, however, is ΔH. Hence, there *is* some heat energy given up or taken in during the ideal reversible working of an electrochemical reactor. It is $\Delta H - \Delta G$ or $T\Delta S$ and is usually negative, i.e., heat is given out. The amounts are small, e.g., in the oxidation of 1 mole of propane into carbon dioxide, $\Delta H = -530.6$ kcal mole^{-1}, but $T\Delta S$ is only -32 kcal mole^{-1}. Electrochemical combustion is *almost* cold.

11.3.11. Making V near V_e Is the Central Problem of Electrochemical Energy Conversion

One cannot change $I_{\max}$ for a given reaction, and ε_f is usually near unity [see Eq. (11.77)]. Consequently, the main efficiency-determining quantity which is subject to variation is V, the actual cell potential. Thus, the overall efficiency of electrochemical energy converters depends on how the overall cell potential varies with the current density which the cell is producing. One might say that it depends on how much of the energy of its own reaction the cell has to use up to get its two electrode reactions to take place at the desired rate, the rest of the energy being available for use outside the cell, i.e., for useful work.

In considering the behavior of electrochemical systems in action when a current is flowing through them, an expression was developed earlier (Section 9.6.7) for the cell potential V as a function of the overall current I. It was shown that, for a very simple electrochemical energy converter having electrodes of the same area A and delivering a current I, one has

$$V = IR_e = (E_{e,\text{so}} - \eta_{a,\text{so}} - \eta_{c,\text{so}}) - (E_{e,\text{si}} + \eta_{a,\text{si}} + \eta_{c,\text{si}}) - IR_i$$
$$= (E_{e,\text{so}} - E_{e,\text{si}}) - \eta_{a,\text{so}} - \eta_{c,\text{so}} - \eta_{a,\text{si}} - \eta_{c,\text{si}} - IR_i \quad (11.79)$$

where E is an equilibrium potential, η is overpotential of the type and location indicated, R_e is the resistance external to the cell, and R_i is the resistance of the space between the electrodes (see also Section 9.6.6). If both the de-electronation and the electronation reactions are running under high-field, or Tafel, conditions, the high-field approximation can be used to relate the two activation overpotentials (that for the source electrode and that for the sink electrode) to the current density

$$\eta_{a,\text{so}} = \frac{RT}{\alpha_{\text{so}} F} \ln \frac{(I/A_{\text{so}})}{i_{0,\text{so}}} \quad (11.80)$$

and

$$\eta_{a,\text{si}} = \frac{RT}{\alpha_{\text{si}} F} \ln \frac{(I/A_{\text{si}})}{i_{0,\text{si}}} \quad (11.81)$$

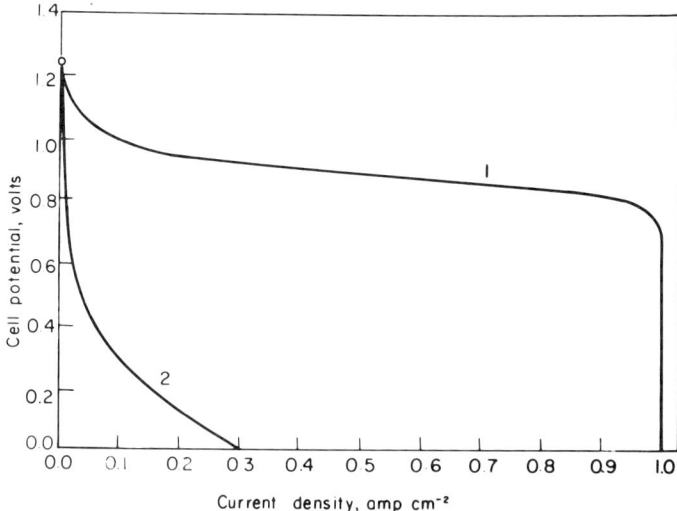

Fig. 11.71. Cell-potential *versus* current-density relations for an idealized electrochemical energy converter with planar, smooth electrodes. Assumed are the following parameters: Curve 1, $i_{0,\text{si}} > 1$ amp cm^{-2}, $i_{0,\text{so}} = 10^{-3}$ amp cm^{-2}; $i_{L,\text{si}} = 1$ amp cm^{-2}, $i_{L,\text{so}} = 1$ amp cm^{-2}; $\alpha_{\text{si}} = \infty$, $\alpha_{\text{so}} = \frac{1}{2}$; $R_i = 10^{-2}$ ohm. Curve 2, $i_{0,\text{si}} = 10^{-3}$ amp cm^{-2}, $i_{0,\text{so}} = 10^{-6}$ amp cm^{-2}; $i_{L,\text{si}} = 1$ amp cm^{-2}, $i_{L,\text{so}} = 1$ amp cm^{-2}; $\alpha_{\text{si}} = \frac{1}{2}$, $\alpha_{\text{so}} = \frac{1}{2}$; $R_i = 1$ ohm.

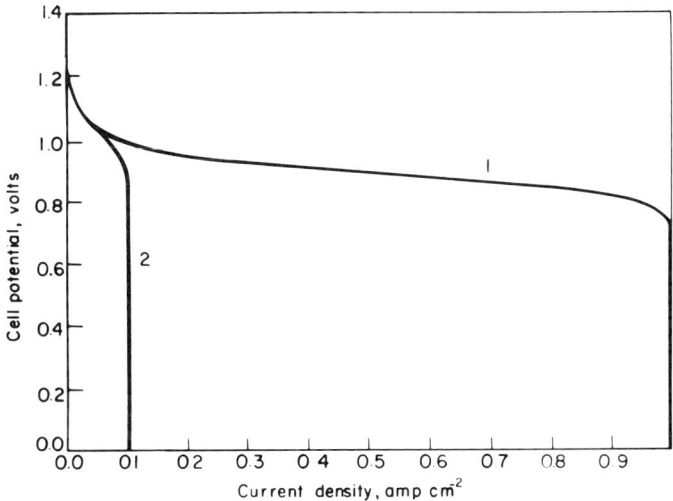

Fig. 11.72. Cell-potential *versus* current-density relations for an electrochemical energy converter for negligible values of *IR*, showing the influence of a limiting current. Curve 1, kinetic parameters correspond to curve 1 in Fig. 11.7; curve 2, the same as curve 1 except $i_{L,\text{si}} = 10^{-1}$ amp cm^{-2} and $i_{L,\text{so}} = 10^{-1}$ amp cm^{-2}.

Further, the expression for the concentration overpotential can be substituted for $\eta_{c,\text{so}}$ and $\eta_{c,\text{si}}$. They are

$$\eta_{c,\text{so}} = \frac{RT}{nF} \ln\left(1 - \frac{I/A_{\text{so}}}{i_{L,\text{so}}}\right) \tag{11.82}$$

and

$$\eta_{c,\text{si}} = \frac{RT}{nF} \ln\left(1 - \frac{I/A_{\text{si}}}{i_{L,\text{si}}}\right) \tag{11.83}$$

Thus, the expression for the cell potential becomes

$$V = IR_e = V_e - \left[\frac{RT}{\alpha_{\text{so}}F} \ln \frac{I/A_{\text{so}}}{i_{0,\text{so}}} + \frac{RT}{nF} \ln\left(1 - \frac{I/A_{\text{so}}}{i_{L,\text{so}}}\right)\right]$$
$$- \left[\frac{RT}{\alpha_{\text{si}}F} \ln \frac{I/A_{\text{si}}}{i_{0,\text{si}}} + \frac{RT}{nF} \ln\left(1 - \frac{I/A_{\text{si}}}{i_{L,\text{si}}}\right)\right] - IR_i \tag{11.84}$$

Equation (11.84) represents a cell-potential–cell-current relation (Fig. 11.71) over a wide range of conditions; that is why it appears complicated even though it is given here for the idealized case of a converter with two planar, smooth electrodes so that the complexities of the current *versus* potential with porous electrodes is avoided.

Under current-density conditions (Fig. 11.72) far below the limiting diffusion values [see Eq. (11.84)] and when the ohmic losses inside the cell are negligible, the activation-overpotential terms dominate the expression for the relation of current to potential, i.e.,

$$V \cong V_e - \frac{RT}{\alpha_{so}F} \ln \frac{I/A_{so}}{i_{0,so}} - \frac{RT}{\alpha_{si}F} \ln \frac{I/A_{si}}{i_{0,si}} \qquad (11.85)$$

At higher current densities than those referred to in Eq. (11.85), the activation-overpotential terms in this equation change much less with current than the IR_i drop term owing to the internal resistance of the cell. Under these conditions, when $(I/A)/i_L$ continues to remain negligible and the variation of V with I (but not its absolute value) is dominated by the IR_i term, one has (Fig. 11.73)

$$V \cong V_e - \text{Constant} - IR_i \qquad (11.86)$$

where the constant represents the activation overpotential which changes more slowly with current density than does the linear ohmic term.[†] At sufficiently high current densities, the I/A of Eq. (11.84) starts becoming comparable with i_L and concentration overpotential starts to reduce the cell potential in a more significant way than the IR_i term, which may now be taken as relatively constant. Thus (Fig. 11.74),

$$V \cong V_e - \text{Constant 2} - \frac{RT}{nF} \ln \left(1 - \frac{I/A_{so}}{i_{L,so}}\right) - \frac{RT}{nF} \ln \left(1 - \frac{I/A_{si}}{i_{L,si}}\right) \qquad (11.87)$$

It is seen, therefore, that the cell potential V and consequently the efficiency of an electrochemical converter (Fig. 11.75) are determined by the activation overpotential, by the electrolyte conductance, and by mass transfer (i.e., the solubility of the reactants) and the factors which dominate the way the efficiency of the conversion of energy *changes* with increase of current density are, respectively, at low current density, the activation overpotential; at medium current density, the electrolyte resistance; and, at the highest current densities, the mass transport. These factors are the ones which *dominate* at a given condition in causing *changes* in the efficiency (or power) in that particular region. But the absolute value of quantities

---

[†] Note how, even in the region in which there is linear behavior of V with respect to I, the *actual value* of the potential which the generator could put out depends on the value of the so-called "constant," i.e., on the activation overpotential and thus correspondingly on the exchange-current densities and the catalytic power of the electrodes.

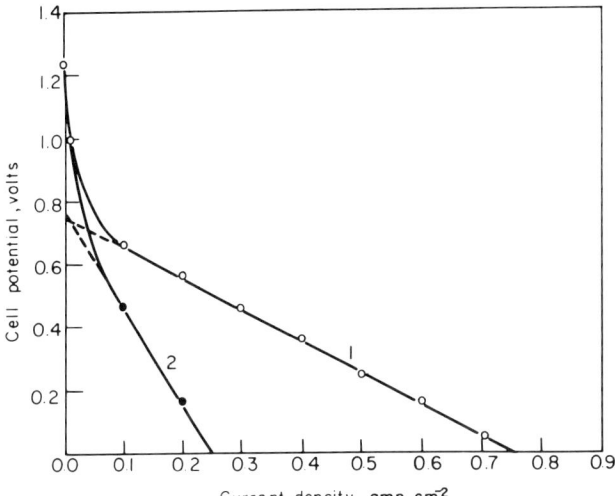

Fig. 11.73. Graphical representation of the influence of the internal resistance of an electrochemical energy converter on the cell potential when mass-transfer polarization is negligible. The early nonlinear part of the curve represents the effect of the activation overpotential on the cell potential before ohmic polarization has become important

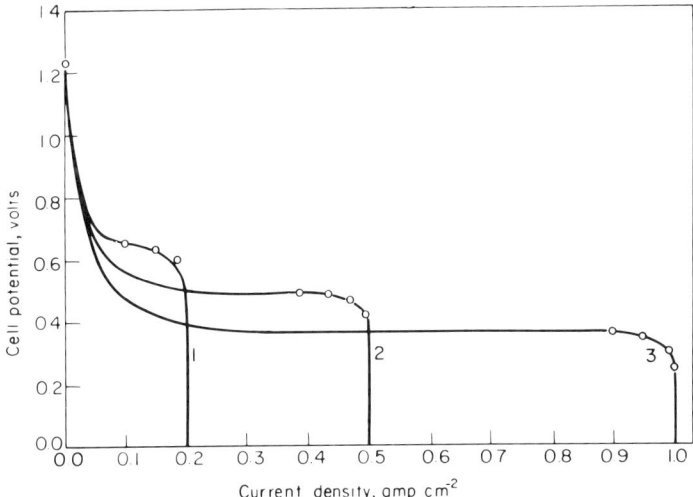

Fig. 11.74. The cell-potential *versus* current-density relations for an electrochemical energy converter (curve 1 of Fig. 11.71), showing the influence of mass-transfer limitations. Curve 1 is for $i_L = 0.2$ amp cm^{-2}; curve 2, for $i_L = 0.5$ amp cm^{-2}; and curve 3, for $i_L = 1.0$ amp cm^{-2}.

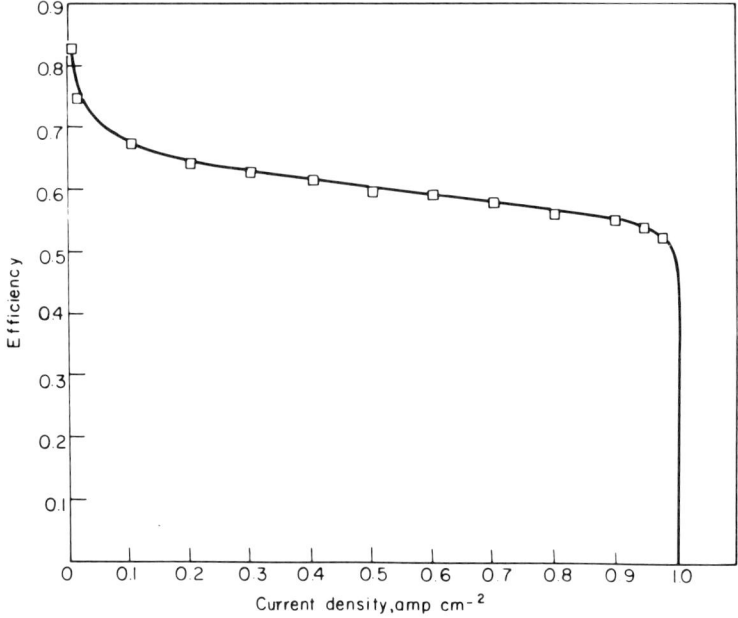

Fig. 11.75. The efficiency *versus* current-density relations for an electrochemical energy converter (calculated for curve 1 of Fig. 11.71).

describing cell behavior is determined by *the sum* of the influences of the activation, ohmic, and diffusional overpotentials. Thus, when mass transport becomes the most important factor in causing the efficiency to change, the actual efficiency of conversion at the point at which diffusion becomes important is influenced also by the value of the activation overpotential, which dominated changes in the situation earlier, and by ohmic influences which are the main causes of increase in overpotential and decrease in cell potential in the middle range of current density (Fig. 11.75).

11.3.12. The Electrochemical Quantities Which Must Be Optimized for Good Energy Conversion

Equation (11.84) also makes another aspect clear, what properties one has to attempt to optimize in electrochemical energy converters. Inspection of Eq. (11.84) and Figs. 11.71 to 11.74 shows that ideal reversible behavior will be approached when i_0 and i_L are very large and the internal electrolyte resistance of the cell is small. The maximization of i_L is a matter of designing and engineering cells to which the diffusion and convection of ions most easily occurs (*cf.* Section 11.3.17). To reduce R_i, electrolytes of as high a conductance as possible are used. The conductance of 1–5M H_2SO_4 and KOH is in the range of 0.5 ohm^{-1} cm^{-1}, while that of most other concentrated aqueous solutions is much smaller. Minimizing R_i is also largely a matter of structural arrangement of the electrodes in the cell, in particular with respect to the type of electrodes (usually *porous*) used.

The main quantity—upon the detailed understanding of which a great deal of the future of electrochemical energy conversion depends—is the exchange-current density i_0. It is through this parameter that electrochemical energy conversion becomes linked up with electrocatalysis (see Section 10.1). The values of i_0 observed for some reactions vary over many orders of magnitude [e.g., $\sim 10^6$ for methanol as a fuel (*cf.* Table 11.12)] at different catalysts, and the aim of fundamental research on electrocatalysis is to understand the phenomenon of this considerable dependence of electrochemical reaction rates on the electronic structure of the substrate so that it is possible to make electrodes with high i values (i.e., small polarization at high rates) for fuels which are cheap and which have, as do the hydro-

TABLE 11.12

Exchange-Current Densities and Tafel Slopes for the Oxidation of Methanol to Carbon Dioxide on Various Catalysts

Catalyst composition	$-\log i_0$	Tafel slope (b)
Pt–Fe	4.0	0.103
Pt–Ni	4.3	0.109
Pt–Rh–Fe	5.4	0.081
Pt–Co	5.5	0.084
Pt–Pb	6.0	0.085
Platinum	7.3	0.066
Pt–Rh	7.9	0.063
Pt–Cu	8.4	0.059
Pt–Au	8.8	0.058
Pt–Co	9.4	0.058
Pt–Ni	9.1	0.060
Pt–Fe	6.5	0.073
Iridium	9.9	0.053

carbons, a large amount of energy per gram and per dollar potentially available if the energy change in their reaction with oxygen can be tapped. Little fundamental research has so far been carried out here, and only the relative order of catalytic activity among various substrates is known for most fuels (*cf.* Table 11.13).[†]

In practice, the maximum-energy-conversion efficiencies obtained in some electrochemical converters are about 75%. Equation (11.77) shows that there is a decrease of efficiency with increase of current density. When the electrochemical converter is working as an engine, therefore, i.e., when it is being used to produce power—energy at a given rate—, the efficiency of the conversion of energy to work will decrease with increase of the power density.

11.3.13. The Power Output of an Electrochemical Energy Converter

When electrochemical energy converters are to be considered for use, e.g., in driving trucks, it is the power output or the rate at which they are able to do work that has to be examined. The power P of an electrochemical energy converter is defined thus

$$P = IV \qquad (11.88)$$

from which it follows that the power is small when I is small even though V is near the maximum V_e. But the power output is also small when I is very large because the sudden growth of concentration overpotential when the current density approaches i_L tends to drive V to zero [see Eq. (11.84)]. Thus, the P versus I curve should pass through a maximum (Fig. 11.76).

The distinction between the situation in which one needs predominantly high efficiency and one in which one wants high power becomes clear when one compares the P versus I and ε versus I curves (Fig. 11.77). When its efficiency is at a maximum, the electrochemical energy converter is a less good power source. As the current density is increased, the power output increases, but the efficiency decreases. Of course, at the highest current drains, both the power and efficiency fall toward zero.

[†] A reason for the restricted knowledge of electrocatalysis arises outside the small *amount* of research so far carried out. It is that much of that work has been done by using porous electrode structures. In these (Section 11.3.17), the control of rate is complicated by contributions from diffusional and ohmic effects. It is difficult to draw clear conclusions concerning electrocatalysis from experiments with porous electrodes.

TABLE 11.13
Relative Efficiencies of Metal Catalysts for Three Reactions[†]

Electrolyte	Catalyst surface	$CH_3OH \rightarrow CO_2$	$HCHO \rightarrow CO_2$	$HCOOH \rightarrow CO_2$
5M KOH	Smooth	Pt > Pd, Ir > Au > Rh		
	Rough	Pd, Pt, Ir > Au > Rh		
1M Na$_2$SO$_4$	Smooth	Ir > Ag, Pt, Pd > Au	Ir > Rh > Pt > Pd > Au	Pd > Ir > Pt, Rh > Au > Ag
	Rough	Au > Pt > Ir > Pd > Rh	Au > Ir, Rh > Pt > Pd	Pd, Ir > Pt, Au > Rh
3.3M KHCO$_3$	Smooth	Pt > Ir > Pd > Au > Ag > Rh	Ag > Rh, Pt, Ir > Au > Pd	Ag > Pt > Pd > Rh, Ir > Au
	Rough	Pt > Ir > Pd > Au > Rh	Au > Ir > Pt, Rh > Pd	Ir > Rh > Pt > Au
1M H$_2$SO$_4$	Smooth	Pt, Ir > Pd	Pt, Ir > Pd > Au	Pt > Ir > Pd > Au, Ag
	Rough	Pt, Ir > Pd, Au > Rh		Ir, Pt > Au > Pd > Rh

[†] K. R. Williams, *An Introduction to Fuel Cells*, Elsevier Publishing Co., New York, 1966.

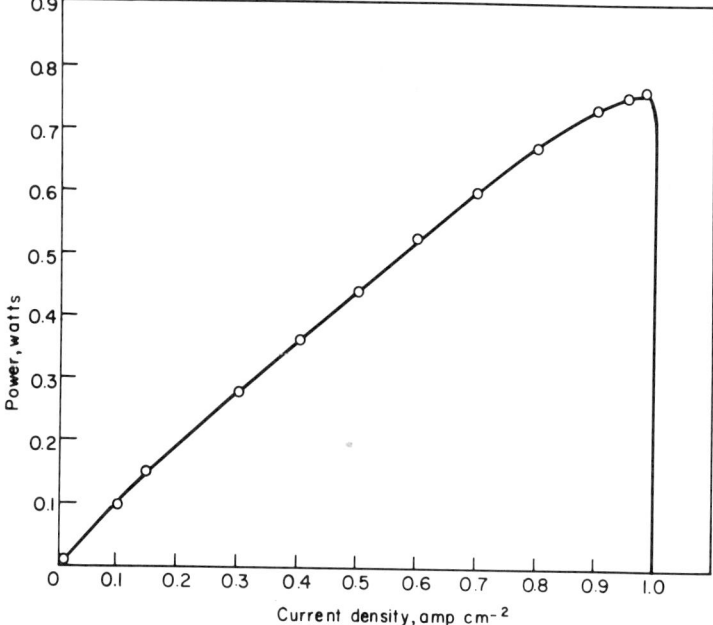

Fig. 11.76. The power *versus* current-density relations for an electrochemical energy converter (calculated for curve 1 of Fig. 11.71).

11.3.14. The Electrochemical Engine

The principal use hitherto of the electrochemical energy converter has been the situation (auxiliary power in space) in which the *weight* of the energy converter plus fuel carried is of primary import. Thus, in any energy-conversion situation where, e.g., in transportation in its most general sense, the system must carry its own fuel with it to provide a certain number of kilowatt hours of energy during a journey of known duration, the main point is the *efficiency* of conversion; the weight of the fuel necessary for a given operation is clearly inversely proportional to it. Hence the use in space exploited the avoidance of the Carnot cycle efficiency loss in electrochemical conversion.

Another probable use of the electrochemical reactor, however, is as a *power* (rather than energy) source and, in combination with an electric motor, as an *electrochemical engine* (Henderson). Here, the stress is on power, the *rate* of delivery of energy per unit weight, and not so much on the efficiency of the conversion though this is, of course, also important. There is at present no advantage to be gained from an electrochemical engine over a thermal-combustion engine in respect to a *power* to weight ratio; in fact, present electrochemical converters tend to be distinctly heavier per unit of power than thermal combustion engines, although the best of them are already better than diesel engines in this respect.

11.3.15. Was the Wrong Path Taken in the Development of Power Sources at the End of the Nineteenth Century?

The extrinsic efficiency losses that may arise in an electrochemical energy converter and their origins have been described in the above sections. It is the *overpotential* associated with electrode reactions which is the general cause for the fall of the cell potential from the reversible value, and the various types of overpotential have been related to the cell potential and hence to its efficiency and its power.

However, an understanding of the fundamentals of electrodics became current in Western Europe and America only after about 1950, and the knowledge was not applied to the theory of electrochemical energy converts until the 1960's. Prior to the 1950's, it was not mechanistically clear why an electrochemical cell did not work at the equilibrium potential V_e

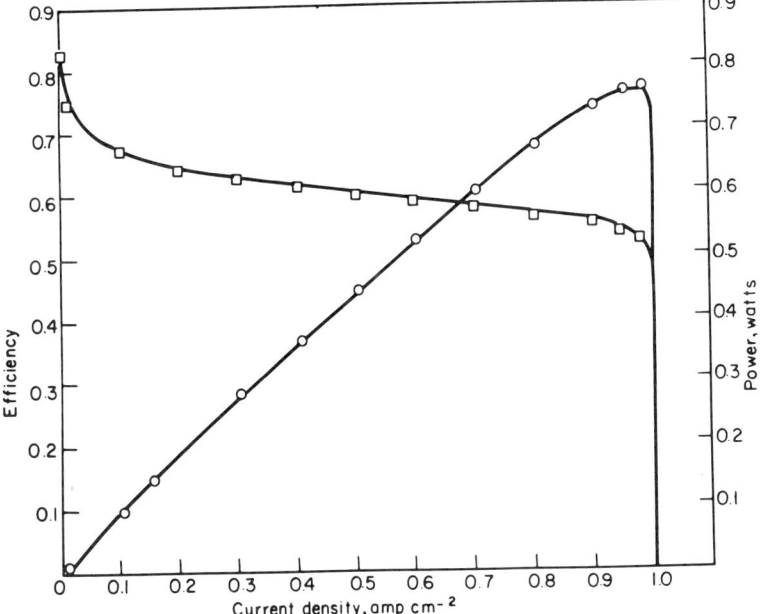

Fig. 11.77. The power *versus* current-density and efficiency *versus* current-density relations for the electrochemical energy converter (calculated for curve 1 of Fig. 11.71).

when the current was small enough for the ohmic and diffusional components (η_0 and η_c, respectively) to be near zero. Indeed, the theoretical situation for converters before about 1955 was buried in obscurity, even for qualitative discussion. A word *polarization* was used to describe the loss of cell potential and hence energy when the cell worked spontaneously. The absence of understanding of the phenomenon of polarization had a great negative influence on the history of energy conversion. The background of this historical point shall be touched upon now.

The first electrochemical energy converter (Fig. 11.78) was described by Grove in 1842. On the basis of developments by Carnot, Thomson, and Clausius in thermodynamics, Ostwald, speaking at the inaugural meeting of the Bunsen Gesellschaft, in 1894, called for the replacement of heat engines by electrochemical energy converters because the latter were not Carnot-limited and thus not susceptible to an *intrinsic* efficiency loss. The internal-combustion engine was developed between about 1860 and 1890, i.e., well after Grove's first exhibition of electrochemical energy conversion and, considering the four-stroke Otto engine, only four years before Ostwald pointed out the fundamental difference in the two approaches to energy conversion. Why, then, did technologists decide to exploit the *chemical* energy of fossil fuels by adopting the wasteful thermal method (Section 11.3.4) of converting it to mechanical and electrical energy rather than the intrinsically more efficient (see Section 11.3.8) electrochemical method? In fact, it is only in the past decade, i.e., more than a century after Grove's work, that the development of electrochemical energy converters has been emphasized. Even this rebirth of interest in non-Carnot-limited converters was prompted by the demands of space technology for carrying

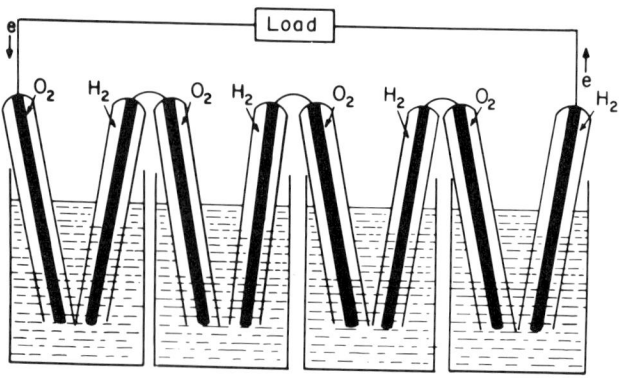

Fig. 11.78. Diagram of the first hydrogen–oxygen fuel cell, constructed by Grove (1842). The dark lines in the axes of the tubes are platinized-foil electrodes.

maximum energy for auxiliary power with minimal weight rather than by the economic use of the reserves of fossil fuels or the desire to replace the pollution-causing internal-combustion machine.

One can hazard a hypothesis on this point of scientific history. The *e*xtrinsic losses of efficiency in the early electrochemical energy converters (which often reduced the overall efficiency of values as low as those observed with the *in*trinsically limited combustion engine) frightened off technology. The fear was based on the virtually *zero understanding of the nature of electrochemical polarization* at interfaces *which existed among scientists and engineers before the* 1950's. What was not realized was that the losses of efficiency in an electrochemical energy converter are *e*xtrinsic and research upon electrodic mechanism could give rise to an understanding which would make possible devices in which polarization was greatly reduced, i.e., the efficiency of conversion was greatly increased (see Section 11.3.9). This substantial diminution falls within the realm of the possible in contrast to the *in*trinsic efficiency loss in combustion engines which can be improved by research only insofar as the temperature of conversion can be increased—an approach which has serious limitations.

Thus, an important hitch in the course of the development of energy converters and what will be eventually regarded as an unfortunate deviation of much of technology as a whole seems rationalized. Toward the last years of the nineteenth century, man could have chosen either the thermal or the electrochemical path (Fig. 11.79) in developing devices to convert the energy

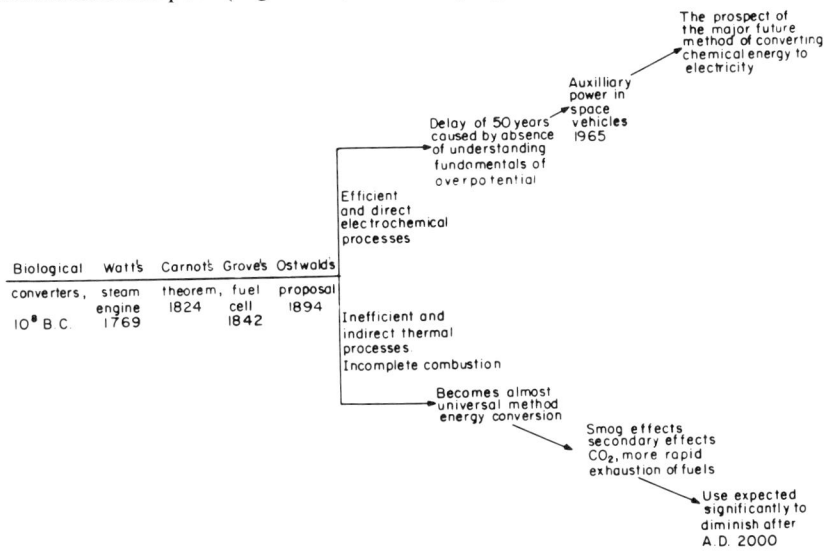

Fig. 11.79. Energy conversion took the wrong path in 1894.

of chemical reactions to mechanical work. The electrochemical method offered a superior prospect because it did not involve an inherent efficiency loss. However, it was the thermal rather than the electrochemical path which was chosen, and the essential reason was the absence of understanding of the mechanism of overpotential, the existence of which in large quantities made the early electrochemical converters have efficiencies which were, in practice, no better than those of good internal-combustion engines.

Thus, the development of chemical kinetics prior to electrochemical kinetics and the stress at the beginning of the century upon the thermodynamics, rather than the kinetics, of electrochemical cells changed the course of technology. The results of that change and the period of the Nernstian hiatus (Section 1.3.4 and Fig. 1.12) are reflected at present in that part of the pollution of the atmosphere which arises from the burning of fossil fuels, the noise and dirt of the cities, and in the relatively high cost of transportation.

11.3.16. Electrodes Burning Oxygen from Air

Different materials can be used as oxidants in fuel cells. Yet it is desirable, if possible, to use O_2 from air at one electrode in every earthbound fuel cell because this avoids the necessity of carrying a second fuel for the cathodic reaction. Hence, the cathodic reduction of O_2 has a special importance in electrochemical reactors. The overall reaction is, in acid solution (see Section 10.4),

$$O_2 + 4H_3O^+ + 4e \to 6H_2O \qquad (11.69)$$

and in alkaline solution,

$$O_2 + 2H_2O + 4e \to 4OH^- \qquad (11.89)$$

There is a grave disadvantage in this important electrode reaction. It has an i_0 value in the region of 10^{-10} amp cm^{-2}, and hence the reaction usually contributes considerably to the overpotential in the functioning of an air-burning electrochemical converter. Electrocatalysis of this reaction is needed more than any other in energy converters.

11.3.17. The Special Configurations of Electrodes in Electrochemical Reactors[†]

In the foregoing material concerning electrochemical reactors (i.e., energy converters and power sources), the explicit assumption is that the

---

[†] The term *electrochemical reactor* is a general one referring to all electrochemical cells,

electronic conductors with which the molecules of the fuels transfer and receive electrons are in the form of planar bodies, e.g., sheet electrodes. The presentation is made in that way because it is possible thereby to present simple equations and make relatively clear deductions concerning the trends arising from changes in the exchange current, conductance, and solubility.

However, were only planar electrodes used, much smaller currents would be observed per geometric (or external) unit area of electrode material than are in fact obtained. Electrochemical converters would in fact have no practical uses. Thus, in all actual electrochemical converters, the electrodes are *porous* structures, e.g., of graphite, in the pores of which is usually the *catalyst material* (e.g., platinum) to and from which the charge transfer occurs.

Earlier workers in this field regarded the higher currents obtained by using porous electrodes as due simply to the fact that the electrolyte was in contact with a larger real area of the electrodic catalyst per geometric area than with planar electrodes. However, this view is now regarded as much less than the whole story. The expression for the power of a generator involves the current, and it may be desirable to work near the maximum value of this, i.e., near the limiting current. Upon what does this limiting current depend? (See Section 9.3.10.) Were it at a planar electrode, the principal variable would be the diffusion-layer thickness and this could be altered by various forms of agitation (but the agitation itself uses up some of the power produced and could hardly make the diffusion layer thinner than $\sim 10^{-2}$ to 10^{-3} cm). Consider, however, a three-phase boundary in a porous electrode at which there exists a gas in contact with a solution and a metal, as shown in Fig. 11.80. The thickness of such menisci vanishes at the tip and increases back towards the bulk of the electrolyte (Cahan).

These variations in meniscus thickness with distance in a pore imply that there is a very large local decrease in δ in the meniscus (of a *wetted* pore) to less than 10^{-6} cm and hence a large increase in the limiting diffusion current. However, if the thickness of the meniscus is too low, the H_3O^+ ions diffusing back to the bulk of the electrolyte from the dissolution of, say,

---

i.e., cells in which the introduction of two substances (fuels) on the electrodes generates power or the injection of electric power synthesizes compounds. The reactor as a converter may thus produce electricity at high efficiency, it may be a power generator in which the construction is mainly aimed at maximizing the power-to-weight ratio, or it may be an electrochemical engine in which the electrochemical generator is connected with an electric motor. *Fuel cell* is the historical and general name for all these devices.

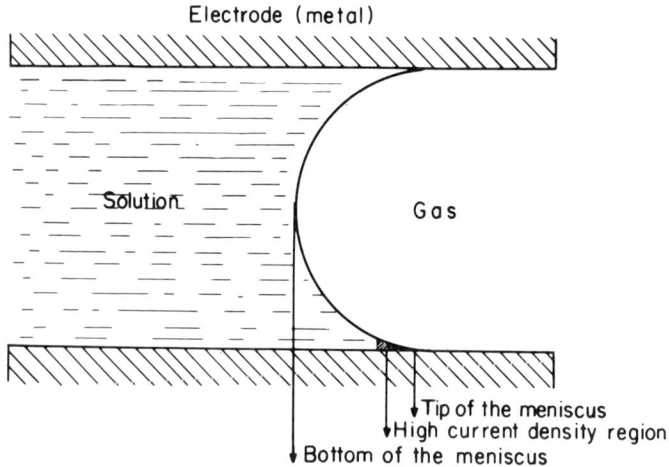

Fig. 11.80. Schematic representation of the three-phase (metal, electrolyte, and gas) interphase in a single pore when the contact angle is a few degrees.

$H_2 + 2H_2O \rightarrow 2H_3O^+ + 2e$ will not be able to escape sufficiently quickly because of the high resistance, due to the low cross section, to the diffusion of ions arising from the very thin section implied by such thin menisci. On the other hand, back where the meniscus is sufficiently thick, say, at 0.01 cm from its beginning, δ is only about the same as for diffusion to a planar surface from dissolved gas in a bulk electrolyte. Hence, the highest currents will not be reached at those parts of a pore where a dissolving gaseous fuel has to diffuse through a thick electrolyte or at the very thinnest end of the meniscus wedge, but at some intermediate distance up the meniscus (Fig. 11.81).

The main reason why a porous gas electrode is so active, therefore, is that it allows particularly large maximum *diffusion* currents by diffusion through (fairly) thin meniscus layers. But this thesis brings a corresponding antithesis because it implies that, farther up the pore where there is no meniscus but bulk solution, the gaseous reactant dissolved therein is reacting at a rate much less per unit length of meniscus, than that in the (moderately thin) meniscus region. Thus, there is little activity in the interior of electrodes of electrochemical reactors; i.e., if the catalyst material present is distributed uniformly throughout the electrode (as in many reactors), much is wasted.

Theory would indicate, therefore, that electrodes should indeed have pores (and hence large numbers of menisci) but they should be thin or the

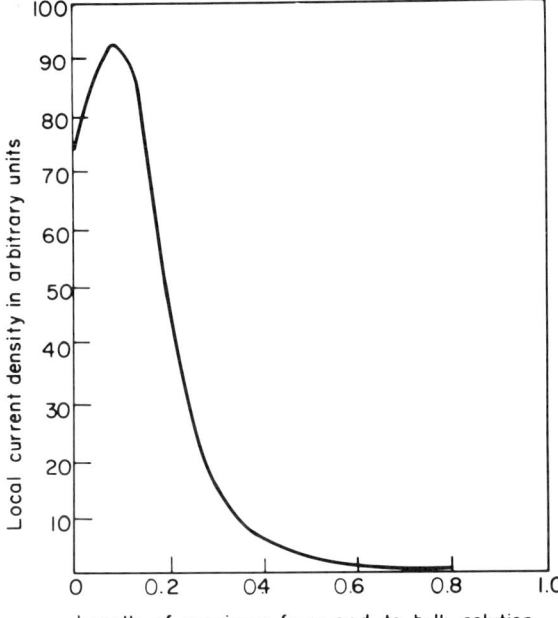

Fig. 11.81. Distribution of current density along the meniscus in a single pore.

catalyst should be concentrated only in a small region of high current density near the three-phase boundary. The attainment of such a model in a practical way is a matter for present research in electrochemical engineering. Such an attainment would mean, because of the reduction in needed catalyst per unit of power, a great (10–100 times?) reduction in the cost of converters for situations in which platinum must at present be used as a catalyst.

11.3.18. Electrochemical Electricity Producers: The Two Basic Types

The life of an electrochemical electricity producer depends on its electrode–electrolyte interfaces' being supplied with the reactants for the spontaneous deelectronation and electronation reactions. The reactants (or fuel) must therefore be stored somewhere. The storage of the fuel can be done either outside the electrochemical system or inside.

If the reactants are stored outside the system, provision must be made for supply lines to feed the fuel to the electrolyte interfaces. The electrochemical system will then be an *open* system exchanging matter with its environment. The reactants, however, can also be incorporated in the cell during manufacture and stored either as electrode material or as part of the electrolyte. In this case, one has a *closed* electrochemical system. Such a system does not need any facilities for maintaining fuel supplies, but, when the stock of reactants is used up by the reactions, the cell's life is over.

The open electrochemical systems with external fuel storage are known, in common parlance, as *fuel cells*, and the closed "one-shot" electrochemical systems with internal stocks of reactants are known as *primary batteries*. These conventional terms, however, (cf. Table 11.18) conceal the fact that the two types of electricity producers are similar except in the location of fuel stocks (and that the reactants can hardly be gaseous in the primary battery).

11.3.19. Examples of Electrochemical Generators

The development and construction of electrochemical generators during the 1960's comprises considerable technology and is already a subject of monographs. Many—several dozen—types of fuel cells have been described in the literature. A representative account of them is outside the scope of this book. Therefore, only a very brief description of some of the systems which seem to have become or likely to become of practical use is given here.

Among fuels already used in fuel cells, besides hydrogen, are many hydrocarbons, several lower alcohols, hydrazine, and ammonia (*cf.* Table 11.14). These fuels are generally used as anodically reacting materials in combination with an oxygen cathode.

A rough division of present electrochemical generators may be made according to the temperature of operation. Thus, if this is below 150°C, the device is called a low-temperature fuel cell;[†] and, above about 500°C, a high-temperature fuel cell. Of the many problems facing research in this new field,[‡] this division reflects the two principal ones.

[†] Within a given temperature range, increasing temperature results in an increased efficiency (*cf.* Table 11.15), but a compromise has to be made with the decreasing stability (i.e., increasing corrodability) of the construction materials.

[‡] Although fuel cells were worked upon sporadically in the nineteenth century (Grove) and to a small extent in the period preceding the war of 1939 (Davtyan, Baur), the large wave of modern work began in the late 1950's, primed by the pioneer and individualistic work of the British engineer, Bacon (and the British electrochemist, Watson). It was funded initially by the United States Navy, later in its major part by the requirements of space vehicles, and now increasingly with the objective of providing pollution-free cities.

TABLE 11.14
Cells Using Fuels Other than Hydrogen[†]

Fuel and type of cell	Electrolyte	Catalysts	Av. power density
Hydrocarbons CH_4, C_2H_6, C_3H_8, etc., temp 100–200°C	Conc H_3PO_4 solution with low vapor pressure	Anode, Pt; Cathode, Pt	10–60 mW cm^{-2} at 0.4–0.5 V
Alcohols CH_3OH, $C_2H_4(OH)_2$, and other alcohols and derivatives, temp 20–80°C	6N KOH	Anode, Pt, Pd, and Ni; Cathode, C, Ag, Pt, and Pd oxides	20–100 mW cm^{-2} at 0.6–0.7 V
	7N H_2SO_4	Anode, Pt; Cathode, Pt	10–50 mW cm^{-2} at 0.4–0.6 V
Nitrogen derivatives NH_2–NH_2, temp 20–60°C	6–12N KOH	Anode, Ni and Co; Cathode, C and Ag	60–200 mW cm^{-2} at 0.8–0.9 V
NH_3, temp 200–400°C	Molten hydroxides	Anode, Pt; Cathode, NiO + Li	100–200 mW cm^{-2} at 0.7–0.8 V
Special fuels and Alkali metals, 20–60°C	6 to 12N KOH	Anode, Steel; Cathode, C, Ag, etc.	100–300 mW cm^{-2} at 1.0–1.2 V

[†] Table based on *Piles à combustible*, Paris, 1965.

Thus, for the low-temperature cells, the principal problem is that of *electrocatalysis*, how to raise the exchange-current density for the oxidation of relatively cheap fuels by using electrode materials which do not make the initial cost of a device too high. The stability of construction materials is not so critical, although a selection is limited if aqueous acid electrolytes are employed. For the high-temperature cells, the temperature alone causes the exchange-current densities to be sufficiently high that there can be

TABLE 11.15
Temperature Dependence of Current Density at a Given Cell Potential for a Typical Hydrogen–Oxygen Fuel Cell

Temperature, °C	Current density, mamp cm^{-2}
20	50
50	120
80	300

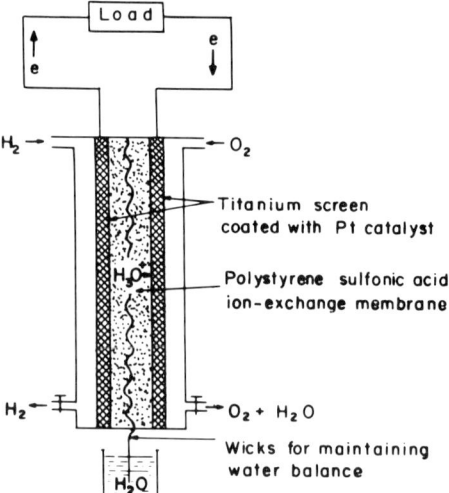

Fig. 11.82. Schematic representation of a single Gemini hydrogen–oxygen fuel cell.

production of energy at a suitable rate without using expensive materials as catalysts. The counter problem then becomes the *stability of the materials*, the containers and electrodes, under the highly corrosive action of the electrolytes.

11.3.19a. *The Hydrogen–Oxygen Cell.* This is the best known of electrochemical generators, having been the first to reach meaningful practical power levels (5 kW, 1959). The practical realization in the 1959 cell by Bacon of achieving considerable power by operating at high gas pressures and somewhat elevated temperatures (up to 150°) was an important stimulus to the present phase of work on electrochemical power.

Several types of hydrogen–oxygen fuel cells have been developed since the accomplishment by Bacon, of which the Gemini and Apollo systems of the American space program are outstanding. A schematic diagram of a single Gemini cell is shown in Fig. 11.82. A unique feature of this cell is the use of a thin cation-exchange membrane as electrolyte (polystyrene sulfonic acid intimately mixed with a Kel-F[†] spine). Each side of this rectangular membrane is covered by a titanium screen coated with a platinum catalyst. The thickness of the entire cell is about $\frac{1}{2}$ mm. Reactions taking place in the cell are

---

[†] *Kel-F* is the commercial name for trifluorochloroethylene resins, plasticizers, etc.

at anode,
$$2H_2 \to 4H^+ + 4e \quad (11.68)$$
at cathode,
$$O_2 + 4H^+ + 4e \to 2H_2O \quad (11.69)$$

Since the conductivity of the membrane is strongly dependent on the water content, the water balance is maintained by wicks' draining or supplying water by capillary action.

The performance of a single cell is shown in Fig. 11.83. An important overpotential loss is, in these cells, due to the membrane resistance and also, as usual, to the oxygen electrode. A storage system built of these cells, having an average power of about 900 W and a maximum power of 2 kW, was used in the first two-man Gemini spacecraft. A noteworthy feature of this storage system is the self-contained provision for collecting water (a by-product of the cell reaction) for drinking purposes in space (1 pt kWh^{-1}). At present, some hydrogen–oxygen fuel cells have attained a power level of 1 W cm^{-2}. Some technical data for hydrogen–oxygen fuel cells in comparison with hydrazine–oxygen fuel cells are given in Table 11.16.

11.3.19b. Reformer-Supplied Hydrogen–Air Cells. Pure hydrogen is expensive. Conversely, hydrogen is an excellent fuel because of its large i_0 value and the resulting possibility of catalyzing its dissolution well (i.e., with small overpotential) on cheap materials such as nickel. To meet this

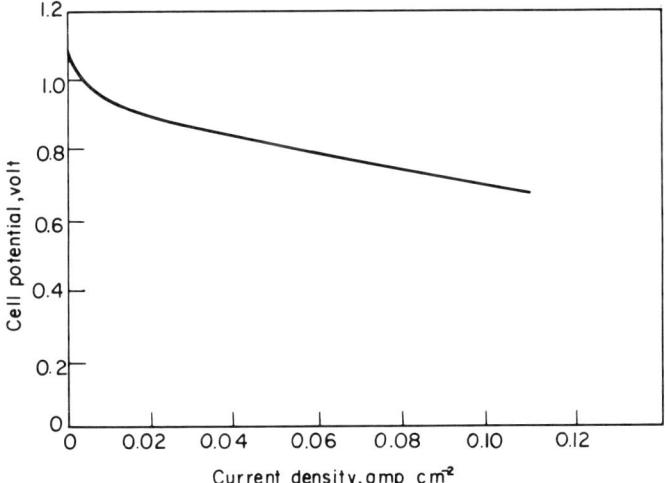

Fig. 11.83. The cell-potential *versus* current-density relations for a Gemini hydrogen–oxygen fuel cell.

TABLE 11.16
Comparison of Performance Characteristics of Hydrogen–Oxygen and Hydrazine–Oxygen Fuel Cells

Basis: 300-W fuel cell; 5-, 14-, and 30-day continuous operation; 60% overall efficiency; cryogenic hydrogen and oxygen storage.

Type of fuel cell	Days of functioning	Anodic fuel			Oxygen			Fuel + O_2 + Tankage	
		Fuel wt, kg	Tankage wt, kg	Total vol, cm^3	Oxigen wt, kg	Tankage wt, kg	Total vol, cm^3	Total wt, kg	Total vol, cm^3
Hydrazine–oxygen	5	11.4	1.1	10.5	11.4	3.1	32.3	27.0	42.8
	14	31.7	3.2	28.9	31.7	5.3	52.6	71.9	81.5
	30	68.2	6.8	62.3	68.2	9.9	86.3	153.0	148.6
Hydrogen–oxygen	5	2.0	18.0	50.4	11.4	3.1	32.3	34.5	82.7
	14	5.0	24.5	133.0	31.7	5.3	52.6	66.5	185.6
	30	12.0	38.0	315.5	68.2	9.9	86.3	128.1	402.0

situation, a series of fuel cells utilize a system in which a cheap hydrocarbon fuel is the origin of the hydrogen, this being produced in an adjoining apparatus, separated from other gases, and fed into the cell.

The *steam hydrocarbon process* is

$$C_nH_{2n+2} + nH_2O \rightarrow (2n+1)H_2 + nCO \tag{11.90}$$

$$CO + H_2O \rightarrow CO_2 + H_2 \tag{11.91}$$

The CO_2 may be removed by absorption in ethylamine or by hydrogen separated by diffusion through palladium or silver–palladium membranes.

Ammonia or metal hydrides may also be used as the chemicals, both being in fact hydrogen carriers. The use of the latter is limited by poor economics, while otherwise they are particularly suitable because of the ease of handling them.

The disadvantage of such a hydrogen supply is, of course, in the weight and size added to the electrochemical device itself and also the cost, the start–stop difficulties, etc.

11.3.19c. Hydrocarbon–Air Cells. Many hydrocarbons, including the main constituents of diesel oil, have been oxidized electrochemically at levels of more than 99% completion. Platinum is the only suitable catalyst material at the present time.

The electrodes are constructed by depositing finely divided platinum on a porous Teflon substrate attached to a base of tantalum (which, after oxidation, resists further corrosion in strong acid).

Unsaturated hydrocarbons can be oxidized at relatively low temperatures. Kinetic data for a number of these compounds are given in Table 11.17. Saturated hydrocarbons, however, can be oxidized at practical rates if the temperature is in the region from 80 to 150°C (Heath and Worsham). Figure 11.84 shows some results for the oxidation of propane (*cf.* Section 9.5.6), for which

$$C_3H_8 + 6H_2O \rightarrow 3CO_2 + 20H^+ + 20e \tag{11.92}$$

The supporting electrolyte is oncentrated H_3PO_4. The power density of such cells is about 0.1 W cm^{-2}. The fact that the power density is about one order of magnitude less than that obtainable with cells which burn hydrogen directly is of course due to the much lower exchange-current densities which are exhibited during the oxidation of the hydrocarbons compared with those exhibited during the oxidation of H_2. This disadvantage (which may be eliminated by future research on novel electrocatalysts, e.g., metal oxides) must be compared with the lesser cost of the hydrocarbon fuel and the lesser

TABLE 11.17
Summary of Results for the Anodic Oxidation of a Number of Unsaturated Hydrocarbons, 1 atm Gas, $1N$ H_2SO_4, 80°C[†]

Hydrocarbon	vs. E_{rev} NHE, mV	vs. i_0, amp cm^{-2}	$\left(\dfrac{\partial V}{\partial \log i}\right)_{pH}$	$\left(\dfrac{\partial \log i}{\partial pH}\right)_V$	$\left(\dfrac{\partial \eta}{\partial pH}\right)_i$ mV	$\dfrac{\partial \log i_0}{\partial pH}$	$\left(\dfrac{\partial V}{\partial pH}\right)_i$ mV	$\left(\dfrac{\partial i}{\partial P}\right)$	Faradaic efficiency, %
Ethylene	30	5.0×10^{-8}	140–160	0.45	~0	~0	−65	<0	100
Allene	0	2.1×10^{-8}	140–160	0.41	~0	~0	−68	<0	93
Propylene	80	2.0×10^{-8}	140–160	0.35	~0	~0	−52	<0	97
1,3-Butadiene	60	0.5×10^{-8}	140–160	0.39	~0	~0	−65	<0	62–88
1-Butene	90	1.2×10^{-8}	140–160	0.47	~0	~0	−67	<0	71
2-Butene	90	1.2×10^{-8}	140–160	0.45	~0	~0	−66	<0	83
Benzene	100	1.2×10^{-8}	140–160	0.40	~0	~0	−51	<0	60–90
Acetylene	−110	4.5×10^{-18}	70–80	0.80	~0	~0	−50	<0	100

[†] Effects of pH were studied at constant ionic strength in mixtures of H_2SO_4, Na_2SO_4, and NaOH over the range of 0.5 to 12.5 pH.

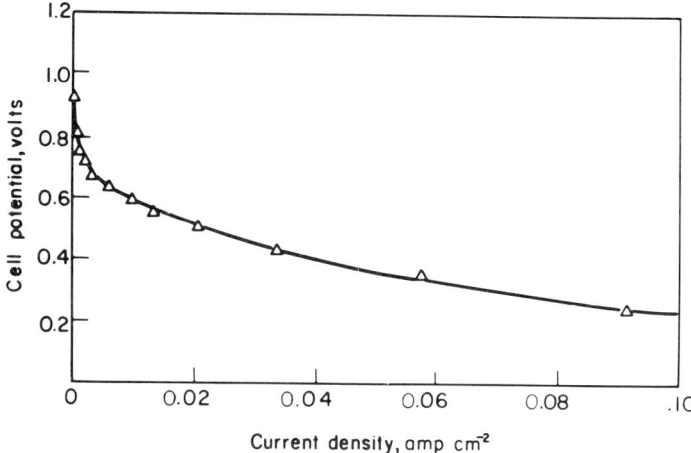

Fig. 11.84. The cell-potential *versus* current-density relations for a propane–air fuel cell in phosphoric acid solution at 150°C.

weight and size which direct hydrocarbon converters have compared with the reformer system (*cf.* Section 11.3.19*b*).

Presently, the basic disadvantages of hydrocarbon-burning systems are the cost of the platinum catalyst and the low exchange-current densities, which cause less-good conversion efficiencies than with, say, hydrogen or even alcohols as fuels.

11.3.19*d*. Dissolved-Fuel Fuel Cells. A line of research on the use of partially oxygenated hydrocarbons and other water-soluble chemicals as fuels in fuel cells has resulted in a number of achievements which place this type of fuel cells close in performance to the hydrogen–oxygen cells. A number of such fuels can render considerable current densities at a reasonable cell potential. The most widely used fuels are methanol and hydrazine.

11.3.19*e*. Natural Gas and CO–Air Cells. The analog to the work of Bacon on hydrogen–oxygen cells in the low-temperature range is the work of Broers in the high-temperature range on cells with CO or natural gas (largely CH_4) for it continues work carried out much before the modern phase of interest in electrochemical generators began. Because methane is so unreactive, direct electrochemical oxidation of it is an inefficient process (high overpotential). Principally, there are two approaches to the use of natural gas as an electrochemical fuel. Mixed with steam, methane is reformed into H_2 and CO with a nickel catalyst either *in situ* or externally.

External reforming makes possible the operation of the cell between 500 and 600°C, whereas *in situ* reforming needs about 750°C.

The electrolyte in the form of a paste with MgO is a molten mixture of Li_2CO_3, Na_2CO_3, and K_2CO_3. A thin layer of porous nickel is the anode, while a finely divided silver constitutes the cathode. A schematic diagram of such a cell is shown in Fig. 11.85.

The reactants at the cathode are CO_2 and air. The reactions occurring in this cell are

at cathode,
$$2CO_2 + O_2 + 4e \rightarrow 2CO_3^{2-} \tag{11.93}$$

at anode,
$$CO + CO_3^{2-} \rightarrow 2CO_2 + 2e \tag{11.94}$$

$$H_2 + CO_3^{2-} \rightarrow CO_2 + H_2O + 2e \tag{11.95}$$

The performance of this cell is shown in Fig. 11.86. It works well over periods of years, and the materials of construction are cheap. It may rather

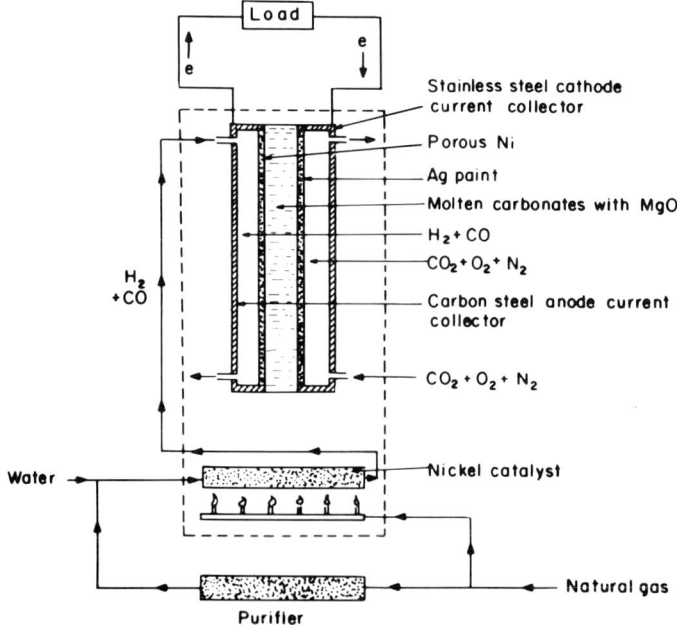

Fig. 11.85. Schematic representation of a high-temperature (500°C) natural-gas fuel cell.

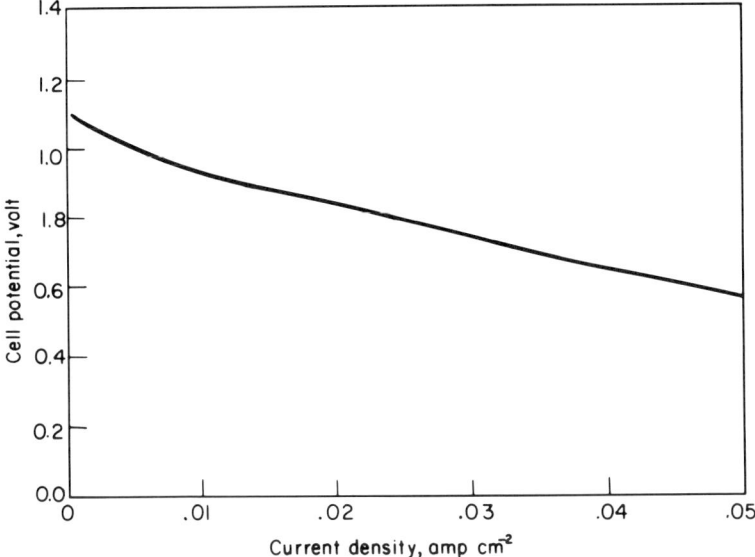

Fig. 11.86. The cell-potential *versus* current-density relations for a high-temperature (500°C) natural-gas fuel cell.

be thought of as an energy converter (perhaps for natural-gas deposits) in which weight is of less importance. However, it may attain a power level of 0.1 W cm^{-2}.

11.3.20. The Relations between Electrochemical Energy Conversion and the Future Dominance of Atomic Energy as the Source of Power

One of the most certain trends of the coming decades is the increasing use of atomic energy as a power source. What effects will this have upon the probable applications of electrochemical power sources? This question can be divided into three further questions.

11.3.20a. Will Atomic Power Sources Compete for Any of the Uses Foreseen for Electrochemical Power Sources? The answer here seems a clear one. Only for the potential large, central fuel-cell stations. Whether it will be more economic to use uranium as a fuel or burn natural gas electrochemically is a question of economics and the local situation in individual countries.† The conversion to electric energy of the stores of natural gas

---

† About half the cost of electricity in a plant or home is that of pushing it through wires. According to calculations, it would be cheaper to pump gases through pipes than pass

and coal must be considered in a different light from the consumption for that purpose of the oil reserve because of the greater use which can be made of oil as raw material for the manufacture of chemicals and food. Some countries have natural-gas supplies which would meet their foreseeable energy needs for more than a century.

In the other developing applications for electrochemical converters, no competition will arise, for all applications except those for central stations fall into classifications where (1) the station is mobile or (2) the power needed is small (100 kW or less).

In the mobile situations (ships, trains, trucks, buses, cars, etc.), atomic reactors are only likely to be used in ships and even then only in relatively large ones. Apart from economic reasons, the balance of which may change with time, there are fundamental aspects of an atomic reactor which make its employment improbable at any time in vehicles of less than several hundred ton weight. Such considerations arise from the necessity for a minimum shielding weight corresponding to a minimum critical mass.

As small power sources, ($\leq$100 kW), atomic reactors are of course even more inapplicable than in most transports.

11.3.20b. Will Electrochemical Means Be Used to Convert Nuclear Power to Electricity? *Atomic energy* is given out as the *heat* of reactions in the controlled fission of U^{235}. The heat is then taken to a heat engine and electricity produced through the same cycle of heating a gas, its impingement on pistons or turbine blades, and the transfer of the resultant mechanical energy to a dynamo. It is desirable to change from this indirect to a more direct method. However, the other direct methods of energy conversion all involve Carnot efficiency losses.

In principle, electrochemical converters could be used to convert nuclear energy to electricity without Carnot losses (Gomberg) and with all the advantages of relative low temperature (compare the materials problems of high-temperature atomic reactors). The method would be to allow the energy of the particles emitted in atomic disintegrations to decompose some substance, e.g., water, and then recombine the products in an electrochemical reactor. Were all the energy of the emitted particles to produce stable molecular species in the radiolysis, conversion efficiencies would be limited only by the fuel-cell's efficiency and could reach toward 75%, nearly double the

---

electricity through wires, and a mode of distribution of nuclear energy to be considered is to have hydrogen produced as the source of atomic power and pumped to sites where electricity is needed, at which the hydrogen would be converted to electric power in fuel cells (or used as heat directly).

best efficiencies in large atomic reactors and much more than double the efficiency of smaller reactors.

When particles from atomic disintegrations pass through a liquid, they momentarily dissociate a large number of particles. The difficulty is that a majority of these recombine. Hence, at the present time, because of the radiolysis losses, there would be an efficiency of only about 10% from this approach. The method is illustrated in Fig. 11.87.

However, such developments are at a very early stage. Could *radical catchers* in the solution (e.g., $Fe^{2+} + OH \rightarrow Fe^{3+} + OH^-$) react with the transient dissociation products before they recombined? Could the radiolytic decomposition be carried out on a substrate which would absorb the radicals before they recombined? There are intriguing possibilities of suitably modifying the radiolysis and making these electronuclear engines of virtually any size prescribed.

11.3.20c. *What Is the Relation between Electricity Storage and Atomic Energy?* To the extent that a substantial fraction of the burning of fossil fuels is carried out electrochemically, the problem of the greenhouse effect and pollution (see Section 11.3.3) is greatly relieved because only one-third to one-half of the CO_2 (and no intermediate products of combustion) is emitted for the same amount of energy obtained. The use of the electrochemical mode of conversion would thus extend the

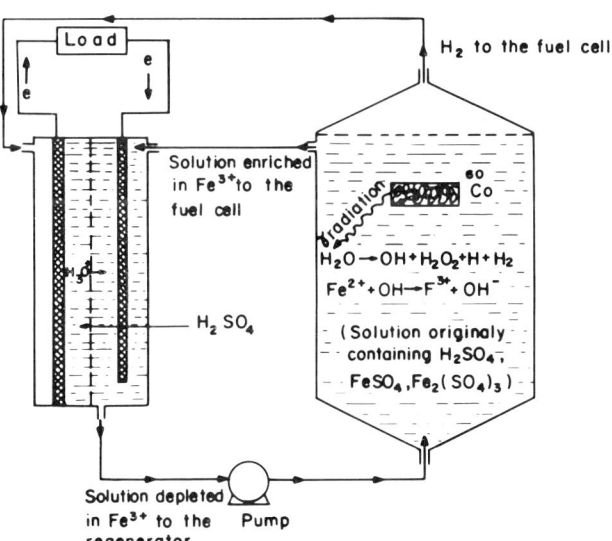

Fig. 11.87. Schematic representation of a radiolytic regeneration fuel cell.

time during which one could inject CO_2 into the atmosphere without significant results from the greenhouse effect from what the present rough calculations indicate as about three decades to about one century. Correspondingly, during this time period, the fraction of electric power which would come from atomic sources would become high and one might assume that raw electricity would become increasingly cheaper, possibly as a result of the introduction of controlled fusion.

When such a situation arises, it will only be possible in conjunction with electrochemical power sources. Thus, abundant cheap electricity would mean an increased economic advantage for all electrical over thermal methods. Transportation would become entirely electric for economic reasons alone. But the electricity from the atomic source has to be stored. The only way electricity can be stored so as to be available on demand is by electrochemical storers, i.e., in batteries.

The nature of such electrochemical electricity storage will be considered in Section 11.4.

11.3.21. A Summary of the Direct Conversion of Chemical Energy to Electricity

At present, most of the energy used by man comes by burning oil. This is a way which further knowledge shows not to be desirable because the oil could be conveniently converted into much-needed food and textiles to meet needs for centuries, whereas its use as an energy source has numerous disadvantages. For one thing, it exhausts a one-time gift of nature at a tremendous rate. Moreover, the CO_2 injection into the atmosphere with which obtaining energy from oil is associated may in about a generation bring far-reaching changes in climate, including significant change in the world sea level.

But there is an even more-urgent economic reason for reappraisal of the mode in which energy is produced. If one must use up fossil fuels (and if natural gas and coal are included, some countries have amounts equivalent to more than a thousand years' needs) to produce energy, then there is a better way to do it than burning it in air and converting only a part of the heat thus produced to energy. This is a way which avoids the intrinsic efficiency loss which occurs in accordance with Carnot's theorem. The only way in which this intrinsic efficiency loss can be avoided is by burning the fuel electrochemically.

Why the Carnot efficiency loss occurs only in heat engines is an interesting story, told with a kinetic model in the text. One finds that electrochemical means of obtaining energy from fuel offers a minimum of double the efficiency in converting chemicals to energy. Another bright point is the directness of the process. One does not go through an intermediate mechanical device (the generator) but obtains the electricity directly and with no moving parts.

Electrodics comes into a very practical phase when one considers electrochemical converters. According to just whether the converter is working at low, intermediate, or high current densities, the efficiency of conversion or the power on the current density depends predominantly on catalysis, electrolyte resistance, and diffusion of the fuel to surfaces, respectively. But these influences *add* to each other, so that as the second and then the third come into play the loss of energy due to the others is present. And as the electrocatalytic loss occurs first (i.e., at the lowest current densities), its effect is always present in the total overpotential, even though the changes in this are largely due to resistance and diffusion at the higher current densities.

Electrocatalysis is the center of electrochemical energy conversion. If one were able to increase it so that rates of operation could be about ten times those on platinum at one-tenth the cost of platinum a major technological happening would occur. Electrocatalysis is one of the great new technological challenges of this time. Progress in it appears likely to take place, if at all, in the fundamental laboratory.

Power, rather than energy-conversion efficiency, again depends upon electrocatalysis. The power per unit weight of electrochemical converters is not yet as good as that of gasoline-burning internal-combustion engines, but it is in the same range as that of diesel engines.

What happened to energy conversion at the end of the nineteenth century? At that time, many of the negative consequences of the use of thermal combustion which are widely discussed today, e.g., pollution, the inefficiency of the combustion engine, etc., were realized among thermodynamicists, but electrochemical kinetics was not at that time realized at all. It was this blockage—no knowledge of what overpotential meant, no model for interfacial electric charge transfer—which made engineers go along the internal-combustion route instead of developing electrochemical engines as they could have done about 1900. They came under the influence of the Great Nernstian Hiatus time of electrochemistry (see Fig. 1.13), when the dead hand of thermodynamics dominated it and its researchers looked blankly at their equations relating potential to interfacial concentration with their vital mistake of assuming that interfacial electron transfer would always be thermodynamically reversible. A century of atmospheric pollution has been the result.

The concept of an electrochemical engine and the development of practical fuel cells had to wait until the 1960's. Man would have had to wait much longer for this development, too, if it were not that special gains could be made by using the properties of porous electrodes. The advantage they give (through much higher limiting current per square centimeter of external surface than with planar electrodes) are not so much because they are porous in the sense of having a very large area within for unit area without as because the menisci there have finite contact angles. Thus, they are thin, *very* thin, and a gaseous fuel dissolves through such thin boundary layers at maximum diffusion-current densities, orders of magnitude above those otherwise possible.

Last of all, electrochemical converters, as well as storers, will be used increasingly as atomic energy becomes more and more the origin of electricity. For one thing, atomic reactors will perhaps always have to be above a limiting weight, some tens of tons, for intrinsic reasons. The very probable future use of fuel cells will be in ships, trains, big trucks, buses, and, of course, space ships and lunar and planetary bases. The development of large fuel-cell stations in competition with atomic power plants depends upon the economics of each country and the price and availability there of fossil fuels.

Further Reading

1. W. R. Grove, "On Voltaic Series and the Combination of Gases by Platinum," *Phil. Mag.*, **14**: 127 (1839).
2. W. Ostwald, *Z. Electrochem.*, **1**: 212 (1894).
3. E. W. Justi and A. W. Winsel, *Kalte Verbrennung Fuel Cells*, Franz Steiner Verlag GMBH, Wiesbaden, Germany, 1962.
4. W. Vielstich, *Brennstoffelemente; Moderne Verfahren zur Electrochemischen Energiegewinnung*, Weinheim, Bergstrasse, Verlag Chemie, Germany, 1965.
5. *Les Piles à combustible*, Publications de L'Institut Français du Pétrole, Éditions Technip, Paris, 1965.
6. "Atmospheric carbon dioxide," Appendix Y4, Report of the Environmental Pollution Panel, PSAC, Washington, D.C. (1965).
7. K. R. Williams, *An Introduction to Fuel Cells*, Elsevier Publishing Co., Amsterdam, London, New York, 1966.
8. L. G. Austin, "Fuel Cells," *Sci. Tech. Infor. Div.*, *NASA*, Washington, D.C. (1967).
9. J. O'M. Bockris and S. Srinivasan, *Fuel Cells: Their Electrochemistry*, McGraw–Hill Book Company, 1969.
10. J. O'M. Bockris and B. Cahan, "Effect of Finite Contact Angle Meniscus on Kinetics in Porous Electrodes," *J. Chem. Phys.* **50**: 1307 (1969).

11.4. ELECTRICITY STORAGE

In the electrochemical converters described hitherto, the two reactants are stored outside the cells and are fed to each electrode, respectively (*cf.*, e.g., Figs. 11.82 and 11.85).

Now, suppose that, instead of feeding the reactants to the electrodes continually, one places the reactants *on* the electrodes. When cathode and anode are joined through an external circuit, the two fuels, each on one inert electrode, will react electrodically at the respective electrode–solution interfaces and a current will flow, just as in an electrochemical converter. For example, suppose that, on one inert electrode substrate, powdered cadmium had been placed (in analogy to the *fuel* of a converter) and, on the other electrode, an oxide of nickel (the other fuel). When the external circuit is closed [Fig. 11.88(a)], the electrode on which the cadmium has been placed undergoes an anodic reaction; it deelectronates with the following overall charge-transfer reaction

$$Cd + 2OH^- \rightarrow Cd(OH)_2 + 2e \qquad (11.96)$$

Correspondingly, at the electrode on which the nickel oxide has been placed, an electronation reaction occurs and can be represented as an overall reaction

$$2NiOOH + 2H_2O + 2e \rightarrow 2OH^- + 2Ni(OH)_2 \qquad (11.97)$$

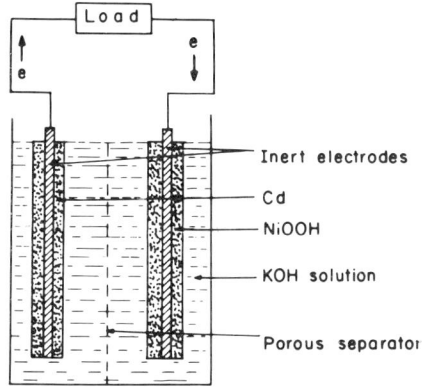
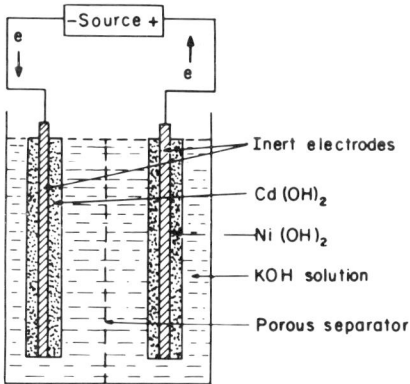

Fig. 11.88(a). A nickel–cadmium cell at the beginning of discharging. At the cathode, $2NiOOH + 2H_2O + 2e \rightarrow 2Ni(OH)_2 + 2OH^-$; at the anode, $Cd + 2OH^- \rightarrow Cd(OH)_2 + 2e$.

Fig. 11.88(b). A nickel–cadmium cell at the beginning of charging. At the cathode, $Cd(OH)_2 + 2e \rightarrow Cd + 2OH^-$; at the anode, $2Ni(OH)_2 + 2OH^- \rightarrow 2NiOOH + 2H_2O + 2e$.

So far, one has an unusual kind of fuel cell with a limited supply of (solid) fuel, which is on each electrode. But what happens if one uses an external source of electric power to push electricity *into* the cell (Fig. 11.88b) in the opposite way to that in which it passes if this cell is allowed to function spontaneously.

If one were dealing with, say, an electrochemical converter with decane as the anodic fuel, the product of the overall cell reaction would be CO_2 and H_3O^+ and, if the pH is acidic, CO_2 is evolved and expelled from the system (in an alkaline solution, the CO_2 would of course form carbonate and eventually precipitate in the cell). Thus, in this decane case, when one tries with the aid of an external power source to push a current in a direction opposite to that in which the *generator* has been functioning spontaneously, there is no chance of reversing the overall reaction of decane oxidation, i.e., of reducing CO_2 to decane at the cathode, while evolving oxygen at the anode (for, among other reasons, the CO_2 has escaped from the system into the atmosphere). But what would happen in a device in which, as with the cell (Fig. 11.88a) whose electrodes contain $Cd(OH)_2$ and $Ni(OH)_2$, the products of the *spontaneous* functioning of the cell are held on the electrodes? One can see this from the above electrode reactions. Nothing escaped from the system during the spontaneous functioning of the electrode reactions (the cell *giving out* electricity, i.e., *discharging*). Hence, on the cadmium electrode, the reaction on the electrode plate covered with cadmium hydroxide can be, when one puts electricity into the device (Fig. 11.88b),

$$Cd(OH)_2 + 2e \rightarrow Cd + 2OH^- \qquad (11.98)$$

and, on the nickel electrode during corresponding circumstances,

$$2Ni(OH)_2 + 2OH^- \rightarrow NiOOH + 2H_2O + 2e \qquad (11.99)$$

One is back to the starting point again, charged up once more.

But, it is clear now what the device represents. It is an *electricity storer*. On the one hand, the substances on its electrodes can react to electricity sent in from an external source by undergoing transformation. On the other hand, upon demand, the storer can allow its electrode reactions to occur spontaneously and send out through an external circuit (Fig. 11.88a) the electricity which was previously put into it. Thus, electricity is stored in the chemicals on the electrodes because they are a certain form; and, in changing over to another, they give out electricity.

11.4.1. Conventional and Descriptive Terminology in Energy Conversion and Storage

One of the difficulties for the student learning electrochemistry is that the terminology of devices in which the energy of chemical reactions is *converted* directly to electricity is confusedly mixed up with that of devices which serve to *store* electricity. The aims and uses of these devices are different in an essential way, and it seems desirable to bring this out in a series of terms which describe the function of the device named.

Another reason for a change in terminology is that many electrochemical terms, particularly those associated with the *storage* of electricity, are fusty. A strange state of affairs existed for a long time in electrochemical energy storage. Electricity storers were known and used in the nineteenth century. They are familiar articles of everyday use, e.g., the car battery. For about four or five decades, their basic design and function, together with their empirical uninformed terminology, hardly changed for they were unchallenged by any rival device. It is only since the beginning of the interest in electrochemical energy converters that the situation in electrochemical electricity storers has begun to respond to modern electrochemical research. But the energy converters belong to the space age, and the newly developing electrochemical electricity storers will belong to the coming atomic age. It would, therefore, be unfortunate if one had to live with a terminology for them belonging to the horseless carriage age.[†]

The essential and considerable difference between the electrochemical energy converters and the electricity storers must, therefore, be reiterated boldly and clearly (see Figs. 11.82, 11.85, and 11.88). The converters take in *substances*, usually stored externally, and produce electric energy directly from the spontaneous reaction between the two substances. The storers take in *electricity* when an electric current is passed through them from an outside source and give back electricity when required by allowing to run spontaneously in the forward direction the reaction which had been made to go backward during charging. In converters (or generators), the cost of the substances undergoing reaction—the fuels—is of vital importance; it determines the cost of the electricity and power produced. The cost of the substrate by which the reactions are catalyzed is also of great importance in determining the initial cost of the converter. In storers, the cost of the substances undergoing the reactions at the electrodes determines the initial cost of the device but it only secondarily influences the cost of the

---

[†] Analogous remarks apply, of course, to other fustyisms such as *electrolysis cells* (substance producers), *electromotive force* (cell potential), and *polarization* (overpotential).

electricity given back from storage (as a result of a contribution to this of the amortization of the materials in the storer).

A summary of the terminology used here is given in Table 11.18. For 20% more words, the descriptive system of terminology gives much more than 100% increase in the indication of function.

11.4.2. The Important Quantities in Electricity Storage

In discussing energy *conversion*, there were two important criteria. If the main purpose is the conversion of chemical energy to provide electricity at minimum cost, conversion *efficiency* is the main point. If the main purpose is the providing of power to a machine, particularly in a transport situation, then *power per unit area* of electrode and *power per unit weight and volume* are important.

In electricity storers, the considerations that determine which are the important criteria differ from those for converters. For the storers, the most often used concepts are those now to be presented.

11.4.2a. Electricity Storage Density. This refers to a measure of the maximum total amount of electricity which can be withdrawn from unit weight of a substance when (of course, in conjunction with another substance in a cell) it undergoes an electrochemical reaction.

Thus, let one formula weight mole of substances (as written on the left-hand side of the stoichiometric equation) enter into an electrodic reaction in which they accept or reject n electrons to produce other substances in one of the two electrode reactions in a cell. Then,

$$M \text{ [g of substance]} = nF \text{ [coulombs of electricity]} \qquad (11.100)$$

$$1 \text{ kg} = \frac{96{,}500 \times 1000\, n}{M} \text{ [C]} \qquad (11.101)$$

or

$$\text{Electricity storage density is } \frac{96.5\, n 10^6}{M} \text{ [C kg}^{-1}\text{]} \qquad (11.102)$$

The electricity storage density is the *amount of electricity* [coulombs] per unit weight which the storer can hold. It states nothing concerning the *energy* [watts or kilowatt hours] which it can store per unit weight. Suppose the potential difference through which the amount of electricity stored passed (the cell potential) was particularly small for a given system, then its electricity storage density could be high but the *energy* it stored (the coulombs flowing times the potential difference passed through) would be small.

TABLE 11.18
Conventional and Descriptive Termonilogy for Electrochemical Reactors

Conventional	Descriptive	Simple representation
Fuel cell	Electrochemical energy producer	Fuel → [Fuel cell] → Electricity
Fuel-cell battery, electric-motor unit	Electrochemical engine	Fuel → [Fuel cell] → [Motor] → Power
Primary battery	Electrochemical energy converter (with internal storage of reactants)	[Battery (fuel contained)] → Electricity
Secondary battery, accumulator	Electricity storer	Electricity → [Secondary battery] [charging]; [Secondary battery] → Electricity [discharging]
Electrolysis cell	Electrochemical substance producer	Electricity → [Electrolysis cell] → Substance

The electricity storage densities for a number of substances are shown in Table 11.19.

Of course, for two substances or reactions sufficiently complementary to make a cell, the coulombs per kilogram of the cell can easily be calculated by using data such as that in Table 11.19 together with a knowledge of n for the reaction concerned.

It is noteworthy that the electricity storage density, e.g., says little of the actual behavior of the substance at an electrode or the possibilities of making an electrode from it. Electricity storage density may be, therefore, only a paper parameter, useful as an indication of what might be possible

TABLE 11.19
Idealized Maximum Electricity-Storage Density for Some Substances

Substance	Reaction	Idealized maximum electricity-storage density, 10^6 C kg^{-1}
H_2	$\frac{1}{2}H_2 \rightarrow H^+ + e$	96.5
Lithium	$Li \rightarrow Li^+ + e$	14.0
Sodium	$Na \rightarrow Na^+ + e$	4.2
Potassium	$K \rightarrow K^+ + e$	2.5
Beryllium	$Be \rightarrow Be^{2+} + 2e$	21.4
Magnesium	$Mg \rightarrow Mg^{2+} + 2e$	8.0
Aluminum	$Al \rightarrow Al^{3+} + 3e$	10.7
Calcium	$Ca \rightarrow Ca^{2+} + 2e$	4.8
Iron	$Fe \rightarrow Fe^{2+} + 2e$	3.5
Nickel	$Ni \rightarrow Ni^{2+} + 2e$	3.3
Zinc	$Zn \rightarrow Zn^{2+} + 2e$	2.3
Cadmium	$Cd \rightarrow Cd^{2+} + 2e$	1.7
Silver	$Ag \rightarrow Ag^+ + e$	0.9
Mercury	$Hg \rightarrow Hg^{++} + e$	1.0
Lead	$Pb \rightarrow Pb^{2+} + 2e$	0.9
Br_2	$\frac{1}{2}Br_2 + e \rightarrow Br^-$	1.2
Cl_2	$\frac{1}{2}Cl_2 + e \rightarrow Cl^-$	2.7
F_2	$\frac{1}{2}F_2 + e \rightarrow F$	5.0
CH_4	$CH_4 + 2H_2O \rightarrow CO_2 + 8H^+ + 8e$	48.1
C_2H_6	$C_2H_6 + 4H_2O \rightarrow 2CO_2 + 14H^+ + 14e$	45.0
C_3H_8	$C_3H_8 + 6H_2O \rightarrow 3CO_2 + 20H^+ + 20e$	43.8
CH_3OH	$CH_3OH + H_2O \rightarrow CO_2 + 6H^+ + 6e$	18.1
C_2H_5OH	$C_2H_5OH + 3H_2O \rightarrow 2CO_2 + 12H^+ + 12e$	25.1
CO	$CO + H_2O \rightarrow CO_2 + 2H^+ + 2e$	6.9
NH_3	$2NH_3 + 6OH^- \rightarrow N_2 + 6H_2O + 6e$	17.0
N_2H_4	$N_2H_4 + 4OH^- \rightarrow N_2 + 4H_2O + 4e$	12.0

with certain substances *if* they are made to react reversibly at electrodes which give the reactions large i_0's. For example, it is possible to write attractive hypothetical electrode reactions with a large number of organic compounds, and many of these have large n values and hence seem to indicate very attractive watt hours per kilogram (see Section 11.10.5). However, these substances often do not have enough electronic conductivity to form part of an electrode structure, or the kinetics of their reactions at room temperature (i_0 factors) are so poor that cells comprising such reactions of organic substances deliver negligible power.

11.4.2b. Energy Density. This refers to the energy which may be extracted from a given weight of a substance or a device per unit weight. Electrical energy is measured by the quantity of electricity (current multiplied by time) multiplied by the potential difference through which such a quantity passes. Thus,

$$\text{Energy density} = \text{Electricity storage density} \times \text{Cell potential} \quad (11.103)$$

Now, by referring again, at first, to an electrode material rather than to two electrode materials making up a cell,

$$\text{Maximum energy density} = \frac{9.65 \times 10^7 n V_e}{\text{Mole}} \quad (11.104)$$

where V_e is the reversible electrode potential of the material in the reaction concerned.[†] The value of Eq. (11.104) is in coulomb volts per kilogram (if "Mole" is expressed in kilograms). The same may be expressed as ampere-second-volt per kilogram = ampere-volt-second per kilogram = = watt seconds per kilogram. Hence, maximum energy density is, in kilowatt hours per kilogram,

$$\frac{9.65 \times 10^7 n V_e}{M\ 60 \times 60 \times 1000} = 26.8 \frac{n V_e}{\text{Mole}} \quad [\text{kWh kg}^{-1}] \quad (11.105)$$

This maximum energy density for an electrode material undergoing an electrode reaction involving n electrons is only an indication of its ability to act as a vessel loaded with realizable electricity, as will be seen below.

[†] Note that the energy density of a *single electrode* is a fictitious quantity since V_e depends on the arbitrarily selected reference electrode (e.g., a normal hydrogen electrode). It reflects the potentialities of an electrode material *if* coupled with the normal hydrogen electrode. Yet, on one side, it gives a basis for comparing different substances, and, on the other side, it gives a rough indication of the energy density which can be expected from coupling two such substances.

TABLE 11.20
Idealized Maximum Energy-Density Values for Some Elements, at 25°C

Element	Reaction	Thermodynamic potential E^0, V	Idealized maximum energy density,[†] kWh kg^{-1}
Lithium	Li → Li$^+$ + e	−3.045	11.76
Sodium	Na → Na$^+$ + e	−2.714	3.16
Potassium	K → K$^+$ + e	−2.925	2.00
Beryllium	Be → Be^{2+} + 2e	−1.847	10.98
Magnesium	Mg → Mg^{2+} + 2e	−2.363	5.21
Calcium	Ca → Ca^{2+} + 2e	−2.866	3.83
Aluminum	Al → Al^{3+} + 3e	−1.662	4.95
Manganese	Mn → Mn^{2+} + 2e	−1.18	1.15
Iron	Fe → Fe^{2+} + 2e	−0.440	0.42
Nickel	Ni → Ni^{2+} + 2e	−0.250	0.23
Zinc	Zn → Zn^{2+} + 2e	−0.763	0.63
Cadmium	Cd → Cd^{2+} + 2e	−0.403	0.19
Lead	Pb → Pb^{2+} + 2e	−0.126	0.03
Silver	Ag$^+$ + e → Ag	+0.799	0.20
Mercury	Hg$_2^{2+}$ + 2e → 2Hg	+0.789	0.11
Bromine	Br$_2$ + 2e → 2Br$^-$	+1.065	0.36
Chlorine	Cl$_2$ + 2e → 2Cl$^-$	+1.359	1.03
Fluorine	F$_2$ + 2e → 2F$^-$	+2.87	4.05

[†] In a hypothetical cell with the normal hydrogen electrode.

TABLE 11.21
Idealized Maximum Energy Density of Hypothetical Electrode Couples at 25°C

Reaction	Cell potential V_e, V	Energy density, kWh kg^{-1}
2Li + F$_2$ → 2LiF	6.05	6.03
2Li + CuCl$_2$ → 2LiCl + Cu	3.06	1.11
2Li + CuF$_2$ → 2LiF + Cu	3.55	1.66
2Li + NiF$_2$ → 2LiF + Ni	2.83	1.34
3Li + CoF$_3$ → 3LiF + Co	3.64	2.12
2Li + CoF$_2$ → 2LiF + Co	2.88	1.40
2Li + CuO → Li$_2$O + Cu	2.25	1.30
Mg + CuF$_2$ → MgF$_2$ + Cu	2.92	1.25
3Mg + 2CoF$_3$ → 3MgF$_2$ + 2Co	2.89	1.52
Mg + AgO → MgO + Ag	2.98	1.08
Ca + CuF$_2$ → CaF$_2$ + Cu	3.51	1.33
Ca + CuO → CaO + Cu	2.47	1.11

Values of such maximum energy densities of certain materials in certain reactions are shown in Table 11.20. The maximum energy density of the best combinations of electrode reactions is shown in Table 11.21.

The real energy densities of cells are less than the values given in Tables 11.20 or 11.21 and sometimes by an unfortunately great fraction, e.g., some one-fifth (Table 11.22). Some of the reasons for this are obvious. The idealized maxima of the calculation of the hypothetical maximum power density take into account the weight of the active materials with no accounting for grid, container, solution, connections, etc. In addition, the V_e value must of course be reduced by the prevailing overpotential for the given conditions of the rate or current density at which it is desired to supply electricity [see Eq. (11.84)]. Nevertheless, these factors do not alone

TABLE 11.22

Realized and Idealized Energy-Storage-Density Parameters for Some Cell Systems

System	Thermodynamic reversible potential V_e, V	Realized energy density (mean), kWh kg^{-1}	Idealized energy density, kWh kg^{-1}	Ratio of realized to idealized energy density
Single discharge				
M-O$_2$-KOH-Zn	1.64	0.15	0.44	0.34
m-DNB-MgBr$_2$-Mg	2.84	0.22	1.54	0.14
Ag$_2$O$_2$-KOH-Zn	1.81	0.11	0.22	0.50
HgO-KOH-Zn	1.34	0.11	0.22	0.50
Hypothetical single discharge				
Li-F$_2$	6.0	–	1.5	
Mg-O$_2$	3.1	–	1.3	
Be-*m*-dinitrobenzene	2.6	–	3.1	
Be-nitrobutanol	2.3	–	3.7	
Multiple charge–discharge				
PbO$_2$-H$_2$SO$_4$-Pb	2.04	0.02	0.18	0.11
NiO(OH)-KOH-Cd	1.48	0.04	0.22	0.18
Ag$_2$O$_2$-KOH-Zn	1.70	0.11	0.44	0.23
Hypothetical multiple charge–discharge				
Be-BeF$_2$-F$_2$	4.5	–	3.30	
Al-AlCl$_3$-Cl$_2$	3.02	–	2.20	

account for the differences between the idealized maxima and the real energy density of electricity storers; this point will be discussed further.

Caution must be exercised even in relating experimentally established energy densities of a cell to a calculation concerning what electricity can be got from it under conditions other than those of the original determination. The quantity referred to as energy is potential × current × time. But these three quantities cannot be varied indiscriminately and the product placed equal to the energy expected for a given weight of cell material taken from an experimentally determined energy density. Thus, potential and current density are not independent variables (Fig. 11.76) and, as stated, the potential which must be substituted in the equation given is not the thermodynamic potential V_e but that diminished by the overpotential for the given rate (i.e., the current density) at which the electricity passes when the storer is applied to a given load. Correspondingly, the current density can be varied only up to a value, for the electrodes concerned, which corresponds to the smaller of the limiting current densities on each electrode. The presence of significant overpotential is equivalent to a real loss in energy available, so that one cannot simply demand from the storer an increased current density to compensate for a fall in cell potential. Thus, the hypothetical increase of current will simply reduce further the effective cell potential because of the increase in overpotential which an increase in current would bring about. The energy density of a storer is hence firmly dependent, by electrode kinetics, on the current density, or the rate at which the cell is discharged (cf. Fig. 11.89).

In summary, calculated maximum energy densities are indicative largely of what system to investigate in considering the development of future electricity storers. Values obtainable in practice (see Table 11.22) are at present one-half to one-fifth the ideal values and must be established empirically. Once measured at a given current density, the energy densities can, even then, be used only as rough guides because the available potential is reduced by the overpotential which arises with increase of current density [see Eq. (11.84)], so that, intrinsically, a variation of energy density with power will tend to occur (Fig. 11.89).

11.4.2c. *Power.* The criteria for a good electricity storer given above relate only to the question: *How much* electricity (or electric energy) can the device hold and deliver per unit weight? As the device is for the storage of electricity, this question is obviously the primary one. But there is another and different question which relates to the *rate* at which the storer would be able to release the energy again. (Rate in the form of current times

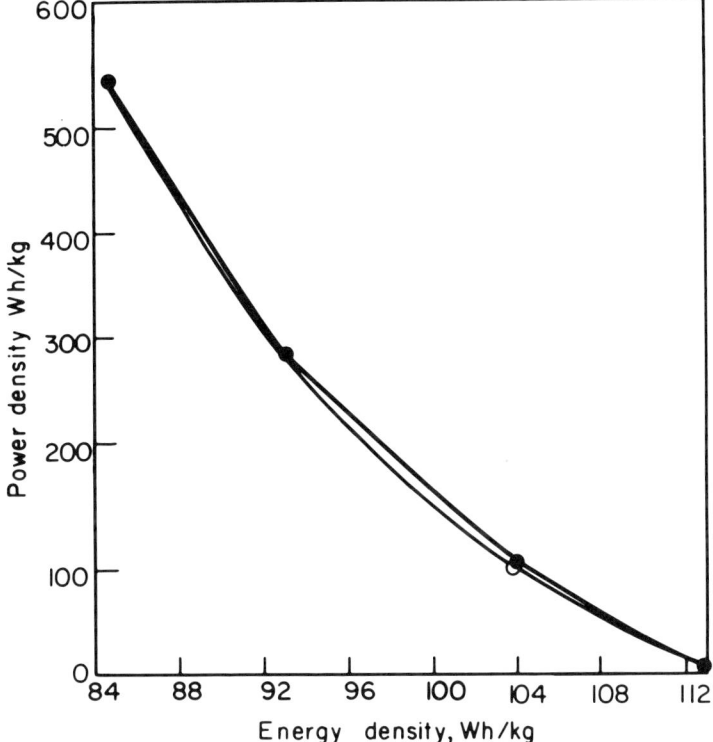

Fig. 11.89. Power-density *versus* energy-density relations for a silver–zinc cell.

potential is the rate of delivering energy, i.e., the power.) One may imagine a device which has an excellent energy storage density but can only release this energy at a slow rate. The disadvantage of such a device would be that only a situation requiring low power could be coped with by it. Hence, the power density of a storer (the number of kilowatts, rather than kilowatt hours, it can put out per unit weight) is also an important index of its merit as a device.

Upon what factors does this power density of an electricity storer depend? The difference in function between electricity storers and electrochemical energy converters has been stressed in Figs. 11.82, 11.85, and 11.88 and in the descriptive terminology which is used here to designate the different functions of fuel cells and secondary batteries (or accumulators) in Table 11.18. But the electricity storer working in the supply mode and the electrochemical generator, or converter, are both electrochemical cells working as energy producers (see Section 11.3.18). Therefore, the same sort of considerations apply to determining the power which can be produced by the electricity storer as apply to the power of an electrochemical generator, and these have already been discussed in principle in Section 11.3.13.

To recapitulate briefly, the power is the product (for each single cell) of the working potential of the cell (its reversible, thermodynamic or maximum potential diminished by the sum of the overpotentials at and between the electrodes in the cell) multiplied by the current of the cell, which is itself a function of overpotential. The power increases with the current density and passes through a maximum as shown in Fig. 11.76.

11.4.2d. Desirable Trends. Thus, finally, let the basic *trends* be summarized, which one would like to see in a storer (some of which [e.g., 7 and 8] may of course be mutually contradictory). In an ideal sense, one wants to maximize the energy density and the power output. Therefore, one would like to incorporate the following in the device:

1. The cell reaction which would have a maximum thermodynamic reversible potential
2. Cell reactions which involve in the overall reaction a maximum number of electrons (n in Eq. 11.105)
3. Reactants of the lowest possible molecular weight
4. Electrode reactions which have the highest exchange-current densities available, at least in the milliampere per square centimeter range
5. A physical structure of the electrodes which gives the highest possible limiting diffusion-current densities
6. Geometric structures which give the minimum distances between the electrodes so that energy-wasting ohmic drops caused by the passage of electricity between the electrodes are minimized
7. The highest possible solution conductance and solubility of dissolved reaction products of discharge [the latter to maximize the limiting currents in the cell reactions, see Eq. (11.84)]
8. Highly insoluble reactants (to avoid diffusion onto the other electrode and self-discharge when the cell is in the charged state)
9. The lightest possible materials of construction compatible with stability and mechanical strength

These criteria represent trends, all of which are desirable, some of which may be mutually contradictory. In real devices, many other factors special to the device have to be considered, and these all make for compromises with the desirable trends which arise from theoretical considerations made only in principle. For example, the basic demand 1 would suggest that an electrode of lithium in a highly conducting aqueous solution together with an electrode of fluorine for the overall reaction $Li + \frac{1}{2}F_2 \rightarrow LiF$ has a very high standard free energy and therefore, through Eq. (11.72), a large thermodynamic reversible potential and, through Eq. (11.105), a large maximum energy density. But the first theorist who tried to assemble such a cell for test would be likely to receive a painful lesson in the need also for a continuing knowledge of the freshman chemistry of the explosive reaction of lithium and water.

Correspondingly, many other factors must be taken into account in an attempt to make new real electric energy storers. If the formation of new, solid phases is involved, will they crystallize stably and adhere to the substrate? Do these solid phases conduct electricity between the grid on which they are laid down and the solution with which they are in contact? Do the solid phases enter into side reactions with the solution and thus corrode, which would change the shape of the electrodes, etc.? And all of these scientific trends must in turn be compromised with the economics of the particular time and place.

11.4.3. Classical Electricity Storers

Besides the already described nickel–cadmium cell and its somewhat older relative, the nickel–iron *Edison cell*, two classical cells of considerable practical interest should be mentioned.

11.4.3a. The Lead–Acid Storage Battery. This electricity storer invented by Planté as early as 1860 is still the one most-used storer at the present time, particularly for driving the electric motor by which internal-combustion engines are made to fire.

It consists basically of two electrodes of lead sheet and the electrolytic solution H_2SO_4 saturated with $PbSO_4$ between them. Upon charging, Pb^{++} is deposited on the lead sheet which is made cathodic by the application of a potential from an outside power source, i.e., electronation occurs. At the other electrode during charging, the Pb^{2+} ions present in solution yield electrons to the anodic lead plate and become Pb^{4+}, after which they undergo hydrolysis to form crystalline PbO_2, which deposits on the plate.

The first attempt at formulating the electrode reactions during charging would therefore be

at cathode,

$$Pb^{2+} + 2e \rightarrow Pb \tag{11.106}$$

at anode,
$$Pb^{2+} + 6H_2O \rightarrow PbO_2 + 4H_3O^+ + 2e \qquad (11.107)$$

During discharge, the reverse of the above reactions occurs. Relatively high current densities can be reached during discharge, i.e., the overpotential developed during the reactions is small. The importance of the fact that the exchange-current density for the discharge of H_3O^+ on lead is low may be seen for, at the cathode during charging, the reaction $Pb^{2+} + 2e \rightarrow Pb$ occurs and not the reaction $H_3O^+ + e \rightarrow \frac{1}{2}H_2 + H_2O$ [i.e., this latter reaction occurs to a negligible extent because of the relation $i = i_0 e^{-\alpha \eta / RT}$ (see Section 8.2.10) and because the exchange-current density for the reaction of hydrogen evolution on lead is particularly low].

There are two principal difficulties which make it desirable to develop electricity storers other than the lead–acid storage system as quickly as possible. The more important one is the weight of the cell per unit of energy produced, i.e., the poor value of the real energy density. The theoretical value for this is 0.18 kWh kg^{-1}, but the practical value is 0.022 to 0.026 kWh kg^{-1}.

The second difficulty is that the plates undergo a process called *sulphation*. When the storer is not supplying electricity, spontaneous discharge takes place. The reactions occurring are
at the lead electrode,

$$Pb + SO_4^{2-} \rightarrow PbSO_4 + 2e \qquad (11.108)$$

$$2H_3O^+ + 2e \rightarrow 2H_2O + H_2 \qquad (11.109)$$

at the PbO_2 electrode,

$$PbO_2 + 2H_2SO_4 + 2e \rightarrow PbSO_4 + 2H_2O + SO_4^{2-} \qquad (11.110)$$

$$Pb + SO_4^{2-} \rightarrow PbSO_4 + 2e \qquad (11.111)$$

Such processes as this gradual conversion of lead and lead dioxide to lead sulfate of larger particle size (than that formed during normal discharge) limit the life of the cell in practice.

Why was such a low energy density cell as the lead–acid battery used for so long without an advance until very recent years to one of the many new electrochemical systems which are possible? One reason has been mentioned; there was no competitive product. But the other reason is a positive one and the reason why the lead–acid cell will not disappear in the next few years as the new electrochemical storage units now being researched are introduced. It is that the battery gives very high current densities per unit area (hence, volume). Its poor power per unit weight is not so important because little total power (or weight) is needed to start an automotive engine. It provides just what is needed for an important application.

11.4.3b. A Dry Cell. *Dry* cells are electrochemical energy storers in which the electrolyte is immobilized in the form of a paste. A typical dry cell is the *Leclanché cell*. A schematic diagram of this cell is shown in Fig. 11.90. The reactions occurring in the cell during discharge are

at anode,
$$Zn \rightarrow Zn^{2+} + 2e \qquad (11.112)$$

at cathode,
$$2MnO_2 + 2H_3O^+ + 2e \rightarrow Mn_2O_3 + 3H_2O \qquad (11.113)$$

Since hydroxide ions are produced during working (because H_3O^+ is consumed), the following irreversible side reactions occur

$$OH^- + NH_4^+ \rightarrow H_2O + NH_3 \qquad (11.114)$$

$$Zn^{2+} + 2NH_3 + 2Cl^- \rightarrow Zn(NH_3)_2Cl_2 \qquad (11.115)$$

$$Zn^{2+} + 2OH^- \rightarrow ZnO + H_2O \qquad (11.116)$$

$$ZnO + Mn_2O_3 \rightarrow Mn_2O_3 \cdot ZnO \qquad (11.117)$$

Owing to the above reactions, the cell is only partially rechargeable and this to such a small extent that it is never done in practice. In spite of this

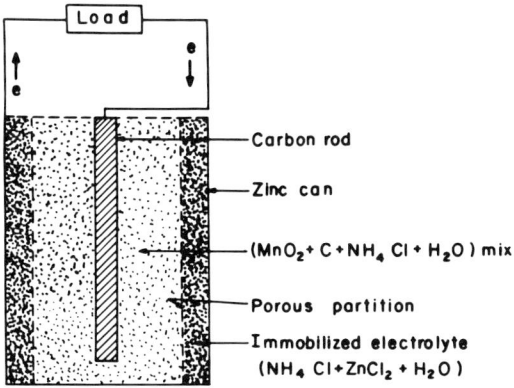

Fig. 11.90. Schematic diagram of a Leclanché cell.

disadvantage, these cells are extensively used as primary batteries (one-shot sources of power). They have an energy density of about 0.14 kWh kg^{-1} if discharged at low current densities.

11.4.3c. Two Relatively New Electricity Storers

(i) *Silver–zinc cell.* An electrochemical energy storer which is particularly noted for high power densities in practice is the silver–zinc cell, which became available in the 1950's (André, Yardney). The materials used as fuels on the electrode substrates are zinc and AgO. Saturated KOH is the electrolyte. A schematic diagram of this cell is shown in Fig. 11.91.

Reactions occurring in the cell during charging may be described as

at anode,
$$Ag + 2OH^- \rightarrow AgO + H_2O + 2e \qquad (11.118)$$
at cathode,
$$ZnO + H_2O + 2e \rightarrow Zn + 2OH^- \qquad (11.119)$$

While discharging, the reverse of the above reactions takes place.

The exchange-current densities of the cell reactions are relatively large; and the side reactions, comparatively slow in rate. These cells have an energy density in the region of 0.1 kWh kg^{-1}. The fairly small number (a few hundred) of charge–discharge cycles which this cell is able to stand before its reactant mass undergoes some form of deterioration is a disadvantage which has to be compared with its advantage in having the highest energy density and power density of commercially available cells. Note that the number of cycles which the cell undergoes before breakdown is a function of the degree to which it is discharged at each cycle (Fig. 11.92).

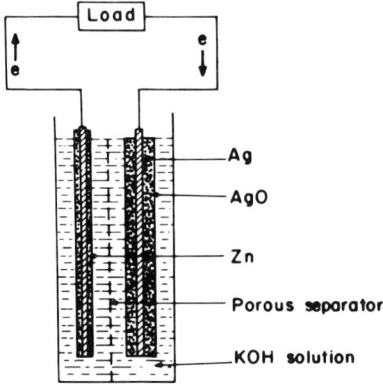

Fig. 11.91. Schematic diagram of a silver–zinc cell.

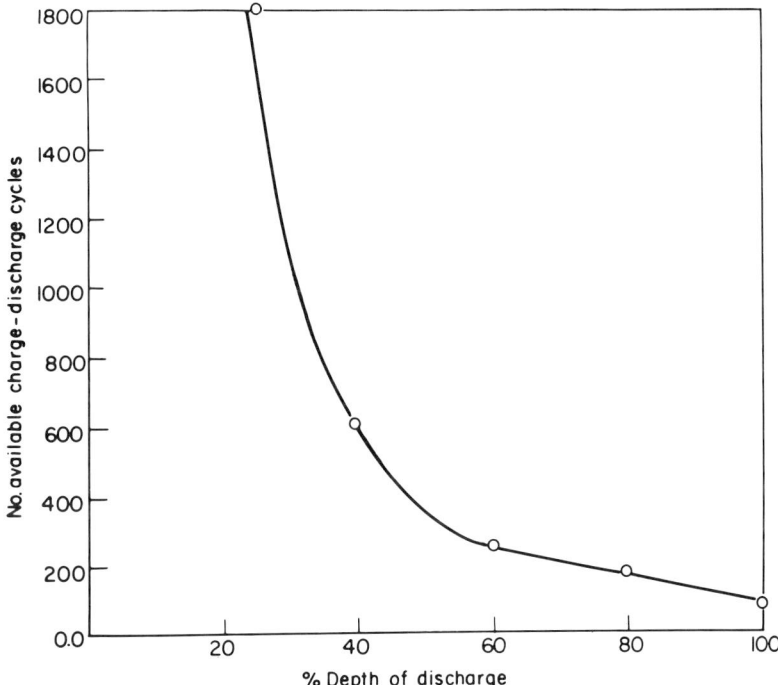

Fig. 11.92. A plot of the number of available charge–discharge cycles *versus* the depth of the discharge for a silver–zinc cell.

(ii) *Sodium–sulfur cell.* It has been pointed out (*cf.* Section 11.4.2) that the hypothetical cell coupling an alkali metal with oxygen with the use of an aqueous solution would be an excellent storer were it not for the violent corrosion reaction with the aqueous solution. One might conceive of separating the alkali metal from the aqueous solution by a membrane through which only alkali metal *ions* would pass, i.e., the reactive atoms of the alkali metal would not contact the aqueous solution. The difficulty with this suggestion is that the conductance of known membranes does not attain a sufficiently high value until temperatures above those at which aqueous solutions boil.

An analogous possibility is to use sulfur (mp $<$ 120°C; bp, 444.6°C) in place of oxygen. The ions corresponding to OH$^-$ ions in aqueous electrolytes are now S^{2-} ions. The advantage is that much higher temperatures can be used than for the aqueous analog, and the conductance of certain membranes (in particular, $Na_2O \cdot 11Al_2O_3$, called β *alumina*) is now sufficient so that passage of current through them does not cause a large and wasteful loss in energy in the form of ohmic heat (Kummer and Weber).

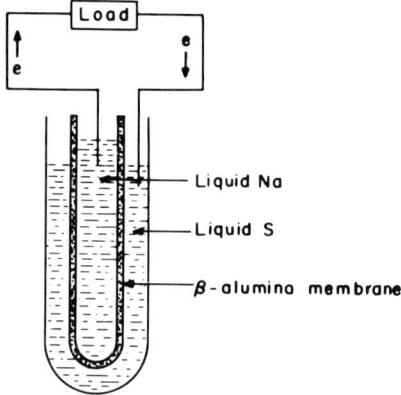

Fig. 11.93. Schematic diagram of sodium–sulfur cell.

The appearance of such a cell is schematically shown in Fig. 11.93. The reactions occurring during charging are

at cathode,
$$2Na^+ + 2e \rightarrow 2Na \tag{11.120}$$
at anode,
$$S^{2-} \rightarrow S + 2e \tag{11.121}$$

While discharging, the above reactions take place in the reverse direction.

The energy density of such a storer is about 0.3 kWh kg^{-1} and about half of this is attained in practice at 300°C.

Such a cell has the advantages of the high current densities associated with the simple alkali-metal reactions Na $\rightarrow$ Na$^+$ + e. (The corresponding reactions for sulfur also are fast, i.e., develop little overpotential.)

Conversely, the cell may suffer some long-time corrosion problems. In this respect, it has only an intermediate temperature—low for a molten-salt cell—, and, hence, the corrosion difficulties are relatively small.

11.4.4. The Large Gap between the Maximum Feasible and the Present Actual Energy Densities of Electricity Storers

The electricity storers available in the late 1960's which give the highest energy densities are the nickel–cadmium and the silver–zinc cells, the energy densities of which are about[†] 0.04 and 0.10 kWh kg^{-1}, respectively. In Table 11.22 are collected the hypothetical maxima under ideal conditions

---

[†] The actual energy densities of cells vary, as explained in Section 11.4.2, depending on the particular rate of charge and discharge to which they are subject.

for some electricity storers. These are certainly unreal conditions for they neglect, e.g., the weight of the containers and electrodes and involve the tacit assumption that it is possible to convert all of the materials on the electrodes into the other form to which it is transformed upon discharge. (Some will be retained in the interior of grains.) Further, the overpotential, which inevitably shows upon working the devices and significantly reduces power densities, is neglected.

What are the feasible goals in energy and power density, respectively, at which research and development should aim for these devices? Such a question would have to be answered separately for each new storer upon which research was proposed. It is, however, the weight of inactive materials on the electrode and in the solution and of materials of construction which adds to unavoidable losses when one tries to obtain from the calculated ideal hypothetical maximum energy density the maximum *feasible* energy density (see Table 11.22). Thus, the latter can be estimated approximately. It turns out that it is reasonable to set as a feasible general aim the attainment of about one-half of the hypothetical maximum.

A great gap exists between that which may now be available in electricity storage and that to which research could lead. Intense interest in electrochemical energy converters commenced in the late 1950's because of the needs of space vehicles. Keen interest in research on electrochemical energy storage began in the middle 1960's because of the pollution difficulties attendent upon burning fossil fuels to power transports. Research on storers is thus in a more rudimentary stage than that on converters. It has only very recently received application of the thinking engendered by the electrode kinetics which was the subject of the research papers of the 1950's.

Attainment of energy densities in electrochemical storers of about half the idealized maximum would have widespread consequences in everyday life, particularly for the possibilities of electrically powered transportation. For example, it can be shown[†] that, were it possible to produce storers

---

[†] The considerations behind this statement are complex, particularly because the theory of the performance of electrochemical engines indicates relative strengths and weaknesses compared with internal-combustion engines which make the balance of the situation depend on the circumstances of operation (e.g., long distance at constant speed or commuter type of transportation). One conclusion on which wide agreement has been reached is that electrochemically powered vehicles would give a decrease in the overall cost of transportation. The gain would increase as the amount of use of the car per year increased and, of course, with the expected continuing fall in the price of electricity with increased use of nuclear power. Thus, taxis and short-haul buses would be, initially, the type of vehicle which would show most gains.

of about two or three times the energy densities available at present (i.e., 100 to 150 Whr lb^{-1}), the powering of motor vehicles for a range of 200 to 250 miles by electrochemical sources would become feasible. To produce such sources is a major aim of electrochemical engineering in the coming years,[†] and the success of this work will depend significantly upon the evaluation of the mechanisms of the electrocrystallization reactions involved in the various storers.

11.4.5. Outlines of Some Possible Future Electricity Storers

A number of possibilities exist by which greatly increased energy density and power density of electricity storers could be realized in certain situations. Some of these will be described, with remarks on the attendant obstacles which have to be overcome by future research.

11.4.5a. Electricity Storage in Hydrogen. The hydrogen–oxygen cell has already been discussed for an electrochemical energy converter, i.e., a cell or series of cells in which hydrogen and oxygen are fed in and produce water, which is evaporated off from the cell.[‡] In practice, such cells can only be economically used if there is a preliminary chemical reactor for obtaining hydrogen cheaply, e.g., from hydrocarbons. However, this consideration would not enter the situation were these cells to be used as electricity storers, because, after the initial supply of hydrogen had been consumed, it could be reproduced by electrolysis. An added advantage, at least in terrestrial use, is the use of air instead of oxygen, i.e., the lack of necessity of storage for the component which functions cathodically during discharge (Fig. 11.94).

Of course, the low molecular weight of hydrogen and the absence of weight of reactants for the oxygen side of the storer makes the energy density of such a hypothetical storage device depend predominantly on the weight of the container material used to store the hydrogen. This in turn depends upon the situation. If volume considerations are not important—a

---

[†] Feasible goals in the research of the next one or two decades are 1 kWh lb^{-1} in storage and 1 kW lb^{-1} in power.

[‡] The use of such water for potable purposes is worthy of consideration as a supplement to natural water supplies, the lack of which limits the viability of much land. Were hydrogen–oxygen fuel cells to be used widely, hydrogen could be produced by electrolysis at central sites with the use of electricity produced by atomic reactors and pumped to consumer sites to be reunited in a fuel cell with oxygen from air. The production of potable water for American space vehicles from the fuels used for the auxiliary energy supply was tested in 1967.

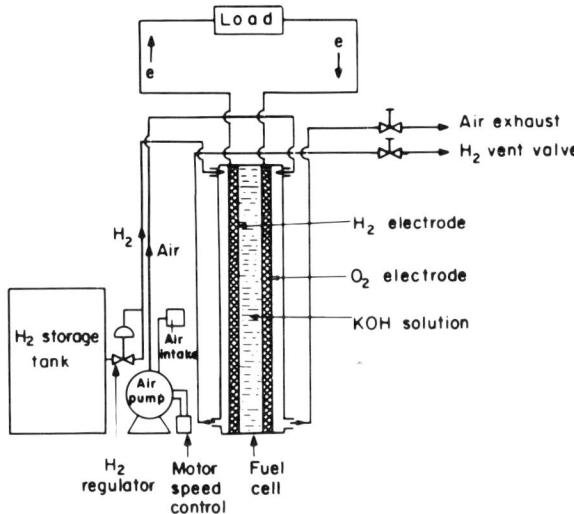

Fig. 11.94. Schematic diagram of a hydrogen–air battery.

rare circumstance—, the hydrogen could be stored at low pressure in relatively light containers. Circumstances might sometimes exist in which balloon storage could be favorable with, therefore, no consumption of ground space.

Storage at economically acceptable volumes for most circumstances presents problems as yet only partially solved. Can the hydrogen be stored cryogenically? To what degree may developments in the plastics field give materials of sufficient strength to contain hydrogen at suitable pressures? What of the properties of materials such as titanium and beryllium in respect to their use in light hydrogen containers? The problem is largely one of the *weight* of a container with sufficient strength if the storer is to be used to power moving vehicles, where the problem of devising a situation so that the containers could withstand collisional shocks becomes important. For very large scale storage, one could contemplate underground storage of hydrogen produced by electrolysis, perhaps of sea water, e.g., from primary energy sources such as the sun, wind, or tidal power. Such questions bring one to some frontiers of research in the engineering aspects of the electrochemical electricity storage situation. The other principal difficulty of the hydrogen–oxygen cell is the slow kinetics of the oxygen electrode. This is a problem which depends upon fundamental work in electrocatalysis and is likely to be solvable (Fig. 11.95); e.g., recently, doped nonstoichiometric oxides of transition metals have been shown to have catalytic properties comparable with those of platinum.

The favorable kinetics of the reaction $2H^+ + 2e \rightarrow H_2$ (with high i_0

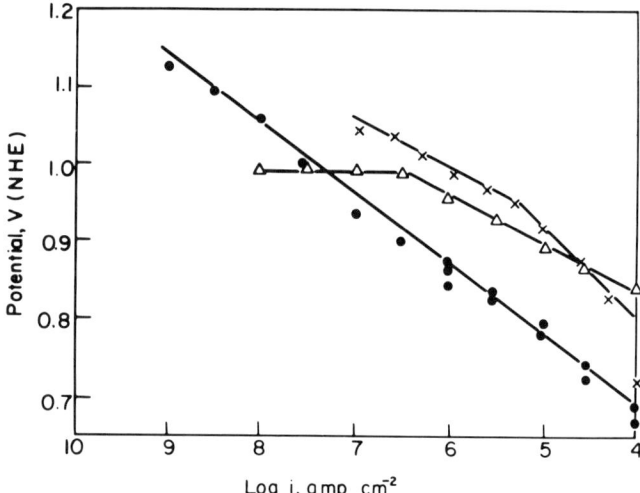

Fig. 11.95. Tafel lines for the reduction of oxygen on platinum and doped sodium–tungsten-bronze electrodes in sulfuric acid solution of pH ~1.5; ×, for sodium-tungsten bronze; △, for oxide-free platinum; and ●, for oxide-covered platinum.

value on cheap substrate material—nickel oxide or nickel—both for anodic dissolution and cathodic evolution), the relatively advanced state of the engineering development of the corresponding fuel cells because of the use in space vehicles, and the absence of solid-state problems in crystallization or corrosion make the prospect for the hydrogen approach to the storage of electricity have rather favorable prospects.

11.4.5b. *Storage by Using Alkali Metals.* It has been pointed out (Section 11.4.2) that it is not only the kilowatt hours per kilogram, the energy density, which an electricity storer can hold which is important in considering its merits in comparison with other storers but also the kilowatts per kilogram which it can produce on discharge. One substantial disadvantage of hydrogen–oxygen cells exists here in the kinetics of the oxygen cathodes which are not, as yet, very favorable (see Section 11.4.5a). Thus, substantial overpotential develops at electrocatalysts such as platinum when it is required to dissolve oxygen on them at significant current densities; and this, of course, reduces the power density and real energy density. Hence, although hydrogen–oxygen cells can attain, at temperatures of less than 100°C, considerable power, substantial improvements would depend on advances in mechanism research on the electrocatalysis of the oxygen dissolution before increase in the velocity of oxygen reduction can be expected.

Among the requirements of a desired energy storer are (see Section 11.4.2) a high reversible cell potential and high i_0 values for both the electrode reactions. An approach to this ideal can be obtained by combining the alkali-metal-dissolution reactions

$$M \rightarrow M^+ + e \qquad (11.122)$$

with the halogen-dissolution reaction

$$X_2 + 2e \rightarrow 2X^- \qquad (11.123)$$

in a molten salt MX.

Such a device not only fulfills the two criteria mentioned [e.g., Li–LiCl–$\frac{1}{2}$Cl$_2$ at 650°C has an open-circuit potential of about 3.5 V and an i_0 value of about 0.1 amp cm^{-2} (on carbon) for the Cl$_2$ electrode] but also has a further unique property. There is no solution in the normal sense in which the diffusion of a solute to and from the electrodes gives limiting mass-transport rates which define the maximum rate of charge and discharge of devices. In pure MX (e.g., liquid LiCl at, say, 650°C), the supply of material to the electrodes is in effect governed by the ionic-conductance parameters and these are about ten times higher in molten salts in the region of the temperatures mentioned than in the corresponding aqueous electrolytes at lower temperatures.

The combination of the high reversible cell potential, exchange-current densities, and lack of transport limitations makes the type of device mentioned in this section have very advantageous characteristics for an energy storer in which high *power* densities are desirable.

What are the problems faced with devices (Fig. 11.96) of this type, e.g., cells in which lithium dissolves into LiCl to form Li$^+$ and chlorine dissolves in it to form Cl$^-$? They are nearly all of one kind, i.e., the instability of the materials of construction of the cell. It is not difficult to find materials on which lithium and pure LiCl do not undergo the electrochemical reactions of corrosion. But Cl$_2$ is difficult to contain, and, when lithium metal is dissolved in LiCl, it provokes electrochemical surface reactions with container materials which break down many materials.

Were research and development able to overcome these difficulties of material instability—corrosion—, electricity storers involving alkali metals and halogens have advantages in power densities which would be difficult to find in any other direction. The field of the stability of materials is a wide one. Solutions to the problems it at present poses may come not only by (much-needed and little-available) fundamental mechanism work by

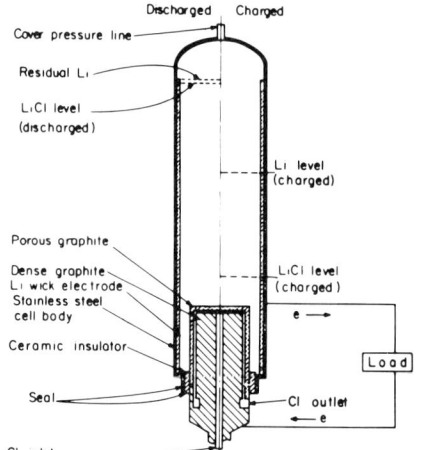

Fig. 11.96. Schematic diagram of lithium–lithium chloride–chlorine molten-salt cell.

which one determines the electrochemical reactions which are the origin of the instability of the materials. It may also come because of the increasing availability of new materials. For example, nonconducting organic materials would probably provide stable substrates in a number of instances. The difficulty has been the *thermal* instability of such substances at temperatures above 250 to 300°C. However, e.g., the polyphenylene type of compounds already give thermal stability into the range of 500 to 600°C.

11.4.5c. Storers Involving Nonaqueous Solutions. Another combination of positive options to aim for in an electricity storer would be, e.g., that its electrodes should have widely different and highly reversible electrode potentials (the latter is equivalent to having high exchange-current densities), it should use a reaction with oxygen as the cathodic discharge step (to avoid the weight of containers for a second fuel), and it should be free of the material instability and breakdown problems associated with the presence of high temperatures.

The first requirement indicates (*cf.* the projected alkali-metal cells) the use of the metals of groups IA or IIA of the periodic Table, in particular, lithium and beryllium (Table 11.20). But these metals corrode violently in aqueous solutions to give hydrogen (from the parallel cathodic reaction $2H^+ + 2e \rightarrow H_2$, which must occur if the central corrosive reaction $M \rightarrow M^+ + e$ is to take place); see Sections 11.2.9 and 11.2.10. The heat engendered together with the presence of oxygen from air causes the well-known explosion. The solution to the problem by using molten salts brings with it the difficulties of instability and breakdown caused by corrosive side reactions not yet understood in a mechanistic way.

Hence, another solution may be obtained, in the use of a nonaqueous solvent, e.g., acetonitrile, in which the corrosion reaction and therefore the presence of evolved H_2 are suppressed toward zero velocity. Series of electrochemical potentials of different metals in water and in a number of nonaqueous solvents are presented in Table 11.23. An aim is to discover a way of causing oxygen to dissolve cathodically in a non-aqueous medium. Corrosion problems at room temperatures are small so long as that of the alkali or alkaline-earth metal is avoided by the use of a nonaqueous solution in which the hydrogen-evolution reaction is absent. An excellent hypothetical electricity storer in this sense would be a rechargeable cell of beryllium with air.

There are many problems facing attainment of a possibility of this kind, certainly more than those of the two types of devices mentioned earlier in this section. For example, no proposal for a satisfactory functioning of an oxygen electrode in a nonaqueous solution has as yet been made. The cathodic reduction of O_2 usually forms water, so that a solvent system would have to be devised to provide an analogous solvent-forming reaction and, then, during charging, the anodic reaction would have to be the evolution of oxygen. The solubility of salts in nonaqueous solution is often

TABLE 11.23

Series of Standard Electrochemical Potentials in Different Solvents with $E^0_{H(H_2O)} = 0$[†]

Element	H_2O	CH_3OH	CH_3CN	HCOOH	$HCONH_2$	N_2H_4	NH_3
Sodium	−2.71	−2.76	−2.73	−2.95	–	−2.74	−2.84
Lithium	−2.96	−3.13	−3.09	−3.01	–	−3.11	−3.23
Calcium	−2.76	–	−2.61	−2.73	–	−2.82	−2.73
Zinc	−0.76	−0.77	−0.60	−0.58	−0.83	−1.32	−1.52
Cadmium	−0.40	−0.46	−0.33	−0.28	−0.48	−1.01	−1.19
Lead	−0.13	−0.23	+0.02	−0.25	−0.26	−0.56	−0.67
Hydrogen	0	−0.03	+0.14	+0.47	−0.07	−0.91	−0.99
Cu–Cu^{2+}	+0.35	+0.31	−0.24	+0.33	+0.21	–	−0.56
Hg–Hg^+	+0.80	+0.71	–	+0.65	–	–	–
Silver	+0.80	+0.73	+0.37	+0.64	–	−0.14	−0.16
Chlorine	+1.36	+1.09	+0.72	+1.24	–	–	+1.04

[†] R. Jasinski, *High-Energy Batteries*, Plenum Press, New York, 1967.

smaller than in aqueous solutions, so that [as the limiting current density is proportional to concentration, see Eq. (9.82)] a relatively small limiting current and hence power would be expected with devices using nonaqueous solutions. Problems of voltage losses due to the resistance of the solution also exist.

However, electricity storers with nonaqueous solutions need not attempt to attain immediately the aim outlined above. Their basic advantage is that they can involve the high electricity-storage capacity of the alkali and alkaline-earth metals and forego problems of the weight of containers for hydrogen or the material instability problems of high temperature cells. Investigations conducted in the past few years have led to the development of liquid-ammonia cells that operate under high pressure and organic electrolyte cells that operate under room temperature and atmospheric pressure and employ solvents such as propylene carbonate, dimethylformamide, dimethyl sulfoxide, acetonitrile, and methyl formate which contain dissolved organic and inorganic salts. A schematic diagram of an organic-electrolyte electricity storer is shown in Fig. 11.97. By employing these electrolytes, it has been possible to operate systems in which the *theoretical* energy densities (see Table 11.24) are some five times higher than those of conventional cells operating in aqueous media. Some possibilities of the use of hybrid situations also exist, e.g., an alkali metal in contact with a nonaqueous solution and an oxygen electrode in contact with an aqueous one, the two being separated by some kind of membrane. Realization of such a device would give the advantages of the large amount of electricity per unit of weight characteristic of the active alkali metals and the avoidance of carrying the cathodic fuel. Correspondingly, the difficulty of finding

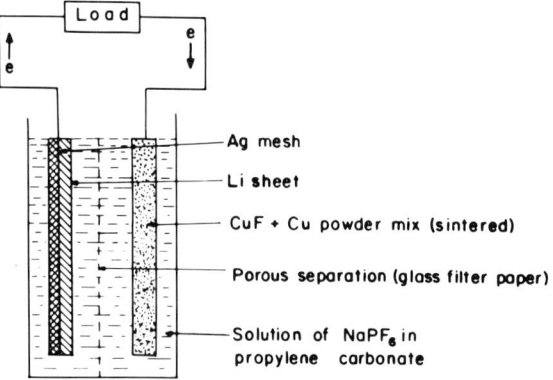

Fig. 11.97. Schematic diagram of an organic-electrolyte energy storer.

TABLE 11.24

Energy-Storage-Density Parameters for Nonaqueous Electrolyte Cell System

System	Open-circuit potential, V	Energy storage density, kWh kg^{-1}, at		
		Zero current	2 mamp cm^{-2}	10 mamp cm^{-2}
Li–LiBr, DMSO†–AgO–Ag	3.14	1.22	1.07	0.90
Li–LiBr, PC‡–KCNS, MnO$_2$–C	3.44	0.98	0.88	0.60
Mg–LiBr, PC–MnO$_2$–C	2.18	0.59	0.45	
Al–KCNS, DMSO–NiCl$_2$–Ni	1.08	0.39		

† DMSO, dimethyl sulfoxide.
‡ PC, propylene carbonate.

a suitable reaction for a nonaqueous situation and oxygen dissolution and the corrosion-inhibition problems of molten-salt cells would be avoided.

11.4.5d. Storers with Zinc in Combination with an Air Electrode. Appreciation of the difficulties facing research in the hypothetical nonaqueous storers brings one to attempt to retain something of the advantages of the absence of corrosion problems—the scourge of the molten-alkali-metal storers—and the reduction in container weight due to the use of air, while abandoning the advantage of using the powerful electricity-storage capacity in lithium or sodium (see Table 11.19). Among the more active metals is zinc ($V_e = -0.76$ V for the reaction Zn → Zn^{++} + 2e), and, in an alkaline solution coupled with an air electrode, the reversible, thermodynamic cell potential is 1.62 V. A schematic diagram of such a cell is shown in Fig. 11.98.

The reactions upon charging would be

at anode,
$$4OH^- \rightarrow 2H_2O + O_2 + 4e \tag{11.124}$$

at cathode,
$$2Zn(OH)_2 + 4e \rightarrow 2Zn + 4OH^- \tag{11.125}$$

and those upon discharging would be the reverse.

The deposition and dissolution of zinc is empirically well understood, largely from studies concerned with the silver–zinc cell (in comparison with which the zinc–air cell would have an obvious economic advantage).

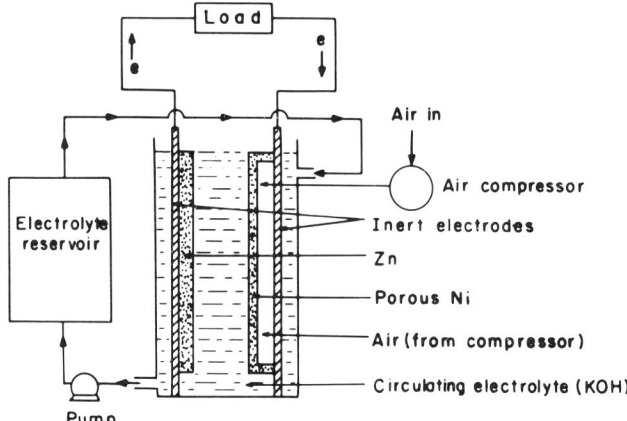

Fig. 11.98. Schematic representation of a zinc–air cell.

In fact, such zinc–air cells have been known for many years but regarded as systems able to give rise only to low power. However, research during the 1960's on oxygen cathodes connected with the functioning of electrochemical converters caused considerable increase in the currents which could be passed through such devices, so that zinc–air cells have fair prospects of attaining the 0.1 to 0.2 kWh kg^{-1} possible as a basis for electrochemically powered transportation.

The difficulties which research on such systems faces—apart from the desired increase in the electrocatalysis of the oxygen-dissolution reaction—is largely concerned with difficulties arising on repeating many times (i.e., each time the cell is charged) the electrocrystallization reaction associated with the cathodic formation of zinc. Dendritic growths (see Section 10.2.15) form and tend to push out from the zinc electrode to form a contact with the anode, which makes the cell short out, and the very high momentary currents cause buckling of the plates, etc. Correspondingly, slow corrosion reactions (arising from the difference in zincate concentration at various parts of the cell and hence currents between these two parts) decrease the stability of the zinc electrodes. These gradually change in shape, growing denser as the charging cycles are repeated.

11.4.6. The Respective Realms of Applicability of Electrochemical Energy Converters and Electricity Storers

It has been pointed out (Section 11.3.20) that, in the decades ahead, the relative usefulness of electrochemical energy converters and electricity storers depends on a variety of circumstances—the availability of atomic reactors, the progress made in research on the control of atomic fusion (which would favor storers), and how the plentiful coal and natural-gas reserves of certain countries (apart from the limited oil reserves, useful also for other purposes) will be used. But there is one situation, an important one, for which some rough prognostications can be made and which, with due allowance for supplies of energy available in each country, seems unlikely to change with time. The situation concerns the type of source which should be used for the onboard powering of transports, namely, vehicles on land and sea, and in space.

Here, an important concept is the mission time in which power must be available without refueling or recharging. Consider two extreme cases, a very short time, say, minutes, and a very long time, say, weeks. Then, for the very short time, it is qualitatively clear that the device which has everything within itself and does not need fuel tanks and auxiliaries outside itself—namely, the storer—will give the most energy for the least weight. Conversely, when the time is sufficiently long, the converter will have the advantage for it remains the same in size however long it has to work, although of course the fuel-container size increases. The storer, on the other hand, would have to have the increased fuel for a lengthening mission time and this fuel would have to be carried on the plates themselves, so that the grid, solution, container, etc., and the total storer itself have all to grow in size. At some time, between the extremes of minutes and weeks in possible time between refueling or recharging, it will become more advantageous for an electrochemically energized transport to use an electrochemical converter of chemical fuels than an electricity storer. (As has been observed above, the direct use of onboard atomic reactors is unlikely for earth use for vehicles of less than several hundred tons (Section 11.4.5)).

How electrochemical power sources compare with other modes of producing energy, when this maximum energy for minimum weight is the criterion, is shown in Fig. 11.99.† A rough, order-of-magnitude result is to say that, for time less than about 10 hr, the storer has less weight for the amount of energy which it gives and, for greater than about 10 hr, the converter weighs less (assuming the power demand is less than about 100 kW). Finally, for a time of about 1000 hr, solar converters become important, and, eventually, for power greater than 100 kW, the enormous intrinsic energy per unit of weight of an unshielded atomic power source

---

† The figure indicates that electrochemical power sources are the favorable ones (*with respect to their energy per unit weight*) for application up to about 100 kW and 1000 hr, a considerable portion of all needed power sources.

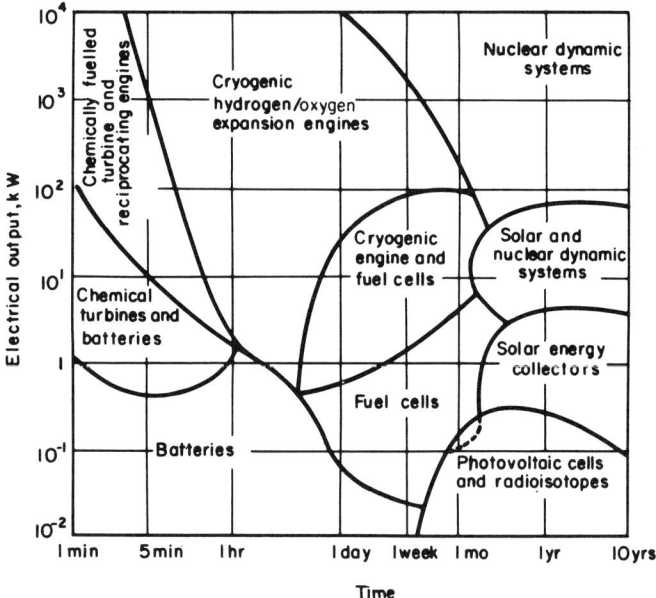

Fig. 11.99. A plot showing regions of applicability of various types of energy converters depending on the mission time and power output. This figure is colloquially referred to as "the map of the Balkans."

begins to overcome the intrinsic great minimum weight which the shielding of an atomic reactor of minimum size requires.

In these remarks, the electricity storers referred to have the material on the plates, as is usual for all past and most future proposed storers. The hydrogen–oxygen and alkali-metal storers would not be subject to the same type of conclusion. They are converters in which the products of the spontaneous energy-producing reactions are not rejected (*cf.* the evolution of CO_2 into the atmosphere in hydrocarbon fuel cells) but re-electrolyzed back to fuels from an outside source of electricity. Yet, they need not remain at the electrode forming them. Such devices might be termed *converter storers*.

11.4.7. Electrochemical Electricity Storage in a Nutshell

Many people, when first introduced to the field of electrochemistry, think that batteries and fuel cells, storers and converters, are only trivially different. In fact, although they both work by electron charge-transfer reactions at interfaces, they have quite different aims. Converters take out the energy of a chemical reaction directly as electricity. Storers store electricity produced elsewhere and recreate it on demand. Avoidance of the confusion that storers and converters are nearly the same, etc., is possible if one sticks to a literal terminology (storer and converter) and not the old-fashioned and arbitrary words (battery and fuel cell).

The important quantities in electricity storage are not the same as those in converters. Here, for the storers, the energy per unit of weight is the most important term, but one may also need to know the power per unit of weight. It depends on circumstance. For powering cars, it is the power density; for auxiliary power in space ships, the energy density.

Electricity storers of the present like the lead–acid storage battery are poor (particularly with respect to energy-density) when looked at in the light of the possibilities which are now clear. The lead–acid storage battery is, however, ideal for the main purpose for which it is used, for starting cars; it gives a high current for a short time, and the energy needed to start is small, so that the total weight of the battery carried can still be negligible compared with the weight of the car.

Two new electricity storers come nearer theoretically indicated possibilities. They give about 0.1 to 0.2 kWh kg^{-1}, but one of them contains expensive materials, and the other, a molten salt at 650°C. Hence, several other types of storers are being looked into. But, first, a look is taken to see which systems simple electrochemical-energy calculations indicate would be of interest if they could be understood well enough to be made in a practical way. It turns out that theory indicates that storers should be feasible which would be just about a whole order of magnitude greater in energy density than is available now, a considerable goal for research. In fact, feasible goals of the next decade in electrochemistry storage research would be 1 kWh kg^{-1} in energy density and 1 kW kg^{-1} in power density.

A number of new electricity storers are incipient in the laboratories; some of them may be within a few years of practicability. Storage in hydrogen has attractions and is already well engineered: were a light-weight method of holding hydrogen to be devised, this path would be a good solution for many needs. Storers using alkaline metals have promise at least in energy and power but require research on corrosion at high temperatures. Non-aqueous solutions would perhaps avoid the corrosion troubles, and alkali metals could be used here, too, with their high potentials. But trouble comes from a low solubility—solute concentration—in such cells, and the limiting currents are not high. A more conservative design, therefore nearer realization, is to be expected for the first of the new generation of storers—metals in combination with the fuel-cell-developed air electrode.

Finally, what realm do these new electrochemical storage devices have for the future? When are electrochemical energy storers relevant; when, fuel cells; when, atomic reactors; etc.? The answers may be calculated from simple considerations based on the energy per unit of weight carried for the times at which recharge or refill would be possible for the application concerned. If this is less than 10 hr, batteries are best; if less than 1000 hr, fuel cells; and, if more, solar converters and atomic reactors. These are results which are limited to engines of less than about 100 kW, but that limitation encompasses quite a large fraction of our mobile-power-source needs.

Further Reading

1. D. H. Collins, ed., *Batteries*, The Macmillan Company, New York, 1963.
2. R. Jasinski, *High-Energy Batteries*, Plenum Press, New York, 1967.
3. J. O'M. Bockris and S. Srinivasan, *Fuel Cells: Their Electrochemistry*, McGraw-Hill Book Company, New York, 1969.

INDEX

absolute mobility, and Einstein relation, 374
absolute potential, impossibility of measurement, 645
absolute potential difference
 attempts to measure, 644
 structured, 660
absolute strength of acid or base, 494
absorption spectroscopy, in study of solutions, 275
acceptor particles, need for in tunneling, 977
acid(s)
 Bronsted's view, 489
 dissociation constant of, 496
 relative strength of, 495
acid–base strength(s), 501, 509, 511
acid strength, 506, 510
acid strength theory, 505
action potential, and nerve impulse, 940
activation, of electrocatalysis, 1170
activation barrier, and rate-determining step, 1003
activation energy, 457
 for diffusion, related to melting point, 547
 electrical contribution to, 872
 for multistep reaction, 1002
 for viscous flow in fused salts, related to melting point, 550
activation-transport control, of electrode reaction, 1054
activity coefficient(s)
 concept, 202
 cube-root law, 269
 Debye–Hückel model, 223
 Debye–Hückel, parameters of, 211, 212
 further reading on, 246
 at high concentration, 245
 and hydration number, 240
 and ideal solutions, 205
 and ion pairs, 260

activity coefficient(s) *(cont.)*
 and ionic strength, 215
 mean, 207, 269
 model for effect of solvent molecule, 240
 of single ionic species, 206
 and solvation, 241
 theory and experiment, 209, 213, 216, 228
 theory of solvent effects on, 242
adatoms, 1177
additives, organic, 1222
adiabatic change, 974, 975
adion(s)
 and adatoms, 1177
 charge on, 1177
 concentration, determination of, 1197
 concentration profile of, 1200
 concentration as function of time, 1189
 and electric field, 1181
 existence of, 1177
 formation of hydrated ions from, 1187
 function in electrodeposition, 1177
 lattices from, and spiral growth, 1203
 surface, deposition from, 1187
 tabulated, 1198
adsorbed intermediates, 1016
adsorption
 of atomic hydrogen, and coverage, 1246
 of cations and anions, at interfaces, 680
 contact (or specific), 748
  definition of, 682
  of desolvated ions, 742
  free energy change in, 742
  variation with potential, 638
 in electrochemistry and chemistry, 683
 organic, 792, 793
  and free energy of flip-flop water molecules, 799
 superequivalent, 638, 748
 and surface excess, distinction, 684

adsorption energy, effect on tunneling, 965
adsorption intermediates, further reading on, 1036
adsorption isotherm, for ions, theory and experiment, 773, 774
adsorption step, in hydrogen evolution, 1153, 1233
aeration, differential, 1302
aggregation of colloidal particles, theory, 838
aims, of this book, vi
air
  containing CO_2, effect on metals, 1268
  electrochemically burned, 1382
air electrode, in storage, 1427
alkali metals, and storage, 1423
alloy(s), 1224, 1225
American convention, 1118
American space program, contributions to fuel cells, 1388
analogy(ies)
  mass transfer and heat transfer, 1041
  semiconductors and electrolytic solutions, 812
angel, Gibbs', 679
anode(s), 1311, 1312
anodic protection, 1318
artificial organs, and electrochemistry, 43
association, of ions, temporary and permanent, 273
atom(s)
  of hydrogen, recombination of, in void, 1334
  labeled, and surface coverage, 1036
atom–atom step, in hydrogen evolution, 1234
atomic energy
  and electricity storage, 1397
  and electrochemical power sources, 1395
  and electrochemistry, 43
atomic power, coming era of, and electrochemistry, 43
autodissociation, proton transfer reactions in, 492
autodissociation constants, 499
automobiles
  necessary electric energy density for driving, 1420
  exhaust products from, 1357
auxiliary electrode, and cathodic protection, 1312
auxiliary electrode, use of in determining overpotential, 890
averaging process, Debye, 141
axon
  as a cell, 937
  potential across, 938
  of squid, 940
Balkans, map of, 1430
bands, bending of, near surface of semiconductor, 816
band picture, 804, 807, 810, 820
barrier(s)
  for consecutive steps, 1002
  electron leak through, 945, 952, 962
  transfer of charge across, further reading on, 946

barrier (s) *(cont.)*
  width of, and probability of tunneling, 954
base, 494, 495
batteries
  in ancient times, 1265
  classical, 1413
  and dendrites, 1221
  mission time in which useful, 1432
benzene, 1098, 1099
biological cells, and charge transfer processes, 841, 1266
biological membranes, 937, 981
biological processes, 4, 840
biological reactions, quantum nature of, 21
biological situation, 941
biology, and electrochemistry, 29, 42, 43
Bjerrum approach, to ion pairs, 253, 257, 260
Bjerrum's integral, 256
blister, bursting, 1334
blood, clotting of, and electrochemistry, 840
bond, stretching of, and electronation reaction, 971
bond strength, effect on electrocatalysis, 1157
Born charging process, 84
Born cycle, 50, 86
Born equation, 57, 69
  and strength of acids, 510
Born model, 49, 50, 70, 71
Bridgman technique, for single crystals, 1218
Bronsted's view of acids, 489
bronzes, 1169
Brownian motion, and movement of piston, 300
bulk properties, and interphase properties, 641
bunching, 1209
butanol
  adsorption of, on mercury, 795, 798
Butler–Volmer equation, 862, 984, 1054
  and biological situation, 941
  and catalysis, 1141
  and contact adsorption, 911
  deduction, 880
  effect of water coverage?, 1015
  and effective potential in catalysis, 1142
  and electric car, 928
  final form for multistep reactions, 1000
  further details, 910
  further reading on, 929
  in galvanostatic transients, 1190
  high-field approximation, 885, 888, 889, 1001
  for iron dissolution and deposition, 1084
  low-field approximation, 882, 892
  for multistep reaction, 998, 1090
    in terms of stoichiometric number, 1006
  and order of reaction, 1009
  in quantum mechanics, 980
  and rate-determining step, 1139
  and structure of interface, 911
  a summary, 928
  and surface coverage, 1014, 1015
  and zeta potential, deduction, 913
cadmium–nickel battery, 1401
calomel electrode, 654

capacitance
  constant, equation for, 760
  and contact adsorption, 751
  dipole, 798
  of diffuse layer, 731, 732
  electrical, determination at interface, 703
  and electrocapillary curves, 721
  and Gouy–Chapman theory, 731
  hump, 754, 761
  at interface, as function of dipole orientation, 788
  of semiconductor-solution interface, 817
  of whole electrode, 790
capacitance–potential curve, lateral adsorption model, 778
capacitance hump, equation for, 761
capacitor(s)
  and dielectric, 134
  and dipoles, 135
  in series, 735
capacity, constant region of, 753
capillary electrometer, 689
cars
  and batteries, 1419
  and fuel cells, 1158
  and pollution, 1419
carbon dioxide
  in atmosphere, as function of time, 1355
  and possible rise in sea level, 1356
carbon monoxide–air cells, 1393
Carnot limitation, 1358, 1360
  and electrochemistry, 1364
  and internal combustion, 1358
  physical interpretation, 1367
catalysis
  Butler–Volmer equations in, 1142
  and electrocatalysis, 987
  for oxygen on doped tungsten bronzes, 1422
  tabulated, 1377
catalyst(s)
  charge transfer, 10
  chemical and electrocatalysts, 1141
  distribution of, and porous electrode, 1384
  distribution of, in porous electrodes, 1172
catalytic activity
  for the oxidation of ethylene, 1161
  oxide-free and oxide surfaces, 1258
cathodic protection, 1309, 1312, 1313
cavity, 150, 1335
cell
  driven, 1128
  electric, potentials in, 649
  electrochemical, discussion of dependence on current, 1128, 1129
  entire, and Nernst equation, 904, 1114
  local, in corrosion, 1270
  and metal–metal potentials in electrocatalysis, 1148
  for the observation of transients, 1184
  relations in, further reading on, 1137
  short-circuited, and stability of metals, 1269
cell model, for liquid electrolytes, 529
cell potential
  and current, in electrochemical energy conversion, 1371

cell potential *(cont.)*
  maximum, 1137
  minimum, 1137
  in self-driving cell, 1131
cellulose, oxidation of, 1169
central ion, excess charge density as function of, 194
charge
  at boundary, 627
  of individual ions, in diffuse layer, 747
  storage of, 703
charge density
  in double layer, components of, 712
  on electrode, determination of, 702
  excess, and ionic atmosphere, 636
charge transfer
  across barriers, further reading on, 946
  and blockage of electrode surface, 1014
  chemical and electrical implications, 846
  and Fermi level, 977
  and formation of intermediates, 1027
  and instability of surfaces, 1268
  and metal–slag equilibrium, 618
  in perspective, 974
  quantification of quantum-mechanical picture, 977
  and rate-determining step, 1185
charge transfer catalysts, 10
charge transfer theory, summary, 893
charged sphere, and Born model, 50, 52
charging process, effect of double layer on, 1190
cheap heat, and electrochemistry, 43
chemical desorption step, in hydrogen evolution, 1234
chemical and electrochemical reactions, further reading on, 989
chemical energy, conversion to electrical, and symmetry factor, 1138
chemical potential
  change arising from ionic cloud, 201
  and computation at interfaces, 672
  and flux of ions, 397
  standard, 203
  and thought experiment, 694
chemical reaction
  and electrical energy, 15
  and electrode reactions, 896
chemistry, and electrochemistry, 28, 38, 869
chi potential, 667
Christmas trees, mini, 1221
chronopotentiometry, 1051
circuitry, and electrochemistry, 43
circuits, 33, 1316
circuits in the body, and electrochemistry, 43
civilization, and surfaces, 1267
cliff, and symmetry factor, 937
closest approach, 225, 741
clotting, of blood, and zeta potential, 841
cloud, near central ion, 193
clusters of ions, 82, 265
coagulation, 839
codeposition, of hydrogen, with metal, 1227
cold combustion, 1369
cold emission, of electrons, 944

colloids, 835, 838
colloid chemistry, further readings on, 841
collodial particle(s)
  energy of interaction for coagulation, 840
  and potential distance relation, 729
  space charge near, 217
colloidal nature of biological processes, 4
combustion
  cold, 1369
  products of, other than carbon dioxide, 1353
competition, between adsorbed water and hydrogen, 1015
complex formation, and mixtures of ionic liquids, 587
complex ion(s)
  concentration as a function of added ligands, 592
  lifetime, 590
  in molten salts, further reading on, 594
  tests for, 589, 593, 594,
  and Raman spectra, in molten salts, 590
compressibility
  calculated from hole model, for molten salts, 585
  and hydration number, 127
concentrated solutions, skepticism on theory of, 281
concentration gradient, 1059
  and chemical potential, 291
  and diffusion flux, 295, 1040
  ion diffusion in, 288
  linear, and Planck–Henderson equation, 419
  in tracer diffusion, 543
concentration overpotential, 1052, 1053
concentration perturbation, Laplace transform of, 329
condenser
  parallel-plate, and double layer, 634
  and storage of charge, 703
conductance
  in migration, further reading on, 367
  in nonaqueous solutions, further readings on, 452
  theory, further reading on, 439
  theory of, in terms of Debye–Hückel–Onsager equations, 435
conductance of true electrolyte, in nonaqueous solution, 452
conduction, 345, 351
conduction band, 809
conductivity
  equivalent, 358
  and ion association, 448
  molar, 357
  of molten salts, 517
  of pure liquid electrolytes, 553
  specific, and current density, 354
conductor(s), 806
  electronic, and passivation, 1320
  ionic, and effect on corrosion, 1273
configuration, of water molecules around proton, 470

conservation of momentum, and radiationless transition, 950
constant-flux diffusion problem, and solution of other problems, 323
constitution of proton, further reading on, 470
contact adsorption, 748
  and Butler-Volmer equation, 911
  and capacitance, 751
  and capacitance of interface, 749
  definition of, 682
  of desolvated ions, 742
  determination of, 743
  free energy change in, 742
  as function of charge, 748, 763
  and image charge, 767
  and ionic radius, 744
  lateral repulsion model, 766
  measurement of, 745
  of negative ions, 637
  and stability of colloids, 840
  and surface state, 821
  tests for isotherm, 769
contact adsorption model, tests for, 775
contact potential difference, 647
convection, 1051–1057
convention
  American, 1118
  international, 1118
conversion, of chemical energy to electricity, 14, 42, 1266, 1358
converters, photogalvanic, and electrochemistry, 43
converter-storers, 1430
coordination number, of proton, 468
copper, deposition of, summary of mechanism, 1202
correspondence principle, 20, 224
corrosion, 1266, 1267
  as affected by solid phases, 1283
  and agitation of solution, 1300
  basic kinetic conditions for, 1274, 1286
  bird's eye view of, 1347
  common examples, 1301
  cost of, 1346
  effect of equilibrium potential, 1294
  effect of transport difficulties, 1294
  effect of purity, 1273
  effect of Tafel slope, 1294
  and electrodics, 1272, 1275, 1285
  embrittlement in, 1347
  electrochemical mechanism, 1268
  and flip-flop water dipole molecule, 790
  and future of fuel cells, 1388
  and hydrogen evolution, 1232
  inhibition of, and electrodics, 1306
  local cell theory of, 1270, 1273
  Nernst equation and potential–pH diagrams in, 1279
  and oxygen electronation, 1252
  potential of, 1285
  rate of , 1276, 1284, 1285
  rate-determining step, 1296
  and reversible potential, 1274
  in sand, 1304
  spontaneous energetics, 1278

corrosion *(cont.)*
  summary of mechanisms, 1345
  thought experiments in, 1272
  through oxygen starvation, 1303
  through paint, 1302
  of ultrapure metals, 1273
  and undevice, 860
  and wastage, 861
  at water line, 1303
  yield-assisted, 1337
corrosion inhibition, and film-forming inhibitors, 1309
corrosion and passivity, further reading on, 1349
costs, reduction of, in fuel cells, 1385
coupled reactions, 1235
coverage
  determination of, 1029
  of electrode, 1235, 1245
    with inhibitors, tabulated, 1309
    and mechanism determination, 1097
    with hydrogen, determination of, 1245
  of organic molecules, on electrodes, 796
  of surface with hydrogen, variation with potential, 1246
crack(s), 1335–1341
crack initiation, testing of, 1343
cracking
  of hydrocarbons, 1158
  stress corrosion, 1335, 1338
crevices, corrosion associated with, 1302
cross coefficient, 828
cryogenics, and storage of hydrogen, 1421
crystal, growth of, and fast-growing face, 1216
crystal faces, rates of deposition on, theory of, 1216
crystal facets, 1212, 1213, 1214
crystal growth, faster at projection under electric field, 1217
crystal growth, morphology of, for copper, 1222
crystal lattice, 65, 1205
crystal lattice plane, kink site on, 1178
crystal plane, 1179, 1213, 1216
crystallization, 1174, 1202, 1204, 1218
crystalline solids, band theory of, 804
cube root law, for activity coefficients, 269
current-centric view, 16
current–distance relation, along meniscus in single pore, 1385
current–potential curve, and internal resistance of fuel cells, 1373
current–potential diagram, for passivation, characteristic, 1317
current–potential laws, 930, 1113
current–potential relation
  and alloy composition, in oxygen reduction, 1259
  and alloy deposition, 1259
  in cells, 1133
  characteristic shape in passivation, 1317
  for Gemini fuel cells, 1389
  and mass transfer limitation, 1373
  at n–p junction, 936

current–potential relation *(cont.)*
  in terms of cell potentials, 1135
current efficiency, 1229, 1232
current transients, and Fick's law, 316
cybernetic organism, 1267
cyborg, 1267
cytochrome
  quantum mechanical tunneling to, 981
  tunneling to, and enzymes as electrodes, 1253
Daniell cell, 858
  sign of voltage of, 1125
d-band character of metals, and oxygen adsorption on, 1163
de Broglie wavelength, 20, 21, 948, 949
Debye charging process, 248, 249
Debye equation, for dielectric constant, 142
Debye theory of dielectric constant of gas, 140
Debye–Hückel activity coefficient, parameters of, 211, 212
Debye–Hückel constant, 190, 197
Debye–Hückel length
  and diffuse charge, 730
  in semiconductors, 816
Debye–Hückel model
  approximations of, 219
  breakdown of, 266
Debye–Hückel radius, of ionic atmosphere, 197
Debye–Hückel theory, 180, 189
  an assessment, 230
  basis of, further reading on, 212
  comparison with experiment, 212
  further reading on, 238
  parentage, 237
  postulates, 236
  summary of derivation, 233
  triumphs and limitations, 212
Debye–Hückel thickness, thickness of ionic cloud in, 220
Debye–Hückel–Onsager equations, 434, 438
  comparison with experiment, 436
  for nonaqueous solutions, 442
decay, 4, 861
decoration, surface, 1266
de-electronation, 352, 847
  definition of, 853
  and desorption of hydrogen, 1246
  effect of field, 875
  and quantum mechanics, 973
dehydration, and electrodeposition, 1176, 1177, 1180
delay time, and electronics, 33
demon, Maxwell's, 679
dendrites, 1220–1221
deposition
  of alloys, 1223
  of alloys, equations for, 1224
  and crystallization, 1202
  metal, advance of growth step in, 1204
  metal, and screw dislocation, 1206
  metal, steps in, 1203
  metal, and nucleation, 1204
  random walk process in, 1186
  on single crystal, 1218

deposition overpotential, and exchange current density, 1215
desorption step, 1153, 1233, 1234
deuterons, mobility of protons and, 472
device
  electrochemical systems as, 851
  electronic, 324
diagnostic coefficients, and propane oxidation, 1107
diagnostic criteria, for de-electronation of ethylene on platinum, 1162
dielectric constant, 55, 152
  of aqueous solutions, 157
  in bulk near ion, 71
  Debye equation for, 142
  and deformation polarization, 145
  and dipoles, 139
  of electrolytic solutions, 157
  gas, Debye theory of, 140
  and ionic solutions, 132
  and ionic solvation, 155
  and oriented dipole layer, 136
  and polarizability, 138
  and relative strength of acids, 507
  and solvation sheath, 156
  of solvent and solution, further reading on, 158
  theory of, for water, 146
  in water, alignment of group, 146
  of water, calculation for, at various temperatures, 154
  of water, in double layer, 756
  of water, theory of, 153
diesel oil, burned electrochemically, 1391
differential equation, for diffusion, integrated, 417
diffuse charge, and Debye–Hückel length, 730
diffuse layer
  charge of individual ions in, 747
  effect on Tafel relation, 915
  and streaming current, 830
diffusion, 27
  of adion to kink site, 1182
  at constant current, quantitative, 1044
  and convection, 1051
  driving force for, 290
  further reading on, 345
  in fused salts, 542
  of hydrogen by interstitial and intergranular paths, 1330
  of hydrogen into metals, 1328
  of hydrogen, into regions of stress, 1330
  linear, independent of time, 1077
  in molten salts, at constant temperature, 548
  of neutral ion pairs, 383
  nonsteady state, gross view, 307
  an overall view, 342
  to peak of crystal, 1219
  as pseudoforce, 306
  spherical, 1002, 1220
  in solution, and electrocrystallization forms, 1219
  on surface, contribution to overpotential, 1199

diffusion *(cont.)*
  surface, deduction of equations for, 1188
  surface, and lattice formation, 1195
  time in, by Einstein–Smoluchowski equation, 333
diffusion coefficient, 296
  and critical concentration for hydrogen, 1328
  and Einstein relation, 374
  and Einstein–Smoluchowski relation, 339
  and holes, in molten salts, 577
  and mobility, 374, 376
  and molecular quantities, 338
  rate process expression for, 342
  and structural properties, 344
  of substances near melting point, 546
  tracer, 545
diffusion control, 1046
diffusion equation, and Nernst-Planck flux equation, 411
diffusion flux
  and Kirchhoff's laws, 1039
  produced by sinusoidal variation of current, 328
diffusion layer, 1055–1058
diffusion layer thickness, and microrough surface, 1219
diffusion potential
  as a function of transport number, 418
  further reading on, 420
  and transport number, 406
diffusion process, boundary conditions, 313
digestion, biochemical, 40
dimerization, in liquid silicates, 610, 612
dipole
  difference of contribution to potential of flip and flop positions, 787
  and electric field, interaction, 784
  interaction with electrode, 784
  orientation of, and capacitor, 135, 143
  water, flip-flop model, 790
  of water, in interior of electrolyte, 624
dipole-covered phase, potential difference through, 667
dipole–dipole interaction potential, at electrodes, 786
dipole orientation, net, at interphase region, 627
dipole potential
  and electrocatalysis, 1145
  at interface, 667
  at electrodes, 783
direct energy conversion, 1358, 1359
  and Carnot limitation, 1360
  by electrochemical means, 1360, 1361
  further reading on, 1400
  summary, 1398
discharge of ions, and dependence on lattice site, 1178
discrete polyanions, 610, 614
dislocation, 1201, 1205, 1206
dispersion forces, 166, 167
dissociation constant, of acid, 496, 497, 498, 500

dissolution
  field-assisted, 1337
  of iron, intermediates in, 1087
  of iron, mechanism for, 1085
  of iron, prediction of various mechanisms, and experiment, 1092
  of kinky surface at bottom of crack, 1337
  of metals under stress, and Miller index, 1335
dissolution–precipitation mechanism, of passivation, 1321, 1324
distribution
  of electrons, among energy levels in metal, 956
  time average, spatial, 182
distribution function, for holes in liquid electrolyte, 537
distribution law
  in Einstein–Smolchowski diffusion, 335
  for size of hole, 534
Doddario Committee, and John Malone's baleful prediction, 1357
donor, 499
doping agent, 819
doping of bronzes, and electrocatalysis of oxygen, 1169
double layer
  charge density components, 712
  charging of, 1027
  concentration of reactants in, and work function, 1150
  constant capacitance of, model for, 758
  "constant" value for capacitance, 753
  effect upon electrocatalysis, 1012
  electrical, becomes trouble layer, 750
  and Gauss's law, 727
  interaction of, and stability of colloid, 837
  ionic cloud at, 722
  isotherm for ions in, 764
  and parallel plate condenser, 634
  and Poisson equation, 724
  potential difference at, 635
  special position of mercury in studies of, 687
  structure of, further reading on, 790
  thickness of, and colloids, 835
double-layer charging, in galvanostatic transients, 1190
double-layer structure, further reading on, 717
double-layer studies at solids, further reading on, 803
double-layer theories, review of, 752
double-layer treatments, history of, 724
drift, ionic, to interface, 391, 845
drift velocity
  calculation of relaxation components, 427
  electrophoretic components of, 427
  and interacting ions, 425
  relaxation component of, 430
driven cell, 1128, 1131
driving force, 343, 412
droplets, corrosion under, 1305
economics, and social importance of overpotential, 16

edge vacancy, and electrodeposition, 1179
effective mass, of hole, 619
efficiency, 1358–1371
e–i junction, law for, 936
Einstein relation, and absolute mobility, 374
Einstein–Smoluchowski equation, 333, 334
  how many ions diffuse?, 333
electric car
  and Butler–Volmer equation, 928
  and electrocatalysis, 1155
electric conduction, and liquid silicates, 606
electric energy storage, needed for automobiles, 442
electric field
  effect upon electrocatalysis, 1168
  effect on rate, 869
  influence on random walk, 350
  local, in polar dielectric, 147
electric reactions, and electrochemistry, 14
electrical energy
  and free energy, 15
  production by thermal and electrochemical means, 40
electricity
  from chemical energy, by electrochemical means, 1362
  conversion of energy to, and electrochemistry, 42
  from heat, 1360
electricity storage
  in alkali metals, 1423
  and atomic energy, 1397
  in hydrogen, 1420
  important quantities in, 1404
electricity storage density, 1404, 1406
electricity storer(s), 1402, 1412, 1413
  future ones, 1420
electrification, 6, 7, 623
electrified interface(s)
  absolute potential difference at, 675
  further reading on, 717
  importance of, in practical situations, 642
  mobile, electrokinetic properties, 826
  retrospect and prospect, 715
  structure of, 718
  thermodynamics of, 688, 698
electrocapillary curves
  basic equation for, 698
  and capacitance, 721
  differentiation of, 704
  facts on, 690
  as perfect parabola, 705
  and surface excess, 710
electrocapillary equation, final general form, 701
electrocapillary maximum, 691
electrocapillary thermodynamics, 713, 714
electrocatalysis, 1141
  activation in, 1170
  and cancellation of thermionic work function, 1147
  and cars, 1155
  and chemical catalysis, 987
  of copper deposition, on various surfaces, 1202

electrocatalysis *(cont.)*
  dependence on electronic properties, 1147
  difference from chemical catalysis, 1143
  and dipole potential, 1145
  effect of double layer on, 1012
  effect of metal–metal potentials, 1148, 1155
  and exchange current density, 1146
  and Galvani potential difference at nonpolarizable electrodes, 1149
  heat of activation in, of nonbonding reactions, 1149
  of hydrocarbons, 1156, 1158
  and hydrocarbon oxidation, 1391
  of the hydrogen evolution reaction, 1155, 1232
  irrelevance to, of experiments with porous electrodes, 1376
  lack of effect of work function on, 1148
  of oxygen, and doping of bronzes, 1169
  potential of, comparison, 1145
  and potential of zero charge, 1142
  rate equations for, compared with those for catalysis, 1143
  in reactions involving adsorbed species, 1153
  reactivity at low temperatures in, 1169
  of redox reactions, 1149
  reference potential for, 1143, 1144
  secondary effects due to double layer, 1151
  secondary effect of work function on, 1151
  and simple redox reactions, 1146
  in space vehicles, 1158
  special position of platinum in, theory of, 1166
  tunneling condition for hydrogen evolution, 1154
  volcano relations in, 1165
electrocatalyst, determination of adsorbed entities at, 1030
electrochemical cells, 1114, 1132
electrochemical converter(s)
  Carnot efficiency limitation avoided in, 1364
  efficiency, 1364, 1366
electrochemical desorption, in hydrogen evolution, 1234
electrochemical device(s)
  as energy producer, 855
  as substance producer, 851
electrochemical electricity storers, 1412
electrochemical energy conversion
  and atomic energy, 1395
  its central problem, 1369
  dominating role of electrocatalysis, 1372
  and Tafel relation, 1370
electrochemical energy converters
  cost of, and porous electrodes, 1385
  deduction of real efficiency, 1366
  and power-rate relation, 1378
electrochemical energy producer, and power density, 859
electrochemical energy storage
  feasible goals, 1431
  summary of, 1430
electrochemical engine, 1157

electrochemical era, 43
electrochemical generators, 1386
  examples of, 1385
electrochemical methods, for surface coverage, 1035
electrochemical model, for slag–metal equilibrium, 618
electrochemical potential(s)
  digression on, 693
  equality of, in different phases, rationalized, 864
  gradient of, 395
  and work of bringing charge particles into material phase, 695
electrochemical producer, 1361
electrochemical reaction(s), 7
  and chemical reactions, 8, 987
  always quantal, 985
electrochemical reactor(s), 9, 1405, 1406
electrochemical system(s)
  as devices, 851
  and metal–metal potential, 1113
  series of potential drops in, 1112
electrochemical vista, 43
electrochemist(s)
  frustrations of, 1042
  training of, 44
electrochemistry
  advances expected, 42
  and artificial limbs, 43
  and atomic reactors, 43
  awakening, 24
  as basic science for advances in postindustrial era, 42
  and biology, 29
  brilliant beginning, 14
  and cheap heat, 43
  and chemistry, 28, 869
  and circuits in the body, 43
  and coming era of atomic power, 43
  and conversion of energy to electricity, 42
  conversion of, to charge transfer orientation, 17
  delay in development of, 18
  and development of circuitry, 43
  and developments in molecular biology, 42, 43
  and direct energy conversion, 1361
  disciplines in, 27
  and electric reactions, 14
  and electronics, 18, 27, 33
  and electron transfer, 25
  future role, 41
  and geology, 29
  an interdisciplinary area, 1, 25, 29, 31
  and interfaces, 22, 23
  interfacial, degree of ubiquity, 39
  involvement of, in many sciences, 28
  and machining, 42, 43
  and medical developments, 43
  need for books, vi
  and new towns, 43
  and other fields, 25
  perspective from afar, 23
  perspective from a medium distance, 24

electrochemistry *(cont.)*
  perspective in time, 39
  and photogalvanic converters, 43
  place in science, 26, 41
  and polluted liquids, 43
  and possible fuel cell heart, 43
  and powering of ships, 43
  and powering of vehicles, 43
  quantum nature of reactions in, 20, 21
  as separate discipline, 38
  and sewage disposal, 43
  sign convention in, 1115
  and stabilization of materials, 42
  and storage of energy, 42
  and time, 38
  and tools, 43
  and transportation, 43
  and urban living, 42
  and water purification, 42
  wider significance?, 38
electrocrystallization, 1129, 1173, 1174, 1219
electrode(s)
  as catalyst, 34, 1139
  de Broglie wavelength at, 20
  porous, 1171
    activity near tip of meniscus, 1384
    diffusion of reactant in, 1384
    and distribution of catalyst, 1384
    vital importantance in fuel cells, 1172
  sick, 1231
electrode kinetics
  and double-layer effects, 916
  and organic reactants, 916
  transfer coefficient as center of, 918
  and zeta potential, 912
electrode processes, quantum-mechanical approach to, 947
electrode reactions
  and heterogeneous reactions, 989
  history of quantum-mechanical developments, 983
  and tunneling, 955
electrode surfaces, 36
electrode–electrolyte potential difference, analysis of, 659
electrodeposits, organic, 1222
electrodeposition
  consecutive stages in, 1180
  and dehydration, 1180
  electronation in, 1176
  function of adions in, 1177
  and hole vacancy, 1179
  influence of potential of zero charge, 1180
  of metals, and tunneling, 1177
  rotation of a spiral in, 1205
  stepwise dehydration in, 1177
  and surface adions in random walk, 1181
  and surfaces which change with time, 1182
  in terms of consecutive reactions, schematic, 1183
elecrodeposition rate, as a function of crystal plane, 1216
electrodics, 19, 846
  and corrosion, 1285
  and electronics, 30, 31, 32

electrodics *(cont.)*
  elementary, further reading in, 909
  elementary, summary, 908, 983
  and inhibition of corrosion, 1306
  and quantum mechanics, 30
  transient techniques in, 34
  transport aspects of, summary, 1076
  in the west before 1950, 1380
electrodissolution, burst of, and Laplace transformation, 330
electrogrowth, 1215
  basic aspects of, 1173
  and kink sites, 1203
  topographical features, 1184
electrokinetic properties, 826
  further reading on, 835
electrolyte(s)
  forces at boundary of, 623
  glasses as, 603
  potential and true, conductance in different media, 179
  pure liquid, 513
  true and potential, 176
electrolytic solutions
  electromagnetic radiation in study of, 274
  and infrared spectroscopy, 278
  and nuclear magnetic resonance spectroscopy, 278
  and Raman spectra, 277
  and semiconductors, 811
electromagnetic methods for investigating solutions, further reading in, 279
electromagnetic radiation, in study of electrolytic solutions, 274
electromagnetic theory of light, 35
electron(s)
  cold emission of, 944
  collision with impurity atom, 956
  distribution among energy levels in metal, 956
  in holes, and the double layer, 825
  image energy of, as function of distance from metal, 945
  leak through barrier, 945
  their mechanics, 947
  near interfaces, potential of, 943
  number of which strike surface of metal, 990
  penetration into forbidden region, 950
  in space region, and wave function, 669
  and tunneling to solution, 959
  in vacuum, and probability of passage through, 952
  which are free to tunnel, 959
electron overlap potential difference, 670
electron sink, 853
  electrode, 1126
electron source electrode 853, 1126
electron transfer
  and electrochemistry, 25
  probability for tunneling, expression for, 978
  type of, in electrochemistry, 20

electron transfer reactions, 494
  and electroneutrality difficulties in conduction, 351
  the 1950's, 17
electron transfer theory, beginnings, 18
electron tunneling, 946
  condition for, 972
electronation, 352, 847
  in corrosion, 1275
  of oxygen, 1251
    and enzymes, 1253
electronation reaction
  and bond stretching, 971
  effect of field, 874
  of hydrogen on platinum, energy terms in, 968
electroneutrality
  conflict with conduction, 351
  and coupling between ionic species, 410
  in fused salts, 543, 566
  principle of, 414
electronics, 18, 27, 30, 31, 32, 33, 43
electro-osmosis, 826
  theory of, 827
electro-osmotic motion, of phases relative to each other, 831
electrophoresis, 832, 834
electrophoretic effect, 424, 425
electrostatic potential
  and charged sphere, 52
  and field strength, 347
  near ion, as function of distance, 193
  variation of, near interface, 664
electrostriction, 126
ellipsometric spectroscopy, 37
ellipsometry, 37, 1319
embrittlement
  in corrosion, 1347
  by hydrogen, 1314, 1338, 1339, 1344
emission
  cold, 944, 952, 955
  hot, of electrons, 941
  thermionic, 953
empty space, in fused salts, 523
energetics, of certain corrosion reactions, 1278
energy(ies)
  of activation, for self-diffusion, 579
  of crack, 1341
  electrical, 11
  free and electrical, 15
  as function of repulsion between water molecules, 970
  of interaction between colloid particles, 838
  of strain, 1340
energy barrier, and rate theory, 341
energy consumption, electrochemical, 1350
energy conversion, 1266, 1358, 1361
  to electricity, and electrochemistry, 42
  as an interdisciplinary field, 31
  and storage, terminology of, 1402
energy conversion efficiency, maximum of, 1376
energy converter(s)
  electrochemical, efficiency of, 1374

energy converter(s) (cont.)
  power and efficiency in, 1379
  thermionic, 1359
  thermoelectric, 1359
energy density, 1407, 1408
  idealized maxima, tabulated, 1408
  and rate of working of cell, 1410
  of stores, feasible values, 1418
  versus power density, 1411
energy gap, 810, 931
energy levels, 805, 807, 956
energy producer, 855
energy-producing device, current and potential in, 1131
energy sources, 1350
  distribution of, 1351
energy states
  discrete, 806
  those accounting for free electrons, 959
energy storage, 1266
  electrochemical, summary, 1412, 1430
  terminology, 1402
energy storage density
  and non-aqueous electrolyte, 1427
  for some realized cells, 1409
energy storers, 1420, 1426, 1428
energy waster, 859
enthalpy
  by Born, and ionic radii, 60, 69
  and electrochemical energy conversion, 1368
  and entropy, of ion–solvent interaction, 59
entropy, and electrochemical energy conversion, 1364, 1368
enzymes, 1253
equilibrium
  difficulty of observing rates near to, 1263
  and electrochemical potential, 696
  at interface, 876
  and Nernst equation, 898
  and steady state, 1018
equilibrium cell potentials, useful ?, 1124
equilibrium potential, 876
  and activity in solution, 905
  limitation in usefulness, 876
equivalent circuit(s)
  for galvanostatic transients, 1026
  for interface involving ideally polarizable electrode, 654
  and pseudocapacitance, 1029
equivalent conductivity, 358
  and concentration, 360
  and ionic mobility, 372
  significance of, 360
era, electrochemical, 43
error function, 321, 1065
Esaki tunnel diode, and tunneling, 956
ethylene
  adsorption of, on platinum, 1160
  rate-determining step in the oxidation of, 1159, 1160
ethylene oxidation
  and catalytic activity, 1161

INDEX xxxix

ethylene oxidation *(cont.)*
  diagnostic criteria for de-electronation on platinum, 1162
  negative pressure effect in, 1160
  radiotracer measurements of coverage, 1160
  rate-determining step, 1160, 1161, 1164
  volcano relation in, 1164
European convention, 1118
evolution, of gases, 1102, 1104
excess charge density, as function of distance from central ion, 194
exchange current density, 876, 877
  catalytic effects due to double layer properties, 1151
  and deposition overpotential, 1215
  determination of, for hydrogen evolution, 1238
  and electrocatalysis, 1146
  and heat of activation, 1150
  for hydrogen evolution, tabulated, 1238
  for metal deposition, 1202
  on noble metals, and platinum–rhodium alloys, 1260
  and polarizability, 895
  and rate-determining step, 997
  and reaction order, 1012
  schematic diagram of, 878
  small, difficulty of measurements with, 1260
  for various crystal faces, 1215
  and work function, 1149
exclusion principle, 963
expansivity
  calculation from hole model, 585
  of liquid silicate models, 609
faces, fast growing, 1217
facets, 1213, 1214, 1217
faradaic rectification, 885
faradaic resistance, 996
fatigue of metals, 1267
Fermi–Dirac distribution, deduced, 958
Fermi energy, 959, 962
Fermi level, 957, 964, 977
Fick's first law, 293, 315, 316, 343, 1056, 1065
Fick's second law, 308, 344, 1040
field
  current produced, 883
  in double layer, 630
  excess, 882
  induced reorientation, 484
  at interphase, 630
  nonlinear, 348
  producing current, 882
  in semiconductors, 814
field strength
  effect on current, 881
  and electrostatic potential, 347
  and flux, 350
film(s)
  electronic conductivity of, and passivation, 1320

film(s) *(cont.)*
  passive, formation of, at the bottom of pits, 1335
  precursor, 1319
film-covered surfaces, determination of properties by ellipsometry, 37
Flade potential, 1316
flash photolysis, 34
flip dipoles, at electrodes, lateral interaction between, 785
flip-flop model, 790, 797
flip-flop water on dipoles, 779
flocculation, 839
flux
  and forces, 894
  as a function of field strength, 350
  of ions, and chemical potential, 397
  of ions, and electrostatic potential, 397
  at low field gradient, 295
  sinusoidally varying, near electrode–solution interface, 327
  time derivitive of, in step function, 330
flux equality condition, 1038
flux equations, and Onsager equations, 411
forbidden region, electron penetration into, 950
forces
  anisotropic, 626
  at boundary of electrolyte, 623
  and fluxes, 894
  in organic adsorption, at electrodes, 792
fossil fuels
  available for centuries if used mainly for food and textiles, 1350
  converted to food and textiles, 1358
  lack of rationality of burning up, 1352
Fourier's law, 894
fraction, of ions, produced in pulse, near electrode, 332
free electrons, energy states for, 959
free energy
  change of, when ion goes from OHP to IHP, 763
  determination by chemical and electrochemical means, 40
  and electrical energy, 15
  of ion–ion interactions, 181
  of organic adsorption, at electrodes, 792, 796
free energy of activation, as function of potential, 390
free energy change
  in contact adsorption, 742
  in ion–solvent interactions, 49, 50
free energy level, of proton, 500
frog, electrical movement of its nerve, 11
Frumkin, effect of leadership in Russia on electrochemistry throughout world, 18
fuel cell
  catalyst, 1383
  first one, 1380
  future of, and corrosion, 1388
  history of, 1386, 1387

fuel cell *(cont.)*
  immediate uses of, 1396
  mission time in which useful, 1432
  practical applications of, 1396
  and production of water, 1420
  as source of drinking water, 1389
  vital importance of porous electrodes in, 1172
fuel cell heart, possible electrochemical development, 43
fuel cell research, funding of, and pollution, 1386
fuels, and electrocatalysis, 1387
Fuoss approach, to ion pairs, 261
Furth approach to work of hole formation, 536
fused oxides, as slags, 616
fused salts *(see also* molten salts)
  and activiation energy for diffusion, 547
  and activation energy for viscous flow, 550
  atomistic theory of transport, 574
  diffusion in, 542
  empty space in, 523
  holes and diffusion coefficients in, 577
  heat of activation for viscous flow, 551
  internuclear distances in, 524
  radiotracer method for transport number determination, 571
  and self-diffusion, 542
  Stokes–Einstein relation in, 552
  transport, and holes, 574
  transport numbers, Stokes' law approach to, 572
  viscous forces and momentum in, 575
Galvani potential, thought experiment synthesis of, 672
Galvani potential difference, 670
  at nonpolarizable electrodes and electrocatalysis, 1149
galvanostatis rise time, 1193
galvanostatic transient, 1021, 1185
  and Butler–Volmer equation, 1190
  and double-layer charging, 1190
  equations for, 1024
galvanostatic transient method, for surface coverage, 1030
Garrett–Brattain space charge region, 812
Gaussian box, 728
Gaussian surface, 137, 151
gels, 839
Gemini fuel cells, used in space, 1388
generators, electrochemical, 1385, 1386
geology, and electrochemistry, 29
Gibbs, his angel, 679
Gibbs' surface excess, 680, 683
glasses, 603
  electrolytic structures of, 603
  as ionic liquids, 603
  liquid silicate as, 603
goals, in storers, 1420
Gouy–Chapman and Helmholdtz–Perrin, relative contributions, 737, 822

Gouy–Chapman model, and potential dependence of capacitance, 732
Gouy–Chapman region, and colloidal stability, 839
Gouy–Chapman theory, and capacitance, 731
grains, 1218
greenhouse effect, 1356
Griffith crack, 1340
group dipole, 147
growth step, 1210, 1211
Guntelberg charging process, 248
happenings, thermal and electrochemical, alternative versions, 40
heart, artificial, possible electrochemical development, 43
heat of activation
  change of, with electrode potential, 924
  in electrocatalysis, of nonbonding reactions, 1149
  and exchange current density, 1150
  for flow in silicate melt, 605
  for proton transport, in aqueous solutions, 473
  and temperature, in proton mobility, 486
  for viscous flow, and melting point, 551
heat to electricity conversion, 1360
heat engine, essence of working of, 1352
heat of hydration
  of hydrogen ion, 105, 467, 468
  by quadrupole model, 101
  relative, as a function of radius, 97
  of transition metal ions, as a function of atomic number, 112
  of transition metal ions, and water stabilization energy, 113
heat of solution, 65, 67
heat of solvation, 63, 66, 88, 96
Helmholtz and Gouy capacities, in series, 736
Helmholtz–Perrin model, 718
Helmholtz–Perrin and Gouy–Chapman, relative contributions, 737
Henderson–Planck equation, solution for, 459
heterogeneity, surface, on solid electrodes, 803
Hiatus, the Great Nernstian, 16
history of double-layer treatments, 724
hole
  average size of, 539
  concept of formation of, 528
  and diffusion coefficients, in fused salts, 577
  effective mass of, 619
  formation of, in valency band, 810
  lifetime of, in fused salts, 576
  in liquid electrolyte, size of, 537
  and transport in fused salts, 574
  viscosity in terms of, 577
hole current, and electron current, 933
hole mobility, values of, 931
hole model, 527
  and compressibility, 584
  and expansivity, 585

hole model *(cont.)*
  for liquid electrolytes, further reading on, 541
  most consistent model at present, 584
  normalizing conditions, 536
  probability of finding hole of radius r, 535
  and rationalization of relation of heat of activation to melting point, 582
hole motion, and electron motion, 811
hole vacany, and electrodeposition, 1179
Hooke's law, 1331, 1334
hot emission of electrons, 941
hump
  of capacitance, 754
  experimental, in capacity–charge curve, 762
hydrated ions
  distance of closet approach to electrodes, 741
  formation from adions, 1187
hydration, 80
  calculations involving quadrupoles, 104
  effect of ligands, 109
  and orbitals, 110
  of transition metal ions, 106
  of transition metal ions, and stabilization of field, 111
hydration number(s)
  activity coefficient as function of, 240
  of alkali and halide ions, by independent methods, 118
  and compressibility, 127
  primary, table of, 131
  and radius, 130
  by various methods, 118
hydration of proton, heat of, 467
hydrazine–oxygen cells, performance tabulated, 1390
hydrazine–oxygen fuel cells, 1389
hydrocarbons
  catalysis of and cars, 1158
  cracking of hydrocarbons, reforming of hydrocarbons, 1158
  electrocatalysis of, 1158
  reformed by steam, 1390
  reforming of, 1391
  saturated, 1391
    mechanism determination of, 1107
    mechanism of oxidation of, 1107
    rate-determining step in the de-electronation of, 1110, 1158
  unsaturated, electrochemical data concerning, 1392
hydrocarbon–air cells, 1391
hydrocarbon fuels, how used?, 1351
hydrocarbon oxidation, 1158–1159
hydrodynamic flow, and convection, 1050
hydrogen
  accumulated, in cracks inside metal, 1339
  accumulation of, at regions of stress, 1333
  adsorbed, and stability of metals, 1328
  adsorption of, and change the path of crystallization, 1129
  and change of mechanical properties, 1129
  diffusion into
    metals, 1328
    regions of stress, 1330
  diffusion by interstitial and intergranular paths, 1330
  effects, and passivation, 1349
  electricity storage in, 1420
  initiation of cracks by, 1335
  and instability of metals, 1338
  kinetics of discharge of, favorable effects on storage of hydrogen, 1421
  partial molar volume of, 1331, 1332
  penetration into bulk of metal, 1329
  pressure of in metals, 1333
  storage of, cryogenics, 1421
hydrogen–air battery, 1421
hydrogen–oxygen cell, 1388, 1390
hydrogen adsorption, atomic, on electrodes, and coverage, 1246
hydrogen bond(s)
  and proton mobility, 468
  in solvation process, 90
  and clusters of water molecules, 88
hydrogen coverage, 1245
  of surface, variation with potential, 1245, 1246
  and various mechanisms, 1247
hydrogen codeposition, 1227, 1229
hydrogen desorption, and de-electronation, 1246
hydrogen embrittlement, 1314, 1338
hydrogen evolution
  adsorption step in, 1153, 1233
  atom–atom step in, 1234
  chemical desorption step in, 1234
  and corrosion, 1232
  and current efficiency, 1232
  deduction of values for transfer coefficient, 1241
  desorption in, 1153, 1234
  determination of path and rate-determining step, 1237
  determination of transfer coefficient by various means, 1238
  and electrocatalysis, 1232
  equations for various mechanisms, 1240
  exchange current density for, tabulated, 1238
  further readings on, 1250
  general, 1231
  history of, 1231
  ion–atom recombination, 1234
  mechanisms, 1235
  on metals, equations for, 1228
  paths, 1233
  reaction paths, 1235
  recombination mechanism in, 1234
  and separation factor, values for, 1249
  tabulated summary of probable mechanisms, 1250
  and transfer coefficient, 1102

xlii   INDEX

hydrogen evolution *(cont.)*
  tunneling conditions for, and electrocatalysis, 1154
hydrogen ion, trigonal pyramid structure of, 466
hydrogen overpotential, experimental characteristics, 1232
hydronium ion, existence of, 487
  proton mobility in, 486
  and water, structure of, 76
iceberg model, for liquid silicates, 615
Ilkovic's equation, 1068
image charge, and contact adsorption, 767
image energy
  of electron, as function of distance from metal, 945
  its place in charge transfer kinetics, 960
  and tunneling, 961
image forces, 661–662
  and quantum mechanical tunneling, 960
image interactions, with charged electrodes, 660
impedance, 896
  high, and measurement of overpotential, 897
  and measuring circuit, 896
impurity atoms, collision of electron with, 956
impurity conduction, in silicon and germanium, 930
inclusions, microscopic, of copper, effect on reaction rate, 1272
indifferent electrolyte, 1060, 1069
induction time, in passivation, 1322
infrared spectroscopy, and electrolytic solutions, 278
inhibition, of corrosion, and electrodics, 1306
inhibitor(s)
  coverage, tabulated, 1309
  adsorption, 1310
  practical examples of, tabulated, 1311
initiation, of cracks, 1335
inner potential, 673
inner potential difference
  between dissimilar phases, 677
  between two identical phases, measurability of, 679
instability
  of metals, 1338, 1343
  of surface, and internal decay, 1335
instantaneous pulse, 329
instruments, high impedance, need for, 896
instrumentation, for potentiostatic transients, 1033
interaction(s)
  ion–ion compared with ion–electrode, 723
  metal–water, 740
  minimal, between dipoles at electrodes, 785
interaction energy, and orientation of dipole near ion, 81
interatomic spacing, 807, 808

interface(s)
  accumulation of substances at, 679
  adsorption of ions at, 738
  affected by image forces, 662
  bird's eye view of structure, 824
  creation by thought experiment, 5
  current–potential laws at, 930
  dipole potential at, 667
  electrified
    importance of, 642
    retrospect and prospect, 715
    review, 716
    structure of, 718, 632
    thermodynamics of, 688
  electron potential near, 943
  examination in terms of transients, 1026
  exchange of electrons through moisture films, 6
  ionic drift to, 845
  at metals, other than mercury, 801
  metal–solution, and Volta potential, 665
  nonpolarizable, 697, 701, 894
  polarizable, 653, 700, 894
  potential differences at
    further readings on, 687
    and surface tension, 688
  profile of concentration at, in adsorption, 681
  semiconductor–electrolyte, 803
  structure of, and Butler–Volmer equation, 911
  thermodynamic deduction of surface excess equations, 700
  two-way electron traffic across, 873
  under transient conditions, 1017
  water molecules in oriented layer at, 633
interfacial tension
  and applied potential, 701
  measurement of, 688
intermediates
  adsorbed, 1016, 1029
  determination of, in benzene oxidation, 1099
  in dissolution of iron, 1087
  and potential–time transients, 1026
  formed by charge transfer, 1027
  and propane oxidation, 1110
internal combustion engine, 1358
  wasting fuel as heat, 1352
internal resistance of fuel cells, effect upon current–potential curve, 1373
internuclear distance, in solid and liquid fused salts, 524
interphase, 2, 630
  surface excess at, 683
interphase properties, and bulk properties, 641
interphase region, 626, 627
ion(s)
  adsorbed, on kink sites, and passivity, 1325
  association of, temporary and permanent, 273
  clusters of, 265

hydrated, distance of closest approach to
electrodes, 741
in sheath of water molecules, 77
ion-atom combination, in hydrogen evolution,
1234
ion-cloud theory, 180
ion-dipole interaction(s), 83
equations deduced, 169
ion-dipole model, of solvent interaction, 80
ion-dipole theory, of solvation, evaluation,
93
ion-electrode interactions, 723
ion-electrode and ion-ion interactions, 723
ion-ion interaction(s), 175
and activity coefficients, 202
and Debye-Hückel theory, 725
free energy of, 181
parentage of theory of, 273
in perspective, 279
ion-quadrupole theory, of solvation, 103
ion-size parameter, 224, 225, 230
ion-solvent interaction, 49
and cavities, 81
effect on activity coefficient calculation,
238
and effect of quadrupole theory, 115
equations for, 59
experiment and theory, 94
free energy change in, 49, 50, 57, 58
further reading on, 116
heat of, and thermodynamic cycle, 66
improvement, by quadrupole model, 100
of individual ions, 114
quadrupole model of, 99
summarizing remarks, 113
thought experiments in, 80
ion-solvent-nonelectrolyte interactions, 158
ion-water interactions, quadrupole theory,
171
ion association, 447
and conductivity, 448
ion association constant, 257, 259
ion migration, as function of electrostatic
potential gradient, 288, 289
ion pairs
and activity coefficients, 260
Bjerrum approach, 260, 263, 264, 265
Fuoss approach, 261
further reading on, 266
and triple ions, 265
ion pair formation, 251, 255
ion pair fraction, 258
ion size parameter, skepticism, 280
ionic association, 251
ionic atmosphere, 428, 636
and ionic migration, 420
radius of, Debye-Hückel, 197, 199
and variation with potential, 636
ionic cloud
asymmetric, 422
catching up with moving ions, 422
chemical potential change arising from,
201

ionic cloud (cont.)
and Debye-Hückel constant, 200
further reading on, 202
relaxation of, 423
shape of, 431
ionic drift
interdependence of, 399
to interface, 845
ionic fluxes, and Onsager equations, 411
ionic groups, in silicates, 596
ionic liquids, 513
further reading on general aspects of, 522
and glasses, 603
lattice-oriented models for, 522
and liquid silicates, 603
mixture of, and complex formation, 587
and Nernst-Einstein relation, 555
slags, 617
ionic migration, 367, 387, 420
ionic mobility, 401
ionic movement, as function of random walk,
299
ionic product, 499
and semiconductors, 819
ionic radius, and contact adsorption, 744
ionic solutions, and dielectric constant, 132,
155
ionic solvation, 155
ionic species, single
activity coefficient of, 206
ambiguity of measurement of properties,
64
ionic strength, definition, 210
ionic transport, and electrochemical potential,
395
ionics, 19, 846
analogy to behavior of electrons in holes,
32
and charge transfer, 1036
definition of, 3
rise and fall, 24
ionization potential, and energy released
from electronation of proton, 967
iridium, 1253, 1255
iron
mechanisms of dissolution of, tabulated,
1091
potential of, as affected by solid phase,
in corrosion, 1283
potential-pH diagram for, 1280
irradiation, ultrasonic, and electrocatalysis,
1170
isotherm
for adsorption of oxygen on platinum, 1257
contact adsorption, tests of the value of,
770
deduction of, for contact adsorption, 768
for ions in the double layer, 764
lateral repulsion model, discussion, 777
and organic adsorption, 797
tests for, 771, 772
isotopes
radioactive, self-diffusion, 542

isotopes *(cont.)*
  substitute, dependence of reaction rate on, 1106
IUPAC sign convention, in detail, 1119
journals, international, in electrochemistry, v
kilowatt hours per kilogram, 1407, 1410
kinetics, of hydrogen discharge, and storage, 1421
kinetic theory, for viscosity, 621
kink sites
  adsorbed ions on, and passivity, 1325
  diffusion of adion to, 1182
  and electrogrowth, 1203
Kirchhoff's law, and diffusion flux, 1039
Laplace transform
  for concentration perturbation, 329
  of constants, 454
  in diffusion problems, 1041
  and Fick's second law, 344
  use in polarography, 1063
  of pulse of flux, 331
Laplace transformation
  and burst of electrodissolution, 330
  definition of, 310
  explanation of, 309
  and Fick's law, 315
  initial and boundary conditions for diffusion process stimulated by constant current, 313
  use in Fick's second law, 312
  and transients, 1041
lattice approach, to concentrated solutions, 266
lattice dislocations, and spiral, 1206
lattice energy, and heat of solution, 67
lattice formation, after surface diffusion, 1195
layer growth, 1208, 1222
leveling, 1222
life, and oxygen electronation, 1252
lifetime
  of complex ions, in molten salts, 590
  of proton in solution, 485
ligands, effects on hydration heat, 109
light
  absorption of, and passivation, 1319
  polarized, application to electrode surfaces, 36
  reflected, and passivation, 1320
  theory of, electromagnetic, 35
  visible, application to electrode surfaces, 36
limiting current
  and hydrogen evolution, 1229
  practical importance of, 1059
  typical experimental values of, 1060
limiting current density, and electric migration, 1075
limiting law, breakdown of, 267
linear diffusion, time-independent, 1077
Lippman equation, 702
liquids, comparison of properties of various kinds, 519

liquid electrolytes
  cell model for, 529
  distribution function for holes in, 537
  further reading on, 533
  gas-oriented models for, 529
  hole model for, further reading, 541
  hole size in, 537
  liquid free volume model for, 530
  summary of models for, 532
  transport numbers in, 566
liquid free volume model
  for pure liquid electrolytes, 530, 531
liquid oxides, 594
liquid silicates
  dimers in, 610, 612
  and electric conduction, 606, 609
  as glasses, 603
  model of discrete anions, 610
  model of icebergs, 615
  network model of, and table, 606
  and polymerization, 611
  structure breaking in, and concentration of metal ions, 604
  ring anions in, 613
  ring formation in, 612
  as slags, 616
local cell, diagrammatic, 1271
local field, 147
  calculation of, 148
logistics, 1036
Los Angeles, and the smoggy state, 1354
Luggin capillary, 891
machining, and electrochemistry, 42, 43, 1267
macrostep(s), 1207, 1208
  and bunching, 1209
  and irregular edge, 1209
magnetohydrodynamic conversion, 1360
map of the Balkans, 1430
mass of hole, effective, 619
materials
  decay of, 4
  dependence of properties on surface, 3
materials science, 3, 5
  definition of, 5
  as an interdisciplinary field, 31
maximum, electrocapillary, 691
maximum cell potential, 1137
maximum efficiency, of electrochemical converter, 1364
maximum energy density, for electrode couples, idealized, 1408
Maxwell, his demon, 679
mean jump distance, 340
mean square distance, time, 453
measurement
  of overpotential, and high impedance, 897
  of Volta potential difference, 841, 842
mechanical properties
  change of, due to hydrogen, 1129
  and hydrogen codeposition, 1229
mechanisms
  biological, 1266

mechanisms *(cont.)*
  of ethylene oxidation, on various metals, 1160
  of hydrogen evolution, 1235
    affected by coverage of hydrogen, 1235
    and coverage, 1247
  of the oxidation of saturated hydrocarbons, 1107
  of oxygen electronation, evaluation, 1253
  of passivation, 1319
  of porous electrode, 1384
  probable, of hydrogen evolution, 1250
  of reaction
    and Faraday's law, 1096
    and surface coverage, 1096
  on surface which change with time, 1182
mechanism determination, 1080
  and coverage of electrodes, 1097
  for ethylene, 1094
  further reading on, 1093, 1110
  and mixed potentials, 1105
  and separation factor, 1106
  on surfaces which change with time, 1182
  techniques for, 1099
mechanistic studies, summarizing remarks, 1090
medical electronics, and electrochemistry, 43
membrane
  effect on gravitational flow, 569
  Tafel relation at, 942
membrane potential, 410, 941
meniscus, 1383, 1384, 1388
mercury
  and double layer studies, 687
  and hydrogen evolution, 1232
metal(s)
  band picture of, with interatomic spacing, 808
  break-up when strained, 4
  of great strength, and electrochemistry, 43
  stretched, dissolution of, 1337
  sudden failure of, 4
metal deposition
  consecutive steps in, 1183
  further reading on, 1230
  with hydrogen, 1227
  and rise time, 1185
  transients in, 1183
metal–metal potential difference, and reaction rates, 1147
metal oxide and silicon atom, interaction in silica network, 602
metal–slag equilibrium, and charge transfer, 618
metal–slag system, as electrode–solution system, 618
metal–solution interface, and Volta potential, 665
metal–solution junction, and semiconductor, 850
metal–water interactions, 740
metallurgical extraction, and electrochemistry, 43

metallurgy, 27, 1266
  and electrochemistry, 28
methanol–water mixtures, abnormal conductivity in, 473
microrough surface, and diffusion layer thickness, 1219
microscopy, 37, 38
microspiral, 1207
microsteps, 1207, 1208, 1209
migration
  conductance in, further reading on, 367
  and diffusion, further reading on, 399
  of ions, as function of electrostatic potential gradient, 289
  ionic, 367
  and radiotracers, 570
Miller index, of metal surface, and cracking under stress, 1335
mixed potential, 1286
  determination of, 1105
  as a function of concentration, 1104
  and mechanism determination, 1105
mobile electrified interfaces, 826
mobility
  absolute, 370
  of charge carriers, in liquids and solids, 471
  conventional, 371
  and diffusion coefficient, 374, 376
  of ions, 369
  of proton, 471, 472, 476
  solvent effect on, at infinite dilutions, 443
mobility method
  in determining solvation number, 125
  for solvation number, 130
  for ionic liquids
    comparison of ability to predict, 583
    facts, 522
    further reading in, 587
    gas-oriented, 529
    lattice-oriented, 522
    summary, 532
    comparison of predictions, 485
model-oriented approach, 1
molecular biology, developments in, and electrochemistry, 42, 43
molten oxides, properties near melting point, 518
molten salts
  average size of hole in, 539
  compressibility calculated from hole model, 585
  detection of ions by Raman spectra, 590
  deviations from Nernst–Einstein relation, 557
  diffusion in, at constant temperature, 548
  distribution law for hole size, 534
  electroneutrality in, near electrode, 566
  future of model, 586
  lifetime of complex ions in, 590
  liquid free volume model, diagrammatic, 531
  models in, further reading, 587

molten salts *(cont.)*
  nonideal behavior due to interactions in, 588
  quasi-lattice model, 527
  Schottky defects in, 527
  tests for complex ions in, 594
  vacancy model for, 526
  volume change on fusion, 525
Morse curve
  and acid-base strength, 501
  and symmetry factor, 922, 926
Morse equation
  a gross approximation, 981
  in proton conductance calculations, 477
moving ion, asymmetric cloud around, 429
multistep reactions, 991
  activation energy for, 1002
  and Butler-Volmer equation in final form, 1000
  further reading on, 1017
  involving stoichiometric number, 1005
n–p junction(s), 930
  diffusion of holes and electrons across, 932
  exponential laws for, 936
  potential difference at, 932
n–p product, 819
n-type semiconductors, 818
  band picture, 820
natural convection, 1050
natural gas, electrochemical burning, 1393 1394
negative potential, superposition of, and stabilization, 1314
Nernst diffusion layer, 1056
Nernst-Einstein equation, limitations, 382
Nernst-Einstein relation, 377, 381
  and diffusion of ion pairs, 583, 586
  gross view of deviations in molten salts, 557
  and heats of activation, 556
  and ionic liquids, 555
  irreversible thermodynamic view of deviations, 558
  and paired vacancy theory of diffusion, 562
Nernstian Hiatus, 16, 23, 24
  and smog, 16
  and pollution, 17
Nernst-Planck equation, and electrolytic transport, 405
Nernst-Planck flux equation, 398
  and diffusion equation, 411
Nernst equation
  dilemmas associated with, 907
  discussion of deduction, 899
  kinetically deduced, 898
  physical significance associated with, 908
  and potential–pH diagrams, in corrosion, 1279
  sphere of relevance, 906
  as zero-current special case, 898
Nernst's relation, and alloys, 1223
nerve cell, diagram of, 937
network, and associated water, 77

network model
  for liquid silicates, 606
  defects of, 609
  table describing, 606
neutral ion pairs, diffusion of, 383
new towns, and electrochemistry, 43
Nobel Prize, given to electrochemist, 1060
nonaqueous solutions, 400
  and electricity storers, 1424
  standard electrochemical potential in, 1425
nuclear magnetic resonance spectroscopy, and electrolytic solutions, 278
nuclear power, direct conversion to electricity, 1396
nucleation
  conditions for, 1204
  and metal deposition, 1204
ohmic resistance, 997
Onsager equations, and ionic fluxes, 411, 828
optical method, of examining passivation, 1319, 1326
optimization, for energy conversion, 1374
orbitals, and hydration, 110
order, of electrodic reaction, 1008
order of reaction in electrodic reactions, 1009
  compared with order of chemical reactions, 1011
organics, effect on smoothing, 1222
organic adsorption, 791
  and electrode charge, 795
  flip-flop model for, 797
  forces in, 792
  further reading on, 800
  maximum, and potential, 798, 799
organic electrolytes, use in electrochemical electricity storers, 1413
organic reactants, and electrode kinetics, 916
organic substances, and morphology, 1222
outer Helmholtz plane, 635
  and Butler–Volmer equation, 911
  locus or reaction, 917
  position of, 757
outer potential, 673
  definition, 663
  diagrammed, 665
  and double-layer studies, 665
overall rate, and rate-determining step, 1002
overpotential, 883
  activation, 1053
  attitude toward, in pre-electrodic days, 1131
  classical picture, 1231
  concentration, 1052
  consequences of lack of understanding, 1381
  definition of, 880
  deposition, and exchange current density, 1215
  and electron queue, 993
  and electronation reaction, 1133
  and exchange current density, 896

INDEX xlvii

overpotential *(cont.)*
  measurement of, and high impedance, 897
  and multistep reaction, a near-equilibrium relation, 994
  and pollution, 1420
  and self-driving cell, 1136
  social importance of knowledge of, 16
  for various steps in multistep reaction, 996
overvoltage of hydrogen, 1231
oxidation of ethylene, dependence of rate upon substrate, 1161
oxide film
  and electrocatalysis, 1258
  electronic conduction of, and passivation, 1320
  and oxygen catalysis, 1258
oxide path, in oxygen evolution, 1104
oxygen
  adsorption on noble metals, related to d-band character, 1163
  catalysis, and doped tungsten bronzes, 1169, 1422
  electronation
    and corrosion, 1252
    evaluation of mechanism, 1253
    and life, 1252
  evolution, oxide path in, 1104
oxygen pressure, and corrosion rate, 1276
oxygen reactions, further reading on, 1263
oxygen reduction
  chemical step in, 1255
  rate-determining step in, 1255
  and stoichiometric number, 1254
oxygen reduction reaction
  catalysis of, 1256
  observation of equilibrium potential, 1263
oxygen starvation, and corrosion, 1303
ozone, connected with pollution, 1353
pH, and current efficiency, 1229
pH-potential diagrams, 1120, 1278
  for iron, 1280
  for lead, 1284
  and Nernst equation, 1279
  and solution concentration, 1280
  uses and abuses, 1281, 1284
paint, corrosion through, 1302
paired vacancies
  diffusion of, 1124
  and Nernst–Einstein relation, 562
parabola
  in electrocapillary curves, 705
  in surface tension–potential relation, 691
parallel plate condenser model, surface, tension–potential relation for, 719
partial molar volume
  of hydrogen, in metal, 1331
  of hydrogen in metals, and distortion of lattices, 1334
particles and formation of holes, 578
passivation, 1315
  and absorption of light, 1319
  characteristic current–potential course, 1317

passivation *(cont.)*
  competition in models, 1324
  dissolution–precipitation, mechanism of, 1321
  electrochemical model, 1325
  and electronic conductivity of films, 1320
  ellipsometry in, 1319
  induction time in, 1322
  mechanism of, 1319
  monolayer model, 1325
  optical method of examining, 1319
  and protection from corrosion, 1348
  and reflected light, 1320
  solid state model, 1325
passivation potential, 1316, 1317
passive films, formation at the bottom of pits, 1335
passivity
  and adsorbed ions on kink sites, 1325
  competing theories of, 1325
  electrodic model of, 1326
  criteria for distinction between models, 1326
  electrodic model of, 1326
  and oxide formation, 1327
  and potential–pH diagram, 1327
  and thickness of oxide in critical region, 1325
passivity and corrosion, further reading on, 1349
paths of hydrogen evolution, 1233
Pauling equation, use in electrocatalytic theory, 1164
penetration of hydrogen, into bulk of metal, 1329, 1334
Pennsylvania, Electrochemistry Laboratory in [University of], vii
permeation
  affected by embrittlement, 1344
  as function of stress, 1331
permeation currents, and evidence for cracking, 1343
perturbation methods, 1019
phases, moving, and the double layer, 826
phase boundary, 624
  double layer at, 630
photochemical reactions, and electrical, 11
photoelectric effect, 674
photogalvanic converters, and electrochemistry, 43
photolysis, flash, 34
pipes, corrosion of, 1304, 1345
Planck's relation, 948
Planck–Henderson equation, 417, 419
platinum
  hydrogen electronation reaction on, 968
  special position in electrocatalysis, reasons for, 1166
platinum electrocatalyst, difficulties, 1158
platinum–gold, catalysis on, 1259
poisons, 1262
Poisson equation
  applied to diffuse charge region inside

INDEX

Poisson equation *(cont.)*
  semiconductor, 812
  deduction of, 282
  solution of, for double-layer situation, 726
Poisson–Boltzmann equation
  linearized form of, 190, 234
  so-called rigorous solution, 247
  tests for validity of solution, further reading on, 250
polarizability, 895
  degree of, at various electrode surfaces, 802
  effect in salting out calculations, 161
  and exchange current density, 895
polarizable interface
  and interfacial thermodynamics, 700
  and lack of equilibrium, 697
polarographic wave, 1066
polarography, 1060, 1062, 1065, 1069
pollution
  of atmosphere, with products of internal combustion, 1352
  and cars, 1419
  and electrochemistry, 43
  and funding of fuel cell research, 1386
  and Nernstian Hiatus, 17
  and overpotential, 1420
  prediction of, 1357
polyanion, discrete, 610, 614
pore
  thin, high limiting current at, 1383
  wetted, and contact angle, 1383
porous electrode(s)
  activity near tip of meniscus, 1384
  and cost of electrochemical energy converters, 1385
  distribution of catalyst in, 1172
  and three-phase boundary, 1383
  use of, 1171
  vital importance in fuel cells, 1172
possible recharge cycles, as function of depth discharge, 1417
potential(s)
  absolute, impossibility of measurement, 645
  anode, 1126
  of average force, 183
  between metals, in cell, 652
  changes during increase of current, as function of type of cell, 1131
  of charged sphere, 52
  chemical, 672
    thought experiment in, 694
  of comparison, in electrocatalysis, 1145
  of corrosion, 1285
  in diffuse layer, as function of distance, 729
  electrochemical, 693, 694
  of electrochemical energy converter, its regions, 1372
  at electrodes, due to dipoles, 783
  electrostatic, variation of near interface, 664
  Flade, 1316
  at membrane, 410

potential(s) *(cont.)*
  of metals, in equilibrium with $1M$ solution of ions, 1282
  metal–solution, measurement of changes in, 650
  mixed, as a function of concentration, 1104
  of oxide formation, 1282
  of passivation, tabulated, 1317
  produced by electrochemical energy converter, and current, 1370
  reversible, of oxygen, determined, 1261
  standard, of certain electrode reactions, tabulated, 1116
  and surface tension at interface, 689
  variation of film thickness with, and passivity, 1325
  variation of, and ionic atmosphere, 636
  Volta, and potential of zero charge, 707
  of zero charge, 691
potential–pH diagram, 1278
potential–distance relation, for colloid particles, 729
potential difference
  absolute
    attempts to measure, 644
    at electrode–electrolyte interface, 670
    across single interface, 648
    structured, 660
  across barrier, 871
  across electrochemical system, 1112
  across interphase, generality, of 631
  between two electrodes, 346
  contact, 647
  of dipole, at electrode–electrolyte interface, 670
  due to electron overlap at interface, 670
  in electro-endosmotic motion, 831
  Galvani, 670
  inside semiconductors, 804
  at interface, 688, 912
  at interfaces, further reading on, 687
  measurability of outer and inner, tabulated, 677
  and n–p junctions, 933
  periodically varying, 1020
  Volta, measurement of, 841, 842
potential drop, in diffuse layer, 728
potential electrolytes, 176, 488
  dissolution of, 491
  and proton transfer reactions, 491
  and true electrolytes, difference between, 178
potential energy
  and Morse curves, 920
  of water system, during rotation, 477
potential energy–distance profile, 866
  for successive motions of ions, 868
potential energy–distance relations, and tunneling through barrier, 962
potential energy–distance theory, and Morse equation, 919
potential energy barrier, for proton transfer, 479

potential energy barrier *(cont.)*
  in acid strength theory, 505
potential energy curves, 921, 925, 971
  and effect of increased M–H bond strength, 1156
  in proton transfer, 969
  and stretching of bonds, 971
  vertical shift under potential change, 924
potential pulse, 1020
potential step, 1020
potential sweep method, 1033
potential of zero charge, 706
  and electrocatalysis, 1142
  and electrodeposition, 1180
  and heat of activation, for metal deposition, 1181
  and organic adsorption, 798
  and potential of maximum adsorption, tabulated, 800
  and surface potential, 707
    tabulated, 864
  and Tafel slope, 915
  in terms of Volta potential, 707
potential–time transients
  effect of intermediates on, 1026
  under diffusion control, 1046
potentiodynamic method, 1033, 1034
potentiostatic circuit, for examination of passivation, 1316
potentiostatic transient(s), 1032
  and adsorption of benzene on platinum, 1033
  instrumentation for, 1033
Pourbaix diagrams, 1121, 1123, 1281
  and spreading of cracks, 1343
power, 1410
  and efficiency, in energy converters, 1379
  and energy density, 1429
power density
  feasible goals for, 1431
  as function of energy density, for silver–zinc cell, 1411
  and lithium–chlorine cell, 1423
power density *versus* energy density, 1411
power output, 1171, 1376
power stations, and fuel cells, 1396
powering of vehicles, and electrochemistry, 43
precipitation, of materials, near electrode, 1322
precursor film, 1319
pre-electrolysis, purification by, 1261
prepassive film, 1319
pressure
  critical, for spreading of crack, 1341
  of hydrogen in metals, 1333
    high, 1333
primary hydration number
  of proton, 469
  table of, 131
primary solvation, 79, 160
probability
  of electron transfer, 978
  of finding one ion near another, 253

probability *(cont.)*
  that hole has radius r, 538
  of tunneling, on both sides of the barrier, 952
probability amplitude, and electron passage, 947
projection, 1266
  and concentrated electric field at, 1217
propane oxidation
  diagnostic coefficients, 1107
  and intermediates, 1110
  possible rate determining steps, 1109
protection
  anodic, diagrammatic, 1318
  cathodic, 1309, 1312
  theory of, 1309
  surface, 1266
protein, quantum mechanical tunneling to, 981
proton(s), 461
  affinity, 466
  conductance and Morse equation, 477
  chain mechanism, 474
  constitution of, further reading on, 470
  free energy level, 500, 503
  heat of hydration of, absolute, 105
  hydrated, and electron tunneling to, 962
  jumping, from water, 475
  lifetime in solution, 485
  mobility of
    abnormally high, 472, 485
    in aqueous solutions, model, 484
    further reading on, 488
    in ice, 486
    rate-determining step in, 485
    and water rotation, detailed scheme, 483
  as nonconforming ion, 461
  in solution, existence of, 465
  solvation, 462
proton jumps
  and classical laws?, 478
  quantum-mechanical, 480
  water re-orientation necessary for, 481
proton transfer
  in autodissociation, 492
  conditions for, 968
  between hydroxonium ions and favorably oriented water molecules, 478
  and potential electrolytes, 491
  potential energy barriers for, 505, 479
  and tunneling conditions, 966
proton transport, 470, 473
pseudocapacitance, equivalent circuit for, 1028, 1029
pseudo-equilibrium, 1236
pseudoforce, and diffusion, 306
pulse
  double, and staircase pulsing, 1019
  square wave, 1019
  with step reversal, 1019
pulse generator, 329
pulsing, and double pulse, 1019
pure liquid electrolytes, 513, 518

purification, 1260, 1261, 1263
purity, effect on corrosion, 1273
quadrupole model, 99, 100, 101, 104
quantum electrochemistry, in 1960's, 19
quantum-mechanical charge transfer, relation to proton transfer, 974
quantum-mechanical proton jumps, 480
quantum-mechanical theory, or radiationless tunneling, 960
quantum-mechanical transfer, basic condition for, 945
quantum mechanics, 27, 41
  deduction of Butler–Volmer equation in, 980
  desirable refinements in electrochemical application to, 981
  of electrode processes, 947
  of electrode processes, further reading on, 985
  of electrode reactions, history of developments, 983
  and electrodics, 30
  symmetry factor theory in, 974
  tunneling conditions for, 972
  of tunneling to proteins, 981
quantum nature of biological reactions, 21
quantum nature of electrochemical reactions, 21
quasi-lattice theory, 271, 272
queues, 992
radicals
  examination of, 1034
  intermediate detection of, 34
  on surface, nature of, and mechanism, 1105
radiotracer method
  and electrochemical method, 1035
  for surface concentration, 1030
radiotracers
  and determination of transport numbers in fused salts, 570, 571
  and electrochemical methods, complementary nature of, 1098
Raman scattering, 277
Raman spectra
  and detection of complex ions in molten salts, 590, 591
  and electrolytic solutions, 277
ramp, various types, 1019
random walk
  in deposition, 1186
  distances moved and time, 306
  influence of electric field on, 350
  ionic movement as a function of, 299
  sum of distance from origin, squared, 303
  and surface adions, in electrodeposition, 1181
rate(s)
  of deposition, on different planes, theory of, 1215
  as a function of substrate, for oxidation of ethylene, 1161
rate-determining step, 1138
  and activation barrier, 1003

rate-determining step *(cont.)*
  and charge transfer, 1185
  with chemical step, 1002
  concept of, 997
  in context, 1138
  in corrosion processes, 1296
  deduction of in terms of barrier height, 1003
  and energy barrier for multistep reaction, 1002
  in ethylene oxidation, 1160
  and exchange current density of partial reactions, 997
  and highest standard free energy, 1003
  and hydrogen evolution, 1237
  in hydrocarbon oxidation, 1159
  and iron deposition and dissolution, 1084
  in some metal depositions, 1195
  in oxygen reduction, 1255
  and path, 1139
  in propane oxidation, 1109
  in proton motion, 485
rate constant
  for adion diffusion, as function of dislocation density, 1201
  determined from transient measurements, for surface diffusion, 1196
  of non-rate-determining reaction, first determination of, 1195
rate equations, for electrocatalysis, compared with catalysis, 1143
rate processes, 340, 341, 342, 387, 391, 455, 867
ratio of currents, in alloy deposition, 1226
rationality, of burning fossil fuels, lack of, 1352
Rayleigh scattering, 277
reaction(s)
  chemical and electrochemical, 8, 40
  coupled, 1235
  on electrodes, several simultaneous, 1274
  electron transfer, in 1950's, 17
  multistep, 991
  photo- and electro-, 11
  thermal, and electrochemical, 14
reaction coordinate, 456
reaction mechanism
  determination of, 1080
  elucidation of its stages, 1095
  and solution entities, 1095
  and surface coverage, 1096
reaction order, 1012, 1013, 1099, 1241
reaction paths, for oxygen electronation, 1255
reaction rate
  its dependence on isotope substitution, 1105
  electrochemical, and relation to metal–metal p.d., 1147
reactivity, at low temperatures, in electrocatalysis, 1169
reciprocity relation, 413
recovery of materials, and electrochemistry, 43

rectification, faradaic, 885
redox reactions, 982, 1046, 1146
reference potential, for electrocatalysis, 1143
reforming, of hydrocarbons, 1158, 1391
relations in cells, further reading on, 1137
relative strength of acid, 495
relative strength of base, 495
relative potential differences, 655, 656
relaxation, of ionic cloud, 423
relaxation methods, and various types of stimuli, 1019
resistance, 996, 997
response
  in concentration, to switch-on of current flux, 317
  of dielectric medium to field, 145
  of system to outside stimulus, 325
reversibility, microscopic, 874
ring anions, in liquid silicates, 613
rise time, 1185–1194
rotating disk electrode, 1058
  development of, 1070
rotation
  potential energy of water system during, 477
  of screw dislocation, schematic, 1206
  of a spiral, in electrodeposition, 1205
  of water and proton mobility: detailed scheme, 483
sacrificial anode, 1312
salting in, 159
  anomalous theory, 164
  normal, 163
salting, out, 158
  further reading on, 168
saturated hydrocarbons, 1107, 1391
scattering, Rayleigh and Raman, 277
scavenger electrolysis, theory of, 1262
Schottky defects, 526
Schrödinger equation, 947
science, advances in, and electrochemistry, 39, 41
screw-thread, and spiral, analogy, 1207
screw dislocations, 1203, 1206, 1210, 1211, 1212
secondary solvation, 79
  effect on solubility, 161
self-diffusion
  coefficients, of various substances, 545
  energies of activation for, 579
  experiments, results of, 544
  experiments with radioactive isotopes, 542
self-driven cell, 1128
self-driving cell, 1131, 1136
semiconductors, 806
  analogy to electrolytes in solution, 32
  conduction of, 809
  and Debye–Hückel length, 816
  doped, 819
  and electrolytic solutions, 811
    analogies tabulated, 812
  impurity additions to, 818

semiconductors (cont.)
  intrinsic, 910
  and metal–solution junction, 850
  and tunneling, 956
  values of mobilities in, 931
semiconductor–electrolyte interfaces, 803
semiconductor electrochemistry
  further reading on, 823
  and nonmetals, 823
semiconductor junctions, 930
separation, of isotopes, and hydrogen evolution, 1246
separation factor, 1106, 1231, 1248, 1249
separators, and dendrites, 1221
servicing center, 992
sewage disposal and electrochemistry, 43
SHE and IUPAC convention, 1117
signs, of electrode potentials, 1120
sign convention
  in electrochemistry, 1115
  IUPAC, in detail, 1119
silica, fused, and liquid water, 596
silicates
  liquid, and transport processes in simple _fused salts, 609
  structure of ions of, 596
  liquid network model of, 606
silicate melt, heat of activation for flow in, 605
silver–zinc cell, 1411, 1416, 1417
single crystals
  Bridgman technique for, 1218
  deposition on, 1218
single ions, 61, 62
sink, for electrons, 853
sinusoidal stimulus, rectified, 1019
slags, 616–618
slow discharge, 1153, 1236
smog
  its formation, 1354
  and Nernstian Hiatus, 16
smoothing, 1222
sodium–sulfur cell, 1417, 1418
sols, 839
solid metals, and mercury, in double-layer studies, 802
solubility, 160, 163
solution(s)
  agitation of, and corrosion, 1300
  entities in, and reaction mechanism, 1095
  ideal, and activity coefficient, 205
  ionic, quasi-lattice approach to, 266
  ionic, standard state in, 204
  visible and ultraviolet absorption spectroscopy in study of, 275
solvated ions, at interface, 633
solvation, 80
  effect on activity coefficient, calculation of 241
  evaluation of ion-dipole of, 93
  as function of orientation time of water molecule, 122
  heats of, for pairs of ions, 63

solvation *(cont.)*
  ion-quadrupole theory of, 103
  primary, 79, 160
  of protons, 462, 469, 470
  secondary, 79
  as time-dependent phenomenon, 121
  of transition metal ions, 106
solvation effects, on activity coefficients, consistency of theory?, 243
solvation number,
  and compressibility, 125, 126
  dynamic model, 124
  further reading on, 132
  and hopping, 122
  and mobility method of determination, 125
  primary, explanation of, 119
  and theory of activity, 244
  ultrasonic method of determination of, 25
  usefulness of concept, 124
solvation theory, electrostatic, further reading on, 72
solvent, its effect upon the strength of an acid, 504
solvent effect(s)
  on activity coefficient, theory of, 242
  on mobility, at infinite dilutions, 443
solvent levels, and strength of acids, 506
solvent properties, fused nonmetallic oxides, 602
solvent sheath, structure of water in, 78
source, for electrons, 853
space charge, 813
  capacitance associated with, 816
  inside semiconductor, 815
  in ionic solution, 217
species, in solution, and mechanism determination
specific adsorption, 748
  and constant adsorption, 638
specific conductivity
  of aqueous and nonaqueous solutions, 446
  and ionic mobility, 372
spectra, Raman, and electrolytic solutions, 277
spectroscopy, 27
  ellipsometric, 37
  infrared, and electrolytic solutions, 278
  nuclear magnetic resonance, and electrolytic solutions, 278
  Raman, and study of solutions, 276
  in study of electrolytic solutions, 272
  visible and ultraviolet absorption, in study of solutions, 275
sphere, 50, 52, 55
spherical cavity
  field in, 148
  with reference dipole, 148
spherical diffusion, 1062
  and growth at surface, 1219
spiral, formation of, from screw dislocation, 1206
spiral growth, 1203
spiral tip, concentration at, 1220

square wave pulse, 1019
stability
  electrodic principles of, 1348
  of interior of metal, and surface hydrogen, 1328
  of metals, 1268
    and adsorbed hydrogen, 1328
    electrodic approach to increase, 1306
  of surface and charge transfer, 1268
stabilization of materials, and electrochemistry, 42, 1314, 1318, 1323
standard electrochemical potentials, in nonaqueous solutions, 1425
standard electrode potentials, table of, problems, 1115
standard free energy of activation, for multistep reactions, 1002
standard free energy of intermediate states, and water coverage, 1016
standard hydrogen electrode, 1011
steady state
  and concentration of intermediates, 1038
  and convection, 1052
  equality of velocity for all steps in, 1038
  and equilibrium, 1018
  and Fick's first law, 343
  and galvanostatic transients, 1025
  how long to establish?, 1018
steam hydrocarbon process, 1390, 1391
steel, and cathodic protection, 1312
step function flux, as function on time, 331
step site, surface diffusion to, 1182
Stern model, 733, 734, 735
stimulation, of interface, to show variation, 1019
stimuli, 1019, 1020, 1021
stoichiometric number, 1004
  and Butler–Volmer equation, 1006
  determination of, 1101
  and hydrogen evolution, 1241
  and multistep reactions, 1005
  and oxygen reduction, 1254
  and symmetry factor, 1007
  tabulated, for hydrogen evolution, 1245
  and transfer coefficient, 1007
Stokes–Einstein relation, 377, 379, 380, 551, 552
Stokes' law
  and proton mobility, 471
  and transport numbers in fused salts, tests, 573
Stokes' law approach, to transport numbers in fused salts, 572
Stokes mobility, 381
storage
  of charge, 703
  of electricity, 40, 1266, 1421
strain, in metals, caused by hydrogen, 1333
streaming current
  and diffuse layer, 830
  relative motion of phases in, 827, 829
streaming current density, equation relating to zeta potential, 831

streaming potential, and specific conductivity, 828
strength
  of acid, absolute, 494
  of acid, and solvent effect, 504
  of base, 494
stress, 1330, 1331
stress corrosion, examples, tabulated, 1338
stress corrosion cracking, 1335, 1336, 1337, 1349
stretching, of bond, in ions at electrodes, 920
structure breaking in liquid silicates, and concentration of metal ions, 604
structure of charged interfaces, birds'-eye view, 824
substance producer, 854, 1127
  as electrochemical device, 851
summarizing remarks, on mechanistic studies, 1090
summary
  charge transfer theory, 893
  of corrosion mechanisms, 1345
  of criteria in good electrochemical energy storage, 1412
  of direct energy conversion, 1398
  of electrochemical electricity storers, 1412
  of electrochemical energy storage, 1430
superequivalent adsorption, 638, 748
superimposition of negative potential, and stabilization, 1314
superposition, of potential of ion and cloud, 199
surface(s)
  and civilization, 1267
  decoration of, 1266
  electrification of, consequences, 7
  electrode, state of, 36
  of metals
    and cracks by hydrogen, 1333
    and instability, 1328
  protection of, 1266
  stability of, and charge transfer, 1268
  and time change, in respect to electrodeposition, 1182
surface adion(s)
  deposition from, 1187
  in random walk, 1181
surface coverage
  during benzene oxidation, 1098
  and Butler–Volmer equation, 1014
  and charge transfer reaction, 1014
  by galvanostatic transients, 1030
  and mechanism of reaction, 1096
surface diffusion
  of adions, in electrodeposition, 1177
  contribution to overpotential, in steady state, 1199
  dependence upon time, 1194
  equations for, 1188
  to kink site, 1182
  and lattice formation, 1195
  to a step site, 1182
  and stepwise dehydration, 1177

surface diffusion parameter, in metal deposition, 1202
surface excess
  and amount adsorbed, 684
  determination, 710
  and distribution of species in the interface region, 683
  electrocapillary equations for, 710
  of individual species, and surface tension, 711
  as macroscopic concept, 684
  and radiotracer measurements, 686
surface potential, 667, 673
  across interface, 677
  between two wires in cell, 675
  measurement of change of, 674
  origin of, 669
  and potential of zero charge, 707
surface state
  and contact adsorption, 821
  and Gouy–Chapman diffuse charge, 822
  in semiconductors, and contact-adsorbed ions, 822
surface tension
  and capacitance, 705
  at interface, and potential, 689
  and potential difference at interface, 688
  and surface excess, 711
  and variation with electrolyte concentration, 693
surface tension–potential relation, from parallel plate model, 719
sweep, triangular, 1019
Swiss cheese, and hole model, 527
symmetry, dependence on hill-shaped potential barrier, 936
symmetry factor, 923
  basic condition for the presence of, 977
  and biological situations, 941
  deduction in terms of potential energy curves, 925
  dependence on potential, 926
  elementary theory, 923
  first model, 871
  and flow of chemical to electrical energy, 1138
  and Morse curves, 922, 923, 926
  potential dependence as a function of of exchange current density, 927
  in quantum mechanics, third model, 674, 977
  second model, 922
  and stoichiometric number, 1007
  and transfer coefficient, 1007
synthesis, 40
Tafel's law, 15
Tafel lines
  for alloy deposition, 1225
  for oxygen reduction on doped bronzes, 1422
Tafel relation
  effect of diffuse layer on, 915

liv   INDEX

Tafel relation *(cont.)*
  and electrochemical energy conversion, 1370
  at membranes, 942
Tafel slope
  effect on corrosion, 1294
  and potential of zero charge, 915
  and reaction order, 1090
tanks, stabilization of, 1318
teflon, use in porous electrodes, 1391
temperature
  of atmosphere, as function of time, 1356
  and dielectric constant, 60
terminology, of electrodes, historical, 1126
terminology, in energy conversion and storage, 1402
test charge, and definition of Galvani potential, 672
textbooks, electrochemical in English, absence of, v
thermal combustion, and waste of energy, 1357
thermionic emission, 953
thermionic energy converter, 1359
thermionic work function, and electrocatalysis, 1147, 1150
thermoelectric energy converter, 1359
thickness of atmosphere, variation of, with concentration, 198
Thompson's hypothesis, 416
thought experiments
  at charged interface, and surface potential, 668
  and definition of chemical potential, 694
  with image forces, 661
three-electrode system, 891
three-phase boundary, at porous electrode, 1383
  importance of, 1384
transients, 32, 1020
  basic equations for, 1027
  cell for measurement, 1184
  and equivalent circuits, 1026
  galvanostatic transients, 1021, 1185
    arrangement for, 1022
    and Butler–Volmer equation, 1190
    and determination for surface radical concentration, 1197
    and double-layer charging, 1190
    solution of equations for, 1191
    and steady state, 1025
  at interfaces, 1017
  and Laplace transformation, 1041
  in metal deposition, 1183
  potential–time
    under diffusion control, 1046
    effect of intermediates on, 1026
  potentiostatic, 1032
    instrumentation for, 1033
transient behavior, and equivalent circuit, 1026
transient methods, 1019
transient state, and Fick's second law, 344

transient techniques, and prospects for electrodic research, 34
thrombosis, electrochemical mechanism of, 840
transfer
  of hydrated ion, to hole site, 1180
  mass transfer and heat transfer, analogies between, tabulated, 1041
  representation of, for ion to kink or edge vacancy, 1180
transfer coefficient
  and Butler–Volmer equation, 1007'
  for cathodic and anodic reactions, 1007
  as center of electrode kinetics, 918
  determination of, 1100
  and hydrogen evolution mechanisms, tabulated, 1235
  and stoichiometric number, 1007
  and symmetry factor, 1007
  tabulated summary of values, 1241
  and Tafel plots, 1103
  for various mechanisms, 1007
transition(s)
  electronic, adiabatic, 974, 977
  radiationless, and conservation of momentum, 950
transition metal ions
  heat of hydration and atomic number, 108
  hydration of, and stabilization of field, 106, 111
  interaction with water, 106
transition time, 1043
and concentration, 1048
  concept of, 1047
  and convection, 1052
  relation to potential, 1048
  and charge transfer, 1036
  to electrode, effect on kinetics, 916
  control, further reading on, 1079
  forces in phenomenological treatment of, 294
  in fused salts, atomistic theory, 574
  in fused salts, further reading on, 573
  of proton, 470
  in solution, work done in, 292
transport aspects of electrodics, summary, 1076
transport numbers, 400, 401, 402, 403, 406, 415, 418, 565, 566, 568, 571, 572
transport phenomena, 541
transport theory, and history, 1042
transport, and electrochemistry, 43
triangular sweep, 1019
triple ions, 265
  and nonaqueous solutions, 449
triple layer, 750
tritium
  use in adsorption measurements, 1030
  use in mechanism determination, 1106
tungsten bronzes, doped, and oxygen catalysis, 1422
tunneling, 952

INDEX lv

tunneling *(cont.)*
  conditions for, effect of adsorption energy on, 965
  condition for, in hydrogen evolution, 1154
  to cytochromes, and enzymes as electrodes, 1253
  and electrode reactions, 955
  and electrodeposition of metals, 1177
  of electron, 946
    conditions for, 972
    and de-electronation reaction, 973
    through barrier, probability of, 952
    and Esaki tunnel diode, 956
    and Fermi level, 964
    forbidden conditions for, 972
    to hydrated proton, 962
    and image energy, 961
    and Morse equation, 981
    need for acceptor particles in, 977
    probability of, and width of barrier, 954
  of protons, 1155
    in ice, 487
    in water, 487
  quantum-mechanical, and biological membranes, 981
  radiationless, 961
  and semiconductors, 956
  simultaneous, of two electrons, unlikelihood of, 1082
  work function in, 963
turbulence, 1055
ultrasonics, and hydration number, 128
ultrasonic irradiation, and electrocatalysis, 1170
ultraviolet and visible adsorption spectroscopy, in study of solutions, 275
undevice, 859, 860
unsaturated hydrocarbons, electrochemical data concerning, 1392
urban living, and electrochemistry, 42
vacancies, types of, 1179
vacancy model for fused salts, 526
vacuum, work function in, 963
virial, definition of, 237
viscosity
  kinetic theory expression for, 621
  of molten salts, 547
visible and ultraviolet absorption spectroscopy, in study of solutions, 275
volcano relation, in electro-organic chemistry, 1164
volcano relation, interpretation of, in electrocatalysis, 1165
Volta potential
  measurability of, 666
  and metal–solution interface, 665
Volta potential difference, theory of measurement, 841
volume change on fusion, molten salts, 525
Wagner–Traud mechanism for corrosion, 1274

waiting lines, 992
water
  adsorption of, and Butler–Volmer equation, 1015
  bound near to ion, and deviations from ideality, 238
  constitution of, in vapor, 464
  de-electronation reaction, and ethylene oxidation, 1160
  de-ionization of, thermal and electrochemical, 40
  as dielectric, 132
  dielectric constant, and association of dipoles, 154
  dielectric unit in, 146
  on electrodes, flip-flop, 779
  immobilized near ion, 77
  interaction, with transition metal ions, 106
  a new form, 76
  orbitals around, 73
  potable, from fuel cells, 1389
  primary region of solvation, 79
  production of, and fuel cells, 1420
  as quadrupole, 98
  in secondary region of solvation, 79
  and silica, 595
  as solvent, 440
  structure of, 72, 80
  theory of its dielectric constant, 1461
water coverage, possible effect on intermediate concentration, 1016
water dipoles
  at electrodes, and capacitance, 788
  flip-flop model, 741, 780
water molecules, adsorption on electrodes, 739
  condensation of, in solvation calculation, 89
  orientation of, at electrodes, 741
water purification, and electrochemistry, 42
water reorientation, 482
  and proton mobility, 482
wave form, sinusoidal, 1052
wave function, of electron, near interface, 670
wavelength, de Broglie, 949
work
  chemical and electrical, separation of, 695
  of transfer, of ion from vacuum to solvent, 50
work done
  in transport in solution, 292
  in transport of unit charge from infinity to interior of phase, 694
work function
  cancellation of, in electrocatalysis, 1147
  and electrocatalysis, 1146
  and exchange current density, 1149
  influence of concentration of reactants in double layer, 1150
  influence upon rate of nonbonding electrochemical reaction, 1149
  of metals, tabulated, 944

work function *(cont.)*
    in tunneling, 963
    and zero charge potential, 1151
water discharge, in ethylene oxidation, 1160
water line, corrosion at, 1303
X-rays, and constitution of protons in
    solution, 462
Young's modulus, 1340
zero charge, potential of, 691, 706
    and electrocatalysis, 1142
zero charge potential, relation to work
    function on, 1151

zero charge situation, 863
zero current, and Nernst's law, 897
zeta potential
    and clotting of blood, 841
    and dependence on concentration, 913
    and electrode kinetics, 912
    relation to concentration, 914
    and streaming current, 831
zinc, and air electrode, storage, 1427
zinc–air cell, and storage, 1428
zinc-containing cells, and dendritic growth, 1428

SONG TITLE INDEX • 679

I Should Have Known You Years Ago 55
I Shoulda Stood in Bed 407
I Simply Adore You 539
I Sold My Heart to the Junkman 395
I Solemnly Swear 175
I Speak to the Stars 531
I Still Do 7
I Still Get a Thrill (Thinking of You) 72, 82
I Still Get Jealous 483
I Still Love to Kiss You Goodnight 43, 471
I Surrender Dear 21, 22, 63
I Take Things Easy 287, 515
I Talk to the Trees 279
I Think of You 117
I Think You'll Like It 301
I Think You're Wonderful 400
I Thought about You 316, 510
I Thrill When They Mention Your Name 468
I Understand 528
I Ups to Her and She Ups to Me 207, 252
I Used to Be Her One and Only 499
I Used to Love Her in the Moonlight 265, 566
I Used to Love You (But It's All Over Now) 38, 514
I Wake Up Smiling 7, 261
I Walk Alone 560
I Walk with Music 55
I Walked In 308
I Wanna Be in Winchell's Column 155
I Wanna Be Loved 162, 191, 192, 417
I Wanna be Loved by You 237, 421, 479
I Wanna Go Places and Do Things 406
I Wanna Sing About You 105, 136
I Want a Girl 309
I Want a Girl (Just Like the Girl That Married Dear Old Dad) 516
I Want a Little Girl 328, 392
I Want It Sweet Like You 503
I Want the World to Know 336, 406
I Want to Be Bad 189
I Want to Be Happy 49, 563
I Want to Be the Guy 407
I Want to Learn to Speak Hawaiian 349
I Want to Live 233
I Want to Ring Bells 72
I Want to Thank Your Folks 27
I Want You for Christmas 497
I Want You from This Day Forward 163
I Want You to Want Me 498, 499
I Was Afraid of That 407
I Was Dancing With Someone 345
I Was Introduced to Heaven (When I Was Introduced to You) 266
I Was Lucky 318, 475
I Was Made to Love You 491
I Wish I Could Tell You 422
I Wish I Had a Girl 233
I Wish I Had My Old Girl Back Again 285, 378
I Wish I Knew the Name (Of the Girl in My Dreams) 401
I Wish I Was Aladdin 154
I Wish I Were Twins 84, 276, 324
I Wish Somebody Knew I Was Lonesome 402
I Wish You Were Here 389
I Wished On the Moon 385
I Woke Up Crying 574
I Woke Up Too Soon 133
I Woke Up With a Teardrop in My Eye 395
I Wonder If She's Waiting 474
I Wonder What My Gal is Doin' Now 371
I Wonder What's Become of Sally 5, 561

I Wonder Where My Baby Is Tonight
  96, 235
I Wonder Who's Kissing Her Now
  212
I Wonder Why   245
I Won't Be Home Anymore When You
  Call   231
I Won't Believe It   409
I Won't Dance   127, 307
I Won't Grow Up   259
(I Would Do) Anything for You   195,
  209
I Wouldn't Change You for the World
  229, 346
I Wrote My Song   469
I'd Be Lost Without You   97, 465
I'd Be Telling a Lie   525, 570
I'd Climb the Highest Mountain   39,
  61
I'd Do It All Over Again   59, 403,
  533
I'd Give a Million Tomorrows   274
I'd Give My Kingdom for a Smile
  244
I'd Know You Anywhere   308, 316
I'd Leave My Happy Home for You
  516
I'd Like to See Samoa of Samoa   471
I'd Like to Set You to Music   531
I'd Love to Call You My Sweetheart
  451
I'd Love to Fall Asleep and Wake Up
  in My Mammy's Arms   7
I'd Love to Make Love to You   119
(I'd Love to Spend) One Hour With
  You   405
I'd Rather Be Blue   132
I'd Rather Be the Girl in Your Arms
  13
I'd Rather Cry over You (Than Smile
  at Somebody Else)   101
I'd Rather Listen to Your Eyes   523
I'd Rather Wake Up By Myself   440
Idaho State Fair   339

If All the Stars Were Pretty Babies
  132
If Dreams Come True   150, 325, 429
If Every Day Would Be Christmas
  444
If He Can Fight Like He Can Love
  322
If He Cared   480
If He Comes In I'm Going Out   287
If He'll Come Back to Me   565
If I Could Be the Sweetheart of a Girl
  Like You   269, 381
If I Could Be With You   226
If I Could Be With You One Hour
  Tonight   76
If I Didn't Already Love You, Baby
  529
If I Didn't Care   6, 256, 442
If I Didn't Have You   6, 181
If I Gave You   304
If I Had a Girl Like You   94, 188, 417
If I Had a Million Dollars   292, 316
If I Had a Ribbon Bow   464
If I Had a Talking Picture of You   92,
  189
If I Had My Life to Live Over Again
  219, 499
If I Had My Way   239
If I Had Only Known   274
If I Had You   266, 448
If I Knew Then   211, 231
If I Knew You Better   435
If I Knew You Were Comin', I'd've
  Baked a Cake   206
If I Love Again   335, 355
If I Love You a Mountain   338
If I Only Had a Match   321
If I Should Lose You   387, 406
If I Thought   312, 315
If I Was a Millionaire   112
If I Were a Bee and You Were a Red
  Red Rose   38
If I Were a Bell   279
If I Were King   406

If I Were You (I'd Fall in Love with Me)   119, 335
If I'd Only Believed in You   83
If I'm Lucky   337
If It Ain't Love   392, 520
If It Isn't Love   48, 220
If It Isn't Pain, Then It Isn't Love   387, 406
If It Wasn't for the Moon   499, 501
If It Wasn't for You   126
If It Were Easy to Do   459
If It's the Last Thing I Do   60
If It's True   392
If It's You   356, 450
If Love Remains   533
If Love Were All   75
If Money Grew On Trees   395
If My Heart Could Only Talk   381, 431, 540
If My Heart Had a Window   275
If the Moon Turns Green   177
If There Is Someone Lovelier Than You   94, 439
If There's Anybody Here (From Out of Town)   447
If This Isn't Love   254
If We Can't Be the Same Old Sweethearts   331
If What You Say Is True   345
If Wishes Were Kisses   104, 477
If You Are but a Dream   139, 219
If You Build a Better Mousetrap   435
If You Can't Get a Girl in the Summertime   493
If You Cared   403
If You Could See Me Now   458
If You Ever Change Your Mind   457, 527
If You Ever Should Leave   60
If You Haven't Got a Girl   83, 206
If You Knew Susie (Like I Know Susie)   90, 92, 323
If You Know the Lord   394
If You Please   510
If You Should Love Me   473
If You Turn Me Down   459
If You Want the Rainbow (You Must Have the Rain)   262
If You Were in My Place   325, 344
If You Were Mine   292, 316
If You Were Only Mine   230, 346
If You Were the Only Girl   38, 165, 322
If You'll Be Mine   304
If You'll Say "Yes," Cherie   351
If Your Aloha Means I Love You   362
If Your Heart's in the Game   493
If Your Kisses Can't Hold the Man You Love   118
If You're Ever Down in Texas Look Me Up   447
If You're in Love You'll Waltz   493
I'll Always Be in Love with You   160, 423, 473
I'll Always Love You   272
I'll Always Think I'm in Heaven When I'm Down in Dixieland   1
I'll Be a Friend With Pleasure   377
I'll Be Around   543
I'll Be Back For More   430
I'll Be Blue Just Thinking of You   534
I'll Be Faithful   525, 559
I'll Be Hanged If They're Gonna Hang Me   339
I'll Be Home For Christmas   142, 242
I'll Be Marching to a Love Song   407
I'll Be Reminded of You   191
I'll Be Seeing You   123, 233
I'll Be With You in Apple Blossom Time   514
I'll Be With You When the Roses Bloom Again   112
I'll Build a Bungalow   480
I'll Buy That Dream   289, 560
I'll Close My Eyes to Everyone Else   494
I'll Dance at Your Wedding   289, 356

I'll Do My Best to Make You Happy 351
I'll Follow My Secret Heart 74
I'll Follow the Boys 310
I'll Follow You 7
I'll Get Along Somehow 126
I'll Get By 7, 504, 505
I'll Keep the Lovelight Burning 499
I'll Make a Ring around Rosy 442
I'll Never Ask for More 7
I'll Never Be the Same 235, 291, 293, 460
I'll Never Fail You 328
I'll Never Forget I Love You 432
I'll Never Have to Dream Again 230, 346
I'll Never Leave You 440
I'll Never Let a Day Pass By 277, 434, 435
I'll Never Let You Cry 327, 379
I'll Never Let You Go 105
I'll Never Make the Same Mistake Again 356
I'll Never Mention Your Name 145
I'll Never Say "Never Again" 557
I'll Never Say No 549, 550
I'll Never See My Baby Any More 102
I'll Never Smile Again 281
I'll Remember 136
I'll Remember April 88, 388
I'll Remember Her 75
I'll Say She Does 227, 228, 234
I'll See You Again 74
I'll See You in My Dreams 229, 235
I'll Sing to You 296
I'll Sing You a Thousand Love Songs 107, 523
I'll Smile Again 398
I'll Stand By 72, 83
I'll String Along with You 107, 523
I'll Take an Option On You 387, 406
I'll Take Care of Your Cares 331
I'll Take Her Back if She Wants to Come Back 331
I'll Take Romance 355
I'll Take Tallulah 254
I'll Understand 142
I'll Wait for You Forever 457
I'll Walk Alone 50, 482, 483
Ill Wind 14, 249
I'm a Big Girl Now 104, 274
I'm a Black Sheep Who Is Blue 387, 406
I'm a Ding Dong Daddy from Dumas 25
I'm a Dreamer—Aren't We All 92, 189
I'm a Fool About My Mama 412
I'm a Fool for Loving You 265
I'm a Hundred Percent for You 356
I'm a Little Blackbird Looking for a Bluebird 320, 505
I'm a Little on the Lonely Side 59
I'm a Lucky Devil (To Find an Angel Like You) 494
I'm a One Man Girl 118, 336, 406
I'm a Stranger Here Myself 341, 533
I'm a Vamp From East Broadway 422
I'm Afraid of You 387, 440
I'm Afraid the Masquerade Is Over 289
I'm Afraid to Come Home in the Dark 509, 546
I'm All A-Twitter over You 406, 407
I'm All Bound 'Round with the Mason-Dixon Line 265, 441, 565
I'm Alone Because I Love You 437, 567
I'm Always Chasing Rainbows 56, 305
I'm an Occidental Woman in an Oriental Mood for Love 17
I'm an Old Cowhand 316
I'm an Unemployed Sweetheart 331
I'm Away from It All 494

I'm Awfully Glad I Met You   322
I'm Beginning to See the Light   116, 145, 204
I'm Bringing a Red, Red, Rose   96
I'm Building a Sailboat of Dreams   134
I'm Building Up to an Awful Letdown   316
I'm Checking Out, Go-om Bye   480
I'm Climbing Up a Rainbow   344
I'm Confessin' That I Love You   342
I'm Counting on You   355
I'm Crazy 'Bout My Baby (And My Baby's Crazy 'Bout Me)   195, 520
I'm Crying Just for You   331
I'm Disappointed in You   282
I'm Falling in Love with Someone   190
I'm Flying   259
I'm Following You   105, 286
I'm Forever Blowing Bubbles   35, 239, 512
I'm Free   171
I'm Getting Sentimental Over You   524, 525
I'm Glad for Your Sake   494
I'm Glad I Waited for You   485
I'm Glad I'm Not Young Anymore   279
I'm Glad There Is You   98, 317
I'm Goin' Home   476
I'm Goin' South   461, 552, 553
I'm Goin' to Dance with the Guy What Brung Me   13
I'm Going Back, Back, Back to Carolina   120
I'm Going Steady with Eddie   184
I'm Going to Lose My Heart Someone   315
I'm Gonna Charleston Back to Charleston   174
I'm Gonna Cross My Fingers   242
I'm Gonna Get You   498
I'm Gonna Lasso a Dream   207

I'm Gonna Live Till I Die   207, 242, 252
I'm Gonna Lock My Heart and Throw Away the Key   447
I'm Gonna Meet My Sweetie Now   82, 164
I'm Gonna Sit Right Down and Write Myself a Letter   7, 569
I'm Good for Nothing but Love   293
I'm Grateful to You   72, 82
I'm Growing Fond of You   148
I'm Growing Fonder of You   321, 534, 568
I'm Happy, Darling, Dancing with You   569
I'm Happy When You're Happy   83
I'm Hatin' This Waitin' Around   497
I'm Hummin', I'm Whistlin', I'm Singin'   154
I'm in a Dancing Mood   149, 206, 456
I'm in a Fog about You   168
I'm in a Happy Frame of Mind   366
I'm in Love   339, 462, 501
I'm in Love All Over Again   127, 307
I'm in Seventh Heaven   227
I'm in the Market for You   178, 179, 306
I'm in the Mood for Love   127, 307
I'm in the Mood for Swing   57
I'm Just a Country Boy at Heart   501
I'm Just a Dancing Sweetheart   497
I'm Just a Lucky So and So   80
I'm Just a Vagabond Lover   507
I'm Just Wild About Animal Crackers   267
I'm Just Wild about Harry   31, 464
I'm Keepin' Company   105, 255
I'm Knee Deep in Daisies (And Head Over Heels in Love)   450
I'm Looking Over a Four-Leaf Clover   94, 458, 551, 554
I'm Lost   395
I'm Madly in Love with You   83
I'm Making a Play for You   407

I'm Making Hay in the Moonlight   446
I'm My Own Grandmaw   219
I'm My Own Grandpaw   219
I'm Nobody's Baby   5, 81, 433
I'm Not Ashamed of You, Molly   475
I'm Not Complainin'   430
I'm Not Good Enough for You   384
I'm Not So Bright   304
I'm on My Way to Mandalay   132
I'm On the Crest of a Wave   189
I'm One of God's Children Who Hasn't Got Wings   10
I'm Only Human After All   108
I'm Painting the Town Red   347, 473, 497
I'm Popeye the Sailor Man   260
I'm Prayin' Humble   171
I'm Satisfied   21
I'm Saving Myself for You   194
I'm Shooting High   249
I'm Sitting on Top of the World   188, 228, 265, 566
I'm Slappin' Seventh Avenue   344
I'm So Afraid of You   421
I'm So Happy I Could Cry   486
I'm Sorry   494
I'm Sorry Dear   442, 498
I'm Sorry I Didn't Say I'm Sorry   402
I'm Stepping Out with a Memory Tonight   289, 560
I'm Still Caring   246, 507
I'm Still Crazy for You   407
I'm Still Not Thru Missing You   252
I'm Still Without a Sweetheart, With Summer Coming On   505
I'm Sure of Everything But You   321
I'm Talking Through My Heart   386, 406
I'm Tellin' the Birds, I'm Tellin' the Bees   39
I'm Telling the World About You   331
I'm the Fellow Who Loves You   188

I'm the Last of the Red-Hot Mamas   562
I'm the Lonesomest Gal in Town   38, 514
I'm the Medicine Man for the Blues   9
I'm the Secretary to the Sultan   407
I'm Thru with Love   235, 269, 290
I'm Tickled Pink With a Blue-Eyed Baby   534
I'm Tired   441
I'm Too Romantic   331
I'm True to the Navy Now   243
I'm Trying So Hard to Forget You   183
I'm Unlucky   441
I'm Waiting for a Wonderful Girl   565
I'm Walkin' on Air   100
I'm Walkin' the Chalk Line   208, 348
I'm Wild About Horns on Automobiles (That Go Ta-Ta-Ta-Ta)   143
I'm Wond'rin'   529
I'm Yours   161, 162, 181
I'm Yours for Tonight   261, 331
The Image of You   8, 569
Imagination   45, 270, 324, 510
Improvisation in Several Keys   168
In a Blue and Pensive Mood   273, 342, 489
In a Boat for Two   423
In a Boat Out to Sea   266
In a Chapel in the Moonlight   197
In a Great Big Way   127
In a Little Bookshop   322
In a Little Hula Heaven   386, 406
In a Little Red Barn   6
In a Little Red Schoolhouse   64
In a Little Secondhand Store   105, 344
In a Little Spanish Town   265, 527, 528, 566
In a Little Waterfront Cafe   412
In a Mist   26
In a Moment of Madness   136, 308
In a Sentimental Mood   116, 252, 325
In a Shady Nook   343

SONG TITLE INDEX • 685

In a Shanty in Old Shanty Town   268, 437, 567
In an Eighteenth-Century Drawing Room   443
In an Old Dutch Garden   170
In Any Language   155
In Between Age   489
In Dixie Land with Dixie Lou   322
In God We Trust   498
In Love in Vain   406
In My Arms   170, 278
In My Bouquet of Memories   265
In My Estimation of You   82, 280
In My Little Red Book   462
In My Merry Oldsmobile   112
In My Solitude   116
In New Orleans   183
In Old New York   190
In Ole Oklahoma   501
In Other Words We're Through   273, 342, 489
In Our Little Studio   378
In the Arms of Love   271, 272
In the Blue of Evening   2
In the Blue of the Night   7, 66
In the Cool Cool Cool of the Evening   55, 316
In the Cool of the Evening   482
In the Dark   26
In the Evening by the Moonlight   474, 516
In the Good Old Summertime   454
In the Hush of the Night   260
In the Land of Beginning Again   320, 321
In the Land of Harmony   237
In the Land of the Shady Palm Trees   135, 283
In the Light of the Stars   451
In the Merry Month of June   454
In the Middle of a Kiss   487
In the Middle of May   8
In the Mood   391
In the Moonlight   49, 422
In the Old Town Hall   344
n the Park in Paree   387, 406
In the Shade of the Old Apple Tree   508, 546
In the Shade of the Pyramids   287
In the Still of the Night   54
In the Sweet Bye and Bye   515
In the Valley of the Moon   496
In the Wee Small Hours of the Morning   199, 296
In the Wink of an Eye   398
In Your Own Little Way   72, 347
In Your Own Town   521
Indian Cradle Song   529
Indian Love Call   138, 173, 180, 479
Indian Summer   107, 190
Indiana   178, 285, 286
Indiana Moon   83, 518
Indiscretion   538
Indoor Outdoor Girl   126
Infatuation   430
Inner Sanctum   494
Inside This Heart of Mine   225
Inspiration   202
Instant Love   104
Interlude   427
Into Each Life Some Rain Must Fall   129, 401
Intrigue   260
Invitation to a Broken Heart   264
Invitation to the Blues   401
Iowa Corn Song   172
Ireland and Someone I Love   120
Ireland Must Be Heaven, For My Mother Came from There   130, 223
Iroquois   352
Irresistible   471
Is I In Love? I Is   409
Is It Just a Summer Romance   72
Is It So?   209
Is It True What They Say about Dixie?   49, 260, 303
Is My Baby Blue Tonight?   175, 502
Is She My Girl Friend?   5, 562

Is That Good?   407
IIs That Religion?   365, 377
Is That the Way to Treat a Sweetheart?   462, 497
Is There Anything Wrong in That?   62
Is There Somebody Else?   338, 403
Is There Still Room for Me   1
Is This Gonna Be My Lucky Summer   33
Isadore the Toreador   133
I'se a Muggin'   469
Isfahan   481
Island Serenade   349
Isle of Capri   169
Isle of Pines   242
Isn't It a Shame   453, 461
Isn't It Heavenly   181, 324
Isn't Love the Strangest Thing?   72, 82
Isn't This a Night for Love?   48, 220
It Ain't All Roses   499
It Ain't Necessarily So   193
It All Begins and Ends With You   245, 364, 444
It All Comes Back to Me Now   251, 573
It All Depends on You   92, 189
It Amazes Me   259
It Can Happen to You   220
It Can't Be Wrong   142
It Couldn't Happen to a Sweeter Girl   44
It Doesn't Cost You Anything to Dream   128
It Don't Mean a Thing   116, 325
It Goes Like This   136
It Had to Be You   228, 235
It Happened in Chicago   486
It Happened in Hawaii   529
It Happened in Monterey   417, 528
It Happened in Sun Valley   523
It Happens Every Spring   338
It Happens to the Best of Friends   366
It Isn't a Dream Any More   347, 431

It Isn't Fair   200
It Looks Like a Big Night Tonight   233
It Looks Like Rain in Cherry Blossom Lane   44, 261
It Might As Well Be Spring   173, 174
It Might Have Been a Diff'rent Story   245, 316, 331
It Must Be Jelly 'Cause Jam Don't Shake Like That   465
It Must Be Love   13, 434, 492
It Must Be True   16, 21, 63
It Never Dawned on Me   72, 265
It Never Rains but It Pours   155
It Never Was You   532
It Only Takes a Minute   333
It Seems Like Old Times   275, 280
It Seems to Be Spring   301, 542
It Shouldn't Happen to a Dream   145, 204
It Started All Over Again   51, 129
It Takes Love to Make a Home   560
It Takes So Long to Say Goodbye   451
It Was a Night in June   154, 396
It Was Meant to Be   422
It Was Only a Sunshower   233
It Was So Beautiful   21, 22, 135
It Was the Dawn of Love   72
It Won't Be Long Now   92
It's a Cute Little Way of My Own   493
It's a Good Day   20, 258
It's a Great Life   144, 406
It's a Great Life If You Don't Weaken   61
It's a Habit of Mine   406
It's a Hap-Hap-Happy Day   343
It's a Hundred to One   231
It's a Lonely Trail   241
It's a Lonesome Old Town   318, 498
It's a Long Way to Tipperary   546
It's a Lot of Idle Gossip   343, 490
It's a Marshmallow World   458

It's a Million to One You're in Love   83
2It's a Most Unusual Day   3
It's a Quiet Town   427
It's a Whole New Thing   331, 347
It's a Wonderful World   3, 247, 433
It's About Time   316, 494
It's All Forgotten Now   351
It's All in the Game   459
It's All Over But the Crying   333
It's All Over Now   466
It's All So New to Me   49, 188, 369, 490
It's All Yours   128, 440
It's Always Goodbye   313
It's Always June in Miami   319
It's Always You   45, 510
It's an Old Southern Custom   324
It's Anybody's Spring   45, 510
It's Been a Long, Long Time   50, 483
It's Been So Long   3, 97
It's Beginning to Look a Lot Like Christmas   549
It's Dark on Observatory Hill   45, 470, 471
It's Easier Said Than Done   275
It's Easter Time   549
It's Easy to Blame the Weather   60
It's Funny to Everyone but Me   133, 230
It's Great to Be Alive   39, 95, 188
It's High Time I Got the Lowdown on You   336
It's in the Stars   62
It's Like Reaching for the Moon   453
It's Love Again   73, 557
It's Love I'm After   327, 378
It's Love, Love, Love   251
It's Love Time   267
It's Magic   50, 484
It's Moonlight All the Time On Broadway   535
It's My Turn Now   60
It's Never Too Late   275, 280

It's No Fun   6, 347
It's No Fun Eating Alone   115
It's Only a Paper Moon   14, 181, 417
It's Raining Sunbeams   209
It's Snowing in Hawaii   384
It's So Nice to Have a Man Around the House   117, 471
It's So Peaceful in the Country   543
It's Somebody Else That You Love   168
It's Sunday Down in Caroline   272, 273, 489
It's Swell of You   155, 397
It's the Beast in Me   427
It's the Darndest Thing   307
It's the Dreamer in Me   98, 510
It's the Girl   19, 358
It's the Going Home Together   254
It's the Last Time I'll Fall in Love   497
It's the Little Things That Count   148, 463
It's the Same Old South   115
It's the Talk of the Town   273, 342, 489
It's the Tune That Counts   434
It's Time to Say Aloha   497
It's Wearing Me Down   187
It's Winter Again   206
It's Within Your Power   154
It's Wonderful   366, 469
It's You I Love   83, 488
It's You or No One   485
I've a Longing in My Heart for You, Louise   183
I've Been Around   319
I've Been Floating Down the Old Green River (On the Good Ship Rock 'n Rye)   237
I've Been in Love Before   277
I've Been Invited to a Party   75
I've Been to a Marvelous Party   75
I've Confessed to the Breeze   564

I've Found a New Baby   363, 364, 548
I've Found the Bluebird   336, 406
I've Got a Cross-Eyed Papa (But He Looks Straight to Me)   217
I've Got a Date With a Dream   155, 397
I've Got a Feelin' You're Foolin'   40, 135
I've Got a Feeling I'm Falling   267, 417, 520
I've Got a New Lease on Love   127, 569
I've Got a Pocketful of Dreams   45, 331
I've Got a Warm Spot in My Heart for You   470, 471
I've Got an Invitation to a Dance   273, 342, 489
I've Got Everything I Want but You   442
I've Got My Eye on You   35
I've Got My Fingers Crossed   250
I've Got My Heart Set on You   397
I've Got Rain in My Eyes   274, 343
I've Got Sand in My Shoes   11
I've Got the World on a String   14, 249
I've Got to Break Myself of You   53
I've Got to Get Hot   188
I've Got to Sing a Torch Song   107
I've Got Two of Everything   315
I've Got You   278
I've Got You All to Myself   407
I've Got Your Number   259, 322
I've Gotta Crow   259
I've Grown Accustomed to Her Face   279
I've Had My Moments   97, 235
I've Heard That Song Before   50, 482
I've Just Come to Say Good-bye   183
I've Just Recieved a Telegram from Baby   516
I've Made a Habit of You   440

I've Never Been Loved by Anyone Like You   370, 371
I've Never Had a Sweetheart Like You   539
I've Only Myself to Blame   296
I've Spent the Evening in Heaven   53
I've Taken a Fancy to You   327, 379
Ivory Towers   140
Ivy   55
Ivy-Covered Arbor   166

J. P. Dolley III   544
Jack, How I Envy You   515
Jack in the Box   68
Jackass Blues   476
Ja-Da   52
The Japanese Sandman   113, 378, 542
Japansy   246
Jasmine and Jade   478
Java Jive   103, 355
Jazz Nocturne   485
Jazznocracy   213
Jealous   268
Jealous Eyes   296
Jealous Lover   50
Jealous Moon   212, 472
Jean   120
Jean, You Ain't Talking to Me   38
Jeanine, I Dream of Lilac Time   147, 455
The Jeep Is Jumpin'   204
Jeepers Creepers   316, 393, 523
Jeep's Blues   204
Jelly Bean (He's a Curbstone Cutie)   38
Jelly Jelly   202
Jelly Roll Blues   333
Jennie Dear   516
Jenny   533
Jericho   336, 404
Jersey Lightning   424
Jesse James   274
The Jet Set   502
Jewels from Cartier Suite   11
Jimmy Had a Nickel   206, 456

SONG TITLE INDEX • 689

Jimmy Valentine   112, 288
Jingle, Jangle, Jingle   278
Jingle Jingle Jingle   303
Jitney Man   202
The Jitterbug Waltz   520
Jitterbug's Lullaby   204
Jodie Man   402
Joe Turner Blues   176
Johannesburg   356
John Hardy's Wife   117
John Silver   99
Johnny   75
Johnny Come Lately   481
Johnny Doughboy Found a Rose in Ireland   149
Johnny Get Your Girl   328
John's Idea   23
Johnson Rag   257
Join the Navy   406, 565
The Joint is Jumpin'   225, 391, 520
The Joint Is Really Jumping   304
The Jones Boy   252, 328
Joobalai   386, 407
Jose O'Neill, the Cuban Heel   222
Joseph! Joseph!   60
Josephine   235, 244
Josephine, My Jo   287
Josephine, Please No Lean on the Bell   343
A Journey to a Star   407
Journey's End   493
Jubilee   55
Judy   55, 260
Judy, Who D'Ya Love?   406
Jukin'   191
Jump for Joy   530
Jump Jump's Here   344
Jumpin' at the Woodside   23
Jumpin' for Joy   430
Jumpin' in a Julep Joint   309
Jumpin' Jive   364
Jumpin' Punkins   117
June in January   385, 406
June, I Love No One But You   195

June Is Bustin' Out All Over   173
June, July and Always   319
June Night   19, 136
Jungle Drums   357
Jungle Shadows   480
Junk Man   276, 324
Just a Baby's Prayer at Twilight   221, 265
Just a Cottage Small   178
Just a Gigolo   49
Just a Kid Named Joe   80, 273
Just a Kiss in the Moonlight   61
Just a Little Bit South of North Carolina   465
Just a Little Closer   324
Just a Little Home for the Old Folks   7
Just a Memory   92, 189
Just a Moon Ago   168
Just a Quiet Evening   316, 542
Just a Simple Melody   60
Just a Word of Sympathy   234
Just an Echo in the Valley   555
Just an Honest Mistake   272
Just an Old Love of Mine   20, 258
Just an Old Sweetheart of Mine   448
Just Another Day Wasted Away (Waiting for You)   496, 504
Just Another Dream of You   66
Just Another Polka   85
Just Around the Corner   516
Just as Though You Were Here   84
Just A-Settin' and A-Rockin'   481
Just Because You're You   136, 506
Just Behind the Times   183
Just For a Thrill   15
Just For Tonight   238, 254, 493
Just Friends   246, 265
Just Imagine   92, 189
Just in Time   67, 484
Just Let Me Look at You   128
Just Like a Butterfly   94
Just Like a Butterfly That's Caught in the Rain   554
Just Like a Melody Out of the Sky   96

Just Like in a Storybook   179, 306
Just My Style   212
Just One Kiss   422
Just One More Chance   73, 226
Just Think of Me Sometime   360
Just to Remind You   230
Just When We're Falling in Love   427
Just You, Just Me   245

Kaigoon   331
Kalamazoo   523
Kalamazoo to Timbuktu   544
Kansas City Kitty   97, 261
Kansas City Stomp(s)   334
Katinka   499
Katrina   279
Keb-lah   495
Keep 'Em Flying   160, 388
Keep It Casual   301
Keep off the Grass   226
Keep on Doin' What You're Doin'   422
Keep Shufflin'   519
Keep Smiling at Trouble   227
Keep Your Promise, Willie Thomas   140
Keep Your Undershirt On   421
Keepin' Myself for You   61, 564
Keepin' Out of Mischief Now   390, 520
Kentucky Days   536
Kentucky Sue   515
Kentucky's Way of Saying Good Morning   509
Kickin' the Cat   511
Kickin' the Gong Around   14, 248
The Kid from Red Bank   185
The Kid from Spain   237
The Kid in the Three-Cornered Pants   275
Killer Diller   334
Kinda Lonesome   54, 55, 393, 406
King for a Day   265
King Kamehaneha   349

King Porter Stomp   333
The Kingdom of Swing   151
The Kinkerjou   493
Kiss by Kiss   245, 318, 420
Kiss Her for Me   554
A Kiss in the Dark   90, 91, 190
The Kiss in Your Eyes   45
Kiss Me Again   190
Kiss Me Once More   266
Kiss Me Sweet   103
Kiss Me with Your Eyes   277
The Kiss That You've Forgotten   266
Kiss the Boys Goodbye   277, 434
A Kiss to Build a Dream On   422
A Kiss to Remember   372
A Kiss to Remind You   115
The Kiss Waltz   44, 106
Kissin' on the Phone   540
The Kissing Bug   467
The Kissing Song   356
The Kissing Trust   521
Kitchy-Koo   1
Kitten on the Keys   67
Kitty from Kansas City   164
Knee Deep in Daisies   157
Knick Knacks on the Mantle   114
Knit One, Purl Two   231
Knock on Wood   222
Knock Wood   516
Kokomo, Indiana   337

La Golodrina   147
La Plume De Ma Tante   206
La Veeda   513
La Vie En Rose   80
The Ladder of Roses   212
The Ladies Who Sing With a Band   301
The Lady from Fifth Avenue   381, 431, 540
The Lady I Love   368, 369
The Lady in Red   94, 560
The Lady in the Tutti-Fruitte Hat   407
Lady, I've Been Kissed Before   427

Lady Love   427
A Lady Loves to Love   338
The Lady Needs a Change   128
Lady, Play Your Mandolin   262
The Lady Who Couldn't Be Kissed   107
The Lady's in Love With You   253, 254, 277
Lake Placid   185
Lament to Love   501
Lamento Gitano   431
L'Amour, Toujours, L'Amour   138
The Lamp Is Low   89, 366
Lamp of Love   442
The Lamp on the Corner   407
The Lamplighter's Serenade   530
The Land of Dreams   194, 300, 476
The Language of Love   529
Lantern of Love   536
Las Vegas Nights   281
The Lass with the Delicate Air   464, 574
The Last Mile Home   242
Last Night I Dreamed of You   175, 203
Last Night I Had That Dream Again   518
Last Night on the Back Porch (I Loved Her Best of All)   39
Last Night Was the End of the World   474
The Last Roundup   196, 197
The Last Time I Saw Paris   14, 173
The Last Two Weeks in July   19, 266
Late Evening Blues   209
Later Than Spring   75
Later Tonight   407
A Latin Tune, A Manhattan Moon and You   107
Laugh and Call It Love   331
Laugh, Clown, Laugh   265
Laugh, You Son-of-a-Gun   387, 406
Laughing at Life   241
Laughing Sailor   533

Laura   316
Lawd, You Made the Night too Long   265, 570
Lazy Afternoon   254
Lazy Bones   316
Lazy Daddy   359
Lazy Mood   52, 87
Lazy River   54
Lazy Weather   233
Lazybones   54, 55
Leanin' on the Old Top Rail   241
Learn to Croon   13, 73, 227, 562
Learn to Love   317
Learning   273, 342, 419, 489
Legend of the Roses   494
Lei Aloha   362
Let a Smile Be Your Umbrella   123, 233
Let It Be Me   252
Let It Rain, Let It Pour   96
Let It Snow! Let It Snow! Let It Snow!   50, 483
Let Me Be the First to Kiss You Good Morning   409
Let Me Call You Sweetheart   410
Let Me Fill Your Day With Music   282
Let Me Give My Happiness to You   380
Let Me Kiss Your Tears Away   471
Let Me Love You Tonight   366
Let Me Off Uptown   121
Let That Be a Lesson to You   230, 316
Let This be a Warning to You, Baby   83, 175
Let's Call It a Day   39, 188
Let's Dream Awhile   501
Let's Fall in Love   14, 249
Let's Get Away from It All   2, 87
Let's Get Friendly   101
Let's Get Lost   278, 308
Let's Give Love a Chance   325, 459, 490

Let's Give Three Cheers for Love   154
Let's Go Back to Where We Started   266, 267
Let's Go for Broke   362
Let's Go Skiing   499
Let's Grow Old Together   437
Let's Have a Party   25
Let's Have a Showdown   377
Let's Have an Old Fashioned Christmas   69
Let's Have Another Cigarette   560
Let's Keep 'Em Flying   389
Let's Make Memories Tonight   497
Let's Pretend There's a Moon   66, 475
Let's Put Out the Lights (And Go to Sleep)   215
Let's Return to God   140
Let's Sing Again   235
Let's Sit and Talk about You   127, 307
Let's Spend an Evening at Home   21, 22
Let's Stay Young Forever   402
Let's Stop the Clock   72
Let's Swing It   310, 347, 497
Let's Take a Walk around the Block   14
Let's Take the Long Way Home   316
Let's Try Again   230, 346
Let's Walk   450
A Letter From Home   241, 499
The Letter That Never Came   104
'Leven Miles from Leavenworth   25
'Leven Thirty Saturday Night   46
Lida Rose   549
Lies   21, 22
Life Begins at Sweet Sixteen   188
Life Begins When You're in Love   400, 434
Life Is a Song (Let's Sing It Together)   568
Life Is Just a Bowl of Cherries   39, 188
Light a Candle in the Chapel   344

Light and Sweet   430
The Lights of Home   384
Lights Out   197
Like a Bolt From the Blue   356
Like He Loves Me   565
Like Someone in Love   45
L'il Abner   356
Li'l Boy Love   209, 277
Lil Darlin'   185
Lila   377
Lilacs in the Rain   89, 366
Lillies of Lorraine   69
Lily of Laguna   530
Limehouse Blues   74
Lindbergh, the Eagle of the U.S.A.   452
Linger Awhile   361, 419
Linger in My Arms a Little Longer, Baby   290
Listen to My Song (Johnny's Song)   532
Listen to That Dixie Band   64
Listen to the German Band   153
A Little Bit Independent   44, 261
A Little Bit Later On   273, 343, 352
Little Black Boy   541
Little Black Dog   69
The Little Boy Blues   304
Little Brown Gal   349
Little Butch   362
Little Buttercup   291, 460
Little by Little   95
A Little China Doll   546
Little Curly Hair in a High Chair   462
Little Did I Dream   253
A Little Door, a Little Lock, a Little Key   557
A Little Dutch Mill   22, 136
Little Genius   381, 431, 540
Little Girl Dressed in Blue   513
Little Girl in Blue   212
Little Gypsy Maid   287
Little Igloo for Two   440
Little Joe from Chicago   547

SONG TITLE INDEX • 693

A Little Kiss at Twilight   386, 407
A Little Kiss Each Morning   555
Little Lady Make Believe   462, 497
Little Log Cabin of Dreams   179
The Little Man Who Wasn't There   177
Little Man You've Had a Busy Day   206, 456, 528
Little Mary Brown   83
Little Old Cathedral in the Pines   240, 241
Little Old Lady   3, 55
Little Old New York   440
A Little On the Lonely Side   403, 533
Little Orphan Annie   432
Little Pal   227
Little Rag Doll   529
The Little Red School   120
A Little Robin Told Me So   83
Little Rose of the Rancho   387
Little Shirley Temple   293
Little Skipper   241
A Little Street Where Old Friends Meet   235
Little Sweetheart   247
The Little White House at the End of Honeymoon Lane   179
Little White Lies   97
Live and Love Today   243
Live and Love Tonight   73, 227
Livery Stable Blues   120, 359
Livin' in the Sunlight, Lovin' in the Moonlight   452
Living Dangerously   491
Living in a Great Big Way   127, 307
Living in Clover   379
Liza   235
Lollipop Ball   52
London On a Rainy Night   473, 526
The London Suite   520
The Lone Grave   104
Lonely   75
Lonely Acres   411
Lonely Little Melody   472

Lonely Man   299
Lonely Melody   73
Lonely Moments   547
Lonely Night   544
Lonely Park   481
Lonesome   322
Lonesome and Sorry   70, 81
A Lonesome Cup of Coffee   427
Lonesome Lover   41, 331
Lonesome Me   66, 71, 390, 520
Lonesome Nights   57
The Lonesome Road   17, 455
Lonesome Tonight   211
The Lonesomest Girl in Town   106, 325
Long About Midnight   195
Long About Sundown   329
Long Ago and Far Away   387, 406
Long Before You Came Along   13
Long Lost Mama   553
The Longest Way 'Round Is the Shortest Way Home   454
Look Around   480, 564
Look at the World and Smile   212
Look for the Girl   536
Look for the Silver Lining   90, 513
Look into Your Baby's Eyes and Say Goo Goo   287
Look Out I'm Romantic   252, 328
Look What I've Got   387, 406
Look What You've Done   9, 49, 422
Look What You've Done to Me   326, 327
(Lookie, Lookie, Lookie) Here Comes Cookie   154
Looking Around   336, 406
Looking Around Corners for You   155, 397
Looking for a Thrill   406
Looks Like a Beautiful Day   537
Looks Like a Cold, Cold Winter   370
Lord and Lady Whoozis   260
Lorraine, My Beautiful Alsace-Lorraine   132

694 • SONG TITLE INDEX

Lost   316
Lost and Found   501
Lost in a Dream   33, 261
Lost in a Fog   127, 307
Lost in Meditation   495
Lost in the Shuffle   119
Lost in the Stars   532
Lotus Blossom   481
Lotus Flower   442
Louise   404, 542
Louise, Louise   25
Louisiana   224, 390, 446
Louisiana Fairy Tale   72, 148, 366
Louisiana Hayride   94, 439
Louisiana Swing   424
Louisville Lady   89, 197
Louisville Lou   5
Lovable   235
Lovable and Sweet   262
Love   304, 460
Love Ain't Nothin' but the Blues   10
Love and a Dime   35
Love and Learn   440
Love and Marriage   510
Love and Rhythm   37, 188
Love at Night (Is Out of Sight)   202
The Love Bug Will Bite You (If You Don't Watch Out)   501
Love Came into My Heart   253
Love Came Out of the Night   419
Love Can Change the Stars   304
Love, Come Take Me Again   550
Love Doesn't Grow on Trees   364
Love Dropped in for Tea   45, 471
Love Grows on the White Oak Tree   500
The Love I Long For   94, 109
Love in Bloom   385, 406
Love Is a Beautiful Thing   217
Love Is a Dancing Thing   94, 440
Love Is a Many-Splendored Thing   123
Love Is a Merry-Go-Round   33
Love Is a Random Thing   301
Love Is All   499, 501

Love Is Eternal   338
Love Is Good for Anything That Ails You   175, 203
Love Is Just Around the Corner   144, 405
Love Is Like a Cigarette   241
Love Is Like a Song   565
Love Is Love Anywhere   14
Love Is the Sweetest Thing   351
Love Is the Thing   525, 571
Love Is Where You Find It   41, 107, 523
Love Letters   191, 193, 570
Love Letters in the Sand   72, 240, 241
Love Lies   136, 325, 457
A Love Like Ours   208, 348
A Love Like This   571
Love Locked Out   351
Love Makes the World Go 'Round   486, 509, 546
Love Marches On   275, 497
Love Me   525, 570
Love Me Forever   435
Love Me or Leave Me   96, 235
Love Me or Leave Me Alone   223
Love Me Sweet and Love Me Long   451
Love Me With All Your Heart   466
Love Moon   195
The Love Nest   180, 202
Love of My Life   381, 450
Love Rules the World   536
Love Sick Blues   137
Love Somebody   251
Love Song   272
Love Song of Renaldo   233
Love Song of Tahiti   235
Love Song of the Nile   40
A Love Tale of Alsace Lorraine   72
Love Tales   419
Love Thy Neighbor   154, 396
Love Turned the Light Out   254
Love Walked In   393
Love, What Are You Doing to My Heart   266

Love Will Find a Way   464
Love Will Tell   379
Love with a Capital "You"   407
Love, You Funny Thing   7
Loveless Love   176
The Loveliest Night of the Year   531
Lovelight   547
Lovelight in the Starlight   136, 208
Loveliness and Love   407
The Loveliness of You   155, 397
Lovely Debutante   78, 499
Lovely Lady   472
Lovely Little Lady   154
Lovely One   453
A Lovely Rainy Afternoon   59, 533
Lovely to Look At   127, 307
A Lovely Way to Spend an Evening   3, 308
Lovely While It Lasted   172
Lover, Come Back to Me   173, 415
A Lover Is Blue   334
Lover of My Dreams   75
A Lover's Fantasy   461
A Lover's Lullaby   52, 391, 517
Lovey Came Back   174, 265, 566
Lovin' Sam (the Sheik of Alabam)   5, 561
Low Down Lullaby   387, 406
Low Down Rhythm   245
Low Down Rhythm in a Top Hat   447
Low Gravy   209
Low Tide   37
Lowdown Blues   464
Lucky   430
Lucky Day   39, 189
Lucky in Love   92, 189
Lucky Lindy   19, 147
Lucky Me, Lovable You   5
Lucky Seven   440
Lullaby in Blue   337
Lullaby in Rhythm   150, 203, 429
Lullaby of the Leaves   368, 567
Lulu's Back in Town   107, 522
Lush Life   480

Ma! (He's Makin' Eyes at Me)   70
Ma Belle   139, 165
Ma, I Miss Your Apple Pie   275
Mack the Knife   532
Mad about the Boy   74
Mad Dogs and Englishmen   74, 508
Mad House   201, 334
Madame, I Love Your Crepe Suzettes   136
Made Up My Mind   122, 296
Mademoiselle de Paree   366
Madly in Love   127, 342
Magic Garden   443
The Magic Kiss   179
Magic Town   502
Magnolia   92, 93
Magnolias in the Moonlight   43, 435
Mahogany Hall Stomp   548
Mahoney's Eleven Arms   322
The Mailman's Got My Letter   431
The Maine Stein Song   507, 508
Mairzy Doats   103, 206, 273
Maisey, Maisey, Fine and Daisey   521
Major and Minor Stomp   99
The Major and the Minor   121
Make a Miracle   278
Make a Wish   11, 531
Make Believe   83, 173
Make Believe Ballroom   86, 391, 545
Make Believe Island   240, 241
Make 'Em Laugh   41
Make Love to Me   142, 297, 300, 309, 475
Make Me a Pallet On the Floor   176
Make My Bed Down in Dixieland   536
Make Someone Happy   67, 484
Make the Man Love Me   440
Make Way for Tomorrow   406
Make with the Music   319
Makin' Faces at the Man in the Moon   206, 398
Makin' Whoopie   96, 235
Ma-lu-na (Bottoms Up)   362

Mama Don't Want No Peas and Rice and Cocoanut Oil   147
Mama Goes Where Papa Goes (Or Papa Don't Go Out Tonight)   5, 561
Mama, I Wanna Make Rhythm   242
Mama Inez   147
Mama Loves Papa   19, 136
Mama Macushla   319
Mama, That Moon Is Here Again   386, 407
Mama's in the Groove   424
Mambo on My Mind   427
Mambo Rock   394
Mammy O' Mine   375, 502
Mammy's Soldier Boy   327
A Man and a Maid   492
A Man and His Dream   331
The Man from Laramie   526
Man From Mars   450
The Man from the South   33, 554
The Man in the Moon   160
A Man Is a Brother to a Mule   402
The Man on the Carousel   59
The Man That Got Away   15
The Man Who Comes Around   161
The Man with the Mandolin   58, 533
The Man with the Weird Beard   104
Manana (Is Soon Enough for Me)   20, 258
Mandalay   16, 46, 282
Mandy, Make Up Your Mind   62, 320, 504
Manhattan Masquerade   11
Manhattan Moonlight   11
Manhattan Serenade   3, 10
Maniac's Ball   146
Many Happy Returns of the Day   44, 106
Many Moons Ago   154, 396
Maple Leaf Rag   300
March of the Grenadiers   434
March of the Musketeers   139, 165
March of the Swing Parade   430

March Winds and April Showers   380, 431, 539, 540
Marcheta   434
Mardi Gras   167
Mardi Gras Parade   25
Margie   70, 80, 360, 408, 566
Maria Elena   425
Marian the Librarian   549
Marianne   415
The Mark Hop   172
Mars   163
Marshmallow Moon   272
Marta (Rambling Rose of the Wild Wood)   147
Martha   63
Mary Ann   83
Mary Dear   222
Mary Had a Little Lamb   490
Mary Lou   283, 360
Mary, Mary Quite Contrary   11
Mary Was a Real Nice Girl   448
Mary's a Grand Old Name   64
Mary's Idea   547
Masquerade   529
The Masquerade Is Over   560
Massachusetts   391
The Matador   427
Matinee   427, 459
Maui Girl   362
Maunaloa   362
May I?   154, 396
May I Have the Next Romance with You?   155, 397
May I Say "I Love You?"   488
May I Still Hold You   427
May the Good Lord Bless and Keep You   549
May the Good Lord Take a Liking to You   494
May Time   420
Maybe   275
Maybe I'm Wrong Again   503
Maybe It's Because   422, 442
Maybe September   272

SONG TITLE INDEX • 697

Maybe, Someday   331
Maybe You'll Be There   33, 141
Me and My Fella and a Big Umbrella   407
Me and My Shadow   105, 227, 417
Me and My Wonderful One   468, 481
Me and the Blues   250
Me and the Boy Friend   331
Me and the Ghost Upstairs   178
Me and the Man in the Moon   261, 330
Me and the Moon   175
Me for You Forever   192, 336
Me, Myself and I   152, 401
Me Queres (Do You Love Me?)   507
Me Too   452, 496, 553
Me Without You   144
Meadowbrook Shuffle   414
Mean to Me   7, 505
Meet Me at No Special Place (And I'll Be There at No Particular Time)   409
Meet Me in St. Louis   474
Meet Me in the Gloaming   206
Meet Me Where They Play the Blues   10
Meet the Beat of My Heart   155
Meet the Sun Halfway   331
Melancholy Blues   179
Melancholy Lullaby   57, 193
Melancholy Rhapsody   185
Melinda   362
Melodies Bring Memories   357
Melody   459
A Melody from the Sky   11, 327
The Melody Has to Be Right   209
Melody in Spring   492
Memories   234, 508
Memories of France   106
Memories of You   31, 390
Memory Lane   400
Memphis Blues   175, 353
Memphis in June   55, 530
Menehene Lullaby   362

Mention My Name in Sheboygan   198, 339
Merrily We Roll Along   310
Merry Ann   317
A Merry Merry Christmas   303
The Merry Monahans   56
The Merry-Go-Round Broke Down   134, 136, 137
Message from Mars   374
Metropolitan Nocturne   11
Mexican Moon   168
Miami   228
Michigan Bank Roll   339
Mickey   546
Mid the Green Fields of Virginia   183
Midnight at the Masquerade   407
Midnight Blue   261, 423
Midnight in a Madhouse   64
Midnight in Paris   290
Midnight Moon   446
Midnight on the Trail   34, 394
Midnight Reflections   460, 461
Midnight Sun   316
Midnight, the Stars, and You   556
Midriff   481
Midway   469
A Mile a Minute   369
Milenberg Joys   299, 300, 309, 333
Milkman, Keep Those Bottles Quiet   389
The Milkman's Matinee   86, 391
The Mill on the Floss   274
The Miller's Daughter Marianne   198
A Million Dreams   235
A Million Dreams Ago   211, 231, 383
A Million Miles Away   462, 497
Milwaukee   544
Mindin' My Business   96, 235
Mine   312, 315
Mine Alone   94, 560
Minnie the Mermaid   90
Minnie the Moocher   143, 325
Minnie the Moocher's Wedding Day   14, 249

Minnie's in the Money   407
Minor Drag   520
Minuet in Jazz   443
Miracle of the Bells   69
Miserlou   425
Miss Annabelle Lee   61, 378, 399
Miss Brown to You   385, 386, 405, 542
Miss Hannah   392
Miss Lulu from Louisville   407
Miss Thing   23
Miss Wonderful   41
Miss You   496, 498, 499
M-I-S-S-I-S-S-I-P-P-I   492
Mississippi Basin   391
Mississippi Mud   21, 58
Mississippi Shiver   68
Mississippi Suite   167
Mississippi Volunteers   64
Missouri Misery   486
Missouri Moon   365, 366
Missouri Scrambler   361
The Missouri Waltz   447, 448
Missy   423
A Mist Is Over the Moon   356
Mister and Mississippi   153
Mister Meadowlark   97, 316
Misty   45
Misty Morning   116
Moanin' for You   101
Moanin' Low   94, 385
Mobile Mud   413
Modern Melody   266
Molly   101
The Moment I Looked in Your Eyes   266
The Moment I Saw You   94, 313, 440, 454
Moments in the Moonlight   152, 200
Moments Like This   11, 253, 276, 393
Mona Lisa   121, 271
A Monday Date   201
Money Is the Root of All Evil   251

The Monkey and the Organ Grinder   494
Monsoon   450
The Mooche   116
Mood Hollywood   98
The Mood I'm In   531
Mood Indigo   116, 325
The Mood That I'm In   453
Mood to be Wooed   204
Moody   140
Moon About Town   486
Moon and Sand   544
The Moon and the Willow Tree   435
Moon at Sea   477
Moon Country   55, 260, 316
The Moon Fell in the River   89, 366
The Moon Got in My Eyes   45
The Moon Is a Silver Dollar   366
The Moon Is Grinning at Me   213
The Moon Is Low   40, 135
Moon Love   79
Moon Mist   117
Moon of Desire   93
Moon of Manakoora   277
Moon on My Pillow   497, 499
The Moon over Madison Square   331
Moon over Miami   44, 261
Moon River   294, 316
Moon Rose   132, 419
Moon Song   73, 227, 244
The Moon Was Yellow   7, 261
Moonbeam   554
Moonburn   55, 193
Moon-Faced and Starry Eyed   533
Moonflowers   45
Moonglow   84, 213, 325, 477
Moonlight and Roses   77
Moonlight and Shadows   208, 406, 472
Moonlight and Violins   310, 497
Moonlight Bay   288, 535
Moonlight Becomes You   45, 510
Moonlight Cocktail   142, 402
Moonlight Fiesta   495

SONG TITLE INDEX • 699

Moonlight Gambler   199
Moonlight in Vermont   483
Moonlight March   347
A Moonlight Memory   489
Moonlight Mississippi (A Whistle Stop)   413
Moonlight Mood   89
Moonlight on the Colorado   328
Moonlight On the Meadow   422
Moonlight on the Mississippi   233
Moonlight Poppies   78
Moonlight Saving Time   233, 399
Moonlight Serenade   366
Moonlight Whispers   52, 343
Moonray   450
Moonrise on the Lowlands   273, 342
Moonshine Over Kentucky   327, 379
Moonstruck   73
The More I Know You   72, 83
More Than Anything in the World   481
More Than Ever   161, 230
More Than You Know   114, 417, 564
Moritat   532
The Morning After   34, 219
The Morning Music of Montmarte   272
Morning, Noon, and Night   11, 488
The Most Wonderful Day of the Year   303
M-O-T-H-E-R   223
Mother, Dixie and You   432
Mother Prairie   494
Mountain High, Valley Low   178
Mr. and Mrs. Is the Name   559, 560
Mr. Crump   175, 353
Mr. Dooley   222
Mr. Five by Five   88, 388
Mr. Fortune Tellin' Man   331
Mr. Ghost Goes to Town   213, 366
Mr. Jelly Lord   334
Mr. Siegel, You Gotta Make It Legal   100, 562
Mrs. Santa Claus   140

Muddy Water   88, 503
Muggin' Lightly   424
"Murder," He Says   278
Muscogee Blue   423
Mush Mouth   334
Music   217
Music Box Rag   402
Music by the Angels   490
Music By the Moon   191
Music for Elizabeth   162
Music for Madame   529
Music for Romance   454
The Music Goes 'Round   434
Music in My Fingers   336
Music in My Heart   11, 127, 307, 531
Music in the Moonlight   166
Music, Maestro, Please   289, 560
Music Makers   388
Music Makes Me   115
Music, Music, Everywhere   14, 250
The Music of a Mountain Stream   413
Must We Say Goodnight So Soon?   266
My Arabian Maid   472
My Arms Are Wide Open   524
My Baby Just Cares for Me   96, 235
My Baby's Arms   493
My Best Girl   96
My Blackbirds are Bluebirds Now   137
My Blue Heaven   17, 96, 541
My Bluebird Was Caught in the Rain   398
My Bluebird's Singin' the Blues   406
My Buddy   96, 234
My Cigarette Lady   508
My Cutey's Due at Two to Two Today   406, 515
My Dancing Lady   127
My Darling   191, 192, 336
My Darling, My Darling   278
My Day Begins and Ends with You   72, 83
My Dog Loves Your Dog   188

My Dream Memory 262
My Dream of the Big Parade 106, 306
My Dream of the South 537
My Dream of Yesterday 448
My Dreams Are Getting Better All the Time 252, 328
My Dreams Have Gone With the Wind 356
My Dreamy China Lady 233, 509
My Extraordinary Gal 447
My Fair Lady 459
My Family Tree 75
My Fate Is in Your Hands 390, 520
My First Impression of You 473, 497
My First Thrill 149, 206, 456
My Foolish Heart 570, 571
My Future Just Passed 300, 542
My Gal 76
My Gal Sal 104
My Galveston Gal 4
My Girl Friday 392
My Greatest Mistake 139
My Gypsy Rhapsody 93
My Heart and I 208, 282, 406
My Heart Cries for You 458
My Heart Is a Hobo 45
My Heart Is an Open Book 154, 396
My Heart Is Taking Lessons 331
My Heart Jumped over the Moon 204
My Heart's At Ease 520, 568
My Heart's in the Middle of May 402
My Honey's Lovin' Arms 322, 423
My Ideal 61, 404, 542
My Impression of You 62
My Inspiration 172
My Irish Daisy 441
My Isle of Golden Dreams 234
My Isle of Love 362
My Kind of Country 277
My Kind of Town 50, 510
My Kinda Love 503
My Lady Loves to Dance 85
My Lady of Japan 441

My Last Goodbye 211
My Lily and My Rose 574
My Little Bimbo Down on the Bamboo Isle 62, 96
My Little Buckeroo 222
My Little Corner of the World 199
My Little Girl 515
My Little Grass Shack in Kealakakua Hawaii 349
My Love 66, 524, 570
My Love for You 69
My Love Is on the Way 254
My Love Is Yours 381
My Love Loves Me 272
My Love Parade 165, 434
My Lucky Star 189
My Mad Moment 301
My Mammy 228, 234, 265, 565
My Man's Gone Now 193
My Mariucca Take a Steamboat 375
My Melancholy Baby 353
My Midnight Frolic Girl 195
My Mom 97
My Moonlight Madonna 529
My Mother Told Me 145
My Mother's Eyes 19
My Mother's Rosary 322
My! My! 277
My! My! Ain't That Somethin' 501
My Ohio Home 96, 235
My Old Flame 73, 227
My Old Gal 403
My Old Man 178, 513
My Old New Hampshire Home 515
My Own 13, 308
My Own True Love 80
My Particular Man 225
My Pearl Is a Bowery Girl 222
My Rainbow 331
My Red Letter Day 149, 456
My Reverie 64
My River Home 369
My Rose of Spain 442
My Secret Love Affair 327

My Sentimental Heart   275
My Shawl   2
My Shining Hour   14, 316
My Ship   532
My Silent Love   191, 192, 485
My Silent Mood   64
My Sin   92, 189
My Sister and I   573
My Song   39, 188
My Song of the Nile   322
My State, My Kansas, My Home   550
My Sugar Plum   324
My Summer Colors   252
My Summer Love   199
My Sunday Girl   423
My Sunny Tennessee   422
My Sweet   310, 312, 313
My Sweeter Than Sweet   300
My Sweetheart   69
My Sweetie Went Away (But She Didn't Say Where, When, or Why)   174, 504
My Tane   349
My Thoughts   469
My Troubles Are Over   261, 331
My Very Good Friend the Milkman   45, 471
My Wife's Gone to the Country-(Hurrah! Hurrah!)   541
My Wild Oat   556
My Window Faces the South   274, 366
My Wish   550
My Wonderful Dream Girl   435
My Wonderful One   167
My Yiddishe Momme   378, 562
Mysterioso   450
Mysterious Mose   102

'N Everything   227, 234
Nagasaki   94, 300, 522
Nasty Man   188, 562
Naughty Hula Eyes   349
Naughty Lola   209

Navajo   509, 546
Navajo Nocturne   477
The Navy Took Them Over, and the Navy Will Bring Them Back   437
The Nearness of You   55, 526
Need I Say?   537
Neglected   303
Neighbors   58, 533
Neither Am I   402
Nesting Time   331
Nevada   97, 163
Never Again   75
Never Feel Too Weary to Pray   549
Never Gonna Dance   127
Never in a Million Years   155, 397
Never Let Me Go   272
Never Let the Same Bee Sting You Twice   287, 466
Never Make a Promise in Vain   339
Never No Lament   116, 426
Never Say Die   61
Never Should Have Told You   134, 137
Never Steal Anything Small   560
Nevermore   74
Nevertheless   237, 421
The New Asmolean Marching Society Student's Conservatory Band   278
A New Moon Is over My Shoulder   135
New Orleans   54, 55, 235
A New Star Is Shining in Heaven   494
New Sun in the Sky   440
New York, New York   67
Next Stop Paradise   105
The Next Time I Care   238
Nice Baby   179
Nickel in the Slot   68
Night after Night   478
The Night before Christmas Song   303
The Night Has a Thousand Eyes   30
Night in Manhattan   386, 406
Night in Sudan   334
The Night Is Young   173, 416

The Night Is Young and You're So
    Beautiful   233, 418, 486
Night Life   547
Night on the Desert   197
Night Owl   215
The Night Ride   374
The Night Shall Be Filled with Music
    126, 301
The Night That Love Was Born   358
The Night They Invented Champagne
    279
The Night Was Made for Love   180
The Night We Called It a Day   2, 87
Night Whispers   464
Nightfall   89, 394
Nightingale   65
A Nightingale Sang in Berkeley Square
    453
Nightmare   449
Nights of Gladness   285
Nightwind   267
Nighty Night Dear   432
Nina Never Knew   11, 103
Nine Little Broken Hearts   153
Nine Little Miles from Ten-Ten-
    Tennessee   452
Nine Old Men   2, 87
Ninety-nine Out of a Hundred Wanna
    Be Loved   263, 452
No, Baby, No   187
No Can Do   462
No Foolin'   179
No Longer   499
No Love, No Nothin'   406
No Man Is an Island   251
No Moon at All   122, 296
No More Love   106
No More Rain   159
No Name Jive   517
No, No, a Thousand Times No   264,
    453
No No Nora   120
No Nothing   175
No One   103
No One But You   459

No Other Arms, No Other Lips   574
No Other Love   427, 538
No Other One   255, 445
No Regrets   217, 498
No Wonder   119
No Wonder I'm Blue   10
No Wonder I'm Happy   8, 83
No Wonder You're Blue   93
Nobody but You   536
Nobody Knows, Nobody Cares   183
Nobody Knows What a Red-Headed
    Mama Can Do   106, 123
Nobody's Darling   56
Nobody's Fault But Your Own   350
Nobody's Love Is Like Mine   344
Nobody's Sweetheart   120, 235, 435,
    436
Nobody's Tears Are Falling but Mine
    402
Nobody's Using It Now   435
Noche De Ronda   465
Nocturne   163
Non-stop Flight   450
Noodlin' Rag   402
Norway   132
Not Mine   435
Not Really the Blues   295
Not So Long Ago   35
Not That I Care   542
Nothing   511
Nothing Can Stop Me Now   471
Nothing Could Be Sweeter   565
Nothing Ever Lasts Forever   319
Nothing's Gonna Stop Me Now   184
Nothing's Too Good For My Baby   83
Nothin's Too Good for My Baby   8
Now and Forever   434
Now I Know   15
Now I Lay Me Down to Dream   211,
    322, 327
Now I'm in Love   449
Now or Never   266
Now That Summer Is Gone   463
Now That You're Gone   235
Now You See It   278

SONG TITLE INDEX • 703

Now You're in My Arms   559
Now You've Got Me Doing It   471
Nowhere Guy   338
Now's the Time to Fall in Love   263, 264, 452
Numb Fumblin'   520

O How I Laugh When I Think How I Cried About You   506
O Marinariello   490
O Sole Mio   206
The Object of My Affection   166, 500
O'Brien Has Gone Hawaiian   362
Ode to an Alligator   281
Of This I'm Sure   299
Oh Baby   90, 96, 562
(Oh Baby) Look What You've Done to Me   71
Oh, Bella Mia   499
Oh, But I Do   440
Oh, By Jingo   38
Oh! By Jingo! Oh, by Gee! (You're the Only Girl for Me)   514
Oh! Frenchy!   70
Oh Gee, Georgie   319
Oh, Gee, Jennie   97
Oh, He Loves Me!   389
Oh, How I Long to Belong to You   565
Oh, How I Love My Darling   553
Oh, How I Miss You Tonight   43, 81, 132
Oh, Johnny   34
Oh Johnny, Oh, Johnny, Oh!   151
Oh, Katherina   147
Oh Look at That Baby   512
Oh! Ma-Ma   508
Oh Me, Oh My, Oh You   564
Oh! Mona   523
Oh, Moon   369
Oh, Mr. Dream Man   329
Oh, My Aching Heart   269, 364
Oh No!   423
Oh Suzanna (Dust Off That Old Pianna)   260

Oh, the Pity of It All   407
Oh, What a Pal Was Mary   237, 261
Oh! What a Thrill (To Hear It From You)   205, 335
Oh What a Pal Was Mary   261
Oh What I Know about You   239
Oh What I'd Do for a Girl Like You   541
Oh, What It Seemed to Be   27, 52
Oh Yeah   372
Oh! You Chicken   132
Oh, You Circus Day   331
Oh You Crazy Moon   510
Oh, You Cutie!   547
Oh You Little Rascal   322
Oh You Sweet One   219
Oh! You Have No Idea   100
Okay, Baby   377
Okay for Sound   128
Okay, Toots   97, 235
Okey Doke   464
Ol' Pappy   273, 342, 489
An Ol' Tin Cup   398
The Old Apple Tree   222
The Old Covered Bridge   197
Old Demon Rum   407
Old Devil Moon   254
Old Faithful   55
Old Fashioned Love   226, 287
An Old Flame Never Dies   440
Old Folks   411, 412, 413
The Old Gypsy Fiddler   93
Old King Tut   222
The Old Lamplighter   462, 497
An Old Lullaby   372
The Old Man of the Mountain   196, 570
Old Man Rhythm   144
Old Man River   173
Old Man Sunshine   94
Old Man Time   137
The Old Music Master   55, 316
Old Pal   235
Old Pal, Why Don't You Answer Me?   221, 265

An Old Piano Plays the Blues   10
Old Playmate   235, 291
Old Sad Eyes   233
The Old Sow Song   508
The Old Spinning Wheel   197
The Old Square Dance Is Back Again   499
An Old Straw Hat   155
An Old Water Mill   497
Ole Buttermilk Sky   55
Ole Miss   176
On a Blue and Moonless Night   360
On a Clear Day   259
On a Clear Day You Can See Forever   254
On a Dew-Dew-Dew-Dewy Day   452, 496
On a Little Bamboo Bridge   453
On a Little Dream Ranch   198
On a Little Street in Singapore   89, 198
On a Little Two-Seat Tandem   338
On a Simmery Summery Day   58, 533
On a Slow Boat to China   278
On a Typical Tropical Night   45
On Accounta I Love You   473
On Green Dolphin Street   238, 526
On Mobile Bay   77
On Revival Day   390
On the Alamo   228, 234
On the Atcheson, Topeka, and the Santa Fe   316, 523
On the Banks of the Wabash   104
On the Beach   10
On the Beach at Bali Bali   319, 453, 461
On the Beach With You   164, 445
On the Boardwalk in Atlantic City   337
On the Bumpy Road to Love   206, 309
On the Eight O' Clock Train   409
On the Gay White Way   407
On the Gin, Gin, Ginny Shore   96, 261
On the Good Ship Lollipop   542
On the Good Ship Mary Ann   234
On the Isle of May   80
On the Little Big Horn   477
On the Loose   37
On the Mississippi   124
On the Old Fall River Line   222, 474, 516
On the Other Side of the Tracks   259
On the Outgoing Tide   40, 528, 529
On the Rio Grande   532
On the Sentimental Side   45, 331
On the Street of Regrets   247
On the Street Where You Live   259, 279
On the Sunny Side of the Street   127, 307
On the Trail   167
On the Waterfall   275
On Top of the World Alone   406
On Treasure Island   44, 261
Once   427
Once and for All   242
Once and for Always   45
Once Around the Moon   458
Once in a Blue Moon   154, 396
Once in a Lifetime   244, 245
Once in Awhile   161
Once in Love With Amy   278
Once Over Lightly   370
Once Upon a Song   275
Once You Find Your Guy   489
One Alone   173, 180, 415
One Day in June   179
One Dozen Roses   231, 524
The One Fish Ball   464, 573
One for My Baby (And One More for the Road)   14, 316
The One Girl   565
One Hour With You   405
The One I Love   235

The One I Love Belongs to Somebody Else   228
The One I Love Can't Be Bothered With Me   235, 463
One Kiss   415
One Life to Live   533
One Little Kiss   422
One Little Kiss Did the Trick   344
One Little Raindrop   319, 400, 442
One Little Word Led to Another   230
One Love   250
The One Man Band   25
One Meat Ball   464, 574
One Minute to One   72, 265
One More Affair   312, 313
One More Time   93
One More Tomorrow   338
One More Vote   337
One Morning in May   55, 365
One Night in Monte Carlo   453
One Night in Trinidad   202
One Night of Love   235, 434
One Night Stand   450
One O'Clock Jump   23
One of Us Was Wrong   370, 372
One Pair of Hands   252
One Step to Heaven   245
One Sunny Day   442
One Sweet Letter from You   39
The One That I Love Loves Me   7, 505
One to Remember   517
One, Two, Button Your Shoe   227
Only a Rose   138
Only Forever   45, 331
Only When You're in My Arms   422
Oodles of Noodles   98
Oo-Goo the Little Worm   102
Ooh, Looka There, Ain't She Pretty?   280
Ooh! That Kiss   94, 523, 567
Ooh, What You Said   55, 316
Ooh! Looka There Ain't She Pretty   500

An Open Book   324
Open Road   492
Orange Blossom Lane   366
Orange Blossom Time   112
Orange Colored Sky   85
An Orchid for the Lady   539
An Orchid to You   154, 396
Orchids for Remembrance   366
Orchids in the Moonlight   115, 564
The Oregon Trail   197
Organ Grinder Blues   545
Organ Grinder's Swing   213, 325, 366
Oriental   65
Oriental Strut   378
Original Dixieland One-Step   408
Ostrich Walk   359
Other Lips   518
Oui, Oui, Marie   132
Our Bungalow of Dreams   512
Our Day Will Come   199
Our Hometown Mountain Band   125
Our Little Ranch House   275
Our Love   30, 64, 119, 369
Our Love Affair   135
Our Love Was Meant to Be   196, 520
Our Old Home Team   451
Our Song   127
Our Waltz   47
Out in the Cold Again   33, 250
Out in the New Mown Hay   100
Out of Breath and Scared to Death of You   315
Out of Nowhere   161, 162, 191, 192
Out of Sight, Out of Mind   127
Out of Space   146
Out of the Blue   220, 435
Out of the Clear Blue Sky   342
Out of the Dawn   96
Out of the Night   203
Out of the Past   95
Out of This World   316, 484
Out the Window   23
Out Where the Blues Begin   127, 307
Outer Drive   99

Outside of Paradise   494
Over and Over   304
Over My Shoulder   556
Over Somebody Else's Shoulder   264, 452
Over the Hill and Far Away   441, 442
Over the Rainbow   14, 181
Over the Rhythm of Raindrops   517
Over the Weekend   331
Over There   65, 222
Overhand   547
Overheard in a Cocktail Lounge   337
Overnight   10, 327, 379, 417

Paddlin' Madeline Home   553
Paducah   407
Pagan Love Song   40
Painting the Clouds with Sunshine   43, 44
Pal of My Cradle Days   375
The Pal That I Loved Stole the Girl That I Loved   344
Palace in Paradise   362
Palesteena   70, 360, 408
Palladium Patrol   294
Pals Just Pals   423
Pancho's Rancho   319
The Panic   98
Papa De-Da-Da   500, 548
Papa Loves Mambo   206, 393
Papa Tree Top Tall   3
Papa, Won't You Dance With Me?   485
Paper Boy   309
Paper Doll   444, 543
Parade of the Jumping Beans   68
Parade of the Milk Bottle Caps   99
Paradise Isle   370, 372
Paradise Valley   241
Paramount on Parade   243
Paramour   146
The Pardon Came Too Late   104
Pardon Me, Pretty Baby   245, 318, 419

Pardon My Love   104, 263
Pardon My Southern Accent   292, 316
Pardon Our French   571
Paree!   406
Paris, France   304
Paris in Spring   154
Park Avenue Fantasy   292, 460
Parsons, Kansas Blues   184
The Party's Over   67, 234, 484
The Party's Over Now   75
Pass Me By   259
Pass That Peace Pipe   32, 304
Passe   325, 457
Passing Fancy   296
Passion Flower   481
Pastel Blue   389, 450
Patty Cake, Patty Cake   225
The Pay Off   98
Peaceful Valley   414
Peanut Vendor   374
The Pearls   334
Pee Wee Squawks   424
Pee Wee's Blues   423
Pee Wee's Song   423
Peekaboo to You   325, 459
Peelin' the Peach   267
Peg O' My Heart   41, 130
Peggy   547
Peggy O'Neil   343
The Penguin   443
Penguin at the Waldorf   517
Penn Beach Blues   511
Pennies from Heaven   44, 45, 227, 522
Pennsylvania 6-5000   159, 160, 457
Pennsylvania Dutch   239
The Penthouse Serenade   48, 220
People Have More Fun Than Anyone   402
People to People   540
Perdido   495
Persian Rug   78, 235
Personality   45, 510
The Pessimistic Character   331

Pete Kelly's Blues 185
Peter Piper 316
Peter Tambourine 443
Pettin' in the Park 107, 523
Piano Man 202
Piano Tuner's Dream 413
Pianology 201
Piccolo Pete 25
Pick Yourself Up 127
Pickin' Cotton 92, 189
Pickin' on Your Mama 407
Picture Me Down Home in Tennessee 222
Picture Me Without You 250
Pig Foot Pete 87, 388
Pigeon Toed Joad 364, 412
Pigeon Walk 331
Pigskin Parade 327
Pine Cones and Holly Berries 550
A Pink Cocktail for a Blue Lady 356
Pink Elephants 554
The Pink Panther 294
Play, Fiddle, Play 12, 93, 256
Play Gypsies, Dance Gypsies 467
Play It Again, Joe 172
Play Me Hearts and Flowers 252
Playmates 102
Please 385, 406
Please Be Kind 60
Please Be There 345
Please Come Back to Me 537
Please Come Out of Your Dream 459
Please Don't Kiss Me 402
Please Don't Take My Lovin' Man Away 515
Please Don't Talk about Me When I'm Gone 61, 473
Please Go 'Way and Let Me Sleep 515
Please Handle With Care 481
Please Keep Me in Your Dreams 255, 445
Please, My Love 560
Please Remember 168

Please Tell Me That You Love Me 267
Plenty of Brass 384
Ploddin' Along 503
Plymouth Rock 185
Poeme 529
Poinciana (Song of the Tree) 29, 462
Polar Bear Strut 440
Polka Dots and Moonbeams 45, 510
The Polka-Dot Polka 407
Pompton Turnpike 361, 414
Poor Butterfly 148, 212
Poor Buttermilk 68
Poor Cinderella 310, 406
Poor Little Angeline 170
Poor Little Rhode Island 483
Poor Little Rich Girl 74
A Poor Man's Roses 85
Poor Mr. Chisholm 177, 178
Poor Old Joe 55
Poor Papa (He Ain't Got Nothin' at All) 417, 553
The Poor People of Paris 257
Poor Robinson Crusoe 4
Poor You 254
Pop! Goes Your Heart 559
Porgy 127
Pork and Beans 402
A Porter's Love Song to a Chambermaid 226, 391
A Portrait of Jennie 409
(Potatoes are Cheaper, Tomatoes are Cheaper) Now's the Time to Fall in Love 452
A Prairie Fairy Tale 252, 473, 497
Praise the Lord and Pass the Ammunition 278
Pray for Sunshine, But Always Be Prepared for Rain 1
Precious Little One 381, 540
Prelude in C Major 450
Prelude to a Kiss 116, 152, 325
Prep Step 301
Pretending 453, 490

Pretending You Care 446
Pretty Baby 234
Pretty Doll 544
A Pretty Girl 234
A Pretty Girl—a Lonely Evening 78
Pretty Kitty Blue Eyes 252, 328
Pretty Kitty Kelly 344
Pretty Little Busybody 477
Pretty Little Cinderella 513
Pretty Little Petticoat 443
Pretty Little Stranger 406
Pretty, Petite, and Sweet 13
Prince of Wails 436
Princess Poo-pooly Has Plenty Papaya 362
Prisoner of Love 66, 143, 404
Private Buckeroo 348
The Prize Waltz 456
Promise 51, 129
Promises 453
Proud of You 301
P.S. I Love You 221, 316
Pucker Up and Whistle 513
Puddin' Head Jones 42, 175
Pu-leeze, Mr. Hemingway 103, 241, 461
Pull Yourself Together 336, 406
Pullman Porter's Parade 1
The Pump Song (It's Hard to Tell the Depth of the Well by the Length of the Handle on the Pump) 125, 260
Puppy Love 356
Pushin' Along 122
Put a Little Rhythm in Everything You Do 154
Put 'Em in a Box, Tie 'Em With a Ribbon, and Throw 'Em in the Deep Blue Sea 484
Put It There, Pal 510
Put On an Old Pair of Shoes 197
Put On Your Old Gray Bonnet 535
Put That Kiss Back Where You Found It 457
Put the Blame on Mame 130, 401

Put Your Arms Around Me, Honey 514
Put Your Arms Around Me Where They Belong 433
Put Your Dream Away for Another Day 281
Put Your Dreams Away 297
Put Your Heart in a Song 531
Puttin' On the Dog 449
Pyramid 152

The Queen of Bohemia 467
Queer Street 334
Quicker Than You Can Say Jack Robinson 80, 322

The Rabbit's Jump 204
Racing With the Moon 247
Ragging the Scales 400
Ragtime Cowboy Joe 1, 62
Railroad Blues 402
Railroad Man 371
Rain 487
The Rain in Spain 279
Rain, Moonlight and Magnolias 89
Rain or Shine 562
Rain, Rain, Go Away 79, 162, 191, 192
Rainbow 13
A Rainbow From the U.S.A. 536
Rainbow of My Dreams 518
Rainbow on the River 11, 530
Rainbow Valley 44, 261
Raincheck 481
Raindrop Serenade 477
Raindrops 269, 381
Raindrops Keep Fallin' on My Head 79
The Rains Came 156
Raise a Little Army of Your Own 448
Rambling Rose 44
Ramona 147, 528
Rampart Street Blues 409
Rampart Street Parade 24

SONG TITLE INDEX • 709

Rancho Pillow   348
The Ranger's Song   493
Reaching for Someone   96, 261
Ready for the River   78, 235
A Real Live Girl   259
Rebecca   546
Reciprocity   142
Reckless Night on Board an Ocean Liner   443
Recollections   478
Red Bank Boogie   23
Red Garters   272
Red Hot and Blue Rhythm   83, 488
Red Hot Mama   418
Red Lips, Kiss My Blues Away   330
Red Rose Rag   536
Red Roses for My Blue Baby   344
Red Sails in the Sunset   169
Red Silk Stockings and Green Perfume   198, 339
Red Sky at Night   315
The Red We Want Is the Red We've Got (In the Old Red, White and Blue)   394
Reflections in the Water   274, 529
Reincarnation   300
Remember   125
Remember Cherie   166
Remember Me   523
Remember Me?   107
Remember Me to Carolina   531
Remember That I Care   533
Remember Tonight   267
Reminiscing   261, 522
Rendezvous Time in Paree   107, 308
Restless   39, 188
Return to the Magic Islands   478
Revolvin' Jones   413
Rhapsody in Blue   167
Rhode Island Is Famous For You   94, 440
Rhumboogie   387
A Rhyme for Love   387, 406
Rhyme Your Name   296

The Rhyming Song   260
Rhythm and Romance   47, 225
Rhythm in My Nursery Rhymes   387
Rhythm Is Our Business   50, 59
Rhythm King   360, 408, 503
Rhythm Man   520, 527
Rhythm of the Rain   318, 475
Rhythm on the River   331
Rhythm Saved the World   59
Rhythm Sundae   201
Riddle Me This   144
Ride Off   478
Ridin' a Riff   202
Ridin' Around in the Rain   17, 280
Riff Interlude   23
The Riff Song   173, 180, 415
Right as the Rain   14
Right at the Start of It   440
The Right Guy for Me   532, 533
The Right Kind of Love   529
The Right Somebody to Love   378
Ring Me Up in the Morning   547
Ringtail Blues   409
Rio Rita   493
Rip Van Winkle Was a Lucky Man   441
Ripples of the Nile   142, 402
Riptide   97
Rise 'n' Shine   565
The River and Me   106
River of Smoke   459
River River   356
The River Seine   402
River, Stay 'Way from My Door   94, 555
Riverboat Shuffle   54, 55, 325
The Road I Didn't Take   170
The Road to Zanzibar   510
Robbins Nest   427
Robins and Roses   44, 261
Rock-a-bye the Boogie   389
Rock-a-bye Your Baby with a Dixie Melody   228, 234, 265, 441, 565
Rockin' and Reelin'   389

Rockin' Around the Christmas Tree   303
Rockin' Chair   54, 55
Rockin' in a Rockin' Chair   37
Rockin' the Bass   461
Roll 'Em   547
Roll Along Prairie Moon   515
Roll On, Mississippi, Roll On   400, 536
Roll Up the Carpet   149, 205, 245
Rolleo Rolling Along   499
Rollin' Down the River   2, 520
Rollin' Plains   381, 431, 540
Rolling Down to Bowling Green   338
Romance   139, 415
Romance Runs in the Family   252
Romany   440
Romany Life   467
Roof Top Serenade   48
A Room with a View   74, 487
The Roots of Heaven   526
Ro-Ro-Rolling Along   310, 329, 399
Rosary of Roses   357
Rose Ann of Charing Cross   529
Rose Marie   173, 180
Rose O'Day (The Filla-ba-dusha-Song)   264, 497
Rose of the Rio Grande   261, 522
Rose of Washington Square   56, 178, 285
Rose Room   194, 546
Rose-Marie   138, 479
Rosemary   275
Roses Are Forget-Me-Nots   356
The Roses Have Made Me Remember   75
Roses in December   290
Roses in the Rain   52
Roses Remind Me of You   453
Rosetta   201
Rosie the Riveter   121, 275
'Round Midnight   178
'Round the Old Deserted Farm   412
Roundabout   342

Row, Row, Row   222, 329
Royal Garden Blues   544, 548
Ruby   366
The Ruby and the Pearl   272
Rudolph, the Red-Nosed Reindeer   303
Rugged But Right   102
Rum and Coca Cola   464, 574
Rum Dum   359
Rumors Are Flying   27
Run for the Roundhouse, Nellie   413
Running Through My Mind   240, 241
Rural Revelations   414
Rural Rhythm   533
Rush Inn Blues   370, 371

Sad Sack   450
Saddest Man in Town   221
Sadie Salome, Go Home!   261
Safara Stomp   209
Sag Mir Darling   266
Saga of the Signal Corps   494
Sagebrush Lullaby   489
Sahara Nights   63
Sail Along Silv'ry Moon   499, 536
Sail Away   75
Sail On, Silv'ry Moon   120
A Sailboat in the Moonlight   275, 280
Sailin' Away on the Henry Clay   234, 509
Sailin' on the Robert E. Lee   400, 537
Sailing at Midnight   44, 261
Sailor Boys Have to Talk to Me in English   199
Sally   165
Sally Doesn't Care   221
Sally, Won't You Come Back   42, 472
Saloon   541
Saltin' Away My Sweet Dreams   402
Sam, the Old Accordion Man   96
The Same Old Moon   136, 180
Sampson Stomp   430
Sam's Song   117, 383

SONG TITLE INDEX • 711

San Antonio   546
San Fernando Valley   221
San Francisco   238
San Sue Street   146
Sand Fiddler   209
Sand in My Shoes   277, 434, 435
The Sandman Cometh   427
Sans-Souci   495
Santa Claus Came in the Spring   292, 316
Santa Claus Is Coming to Town   72, 148
Santa Claus Is Riding the Trail   319
Santa Fe Trail   170
Santa Rosa   309
Sapho Rag   407
Sapphire   33
Saratoga Shout   424
Sarong   459
Saskatchewan   260
Satan Takes a Holiday   63
Satanic Blues   360
Satan's Holiday   511
Satin Doll   116, 316, 480
Saturday   36
Saturday Afternoon Before the Game   338
Saturday Night at the Nobles   352
Saturday Night Is the Loneliest Night of the Week   50, 483
Savannah   132
Save a Little Sunbeam   153
Save It Pretty Mama   392
Save Me a Dream   241
Save the Last Dance for Me   203
Save Your Sorrow (For Tomorrow)   452
Saving Myself for You   60
Savoy Shout   424
Saw Mill River Road   493
Say a Prayer Every Day   140
Say It   277
Say It!   4, 47
Say It Again   399

Say It While Dancing   83
Say It With a Kiss   316, 523
Say It With a Solitaire   331
Say It With a Uke   336, 406
Say It With Songs   105
Say "Oui" Cherie   565
Say That You Love Me   336, 406
Say the Word and It's Yours   149, 206, 456
Say When   188
Say "Yes" Today   96
Says My Heart   253, 276
Scalawag   325
Scandinavia   368
The Scat Song   365, 367
Scatter-brain   45
Scattered Toys   240
The Scene Changes   197
School Days   64, 111
The Scissors Grinder's Song   537
Scratchin' the Gravel   547
Scrub Me Mama With a Boogie Beat   388
Scuttlebutt   450
Seal It With a Kiss   440
The Second Time Around   50, 51, 510
Secondhand Rose   62, 178
Secret Love   123, 531
The Secret of My Success   47
See Dixie First   64
See the Monkey   178
See What the Boys in the Back Room Will Have   208, 277
Seein' Is Believin'   3, 6
Seems Like Yesterday   473
Seems to Me   387
Seminole   352, 509, 546
Send My Baby Back to Me   85
Sensation   359
Sensation Rag   359
Sent for You Yesterday   23
Sentimental and Melancholy   316, 542
Sentimental Baby   364

Sentimental Gentleman from Georgia   365, 367
Sentimental Journey   161
Sentimental Lady   116, 426
September in the Rain   107, 522
September Song   532
Serenade   319
Serenade of the Bells   149
Serenade to a Sleeping Beauty   430
Serenade to Nobody in Particular   99
Serenade to the Stars   308
A Serenade to You   461
Serenata   33
Setting the Woods on Fire   344
Seven Come Eleven   151
Seven Little Girls   199
Seventy-Six Trombones   549
The Shabby Old Cabby   462
The Shadow of Your Smile   295, 531
The Shadow Waltz   107
Shadows   52
Shadows on the Moon   235
Shadows on the Swanee   45, 470, 568
Shady Lady Bird   32, 304
A Shady Tree   96
Shake Down the Stars   84, 510
Shake Hands With a Millionaire   398
Shake Your Feet   472
Shakin' the African   392
Shanghai   85, 199
Shanghai Shuffle   68
Shangri-La   293, 458
Sharecroppin' Blues   413
She Broke My Heart in Three Places   104
She Came Rolling Down the Mountain   400, 454
She Didn't Say "Yes"   180
She Loves Me Not   396
She Reminds Me of You   154, 396
She Shall Have Music   149, 206, 456
She Told Him Emphatically, No   102
She Was a China Teacup and He Was Just a Mug   387, 406
She Wore a Yellow Ribbon   353
The Sheik of Araby   467, 469
She'll Always Remember   303
She'll Love Me and Like It   406
Sheltered by the Stars, Cradled by the Moon   520
Shenanigans   494
Shepherd's Serenade   480
She's a Great, Great Girl   554
She's from Missouri   510
She's Funny That Way   78, 542
She's Mine, All Mine   422
She's Such a Comfort to Me   438
Shim Sham Shimmy   98
Shim-Me-Sha-Wabble   548
Shine   39, 287
A Shine on Your Shoes   94, 439
The Shining Sea   295
Ship of Fools   526
Shoe Shine Boy   59
Shoe Shiner's Drag   334
Shoo Fly Pie and Apple Pan Dowdy   141
Shoot the Works   57, 396
Short & Sweet   100
A Short Short Story   281
Shorty George   23
Should I?   40, 135
Should I Be Sweet?   565
Shout, Sister, Shout   545
Show Me the Way to Get Out of This World   87
Show Your Linen, Miss Richardson   177
Shreveport Stomp   334
Shrimp Boats   538
Shuffle Along   464
Shuffle off to Buffalo   106, 523
Shy and Sly   402
Siam   132
Siberian Sleigh Ride   443
Side by Side   554, 558
Side Street in Gotham   11
Sidewalk Blues   334

SONG TITLE INDEX • 713

The Sidewalk Serenade   58, 533
Sidewalks of Cuba   356, 366
Signs of a Honeymoon   546
Silence Is Golden   140
Silent Senorita   407
Silhouette   33
Silhouetted in the Moonlight   316
Silhouettes Under the Stars   497
Silver and Gold   303
Silver Bells   271, 536
Silver Dollar   363, 364
Silver Heels   78
Silver on the Sage   407
Silver River   166
Silvery Moonlight   239
Simple and Sweet   233
The Simple Things in Life   49, 188, 250
Since I Found You   423, 554
Since You Went Away   170
Sincerely Yours   220
Sing a Little Love Song   326
Sing a Little Lowdown Tune   496
Sing a Little Tune   208, 348
Sing a New Song   538
Sing a Song of Sunbeams   331
Sing an Old-Fashioned Song (To a Young Sophisticated Lady)   569
Sing, and Let Your Hair Down   471
Sing, Baby, Sing   378, 379
Sing, Brother, Sing   227
Sing It Way Down Low   55
Sing, Katie (But Leave the Piano Alone)   100
Sing Me a Song of the Islands   362
Sing Me a Song of the South   354
Sing Me a Swing Song   55
Sing Me an Old Fashioned Song   471
Sing Our Song of Love   544
Sing Something Simple   215
Sing Song Girl   178, 306
Sing UCLA   31
Sing You Sinners   181
Sing Your Worries Away   163

Singin' in the Bathtub   62, 288, 524
Singin' in the Rain   40, 135
Singin' in the Saddle   381, 540
Singin' the Blues (Till My Daddy Comes Home)   70, 265, 270, 360, 408, 566
Singing a Happy Song   318, 475
Singing a Song to the Stars   324
Singing a Vagabond Song   48, 399
The Singing Hills   79, 338
Singing River   362
The Singing Sands of Alamosa   142
Single Little Tingle of My Heart   402
A Sinner Kissed an Angel   79
Sioux City Sue   352, 460
A Siren Dream   452
The Siren's Song   550
Sit Down, You're Rocking the Boat   442
Sittin' in a Corner   235, 320
Sittin' in the Sand A-Sunnin'   260, 448
Sittin' on a Backyard Fence   233
Sittin' on a Log (Pettin' My Dog)   68
Sittin' on a Rainbow   101
Six Lessons from Madam LaZonga   331, 347
Six Studies in Modern Syncopation   414
Six Women   188
Sixty Seconds Got Together   80, 274
The Skeleton in the Closet   227
Skeleton Jangle   359
Skip It   469
Skrontch   344
The Sky Fell Down   11, 193
Skylark   54, 55
Slave of Love   464
Sleeping Beauty   10
Sleepy Head   97, 164, 235
Sleepy Lagoon   257
Sleepy Moon   407
Sleepy Serenade   163, 464
Sleepy Time Gal   113, 542

Sleepy Valley   179, 474
Sleigh Ride in July   45
A Slight Case of Ivory   168
Slow Mood   52
Slowly but Surely   264
Sluefoot   78, 432
Slumbertime Along the Suwanee   468
Small Fry   55, 276
Small World   299, 444
Smarty   136, 514
Smashing Thirds   520
Smile, Darn Ya, Smile   318, 357, 398
Smile for Me   25
A Smile Will Go a Long, Long Way   8, 9, 81
Smiling at Trouble   144
Smiling Irish Eyes   368
Smoke Dreams   40, 135, 247
Smoke Gets in Your Eyes   180
Smoke Rings   146, 525
Smokey Mary   24
Snag It   358
Snake Charmer   380, 540
Snake Hips   548
Snowball   55
Snowy Morning Blues   226
Snug as a Bug in a Rug   277, 292, 293
Snuggle on Your Shoulder   280, 568
So Ashamed   6, 83
So at Last It's Come to This   235, 291, 460
So Beats My Heart for You   186
So Close to Me   244
So Dear to Me   327
So Divine   499
So Do I   565
So Far So Good   257, 334
So Help Me   84, 510
So I Married the Girl   290
So Little Time   89
So Long   327, 332
So Long for Now   211
So Long, Oolong   422
So Long Train Whistle   384

So Lovely   119
So Madly in Love   529
So Many Memories   557
So Nice   188
So Nice Seeing You Again   94
So the Bluebirds and the Blackbirds Got Together   21
So This Is Heaven   45, 471
So This Is Love   206, 273
So Tired   332
So What?   387, 406
So What You Wanna Do?   254
So Would I   45
So You're the One   250, 573
Sobbin' Blues   238
Soft As Spring   543
Soft Hearted   278
Soft Summer Breeze   194
Softly, as in a Morning Sunrise   173, 415
Soliloquy   33
Solitude   84, 325
Solo Flight   334
Some Boy   472
Some Day   138
Some Enchanted Evening   174
Some Like It Hot   253
Some of These Days   37, 346
Some Other Bird Whistled a Tune   446
Some Other Time   484
Some Rainy Day   275
Some Sunday Morning   114, 222, 234, 250
Some Sunny Day   185
Some Sweet Day   378, 455
Somebody   314
Somebody Bad Stole de Wedding Bell (Who's Got de Ding Dong?)   199, 296
Somebody Cares for You   453
Somebody Else Is Taking My Place   332
Somebody Else's Picture   275

SONG TITLE INDEX • 715

Somebody Else's Roses   339
Somebody Loses, Somebody Wins   47, 225
Somebody Loves Me   90, 91, 286
Somebody Loves You   89
Somebody Ought to Be Told   416
Somebody Stole Gabriel's Horn   525
Somebody Told Me   345
Somebody Up There Likes Me   238
Somebody's Birthday   63
Somebody's Lonely   83
Somebody's Thinking of You Tonight   381
Somebody's Wrong   114
Someday I'll Find You   74
Somehow Days Go By   444
Someone Is Losin' Susan   320, 321
Someone Loves Someone   402
Someone Loves You After All   493
Someone Stole Gabriel's Horn   525
Someone to Share My Dreams   314
Someone's Rocking My Dream Boat   395
Something   511
Something Happened to Me   34
Something Has Happened to Me   219
Something in Here   245
Something Old—Something New   211
Something Seems Tingle-ingling   139
Something Tells Me   316, 523
Something to Live For   481
Something to Remember You By   94, 438
Something to Sing About   435
Something Very Strange   75
Something's Gotta Give   316
Sometimes I'm Happy   49, 563
Somewhere Beyond the Sunset   398
Somewhere in Old Wyoming   497
Somewhere in the Night   337
Somewhere My Love   531
Somewhere with Somebody Else   44, 261

Sompin' at the Savoy   429
A Song Is Born   389
The Song Is the Thing   494
Song of Delilah   272, 571
The Song of Surrender   107, 121, 271, 570
Song of the Blues   229
Song of the Cotton Fields   36, 520
Song of the Crusades   406
Song of the Fool   265
Song of the Highwayman   95
Song of the Moonbeams   497, 499
The Song of the Mounties   479
Song of the Swanee   424
Song of the Vagabonds   138
The Song of the Victory Fleet   540
Song of the Wanderer   77
Song without Words   314
Songs of Yesterday   183
Sonny Boy   92, 227
Sonya   446
So-o-o-o-o in Love   407
Sophisticated Lady   116, 325, 365
Sophisticated Swing   213, 366
Sous le Ciel de Paris   142
South American Way   107
South Rampart Street Parade   171, 172
South Wind   529
Spain   228, 235
Spanish Dream   370
Spanish Rose   448
Spanish Shawl   309, 436
Speak Low   341, 532
Speak Well of Me   517
Speaking Confidentially   127, 307
The Sphinx   381, 495
The Spider and the Fly   225
Spin a Little Web of Dreams   233
Spinner   491
The Spirit Is Willing   159, 160
The Spirit of St. Louis   487
Spooky Takes a Holiday   63
S'posin'   86

Spread a Little Happiness   118
Spread a Little Happiness As You Go   553
Spring Again   532
Spring Cleaning (Getting Ready for Love)   381, 431, 540
Spring Fever   33
Spring Will Be a Little Late This Year   278
A Square in the Social Circle   272
Squeeze Me   519, 520, 545
St. Andrews by the Sea   319
St. Louis Blues   175, 176
St. Louis Gal   409
St. Louis Shuffle   370, 520
Stairway to the Stars   292, 293, 366, 460
Stampede   187
Stand Up and Sing for Your Father   368
The Stanford Scalp Song   31
Stanley Steamer   202
Star Eyes   88, 388
A Star Fell Out of Heaven   154, 397
A Star Is Born   266
Stardust   53, 54, 55, 285, 365, 569
Stardust on the Moon   93
Stargazing   273, 342, 489
Starlight   368, 568
Starlight Souvenirs   449
Starlight, Starbright   88, 388
Starlit Hour   89, 366
Stars and Soft Guitars   93
Stars Fell on Alabama   366, 367, 419
Stars in My Eyes   127
Stars in Your Eyes   163
Stars over the Desert   217
The Stars Remain   157
The State of My Heart   471
The Stately Homes of England   75
Stay As Sweet As You Are   154, 396
Stay on the Right Side of the Road   250
Stay on the Right Side, Sister   33
Stay Out of My Dreams   369, 525
Stay With Me   259
Stealin' Apples   391, 520
The Steam Is on the Beam   301
Steamboat Bill   454
Steamboat Stomp   334
Stella   228
Stella by Starlight   526, 570
Step to the Rear   259
Steppin' It Off   370
Stevedore's Serenade   153
Still I Love Her   143
Still the Bluebird Sings   331
Stockholm Stomp   370
Stolen Kisses   423
Stomp Off, Let's Go   159
Stompin' at the Savoy   150, 391
Stompin' at the Stadium   34
Stop Beatin' around the Mulberry Bush   34, 393
Stop It's Wonderful   394
Stop! It's Wonderful   34
Stop, Look   469
Stop, Look, and Listen   136
Stop! You're Breakin' My Heart   249
Stop That Dancin' Up There   544
Stormy Monday Blues   202
Stormy Weather   14, 249
The Story Adam Told to Eve   441
The Story of a Starry Night   206, 252, 273
The Story of My Life   427, 459
Stout Hearted Men   415
Straight from the Shoulder   154
Straight Life   295
Straight to Love   202
Straighten Up and Fly Right   325
Strange   254
Strange Enchantment   209, 277
Strange Interlude   28, 29, 203
Strange Interval   464
A Strange Loneliness   338
A Stranger Called the Blues   502
Stranger in Town   501

SONG TITLE INDEX • 717

Strangers   72, 357
A Strawberry Moon   199, 339
The Strawberry Roan   513
Street of Dreams   265, 570
The Streets of Laredo   121, 272
Strictly for the Persians   64
Strictly from Dixie   37
Strictly in the Groove   332
Strictly Instrumental   27, 298, 444
Strike Me Pink   39, 188
Strike Up the Band, Here Comes a Sailor   474, 521
A String of Pearls   84, 159
Stringin' the Blues   511
The Strip Polka   316
Strollers We   467
Strolling Through the Park One Day   112, 541
Strut Miss Lizzie   257
Struttin' with Some Barbecue   15
Stud Polka   537
A Study in Blue   63
A Study in Brown   63
A Study in Green   63
A Study in Red   63
Stuff Like That There   271
Stumbling   67
Such Stuff as Dreams Are Made Of   233
Suddenly It's Spring   45
Suez   167
Sugar   326, 376
Sugar Blues   545
Sugar Foot Stamp   309, 357
Sugar Moon   536
Sugar Plum   235
Suicide Blues   354
Suicide Is Painless   295
Summer Holiday   69, 303
Summer Is A-comin' In   254
Summer Love   51
Summer Rain   251
Summer Sequence   191
Summer Souvenirs   72, 347

Summer Vacation   117, 356
Summertime   193
Summit Ridge Drive   450
The Sun Forgot to Shine This Morning   51
Sun Valley Jump   159, 160
Sunbonnet Sue   64, 111
Sunday   165
Sunday Go To Meetin' Time   446
Sunday, Monday or Always   45, 510
Sunny Side of the Rockies   217
The Sunny Side of Things   531
Sunny Side Up   189
Sunny, Who?   173, 180
Sunrise   440
The Sunrise   433
Sunrise Boogie   52
Sunrise in Napoli   52
Sunrise Serenade   52, 256
Sunshine and Roses   234
Suppose I Had Never Met You   13, 492
Sure As You're Born   451
Surprise   272
The Surrey with the Fringe on Top   173
Susie   341
Swanee   49, 228, 234
Swanee River Blues   472
Swanee River Rhapsody   143
Swan's Serenade   487
Sweet and Hot   13, 562
Sweet and Lovely   16, 166, 498
Sweet and Simple   188
Sweet and Tender   500
Sweet As a Song   155, 397
Sweet Child   264
Sweet Dixie Lady   103
Sweet Dreams   83, 142
Sweet Dreams, Sweetheart   6, 222, 250
Sweet Eloise   79, 332
Sweet Emalina   76

Sweet Georgia Brown 28, 29, 346, 376
Sweet Heartache 473
Sweet Heartaches 462
Sweet Horn 209
A Sweet Irish Sweetheart of Mine 494
Sweet is the Word for You 386, 406
Sweet Jennie Lee 97
Sweet Lady 223
Sweet Leilani 361, 362, 386
Sweet Like You 61
Sweet Lil 21
Sweet Lorraine 365
Sweet Lucy Brown 395
Sweet Madness 525, 570
Sweet Man 377
Sweet Man O' Mine 408
Sweet Misery of Love 463
Sweet Mumtaz 424
Sweet Music 94, 107, 523
Sweet Nevada 440
The Sweet Potato Piper 331
Sweet Savannah Sue 37, 520
Sweet So and So 324
Sweet Someone 155, 397, 422
Sweet Stranger 6, 274, 539
Sweet Substitute 334
Sweet Sue 243, 570
Sweet Thing 7, 19
Sweet Varsity Sue 310, 497
Sweeter Than You 422
The Sweetest Girl This Side of Heaven 13
Sweetest Melody 370, 371
The Sweetest Music This Side of Heaven 280
Sweetheart 83, 142
Sweetheart Darlin' 235, 479
Sweetheart of My Student Days 235, 463
Sweetheart Time 324
The Sweetheart Tree 294
Sweetheart, We Need Each Other 493

Sweethearts Forever 49
Sweethearts Holiday 402
Sweethearts on Parade 280, 346
Sweetie, Be Kind to Me 354
Sweetie Pie 275
The Sweetness of You 430
Swing, Mr. Charlie 37, 409
Swing Session in Siberia 372
Swing Tap 387
Swingin' at the Sugar 172
Swingin' Doors 183
Swingin' Down 201
Swingin' Down the Lane 228, 235
Swingin' in a Hammock 357, 444, 534
Swingin' in the Dell 204
Swingin' on a Star 510
Swingin' the Blues 23
Swinging on a Star 45
Swingtime in Honolulu 344
Swingtime in the Rockies 334
Swingy Little Thingy 473
The Swiss Bellringer 431
Sympathy 139
Symphony 257
Syncopated Heart 480
Syncopated Hula Love Song 362

T'Ain't Nobody's Bizness if I Do 545
A Table in a Corner 486
Taboo 425
Tabu 425
Tain't No Sin 97
'Tain't No Sin (To Take Off Your Skin and Dance Around in Your Bones) 261
'Tain't No Use 253, 290
'Tain't So, Honey, 'Tain't So 412
Take a Lesson From the Lark 387, 406
Take a Little One-Step 564
Take a Number from One to Ten 154

SONG TITLE INDEX • 719

Take a Tip From the Whippoorwill 274
Take Another Guess 310, 347, 453
Take In the Sun, Hang Out the Moon 265, 553, 566
Take It Easy 127, 307, 328, 334
Take It Easy by Slow 362
Take It from Me 520
Take It From There 407
Take Me 79
Take Me Away 494
Take Me Back to the Garden of Roses 133
Take Me Back to My Boots and Saddle 380, 431, 540
Take Me Back to New York Town 516
Take Me Back to Those Wide Open Spaces 499
Take Me in Your Arms 365
Take Me Out to the Ballgame 513
Take Me to My Alabam' 498
Take Me to That Midnight Cakewalk Ball 1
Take Me to That Swannee Shore 147
Take Me to the Land of Jazz 535
Take My Heart 569
Take My Love 238
Take the "A" Train 480
Take the High Ground 526
Take This Ring 486
Take Your Girlie to the Movies 261, 534
Take Your Shoes Off, Baby (And Start Running Through My Mind) 17
Take Your Sins to the River 281
Take Your Tomorrow 224, 390
Takes Two to Make a Bargain 154, 396
Takes Two to Tango 206
Takes You 83
Taking a Chance On Love 108, 254
Talk to Me, Baby 95
Talking to Myself 71, 290

Talking to Myself About You 478, 538
Tamiami Trail 136, 432
Tammy 121
Tampico 130, 181, 401, 413
Tangerine 316, 435
Te Amo 537
Tea for Two 49, 563
Teach Me Tonight 50, 88
Teacher's Pet 402
Tears from My Inkwell 94
Tears in My Heart 540
Tears on My Pillow 264
Teasin' 53
Teasing 287, 515
Teddy Bear Boogie 381
Teenage Prayer 394
Tell Me a Story 457, 477
Tell Me at Midnight 35
Tell Me How Long's the Train Been Gone 345
Tell Me Why 65
Tell Me Why You and I Should Be Strangers 446
Tell Me Why You Smile, Mona Lisa 114
Tell Me With a Love Song 250
Tell Me With Your Eyes 515
Tell Me You Love Me 333
Tell the Truth 369
Telling All My Troubles to the Daisies 319
Telling It to the Daisies 265, 522, 567
Temptation 40, 135
Ten Little Fingers and Ten Little Toes 343, 437
Ten Little Miles from Town 436
Ten O'Clock Town 62, 488
Ten to One It's Tennessee 183
Tender Is the Night 97, 123
The Tender Trap 50, 510
A Tender Word Will Mend It All 402
Tenderly 168, 257
Tennessee Fish Fry 440

Tennessee Lazy   432
Tenting Down in Tennessee   553
Tessie, Stop Teasin' Me   367
Texas Li'l Darlin'   95
Texas Tornado   327
Thank Heaven for Little Girls   259, 279
Thank Heaven for You   387, 406
Thank You for a Lovely Evening   127, 307
Thank You, Mr. Moon   358
Thank Your Lucky Stars   440
Thanks   73, 227
Thanks a Million   227, 235
Thanks for Everything   155, 230, 397
Thanks for the Memory   386, 406
Thanks To You   62, 534
Thanksgivin'   55, 316
That Certain Party   96, 235
That Dreamy Italian Waltz   305
That Face!   304
That Feeling in the Moonlight   59
That Girl of Mine   498
That Lindy Hop   390
That Little Boy of Mine   244
That Man's Here Again   430
That Mellow Melody   265, 322
That Minor Strain   287
That Never-to-Be-Forgotten Night   123, 497
That Old Black Magic   14, 316
That Old Devil Moon   181
That Old Feeling   40, 123
That Old Gang of Mine   94, 188, 416
That Old Girl of Mine   509
That Old Irish Mother of Mine   222, 516
That Particular Friend of Mine   209
That Railroad Rag   513
That Same Old Way   309
That Sentimental Sandwich   209, 277
That Sly Old Gentleman   331
That Was Before I Met You   322
That Was My Heart   225

That Week in Paris   356
That Wonderful Boy Friend of Mine   377
That Wonderful Worrisome Feeling   274, 402
That's Amore   523
That's A-Plenty   75, 377, 378
That's Entertainment   94, 440
That's for Me   45, 331
That's Georgia   451
That's Good   547
That's Good Enough for Me   402
That's Gratitude   354
That's Him   341
That's How I Feel about You   83
That's How I Love the Blues   304
That's How It Goes   343
That's My Desire   213
That's My Girl   83
That's My Home   394
That's My Mammy   344
That's My Weakness Now   160, 473
That's No Dream   474, 516
That's the Beginning of the End   251
That's What a Rainy Day Is For   328
That's What I Like About the South   391
That's What I Want for Christmas   303
That's What Life Is Made Of   351
That's What Puts the "Sweet" in Home Sweet Home   347
That's What You Mean to Me   83
That's What You Think   501
That's When I Learned to Love You   24
That's Where I Came In   497
That's Where the South Begins   118
That's Why Darkies Were Born   39, 188
That's You, Baby   326
That's You Sweetheart   369
Them Hillbillies are Mountain Williams Now   339

Them There Eyes   377, 491, 502
The Theme from Picnic   213
Then I'll Be Happy   39, 61, 136
Then I'll Be Tired of You   440
Then You've Never Been Blue   265, 566
There Ain't No Sweet Man Worth the Salt of My Tears   131
There Are Such Things   3, 19, 20, 321
There Goes My Attraction   273, 343
There Goes My Heart   83, 461
There Goes That Song Again   482, 560
There I Go   573
There I Go Dreaming Again   39, 188
There Is a Red Bordered Flag in the Window   448
There Is No Breeze   267
There Is No Greater Love   230, 490
There Isn't Any Limit to My Love   149, 206, 456
There Isn't Any Special Reason   502
There I've Said It Again   121, 122, 296
There Must Be a Way   141
There Must Be Somebody Else   377
There Never Was a Girl Like You   509, 546
There Never Was a Town Like Paris   207, 348
There Nothing Else to Do in Na-La-Ka-Mo-Ka-Lu   327
There Once Was an Owl   467
(There Oughtta Be a) Moonlight Saving Time   399
There Shall Be No More Tears   125
There Won't Be a Shortage of Love   275
(There'll Be Bluebirds Over) The White Cliffs of Dover   241
There'll Come a Time   183
There'll Never Be Another You   344
There's a Blue Sky 'Way Out Yonder   124, 125

There's a Boat Dat's Leavin' Soon for New York   193
There's a Brand-New Picture in My Picture Frame   134
There's a Broken Heart for Every Light on Broadway   132, 223
There's a Cabin in the Pines   197
There's a Cradle in Caroline   7, 265
There's a Different "You" in Your Heart   233
There's a Faraway Look in Your Eye   328
There's a Girl in the Heart of Maryland   56, 285
There's a Gold Mine in the Sky   240, 241
There's a Hole in the Old Oaken Bucket   475
There's a Little Bit of Bad in Every Good Little Girl   62, 132
There's a Little Lane Without a Turning   322
There's a Lot of Moonlight Being Wasted   462
There's a Lull in My Life   155, 397
There's a Man in My Life   301
There's a Platinum Star in Heaven Tonight (Jean Harlow)   462
There's a Rainbow 'Round My Shoulder   105, 227
There's a Ranch in the Sky   451
There's a Wah Wah Girl in Agua Caliente   97
There's Always a Happy Ending   149, 457
There's Always My Heart   459
There's Danger in a Dance   407
There's Danger in Your Eyes, Cherie   317, 399, 534
There's Frost on the Moon   8, 569
There's Honey on the Moon Tonight   72, 79, 148
There's Life in the Old Girl Yet   75

There's Music in My Heart Cherie 179
There's Never Been a Love Like Ours 172
There's No Depression in Love 101
There's No Other Girl 360
There's No Place Like Your Arms 34, 393
There's No Tomorrow 206
There's Nothing Like a Song 338
There's Nothing the Matter With Me 348, 431
There's Nowhere to Go but Up 532
There's Rain in My Eyes 6, 442
There's Something in That 324
There's That Look in Your Eyes Again 155, 397
There's Two Sides to Every Story 569
There's Yes Yes in Your Eyes 136, 432
These Foolish Things 267, 268
These Things Are Known 219
These Will Be the Best Years of Our Lives 296
They Call the Wind Maria 259, 279
They Called It Dixieland 114
They Can't Convince Me 402
They Cut Down the Old Pine Tree 115, 196
They Go Wild, Simply Wild, Over Me 130
They Say 191, 296
They Talk a Different Language 95
They're Either Too Young or Too Old 278, 440
Thine Alone 190
Things Ain't What They Used to Be 117
Things Have Changed 531
Things Might Have Been So Different 72, 265
The Things We Did Last Summer 485
Think Well of Me 412
Thinking of You 96, 420
This Changing World 486
This Could Be the Start of Something Big 9
This Heart of Mine 135, 523
This Is a Chance of a Lifetime 186
This Is All I Ask 221
This Is Always 337
This Is It 128, 424, 440
This Is Madness 84, 510
This Is My Country 388
This Is My Favorite City 338
This Is My Night to Dream 45, 331
This Is New 533
This Is No Dream 83, 99, 448, 449
This Is Our Last Night Together 157
This Is Romance 108, 191, 192
This Is the Missus 39, 188
This Is the Night 122
This Is Worth Fighting For 473
This Little Ripple Had Rhythm 387, 407
This May Be the Night 155
This May Last Forever 442
This Nearly Was Mine 174
This Night 235
This Time 538
This Time It's Love 72, 265
This Time It's Real 30, 119
This Time It's True Love 75
This Time the Dream's on Me 14, 316
Those Lazy-Hazy-Crazy Days of Summer 497
A Thousand Dreams of You 11, 530
A Thousand Goodnights 97
The Thousand Islands Song (I Left My Love on One of the Thousand Islands) 198, 458
A Thousand Violins 272
Three Dreams Are One Too Many 142, 484
Three Little Fishies 101
Three Little Sisters 328

SONG TITLE INDEX • 723

Three Little Words   237, 421
Three O'Clock Jump   151
Three on a Match   113, 114
The Three Rivers   531
Three Times a Day   406
Three Wishes   380
Three's a Crowd   106, 233
The Thrill Is Gone   39, 188
Thrill Me   181
Thrilled   21, 22, 163, 186
Through (How Can You Say We're Through)   305, 330
Through Children's Eyes   272
Through the Years   192
Throw Another Log On the Fire   310, 496
Throwin' Stones at the Sun   338, 462
Thru the Courtesy of Love   221
Thru With Love Affairs   209
Thunder over Paradise   387, 406
Thy Will Be Done   498
Tia Juana   68
The Tickle Toe   202
Tie a Little String Around Your Finger   463, 564
Tie Me to Your Apron String Again   450
Tiger Rag   359
Till All the Stars Fall into the Ocean   471
Till the Clock Strikes Three   198
Till the Clouds Roll By   550
Till Then   298, 443, 444
Till There Was You   549
Till Tomorrow   291
Till We Meet Again   112, 542
Timber   198
Timbuctu   422
Time   449
Time After Time   50, 485
Time Alone   339
Time and Again   469
A Time for Love   295, 531
Time on My Hands   3, 153, 564

Time to Sing   422
Time Waits for No One   137, 497
Time Was   425, 427
Time Will Tell   200, 245
Timmy   362
Tin Roof Blues   299, 300, 309, 475, 476
Tina   170
Ting-a-Ling (The Song of the Bells)   268
Tiny Little Fingerprints   347, 497
Tiny Room   304
Tiny Tim   529
Tiptoe Through the Tulips with Me   43, 106
Tired   282, 401
Tired of It All   422
Tired Teddy Bear   178
'Tis Autumn   345
Tishomingo Blues   548
To a Sweet Pretty Thing   8, 569
To Be in Love   7
To Be Loved by You   574
To Be Worthy of You   168
To Each His Own   121
To Get Along with the Beautiful Girls   126
To Know You Is to Love You   51, 106, 189, 401
To Love You and Lose You   532
To Remind Me of You   444
To Whom It May Concern   158, 321, 327
To You   83, 99, 100, 448
To You, Sweetheart, Aloha   361
Tobacco Auctioneer   443
Today and Tomorrow   282
Together   92
Tomorrow   221
Tomorrow Is Another Day   235, 238
Tomorrow Mountain   254
Tomorrow Night   170
Tonight I'm Thinking of You   102
Tonight Is Mine   235

Tonight, Lover, Tonight   475
Tonight or Never   245, 420
Tonight Will Never Come Again   212
Tonight You Belong to Me   417
Tonight's My Night   310
Tony's Wife   253
Too Beautiful for Words   66, 475
Too Beautiful to Last   281
Too Good to Be True   34, 250
Too Late   265
Too Late Now   254
Too Many Rings Around Rosie   564
Too Many Tears   106
Too Marvelous for Words   256, 316, 542
Too Much Imagination   45, 471
Too Much in Love   242
Too Tired   451
Too Wonderful for Words   472
Too-Ra-Loo-Ra-Loo-Ral (That's an Irish Lullaby)   447
Toodle-oo   275, 564
Toot Toot Tootsie, Goodbye   120, 228, 234
Topic of the Tropics   343
Topper   217
Toreador Song   480
Tormented   213
Totem Tom-tom   480
Touch and Go   527
A Touch of Love   281
A Touch of Texas   278
The Touch of Your Hand   180
The Touch of Your Lips   352
Town Without Pity   526
The Toy Trumpet   443
Toyland   190
The Toyland Band   102
Trade Winds   137, 497
Traffic Jam   308, 450
The Trail of Dreams   487
The Trail of the Lonesome Pine   56, 285
Train of Love   574

Transatlantic Rhapsody   380
Trav'lin' All Alone   225
Trav'lin' Light   316, 334
Travellin' Shoes   183
A Tree Was a Tree   154, 396
Trinidad   427
The Trolley Song   32, 304
Tropic Trade Winds   349
Trouble   549
Trouble In Paradise   6, 439, 538
The Trouble with Me Is You   499, 501
Truckin'   33
True   430, 539
True Blue Lou   73, 404, 542
True Confession   73, 208
True to Two   336, 406
Truly Wonderful   204
Trumpet Rag   384
Trust In Me   6, 539
Try a Little Tenderness   551, 555, 556
Try Getting a Good Night's Sleep   392
Try to Forget   180
T.S.U. Alma Mater   327
Tuck Me to Sleep in My Old 'Tucky Home   265, 320, 566
Tulip Time   472
Turkish Tom Tom   28, 131
Turn Back the Hands of Time   517
Turn Off the Moon   73
Turn On the Heat   189
Turn On the Moon   253
The Twelfth of Never   274
Twelve O'Clock at Night   175
Twentieth Century Blues   74
Twenty Million People   244
Twenty-four Hours a Day   488
Twenty-Four Hours of Sunshine   459
Twenty-one Dollars a Day—Once a Month   28, 245
Twentyeighth and Eighth   423
Twilight in Turkey   443
Twilight Interlude   494

SONG TITLE INDEX • 725

Twilight on the Trail   11, 327
Twinkle Twinkle Little Star   356
Two Buck Tim from Timbuctoo   206
Two Cigarettes in the Dark   378, 530
Two Dreams Got Together   134
Two Fools in Love   132
Two for the Blues   185
Two for the Record   543
Two for Tonight   154
Two Hearts in Three-Quarter Time   567
Two Heavens   170
Two in Love   549
Two Irish Fairy Tales   486
Two Little Blue Little Eyes   274, 508, 529
Two Little Fishes and Five Loaves of Bread   178
Two Little, True Little Eyes   354
Two Loves Have I   335
Two Seats in the Balcony   47
Two Sleepy People   55, 276
Two Thirds of the Tennessee River   145
Two Tickets to Georgia   72, 496, 568
Tzena, Tzena, Tzena   366

Ugly Chile   544
Ukulele Lady   235
The Umbrella Man   58, 419, 477
Unbelievable   274
Unchained Melody   574
Uncle Joe's Music Store   25
Unconditional Surrender   122
Under a Blanket of Blue   272, 273, 342, 489
Under a Ceiling of Stars   468
Under a Strawberry Moon   529
Under a Texas Moon   368
Under a Wurzburger Tree   516
Under Paris Skies   142
Under the Anheuser Bush   516
Under the Ukulele Tree   94
Under the Yum Yum Tree   474, 516

Under Your Spell   440
Under Your Window Tonight   309
Underneath Hawaiian Skies   120
Underneath the Arches   306
Underneath the Harlem Moon   153
Unforgettable   152
Unsuspecting Heart   381
Until   139
Until the Real Thing Comes Along   59, 207, 348
Until Today   83
Until We Meet Again, Sweetheart   266
Up Among the Chimney Pots   489
Up and At 'Em   370
Up in a Balloon   454, 535
Up in My Aeroplane   546, 547
Up in New Hampshire   119
Up in the Clouds   422
Up Jumped the Devil   343
Upon the Hudson Shore   442
Us on a Bus   255, 445
The U.S.A. and You   407
Used to You   227

V for Victory   102
Vagabond Dreams   54, 55, 256
Valse Moderne   356
Vamping a Coed   209
Varsity Drag   92, 189
The Velvet Glove   471
Velvet Moon   84, 337
A Very Precious Love   123
The Very Thought of You   351
Vieni, Vieni   508
Violets for Your Furs   2, 87
Violins Are Playing   240
Violins Were Playing   241
Viper's Drag   520
Vo-Do-Do-De-O Blues   562
Voice of the Tradewinds   362
Volare   366
Vote for Mr. Rhythm   386, 406

Wabash Blues   400
Wabash Moon   105
Wacky Dust   263
Waddlin' at the Waldorf   99
Wagon Wheels   89, 197
Wah-hoo   136
Waikiki Wedding   361
The Wail   432
Wait for Me, Mary   462, 499
Wait Till the Sun Shines, Nellie   474, 516
Wait Till You See Ma Cherie   406
Wait Until Dark   272
The Waiter and the Porter and the Upstairs Maid   316
Waitin' at the Gate for Katy   235
Waitin' for the Evenin' Mail   24
Waitin' for the Evening Train   440
Waitin' for the Train to Come In   465
Waiting for the Robert E. Lee   146
Waiting for You   564
Wake the Town and Tell the People   274
Wake Up and Dream   104
Wake Up and Live   155, 397
Wake Up and Sing   497
Wake Up, Chillun, Wake Up   414, 503
Wake Up, Sleepy Moon   398
Walk On By   79
Walk on the Wild Side   80
Walk, Jennie, Walk   446
Walkin' and Swingin'   547
Walkin' My Baby Back Home   7, 505
Walkin' the Dog   38
Walking with Susie   70, 157, 158, 326
Wall Street Blues   176
The Wall   105
Waltz in Swing Time   127
The Waltz Lived On   407, 387
Waltz Me Around Again   64
Waltz Me Around Again, Willie ('Round, 'Round, 'Round)   454
The Waltz of Love   512

The Waltz That Made You Mine   120
A Waltz Was Born in Vienna   279
The Waltz You Saved for Me   235, 244
Waltzing in a Dream   524, 571
Waltzing in the Clouds   235
Waltzing With a Dream   461
Wanderlust   204
Wang Wang Blues   48
Wanted   140
Wanting You   140, 415
War Dance for Wooden Indians   443
Warm and Willing   272
Warm Valley   116
Was I To Blame? (For Falling in Love With You)   235, 346, 570
Was It Rain?   175, 203
Was Last Night the Last Night   29
Was She Prettier Than I?   304
Was That the Human Thing to Do?   123, 568
Washboard Blues   54, 55
Washington Squabble   209
Washington Wobble   116
Wasn't It Nice   463
Watch That First Step   219
Watchin' the Trains Go By   446
Watching the Clouds Roll By   421
Water Under the Bridge   378, 530
Watermelon Weather   531
Way Back Home   264
'Way Down in Cotton Town   261
'Way Down Yonder in New Orleans   75, 257
The Way I Feel Today   392
The Way It Might Have Been   304
The Way That the Wind Blows   251
The Way You Look Tonight   127
We Always Get Our Girl   407
We Did It Before   497
We Did It Before and We Can Do It Again   137
We Just Couldn't Say Goodbye   555
We May Never Meet Again   495

SONG TITLE INDEX • 727

We Three (My Echo, My Shadow, and Me)   338, 403
We Were the Best of Friends   265, 321
We Will Always Be Sweethearts   406
We Won't Make a Song and Dance   312, 315
Wear Your Sunday Smile   406
Weary of It All   75
Weather Man   61
A Weaver of Dreams   117, 571
Wedding Bells Are Breaking Up That Old Gang of Mine   123, 233
Wedding Invitations   328
The Wedding of the Painted Doll   40, 135
The Weekend of a Private Secretary   177, 316
Weep No More, My Baby   162, 192
Weep They Will   51
Welcome Stranger   316
Welcome to My Dream   45
Welcome to My Dreams   510
Welcome to the Club   502
Well, All Right   387
We'll All Go Home   546
We'll All Go Riding On a Rainbow   556
Well, I'll Be Switched   338
We'll Be Together Again   129
We'll Sing the Old Songs   524
Were You Sincere?   318, 419
We're Back Together Again   331
We're Going Over   474
We're in the Money   523
We're in the Navy   389
We're Not Dressing   396
We're Not Getting Any Younger, Baby   356
We're Off to See the Wizard   14
We're All Together Now   407
West End Blues   358, 545
West Indies Blues   545
West of the Great Divide   541

West Point Forever   181
West Wind   341, 347, 565
We've Got the Moon and Sixpence   262
We've Got the Stage Door Blues   422
What a Beautiful Night   422
What a Day!   554
What a Difference a Day Made   2
What a Girl Can Do   422
What a Girl, What a Night   432
What a Life   10
What a Little Moonlight Can Do   556, 557
What a Night   371, 371
What a Perfect Combination   9, 49, 422
What a Sweet Surprise   352
What Am I Gonna Do About You?   485
What Am I Supposed to Do   105
What Are You Doin' Tonight?   501
What Are You Doing New Year's Eve   278
What Are You Doing the Rest of Your Life?   497, 499
What Can You Say in a Love Song?   14
What Did I Do?   338
What Do the Animals Do (When They Wanna Say I Love You?)   372
What Do We Do on a Dew Dew Dewy Day?   223
What Do You Know About Love?   80, 155, 274
What Do You See in Her?   533
What Do You Want to Make Those Eyes at Me For?   223, 305, 330
What Goes On Here in My Heart?   386, 407
What Goes Up Must Come Down   33
What Good Is Good Morning   433
What Good Would It Do?   175
What Have We Got to Lose   11

What Have You Got That Gets Me?
    387, 406
What Have You Got to Lose (But Your
    Heart)?   303
What I Was Warned About   304
What Is It?   498
What Is There to Say   108, 181
What Is This That I Feel?   338
What More Can a Woman Do?   20,
    258
What More Can I Ask?   351
What More Do You Want?   574
What, No Women   319
What Price Lyrics   21
What the World Needs Now   79
What Will I Do in the Morning?   520
What Will I Tell My Heart?   152, 256,
    494
What Would I Care?   421
What Would You Do?   406
What Wouldn't I Do for That Man?
    156, 180, 181
What You Goin' to Do When the Rent
    Comes 'Round?   474, 516
Whatcha Gonna Do When There Ain't
    No Jazz?   261, 534
Whatcha Gonna Do When There Ain't
    No Swing?   273, 342
What's a Fellow Gonna Do?   553
What's Good about Good-Night   128
What's It Gonna Be   51
What's Keeping My Prince Charming?
    208, 348
What's New?   45, 171
What's the Good Word, Mr. Bluebird
    402
What's the Matter With Father?   508,
    546
What's the Matter with Me?   447
What's the Name of That Song?   255,
    445
What's the Pitch   423
What's the Reason I'm Not Pleasin'
    You   166, 500

What's the Use   229, 346
What's This World A-Comin' To   414
What's Your Story, Mornin' Glory?
    547
Wheel of Fortune   27
When   224, 390
When a Blackbird Is Blue   366
When a Gypsy Makes His Violin Cry
    93, 468
When a Lady Meets a Gentleman
    Down South   62, 359
When a Prince of a Fella Meets a Cin-
    derella   510
When a Woman Loves a Man   177,
    221, 316, 385, 417
When Am I Gonna Kiss You Good
    Morning?   170
When April Comes Again   273, 343,
    490, 538
When Buddha Smiles   40, 134, 378
When Day Is Done   90
When Did You Leave Heaven?   43,
    542
When Dixie Stars Are Playing Peek-a-
    boo   409
When Does This Feeling Go Away?
    304
When Hearts Are Young   415
When I Climb Down from My Saddle
    35
When I Come Back to You   475
When I Dream of You   201
When I Fall in Love   571
When I First Met Mary   512
When I Get You Alone Tonight   131,
    151
When I Go A-Dreamin'   34, 393
When I Grow Too Old to Dream   173,
    416
When I Grow Up   188
When I See An Elephant Fly   518
When I Take My Sugar to Tea   69,
    123, 233
When I Think of You   103

When I Was the Dandy and You Were the Belle 175
When I'm Not Near the Girl I Love 181, 254
When I'm With You 154, 397
When It's Darkness On the Delta 342
When It's Harvest Time 499
When It's Nightime in Italy, It's Wednesday Over Here 240
When It's Sleeptime in Hawaii 395
When It's Sleepy Time Down South 394
When It's Sunday Down in Caroline 342
(When It's) Darkness on the Delta 489
When Kentucky Bids the World "Good Morning" 528
When Lights Are Low 57, 235, 248, 548
When Love Comes Swinging Along 188, 250
When Love Knocks at Your Heart 197
When My Baby Smiles at Me 474
When My Dream Boat Comes Home 133, 136
When My Ship Comes In 97
When My Sugar Walks Down the Street 17, 306, 325
When Private Brown Becomes a Captain 389
When Santa Gets Your Letter 303
When Somebody Things You're Wonderful 557
When the Bees Are in the Hive 41
When the Bloom Is on the Sage 513
When the Grown-up Ladies Act Like Babies 1
When the Harvest Days Are Over 516
When the Leaves Bid the Trees Goodbye 255, 445

When the Lights Are Soft and Low 494
When the Lights Go On Again All Over the World 27, 298, 345, 443
When the Moon Comes Over the Mountain 223, 398, 555
When the Moon Hangs High 536
When the One You Love Is Gone 183
When the Organ Played "O Promise Me" 319, 453
When the Pussy Willow Whispers to the Catnip 137
When the Red, Red Robin Comes Bob-Bob Bobbin' Along 553, 558
When the Rest of the Crown Goes Home 106
When the Roses Bloom Again 47, 242
When the Sandman Rides the Trail 529
When the Sun Comes Out 250
When the Sun Sets South 37
When the Swallows Come Back to Capistrano 395
When There's a Breeze on Lake Louise 163
When There's No One to Love 442
When They Ask About You 473
When Tomorrow Comes 123, 233
When Twilight Comes 198
When We're Alone 48, 220
When We're Young 543
When Yankee Doodle Learns to Parlez Vous Francais 344
When You Are in My Arms 144
When You Climb Those Golden Stairs 221
When You Dream About Hawaii 422
When You Look in Your Looking Glass 297
When You Love Only One 440
When You Trim Your Xmas Tree 51
When You Walk in the Room 128

When You Walk With the One You Love   357
When You Were Mine   535
When You Wish Upon a Star   526
When You Wore a Tulip (And I Wore a Big Red Rose)   535
When Your Hair Has Turned to Silver   89, 496
When Your Lover Has Gone   487
When You're a Long Long Way From Home   265, 319
When You're Alone   537
When You're in Love   66, 475
When You're in the Room   356
When You're Smiling (The Whole World Smiles with You)   132, 152, 451
When You're Traveling All Alone   241
When You've Got a Little Springtime in Your Heart   556
When Yuba Plays the Rhumba on the Tuba   215
Where Am I?   107, 523
Where Are You   308
Where Can He Be?   440
Where Can I Go Without You?   571
Where Did Robinson Crusoe Go With Friday on a Saturday Night?   265, 322, 565
Where Did You Get That Girl?   237
Where Did You Learn to Dance?   338
Where Do I Go from You?   43, 560
Where Do We Go from Here, Boys?   223
Where Do We Go From Here?   536
Where Do You Work-a, John   522
Where Has My Hubby Gone Blues   564
Where Have You Been All My Life   336, 356, 406
Where In Central Park?   331
Where in the World   155

Where Is My Bess?   193
Where Is My Boy Tonight?   354
Where Is My Heart?   387, 406
Where Is My Old Girl Tonight?   553
Where Is the Man I Married?   304
Where Shall I Find Him?   75
Where the Blue of the Night Meets the Gold of the Day   442, 505
Where the Blues Were Born in New Orleans   53
Where the Lazy Daisies Grow   137
Where the Lazy River Goes By   3
Where the Morning Glories Twine 'Round the Door   114, 234, 516
Where the Shy Little Violets Grow   235
Where the Sweet Magnolias Grow   474, 516
Where There's Smoke There's Fire   273, 489, 342
Where There's You There's Me   149, 457
Where Was I?   107, 182
Where Was Moses When the Lights Went Out?   262
Where You Are   94, 178
Where You Gonna Be When the Moon Shines   275
Wherever You Are   207, 422
Whiffenpoof Song   508
While a Cigarette Was Burning   241
While the Band Is Playing Dixie   521
While We Danced at the Mardi Gras   316
Whisper   333
Whisper in My Ear   442
Whisper Not   502
Whispering   65, 419
Whispering Grass   129, 130, 132
Whispers in the Dark   208, 406
Whispers in the Night   450
Whistle   422
Whistle Away Your Blues   336
Whistle Blues   547

SONG TITLE INDEX • 731

Whistle Stop   191
Whistler's Mother-in-Law   517
The Whistling Boy   127, 128
Whistling for a Kiss   336
Whistling in the Dark   485
Whistling in the Light   407
White Christmas   17, 482, 541
The White Cliffs of Dover   47
White Heat   94, 213, 440
White Jazz   146
White Sails   13
Who Am I?   43, 63, 484
Who Are We to Say?   235
Who Are We?   274
Who Are You With Tonight?   509, 546
Who Blew Out the Flame?   123, 366
Who Calls?   303
Who Can I Turn To?   543
Who Cares?   314, 561
Who Do You Know in Heaven?   89
Who Do You Think I Am?   304
Who Else but God   125
Who Loves You as I Do?   324
Who Loves You?   83
Who Threw Confetti in Angelo's Spaghetti?   357
Who Told You I Cared?   541
Who Walks In When I Walk Out?   136, 149, 206
Who Wants a Bad Little Boy   133
Who Wants to Sing My Love Song   392
Who Wouldn't Be Blue   83
Who Wouldn't Be Jealous of You   451
Who Wouldn't Love You?   51, 129
Whoa, Tillie, Take Your Time   76
Who'd Be Blue   553
The Whole World Is Singing My Song   252, 328
Who'll Be the One?   372
Who's Afraid   491
Who's Afraid of Love   379
Who's Excited   450
Who's Sorry Now   237, 420, 469
Who's Taking You Home Tonight?   454
Who's Your Little Whosis?   29, 203, 369, 370, 372
Whose Heart Are You Breakin' Tonight   310
Whose Honey Are You?   72, 152
Whose Izzy Is He?   160
Why?   488
Why Am I So Romantic?   422
Why Begin Again   389, 450
Why Can't This Night Go On Forever   230, 347
Why Can't You?   228
Why Couldn't It Be Poor Little Me   229, 235
Why Did I Kiss That Girl?   39, 188
Why Did It Have to End So Soon?   490, 533
Why Do I Dream Those Dreams   107, 523
Why Do I Lie to Myself About You?   72, 82
Why Do I Love You   173
Why Do They All Take the Night Boat to Albany?   442
Why Do They Always Say No?   344
Why Does It Have to Be Me?   459
Why Doesn't Somebody Tell Me These Things   447
Why Don't We Do This More Often?   347, 560
Why Don't You Do Right?   258
Why Don't You Fall in Love With Me?   264, 529
Why Don't You Practice What You Preach?   149, 206, 456
Why Dream?   385, 405
Why Have a Falling Out?   58, 533
Why Is It?   251
Why Not?   185
Why, Oh Why?   406, 565

Why Remind Me   491
Why Shouldn't It Happen to Us?   208, 348
Why Stars Come Out at Night   352
Why Talk about Love?   327
Why Was I Born   173
Why'd Ya Make Me Fall in Love?   97
Wild and Woolly Willie   371
Wild Cat   511
The Wild Dog   511
Wild Honey   78, 172, 498
Wild Man Blues   334
Wild Root   185
Wild Rose   165
Wildflower   180, 478, 563
Wilhelmina Down on Wabash Avenue   338
Will I Ever Know?   387, 406
Will Love Find a Way   35
Will You Remember   144
Will You Remember Me?   170, 433
Will You Still Be Mine?   2, 87
Willie   64
Willie Off the Yacht   148
Willow Road   502
Willow Tree   389, 519
Wimmin! Aaa!   100
The Wind at My Window   407
Wind Flowers   254
The Wind in the West   556
The Wind in the Willows   118
Windmill Willie   468
Window of Dreams   247
Wine Song   235
Winin' Boy Blues   334
Winsocki   32
Winter Sun   356
Winter Weather   449
Winter Wonderland   28, 468
Wintertime   407
Wintertime Dreams   42
Wise Guy   333
Wish Me a Rainbow   272
The Wish That I Wish Tonight   222
Wishful Thinking   407

Wishing   90
Witchcraft   259
With a Little Bit of Luck   279
With All My Heart   235
With All My Heart and Soul   477
With Every Breath I Take   385, 406
With Faith in Your Heart   381
With Her Head Tucked Underneath Her Arm   508
With Love in My Heart   246
With My Eyes Wide Open I'm Dreaming   154, 396
With Plenty of Money and You   107, 522
With the Wind and the Rain in Your Hair   257
With Thee I Swing   4
With You Beside Me   244
With You on My Mind (I Can't Write the Words)   125
Without a Song   114, 417, 564
Without a Word of Warning   154, 396
Without That Gal   97
Without You   195
Woe Is Me   119
Wolverine Blues   333
A Woman Alone with the Blues   414
A Woman Is a Sometime Thing   193
Wonderful Pal, Dawning   377
Wonderful You   398
Won't Somebody Please Write a Song   281
Won't You Come Over to My House?   508, 546
Won't You Say You Love Me?   338
Wood and Ivory   374
The Woodchuck Song   297
The Wooden Soldier and the China Doll   230, 346
Wooden Wedding   342
Woodland Reverie   453
The World is Mine   380, 571
Worried Over You   344
Would I Love You?   427, 471

SONG TITLE INDEX • 733

Would I Mind?   269
Would There Be Love?   154
Would You?   135
Would You Believe Me?   222
Would You Care?   183
Would You Like to Take a Walk?   94, 417, 522
Would You Object?   469
Would You Rather Be a Colonel With an Eagle on Your Shoulder or a Private With a Chicken on Your Knee?   157, 326
Wouldn't I Be a Wonder?   557
Wouldn't It Be Loverly   279
Wouldst Could I But Kiss Thy Hand, Oh Babe   361, 414
Wrap Your Troubles in Dreams   21, 22, 249, 329
Wrap Yourself in Cotton Wool   454
Wrappin' It Up   187
Write Me a Love Song, Baby   424

Ya Got Me   34, 393
Ya Gotta Know How to Love   160
Yaaka Hula Hickey Dula   534, 565
Yacht Club Swing   520
The Yankee Doodle Boy   65
Yankee Doodle Never Went to Town   177
Yankee Doodle Tan   225
Yeah Man   409, 464
A Year From Today   228
Yearning   43, 81
Yellow Dog Blues   176
Yellow Flower   254
The Yellow Rose of Texas   145
Yes, My Darling Daughter   257
Yes Sir, That's My Baby   96, 235
Yesterdays   180
Yesterday's Gardenias   339, 403
Yesterthoughts   190, 443
Yip-I-Addy-I-Ay   65
Yodelin' Jive   387
Yoo Hoo   227
You   83, 338

You Ain't Been Livin' Right   527
You Ain't Heard Nothin' Yet   228, 234
You Ain't Nowhere   392
You Always Hurt the One You Love   129, 401
You and I   13, 549
You and I Know   440
You and Me   136, 209
The You and Me That Used to Be   43
You and the Night and the Music   94, 439
You and Your Kiss   128
You and Your Love   316
You Answer "No," Marie   315
You Appeal to Me   471
You Are Music   293
You Are My Lucky Star   40, 135
You Are There   75
You Belong to Me   467
You Better Go Now   394
You Broke the Only Heart That Ever Loved You   269
You Brought a New Kind of Love to Me   69, 123, 233
You Call It Madness   65, 66, 71
You Came a Long Way From St. Louis   426, 427
You Came Along   324, 406
You Came to My Rescue   386, 406
You Can Be Kissed   107, 523
You Can Cry on Somebody Else's Shoulder   403
You Can Dance With Any Girl at All   564
You Can Depend On Me   201
You Can Have Him, I Don't Want Him, Didn't Love Him Anyhow Blues   100
You Can Make My Life a Bed of Roses   39, 188
You Can Say That Again   172
You Can Tell She Comes from Dixie   6, 490
You Can't Be True, Dear   105

You Can't Cheat a Cheater   460
You Can't Do What My Last Man Did   224
You Can't Have Everything   155
You Can't Hide Your Heart   444
You Can't Hold a Memory in Your Arms   574
You Can't Lose Me   475
You Can't Play Every Instrument in the Band   148
You Can't Pull the Wool over My Eyes   6, 310, 347
You Can't See the Sun When You're Crying   130, 401
You Can't Stop Me from Dreaming   134
You Can't Stop Me From Loving You   207, 348
You Couldn't Be Cuter   128
You Crazy Moon   45
You Danced into My Heart   315
You Danced Your Way into My Heart   312
You Darlin'   555
You Discover You're in New York   407
You Do   337
You Do the Darndest Things, Baby   327, 378
You Don't Have to Be a Baby to Cry   447
You Don't Have to Know the Language   510
You Don't Know How Much You Can Suffer   134
You Don't Know What Lonesome Is   524
You Don't Know What Love Is   87, 388
You Don't Love Right   445
You Don't Need Glasses to See   400
You Dreamer You   490
You Dropped Me Like a Red Hot Penny   569
You for Me, Me for You   287
You Forgot About Me   179, 339
You Forgot Your Gloves   114
You Gave Me Everything but Love   235
You Go to My Head   72, 148
You Gorgeous Dancing Doll   333
You Gotta Be a Football Hero   126, 264, 452
You Grew Up to Be Some Baby   328
You Guess Is As Good As Mine   206
You Have Everything   94, 440
You Have Taken My Heart   221, 316
You Have to Live a Little   486
You Hit the Spot   154, 397
You Intrigue Me   343
You Know You Belong to Somebody Else   330, 536
You Leave Me Breathless   136, 208
You Lucky People You   45, 510
You Made Me Love You   305, 329
You Make Me Dream Too Much   484
You Make Me Feel So Young   337
You Meet the Nicest People in Your Dreams   252
You Missed the Boat   137
You Must Have Been a Beautiful Baby   316, 523
You Need Someone   144
You Never Looked So Beautiful   3, 97
You Never Say Yes   239
You Only Live Once   11, 530
You Opened My Eyes   402
You or No One   494
You Ought to See Sally on Sunday   556
You Oughtta Be in Pictures   191, 193, 486
You Said It   562
You Say the Nicest Things, Baby   308
You Showed Me the Way   161, 308
You Smiled at Me   422
You Started Me Dreaming   72, 82
You Started Something   407
You Stayed Away Too Long   47
You Stepped Out of a Dream   41, 235
You Still Belong to Me   243

SONG TITLE INDEX • 735

You Taught Me How to Love You, Now Teach Me to Forget   322
You Taught Me to Love Again   100
You Tell Her, I Stutter   136, 416
You Tell Me Your Dream (And I'll Tell You Mine)   77
You That I Love   352
You Took My Breath Away   73
You Took the Words Right Out of My Heart   386, 406, 407
You Turned the Tables on Me   11, 327
You Was   531
You Was Right, Baby   20, 258
You Went to My Head   119
You Were Meant for Me   40, 135
You Were There   389
You Won't Be Satisfied Until You Break My Heart   477
You Wonderful You   523
You Wouldn't Fool Me, Would You?   189
You, You Darlin'   222
You'd Be a Vision in Television   319
You'd Better Love Me   304
You'll Always Be the One I Love   466
You'll Always Be the Same Sweet Girl   474
You'll Be Mine in Apple Blossom Time   496
You'll Be Reminded of Me   449
You'll Be Reminded of Me, My Fantasy   319
You'll Know When It Happens   275
You'll Never Be Lonely   517
You'll Never Get Away   251, 574
You'll Never Get Up to Heaven That Way   260
Young and Warm and Wonderful   574
Young at Heart   258
Young Emotions   274
Young Man with a Horn   136
A Young Man's Fancy   513, 561
The Younger Generation   75
Younger Than Springtime   174
Your Eyes Have Told Me So   234

Your Feet's Too Big   132
Your Guess is Just as Good as Mine   149, 456
Your Head on My Shoulder   235
Your Heart Will Tell You So   95
Your Love   168, 526
Your Mother's Son-in-Law   207, 348
Your Red Wagon   87, 388
Your Wish Is My Command   494
You're a Builder Upper   14, 181
You're a Doggone Dangerous Girl   331
You're a Grand Old Flag   65
You're a Heavenly Thing   269, 568
You're a Lucky Fellow, Mr. Smith   389
You're a Million Miles from Nowhere   265
You're a Sweet Little Headache   386, 407
You're a Sweetheart   3, 308
You're All I Need   235, 237
You're All the World to Me   254, 259
You're Always in My Arms   493
You're an Angel   127, 307
You're an Old Smoothie   40, 90, 542
You're Beautiful Tonight, My Dear   280, 568
You're Dangerous   510
You're Devastating   180
You're Driving Me Crazy   96
You're Far from Wonderful   342
You're Getting to Be a Habit with me   106, 523
You're Giving Me a Song and a Dance   6, 490
You're Gonna Fall and Break Your Heart   299, 444
You're Gonna Lose Your Gal   330, 568
You're Here, You're There   275
You're in Kentucky   451
You're in Kentucky, Sure as You're Born   147
You're in Love   137, 139
You're in Love with Someone Else   278

You're in My Power (Ha! Ha! Ha!) 58, 206
You're in the Right Church but the Wrong Pew 287, 466
You're Irish and You're Beautiful 264
You're Just a Dream Come True 229, 346
You're Looking for Romance 274, 343
You're Lovely, Madame 387, 406
You're Lucky to Me 31, 390
You're Mine You 161, 191, 192
You're My Desire 325
You're My Everything 94, 523, 567
You're My Girl 485
You're My Past, Present and Future 396, 153
You're My Thrill 61, 157
You're Never Too Old to Learn 480, 564
You're Nobody 'Til Somebody Loves You 59, 332, 477
You're Not Pretty, but You're Mine 253
You're Not So Easy to Forget 356
You're Not the Kind 213, 325
You're On My Mind 422
You're Only in My Arms to Cry on My Shoulder 344
You're Out of This World to Me 267
You're Perfect 115, 324
You're Simply Delish 135, 324
You're Slightly Terrific 327
You're So Darn Charming 471
You're So Desirable 352
You're So Different 366
You're So Easy to Remember 554
You're So Good to Me 484
You're So Right for Me 272
You're So Sweet 324
You're Still in My Heart 101
You're Such a Comfort to Me 154, 396

You're the Cream in My Coffee 92, 189
You're the Dream, I'm the Dreamer 47
You're the Moment in My Life 345
You're the One 302, 440, 565
You're the One (You Beautiful Son-of-a-Gun) 126
You're the One I Care For 158, 159, 266, 267, 282
You're the Rainbow 407
You're the Sweetest Girl 25
You're There in a Dream 424
You're Too Dangerous 80
You're Too Lovely to Last 309
You're Too Much 133
You're Too Sharp to Be Flat 309
You're Twice as Nice as the Girl in My Dreams 534
You're Welcome 230
You're Wonderful 571
Yours and Mine 40, 45, 135
Yours for a Song 486
Yours Is My Heart Alone 467
Yours Truly is Truly Yours 83
You's a Viper 469
You've Changed 51, 128
You've Got Everything 97, 235
You've Got Me Crying Again 230, 347
You've Got Me in the Palm of Your Hand 261, 330
You've Got to See Mamma Ev'ry Night (Or You Can't See Mama at All) 70, 416

Ziegfeld Girl 56
Zigeuner 74
Zilch's Hats 356
Zing! Went the Strings of My Heart 178, 179
Zip-a-Dee-Doo-Dah 560
Zonky 520
Zwei Herzen im Dreivierteltakt 567

About the Author

Warren Vaché Sr. is a well-known personage in American music. As a string bass player, he has worked with many outstanding musicians, including Bobby Hackett, Pee Wee Russell, Pee Wee Erwin, Wild Bill Davison, among others, and is on personal and friendly terms with the people he writes about.

As the leader of a jazz band, the Syncopatin' Seven, he is represented on three current CDs: *Swingin' and Singin'* (Jazzology JCD 202), *The Syncopatin' Seven Celebrate the Music of Harry Barris* (Circle CCD 57), and *The Syncopatin' Seven Celebrate the Music of Isham Jones* (Jazzology JCD 296).

Warren has authored four other books: *This Horn for Hire,* the biography of trumpeter Pee Wee Erwin; *Crazy Fingers,* the life and career of pianist-bandleader Claude Hopkins: *Back Beats and Rim Shots,* the story of drummer Johnny Blowers: *Jazz Gentry: Aristocrats of the Music World.*

Vaché is the father of two internationally famous jazz musicians, Warren Vaché Jr. and Allan Vaché. He lives with his wife, Madeline, in Rahway, New Jersey.